AF895565

MEDICINAL PLANTS:
Phytochemistry, Pharmacology and Therapeutics

– Volume 3 –

The Editor

Dr. Vijay Kumar Gupta, Ph.D., FLS, London (born 1953-) Chief Scientist, CSIR- Indian Institute of Integrative Medicine, Jammu, India. He did his M.Sc. (1975) and Ph.D. (1979) in Zoology both from University of Jammu, Jammu-India. His research capabilities are substantiated by his excellent work on histopathology, ecology and reproductive biology of fishes, turtles, birds and mammals, which has already got recognition in India and abroad.

Dr. Gupta has to his credit more than 100 scientific publications and review articles which have appeared in internationally recognized Indian and foreign journals. Founder fellow, life member and office bearer of many national societies, academies and associations. He has successfully completed a number of research/consultancy projects funded by government, private and multinational agencies. His current areas of interest are histopathology, toxicology, pre-clinical safety pharmacology, reproductive efficacy studies of laboratory animals and biodiversity.

He is the Series Editor of the recently published multi-volume set of books, "**Comprehensive Bioactive Natural Products (Vols. 1-8)**", published by M/S Studium Press, LLC, USA. He is also Editor-in-Chief of the books, "**Utilisation and Management of Medicinal Plants**", "**Medicinal Plants: Phytochemistry, Pharmacology and Therapeutics (Vols. 1 and 2)**", "**Traditional and Folk Herbal Medicine**", "**Natural Products: Research Reviews**", "**Bioactive Phytochemicals: Perspectives for Modern Medicine**", "**Perspectives in Animal Ecology and Reproduction (Vols. 1-9)**" and "**Animal Diversity, Natural History and Conservation (Vols. 1-3)**". The Editor-in-Chief of the American Biographical Institute, USA, has appointed him as *Consulting Editor* of *The Contemporary Who's Who*. Dr. Gupta also appointed as Nominee for the *Committee for the Purpose of Control and Supervision of Experiments on Animals* (CPCSEA, Govt. of India). The *Linnaean Society of London, U.K.* has awarded fellowship to him in November 2009 in recognition of his contribution towards the cultivation of knowledge in Science of Natural History. Recently, Modern Scientific Press, USA has nominated Dr. Gupta as the Editor of the *International Journal of Traditional and Natural Medicine*.

Dr. Gurdarshan Singh (born 1962–), working as Senior Scientist in the PK-PD-Toxicology Division of Indian Institute of Integrative Medicine (CSIR), Jammu. He did his M.Sc. (1987), M.Phil. (1989) and Ph.D. (1993) in Zoology from University of Jammu, Jammu-India.

Dr. Singh has published 26 research papers, both in national and international journals of high repute and impact factor. He is also actively involved in both in-house and sponsored research projects for the generation of pre-clinical safety pharmacological data for IND (investigational New Drug) application. His main areas of research are toxicity and regulatory pharmacology of new chemical entities for drug development programme in addition to anti-allergy and anti-asthmatic screening of herbal as well as synthetic molecules.

Dr. Surjeet Singh (born 1958-) did his M.Sc. (1990) in Zoology and Ph.D. (2003) in the field of Inflammation from University of Jammu, Jammu-India and presently working as senior pharmacologist in the Department of Pharmacology, Indian Institute of Integrative Medicine(CSIR), Jammu.

He has an experience of 28 years in the field of inflammatory pharmacology, preclinical toxicology and safety and has published about 50 research papers and review articles in national and international journals. He is the life member of Indian Pharmacological Society and honour of the member *UNESCO Workshop* in 1992.

Dr. Anpurna Kaul (born 1956-) working as Scientist in Pharmacology Division of, Indian Institute of Integrative Medicine(CSIR), Jammu- India. She has completed her Ph.D. in 1996 from University of Jammu, Jammu-India. Her field of research is immune-pharmacology.

Dr. Kaul has 35 research publications in national and international journals, 12 patents and also presented several research papers in various symposia/conferences. She has screened more than 4000 plant extracts/fractions and pure compounds on immune system by *in vivo* and *in vitro* methods. She has also actively participated in Indo-Malaysian research project as member and visited Malaysia in 2007 to attend *Women's Health and Asian Traditional Medicine* Conference.

MEDICINAL PLANTS:
Phytochemistry, Pharmacology and Therapeutics

– Volume 3 –

Editor-in-Chief
V.K. Gupta
Chief Scientist
CSIR-Indian Institute of Integrative Medicine
Canal Road, Jammu – 180 001, India

Editors
G.D. Singh
Senior Scientist
PK-PD-Toxicology Division

Surjeet Singh
Scientist
Department of Pharmacology

A. Kaul
Scientist,
Department of Pharmacology

CSIR-Indian Institute of Integrative Medicine
Canal Road, Jammu – 180 001, India

2014
Daya Publishing House®
A Division of
Astral International Pvt. Ltd.
New Delhi – 110 002

© 2014 AUTHORS
ISBN 9789351300977

Publisher's note:
Every possible effort has been made to ensure that the information contained in this book is accurate at the time of going to press, and the publisher and author cannot accept responsibility for any errors or omissions, however caused. No responsibility for loss or damage occasioned to any person acting, or refraining from action, as a result of the material in this publication can be accepted by the editor, the publisher or the author. The Publisher is not associated with any product or vendor mentioned in the book. The contents of this work are intended to further general scientific research, understanding and discussion only. Readers should consult with a specialist where appropriate.
Every effort has been made to trace the owners of copyright material used in this book, if any. The author and the publisher will be grateful for any omission brought to their notice for acknowledgement in the future editions of the book.
All Rights reserved under International Copyright Conventions. No part of this publication may be reproduced, stored in a retrieval system, or transmitted in any form or by any means, electronic, mechanical, photocopying, recording or otherwise without the prior written consent of the publisher and the copyright owner.

Published by : **Daya Publishing House**®
A Division of
Astral International Pvt. Ltd.
– ISO 9001:2008 Certified Company –
4760-61/23, Ansari Road, Darya Ganj
New Delhi-110 002
Ph. 011-43549197, 23278134
E-mail: info@astralint.com
Website: www.astralint.com

Laser Typesetting : **Classic Computer Services**, Delhi - 110 035

Printed at : **Replika Press Pvt. Ltd.**

PRINTED IN INDIA

Editorial Board

– Editor-in-Chief –

Dr. V.K. Gupta

Chief Scientist
CSIR-Indian Institute of Integrative Medicine,
Canal Road, Jammu – 180 001, INDIA
Tel: +91-0191-2520039; 09419228396 (M)
Fax: +91-0191-2569017, 2569023
E-mail: vgupta_rrl@yahoo.com; vkguptaiiim@gmail.com

– Editors –

Dr. G.D.Singh

PK-PD-Toxicology Division,
CSIR- Indian Institute of Integrative Medicine,
Canal Road, Jammu – 180 001, India
Fax: +91-0191-2569017, 2569023
E-mail: singh_gd@rediffmail.com

Dr. Surjeet Singh

Department of Pharmacology,
CSIR- Indian Institute of Integrative Medicine,
Canal Road, Jammu – 180 001, India
Fax: +91-0191-2569017, 2569023
E-mail: ssingh@iiim.ac.in

Dr. A. Kaul

Department of Pharmacology,
CSIR- Indian Institute of Integrative Medicine,
Canal Road, Jammu – 180 001, India
Fax: +91-0191-2569017, 2569023
E-mail: akoul@iiim.ac.in

– Members –

Prof. Yu ZHAO

Cheung-Kung Chief Professor of Chinese Ministry of Education,
Assistant to President of Dali University,
Honorary Dean of College of Pharmacy,
Dali University, Wanhua Road, Yunnan Province, China
Tel: +86-0872-2214251; Fax: +86-0872-2257401
E-mail: dryuzhao@126.com; dryuzhao@hotmail.com

Prof. Yukihiro Shoyama

Faculty of Pharmaceutical Science,
Nagasaki International University,
His Ten Bosh, Sasebo, Nagasaki, Japan
Tel: +81-956-20-5653
E-mail: shoyama@niu.ac.jp

Prof. Ian Fraser Pryme

Department of Biomedicine,
University of Bergen,
Jonas Lies vei 91, N-5009 Bergen, Norway
Tel: +47-55-586438 Fax: +47-55-586360
E-mail: ian.pryme@biomed.uib.no

Dr. George Qian Li

Herbal Medicines Research and Education Centre,
Faculty of Pharmacy,
The University of Sydney,
NSW 2006, Australia
Tel: 612 9351 Fax: 9351 8638
E-mail: george.li@sydney.edu.au

Dr. Julia Serkedjieva

Institute of Microbiology,
Bulgarian Academy of Sciences
26, Academician Georgy Bonchev St.
1113 Sofia, Bulgaria
Tel: +359 2 979 31 85 Fax: +359 2 870 01 09
E-mail: jserkedjieva@microbio.bas.bg

T.C.
YEDİTEPE ÜNİVERSİTESİ
ECZACILIK FAKÜLTESİ

Foreword

Plants have been used for human health since thousands of years. Firstly they were used in simple and multi-component prescriptions by the people. Then they had become a primary source for the discovery of new medicines. Particularly after the second world war, with the improvements in chemical techniques, herbal medicines came to be replaced by synthetic drugs and popularity of the medicinal plants had gradually been declined. However, in spite of continuous and enormous developments in chemical, biological and medical sciences, contrary to expectations, this trend has been reversed in favor of herbal medicines in the last two or three decades.

This current worldwide growing interest for herbal medicines cannot be interpreted only with the trend towards "Green Revolution" by the people. This might be a reflection of improved techniques in chemical, biological and medical sciences. Especially after 1990's, such technical improvements have provided achievements in deeper investigations on medicinal plants; their efficiency have been proven in scientific terms, the chemical profile and active principles have been defined, the mechanism of activity have been explained and their safety profiles have been betrayed. Moreover, etiological discoveries in medicine also have revealed the contribution of plant remedies with their multi-component compositions in the treatment of diseases. Therefore, contributions of the qualified investigations on the phytochemical, pharmacological and toxicological profiles of the medicinal plants are extremely important.

It is my privilege to write the foreword to the Volume 3 entitled "***Medicinal Plants: Phytochemistry, Pharmacology and Therapeutics***" edited by Dr. Vijay Kumar Gupta.

I would like to congratulate him for his painstaking work in preparing this serial book. It is hoped this serial book will attract the attention of scientific community particularly of ethnopharmacologists, ethnobotanists, phytochemists, biochemists and others engaged in the natural products research.

Prof. Dr. Erdem Yesilada

Head, Department of Pharmacognosy and Phytotherapy,
Faculty of Pharmacy, Yeditepe University, Istanbul, Turkey
Associate Editor, Journal of Ethnopharmacology (Elsevier)

Preface

Since the beginning of human civilization, medicinal plants have been used by mankind for its therapeutic value. Nature has been a source of medicinal agents for thousands of years and an impressive number of modern drugs have been isolated from natural sources. Many of these isolations were based on the uses of the agents in traditional medicine. The plant-based medicine is still the mainstay of about 75 - 80% of the world population, mainly in the developing countries, for primary health care. This is primarily because of the general belief that herbal drugs are without any side effects besides being cheap and locally available. According to the World Health Organization (WHO), the use of herbal remedies throughout the world exceeds that of the conventional drugs by two to three times.

Written records about medicinal plants date back at least 5000 years to the Sumerians, and archeological records suggest even earlier use of medicinal plants. Some of the oldest known medicinal systems of the world such as Ayurveda of the Indus civilization, Arabian medicine of Mesopotamia, Chinese and Tibetan medicine of the Yellow River civilization of China and Kempo of the Japanese are all based mostly on plants. The ancient cultures are known for their systematic collection of information on herbs and their rich and well-defined herbal pharmacopoeias. Although some of the therapeutic properties attributed to plants have proven to be erroneous, medicinal plant therapy is based on the empirical findings of hundreds and thousands of years. The strong historic bond between plants and human health began to unwind in 1897, when Friedrich Bayer and Co. introduced synthetic acetyl salicylic acid (aspirin) to the world. About 25% of the drugs prescribed worldwide come from plants, 121 such active compounds being in current use. Of the 252 drugs considered as basic and essential by the World Health Organisation, 11% are exclusively of plant origin and a significant number are synthetic drugs obtained from natural precursors. Examples of important drugs obtained from plants are digoxin from *Digitalis* spp., quinine and quinidine from *Cinchona* spp., vincristrine and vinblastine from *Catharanthus roseus*, atropine from *Atropa belladonna* and morphine and codeine from *Papaver somniferum*.

With the advances in synthetic methodology and the development of more sophisticated isolation and analytical techniques, many more of these phytochemicals should be identified as effective as synthetic medicines with fewer or no side effects and contraindications. There is, however, much scope for further systematic research in screening medicinal plants for these phytochemicals and assessing their potential in protecting against different types of diseases. It has been proved that although the effects of natural remedies may seem slower, the results are sometimes better on the long run especially in chronic diseases.

The present volume of the book series entitled, "***Medicinal Plants: Phytochemistry, Pharmacology and Therapeutics - Vol. 3***" illustrates the types of critical discoveries that emerge from the interface of chemistry and biology, containing 25 research and review chapters written by leading authorities in their respective fields of research. The aim is to enthuse the reader with this active and exciting area of research and to lay a solid foundation on which further study of its various facts may be based and the editors will consider themselves to be amply rewarded if this humble piece of work proves to be useful for those it is meant. The book series shall hopefully provide an outlet for research work carried out in these fields. I am greatly indebted to the scientists and researchers who have contributed for this volume. Finally, I am thankful to the publisher for their unstinted support for maintaing the high publishing standards of this serial book.

Dr. V.K. Gupta

Contents

Medicinal Plants: Phytochemistry, Pharmacology and Therapeutics Vol. 3 (2014) *Pages* **1–38**
Editor-in-Chief: **V.K. Gupta**
Published by: **DAYA PUBLISHING HOUSE, NEW DELHI**

1

Phytochemistry, Pharmacology and Therapeutics of *Hemidesmus indicus* (L.) R. Br.

K. Manjulatha[1], K. Saritha[1] and O.H.Setty[1*]

ABSTRACT

Hemidesmus indicus (L.) R. Br. has been revered for its medicinal properties for nearly a thousand years in Indian traditional medicine, Ayurveda and commonly known as Indian Sarsaparilla/Anantamool/Sariva. It is used as a vital herb for healing many ailments and considered as magical spiritual dream herb in Ayurvedic medication. It has long been used in Indian folk medicine to treat diversified diseases such as liver diseases, asthma, arthritis, bronchitis, syphilis, epileptic seizures, high blood pressure, skin diseases (eczema and psoriasis), rheumatism, chronic nervous diseases, impotence and immune disorders. Mainly the roots of this plant are endowed with many pharmacological properties and since long enjoyed a reputation as tonic, alterative, diaphoretic, diuretic, depurative. It is also known to help relieving stress by inducing overwhelming sensation of relaxation, euphoria and tranquility. It forms an important ingredient in "Sarivadyasava" an Ayurvedic preparation used in treatment of gout, diabetes and related skin complications, skin infections. Experimental evidences exists for anticancer, anti hyperlipidaemic, renoprotective, hepatoprotective, anti-diarrhoeal, anti-ulcerogenic, antiatherogenic, antileprotic, antioxidant, anti-inflammatory, analgesic, antipyretic, antimicrobial, immunomodulatory activities. Plant is rich in wide variety of biologically active compounds of essential oils, glycosides, resins, saponins, steroids and tannins. Anantamool has a history of safe use in clinical practices. In this review we have

1 Department of Biochemistry, University of Hyderabad, Hyderabad – 500 046, A.P., India

* *Corresponding author*: E-mail: ohssl@uohyd.ernet.in

put our effort to gather and compile updated information regarding the recent work and current trends of medicinally important herb – Hemidesmus indicus in the field of modern phytomedicine.

Keywords: Anantamool, *Hemidesmus indicus*, Phytochemistry, Pharmacology, Therapeutics.

Introduction

Natural products are being explored by mankind throughout their existence for nutrition and many more applications. Indigenous cultures have learnt to exploit the properties of secondary metabolites in many ways *e.g.*, specific plants or parts of them have been used as poisons, analgesics, stimulants, preservatives, colorants, tanning agents for tanning leather etc. (De Pasquale, 1984). Natural products have been the major sources of chemical diversity for starting materials while driving pharmaceutical discovery over the past century. Historically, pharmaceutical companies have utilized plant extracts to produce relatively crude therapeutic formulations (Firn and Jones, 2003). Natural products have proven to be the richest source of medicinal compounds. Novel natural products and proprietary chemical scaffolds are in high demand by the pharmaceutical and agrochemical industries because compounds with novel chemical entities can be used to generate chemical libraries by parallel synthesis. In the present review, we have summarized the information concerning the geographical distribution, botanical description, medicinal uses, ethnobotanical studies, biological activities, toxicological studies and formulations of this plant.

Figure 1.1: *Hemidesmus indicus* (L.) R. Br. Plant, phytochemicals and roots.

Geographical Distribution

Hemidesmus indicus R. Br. (*H. indicus*) from the family Asclepiadaceae is a well known drug ingredient with many medicinal applications in Ayurvedic system of medicine. *H. indicus* is originated from India where it is still primarily found growing wildly. *H. indicus* is a prostrate or semi-erect shrub found throughout India from upper Gangetic plains, east-wards to Assam, throughout Central, Western and Southern India upto an elevation of 600 m. It is also known to grow in Malaysia, Indonesia, Pakistan, Bangladesh and Sri Lanka. (Sasidharan, 2004; Siddique *et al.*, 2004; Anonymous, 2005; Nayar *et al.*, 2006). The different vernacular names known are Ananta-mula, Ananthamoola, *Asclepias pseudosarsa*, Country Sarasaparilla, Durivel, East Indian Sarsaparilla, Eternal root, False Sarsaparilla, Fragrant one, Gadisugandhi, Gopakanya, *Hemidesmus pubescens*, *Hemidismus Indica-Radix*, Kapuri, Karibandha, Magrabu, Muttavapulagamu, Naga-jihva, Naruninti, Nunnari, Nunnery root, Onontomulo, *Periploca indica*, Sariva, *Smilax aspera*, Sogade, Sugandhi-pala, Sugandi root, Upalasari, White Sariva (Satheesh *et al.*, 2006).

Botanical Description and Medicinal Uses

H. indicus is a perennial, fast-growing thin creeper vine; that sends tendrils out at every node to cling to the surrounding vegetation for stability and support. It is a climbing slender plant with twining woody stems and rust-coloured bark, leaves opposite, petiolate, entire, smooth, shiny and firm, varying in shape and size according to their age. Flowers are small green outside, deep purple inside, in axillary, sessile racemes, imbricated with flowers, followed with scale like bracts and produced all through the year. Fruits are long slender spreading follicles. The seeds are white and covered in tiny silvery white hairs. The dried Indian sarsaparilla roots are medicinally important and officially included in Indian Pharmacopoeia. Roots occur in pieces, about 30 cm long and 3-8 mm in diameter, cylindrical, thick, hard, somewhat tortuous, sparsely branched, provided with few thick rootlets and secondary roots, external appearance dark brown, sometimes with violet grey tinge, centre yellow, woody, surrounded by a mealy white cortical layer, bark brownish, corky, marked with transverse cracks and longitudinal fissures and easily detachable from the hard central core, odour is characteristic, taste sweetish, slightly acrid and aromatic. The root is harvested in the autumn and dried for the later use. Huge quantities of plant materials and extracts are imported for manufacture of Ayurvedic, Unani, Homeopathic preparations (Aiyer, 1951; Prasad and Wahi, 1965; Warrier *et al.*, 2000).

The known medicinal properties of roots are antipyretic, anti-diarrhoeal, astringent, blood purifier, diaphoretic, diuretic, refrigerant and tonic, respiratory disorders (bronchitis, asthma), skin diseases (leprosy, itching), eye diseases, syphilis, epileptic fits in children, kidney and urinary disorders, loss of appetite, burning sensation and rheumatism (Nadkarni 1989; Chopra *et al.*, 1956; Anonymous, 1986, 1997; Kirthikar and Basu, 1980; Murugesa Mudhaliar, 1988; Mukherjee, 1953). Root bark is used to cure dyspepsia, loss of appetite, nutritional disorders, fever, skin diseases, ulcer, syphilis, rheumatism and leucorrhoea (Nadkarni, 1989). Stem of *H. indicus* is used as diaphoretic, diuretic, laxative and in treating brain, liver and kidney diseases, syphilis, urinary discharges, uterine complaints, leucoderma, cough and asthma (Kirthikar and Basu, 1980).

Ayurvedic properties of the *H. indicus* are Rasa-*tikta, Madhura*; guna-*guru, Snigdha*; Veerya-*sheeta*; Vipaka-*madhura*. Doshaghnata: Tridoshashamaka; Rogaghnata: Daha, Shotha, Netrabhisyanda, Aruchi, Agnimandya, Atisara, Pravahika, Vatarakta, Phiranga, Upadansha, Amvata, Gandmala, Pradara, Garbhasrava, Stanyavikara, Shukradaurbalya, Mootrakrichchhra, Paittika prameha, Kushtha, Visarpa, Visphota, Jwara, Daurbalya, Pandu, Visha, Kasa, Shwasa; Karma: Rochana, Deepana, Pachana, Anulomana, Raktashodhaka, Shothahara, Kaphaghna, Vrishya, Stanyashodhana, Garbhasthapana, Mootrajanand, Mootravirajaniuhya, Kushthagna, Jwaraghna, Dahaprashamana, Rasayana and Vishaghna (Sharma *et al.*, 2000).

Ethnobotanical Studies

Some of the ethnobotanical studies on *H. indicus* were carried out and observed for many useful benefits that are followed in traditional medicine.

Entire Plant

The dried entire plant is used for skin diseases (Shah and Gopal, 1985; Pushpalingadan and Atal, 1984). Along with *Piper longum* and *Piper nigrum*, it is used for postpartum recovery, diarrhoea, to improve appetite (Girach *et al.*, 1994; Reddy *et al.*, 1988), for impotency (Paul, 1979), to reduce body heat (Ramachandran and Nair, 1981), to increase blood circulation, cure stomach ulcers (Jain and Singh, 1984, 1996), treat scorpion sting, snake bite, cure fever (Sharma *et al.*, 1979), as blood purifier (Malhotra and Murthy, 1973; Sharma *et al.*, 1979; Pullaiah and Kumar, 1996), it has cooling effect, cures venereal diseases (Singh and Maheswari, 1983), diabetes and fever (Khanna *et al.*, 1996).

Roots

The roots are used in preparation of sweet smelling drink (Prasad *et al.*, 1996). Roots decoction is used for curing high fever and skin diseases (Sudhakar and Rao, 1985; Vyas, 1993), menorrhagia (Nagaraju and Rao, 1990), used as anti-rheumatic, diuretic, anti-inflammatory, to treat snake bite (Alam *et al.*, 1994), used as blood purifier (Reddy *et al.*, 1988), to cure stomach pain, diarrhea (Sabnis and Bedi, 1983), and venereal diseases (Shah and Gopal, 1985), as an anti-venom (Selvanayahgam *et al.*, 1994). Externally it is applied to provide relief from scorpion stings (Singh and Ali, 1992). The roots are peeled and eaten raw as a blood purifier, cooling beverage (John, 1984) and for treating skin diseases (Arseculratne *et al.*, 1985). The rind of the root is chewed for sore mouth (Prasad *et al.*, 1964, 1996). Fresh roots paste in combination with neem oil is applied on scalp of children for development of skull bones to enable carrying bead-loads in adult age (Banerjee and Pal, 1996). Roots along with *Mimosa pudica* when taken orally during the menstrual period, treat leucorrhoea and also acts as body tonic (Bhandary *et al.*, 1995). Along with black pepper, it is used for fevers of long duration and with milk it is taken for anemia (Singh and Prakash, 1996). A decoction of *Elephantopus scaber, H. indicus* and *Piper nigrum* is used for gonorrhoea (Sahu, 1984), skin affections, syphilis and as a tonic (Atal *et al.*, 1986).

With honey, the fresh roots are used for health and vitality. Along with the woods of *Acacia sundra* and *Cinnamomum zeylanicum*, it is used to make a soft, nourishing beverage to promote youthfulness, health and vitality (Pushpangadan and Atal, 1984; Gupta *et al.*, 1992).

Phytochemistry

Phytochemicals are secondary metabolites found in plants which are responsible for colour, organoleptic properties, provide protection against insect attacks, plant diseases and for its consumers exhibit a number of protective functions (David and Emma, 2011). Phytochemical studies have been carried out on roots, stem, leaves and entire plant. Different phytochemicals were found in preliminary chemical tests and same on isolation from the different parts and whole of plant, the studies revealed the presence of steroids, terpenoids, flavonoids, coumarins, aldehydes, pregnane glycosides and others. The root oil constituents were found to be rich with terpenoids, aldehyde and aliphatic acids. The frequently occurring active constituents of *H. indicus* are benzoids, pregnane glycosides, terpenoids and others (Murti *et al.*, 1941; Michael, 2004; Kumar *et al.*, 2007; Austin, 2008; Aneja, 2008).

Steroids

Steroids are a class of organic compounds with a chemical structure that contains the core of gonane or a skeleton derived from three cyclohexane rings and one cyclopentane ring. Hemidosterol, hemidesmol and steriods were isolated from the plant (Dutta, 1938). β-sitosterol (Figure 1.2: **1)** was reported from the roots (Chatterjee and Bhattacharya, 1955 and Roy *et al.*, 2002) and also from the stem (Gupta *et al.*, 1992), β-sitosteryl glucuronate was isolated from the roots (Roy *et al.*, 2002).

Pregnane Glycosides

Pregnanes are steroidal derivatives with carbons present at positions 1 through 21, pregnene with one double bond, pregnadienes with two double bonds and pregnenolone contains ketone and hydroxyl functional groups. Pregnane oligoglycoside, pregnane ester diglycoside and pregnane glycoside were isolated from the dried stem parts. **Pregnane oligoglycosides** isolated are Indicusin is 11α,12β-di-O-acetyl-orgogenin-3-O-β-D-cymaropyranosyl (1→4)-O-β-D-cymaropyranosyl (1→4)-O-β-D-cymaropyranoside, medidesmine (Figure 1.2: **13)** is sarcostin-3-O-α-D-glucopyranosyl(1→4)-O-β-D-digitoxopyran osyl(1→4)-O-β-D-oleandropyranoside, hemisine is calogenin-3-O-β-D-cymaropyranosyl (1→4)-O[3-O-methyl]β-D-glucopyranosyl(1→4)-O-β-D-glucopyranosyl(1→4)-O-β-D-cyma ropyranoside and desmisine (Figure 1.2: **12)** is calogenin-3-O-β-D-xylopyranosyl(1→4)-O-β-D-digitoxopyranosyl(1→4)-O-β-D-xylopyranosyl(1→4)-O-β-D-digitoxopyranoside were isolated from $CHCl_3$-EtOH (3: 2) fraction of dried stem (Deepak *et al.*, 1995a, 1995b, 1997). **Pregnane ester diglycoside** isolated was desinine is β-3-O-β-D-oleandropyranosyl-(1→4)-β-D-oleandropyranoside (Figure 1.2: **8)** (Oberoi *et al.*, 1985). **Pregnane glycosides** isolated are denicunine is calogenin-3-O-3-O-methyl-β-D-fucopyranosyl-(1→4)-O-β-D-oleandropyranosi de (Figure 1.2: **11a)**, heminine is calogenin-3-O-β-D-cymaropyranosyl-(1→4)-O-β-D-digitoxopyra noside (Figure 1.2: **11b)** (Sigler *et al.*, 2000), hemindicusin, is calogenin-3-O-methyl-α-L-

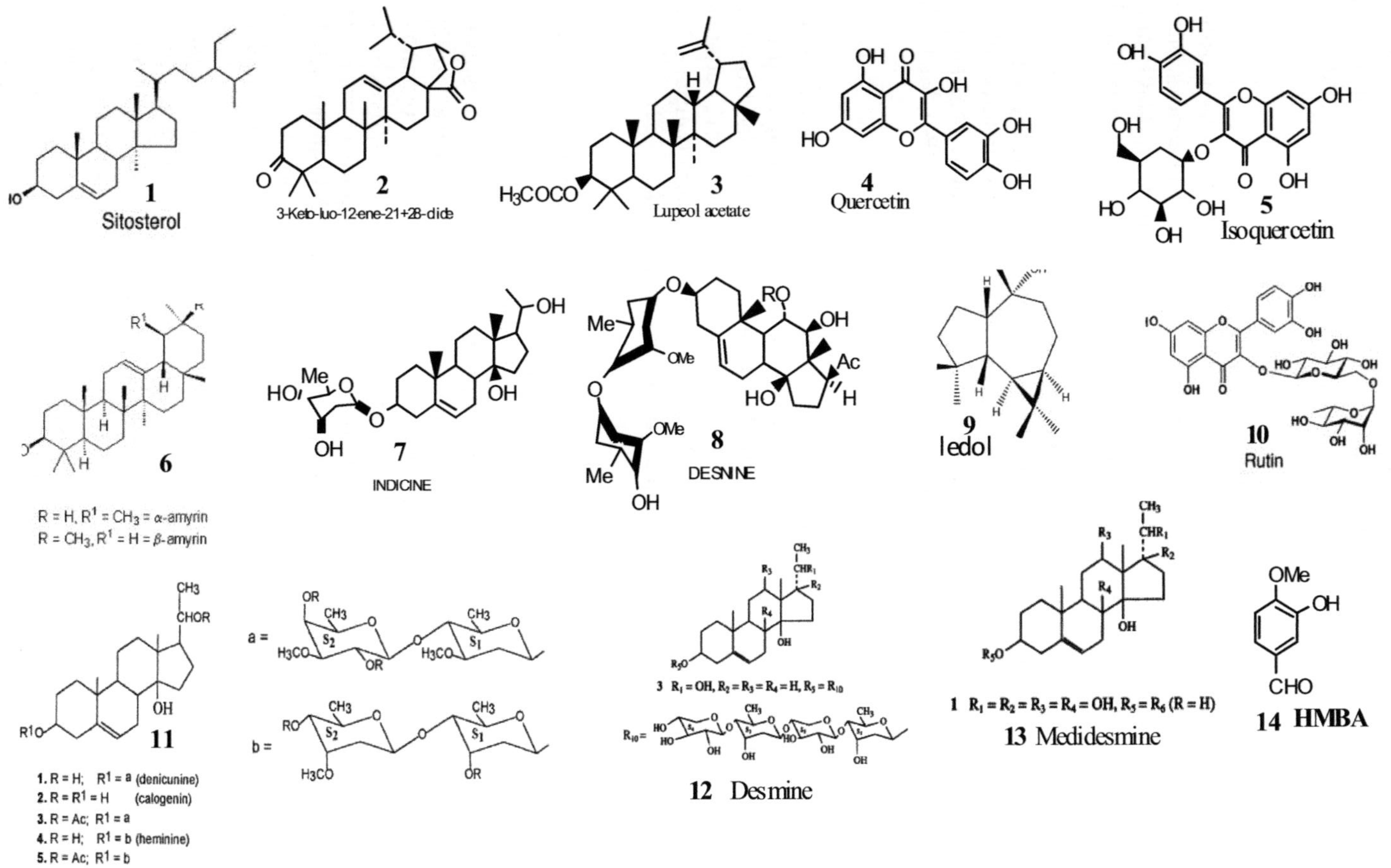

Figure 1.2: Secondary metabolites isolated from plant *Hemidesmus indicus.*

rhamnopyranoside, hindicusine is 20-O-benzoyl calogenin-3-O-β -D-digitoxopyranoside from the $CHCl_3$-EtOH (3: 2) fraction of methanolic extract (Sethi *et al.*, 2006, 2010), indicine (Figure 1.2: **7**) is calogenin-3-O-β-D-digitoxopyranoside and hemidine calogenin-3-O-β-D-boivinopyranoside (Prakash *et al.*, 1991), hemidescine is 20-O-acetyl calogenin 3-O-β-D-digitoxopyranosyl(1→4)-O-β-D-oleandropyranoside, emidine is calogenin-3-O-β-D-digitoxopyranosyl(1→4)-O-β-D-digitoxopyranosyl(1→4)-O-β-D-digitox opyranoside (Chandra *et al.*, 1994). **Pregnenolone glycosides** isolated from roots are hemidesmosides A-C and plocoside A (Zhao *et al.*, 2012).

Terpenoids

Terpenes are hydrocarbons resulting from the combination of several isoprene units. Terpenoids are modified terpenes, wherein methyl groups have been moved or removed, or oxygen atoms are added. The essential oil of roots that possess many pharmacological activities has been studied and analysed the presence of following terpenes using GCMS are 1,8cineol, camphor, pinocarveol, β-pinenoxide, pinocarvone, borneol, 4-terpenenol, bornyl acetate, myrtenal, α-terpineol, verbenone, myrtenol, linalyl acetate, isobornyl acetate, dihydrocarvyl acetate, α-terpinyl acetate, β-elemene, cis-caryophyllene, Isocaryophyllene, β-selinene, nerolidol, ledol (Figure 1.2: **9**) (Nagarajan *et al.*, 2001).

The terpenoids isolated from the *H. indicus* were lupeol, α-amyrin, β-amyrin (Figure 1.2: **6**) from roots and stem (Murthi *et al.*, 1941; Gupta *et al.*, 1981; Chatterjee and Bhattacharya, 1955; Padhy *et al.*, 1973). The derivatives of above reported terpenoid isolated were lupeol acetate (Figure 1.2: **3**), β-amyrin acetate, hexatriacontane, lupeol octacosonate from roots (Murthi *et al.*, 1941; Padhy *et al.*, 1973; Roy *et al.*, 2001). Terpenoid lactone (3-Oxo-12-lupen-28, 21-olide) (Figure 1.2: **2**), terpenoid ketone, lupanone, and derivatives such as δ^{12}–dehydrolupanyl-3-β-acetate, δ^{12}–dehydro lupeol acetate were isolated from hexane soluble portion of ethanolic extract of stem (Gupta *et al.*, 1992).

Six new pentacyclic triterpenes including two oleanenes were identified as olean-12-en-21 β-yl acetate, and olean-12-en-3α-yl acetate, three ursenes characterized as 16(17)-seco-urs-12,20(30)-dien-18α-H-3β-yl actetate, urs-20(30)-en-18β-H-3β-yl acetate and 16(17)-seco-urs-12,20(30)dien-18- α-H-3β-ol and a lupene formulated as lup-1,12-dien-3-on-21-ol, known compound β-amyrin acetate from the roots (Roy *et al.*, 2001). Acyclic triterpenic acid; acyclic diterpenic ester and monocyclic sesterterpene esters isolated and their structures were established as 2,6,10,14,18,22-hexamethyltetracos-1-oic acid; n-octyl-2,6,10,14-tetramethyl hexadec-7-ol-10-en-13-on-1-octanoate and n-non-2'-en-1'-yl-13(15,19,19-trimethyl-cyclohex-14, 16-dienyl)-2, 6, 10-trimethyl-tetradec-6-ol-13-on-1-oate (Roy *et al.*, 2002).

Lupeol octacosanoate is a marker compound that was isolated from petroleum ether extract of roots and designated as a unique compound. The limit of detection and limit of quantitation were 159.5 and 483.5 ng/spot of root powder by HPTLC method and with an average amount of Lupeol octacosanoate found in the root powder was 36.5 mg/gm (Darekar *et al.*, 2008).

Flavonoids

Flavonoids or bioflavonoids, chemically are 2-phenyl-1,4-benzopyrone.These belong to ubiquitous group of polyphenolic substances that are present in most plants, concentrated in the seeds, fruit skin or peel, bark and flowers fulfilling many functions (J. B. Harborne, 1998). Flavonoids isolated from *H. indicus* were rutin (Figure 1.2: **10)**, hyperoside, quercetin (Figure 1.2: **4)**, iso-quercitrin (Figure 1.2: **5)** from leaves and their glycosides from flowers (Subramanian and Nair, 1968).

Benzoid Derivatives

Benzoid derivatives are aromatic compounds that constitute benzene ring as its nucleus. Some of the benzoid derivatives reported from this plant are 2-hydroxy-4-methoxybenzaldehyde (HMBA, Figure 1.2: **14)**, salicylaldehyde, methyl salicylate as constituent of oil from roots (Nagarajan *et al.*, 2001). 4-Hydroxy-3-methoxybenzaldehyde (vanillin), benzaldehyde, HMBA, 3-Hydroxy-4-methoxybenzaldehyde are reported from the stem (Gupta *et al.*, 1992).

HMBA is a marker compound in the plant and in roots. Chakraborty *et al.* (2008) suggested that biosynthesis of HMBA is mediated by phenylalanine ammonia-lyase in roots. By GCMS method of quantification, HMBA was found as a major compound with >90 per cent in their essential volatile oils with variation from 0.03 to 0.54 per cent (Nagarajan and Lingamallu, 2003) and about 97.89 per cent (Sreelekha *et al.*, 2007). Using densitometric HPTLC, HMBA was found as marker compound in hexane extract of the dry root powder and method for quantification was developed, established and validated (Darekar *et al.*, 2009). On reverse phase HPLC, a method was developed and validated for HMBA and 2-hydroxy-4-methoxybenzoic acid (HMBAi) for quantification in the roots. In this method linear ($r > 0.998$) was found in the range of 5-350 µg/mL for HMBA and for HMBAi ($r > 0.999$) in the range of 10-300 µg/mL. The limits of detection (LODs) for HMBA and HMBAi were found to be 0.84 and 2.34 µg/mL (Sircar *et al.*, 2007).

Coumarinolignoids

Coumarin is a phenolic chemical class with benzopyr-2-one as basic skeleton found in many plants. Lignan is also a phenolic known as phenylpropanoid, that is built up of C_6-C_3 basic unit. When both coumarin and lignan phenolics are condensed they form a Coumarinolignoid. From leaves Hemidesminine was isolated (Mandal *et al.*, 1991). Two new coumarinolignoids, hemidesmin-1, hemidesmin-2 were isolated from the roots (Das *et al.*, 1992).

Other

Tannins, resins isolated from leaves (Dutta, 1938), dodecanoic acid, hexadeconoic acid from roots (Nagarajan *et al.*, 2001) and one new condensed phenylpropanoid glucoside from roots (Zhao *et al.*, 2012).

Microbiological Studies

Against *Streptococcus mutants* plant extracts exhibited minimum inhibitory concentration (MIC) at 410 µg/mL for aqueous extract, at 200 µg/mL for methanolic and at 320 µg/mL for 50 per cent methanolic extract. 50 per cent methanolic and aqueous extracts were effective against *Corynebacterium diphtheria, Diplococcus pneumonia, Staphylococccus aureus, Streptococcus pyogens* and *Streptococcus viridans*. The antimicrobial activity of plant extracts against *Helicobacter pylori* from human isolates was observed at MIC of 75 µg for flowering and vegetative season and minimum lethal concentration (MLC) at 100 µg for vegetative and 75 µg for flowering periods (Austin *et al.*, 2003). The antimicrobial study found that roots aqueous extract showed inhibition at the range of 0.04 mg to 0.1 mg against the pathogenic bacteria *Staphylococcus aureus, Pseudomonas aeruginosa* and *Klebsiella pneumoniae* in an *in vitro* condition (Gayathri and Kannabiran, 2009b). Saponin fraction from the roots extract exhibited remarkable antimicrobial activity against *Staphylococcus aureus, Salmonella typhi, Klebsiella pneumoniae, Aspergillus flavus, Aspergillus fumigatus* and *Aspergillus niger* (Khanna *et al.*, 2008). Activity studies of extracts tested against the MDR isolates of *Klebsiella pneumoniae* from Chennai found to be significant (Kamatchi *et al.*, 2009). A study on steam distillates of roots alone and in combination with cow urine revealed antimicrobial and antihelminthic potential of the distillates against bacteria, fungi and worms, wherein roots sample alone was superior over combination samples (Kekuda *et al.*, 2009). The plant extracts could not inhibit the activity of anaerobic pathogen *Propionibacterium acne* (Jain and Bansal, 2003).

Antidiarrhoeal studies of the chloroform and methanolic extracts of roots were screened for antienterobacterial potential by disc diffusion, modified agar well, swab method wherein the fatty substance separated from methanolic extract found to be more active over other extracts. Copper and zinc were detected as trace elements in the active extracts thus, Das *et al.*, suggested that presence of these trace elements along with other active constituents may be contributing to the antienterobacterial activity (Das *et al.*, 2006a). Mimicking of host cell receptors, thereby blocking its attachment to host cell and pathological effects by glycosides in root extract was shown by pre-coating the cells (both Int 407 and P388D1) with these glycosides prior to infection, and by neutralizing the wild-type bacteria (Das and Devraj, 2006). In *Salmonella enterica serovar Typhimurium* (*S. typhimurium*)-induced cytotoxicity in a human intestinal epithelial cell line (Int 407), the roots chloroform extract composed of sterols and fatty acids at dose of 100 µg/mL abrogated the invasion ability forming a coat of adherent and invaded bacteria which is completely occupying the cytoplasm (Das and Devraj, 2007). Later in another investigation the roots chloroform extract blocked the cytotoxic activity of *S. typhimurium* in a macrophage cell line supported with ultrastructures of P388D1 cells infected with wild bacteria showed many ingested bacteria and characteristic *Salmonella*-containing vacuoles (Das and Devraj, 2008).

The anti-methicillin resistant *Staphylococcus aureus* (MRSA) activity of stem ethanolic extracts was detected with inhibition zone size ranged from 11 to 44 mm and MIC varied from 0.32 to 3.25 mg/mL (Aqil *et al.*, 2006), 0.31-6.25 mg/mL against β-lactamases producing enteric bacteria (Ahmad and Aqil, 2007). Dose dependent antimutagenic activity of the stem extract at 100 µg/plate exhibited the inhibition of His+ revertants against direct acting mutagens, methyl methanesulphonate (MMS) and sodium azide (NaN_3) induced mutagenicity in *Salmonella* tester strains TA 97a, TA 100, TA 102 and TA 104 evident from linear regression analysis of the data.

Antifungal activity of plant was studied against *Microsporum canis, M. gypseum, Phialophora jeanselmei, Piedraia hortae* and *Trichophyton mentagrophytes.* 95 per cent ethanolic and aqueous extracts were studied against *Candida albicans* and *C. tropicalis.* Against Ranikhet, antiviral activity (Dhar *et al.*, 1968), antiplaque formation (Namba *et al.*, 1985), antifilarial activities (Suresh and Rai, 1990) were carried out. Weak activity was found against *Setaria digitata*, nematocidal activity against *Toxocara canis* (Ali *et al.*, 1991; Kiuchi *et al.*, 1989) and no activity against keratinophilic fungi (Qureshi *et al.*, 1997). *Mycobacterium leprae* sample was collected from the leprosy patients, was infected in mice and were orally treated with aqueous extract at dose level of 100 mg/kg BW. They found delay in cutaneous hypersensitivity stimulation, decrease in phagocytosis, as immune-modulator and immuno-suppressant (Gupta, 1981, Atal *et al.*, 1986). Water treatment with plant extracts found larvicidal activities at LC_{50} of 0.59 per cent statistically at $p < 0.05$ level, water colour with 13.8 HU, turbidity 33 mg/L and pH 7.1 when compared to controls, these reduced values indicate water sedimentation property (Arjunan *et al.*, 2012).

Pharmacological Activities

Antiulcer

Study on antiulcer property of the root extracts found that extracts of flowering season possess antiulcer properties that exhibited significant reduction in the formation of gastric and duodenal lesions in rats induced by pylorus ligation, aspirin induced peptic and cysteamine induced duodenal ulcers. They found decrease in the aggressive factors like pepsin and proteins and an increase in the resistance factors like pH, hexose, hexosamine, fucose and sialic acid. Increase in hexosamine and carbohydrate/protein ratio and decreased pepsin content that supports for the increase in mucous secretion and ulcer protection that is comparable with standard drug Ranitidine and Omeprazole. These results suggested that extract may be selectively inhibiting $PGF_2\alpha$ (Anoop and Jegadeesan, 2003).

Anti inflammatory

The ethylacetate extract of roots exhibited significant inhibition of inflammation in both acute and subacute inflammation induced by carrageenan, bradykinin, S-hydroxy tryptamine, but less active in granuloma pouch and cotton pellet implantation and ineffective in dextran induced inflammation methods in rats (Dutta *et al.*, 1982). The treatment with the hydroalcoholic root extract (100, 200 and 300 mg/kg BW, p.o.) significantly prevented increase in volume of paw edema and formation of granulation tissue in dose dependent manner and maximal effect was observed at 300 mg/kg BW which was comparable to phenylbutazone 100 mg/kg BW i.p. (Farook *et al.*, 2011). In carragenan induced paw oedema also exhibited significant reduction in volume between 2-4 h on treatment with roots methanolic extract (Lakshman *et al.*, 2006).

Antinociception and Antipyretic

The antinociception property was observed in acetic acid induced writhing, tail flick and hot plate method of alcoholic extract of roots in dose dependent manner

(Verma *et al.*, 2005; Farook *et al.*, 2011). At dose of 300 mg/kg BW caused decrease in body temperature of rats induced with yeast and treated with hydroalcoholic extract of roots (Farook *et al.*, 2011). Standard drug paracetamol 100 mg/kg BW and test *H. indicus* extract at a dose of 100, 200 and 400 mg/kg BW reduced the yeast-elevated rectal temperature compared to control group (Lakshman *et al.*, 2006). A traditional medicine Jwarhar mahakashayan an ayurvedic preparation of the roots of *H. indicus* has been used to cure antipyretic-analgesic effect. Studies of the aqueous extract of above preparation found to possess antipyretic-analgesic property with very low ulcerogenicity and toxicity in animal model (Gupta *et al.*, 2010).

Hepatotonic and Hepatotoxic

Oral treatment of ethanol extract of roots (100 mg/kg BW, for 15 days) significantly prevented rifampicin and isoniazid-induced hepatotoxicity in rats with decrease in level of liver mitochondrial protein and the activities of isocitrate dehydrogenase, α-ketoglutarate dehydrogenase, succinate dehydrogenase, malate dehydrogenase, NADH dehydrogenase and cytochrome c. There was an increase in mitochondrial lipid peroxidation with a significant decrease in the activities of antiperoxidative enzymes such as catalase (CAT) and superoxide dismutase (SOD) (Prabakan *et al.*, 2000). The rats pretreated with methanolic extract of roots (100-500 mg/kg BW, p.o.) exhibited rise in the levels of enzymes namely serum glutamate pyruvate transaminase (SGPT), serum glutamate oxaloacetate transaminase (SGOT), alkaline phosphatase (ALP) but it was significantly less as compared to those treated with paracetamol or CCl_4 alone (Baheti *et al.*, 2006). The hepatoprotective effect of roots methanolic extract was comparable with the standard silymarin (100 mg/kg, BW) at a dose of 250 mg/kg BW in CCl_4 induced damage while 500 mg/kg BW in case of paracetamol induced hepatic damage with altered serum enzyme levels. The studies were supported with histopathological changes which were near to normal (Asha *et al.*, 2011).

Studies on roots ethanolic extract treated on rats with ethanol-induced liver injury was found to be protective against free radical-mediated oxidative stress in plasma, erythrocytes and liver at a dose of 500 mg/kg BW/day administered for 30 days. With effect of treatment, there was significantly increase in body weight, decrease in liver/BW ratio and decrease in activities of liver marker enzymes such as aspartate transaminase (AST), alanine transaminase (ALT), alkaline phosphatae (ALP), γ-glutamyl transpeptidase (GGT) and lactate dehydrogenase (LDH). In addition there was an increase in the levels of plasma, erythrocyte, and hepatic thiobarbituric acid-reactive substances (TBARS), hydroperoxides (LOOH), and conjugated dienes (CD) and decreased activities of superoxide dismutase (SOD), catalase (CAT), glutathione peroxidase (GPx), reduced glutathione (GSH), vitamin C, and alpha-tocopherol (vitamin E) (Saravanan *et al.*, 2007). Further *in vivo* studies were continued (Saravanan and Nalini, 2008) on rats at a dose of 5 g/kg BW administered for 60 days and found an increase in food intake and weight, decreased TBARS levels, near normal levels of glutathione-dependent enzymes, increased total protein, albumin, globulin and liver glycogen contents, an increased A/G ratio and decreased liver marker enzyme activities and ceruloplasmin levels along with the intensity of the bands was near normal in FTIR.

HMBA isolated from roots at a dose of 200 mg/kg BW when co-administered in alcohol injured rats, treated for 30 days observed reduction in lipidemia and hepatoprotective potential, supported with biochemical and histopathological findings (Sarvanan and Nalini, 2007b). There was a significant increase in body weight, activity of enzymic and non-enzymic antioxidants in plasma, erythrocytes and also in the levels of plasma and liver vitamin C and α-tocopherol. There was decrease in the liver-body weight ratio, AST, ALT, ALP, GGT, LDHS, lipid peroxidative marker levels on comparison with untreated ethanol-administered rats. Matrix metalloproteinases (MMPs) are involved in the extracellular matrix degradation during ethanol intoxication, both root extract (500 mg/kg BW) and HMBA (200 μg/kg BW) exhibited strong inhibitory activity on the quantitative and qualitative properties of hepatic collagen and MMPs. There was significant reduction in liver fibrotic markers, collagen content, MMP2, MMP9, liver collagen, hydroxyproline, cross-linked fluorescence, shrinkage temperature and lipid peroxidation. There was enhanced solubility of liver collagen and ascorbic acid levels when compared with untreated ethanol-fed rats. In ethanol induced hyperlipidemia rats, they observed increase in food intake, body weight, hepatic and plasma lipids, lipoproteins and decrease in lipids and lipoprotein concentrations (Saravanan and Nalini, 2007a, 2007c, 2007d).

Studies from our research laboratory observed the effect of aqueous roots extract of *H. indicus* at dose of 100 mg/kg BW against bromobenzene at a dose level of 10 mmol/kg BW induced oxidative damage in rat liver mitochondria and found significant beneficial effects like, stimulation in respiration, prevented the rise in lipid peroxides and protein carbonyls, increased the level of sulphydryl groups back to control level. Administration of vitamin E could not reverse as effectively as extract (Gopi and Setty, 2010).

Diuretic and Effect on Renal Injury

Alcoholic and steam distilled root extracts had no diuretic activity but aqueous extract caused a slight increase in urinary flow in rats but not in dogs (Satoskar *et al.*, 1962). Investigations on roots aqueous and ethanolic extracts at a dose of 400 mg/kg BW was carried out. There was a significant increase in urine output with onset of this diuretic action was gradually within 5 h and lasted upto 24 h and the aqueous extract caused marked increase in urinary Na^+ and K^+ levels but the routine urinalysis showed non-significant alterations in pH and specific gravity (Gadge and Jalalpure, 2011). Treatment with root powder helped in the management of renal impairment, which was induced by gentamicin exhibited as a promising adjunct therapy as it reduces nephrotoxicity caused by aminoglycosides in rats. The findings were supported with histological examination of kidneys and haematological evaluations (Kotnis *et al.*, 2004, 2005).

The ethanolic extract of plant at a dose of 500 mg/kg BW was administered for 30 days in rats was against ethanol-induced kidney injury. There was a significant decrease in the levels of serum urea, uric acid and creatinine as well as kidney TBARS, LOOH and CD. There was an elevation in the activities of SOD, CAT, GPx, GSH, vitamin C and vitamin E in kidney as compared to that of untreated ethanol-administered rats. Findings were supported with histopathological observations. These findings suggest that treatment with ethanolic extract offers protection against free radical-mediated oxidative stress in kidney of animals treated with ethanol-induced nephrotoxicity (Saravanan and Nalini, 2007c).

Amelioration of renal tissue was evidenced with protection from oxidative damage by decreased levels of lipid peroxidation, serum urea level from 41.3 +/- 2.86 mg/dL to 34.54 +/- 0.37, creatinine levels from 1.1 +/- 0.02 to 0.76 +/- 0.09 mg/dL. Histopathological observations also support restoration in renal tissue architecture on cisplatin induced structural alteration (Sandeep *et al.*, 2010).

Antioxidant effect of the ethanolic extract of roots in rats with ethanol-induced nephrotoxicity was observed with protection against free radical-mediated oxidative stress in kidney of animals. Extract was administered at a dose of 5 g/kg BW of ethanol for 60 days to male Wistar rats. Protection against ethanol-induced liver injury in animals with *H. indicus* extract was also observed against free radical-mediated oxidative stress in plasma, erythrocytes, and liver of animals (Sarvanan and Nalini *et al.*, 2007c).

Antidiarrhoeal

In vitro and *in vivo* study of antidiarrhoeal effects of methanolic root extract possesses antienterobacterial effect. There was decrease in the amount of wet faeces in castor oil induced diarrhoea in rats and also inhibition by agar well diffusion method. They suggested that antidiarrhoeal effect may be by inhibition of intestinal motility and by bacteriocidal activity of root extracts (Das *et al.*, 2003). The root powder or its water extract can be incorporated in oral rehydrating salt solution (ORS) to increase its antidiarrhoeal efficacy by increasing the absorption of water, Na^+ and K^+ (but not glucose) from the sac and intestinal motility was not affected (Evans *et al.*, 2004).

Antivenom

The methanol root extract was explored for the first time for neutralization of snake venom (*Vipera russellii*) activity and the extract significantly neutralized the viper venom-induced lethality and hemorrhagic activity in albino rat and mouse (Alam *et al.*, 1996). Further the studies were carried out with the major constituent from the root extract HMBA and found it antagonized viper venom-induced lethal, haemorrhagic, coagulant and anticoagulant activities in experimental rodents with reduction in cotton pellet-induced granuloma, fall in body temperature in yeast-induced pyrexia, neutralized serum phosphatase, transaminase activity as well neutralised the free radicals formed by TBARS and SOD in viper venom-induced changes. HMBA when explored on rabbits which were immunized with *Vipera russellii* venom found to have with positive adjuvant effects and higher neutralization capacity that were compared with the antiserum raised with venom alone (Alam and Gomes, 1998a, 1998b). Protective effect of root extract was administered orally (7.2 mg/kg BW) to mice experimentally envenomed with rattlesnake venom subcutaneous injection (2.5-15 μg/kg BW) (Samy *et al.*, 2008).

Lupeol acetate isolated from roots methanolic extract could significantly neutralize lethality, haemorrhage, defibrinogenation, edema, PLA(2) activity induced by *Daboia russellii* venom and found protection by antiserum action. It also neutralized *Naja kaouthia* venom induced lethality, cardiotoxicity, neurotoxicity and respiratory changes in experimental animals. Lipid peroxidation and SOD were antagonized by lupeol acetate on inducing venom (Chatterjee *et al.*, 2006).

Anti-epileptic Activity

The anti-epileptic efficacy of the root extract was observed in Maximal electroshock (MES) and Isoniazid (INH) convulsion induced methods. In the investigation with MES induced convulsion method, the extract at different concentrations (100, 300 and 500 mg/kg BW) significantly reduced the time spent in hind limb extensor phase. In INH method, the convulsions were significantly reduced with onset of convulsions. (Madhu *et al.*, 2009)

Antidiabetic

Diabetes mellitus is a chronic metabolic disorder associated with hyperglycemia, oxidative stress and dyslipidemia. In streptozotocin-induced diabetic rats roots aqueous extract was administered through oral route and observed decrease in blood glucose level significantly at 5 h and restored serum electrolytes, glycolytic enzymes and hepatic cytochrome P-450-dependent enzyme systems by preventing the formation of liver and kidney lipid peroxides at the end of 12 weeks from the study period (Gayathri and Kannabiran, 2008). Further studies were carried out on isolated HMBA (500 µg/kg BW) from roots and found it as a major active principle possessing antidiabetic potential in streptozotocin-induced diabetic rats administered for 7 weeks by oral intubation method. There was restoration in the levels of low affinity Ca^{2+}-ATPase, indicating insulin secretory effect. The increased levels of plasma glucose, plasma, erythrocyte and erythrocyte membrane lipid peroxidation and decrease in level of plasma insulin, low affinity towards Ca^{+2}-ATPase demonstrate that HMBA exhibits promising antidiabetic activity and also helps to maintain glycemic control by curbing ROS (Gayathri and Kannabiran, 2009c). In another investigation, the hypoglycemic effect of HMBA at a dose of 500 µg/kg BW observed reduced levels of glucose, decreased activities in glucose-6-phosphatase and fructose-1,6-bisphosphatase indicating suppression in the gluconeogenesis and normalization of hexokinase and phosphoglucoisomerase activities indicating its role in inducing glycolysis on streptozotocin-induced diabetic rats (Gayathri and Kannabiran, 2009e). The HMBA (500 µg/kg BW) also have protective effect on liver, kidney and pancreas apart from hypoglycemic effect which was observed in streptozotocin-induced oxidative stress administered for 7 weeks that significantly decreased blood glucose levels, malondialdehyde concentrations in the liver, kidney and pancreas and activities of α-amylase and α-glucosidase ($F>0.05$; $P<0.001$). Its antioxidant activity was supported by quenching the lipid peroxidation and restoring the GSH levels (Gayathri and Kannabiran, 2009a, d). Studies of HMBA oral treatment showed ameliorative effect on liver, kidney and pancreas injury in streptozotocin-induced diabetic rats on correlating with biochemical and histopathological changes (Gayathri and Kannabiran, 2010).

The roots were investigated for antidiabetic and antihyperlipidemic effect at a dose of 400 mg/kg BW administered for 4 weeks in alloxan-induced diabetic rats and there was significant decrease in the serum cholesterol, triglyceride, free fatty acids and phospholipid which indicates the hypoglycemic and hypocholesterolemic effect (Sowmia and Kokilavani, 2007). Antidiabetic effect of roots extract (400 mg/kg BW) on oral administration for 30 days showed altered levels of blood glucose, Hb, glycosylated Hb, plasma insulin, protein, lipid peroxides, enzymic and non-enzymic antioxidants, lipid profile, muscle glycogen content in diabetic rats were significantly reverted back to near basal values. The results were comparable with glyclazide, an oral standard hypoglycemic drug (Subramanian *et al.*, 2012).

Anticancer

The roots decoction was found as cytotoxic on HepG2 cells (Thabrew *et al.*, 2005). The roots methanolic extract showed inhibition on colon adenocarcinoma cell line with IC_{50} 60 µg/mL by MTT assay and suggested that this may be due to the presence of saponins, tannins and steroids (Pasumarthi *et al.*, 2011). There was a promising inhibition of diethylnitrosamine mediated expression of Glutathione S-transferase P form in rat liver by decoction of *Nigella sativa* seeds, *H. indicus* roots, and *Smilax glabra* rhizome treated for 10 weeks in rats. Further studies revealed antihepatocarcinogenic potential by inhibition as above but also the carcinogen mediated development of overt tumors and histopathological changes were leading to tumor development (Iddamaldeniya *et al.*, 2003, 2006). In HepG2 cells, poly-herbal decoction of *Nigella sativa* seeds, *H. indicus* roots and *Smilax glabra* rhizomes can induce DNA fragmentation, characteristic morphological changes associated with apoptosis, in a time and dose dependent manner, upregulate the expression of the pro-apoptotic gene Bax and down regulate expression of anti-apoptotic Bcl-2 and enhanced the activities of caspase-3 and caspase-9 (Samarakoon *et al.*, 2010, 2012).

Plant extract on human leukemic cell lines CEM (acute T lymphoblastic leukemia), HL-60 (acute promyelocytic leukemia), REH (non-T, non-B lymphoblastic leukemia) and KU812F, modulated many components of intracellular signaling pathways involved in cell viability and proliferation and altered the protein expression, eventually leading to tumor cell death, mediated by a loss of mitochondrial transmembrane potential and increased Bax/Bcl-2 ratio. ADP, adenine nucleotide translocator and mitochondrial permeability transition pore inhibitors did not reverse extract-induced mitochondrial depolarization. Extract induced a significant Ca^{2+} raise through the mobilization of intracellular Ca^{2+} stores. Moreover, the extract significantly enhanced the antitumor activity of three commonly used chemotherapeutic drugs (methotrexate, 6-thioguanine, cytarabine). A clinically relevant observation is that its cytotoxic activity was also recorded in primary cells from acute myeloid leukemic patients (Fimognari *et al.*, 2011).

Protection against DEN-mediated carcinogenic changes in rat liver can be achieved by long term treatment with the decoction comprised of *Nigella sativa* seeds, *Smilax glabra* rhizome and *H. indicus* root bark (Iddamaldeniya *et al.*, 2006).

Antioxidant and Free Radical Scavenger

The dried plant material extract at a dose level of 1.5 and 3.0 mg/kg BW was effective as chemopreventive agent, capable of ameliorating cumene-hydroperoxide induced cutaneous oxidative stress and tumor promotion with significant inhibition of cutaneous oxidative stress, epidermal ornithine decarboxylase activity and enhanced DNA synthesis (Sultana *et al.*, 2003a).

The root bark extract was evaluated for antioxidant property and found that scavenging of DPPH and superoxide radicals was intense (EC_{50} = 18.87 and 19.9 µg/mL respectively) while in NO radical was moderate, inhibited lipid peroxidation of liver homogenate (EC_{50} = 43.8 µg/mL) and the hemolysis induced by phenylhydrazine (EC_{50}=9.74 µg/mL) confirming the membrane stabilization activity (Ravishankara *et al.*, 2002).

The methanolic extract of roots was found with 50 per cent inhibition of lipid peroxidation at a dose of 217.5µg/mL and for scavenging of hydroxyl and superoxide radicals were at 73.5 and 287.5 µg/mL *in vitro*. Intravenous administration at a dose of 5 mg/kg BW for rabbits delayed plasma recalcification time and enhanced the release of lipoprotein lipase enzyme significantly. The extract also inhibited ADP-induced platelet aggregation *in vitro* (50-250 µg), which was comparable with commercial heparin (Mary *et al.*, 2003a). The methanolic and aqueous extracts of roots *in vitro* antioxidant studies were carried out and both extracts exhibited similar scavenging effects against ABTS and superoxide radicals whereas against DPPH and nitric oxide, the methanol extract was more effective (Murali *et al.*, 2010). Antioxidant potential of roots methanolic extract was observed at a dose of 19.52 mg ascorbic acid/g and NO Scavenging Capacity Assay at a dose of 2.334 ± 0.145 µg/mL methanolic extract of roots (Zulfiker *et al.*, 2011). In studies of Zulfiker *et al.*, they observed more potent NO Scavenging activity compared to moderate results of earlier studies (Ravishankara *et al.*, 2002). They explained the difference in the antioxidant activity may be due to the climatic conditions, the geographical environment of the sites harvest, the period of the collect, the enzymatic content responsible for the biosynthesis and the regulation of gene expression by environmental factors. The antioxidant potential of free phenolic fraction (HDFP) was more than the bound phenolic fraction (HDBP) of roots of *H. indicus*. The HDFP and HDBP fractions could inhibit lipid peroxidation (IC_{50}–19.5 ± 0.5 and 21.7 ± 0.5 µg GAE/mL), protect DNA, free radical scavenging ((IC_{50}–7 ± 0.2 and 8.6 ± 0.2 µg GAE/mL), have reducing power (110.3 ± 2 and 33.5 ± 1 U/g) and cytoprotective activities (14.8 ± 0.4 and 14.5 ± 0.5 µg GAE/mL). HDFP was constituted by gallic (18 per cent), caffeic (17 per cent), ferulic acids (16 per cent) and HDBP by syringic acid (35 per cent) as major phenolic acids. The results indicated a 34- and 27-folds better contribution to the antioxidant activity by HDFP and HDBP respectively than that of standard HMBA (Jayaram *et al.*, 2011). Antioxidant activity of roots extract was assessed by ABTS, DPPH and FRAPS assays and their total phenolic contents measured by Folin-Ciocalteu assay. The highest hydroxyl radical scavenging activity was observed in comparison to other plants was studied (Surveswaran *et al.*, 2007, 2010). The methanolic extract of stem showed 77 per cent decrease in DPPH scavenging which was measured as decolorizing activity followed by the trapping of the unpaired electron of DPPH (Zahin *et al.*, 2009).

Antithrombitic

In antithrombotic activity, the intravenous administration of roots extract (5mg/kg BW) in rabbits delayed the plasma recalcification time and enhanced the release of lipoprotein lipase enzyme significantly. The extract also inhibited ADP-induced platelet aggregation *in vitro* (50-250 μg), which was comparable to commercial heparin (Mary *et al.*, 2003a). Antithrombolysis effect of *H. indicus* showed 20.3 per cent clot lysis with reference to Streptokinase (86.2 per cent) in an *in vitro* thrombolytic model (Prasad *et al.*, 2007).

Anti-arthritic Activity

The protective effects of hydroalcoholic (450 mg/kg BW, p.o.), ethyl acetate (75 mg/kg BW, p.o.), chloroform (60 mg/kg BW, p.o) extract and residual fractions (270 mg/kg BW, p.o.) from roots of *H. indicus* on arthritis *in vitro* model in rats were studied. There was significant decrease in physical and biochemical parameters supported by good tissue architecture in histopathological analysis. The hydroalcoholic extract and ethyl acetate fraction protective effects were comparable with methotrexates which were used as positive control (Mehta *et al.*, 2012).

Toxicity Studies

In cisplatin-induced genotoxicity on bone marrow cells treated with root aqueous extracts at doses 50, 100 and 200 mg/kg BW found protective effect in an inverse dose-dependent manner. But, under split dose doses (10, 20 and 40 mg/kg BW/day) regime it conferred a higher level of genoprotection and was not cytotoxic at the lower two doses. In cisplatin-induced genotoxicity experiments on lymphocytes when treated with root extract found reduction in frequencies of sister chromatid exchanges, chromosome aberrations and micronucleated binucleate cells at the lower concentrations of 4 and 8 μg/mL ($p<0.05$), proliferative rate index, mitotic index, cytokinesis-block proliferative index ($p<0.05$) and significant increase in the percentage of chromosome aberrations at higher concentrations (Ananthi *et al.*, 2010a, 2010b).

Plant Tissue Culture Studies

In vitro culture of this species might offer an alternative method for production of these important pharmaceuticals, which would reduce the collection pressure on this rare plant. Tissue cultures were established from shoot tip, stem, and roots (Heble and Chadha, 1978) and found that all the cultures contained phytosterols but 16-dehydropregnenolone only in stem callus. Leaves stem, and roots of the plant *H. indicus* contained cholesterol, campesterol, sitosterol and 16-dehydropregnenolone. Roots contained the maximum concentration of 16-dehydropregnenolone (0.04 per cent), stem (0.006 per cent) and leaves (traces). The tissue cultured roots, <"20 mg dry weight (DW), in the dark yielded 550 mg root DW/flask and 0.18 per cent HMBA in the roots after 30 days. The concentration of HMBA per plant was 2-3 fold higher in micropropagated plants though on unit dry root biomass (0.12 per cent per g DW) (Sreekumar *et al.*, 1998, 2000). *In vitro* biosynthesis of few antioxidants (lupeol, vanillin, and rutin) by shoot cultures and callus cultures from roots and leaves was carried out. Shoot cultures and leaf callus cultures produced more antioxidants than root callus cultures (Misra *et al.*, 2005). Further studies (Misra and Mehrotra, 2006) established callus culture and attempted to increase the yield of three medicinally important secondary metabolites lupeol, vanillin, and rutin by treatment with γ-rays, ethylmethane sulfonate, colchicines, and sodium azide (NaN_3). Vanillin content was found to increase with all the treatments whereas lupeol and rutin specifically increased when treated with 3 Krads of γ-rays and 0.05 per cent NaN_3, respectively. Biomass was generally found to decrease on treatment with the mutagens.

Saha *et al.*(2003) studied *in vitro* propagation through bud multiplication which was obtained in Murashige and Skoog's (MS) basal medium supplemented with 0.1 mg/L NAA and 2.0 mg/L BAP (7-8 shoots per explant) and the bud break time was only 4 days after inoculation, subsequent plant regeneration shootlets within 3-4 weeks and shoots within 2 weeks were observed. Finally the rooted plants were successfully established in the soil with 85 per cent survival rate. Callus cultures were obtained from axillary shoots on (MS) medium supplemented with 2, 4-D, L-1 α-naphthalene acetic acid (NAA), IAA, and Kn (Siddique and Bari, 2006). *In vitro* propagation which was rapid from axillary shoots and found highest percentage (90 per cent) of shoot regeneration in 1.0 mg NAA, regeneration best on MS medium containing. 2.5 mg/L Kn and 2.0 mg/L IBA. The highest frequency (85.00 per cent) of organogenic callus induction was observed in MS medium containing 1.0 mg/L NAA and 2.0 mg/L Kn. Development of adventitious shoots occurred when the calli were subcultured in MS medium supplemented in the BAP, NAA, IAA and Kn. Shoots differentiation was best (90.00 per cent) from axillary shoot derived callus on MS medium containing 1.0 mg/L NAA and 2.5 mg/L Kn. Regenerated shoots were rooted best in MS supplemented with IBA and Kn (Siddique and Bari, 2010).

Method was standardized for vegetative propagation of *H. indicus* by leaves and stem cuttings. Leaves on stem cuttings turned pale and shed after 3-4 days and axillary buds sprouted after 5-7 days which grew into young shoots. Rooting was not observed in cuttings treated with water that externally supplied auxins that are necessary for rooting about 2000 ppm of IAA (Ramulu *et al.*, 2005).

Protocol for *in vitro* plant regeneration from callus *via* somatic embryogenesis was developed for this plant using leaf explants. Nodular embryogenic callus was initiated from leaf explants when inoculated onto MS medium supplemented with 2, 4-D (1 mg/L). These calli when sub-cultured on hormone free N, N-basal medium developed into somatic embryos within 8 weeks in dark condition. Subsequently, embryos germinated into complete plantlets on the same medium with well developed shoot and root system. *In vitro* derived plants were hardened in the soil, sand and compostion mixture (2: 1: 1) (Schult *et al.*, 2008; Ghatge *et al.*, 2008). Use of genetic and protein markers for characterisation of medicinal ayurvedic plants (Matharu *et al.*, 2010). Micropropagation of *H. indicus* thriug axillary bud culture (Patnaik and Debata, 1996).

Ecofriendly Uses

The bark of *H. indicus*, an extensively available plant biomass was used as biomaterial for removal of lead metal (Pb) from aqueous streams (Chandra Sekhar *et al.,* 2004, 2005). The roots were used to remove the color from synthetic wastes in water (Manogaran *et al.,* 2005). Phenol has been notorious for living being, which is one of the most toxic industrial pollutants; even at low concentration it affects the water quality and is fatal at high concentrations. Activated carbon is generally used to detoxify water by adsorption of phenol. The residue of roots can be used as an alternative to detoxify the phenol in water by adsorption, but when compared to activated carbon it was found that the adsorption capacity of residue was marginal. With respect to economical and ecofriendly point of view, the roots residue justifies its use as it is easily accessible and inexpensive. A special consideration justifies for the use of roots residue for adsorption of phenol as it is a byproduct after the extraction of juices with medicinally value from the same (Srihari *et al.,* 2005).

Chemoprotective

Co-administration of gentamicin and ethanolic extract of *H. indicus* to organotypic cultures, the extract was able to significantly counteract the toxic effect of gentamicin on hair cells, at concentrations of 25 and 50 µg/mL and suggested that otoprotective action derives from specific inhibition of the apoptotic routine induced by gentamicin treatment (Previati *et al.,* 2007).

Miscellaneous

Pectic polysaccharides from roots inhibited hemagglutination at 40µg/mL of MIC (Satish *et al.,* 2007). The wound healing activity of roots methanolic extract was significant with cellular proliferation, promoting the formation of granulation tissue and improving the healing index (Ganesan *et al.,* 2012). The effect of ethanolic extract was studied on delayed type hypersensitivity, humoral responses to sheep red blood cells, skin allograft rejection and phagocytic activity of the reticuloendothelial system in mice and observed that there was suppression in both the cell-mediated and humoral components of the immune system (Atal *et al.,* 1986). Plant extract has been found to interact with cyclosporine A in a way to reduce its immunosuppressive activity at 1000 µg/mL *in vitro* culture lymphocytes by XTT assay for cell proliferation after 24 h in culture (Kainthla *et al.,* 2007). Radioprotective effect of root extract was found to protect microsomal membranes as evident from reduction in lipid peroxidation values and also protect DNA from radiation induced strand breaks (Shetty *et al.,* 2005). The hydroalcoholic extract of roots was investigated for possible sunscreen activity by using *in vitro* SPF method and extract was prepared according to methodology of Indian Pharmacopoeia by soxhlet apparatus. Sunscreen activity was studied using optometrics LLC, SPF-290S. SPF formulation was found 1.3 with ultra boot star rating 1 which approaches toward sunscreen activity. Studies found that at low concentrations, *H. indicus* extracts inhibit the interactions between nuclear factors and target DNA elements mimicking sequences recognized by the nuclear factor kappaB (NF-kappaB) (Lampronti *et al.,* 2005). The root extract can protect ischemic myocardium against contractile dysfunction and reperfusion-induced arrhythmias and reduce the extent of irreversible tissue damage following I/R in rat hearts (Khandelwal *et al.,* 2010, 2011).

Formulations

H. indicus forms an ingredient of about 46 Ayurvedic preparations either alone or in combination with other drugs (Iyer, 1983). The lists of important Ayurvedic preparations are Dasamoolarishta, Dhanwamthararishta, Balamritham, Saribadyasavam, Anuthaila, Amrithadi enna, Aswagandhadi yamaka, Gandha taila, Chandanadi taila, Triphaladi taila, Dhanwamthara taila, Neeleedaladi taila, Pinda taila, Balaswagandhadi taila, Manjishtadi taila, Madhuyashtyadi taila, Mahabala taila, Lakshadi taila, Sanni enna, Sidharthadi taila, Agragrahyadi kashaya, Jeevanthyadi kashaya, Triphalamarichadi mahakashaya, Dasamoolabaladi maha kashaya, Drakshadi kashaya, Dhanwamthara kashaya, Mahathiktha kashaya, Mridweekadi kashaya, Vidaryadi kashaya, Satavaryadi kashaya, Saribadi kashaya, Marmagudika, Manasamithra vataka, Kalyanaka ghritha, Jathyadi ghritha, Dadhika ghritha, Naladadi ghritha, Panchagavya ghritha, Pippalyadi ghritha, Brihachagaladi ghritha, Mahakalyanaka ghritha, Mahakooshmandaka ghritha, Mahathiktha ghritha, Vasthyamayanthaka ghritha, Varahyadi ghritha and Madhusnuhi rasayana.

A process is developed for manufacture of medicated talcum powder from biotechnologically derived roots extract (Sivanandan *et al.*, 2010). Melanin generation inhibitor (Hashimoto *et al.*, 2010) makes it possible to prevent or control safe development of pigmentation, such as a pigmented spot by aging or sun tanning and freckles. Formulation with roots as one along other is effective in inhibiting elastase, collagenase, tyrosinase activity, preventing and improving skin aging, pigmentation due to insolation, senile plaque, freckle, chloasma, facial mottle, etc. (Yaginuma *et al.*, 2004a b). Plant is used as lipase inhibitor for skin/hair external preparation (Takahashi *et al.*, 2006b). Topical creams for alleviating chronic skin disorders as psoriasis, eczema and lichen planus and/or promoting general health comprise mixtures of hydrolyzed ghee (butterfat) and herbal extracts where roots are used as a component (Shah, 1994, 1997). Polyherbal formulation with multiple therapeutic uses which provides healthy skin and other health benefits specially as skin nourisher, anti-acne, anti-microbial, anti-oxidant, immune-modulation, antiinflammatory, bowel cleanser, fairness for skin and protection of skin from UV radiation from sun rays (Singh *et al.*, 2008). Antipsoriatic and antidandruff uses of roots are also reported (Pushpangandan *et al.*, 2009). Polyherbal composition contains the root which is used as antiinflammatory agent (Chatterjee *et al.*, 2005, 2011). A herbal composition is reported for relieving pain (Bhawarlal, 2008, Biswazit *et al.*, 2010). The cosmetic formulation has antioxidant effect, and can be used for preventing pachylosis and improving skin luster and elasticity (Nanba *et al.*, 1995). Plant extracts are used as hair growth stimulators and compositions of hair care products (Ayuzawa *et al.*, 2010). Day-night cosmetic composition functions to impart protection to the skin from the harmful rays of the sun and as a skin lighter. The invention relates to a skin composition having skin-lightening and rough skin-improving effects, wherein the composition contains plant extract, conchiolin hydrolyzate or reaction product of conchiolin hydrolyzate with succinic acid anhydride (Narayana *et al.*, 2011a, 2011b; Shimomura *et al.*, 2001).

Caps HT2, an ayurvedic formulation is used as antiatherogenic with *H. indicus* as an ingredient (Mary *et al.*, 2003b). The extract of plant is used as testosterone 5α-

reductase inhibitor (Takahashi *et al.*, 2006). Capsule Sujat, herbal formulation with *H. indicus* roots along with other 8 plants is prepared based on the Ayurvedic text Sushruta Samhita to improve clinical conditions, such as pregnancy-induced hypertension (PIH) and intrauterine growth retardation (IUGR) of the fetus associated with low birth weights, increased perinatal morbidity and mortality. This herbal preparation showed antioxidant potential (Palep, 2003). A novel herbal chocolate was formulated for antistress which is in solid or semisolid form with nutritious composition rich in proteins, minerals fortified with standardized herbs. This chocolate is proven with pharmacological activities such as anti-oxidant, antistress and adaptogenic property and relieves the anxiety (Sreerama, 2006). Roots are used in formulations as antioxidant preservative (Nagarajan, 2006). Carctol formulation is used for the treatment of cancer with *H. indicus* plant as ingredient (Tiwari, 2008). Herbal formulations with nutritional value are used for treating metabolic disorders (Shankar *et al.*, 2009) and as a sunscreen cream (Waje *et al.*, 2010).

Rhizome forms an ingredient for treating viral infection, particularly infections with hepatitis B virus or hepatitis C virus (Shalaby *et al.*, 2002). Plant is used in herbal composition which is effective in treatment of AIDS (Ayare *et al.*, 2009) Herbal composition is used for the treatment of HIV (Samuel *et al.*, 2006; Pandey *et al.*, 2009). Herbal tablet and powder are used as blood energiser, antimicrobial (Parikh *et al.*, 2009) and diuretic herbal syrup (Waghulkar *et al.*, 2009).

Anticigarette herbal formulation is used as an antidote to tobacco (Karerat *et al.*, 2006a, 2006b). Synergistic combination of fiber or powder or granules of the aerial parts of plants are used for controlling body weight and composition (Palpu *et al.*, 2006). Herbofizz, a revitalizing soft drink that provides body coolants, thirst controllers, energy provider, appetisers, nourishers, and dietary supplements (Raghuraman, 2007). For quenching thirst locally roots are used by boiling them in drinking water and known as Dahasamani (Vijayalakshmi and Prakasam, 2007). This plant is used in panchakavya, an ayurvedic preparation, to enhance its shelf life to two years and make it marketable as packaged commodity (Sundaresan, 2007). It also forms as an ingredient in incense products (Kesari *et al.*, 2009). In Chinese medicine, its composition is used for the treatment of stress and peripheral vascular disease (Badmajew *et al.*, 1988). Panamrutham a herbal drink from the *H. indicus* plant (Liu *et al.*, 2008).

Conclusions

Hemidesmus indicus has been traditionally claimed for a large number of pharmacological actions, ayurvedic uses, documented for medicinal uses. Indiscriminate and imprudent exploitation of the natural resources by mankind is accountable for the current status of this plant. It is thus a matter of utmost concern to human life that urgent action is to be taken to prevent further diminution in availability of medicinal plants. Scientific studies conducted have verified many of the traditional uses of this plant. More recently, the cosmetic industry has shown interest for sunscreen properties and healing many skin diseases. In many of the herbal formulations plant especially roots are used as an ingredient for antiviral, anti-cancer, anti-inflammatory, body heat reducing agent, etc. Roots decoction is replaced with coffee/tea as refreshing drink and its odour is due to benzoid derivatives in it. The phytochemistry of *H. indicus* has been investigated revealing that it contains pregnane glycosides, terpenes, terpenoids, benzoids and other. Its medicinal properties can be further exploited in future by pharmaceutical industries to treat intractable diseases.

Acknowledgements

KM and KS thank UoH-DBT-CREBB for financial support and University of Hyderabad for Libarary facilities and e-search engines, journals resources for collection and compilation of literature.

References

Aiyer, K.N. (1951). Pharmacognosy of Ayurvedic Drugs of Travancore, Cochin. Central Research Institute, Trivandrum.

Alam, M.I., Auddy, B., and Gomes, A. (1994). Isolation, purification and partial characterization of viper venom inhibiting factor from the root extract of the Indian medicinal plant Sarsaparilla (*Hemidesmus indicus* R. Br.). *Toxicon: official Journal of the International Society on Toxicology*, **32(12):** 1551-1557.

Alam, M.I., Auddy, B., and Gomes, A. (1996). Viper venom neutralization by Indian medicinal plant (*Hemidesmus indicus* and *Pluchea indica*) root extracts. *Phytotherapy Research*, **10(1):** 58-61.

Alam, M.I., and Gomes, A. (1998a). Viper venom-induced inflammation and inhibition of free radical formation by pure compound (2-hydroxy-4-methoxy benzoic acid) isolated and purified from anantamul (*Hemidesmus indicus* R.Br) root extract. *Toxicon, official Journal of the International Society on Toxicology*, **36(1):** 207-215.

Alam, M.I., and Gomes, A. (1998b). Adjuvant effects and antiserum action potentiation by a (herbal) compound 2-hydroxy-4-methoxybenzoic acid isolated from the root extract of the Indian medicinal plant 'sarsaparilla' (*Hemidesmus indicus* R. Br.). *Toxicon, official Journal of the International Society on Toxicology*, **36(10):** 1423-1431.

Ali, M.A., Mikage, M., Kiuchi, F., Tsuda, Y., and Kondo, L. (1991). Screening of crude drugs used in Bangladesh for nematodical activity on the larva of *Toxocara canis*. *Shoyakugaku Zasshi*, **45:** 206-214.

Ananthi, R., Chandra, N., and Santhiya, S.T. (2010a). Protective effect of *Hemidesmus indicus* R.Br. root extract against cisplatin-induced cytogenetic damage in mouse bone marrow cells. *Genetics and molecular biology*, **33(1):** 182-185.

Ananthi, R., Chandra, N., Santhiya, S.T., and Ramesh, A. (2010b). Genotoxic and antigenotoxic effects of *Hemidesmus indicus* R. Br. root extract in cultured lymphocytes. *Journal of Ethnopharmacology*, **127(2):** 558-560.

Aneja, V., Suthar, A., Verma, S., and Kalkunte, S. (2008). Phyto-pharmacology of *Hemidesmus indicus*. *Pharmacognosy Reviews*, **2(3):** 143-150.

Anoop, A., and Jegadeesan, M. (2003). Biochemical studies on the anti-ulcerogenic potential of *Hemidesmus indicus* R.Br. *var. indicus*. *Journal of Ethnopharmacology*, **84(2-3):** 149-56.

Anoop, A., Jegadeesan, M., and Gowrishankar, R. (2003). Antimicrobial activity of *Hemidesmus indicus* var. *indicus* R.Br. Against human isolates of *Helicobacter pylori*. *Natural Product Sciences*, **9(1):** 1-3.

Anonymous. (1986). The useful plants of India. CSIR, New Delhi, India.

Anonymous. (1997). The Wealth of India. Raw materials, Vol. III, V and X, CSIR, New Delhi, India.

Anonymous. (2005). Quality Standards of Indian Medicinal Plants. Indian Council of Medical Research, New Delhi, **2:** 119-128.

Aqil, F., Ahmad, I., and Owais, M. (2006). Evaluation of anti-methicillin-resistant *Staphylococcus aureus* (MRSA) activity and synergy of some bioactive plant extracts. *Biotechnology journal*, **1(10):** 1093-1102.

Aqil, F., and Ahmad, I. (2007). Antibacterial properties of traditionally used Indian medicinal plants. *Methods and Findings in Experimental and Clinical Pharmacology*, **29(2):** 79-92.

Aqil, F., Zahin M., and Ahmad, I. (2008). Antimutagenic activity of methanolic extracts of four ayurvedic medicinal plants. *Indian journal of experimental biology*, **46(9):** 668-672.

Arjunan, N., Murugan K., Madhiyazhagan, P., Kovendan, K., Prasannakumar, K., Thangamani, S., and Barnard, D.R. (2012). Mosquitocidal and water purification properties of *Cynodon dactylon*, *Aloe vera*, *Hemidesmus indicus* and *Coleus amboinicus* leaf extracts against the mosquito vectors. *Parasitology research*, **110(4):** 1435-1443.

Arseculeratne, S.N., Gunatilaka, A.A.L., and Panabokke. (1985). Studies on medicinal plant of Srilanka. Part 14: Toxicity of some traditional medicinal herbs. *Journal of Ethnopharmacology*, **13:** 323-335.

Asha, S., Taju, G., and Jayanthi, M. (2011). Study of hepatoprotective effect of *Hemidesmus indicus* on paracetamol induced liver damage in rats. *Journal of Pharmacy Research*, **4(3):** 624-626.

Atal, C.K., Sharma, M.L., Kaul, A., and Khajuria, A. (1986). Immunomodulating agents of plant origin I: Preliminary screening. *Journal of Ethnopharmacology*, **18(2):** 133-141.

Austin, A. (2008). A review on Indian Sarsaparilla, *Hemidesmus indicus* (L.) R. Br. *Journal of Biological Sciences*, **8:** 1-12.

Austin, A., Jegadeesan, M., and Gowrishankar, R. (2003). Antimicrobial activity of *Hemidesmus indicus* var. *indicus* R.Br. against human isolates of *Helicobacter pylori*. *Natural Product Sciences*, **9(1):** 1-3.

Ayare, S. (2005). Herbal compositions for effective treatment of AIDS, and preparation thereof. *PCT Int. Appl.*, WO 2005030232 A2 20050407.

Ayuzawa, H., and Mohamado, Yamashita, R. (2009). Collagen production stimulators [machine translation]. *Jpn. Kokai Tokkyo Koho.*, JP 2009184997 A 20090820.

Ayuzawa, D., Hossain, M.N., Nagano, E., and Yamashita, R. (2010). Hair matrix cell growth promoters and hair papilla cell growth promoters containing plant extracts, hair growth stimulating compositions containing the agents, and hair care method using the compositions. *Jpn. Kokai Tokkyo Koho.*, JP 2010222273 A 20101007.

Badmajew, V. (1988). A Chinese medicinal composition for the treatment of stress and peripheral vascular disease. *Eur. Pat. Appl.*, EP 265662 A2 19880504.

Baheti, J.R., Goyal, R.K., and Shah, G.B. (2006). Hepatoprotective activity of *Hemidesmus indicus* R. br. in rats. *Indian journal of experimental biology*, **44(5):** 399-402.

Banerjee, D.K., and Pal, D.C. (1996). Plant used by the tribals of North India Plains for Hair and Scalp Preparation. In: Ethnobiology in Human Welfare, Jain, S.K. (Edition). Deep Publications, New Delhi, India.

Bhandary, M.J.M., Chandrasekhar, K.R., and Kaveriappa, K.M. (1995). Medical ethnobotany of the siddis of Uttara Kannada district, Karnataka. India. *Journal of Ethno-pharmacology*, **47:** 149-158.

Bhawarlal, J.B. (2008). A herbal composition for relieving pain from joints and bones and process thereof. *PCT Int. Appl.*, WO 2008015697 A2 20080207.

Biswazit, B. (2010). A process for preparing an Ayurvedic herbal composition for pain treatment. *Indian Pat. Appl.*, IN 2009KO01436 A 20100205.

Chakraborty, D., Sircar, D., and Mitra, A. (2008). Phenylalanine ammonia-lyase-mediated biosynthesis of 2-hydroxy-4-methoxybenzaldehyde in roots of *Hemidesmus indicus*. *Journal of Plant Physiology*, **165(10):** 1033-1040.

Chandra, R., Deepak, D., and Khare, A. (1994). Pregnane glycosides from *Hemidesmus indicus*. *Phytochemistry*, **35(6):** 1545-1548.

Chandra, S.K., Kamala, C.T., Chary, N.S., Balaram, V., and Garcia, G. (2005). Potential of *Hemidesmus indicus* for phytoextraction of lead from industrially contaminated soils. *Chemosphere*, **58(4):** 507-514.

Chandra, S.K., Kamala, C.T., Chary, N.S., Sastry, A.R.K., Nageswara, R.T., and Vairamani, M. (2004). Removal of lead from aqueous solutions using an immobilized biomaterial derived from a plant biomass. *Journal of hazardous materials*, **108(1-2):** 111-117.

Chatterjee, I., Chakravarty, A.K., and Gomes, A. (2006). *Daboia russellii* and *Naja kaouthia* venom neutralization by lupeol acetate isolated from the root extract of Indian sarsaparilla *Hemidesmus indicus* R.Br. *Journal of Ethnopharmacology*, **106(1):** 38-43.

Chatterjee, R.C., and Bhattacharya, B.K. (1955). A note on the isolation of (β-sitoserol from *Hemidesmus indicus*. *Journal of Indian Chemical Society*, **32:** 485-486.

Chatterjee, T.K. (2005). Polyherbal composition as antiinflammatory agent. *PCT Int. Appl.*, WO 2005120529 A1 20051222.

Chaterjee, T. K. (2011). Anti-inflammatory tablet (ostranil plus) compositions. *Indian Pat. Appl.*, IN 2011DE00126 A 20110204.

Chopra, R.N., Nayar, S.L., and Chopra, I.C. (1956). Glossary of Indian Medicinal Plants. CSIR, New Delhi, India.

Darekar, R.S., Khetre, A.B., Sinha, P.K., Jeswani, R.M., and Damle, M.C. (2008). Quantitation of lupeol octacosanoate in *Hemidesmus indicus* R. Br. root powder by HPTLC. *Rasayan Journal of Chemistry*, **1(3):** 526-531.

Darekar, R.S., Khetre, A.B., Singh, S.M., and Damle, M.C. (2009). HPTLC quantitation of 2-hydroxy-4-methoxybenzaldehyde *in Hemidesmus indicus* R.Br. root powder and extract. *Journal of Planar Chromatography—Modern TLC.*, **22(6):** 453-456.

Das, S., Prakash, R., and Devaraj, S.N. (2003). Antidiarrhoeal effects of methanolic root extract of *Hemidesmus indicus* (Indian sarsaparilla)—an *in vitro* and *in vivo* study. *Indian journal of experimental biology*, **41(4):** 363-366.

Das, S., and Devaraj, S.N. (2006a). Antienterobacterial activity of *Hemidesmus indicus* R. Br. root extract. *Phytotherapy research*, **20(5):** 416-421.

Das, S., and Devaraj, S.N. (2006b). Glycosides derived from *Hemidesmus indicus* R. Br. root inhibit adherence of *Salmonella typhimurium* to host cells: receptor mimicry. *Phytotherapy research*, **20(9):** 784-793.

Das, S., and Devaraj, S.N. (2007). Protective role of *Hemidesmus indicus* R. Br. root extract against *Salmonella typhimurium*-induced cytotoxicity in Int 407 cell line. *Phytotherapy research*, **21(12):** 1209-1216.

Das, S., and Devaraj, S.N. (2008). Effect of *Hemidesmus indicus* R.Br. root extract against *Salmonella enterica serovar Typhimurium*-induced apoptosis in murine macrophage cell line (P388D1). *The Indian journal of Medical Research*, **128(5):** 647-657.

Das, P. C., Joshi, P. C., Mandal, S., Das, A., Chatterjee, A., and Banerji, A. (1992). New coumarino-lignoids from *Hemidesmus indicus* R. Br. *Indian Journal of Chemistry, Section B: Organic Chemistry Including Medicinal Chemistry*, **31B (6):** 342-5.

David, O. K., and Emma, L.W. (2011). Herbal extracts and phytochemicals: Plant secondary metabolites and the enhancement of human brain function. *Advances in Nutrition*, **2:** 32–50.

De Pasquale, A. (1984). Pharmacognosy: The oldest modern science. *Journal of Ethnopharmacology*, **11:** 1–16.

Deepak, D., Srivastava, S., and Khare, A. (1995a). Indicusin–a pregnane diester triglycoside from *Hemidesmus indicus* R. Br. *Natural Product Letters*, **6(2):** 81-86.

Deepak, D., Srivastava, S., and Khare, A. (1995b). Structural studies of oligosaccharide segment of hemisine and desmisine. *Trends in Carbohydrate Chemistry developed from a Carbohydrate* Conference, 9th, Lucknow, India, 69-74.

Deepak, D., Srivastava, S., and Khare, A. (1997). Pregnane glycosides from *Hemidesmus indicus.Phytochemistry*, **44(1):** 145-151.

Dhar, M.L., Dhar, M.M., Dhawan, N.B., Mehrotra, N.B., and Ray, C. (1968). Screening of Indian plants for biological activity: Part I. *Indian Journal of Experimental Biology*, **6:** 232-247.

Dutta, A.T., Ghosh, S., and Chopra, R.N. (1938). Chemische Untersuchung der Wurzeln von *Hemidesmus Indicus*. I. Teil. *Archiv der Pharmazie und Berichte der Deutschen Pharmazeutischen Gesellschaft*, **276:** 333-340.

Dutta, M.K., Sen, T.K., and Sikdar, S. (1982). Some preliminary observations on the anti-inflammatory properties *Hemidesmus indicus* in rat. *Indian Journal of Pharmacology*, **14:** 78.

Evans, D.A., Rajasekharan, S., and Subramaniam, A. (2004). Enhancement in the absorption of water and electrolytes from rat intestine by *Hemidesmus indicus* R. Br. root (water extract). *Phytotherapy research*, **18(7):** 511-515.

Farook, S.M., Atlee, W.C., Kannan, S., Kumar, S.D., and Murnal, S. (2011). Assessment of analgesic, anti-pyretic and anti-inflammatory activity of hydro-alcoholic fraction of *Hemidesmus indicus* root in experimental animals. *Pharmacia Lettre*, **3(1):** 442-447.

Fimognari, C., Lenzi, M., Ferruzzi, L., Turrini, E., Scartezzini, P., Poli, F., Gotti, R., Guerrini A., Carulli, G., Ottaviano, V. Cantelli, F.G., and Hrelia P. (2011). Mitochondrial pathway mediates the antileukemic effects of *Hemidesmus indicus*, a promising botanical drug. *PloS one*, **6(6):** e21544.

Firn, R.D., and Jones, C.G. (2003). Natural products – a simple model to explain chemical diversity. *Natural Product Report*, **20**: 382-391.

Gadge, N.B., and Jalalpure, S.S. (2011). Natriuretic and saluretic effects of *Hemidesmus indicus* R. Br. root extracts in rats. *Indian journal of pharmacology*, **43(6):** 714-7.

Ganesan, S., Parasuraman, S., Maheswaran, S.U., and Gnanasekar, N. (2012). Wound healing activity of *Hemidesmus indicus* formulation. *Journal of pharmacology and pharmacotherapeutics*, **3(1):** 66-67.

Gayathri, M., and Kannabiran, K. (2008). Hypoglycemic activity of *Hemidesmus indicus* R. Br. on streptozotocin-induced diabetic rats. *International Journal of Diabetes in Developing Countries*, **28(1):** 6-10.

Gayathri, M., and Kannabiran, K. (2009a). Effect of 2-hydroxy 4-methoxy benzoic acid isolated from the roots of *Hemidesmus indicus* on blood glucose and tissue lipid peroxidation in streptozotocin-induced diabetic rats. *Journal of Pharmacy Research*, **2(4):** 627-629.

Gayathri, M., and Kannabiran, K. (2009b). Antimicrobial activity of *Hemidesmus indicus*, *Ficus bengalensis* and *Pterocarpus marsupium* Roxb. *Indian Journal of Pharmaceutical Sciences*, **71(5):** 578-581.

Gayathri, M., and Kannabiran, K. (2009c). Effect of 2-hydroxy-4-methoxy benzoic acid from the roots of *Hemidesmus indicus* on streptozotocin-induced diabetic rats. *Indian Journal of Pharmaceutical Sciences*, **71(5):** 581-585.

Gayathri, M., and Kannabiran, K. (2009d). Antidiabetic activity of 2-hydroxy 4-methoxy benzoic acid isolated from the roots of *Hemidesmus indicus* on streptozotocin-induced diabetic rats. *International Journal of Diabetes and Metabolism*, **17(2):** 53-57.

Gayathri, M., and Kannabiran, K. (2009e). Hypoglycemic effect of 2-hydroxy 4-methoxy benzoic acid isolated from the roots of *Hemidesmus indicus* on streptozotocin-induced diabetic rats. *Pharmacologyonline,* **1:** 144-154.

Gayathri, M., and Kannabiran, K. (2010). 2-hydroxy 4-methoxy benzoic acid isolated from roots of *Hemidesmus indicus* ameliorates liver, kidney and pancreas injury due to streptozotocin-induced diabetes in rats. *Indian journal of experimental biology,* **48(2):** 159-64.

Ghatge, S.R., Kedage, V.V., and Dixit, G.B. (2008). Somatic embryogenesis and plant regeneration of a rare and multipurpose medicinal plant *Hemidesmus indicus* (L.) Schult. *Plant Cell Biotechnology and Molecular Biology,* **9(1 and 2):** 71-74.

Giriach, R.D., Aminuddi, P.A., and Siddiqui, Khan, S.A. (1994). Traditional plant remedies among the Kondh of district Dhenkanal (Orissa). *International Journal of Pharmacognosy,* **32:** 274-283.

Gopi, S., and Setty, O.H. (2010). Beneficial effect of the administration of *Hemidesmus indicus* against bromobenzene induced oxidative stress in rat liver mitochondria. *Journal of Ethnopharmacology,* **127(1):** 200-203.

Gupta, P.N. (1981). Antileprotic action of an extract from "Anantamul' (*Hemidesmus indicus* R. Br.). *Leprosy in India,* **53(3):** 354-9.

Gupta, M.M., Verma, R.K., and Misra, L.N. (1992). Terpenoids from *Hemidesmus indicus. Phytochemistry,* **31(11):** 4036-4037.

Gupta, M., Shaw, B.P., and Mukherjee, A. (2010). A new glycosidic flavonoid from Jwarhar mahakashay (antipyretic) Ayurvedic preparation. *International journal of Ayurveda research,* **1(2):** 106-11.

Harborne, J.B. (1998) Phytochemical Methods: A Guide to Modern Techniques of Plant Analysis. **Edition III**. London: Chapman and Hall.

Hashimoto, H., Tani, K., Yamashita, R., and Ayuzawa, H. (2010). Melanin generation inhibitor [machine translation]. *Jpn. Kokai Tokkyo Koho.,* JP 2010100554 A 20100506.

Heble, M.R., and Chadha, M.S. (1978). Steroids in cultured tissues and mature plant of *Hemidesmus indicus* RBr. (Asclepiadiaceae). *Zeitschrift fuer Pflanzenphysiologie,* **89(5):** 401-6.

Iddamaldeniya, S.S., Wickramasinghe, N., Thabrew, I., Ratnatunge, N., and Thammitiyagodage, M.G. Protection against diethylnitrosoamine-induced hepatocarcinogenesis by an indigenous medicine comprised of *Nigella sativa, Hemidesmus indicus* and *Smilax glabra*: a preliminary study. *Journal of Carcinogenesis,* **2(1):** 6.

Iddamaldeniya, S.S., Thabrew, M.I., Wickramasinghe, S.M.D.N., Ratnatunge, N., and Thammitiyagodage, M.G. (2006). A long-term investigation of the anti-hepatocarcinogenic potential of an indigenous medicine comprised of *Nigella sativa, Hemidesmus indicus* and *Smilax glabra. Journal of Carcinogenesis,* **5:** 11-15.

Iyer, S.R. (1983). Ayurveda Yogasamgraham, Vaidyaratnam P.S. Varier's Arya Vaidya Sala, Kottakkal.

Jain, S.P., and Singh, S.C. (1984). Ethano-medico-botanical survey of Ambikapur Dist., M. P. 4th *International Congress of Ethnobiology*. NBRI, Lucknow, U.P., India.

Jain, S.P. (1996). Ethno-Medico-Botanical Survey of Chaibasa, Singhbhum District, Bihar. In: Ethnaobotany in South Asia, Maheswari, J.K. (Edition), Scientific Publishers, Jodhpur, India.

Jain, A., and Basal, E. (2003). Inhibition of *Propionibacterium acnes*-induced mediators of inflammation by Indian herbs. *Phytomedicine: International Journal of Phytotherapy and Phytopharmacology*, **10(1):** 34-38.

Jayaram, S., and Dharmesh, S.M. (2011). Assessment of antioxidant potentials of free and bound phenolics of *Hemidesmus indicus* (L) R.Br against oxidative damage. *Pharmacognosy research*, **3(4):** 225-231.

John, D. (1984). One hundred useful raw drugs of the Kani tribes of Trivandrum forest division, Kerala, India. *International Journal of Crude Drug Research*, **22:** 17-39.

Kainthla, R.P., Kashyap, R.S., Purohit, H.J., Deopujari, J.Y., Taori, G.M., and Daginawala, H.F. (2007). Study of interaction between cyclosporine A and herbal extracts on *in vitro* cultured human T-cells. *Journal of Cell and Tissue Research*, **7(1):** 877-880.

Kamatchi, C., Magesh, H., Sekhar, U., and Vaidyanathan, R. (2009). Identification of clonal clusters of *Klebsiella pneumoniae* isolates from Chennai by extended spectrum β-lactamase genotyping and antibiotic resistance phenotyping analysis. *American Journal of Infectious Diseases*, **5(2):** 74-82.

Karerat, A.K., Ifthikar, O.R.M., Varghese, J., Venugopal, V.A., Pushpangadan, P., Rawat, A.K.S., Rao, C.V., and Govindarajan, R. (2006). Anticigarette herbal formulation as an antidote to tobacco. *Indian Pat. Appl.*, IN 2004DE02504 A 20061117.

Karerat, A.K., Ifthikar, O., Rowther, M., Varghese, J., Venugopal, V.A., Pushpangadan, P., Rawat, A.K.S., Rao, Chandana, V., and Govindarajan, R. (2006). Anticigarette herbal formulation as an anti-dote to tobacco. *PCT Int. Appl.*, WO 2006061848 A1 20060615,

Kekuda, T.R., Prashith, M., Sunitha, C., Mukunda, S., Nayana, K.V., Aiyer, M., and Rohini, T.R. (2009). *In vitro* antimicrobial and anthelmintic activity of steam distillates of *Hemidesmus indicus* and *Swertia chirata* alone and in combination with Cow urine. *BioTechnology: An Indian Journal*, **3(2):** 81-86.

Kesari, K.N., and Kesari, M. (2009). Preparation of incense composition. *Indian Pat. Appl.* IN 2007MU01401 A 20090403.

Khandelwal, V.K.M., Balaraman, R., Ondrejcakova, M., Pancza, D., and Ravingerova, T. (2010). Effect of *Hemidesmus indicus* on ischemia-reperfusion injury in the isolated rat heart. *Pharmaceutical biology*, **48(6):** 611-614.

Khandelwal, V.K.M., Balaraman, R., Pancza, D., and Ravingerova, T. (2011). *Hemidesmus indicus* and *Hibiscus rosa-sinensis* affect ischemia reperfusion injury in isolated rat hearts. *Evidence-based complementary and alternative medicine* : eCAM 2011.

Khanna, K.K., Mudgi, V., Shukla, G., and Srivastava, P.K. (1996). Unreported Ethnomedicinal uses of Plants from Mirzapur Districy. U.P. In: Ethnaobotany in South Asia, Maheswari, J.K. (Edition), Scientific Publishers, Jodhpur, India.

Khanna, V.G., and Kannabiran, K. (2008). Antimicrobial activity of saponin fraction from the roots of *Hemidesmus indicus*. *Research Journal of Medicinal Plant*, **2(1):** 39-42.

Kirthikar, K.R., and Basu, B.D. (1980). Indian Medicinal Plants. Vol. 1-4. Bishen Singh Mahendra Pal Singh, Dehradun, India.

Kiuchi, F., Hioki, M., Nakamura, N., Miyashita, N., Tsuda, Y., and Kondo, K. (1989). Screening of crude drugs used in Srilanka for nematocidal activity on the larva of *Toxocaria canis*. *Shoyakugaku Zasshi*, **43:** 288-293.

Kotnis, M.S., Patel, P., Menon, S.N., and Sane, R.T. (2004). Renoprotective effect of *Hemidesmus indicus*, a herbal drug used in gentamicin-induced renal toxicity. *Nephrology* (Carlton, Vic.), **9(3):** 142-152.

Kotnis, M.S., and Sane, R.T. (2005). Renoprotective and lipid lowering oral compositions including *Hemidesmus indicus*. PCT Int. Appl., WO 2005048915 A2 20050602.

Kumar, A.R., Rathinam, K.M.S., and Prabhakar, G. (2007). Phytochemical screening of selected medicinal plants of asclepiadaceae family. *Asian Journal of Microbiology, Biotechnology and Environmental Sciences*, **9(1):** 177-180.

Lakshman, K., Shivaprasad, H.N., Jaiprakash, B., and Mohan, S. (2006). Anti-inflammatory and antipyretic activities of *Hemidesmus indicus* root extract. *African Journal of Traditional complementary and alternative medicine*, **3(1):** 90 – 94.

Lampronti, I., Khan, M.T.H., Bianchi, N., Ather, A., Borgatti, M., Vizziello, L., Fabbri, E., and Gambari, R.B. (2005). Medicinal plant extracts inhibiting molecular interactions between nuclear factors and target DNA sequences mimicking NF-kappaB binding sites. *Medicinal chemistry* (Shariqah, United Arab Emirates), **1(4):** 327-333.

Liu, Y., Mulabagal, V., Bowen-Forbes, C.S., and Nair, M.G. (2008). Bioactive Compounds Present in the Indian Herbal Drink Prepared by Boiling the Commercial Product "Panamrutham". *Abstracts, 42nd Western Regional Meeting of the American Chemical Society*, Las Vegas, NV, United States, September 23-27 WRM-194.

Madhu, A., Keerthi, P.H.V., Singh, J., and Shivalinge, G.K.P. (2009). To evaluate the anti-epileptic activity of aqueous root extract of *Hemidesmus indicus* in rats. *Archives of Pharmaceutical Sciences and Research*, **1(1):** 43-47.

Malhotra, S.K., and Murthy, S. (1973). Some useful and medicinal plants of Chandrapur district, Maharashtra state. *Bulletin of Botanical Survey of India,* **15:** 13-21.

Mandal, S., Das, P.C., Joshi, P.C., and Chatterjee, A. (1991). Hemidesminine–a new coumarinolignoid from *Hemidesmus indicus* R. Br. *Indian Journal of Chemistry, Section B: Organic Chemistry Including Medicinal Chemistry,* **30B (7):** 712-713.

Manogaran, S., Sulochana, N., and Chitravel, M. (2005). Colour removal of synthetic wastewater by using biomaterials. *Asian Journal of Chemistry,* **17(4):** 2860-2862.

Mary, N.K., Achuthan, C.R., Babu, B. H., and Padikkala, J. (2003a). *In vitro* antioxidant and antithrombotic activity of *Hemidesmus indicus* (L) R.Br. *Journal of Ethnopharmacology,* **87(2-3):** 187-191.

Mary, N.K., Babu, B.H., and Padikkala, J. (2003b). Antiatherogenic effect of Caps HT2, a herbal Ayurvedic medicine formulation. *Phytomedicine: international journal of phytotherapy and phytopharmacology,* **10(6-7):** 474-482.

Matharu, G.K., Thappa, D., Kamble, V., Krishnan, A., and Sane, R.T. (2010). Use of genetic and protein markers for characterization of medicinal Ayurvedic plants. *International Journal of Pharmaceutical Sciences and Research,* **1(12):** 95-110.

Mehta, A., Sethiya, N.K., Mehta C., and Shah, G.B. (2012). Anti-arthritis activity of roots of *Hemidesmus indicus* R.Br. (Anantmul) in rats. *Asian Pacific journal of tropical medicine,* **5(2):** 130-135.

Michael, W. (2004). Phytochemical Diversity of Secondary Metabolites. *Encyclopedia of Plant and Crop Science,* New York, 915.

Misra, N., Misra, P., Datta, S.K., and Mehrotra, S. (2005). *In vitro* biosynthesis of antioxidants from *Hemidesmus indicus* R. Br. Cultures. *In Vitro Cellular and Developmental Biology: Plant,* **41(3):** 285-290.

Misra, N., and Mehrotra, S. (2006). Effect of mutagens on production of secondary metabolites in callus culture of Indian sarsaparilla (*Hemidesmus indicus*). *Horticulture, Environment and Biotechnology,* **47(1):** 23-27.

Mukherjee, B. (1953). The Indian Pharmaceutical Codex, Vol. II, Indigenous drugs. CSIR, New Delhi, India.

Murali, A., Ashok., Purnima., and Madhavan, V. (2010). Antioxidant activity of roots of *Hemidesmus indicus* var. *pubescens*–an *in vitro* study. *Pharmacologyonline,* **(3):** 121-129.

Murti, P., Bhaskara R., and Seshadri, T.R. (1941). Chemical components of roots of *Decalepis hamiltonii* (Makali veru). III. Comparison with *Hemidesmus indicus* (Indian sarsaparilla). *Proceedings–Indian Academy of Sciences,* **13A:** 399-403.

Murti, P.B.R., and Seshadri, T.R. (1941). Chemical components of roots of *Decalepis hamiltonii* (Makali veru). IV. Resinols of *Decalepis hamiltonii* and *Hemidesmus indicus. Proceedings–Indian Academy of Sciences,* **14A:** 93-99.

Murugesa, M.K.S. (1988). Gunapadam. Vol I (In Tamil), Tamil Nadu Siddha Medical Board, Chennai, India.

Nadkarni, A.N. (1989). Indian Material Medica, Vol. I, Popular Book Depot, Bombay, India.

Nagaraju, N., and Rao, K.N. (1990). A survey of plant crude drugs of Rayalseema, Andhra Pradesh, India. *Journal of Ethnaopharmacology*, **29:** 137-158.

Nagarajan, S., Rao, L.J.M., and Gurudutt, K.N. (2001). Chemical composition of the volatiles of *Hemidesmus indicus* R. Br. *Flavour and Fragrance Journal,* **16(3):** 212-214.

Nagarajan, S., and Rao, L.J.M. (2003). Determination of 2-hydroxy-4-methoxybenzaldehyde in roots of *Decalepis hamiltonii* (Wight and Arn.) and *Hemidesmus indicus* R.Br. *Journal of AOAC International*, **86(3):** 564-567.

Nagarajan, S., and Rao, L.J.M. (2006). Preparation of antioxidant preservative from the roots of Indian sarsaparilla (*Hemidesmus indicus*). *Indian Pat. Appl.*, IN 2004DE00595 A 20060602.

Namba, T.M., Tsunezuka, D.M.R.R.B., Dissanayake, P.U., Satio, K., Kakiuci, N., and Hattori, M. (1985). Studies on dental caries prevention by traditional medicines (Part VII). Screening of Ayurvedic medicines for antiplaque action. *Shoyakugaku Zasshi,* **39:** 146-153.

Nanba,T., Hattori, Y., Shimomura, K., and Takamatsu, S. (1995). Cosmetic containing plant extract. *Jpn. Kokai Tokkyo Koho.*, JP 07061918 A 19950307.

Narayana, D.B.A., Kattiyar, C.K., Goel, R., Krishnan, K.H.R., Dobriyal, R.M., and Kumar, L. (2011a). Novel cosmetic compositions comprising herbal extracts. *Indian Pat. ApplIN* 2000DE00997 A 20110624.

Narayana, D.B.A., Kattiyar, C.K., Goel, R., and Krishnan, K.H.R. (2011b). Day and night cosmetic composition. *Indian Pat. Appl.* IN 2000DE00998 A 20110624.

Nayar, T.S., Beegam, A.R., Mohanan, N., and Rajkumar, G. (2006). Flowering Plants of Kerala: A Handbook of Tropical Botanic Garden and Research Institute, Thiruvanathapuram, Kerala, India.

Oberai, K., Maheshwari, P. K., and Anakshi, K. (1985). A pregnane ester diglycoside from *Hemidesmus indicus*. *Phytochemistry,* **24(10):** 2395-2397.

Padhy, S. N., Mahato, S. B., and Dutta, N. L. (1973). Triterpenoids from the roots of *Hemidesmus indicus. Phytochemistry,* **12(1):** 217-218.

Palpu, P., Rao, C.V., Govindarajan, R., Rawat, A.K.S., Srivastava, S.K., Nair, K.N., Khatoon, S., Rastogi, S., Ojha, S.K., and Deb, B. (2006). Functional herbal cigarette in controlling body weight and composition. *PCT Int. Appl.*, WO 2006111833 A1 20061026.

Palep, H.S. (2003). Sujat: An herbal antioxidant for superior fetal outcome. *Research Communications in Pharmacology and Toxicology*, **8(1 and 2):** IV/85-IV/102.

Pandey, S.K. (2009). A herbal composition for the treatment of HIV and a process of preparing the same. *Indian Pat. Appl.*, IN 2004DE01330 A 20090619.

Parikh, B.M. (2009). Herbal tablet and powder as blood energizer. *Indian Pat. Appl.*, IN 2007MU02469 A 20090703.

Pasumarthi, S., Chimata, M.K., Chetty, C.S., and Challa, S. (2011). Screening of phytochemical compounds in selected medicinal plants of Deccan Plateau and their viability effects on Caco-2 cells. *Journal of Medicinal Plants Research*, **5(32):** 6955-6962.

Patnaik, J., and Debata, B.K. (1996). Micropropagation of *Hemidesmus indicus* (L.) R. Br. through axillary bud culture. *Plant Cell Reports*, **15(6):** 427-430.

Paul, S. R. (1979). Medicinal plants of Netarhat, Bihar, India. *International Journal of Crude Drug Research*, **15:** 79-97.

Previati M., Corbacella E., Astolfi L., Catozzi M., Khan, M.T.H., Lampronti, I., Gambari, R., Capitani, S., and Martini, A. (2007). Ethanolic extract from *Hemidesmus indicus* (Linn) displays otoprotectant activities on organotypic cultures without interfering on gentamicin uptake. *Journal of Chemical Neuroanatomy*, **34(3-4):** 128-133.

Prabakan, M., Anandan, R., and Devaki, T. (2000). Protective effect of *Hemidesmus indicus* against rifampicin and isoniazid-induced hepatotoxicity in rats. *Fitoterapia*, **71(1):** 55-59.

Prakash, K., Sethi, A., Deepak, D., Khare, A., and Khare, M. P. (1991). Two pregnane glycosides from *Hemidesmus indicus. Phytochemistry* **30(1):** 297-299.

Prasad, S., Wahi, S.P., and Mishra, L.C. (1964). Pharmacognostical investigation on Indian Sarasaparilla, its substitutes and adulterants, Part I, *Hemidesmus indicus* R. Br. *Indian Journal Pharmacy*, **36:** 81.

Prasad, S., and Wahi, S.P. (1965). Pharmacognostical investigation on Indian Sarasaparilla Part II. Root and root-stock of *Hemidesmus indicus* R. Br. *Indian Journal of Pharmacy*, **37:** 35-39.

Prasad, N.P., Ranjit Singh, Narayanan, L.M., and Natarajan, C.R. (1996). Ethnobiology of the Kanikkars of South Tamil Nadu, In: Ethnaobotany in South Asia, Maheswari, J. K. (Edition), Scientific Publishers, Jodhpur, India.

Prasad, S., Kashyap, R.S., Deopujari J.Y., Purohit, H.J., Taori, G.M., and Daginawala, H.F. (2007). Effect of *Fagonia Arabica* (Dhamasa) on *in vitro* thrombolysis. *BMC complementary and alternative medicine*, **7:** 36.

Pullaiah, T., and Kumar, D.C.T. (1996). Herbal Plants in Mannanur Forest Mahaboob Nagar Dist., A.P. In: Ethnaobotany in South Asia, Maheswari, J. K. (Edition), Scientific Publishers, Jodhpur, India.

Pushpangadan, P., and Atal. (1984). Ethano-medico-botanical investigations in Kerala I. Some primitive tribals of Western Ghats and their herbal medicine. *Journal of Ethnopharmacology*, **11:** 59-77.

Pushpangandan, P., Rajasekharan, S., George, V., and Singh, O.V. (2009). A process of manufacture of antipsoriatic and antidandruff drug from plant extracts. *Indian Pat. Appl*. IN 1999MA00257 A 20090306.

Qureshi, S., Rai, M.K., and Agarwal, S.C. (1997). *In vitro* evaluation of inhibitory nature of extracts of 18-plant species of Chhindwara against 3-keratinophilic fungi. *Hindustan Antibiotic Bulletin,* **39:** 56-60.

Raghuraman, M.K. (2007). Herbofizz. *Indian Pat. Appl.*, IN 2006CH01797 A 20070511.

Ramachandran, V.S.S., and Nair, N.C. (1981). Ethnobotanica observations on Irulars of Tamil Nadu, India. *Journal of Econ. Taxon. Botany*, **2:** 183-190.

Ramulu, D.R., Murthy, K.S.R., and Pullaiah, T. (2005). Vegetative propagation of *Hemidesmus indicus* R. BR. by stem cuttings. *Indian Forester*, **131(11):** 1505-1508.

Ravishankara, M.N., Shrivastava, N., Padh, H., and Rajani, M. (2002). Evaluation of antioxidant properties of root bark of *Hemidesmus indicus* R. Br. (Anantmul). *Phytomedicine: International Journal of Phytotherapy and Phytopharmacology*, **9(2):** 153-160.

Reddy, M.B., Reddy, K.R., and Reddy, M.N. (1988). A Survey of medicinal plants of Chenchu tribes of Andhra Pradesh, India. *International Journal of Crude Drug Research,* **26:** 189-196.

Roy, S.K., Ali, M., Sharma, M.P., and Ramachandram, R. (2001). New pentacyclic triterpenes from the roots of *Hemidesmus indicus. Die Pharmazie,* **56(3):** 244-246.

Roy, S.K., Ali, M., Sharma, M.P., and Ramachandram, R. (2002). Phytochemical investigation of *Hemidesmus indicus* R. Br. Roots. *Indian Journal of Chemistry, Section B: Organic Chemistry Including Medicinal Chemistry,* **41B (11):** 2390-2394.

Sabnis, S.D., and Bedi, S.J. (1983). Ethnobotanical studies in Dadra-Nagar Haveli and Daman. *Indian Journal of Forestry*, **6:** 65-69.

Saha, S., Mukhopadhyay, M.J., and Mukhopadhyay, S. (2003). *In vitro* clonal propagation through bud culture of *Hemidesmus indicus* (L) R Br: an important medicinal herb. *Journal of Plant Biochemistry and Biotechnology,* **12(1):** 61-64.

Sahu, T.R. (1984). Less known uses of weeds as medicinal plants. *Anc. Sci. Life*, **3:** 245-249.

Samarakoon, S.R., Thabrew, I., Galhena, P.B., and Tennekoon, K.H. (2010). A comparison of the cytotoxic potential of standardized aqueous and ethanolic extracts of a polyherbal mixture comprised of *Nigella sativa* (seeds), *Hemidesmus indicus* (roots) and *Smilax glabra* (rhizome). *Pharmacognosy Research*, **2(6):** 335-342,

Samarakoon, S.R., Thabrew, I., Galhena, P.B., and Tennekoon, K.H. (2012). Modulation of apoptosis in human hepatocellular carcinoma (HepG2 cells) by a standardized herbal decoction of *Nigella sativa* seeds, *Hemidesmus indicus* roots and *Smilax glabra* rhizomes with anti- hepatocarcinogenic effects. *BMC complementary and alternative medicine*, **12(1):** 25.

Samuel, M. (2006). Side effect less and life prolonging 49 herbs herbal medicine for aids patients, by immunepathy. *Indian Pat. Appl.*, IN 2004CH00681 A 20060616.

Samy, R.P., Thwin, M.M., Gopalakrishnakone, P., and Ignacimuthu, S. (2008). Ethnobotanical survey of folk plants for the treatment of snakebites in Southern part of Tamilnadu, India. *Journal of Ethnopharmacology*, **115(2):** 302-312.

Sandeep, D., Krishnan, N., and Cherupally, K. (2010). Amelioration of cisplatin-induced nephrotoxicity by extracts of *Hemidesmus indicus* and *Acorus calamus*. *Pharmaceutical Biology*, **48(3):** 290-295.

Saravanan, N., and Nalini, N. (2007a). Antioxidant effect of *Hemidesmus indicus* on ethanol-induced hepatotoxicity in rats. *Journal of Medicinal Food*, **10(4):** 675-682.

Saravanan, N., Rajasankar, S., and Nalini, N. (2007b). Antioxidant effect of 2-hydroxy-4-methoxy benzoic acid on ethanol-induced hepatotoxicity in rats. *The Journal of Pharmacy and Pharmacology*, **59(3):** 445-453.

Saravanan, N., and Nalini, N. (2007c). Effect of 2-hydroxy 4-methoxy benzoic acid on an experimental model of hyperlipidaemia, induced by chronic ethanol treatment. *The Journal of Pharmacy and Pharmacology*, **59(11):** 1537-1542.

Saravanan, N., and Nalini, N. (2007d). Impact of *Hemidesmus indicus* R.Br. extract on ethanol-mediated oxidative damage in rat kidney. *Redox report: communications in free radical research*, **12(5):** 229-235.

Saravanan, N., and Nalini, N. (2007e). Inhibitory effect of *Hemidesmus indicus* and its active principle 2-hydroxy 4-methoxy benzoic acid on ethanol-induced liver injury. *Fundamental and Clinical Pharmacology*, **21(5):** 507-514.

Saravanan, N., and Nalini, N. (2008). *Hemidesmus indicus* protects against ethanol-induced liver toxicity. *Cellular and Molecular Biology Letters*, **13(1):** 20-37.

Sasidharan, N. (2004). Biodiversity Documentation for Kerala. Flowering Plants. *Kerala Forest Research Institute, Peechi*, Kerala, India.

Satheesh, G., Tushar, K.V., Unnikrishnan, K.P., Hashim, K.M., and Indira, B. (2006). *Hemidesmus indicus* (L.) R. Br.: A review. *Journal of Plant Sciences*, **3:** 146-156.

Sathisha, U.V., Jayaram, S., Harish, N.M.A., and Dharmesh, S,M. (2007). Inhibition of galectin-3 mediated cellular interactions by pectic polysaccharides from dietary sources. *Glycoconjugate Journal*, **24(8):** 497-507.

Satoskar, R.S., Shah, L.G., Bhatt, K., and Sheth, U.K. (1962). Preliminary study of pharmacologic properties of Anantmul (*Hemidesmus indicus*). *Indian Journal of Physiology and Pharmacology*, 668-676.

Schult., Ghatge., Swaroopa, R., Kedage., Vinayak, V., Dixit., and Ghansham, B. (2008). Somatic embryogenesis and plant regeneration of a rare and multipurpose medicinal plant *Hemidesmus indicus* (L.). *Plant Cell Biotechnology and Molecular Biology*, **9(1 and 2):** 71-74.

Selvanayahgam, Z.E., Gnanevendhan, S.G., Balakrishna, K., and Rao, R.B. (1994). Antisnake venom botanicals from ethnomedicine. *Journal Herbs Spices Medicinal Plants*, **2:** 45-100.

Sethi, A., Srivastava, S.S., and Srivastav, S. (2006). Pregnane glycoside from *Hemidesmus indicus* R. Br. *Indian Journal of Heterocyclic Chemistry*, **16(2):** 191-192.

Sethi, A., Bhatia, A., Srivastava, S., Bhatia, G., Khan, M.M., Khanna, A.K., and Saxena, J. K. (2010). Pregnane glycoside from *Hemidesmus indicus* as a potential antioxidant and anti-dyslipidemic agent. *Natural Product Research*, **24(15):** 1371-1378.

Shah, E. (1994). Creams useful for treating psoriasis, eczema and lichen planus. *Brit. UK Pat. Appl.*, GB 2274058 A 19940713.

Shah, E. (1997). Herbal compositions for the treatment of skin disorders such as psoriasis, eczema and lichen planus. *U.S.patent, US* 5693327 A 19971202.

Shah, G.L., and Gopal, G.V. (1985). Ethanomedical notes from the tribal inhabitants of the North Gujarat. *Journal of Econ. Taxon. Botany*, **6:** 193-201.

Shalaby, S.I.A., and Allah, E.M.A. (2002). Herbal compositions and treatment methods/Herbal composition useful for treating viral infection with hepatitis B virus or hepatitis C virus. *U.S. Pat. Appl. Publ.* US 20020160065 A1 20021031.

Shankar, T.S. (2009). Herb-based nutritional composition for correction of metabolic disorders. *Indian Pat. Appl.*, IN 2009KO01297 A 20091127.

Sharma, P.K., Dhyani, S.K., and Shankar. (1979). Some useful and medicinal plants of the district Dehradun and Siwalik. *Journal of Science Research Plant Medicine*, **1:** 17-43.

Sharma, P.C., Yelne, M.B., and Dennis, T.J. (2000). Database on Medicinal Plants used in Ayurveda. Central Council for Research in Ayurveda and Siddha. Edition I, Department of Indian System of Medicine, Govt. of India, New Delhi.

Shetty, T.K., Satav, J.G., and Nair, C.K.K. (2005). Radiation protection of DNA and membrane *in vitro* by extract of *Hemidesmus indicus*. *Phytotherapy Research*, **19(5):** 387-390.

Shimomura, K., and Hattori, F. (2001). Skin compositions containing conchiolin hydrolyzates and plant extracts. *Jpn. Kokai Tokkyo Koho.*, JP 2001316239 A 20011113.

Siddique, N.A., Bari, M.A., Naderuzzaman, N., Khatun, M.H., and Rahman. (2004). Collection of indigenous knowledge and identification of endangered medicinal plants by questionnaire survey in Barind Tract of Bangladesh. *Journal of Biological Science*, **4:** 72-80.

Siddique, N.A., and Bari, M. (2010). Plant regeneration from axillary shoot segments derived callus in *Hemidesmus indicus* (L.) R. Br. (Anantamul) an endangered medicinal plant in Bangladesh. *Journal of Plant Sciences*, **1(1):** 42-48.

Siddique, N.A., Teixeira, D. S., Jaime, A., Anisuzzaman, M., and Bari, M.A. (2006). Comparative *in vitro* study of plant regeneration from axillary shoot derived callus in *Aristolochia indica* Linn., and *Hemidesmus indicus* (L.) R. Br endangered medicinal plants in Bangladesh. *Journal of Plant Sciences*, **1(4):** 286-296.

Siddique, N.A., and Bari, M. (2010). A Plant regeneration from axillary shoot segments derived callus in *Hemidesmus indicus* (L.) R. Br. (Anantamul) an endangered medicinal plant in Bangladesh. *Journal of Plant Sciences*, **5(1):** 61-67.

Sigler, P., Saksena, R., Deepak, D., and Khare, A. (2000). C_{21} steroidal glycosides from *Hemidesmus indicus*. *Phytochemistry*, **54(8):** 983-987.

Singh, K.K., and Prakash, A. (1996). Observations on Ethnobotany of the Koltribe of Varanasi District, U.P., India. In: Ethnobotany in South Asia, Maheswari, J.K. (Edition), Scientific Publishers, Jodhpur, India.

Singh, R., Katiyar, C.K., Upadhyay, D.J., Saini, K.S., and Sharma, N. (2008). Polyherbal formulation with multiple therapeutic uses. *Indian Pat. Appl.*, IN 2006DE02757 A 20080905.

Singh, V.K., and Ali, Z.A. (1992). A contribution to the ethno-pharmacological study of the Udaipur forests of Rajasthan, India. *Fitoterapia,* **63:** 136-144.

Singh, K.K., and Maheshwari, J.K. (1983). Traditional phytotherapy among the tribals of Varanasi district., U.P. *Journal of Econ Taxa. Botany,* **4:** 803-829.

Sircar, D., Dey, G., and Mitra, A. (2007). A validated HPLC method for simultaneous determination of 2-hydroxy-4-methoxybenzaldehyde and 2-hydroxy-4-methoxybenzoic acid in root organs of *Hemidesmus indicus. Chromatographia,* **65(5/6):** 349-353.

Sivanandan, S., Seeni, S., and Nair, G.P.S. (2010). A process for manufacture of medicated talcum powder from biotechnologically derived root extract of *Hemidesmus indicus. Indian Pat. Appl.*, IN 2001MA00068 A 20100618.

Soumia, C., and Kokilavani, R. (2007). Antidiabetic and antihypercholesterolemic effect of *Hemidesmus indicus* Linn.R. root in alloxan-induced diabetic rats. *Anc Sci Life,* **26(4):** 4-10.

Sreekumar, S., Seeni, S., and Pushpangadan, P. (1998). Production of 2-hydroxy-4-methoxybenzaldehyde using root cultures of *Hemidesmus indicus. Biotechnology Letters,* **20(7):** 631-635.

Sreekumar, S., Seeni, S., and Pushpangadan, P. (2000). Micropropagation of *Hemidesmus indicus* for cultivation and production of 2-hydroxy 4-methoxy benzaldehyde. *Plant Cell, Tissue and Organ Culture,* **62(3):** 211-218.

Sreerama, M.D.V. (2006). Anti-stress herbal formulation. *Indian Pat. Appl.* IN 2005CH01469 A 20060317.

Sreelekha, M., Jirovetz, L., and Shafi, P.M. (2007). Comparative study of the essential oils from *Hemidesmus indicus* and *Decalepis hamiltonii. Asian Journal of Chemistry,* **19(6):** 4942-4944.

Srihari, V., Das, P., Meera, A., and Devi, G.V. (2005). Study on adsorption of phenol from aqueous media using extracted residue of *Hemidesmus indicus. Asian Journal of Microbiology, Biotechnology and Environmental Sciences,* **7(3):** 469-472.

Subramanian, S.S., and Nair, A.G.R. (1968). Flavonoids of some asclepiadaceous plants. *Phytochemistry,* **7(9):** 1703-4.

Subramanian, S.A., and Thamizhiniyan, V. (2012). Antihyperglycemic, antioxidant and antidyslipidemic properties of *Hemidesmus indicus* root extract studied in alloxan-induced experimental diabetes in rats. *International Journal of Pharmaceutical Sciences and Research,* **3(1):** 227-234.

Sudhakar, S., and Rao, R.S. (1985). Medicinal plants of upper East Godavari district, A.P., and need for establishment of Medicnal farm. *Journal of Econ. Taxon. Botany*, **7:** 399-405.

Sultana, S., Khan, N., Sharma, S., and Alam, A. (2003). Modulation of biochemical parameters by *Hemidesmus indicus* in cumene hydroperoxide-induced murine skin carcinoma: possible role in protection against free radical -induced cutaneous oxidative stress and tumour promotion. *Journal of Ethnopharmacology*, **85:** 33-41.

Sultana, S., Alam, A., Khan, N., and Sharma, S. (2003). Inhibition of cutaneous oxidative stress and two-stage skin carcinogenesis by *Hemidesmus indicus* (L.) in Swiss albino mice. *Indian journal of experimental biology*, **41(12):** 1416-1423.

Sundaresan, S. (2007). Method of enhancing the shelf life of panchakavya and method of preparing the same. *Indian Pat. Appl.*, IN 2003MA00061 A 20070511.

Suresh, M., and Rai, R.K. (1990). Cardol the antifilarial principle from *Anacardium occidentale*. *Current Science*, **59:** 477-479.

Surveswaran, S., Cai, Yi-Z., Corke, H., and Sun, M. (2007). Systematic evaluation of natural phenolic antioxidants from 133 Indian medicinal plants. *Food Chemistry*, **102(3):** 938-953.

Surveswaran, S., Cai, Y.Z., Xing, J., Corke, H., and Sun, M. (2010). Antioxidant properties and principal phenolic phytochemicals of Indian medicinal plants from Asclepiadoideae and Periplocoideae. *Natural Product Research*, **24(3):** 206-221.

Takahashi, H., Suzuki, T., and Murata, K. (2006). A testosterone 5 α-reductase inhibitor containing extracts of plants such as *Bergenia ligulata* (Wall.) Engl., *Hemidesmus indicus* (L.) R. Br., and *Adiantum capillus-veneris* L. *Jpn. Kokai Tokkyo Koho* JP 2006257060 A 20060928.

Takahashi, H., Suzuki, T., and Murata, K. (2006). A lipase inhibitor containing extracts of plants such as *Bergenia ligulata* (wall.), *Hemidesmus indicus* (L.) r. br., and *Psoralea corylifolia* L. *Jpn. Kokai Tokkyo Koho*, JP 2006257058 A 20060928.

Thabrew, M.I., Mitry, R.R., Morsy, M.A., and Hughes, R.D. (2005). Cytotoxic effects of a decoction of *Nigella sativa*, *Hemidesmus indicus* and *Smilax glabra* on human hepatoma HepG2 cells. *Life Sciences*, **77(12):** 1319-1330.

Tiwari, N.L. (2008). Carctol. *Indian Pat. Appl.*, IN 2007DE00421 A 20080912.

Verma, P.R., Joharapurkar, A.A., Chatpalliwar V.A., and Asnani A.J. (2005). Antinociceptiv e activity of alcoholic extract of *Hemidesmus indicus* R.Br. in mice. *Journal of Ethnopharmacology*, **102(2):** 298-301.

Vijayalakshmi, L.J., and Prakasam, V.R. (2007). Changes in total hardness of water when mixed with thirst quenching herbal product called Dehasamani. *Indian Journal of Environment and Ecoplanning*, **14(3):** 447-452.

Vyas, Y. (1993). Utility of Juvardhan dashomani in the treatment of fever. *Sachitra Ayurveda*, **45:** 748-755.

Waghulkar, V.M. (2009). Formulation and evaluation of diuretic herbal liquid syrup from *Hemidesmus indicus*. *Research Journal of Pharmacy and Technology*, **2(4):** 874-875.

Waje, A., Kale, S., Ammar, A., Ghoge, P., and Sonawane, A. (2010). Formulation and *in vitro* evaluation of sun protection factor of *Hemidesmus indicus* (L.) Fam. Periplocacae Hydro-alcoholic extracts sunscreen cream. *International Journal of Pharmaceutical Sciences* (Bangalore, India), **2(3):** 664-668.

Warrier, P.K., Nambiar, V.P.K., and Ganapathy, P.M. (2000). Some important medicinal plants of the Western Ghats, India-A profile. *International Development Research Centre*, 159-174.

Yaginuma, A., Honda, H., Murata, K., Suzuki, T., and Matsuda, H. (2004a). Topical skin preparation as active oxygen scavenger. *Jpn. Kokai Tokkyo Koho.*, JP 2004182712 A 20040702.

Yaginuma, A., Honda, H., Murata, K., Suzuki, T., and Matsuda, H. (2004b). Skin whitening external preparation comprising *Myrica nagi* hook.f.non thunb., *Saraca asoca* de wilde, *Hemidesmus indicus* (L.) R.BR., and *Onosma bracteatum* Wall. Used for preventing skin aging and pigmentation. *Jpn. Kokai Tokkyo Koho.*, JP 2004182711 A 20040702.

Zahin, M., Aqil, F., and Ahmad, I. (2009). *In vitro* antioxidant activity and total phenolic content of four Indian medicinal plants. *International Journal of Pharmacy and Pharmaceutical Sciences*, **1(Suppl. 1):** 88-89.

Zhao, Z., Matsunami, K., Otsuka, H., Negi, N., Kumar, A., and Negi, D.S. (2012). A condensed phenylpropanoid glucoside and pregnane saponins from the roots of *Hemidesmus indicus*. *Journal of Natural Medicines*, DOI 10.1007/s11418-012-0659-6.

Zulfiker, A.H.M., Ahmed, D., Alam, M.B., Sana, M.R., Saha, S.K., Khalil, M.I., Menon, T.M.A., and Rana, M.S.(2011). Phenolic content and *in vitro* antioxidant potential of selected medicinal plants of Bangladesh. *Journal of Pharmacy Research*, **4(7):** 1991-1998.

Medicinal Plants: Phytochemistry, Pharmacology and Therapeutics Vol. 3 (2014) *Pages* **39–77**
Editor-in-Chief: **V.K. Gupta**
Published by: **DAYA PUBLISHING HOUSE, NEW DELHI**

2

Review on Phyto-pharmacological and Ethnobotanical Aspects of *Cassia alata* L.

Taufiq Rahman[1], Rownak Jahan[2] and Mohammed Rahmatullah[2*]

ABSTRACT

Cassia alata L. (Fabaceae), commonly known as the ringworm senna, candlestick senna or candlebush, is a beautiful flowering shrub that possesses immense ethnomedicinal values almost all over the world including countries of Southeast Asia, Africa, Northern Australia and Latin America. It is globally recognized to be effective against ringworm (hence the name, 'ringworm shrub') and some other skin diseases of allergic, inflammatory and microbial origin. The plant finds its second most notable traditional use as an effective laxative. In addition to these, the uses of the plant as diuretic, abortifacient, diaphoretic, insecticide, tonic and antidote to snakebite as well as in fever, menstrual disorders, rheumatism, asthma, bronchitis, syphilis and herpes are also prevalent among various countries of the world. Many of these traditional uses of C.alata have been scientifically justified by studies using crude extracts and isolated compounds from various parts of this plant. For example, there are evidences in favor of significant antimicrobial (including both bacteria and fungi, some with antibiotic resistance), anti-parasitic, analgesic, anti-inflammatory, hypoglycemic, hepatoprotective, anti-platelet aggregator, anti-oxidant, anti-allergic, anti-asthmatic, purgative and anti-tumor activities. Various phytochemical assays have suggested that Cassia alata, like most Cassia and Senna plants, is notably enriched in bioactive anthraquinones (e.g., rhein, physicon, aloe-emodin, emodin, chrysophanol, alquinone, sennosides) and flavonoids (e.g., kaempferol, chrysoeriol, naringenin, apigenin, quercetin etc. and their glycosides such

1 Department of Pharmacology, University of Cambridge, Tennis Court Road, CB2 1PD, Cambridge, UK

2 Faculty of Life Sciences, University of Development Alternative, House No. 78, Road No. 11A (new), Dhanmondi R/A, Dhaka-1205, Bangladesh

* *Corresponding author*: E-mail: rahamatm@hotmail.com

as astragalin). There are also reports of presence of terpenoids (e.g., ursolic acid), alkaloids (cannabinoid and indole type in particular), fatty acids and sterols, saponins and tannins in different parts of the plant. In this chapter, we present a comprehensive review on hitherto known biological activities and phytochemical entities in Cassia alata along with potential therapeutic significances of the various plant constituents.

Keywords: *Cassia alata*, Phytochemicals, Pharmacological activity, Review.

Introduction

Cassia alata L. is a shrub belonging to the Fabaceae family in the Plantae kingdom. Various synonyms of this plant include *Cassia alata* L. var. *perennis* Pamp., *Cassia alata* L. var. *rumphiana* DC., *Cassia bracteata* L.f., *Cassia herpetica* Jacq., *Cassia rumphiana* (DC.) Bojer, *Herpetica alata* (L.) Raf., and *Senna alata* (L.) Roxb. In English, it is known by Candelabra bush, Candlebush, Candlestick plant, Candlestick senna, Christmas-candle, Empress-candleplant, Popcorn senna, Ringworm senna, Ringworm bush, Ringworm shrub, Seven golden candlesticks, and Wild senna. In French, the plant is variously called as Bois dartre, Casse ailée, and Dartrier, Épis d'or. In German the plant is known as Kerzenstrauch. The Hindi names of the plant include Dadmurdan, Datkapat and Senamukhi. The plant is known as Dardmordon in Bengali.

The plant can be found in much of the Tropics, including Africa, Southeast Asia, the Pacific Islands and tropical America. It is a heliophilous and hygrophilous plant and can often be found growing besides rivers and streams. It can grow up to 10 meter tall, and 3 meter across with its candelabra-shaped branching, but more typically grows as a shrub, about 1-2 meters high. The leaves are paripinnate and can be up to 75 cm long and are composed of 7-14 pairs of large, oblong leaflets, each around 8-20 cm long and 3-10 cm in width. Flowers are cup-shaped with a bright yellow color, and carried in erect terminal clusters arising from leaf axils. The flowering and fruiting period of this plant is during October-December in tropical countries.

The leaves and leaf-bearing branches have found major uses worldwide traditionally as a laxative and for the treatment of ringworm infections. Because of the scientific approval of its traditional usages, this plant has been accepted by the World Health Organization to be incorporated in primary health care programs, and also included in the Thai National List of Essential Drugs as herbal medicines according to Drug Act B.E. 2510 (Wuthi-udomlert *et al.*, 2010). The *Cassia* genus has been well known to physicians from ancient times. There is some confusion among botanists as to whether the *Cassia* or *Senna* genera are different or the same. However, at present the genus *Cassia* is thought to include some 600 species, including herbs, shrubs and trees. The more important species among this genus include *Cassia alata*, *Cassia angustifolia* Vahl., *Cassia acutifolia* Delile, *Cassia fistula* L., *Cassia nodosa* Buch.-Ham. ex Roxb., *Cassia obtusifolia* L., *Cassia sophera* L., and *Cassia tora* L. (Agarkar and Jadge, 1999).

Ethnomedicine

The Carib population of Guatemala uses the plant for medicinal purposes for the treatment of external skin infections (Giron *et al.*, 1991). The plant is also used in traditional medicine of Nigeria for treatment of skin infections caused by *Trichophyton tonsurans* and *Microsporum soudanense*, and which has been validated by scientific experiments (Eja *et al.*, 2009). Extracts of the plant has been reported to be used traditionally in Ivory Coast, West Africa to treat bacterial infections caused by *Escherichia coli* (*E. coli*), and fungal infections caused by *Candida albicans* (*C. albicans*) and dermatophytes (Crockett *et al.*, 1992). Leaves of the plant are used as anthelmintics for treatment of dogs in ethnoveterinary practices of Trinidad and Tobago (Lans *et al.*, 2000). The plant is considered of dermatologic importance in traditional medicine of Nigeria (Ajose, 2007). In Southwestern Nigeria, the plant is widely used in ethnomedicine practices to treat hypertension, sickle cell anaemia, and diabetes (Okpuzor *et al.*, 2009).

In Cameroon, the leaves and stem bark of the plant are used to treat hepatitis, skin diseases, jaundice, gastroenteritis, intestinal helminthiasis, eczema, tryphoenteritis and ring worm. The young leaves are used in rural areas to treat constipation and food poisoning (Adjanohoum *et al.*, 1996). In Bangladesh, various parts of the plant are used for treatment of boils, carbuncles, cuts and wounds, foul ulcer, dysentery, ringworm, itches, eczema, helminthiasis, and various intestinal disorders (Awal *et al.*, 2004; Ghani, 1998).

In Indian systems of medicine (Ayurveda, Unani, Siddha), the various parts of the plant are used as vermicide, astringent, purgative, expectorant, and to treat skin diseases (Mohideen *et al.*, 2005). The Ayurvedic name of the plant is dadrughna, while in Siddha, the Tamil name of the plant is seemai agathi. Skin diseases treated by the leaves of the plant include herpes, blotch, eczema, and mycosis (washerman's itch). Dried leaves are used in leprosy (Khare, 2007).

Pharmacological Findings

Hypoglycemic Activity

The oral effectiveness of leaf extract has been studied in streptozotocin (STZ)-induced hyperglycemia in rats. While the extract had no effect on glucose levels in normoglycemic animals, it reduced the blood sugar in STZ-induced hyperglycemic animals (Palanichamy *et al.*, 1988).

The EtOAc extract of leaves have been reported to possess hypoglycemic activity; at a dosage of 5 mg per 20g mouse body weight, it decreased the blood sugar level of mice by 58.3 per cent (Villasenor *et al.*, 2002).

Analgesic and Anti-inflammatory Activities

The analgesic activity of leaf extract and a leaf constituent, namely, kaempferol 3-O-sophoroside has been reported in mice and rats using the tail clip, tail flick, tail immersion and acetic acid-induced writhing methods (Palanichamy and Nagarajan, 1990a). Alcoholic extract of leaves and a component present in the extract, kaempferol 3-O-sophoroside, both showed significant anti-inflammatory activity compared with a standard anti-inflammatory drug, phenylbutazone (Palanichamy and Nagarajan, 1990b).

Hexane extract of leaves showed analgesic activity as determined in acetic acid-induced gastric pain models; both hexane and EtOAc extracts exhibited anti-inflammatory activity in carrageenan-induced inflammation in mice (Villasenor *et al.*, 2002).

Hexane extract of leaves exhibited anti-inflammatory effect in complete Freund's adjuvant (CFA) arthritis in rats through showing protective effects against cartilage degradation on the femoral head of the knee joint (Lewis and Levy, 2011).

Cassiaindoline (a dimeric indole alkaloid) has been isolated from leaves, which exhibited analgesic and anti-inflammatory activities in mouse models (Villasenor and Sanchez, 2009).

Heat-treated leaf extract showed strong inhibitory effects on Con A-induced histamine release from rat peritoneal exudates cells. The extract also inhibited 5-lipoxygenase and cyclooxygenases, COX-1 and COX-2 (Moriyama *et al.*, 2003a).

Ethyl acetae and butanol partitioned fractions of crude methanolic fraction of leaf were able to significantly reduce paw edema using carrageenan-induced rat paw edema model (Patrick-Iwuanyanwu *et al.*, 2011).

Alleviation of Gastrointestinal Disorders

Leaves of the plant have been observed to relieve constipation significantly when orally administered to constipated human patients (Thamlikitkul *et al.*, 1990). The leaves also showed laxative properties in male albino rats and has been proposed as a candidate for drug development in Nigeria (Elujoba *et al.*, 1989).

The hexane, chloroform and EtOAc extracts of leaves caused an immediate decrease in mice with diarrhea following administration (Villasenor *et al.*, 2002).

Anti-microbial Activity

Water extract of the plant has been used successfully for treatment of opportunistic infections in AIDS patients, where the extract has been found to significantly inhibit *Escherichia coli* and *Candida albicans* infections. It has been suggested that the water extract may be of therapeutic potential for treatment of such opportunistic infections (Crockett *et al.*, 1992).

The therapeutic efficacy of leaf extract against *Pityriasis versicolor* (a fungus causing flaky discolored patches to appear mainly on the chest and back skin) has been reported for human infections, and it has been suggested that the leaf extract can be reliably used as a herbal medicine (Damodaran and Venkataraman, 1994).

Ethanolic extract of leaves has been reported to be active against various species of dermatophytic fungi. The minimum inhibitory concentration (MIC) values of the extract showed that *Trichophyton mentagorphytes* var. *interdigitale*, *Trichophyton mentagrophytes* var. *mentagrophytes*, *Trichophyton rubrum* and *Microsporum gypseum* had MIC of 125 mg/ml, whereas *Microsporum canis* had a MIC of 62.5 mg/ml. Additionally, inhibition was observed on the macroconidia of *M. gypseum*, which resulted in structural degeneration beyond repair (Ibrahim and Osman, 1995). A high anthraquinone content has been reported to be effective for anti-fungal activity against dermatophytes like *T. rubrum*, *T. mentagrophytes*, and *M. gypseum* (Sakunpak *et al.*, 2009).

Chloroform extract of leaves showed activity against *T. mentagrophytes* but not against *C. albicans*. The hexane and EtOAc extracts showed mild inhibition of both organisms, with the EtOAc extract being more active against *C. albicans* (Villasenor *et al.*, 2002).

Extracts of leaf powder containing anthraquinone compounds in different forms like anthraquinone aglycon extract, anthraquinone glycoside extract, anthraquinone aglycons from glycosidic fraction, crude ethanol extract, and anthraquinone aglycon from crude ethanol extract showed best results with anthraquinone aglycons from glycosidic fraction against clinical strains of various dermatophytes, namely, *T. rubrum*, *T. mentagrophytes*, *Epidermophyton floccosum* and *M. gypseum* (Wuthi-udomlert *et al.*, 2010).

In disc diffusion assays, extract of the plant showed strong inhibitory effect against the acne-inducing bacteria, *Propionibacterium acnes* (Chomnawang *et al.*, 2005).

Use of ointments made with ethanolic extracts of leaves as topical treatments of chronic crusty or acute lesions of dermatophilosis in bovines (usually caused by *Dermatophilus congolensis*), induced healing of the disease without recurrence (Ali-Emmanuel *et al.*, 2003).

Ethanolic extract of a combination of leaves of *C. alata* and *Ocimum sanctum* have been reported to have anti-*Cryptococcus* (a pathogenic fungus genera) activity (Ranganathan and Balajee, 2000).

The methanol extracts of leaves, flowers, stem and root barks of the plant has been reported to show a broad spectrum of anti-bacterial activity, with the dichloromethane fraction of the flower extract being the most effective (Khan *et al.*, 2001). Aqueous flower extract has been reported to be inhibitory for various fungi like *Aspergillus flavus* (NCBT 101), *Aspergillus parasiticus* (NCBT 128), *Fusarium oxysporum* (NCBT 156), *Candida albicans* (NCBT 140), and *Microsporum audouinni* (NCBT 173) (Abubacker *et al.*, 2008).

Ethanolic extract of the plant showed inhibition against adherence of *Streptococcus mutans* ATCC 25175 and TPF-1 *in vitro* (note that *S. mutans* is a facultatively anaerobic, Gram-positive coccus-shaped bacterium commonly found in the human oral cavity and is a significant contributor to tooth decay) (Limsong *et al.*, 2004).

Aqueous extract of leaves demonstrated inhibitory activities against *Pseudomonas aeruginosa* (a frequent cause of nosocomial infections such as pneumonia, urinary tract infections (UTIs), and bacteremia), *Staphylococcus epidermis* (cause of nosocomial infections), *Staphylococcus aureus* (causing skin infections), and *Bacillus subtilis*. Six active compounds were isolated from the aqueous fraction, namely, kaempferol, kaempferol-O-diglucoside, kaempferol-O-glucoside, quercetin-O-glucoside, rhein, and danthron (Saito *et al.*, 2012a).

Aqueous, methanol, chloroform and petroleum extracts of leaves inhibited clinical isolates of *S. aureus*, *C. albicans*, *E. coli*, *Proteus vulgaris*, *P. aeruginosa*, and *B. subtilis*; the anti-microbial activity has been attributed to the presence of phenols, tannins, anthraquinones, saponins and flavonoids in the extracts (Idu *et al.*, 2006). Aqueous, methanol, chloroform and petroleum ether extracts of flowers exhibited anti-microbial activity against clinical isolates of *S. aureus*, *C. albicans*, *E. coli*, *P. vulgaris*, *P. aeruginosa* and *B. subtilis* (Idu *et al.*, 2007).

C. albicans has been reported to show concentration dependent susceptibility towards both ethanol and water extracts of barks, while ethanol and water extracts of leaves inhibited *S. aureus* (Somchit *et al.*, 2003).

Aloe-emodin, isolated from leaves has been shown to be active against *B. subtilis*, *P. aeruginosa*, *C. albicans*, *T. mentagrophytes* and *Aspergillus niger* (Hofilena *et al.*, 2000).

Methanol extract of leaves showed inhibitory activity against methicillin-resistant *S. aureus* with the active components identified as kaempferol, kaempferol 3-O-β-glucopyranoside, kaempferol 3-O-gentiobioside, and aloe-emodin (Hazni *et al.*, 2008).

Methanolic extract of stems and leaves showed inhibitory activity against *B. subtilis* and *Salmonella typhi* with corresponding MIC values of the leaf extract being 1.25 and 1.5 mg/ml, respectively. Ethanolic extract of both stems and leaves were effective against *S. aureus* (Alam *et al.*, 2009).

A cannabinoid dronabinol alkaloid has been reported from roots with activity against *P. aeruginosa*, *Klebsiella pneumoniae*, *E. coli*, *S. aureus*, *C. albicans*, and *A. niger* (Okwu and Nnamdi, 2011).

Mutagenicity Studies

The chloroform extract of leaves was found to be anti-mutagenic, at a dosage of 2 mg/20g mouse, with a 65.8 per cent inhibition in the mutagenicity of tetracycline (Villasenor *et al.*, 2002). Powder, leaves and a lotion from Akapulko (*C. alata*) reportedly decreased the genotoxic activity of methylnitrosoamine, Me-methanesulfonate, and tetracycline by decreasing the chromosome breaking effects of these genotoxins (Balboa and Lim-Sylianco, 1992). On the other hand, genotoxicity tests carried out with *Salmonella typhimurium* TA98, TA100, TA1535, and TA1537, and *E. coli* WP2uvrA, with or without the addition of S9 mixture (Ames test), suggested that the plant is a potent mutagen (Hong and Lyu, 2011). Aloe-emodin, isolated from leaves has been shown to be an anti-mutagen; a micronucleus test indicated a 71 per cent reduction in the number of micronucleated polychromatic erythrocytes induced by Mitomycin C (Hofilena *et al.*, 2000).

Anti-malarial Activity

'Saye', a traditional medicine used in Burkina Faso, containing extracts of the plants *Cochlospermum planchonii* (rhizome), *Cassia alata* (leaf), and *Phyllanthus amarus* (whole plant) reportedly showed significant effect against *Plasmodium falciparum* and *Plasmodium berghei* parasites grown *in vivo* (Traore *et al.*, 2008; Yerbanga *et al.*, 2012). Three quinine compounds isolated from the plant demonstrated anti-plasmodial activity with IC_{50} values below 1 microg/ml (Kayembe *et al.*, 2010).

Anti-parasitic Activity

Acaricidal activity has been reported for ethanolic extract of leaves against *Rhipicephalus* (*Boophilus*) *annulatus* (Ravindran *et al.*, 2012).

The plant can prove useful in control of helminthes in livestock; acetone extract of leaves demonstrated significant inhibitory activity against egg hatchability and larvae of *Haemonchus contortus* (also known as red stomach worm and one of the most pathogenic nematodes of ruminants) (Ademola and Eloff, 2011).

Cestocidal activity has been reported for crude alcoholic extract of the plant, as demonstrated by early paralysis and mortality of the fowl tapeworm, *Raillietina tetragona*, when exposed to the extract (Kundu and Lyndem, 2012). Ethanol extract of air-dried leaves demonstrated inhibitory activity against the cestode *Hymenolepis diminuta* (rat tape worm causing hymenolepiasis) (Kundu *et al.*, 2012).

Acetone, chloroform, peteroleum ether, methanol and hexane extracts of fruits demonstrated high lethality when tested on adults of *Callosobruchus chinensis* (bean weevil, pulse beetle) (Upadhyay *et al.*, 2011).

Cytotoxic Activity

Hexane extract of leaves showed cytotoxic activity against parental A549 lung cancer cells with IC_{50} value of 143 microg/ml (Levy and Lewis, 2011). Extract of the plant has been reported to inhibit proliferation of leukemia cells (L1210) through inhibition of polyamine synthesis (Pieme *et al.*, 2009).

Other Pharmacological Activities

The EtOAc extract of leaves, when administered, caused an immediate decrease in motor activity, enophthalmus (recession of the eyeball within the orbit), hyperemia (increase of blood flow to different tissues of the body), and micturition (urination). At a dose of 150 mg per 20g mouse weight, the extract caused paralysis, screen grip loss and enophthalmus accompanied by drooping and closure of the eyelids (Villasenor *et al.*, 2002).

Acetylcholinesterase inhibition has been observed with extract of the plant (Feitosa *et al.*, 2011), suggesting that active components may be present, which can prove useful in the treatment of myasthenia gravis, glaucoma, Lewy Body Dementia, and Postural Tachycardia Syndrome.

Anti-oxidant activity, as measured using DPPH radical scavenging assay has been observed with crude methanol extract of leaves, flowers and pods; the active ingredient has been shown to be kaempferol (Panichayupakaranant and Kaewsuwan, 2004). Ethyl acetate extract of flowers also demonstrated good anti-oxidant activity; the observed activity has been attributed to presence of flavonoids and phenols in the extract (Subramanian and Venugopal, 2011).

Kaempferol 3-O-sophoroside, isolated from the plant and used as an active ingredient in a cosmetic preparation was shown to induce increased expression of GADD45α and GADD45γ proteins in human keratinocyte with or without UV-B irradiation, and thus increasing the efficiency of the DNA repair systems of the cells (Bardey *et al.*, 2005).

Ethanolic extract of leaves demonstrated hepatoprotective activity by protecting albino rats against carbon tetrachloride-induced liver damage (Anandan *et al.*, 2008). Extract of the plant has also been shown to exhibit hepatoprotective activity in rifampicin-induced hepatotoxicity through lowering the levels of serum marker enzymes (sGOT, sGPT, sALP) and elevating the levels of GSH, SOD, CAT, GGTP and TP in a dose-dependent manner in rats (Neharkar *et al.*, 2010).

Aqueous extract of the plant has been shown to inhibit the cytochrome P450 (CYP) enzymes CYP2D6 and CYP3A4, suggesting that the aqueous extract of this plant can cause adverse drug-drug interactions and therefore should be administered with caution (Appiah-Opong *et al.*, 2008).

Leaf extract of the plant has been shown to be a promising candidate for treatment of vitiligo by increasing melanin content in melb-a mouse melanoblast cells in a dose-dependent manner; in addition, the extract induced tyrosinase activity and altered melb-a cell morphology and induced direct migration of treated cells (Babitha *et al.*, 2010).

Astragalin (kaempferol-3-O-β-glucoside), extracted from leaves, has been shown to induce DNA adducts *in vitro* and repairable DNA damage in the yeast, *Saccharomyces cereviseae* (Saito *et al.*, 2012b).

The hydro-alcoholic extract of leaves showed anti-allergic activity by significantly inhibiting triple antigen/sheep serum-induced mast-cell degranulation in rats, as well as inhibiting lipoxygenase (LOX) activity; two isolated bio-active constituents were identified as rhein and kaempferol and both constituents showed potent inhibition of mast-cell degranulation, but rhain also inhibited LOX enzyme (Singh *et al.*, 2012).

Toxicity Studies

Leaves have been reported to cause marked toxic effects on rats. Furthermore, ethanol extract of leaves and compounds isolated from the leaves like emodin, kaempferol, aloe-emodin and rhein demonstrated subtle hepatorenal toxicity in rats (Yagi *et al.*, 1998). An aqueous extract of leaves, when administered orally to rats resulted in loss of appetite, emaciation and loss of weight, as well as significant dose-dependent decreases in the levels of hemoglobin and erythrocyte count. In addition, increased packed cell volume, mean corpuscular volume, and mean corpuscular hemoglobin concentration were observed (Sodipo *et al.*, 1998).

Reported Phytochemical Constituents

Ethanol extract of whole plant contained chrysophanol, emodin, rhein, and aloe-emodin (Villaroya and Bernal-Santos, 1976). An anthraquinone, alquinone has been reported (Yadav and Kalidhar, 1994).

Kaempferol, rhein, aloe-emodin, and β-sitosterol have been reported to be present in leaves (Rao *et al.*, 1975) (Figures 2.1–2.3). Aloe-emodin in the free form as well as glycosides of rhein and aloe-emodin has been reported for leaves (Rai, 1978). Another study with leaf extracts has identified aloe-emodin, sitosterol, and stigmasterol as being present (Hofilena *et al.*, 2000). Kaempferol 3-O-sophoroside has been found in leaves of the plant (Palanichamy and Nagarajan, 1990). Leaves have also been shown to possess kaempferol 3-gentiobioside (Moriyama *et al.*, 2001; Moriyama *et al.*, 2003b). A new dimeric indole alkaloid, cassiaindoline, has been reported from leaves

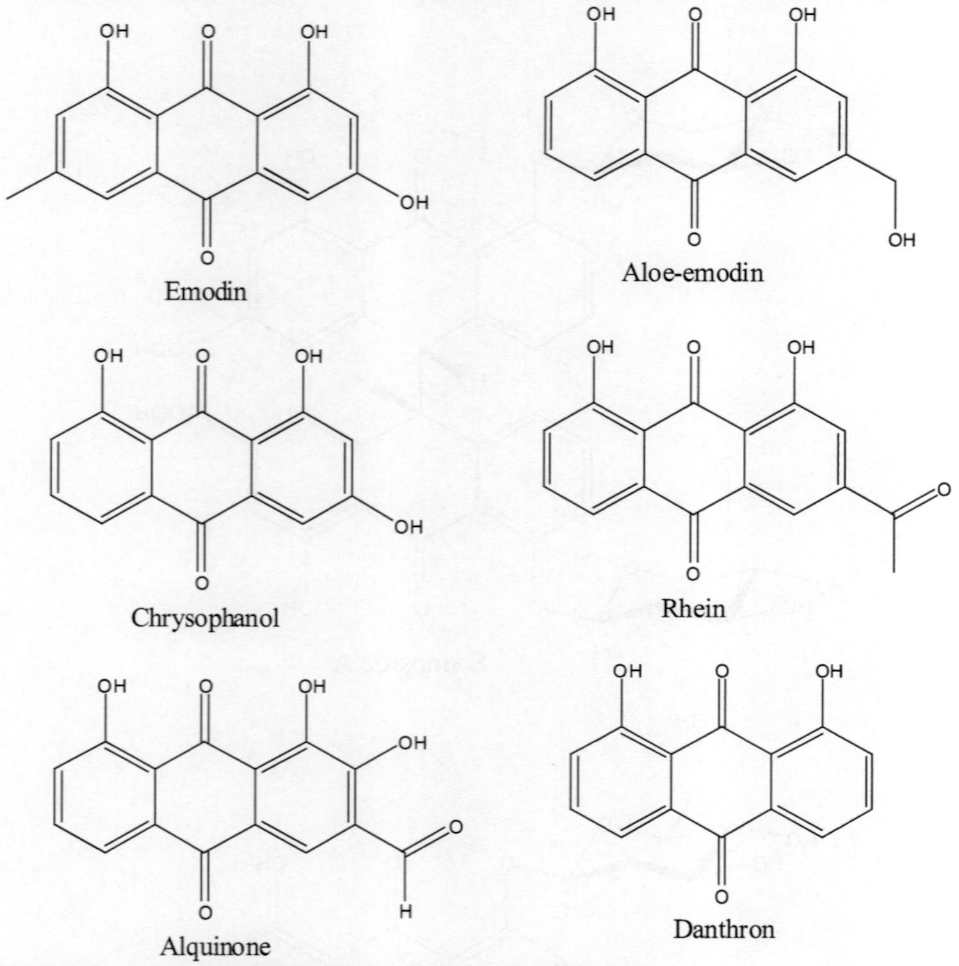

Figure 2.1: Common anthraquinones present in *Cassia alata*.

(Villasenor and Sanchez, 2009). Six compounds were isolated from the aqueous fraction of leaves, namely, kaempferol, kaempferol-O-diglucoside, kaempferol-O-glucoside, quercetin-O-glucoside, rhein, and danthron (Saito *et al.*, 2012) (Figure 2.1). Two other anthraquinones isolated from leaves included isochrysophanol and physcion-L-glucoside (Smith and Ali, 1979). The compound, 2,3,7-tri-O-methylellagic acid has been isolated from leaves (Alam *et al.*, 2003). Adenine, a platelet aggregating inhibitor, has been reported to be present in leaves (Moriyama *et al.*, 2003c). A study on fatty acid composition of leaves indicated that the major acids were stearic acid, palmitic acid, lauric acid, myristic acid, oleic acid, and linoleic acid (Rahman *et al.*, 2005a). A new flavanone, 5,7,4′-trihydroxy flavanone and an isoflavone, 6,8,4′-trihydroxy isoflavone has also been isolated from leaves (Rahman *et al.*, 2005b; Rahman

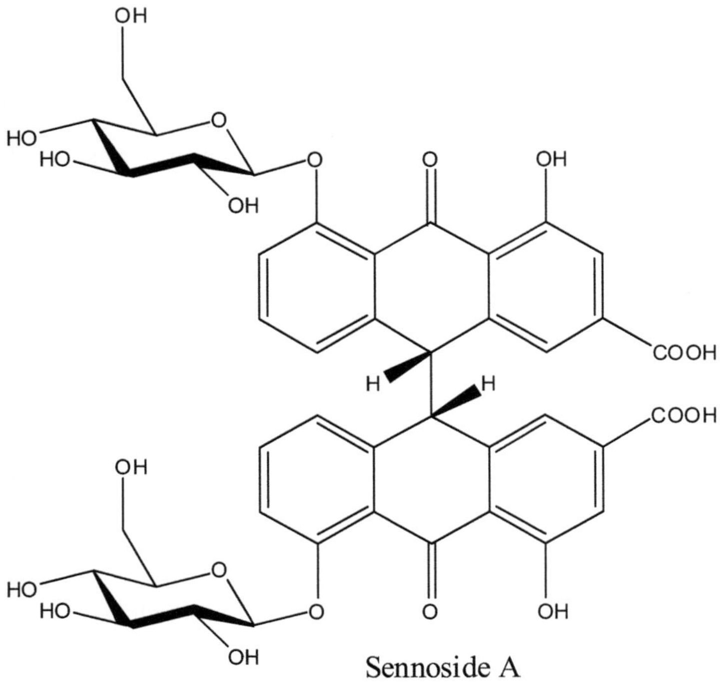

Sennoside A

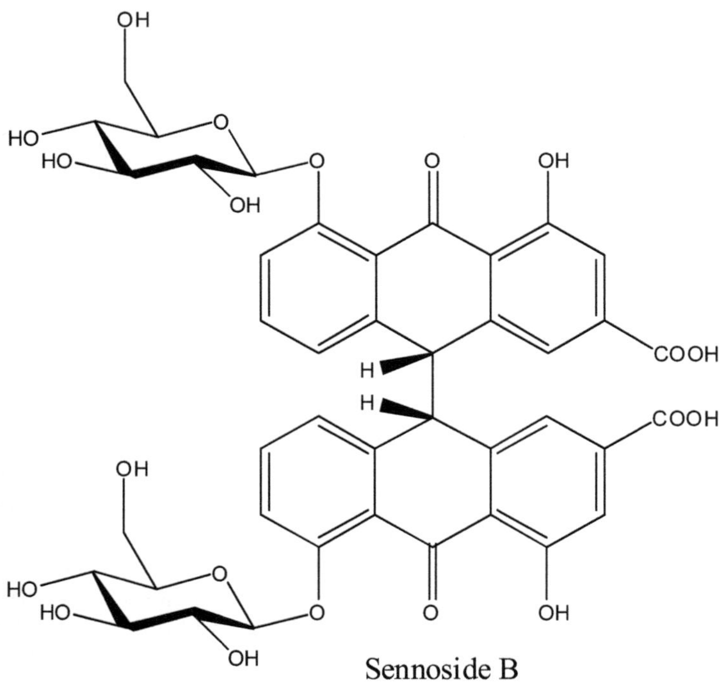

Sennoside B

Figure 2.2: Common anthraquinone glycosides from *Cassia alata*.

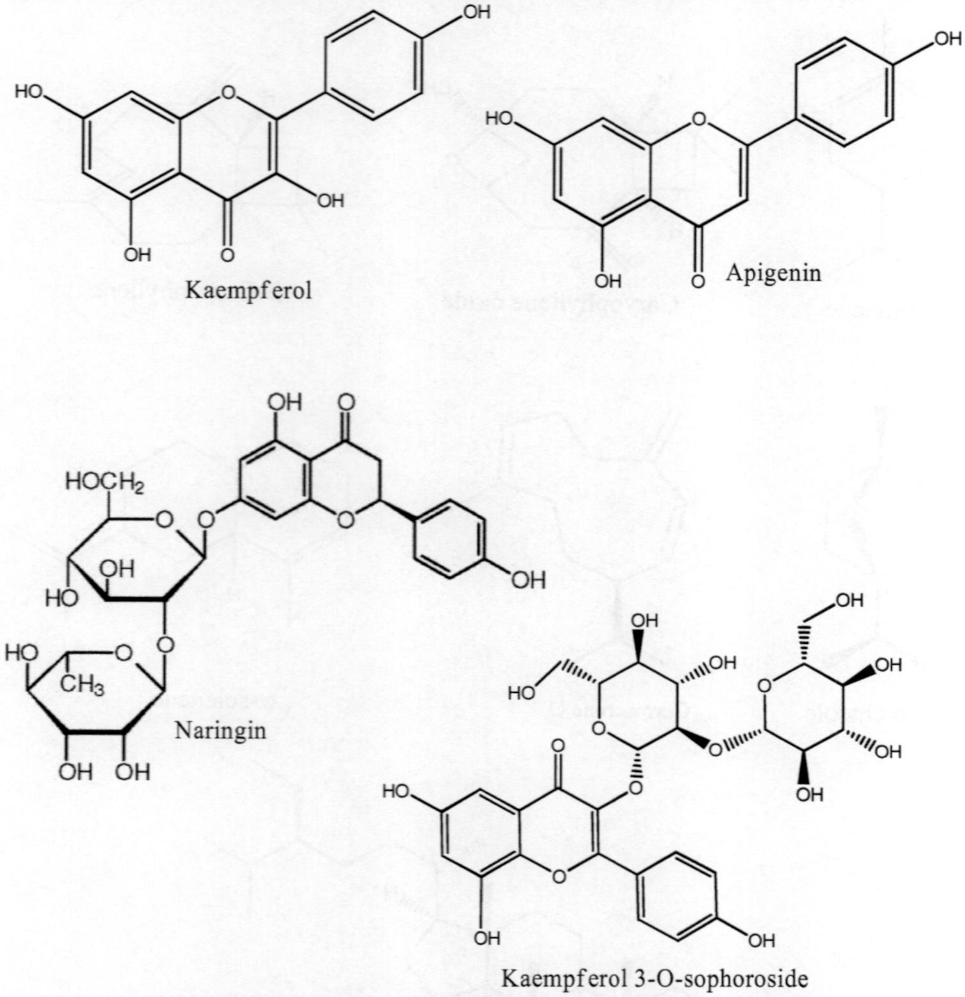

Figure 2.3: Some major flavonoid and related compounds present in *Cassia alata*.

et al., 2005c). Parahydroxy benzoic acid and a flavones – 3,5,7,4′-tetrahydroxy flavones has been shown to be present in leaves (Rahman *et al.*, 2006a,b).

The essential oil obtained by hydrodistillation of leaves has been found to contain forty four compounds, the major constituents being linalool (23.0 per cent), borneol (8.6 per cent), and pentadecanal (9.3 per cent) (Agnaniet *et al.*, 2005). Analysis of fatty acid composition of seed oil indicated the presence of oleic acid, linoleic acid, palmitic acid, stearic acid, arachidic acid, and behenic acid (Rahman *et al.*, 2006). Additionally, volatile constituents reported for oil obtained from leaves included 1,8 cineole, β-caryophyllene, caryophyllene oxide, limonene, germacrene D, and α-selinene (Ogunwande *et al.*, 2010) (Figure 2.3).

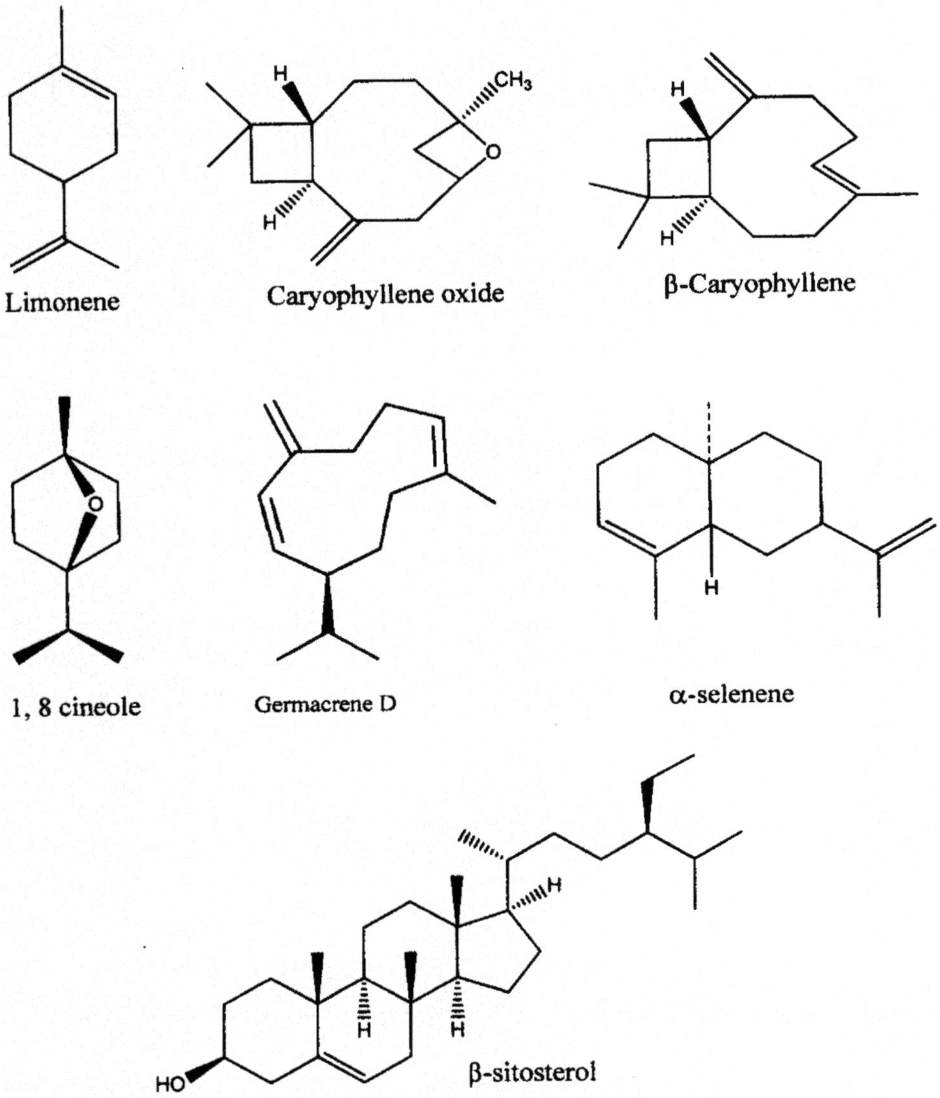

Figure 2.4: Some major triterpenoids present in *Cassia alata*.

Anthraquinone derivatives like sennosides A and B ($C_{42}H_{38}O_{20}$) have been isolated from hairy roots (Putalun *et al.*, 2006) (Figure 2.4). Two anthraquinones, I (R1 = R2 = R3 = H) and II (R1 = Me, R2 = HO, R3 = β-D-glucopyranosyl) has been isolated from roots (Tiwari and Yadava, 1971). HPLC analysis has shown the presence of the anthraquinones – rhein, aloe-emodin, emodin, chrysophanol and physcion, and a flavonoid, kaempferol in roots (Fernand *et al.*, 2008) (Figure 2.2). It has been reported that the roots contain the anthraquinones, 1,5-dihydroxy-2-methylanthraquinone, physcion and alatinone, in addition to n-octacosane, α-amyrin arachidate, tetracosanol, β-sitosterol, ursolic acid, and β-sitosterol-β-D-glucoside (Jain *et al.*, 2010) (Figure 2.3).

Fruits have been reported to contain higher concentration of anthraquinones in the form of free rhein, aloe-emodin, emodin, as well as their glycosides (Rai, 1978).

From the seeds of the plant, the flavonoid glycosides, chrysoeriol-7-O-(2"-O-β-D-mannopyranosyl)-β-D-allopyranoside and rhamnetin-3-O-(2"-O-β-D-mannopyranosyl)-β-D-allopyranoside has been isolated (Gupta and Singh, 1991). A cannabinoid dronabinol alkaloid, 4-(butylamino)-10-methyl-6-hydroxy cannabinoid dronabinol has been reported from seeds (Okwu and Nnamdi, 2011).

An anthraquinone, alatinone (1,5,7-trihydroxy-3-methylanthraquinone) has been reported in stems (Hemlata and Suraj, 1993). This compound has later been shown to be actually emodin (1,6,8-trihydroxy-3-methylanthraquinone) (Kelly *et al.*, 1994). Stems also contained an anthraquinone derivative, alatonal (2-formyl-1,3,8-trihydroxyanthraquinone) (Hemlata and Kalidhar, 1994a). An anthrone (alarone), has been reported from stems and has been characterized as 3-formyl-1,6,8,10-tetrahydroxyanthrone (Hemlata and Kalidhar, 1994b). The anthraquinone, 1,5-dihydroxy-2-methylanthraquinone has been isolated from stems along with β-sitosterol and 5-hydroxy-2-methylanthraquinone-1-O-rutinoside (Rai and Prasad, 1994). The anti-oxidant flavonoids, naringin and apigenin have been reported in aerial parts of the plant (Okpuzor *et al.*, 2009) (Figure 2.2).

Therapeutic Potential and Future Studies

It is quite evident from aforementioned sections that various parts of *Cassia alata* offer an impressive range of pharmacological activities. The latter can be attributed to the presence of several classes of bioactive constituents including notably anthraquinones, flavonoids and terpenoids. To assess therapeutic potentials of *Cassia alata* or (any bioactive plant), it is important that we remain aware of the pharmacodynamics of individual phytochemicals present in *C.alata*. Below we attempt to present comprehensive reviews on hitherto known bioactivity profiles of some major chemical constituents of *C.alata* belonging to the anthraquinone and flavonoids family since these represent the major bioactive constituents that are abundantly present in *C.alata*.

Anthraquinones

A number of phytochemical investigations of *Cassia alata* have led to the identification of several anthraquinones (both the aglycones as well as corresponding glycosides) in various parts of the plant including stems, leaves, roots, fruits as well as the whole plant (section 4). Of these, notable aglycones include emodin, aloe-emodin, rhein, chrysophanol, alatinone, physcion, 1,5-dihydroxy-2-methylanthraquinone and sennosides. Bioactivity profiles of few chosen anthraquinone aglycones (isolated from various plants including *Cassia* sp.) are given below. Emodin (1, 3, 8-trihydroxy-6-methylanthraquinone) and aloe-emodin are the most bioactive ones among the *Cassia* anthraquinones.

Anticancer Activity

Anthraquinones, notably emodin and to a lesser extent, aloe-emodin are well-known to have potent anti-cancer activities revealed through a number of studies that involved various cancer or tumour cells notably related to pancreas, lungs, prostrate, colon, leucocytes, lymphocytes, melanocytes, stomach and breast. There appear to be several ways these compounds exert their potential anti-cancer activities. For example, they have been found to inhibit proliferation, invasion, adhesion, metastasis and angiogenesis in one hand (Liu *et al.*, 2011; He *et al.*, 2008; Suboj *et al.*, 2012; Lin *et al.*, 2012; He *et al.*, 2009; Kaneshiro *et al.*, 2006; Huang *et al.*, 2004). On the other hand, emodin and aloe-emodin also seem to induce or promote cell cycle arrest and apoptosis (both the death receptor as well as mitochondrial pathways) (Chen *et al.*, 2007; Lin *et al.*, 2012; Guo *et al.*, 2012; Wei *et al.*, 2011; Chang *et al.*, 2011; Guo *et al.*, 2007a) as well as promoting cellular differentiation (Guo *et al.*, 2007; Tabolacci *et al.*, 2011; Guo *et al.*, 2007b; Tabolacci *et al.*, 2010). There are evidences all these activities of anthraquinones led to the sensitization of cancer cells to few other anticancer drugs such as gemcitabine (Guo *et al.*, 2012; Wei *et al.*, 2011), cisplatin (Kurokawa *et al.*, 2010) and paclitaxel (Zhang *et al.*, 1999). These helped in improving the efficacy of these chemotherapeutic agents as well as reducing their toxicities.

The molecular mechanisms underlying all the aforementioned activities of emodin and aloe-emodin that contribute towards their anti-cancer effects are emerging but these mechanisms seem to vary depending on types of cells or cancers. Since multiple signal transduction pathways are often implicated in various forms of carcinogenesis, it is unsurprising that emodin and aloe-emodin can intervene in critical steps in one or several cancerous pathways. Due to their planar chemical structure, several anthraquinones were postulated to be able to intercalate into DNA (Swanbeck, 1966). Indeed there are few reports available where emodin caused DNA-double strand breaks by stabilizing topoisomerase II-DNA cleavage complexes and by inhibiting ATP hydrolysis of topoisomerase II (Li *et al.*, 2010) whilst aloe-emodin indirectly triggered oxidative damage to DNA through generation of reactive oxygen species (Lee *et al.*, 2006; Singh and Trigun, 2011). The latter is intriguing as anthraquinones are also known to have potential antioxidant effects (Malterud *et al.*, 1993). Presumably anthraquinone-induced DNA-damage could account for their ability to trigger intrinsic pathway of apoptosis.

Perhaps more common anticancer mechanisms of emodin and aloe-emodin, as with most other chemopreventive phytochemicals, stem from their ability to modulate the expression level and activity of various proteins, many of which are products of oncogenes and tumour suppressor genes and thus related to growth, proliferation, cell death and survival pathways, *e.g.*, increased expression of p53 (Kuo *et al.*, 2002; Lin *et al.*, 2006), decreased expression of pro-survival proteins such as inhibitors of apoptosis (IAPs, pro-survival factors) (Guo *et al.*, 2012), survivin and beta catenin (Guo *et al.*, 2009; Liu *et al.*, 2011). Both emodin and aloe-emodin seem to be able to inhibit various kinases that are implicated in cellular growth, proliferation, differentiation, migration and angiogenesis. For example, direct inhibition of pro-survival/proliferative kinases such as Akt (also known as PKB) (Wei *et al.*, 2011), mammalian target of Rapamycin 2 (mTOR2) (Liu *et al.*, 2012), casein kinase II (Yim *et al.*, 1999) and protein kinase C (Acevedo-Duncan *et al.*, 2004). Last but not the least, anthraquinones have been reported to inhibit several components of the MAP kinase pathways. For example, emodin inhibited interleukin-6-induced activation of Janus-activated kinase 2 (JAK2) and phosphorylation of signal transducer and activator of transcription 3 (STAT3), followed by the decreased expression of myeloid cell leukemia 1 (Mcl-1) protein (Muto *et al.*, 2007). In some other studies, emodin inhibited tumor-associated angiogenesis through inhibition of extracellular signal-regulated kinase 1/2 phosphorylation (Kaneshiro *et al.*, 2006; Zhou *et al.*, 2006).

Apart from enzymes, anthraquinones have been reported to directly inhibit various various growth factor receptors (receptor tyrosine kinases) that are hyperactive in various cancers. These include epithelial growth factor receptor (EGFR), vascular endothelial growth factor receptor 2 (VEGFR-2), fibroblast growth factor receptor 1 (FGFR-1), platelet-derived growth factor receptor b (PDGFR-b), HER-2/neu and Kit (stem cell factor receptor) (Zhang *et al.*, 1999; Lee *et al.*, 2010a; Kwak *et al.*, 2006; Lu *et al.*, 2008). In all these cases, emodin is the active anthraquinone. While inhibition of these growth factor receptors contributed towards antiproliferative activity of emodin in general, there were some additional specific effects. For example, inhibition of VEGFR-2 and FGFR-1 contributed to the observed anti-angiogenic activity of emodin while its ability to inhibit of Kit (stem cell factor receptor) signalling could make it a good candidate for the the development of antipigmentation agents since it can radically block the differentiation and proliferation of pigment cells.

In addition to various kinases, emodin has been reported to directly inhibit some phosphatases that contributes to its anti cancer activity in some cells. For example, Han *et al.* (2012) reported that emodin strongly inhibited the activity of the phosphatase of regenerating liver-3 (PRL-3) with IC_{50} values of 3.5μM and blocked PRL-3-induced tumor cell migration and invasion in a dose-dependent manner (Han *et al.*, 2012). Choi *et al.* (2007) observed that anthraquinones such as emodin, physcion, chrysophanol and rhein can inhibit Cdc25B phosphatase that led to inhibition of growth of human colon cancer cells (Choi *et al.*, 2007).

Apart from directly inhibiting various enzymes and growth factors, anthraquinones also are known to modulate the expression of various proteins that are involved in different stages of cancer and are involved in apoptosis and such activity involves their ability to interact with some transcription factors, notably AP-1 and NF-κB (Huang *et al.*, 2004). For example, emodin was shown to exhibit antiproliferative and antimetastatic activity on pancreatic cancer both *in vitro* and *in vivo*, which may be related to down-regulation of NF-κB and its regulated molecules such as survivin and MMP-9 proteins (Liu *et al.*, 2011). Aloe-emodin significantly inhibited invasion, migration, and adhesion capacities of human carcinoma cells (HO-8910PM) cells, and, furthermore, reduced the protein and mRNA expression of focal adhesion kinase (FAK) (He *et al.*, 2008). Aloe-emodin inhibits colon cancer cell migration/angiogenesis by downregulating MMP-2/9, RhoB and VEGF via reduced DNA binding activity of NF-κB (Suboj *et al.*, 2012). Aloe-emodin has been found to inhibit invasion of nasopharyngeal carcinoma cell by suppressing the expression of MMP-2 via the p38 MAPK- NF-κB signaling pathway (Lin *et al.*, 2010a). Emodin has been reported to downregulate glutathione S-transferase P1-1 gene expression via activating protein 1 and nuclear factor kappaB inhibition (Duvoix *et al.*, 2004). Glutathione S-transferase P1-1 (GSTP1-1) is a phase II drug metabolism enzyme implicated in carcinogenesis and development of resistance to anti-cancer drugs. Aloe-emodin-mediated inhibition of tumor cell nitric oxide (NO) release coincided with a reduction in cytokine-induced accumulation of transcription and translation products of genes encoding inducible NOS and its transcription factor IRF-1 (Mijatovic *et al.*, 2004). In glioma cells, emodin can effectively inhibit hyalouronic acid-induced MMP secretion and invasion of glioma through inhibition of FAK, ERK1/2 and Akt/PKB activation and partial inhibition of AP-1 and NF-κB transcriptional activities. Consequently, these results provide important insights into emodin as an anti-invasive agent for the therapy of human glioma (Kim *et al.*, 2005). Emodin has been reported to have potential anti-tumor effect on pancreatic cancer via its dual role in the promotion of apoptosis and suppression of angiogenesis, probably through regulating the expression of NF-κB and NF-κB-regulated angiogenesis-associated factors (Lin *et al.*, 2012).

Last but not the least, anthraquinones also can also promote differentiation of cancer cells. Tabolacci *et al.* (2011) have shown that aloe-emodin can promote the macrophage differentiation of human U937 monoblastic leukemia cells, suggesting that this anthraquinone could be a potential candidate as a differentiation-inducing selective agent for therapeutic treatment of leukemia. Same group also reported similar activity of aloe-emodin in B16-F10 melanoma cells (Tabolacci *et al.*, 2010). Anticancer effect of aloe-emodin on cervical cancer cells involves G2/M arrest and induction of differentiation (Guo *et al.*, 2007a). Lee *et al.* (2008) identified emodin as one of compounds activating the mRNA expression of bone morphogenetic protein (BMP)-2 in the differentiation process of mouse osteoblastic MC3T3-E1 subclone 4 cells possibly though the activation of both Akt and MAP kinases by activating phosphatidylinositol 3-kinase (PI3K) (Lee *et al.*, 2008).

There are few reports where some attempts have been made to conjugate anthraquinones with other anti-cancer drugs with a view to developing more efficacious and less toxic agent. Yuan *et al.* (2012) synthesized a rhein-aloe-emodin hybrid molecule and found that such hybridization significantly improved anti-tumor activity against human hepatoma HepG2, human nasopharyngeal carcinoma CNE, human lung cancer NCI-H460, human ovarian cancer SK-OV-3, and human cervical cancer Hela cells (Yuan *et al.*, 2012). Zhalo *et al.* (2012) synthesized a series of conjugates of 5-Fluorouracil (5-FU) and emodin and observed that some of conjugates exhibited better or comparable *in vitro* antitumor activity to 5-FU and emodin and low toxicity in the normal cell (Zhao *et al.*, 2012).

Anti-inflammatory and Related Activities

There are several reports suggesting some anthraquinones, notably emodin possess prominent anti-inflammatory activities through diverse mechanisms and this may contribute to cancer chemoprevention as well. For example, in a murine model of asthma, emodin has been found to delay the progression of airway inflammation with significant reduction of ovalbumin-induced eosinophilia in lung tissue and mucus hyper-secretion by goblet cells (Chu *et al.*, 2012). Emodin has also been shown to inhibit mast cell activation and thereby the anaphylactic reaction through suppression of the receptor-proximal Syk-dependent signaling pathways and thus provided a basis for development of a novel anti-allergic drug (Lu *et al.*, 2011). Lin *et al.* (2010b) found that emodin attenuates acute rejection of liver allografts by inhibiting hepatocellular apoptosis and modulating the Th1/Th2 balance in rats. Emodin and physicon were reported to cause dose-dependent inhibition of NO production through suppressing inducible nitric oxide synthase (iNOS) and also found to possess significant anti-inflammatory activity when tested through the carrageenan-induced rat-paw oedema model (Ghosh *et al.*, 2010). Park *et al.* (2009) using murine macrophages reported that aloin and aloe-emodin suppress inflammation possibly via downregulating expression levels of inducible NO synthase (iNOS) and cyclo-oxygenase 2 (COX-2) enzymes. Li *et al.* (2009) have found that emodin could ameliorate pancreatic inflammation partly through reducing the expression of stromal derived factor-1 (SDF-1) which is an efficacious leukocyte chemoattractant. Apart from these, there are many other studies reporting potential immunosuppressive action of emodin both *in vitro* and *in vivo*, possibly through downregulation of cytokine production, suppression of lymphocyte proliferation and generation of free radical (Huang *et al.*, 1992). Ha *et al.* (2011) reported that emodin can inhibit proinflammatory cytokines and vascular endothelial growth factor (VEGF) productions, and histone deacetylase 1 (HDAC1) activity in hypoxic rheumatoid synoviocyte. Emodin was isolated from Rhubarb as the most active anti-allergic ingredient that inhibited 5-lipooxygenase (5-LOX) –mediated leucotriene production in rat basophilic leukemia (RBL -1) cells (Jin *et al.*, 2011). Aloin and aloe-emodin were reported to possibly suppress the inflammatory responses by blocking iNOS and COX-2 mRNA expression (Park *et al.*, 2009).

Activity on Glucose and Fat Metabolism

There a good number of studies revealing potential role of emodin and aloe-emodin in glucose and fat metabolism. For example, emodin has been reported found to promote multimerisation of adiponectin which is important for adiponectin's role as an insulin-sensitizing adipokine. Such activity of emodin appears to be due to its ability to activate AMP kinase (AMPK) as well as PPARγ (Chen *et al.*, 2012). Emodin (IC_{50} 4.2 μM) (and also kaempferol (IC_{50} 4.2 μM)) had significant inhibitory action on DYRK1A, a dual specificity tyrosine-phosphorylation-regulated kinase that have diverse roles in neuronal development and adult brain physiology, and increased activity has been linked to neurodegenerative diseases (Grabher *et al.*, 2012). Aloe-emodin has been reported to inhibit adipocyte differentiation and maturation during *in vitro* human mesenchymal stem cell adipogenesis, possibly by downregulating PPARγ and C/EBPα expressions (Subash-Babu *et al.*, 2012). Like aloe-emodin, emodin also has been shown to activate PPARγ and thus effectively ameliorated p38 over-activation and hypocontractility in high-glucose induced mesangial cells (Liu *et al.*, 2009). Emodin with PPARγ ligand-binding activity has been shown to promote adipocyte differentiation and increases glucose uptake in 3T3-Ll cells (Yang *et al.*, 2007). Emodin has been reported as a potent and selective inhibitor 11beta-hydroxysteroid dehydrogenase type 1 (11β-HSD1) in diet-induced obese mice, thereby highlighting its potential as a lead for developing drugs against metabolic syndrome or type 2 diabetes (Fend *et al.*, 2010). Emodin has been shown to prevent intrahepatic fat accumulation, inflammation and redox status imbalance during high-fat/high-fructose diet-induced hepatosteatosis in rats (Alisi *et al.*, 2012). Similar role of emodin was reported by another research group (Wang *et al.*, 2012). Aloe emodin glycosides have been reported to stimulate glucose transport and glycogen storage through PI3K dependent mechanism in L6 myotubes and inhibit adipocyte differentiation in 3T3L1 adipocytes (Anand S, Muthusamy *et al.*, 2010). Recent studies have identified high-glucose induced p38 MAPK (p38) over-activation in mesangial cells. Mesangial hypocontractility is the major underlying mechanism; however, no ameliorating agents are currently available. Emodin effectively ameliorated p38 over-activation and hypocontractility in high-glucose induced mesangial cells, possibly via activation of PPARgamma (Liu *et al.*, 2009).

Antimicrobial Activities

Emodin was found to be effective against herpes simplex virus *in vitro* and *in vivo* (Xiong *et al.*, 2011). Aloe-emodin has been reported to be an interferon-inducing agent with antiviral activity against Japanese encephalitis virus and enterovirus 71 (Lin *et al.*, 2008). Emodin has been reported as a potent anti-HSV (herpes simplex virus type 1) agent that inhibits the yields of HSV-1 via the suppression of a novel target, UL12. Emodin specifically inhibited the nuclease activity of HSV-1 UL12 alkaline nuclease in a biochemical assay (Hsiang *et al.*, 2008). Emodin and physicon (emodin 3-methyl ether) have been reported to have potent and specific neuraminidase inhibitory activity (Lee *et al.*, 2003). Alves *et al.* (2004) suggested that the antiviral activity of emodin may also stem from its ability to disrupt the two major phospholipids present in bacterial membranes, phosphatidylethanolamine and phosphatidyl-glycerol.

In a cell based screening assay, emodin was one of the four compounds that exhibited interferon-mimetic activity through activation of the JAK-STAT pathway and thus proved its potential to be a novel anti-viral adjuvant (Tai *et al.*, 2012).

Guided by an *in silico* approach, Ding *et al.* (2011) observed that emodin (emodin) significantly inhibited biofilm formation of *Pseudomonas aeruginosa* at 20 µM and induced proteolysis of the quorum-sensing signal receptor TraR in *Escherichia coli* at a concentration of 3-30 mM. Emodin also increased the activity of ampicillin against *P. aeruginosa*. Therefore, emodin was proposed to be suitable for development into an antivirulence and antibacterial agent. In another study, emodin showed inhibitory effects on glycolytic acid production and glucosyltransferase activity of *S. mutans* and *S. sobrinus* in addition to antibacterial activities (Ban *et al.*, 2010). Synergistic effect was observed with emodin in combination with ampicillin or oxacillin against methicillin-resistant *Staphylococcus aureus* (Lee *et al.*, 2010b).

Rhein, physcion, aloe-emodin and chrysophanol isolated from *Rheum emodi* rhizomes exhibited antifungal activity against *Candida albicans*, *Cryptococcus neoformans*, *Trichophyton mentagrophytes* and *Aspergillus fumigatus* (Agarwal *et al.*, 2000). Kim *et al.* (2004) also reported antifungal activity of emodin, physcion, and rhein. Emodin was also reported to have antiplasmodial activity (Ajaiyeoba *et al.*, 2008).

Neurological Effects

Mizuno *et al.* (2008) showed that application of emodin to neocortical cultures attenuated the phosphorylation of ErbB1 and ErbB2 and concluded that emodin can both attenuate EGF receptor signaling and ameliorate behavioral deficits associated with psychosis. Emodin-8-O-β-D-glucoside is able to provide neuroprotection against cerebral ischemia-reperfused injury and glutamate induced neuronal damage through exerting antioxidative effects and inhibiting glutamate neurotoxicity (Wang *et al.*, 2007). Similar activity was observed for aloe-emodin (Lin *et al.*, 2007). Emodin was reported to inhibit monoamine oxidase B (MAO-B) in a dose-dependent way, suggesting possible anti-depressant activity of this compound as well the plants enriched in this compound (Kong *et al.*, 2004). Similar inhibition on MAO enzymes were also observed with the sennosides (Hiraoka *et al.*, 1989). Pickhardt *et al.* (2005) found that emodin could inhibit abnormal aggregation of tau protein into paired helical filaments which is one of the hallmarks of Alzheimer's disease.

Various Other Activities

There are reports suggesting that some anthraquinones including emodin, aloe-emodin and chrysophanol possess significant hepatoprotective activity (Arosio *et al.*, 2000; Lee *et al.*, 2012; Qian *et al.*, 2011). On the contrary, emodin and rhein showed nephrotoxicity in Sprague Dawley rats partially through induction of cathepsin B (CB) protein and activation of CB protease (Wang *et al.*, 2008). Emodin has been shown to lessen hypoxia-induced embryotoxicities in cultured mouse whole embryos via upregulation of hypoxia-inducible factor 1α and intracellular superoxide dismutases (Yon *et al.*, 2011). *In silico* docking studies have shown that emodin can potentially inhibit certain cytochrome P-450 enzyme isoforms (1A1, 1A2 and 2B1) (Sridhar *et al.*, 2012) and therefore may potentially lead to drug-drug interaction. Aloe-emodin and rhein stimulated colonic chloride and fluid secretion by directly potentiating cystic fibrosis transmembrane regulator (CFTR) channels (Yang *et al.*, 2011). Emodin has been reported to have potent inhibitory activity on 5a-reductase and this could contribute to the anti-androgenic activity of various plants rich in emodin and few other anthraquinones (Cho *et al.*, 2010). Emodin had inhibitory effect of emodin on bleomycin-induced pulmonary fibrosis in mice (Chen *et al.*, 2008). Some anthraquiones including emodin and physicon were found to possess significant tyrosinase inhibitory activity. Physicon in particular had good dermal penetration profile and therefore suggested to be a useful for skin-whitening purpose through inhibition of tyrosinase (Leu *et al.*, 2008). *In vitro* experiments have revealed that emodin is a potent profibrinolytic and wound healing agent (Radha *et al.*, 2008). Emodin has been reported to have larvicidal activity against three mosquito species (Yang *et al.*, 2003).

Flavonoids

As mentioned in section 4, the major flavonoids present in the plant (particularly in the leaves) include kaempferol both as free aglycone and as various glycosides. We therefore focus on therapeutic potential of kaempferol and one of its glycoside, astragalin (kaempferol-O-glucoside).

Antioxidant Activity

Flavonoids are well known to possess anti-oxidant activity. Both kaempferol and few of its glycosides including astragalin have been reported to have significant anti-oxidant activity *in vitro* and *in vivo* (Aniya *et al.*, 2005; Verma *et al.*, 2009; Sanz *et al.*, 1994; Hibatallah *et al.*, 1999). From a mechanistic point of view, the observed anti-oxidant activity of these compounds is mainly due to their ability to decrease reactive oxygen species (ROS) such as superoxide anion, hydroxyl radical and peroxinitrite levels at submicromolar concentrations (Wang *et al.*, 2006; Heijnen *et al.*, 2001). Additionally at higher concentrations, kaempferol activate enzymes that generate ROS, such as xanthine oxidase (Nagao *et al.*, 1999; Ozyurek *et al.*, 2009).

Anti-inflammatory Activity

Like many other flavonoids, kaempferol (and few of its glycosides) have been reported to possess significant anti-inflammatory activity (Calderón-Montaño *et al.*, 2011). For example, kaempferol suppresses eosionphil infiltration and airway inflammation in airway epithelial cells and in mice with allergic asthma (Gong *et al.*, 2012). Such anti-inflammatory activity of kaempferol may stem from its ability to inhibit the activity as well as to suppress the expression (through modulation of transcription factors like NFkB, AP-1) of some pro-inflammatory enzymes or cytokines (COX-2, iNOS, IL-8 etc.) and thereby downregulating the production of inflammatory mediators (Mahat *et al.*, 2010; Kim *et al.*, 2010; Lee *et al.*, 2009; Hirose *et al.*, 2009; Park *et al.*, 2008; Hämäläinen *et al.*, 2007; Kempuraj *et al.*, 2005; Liang *et al.*, 1999).

Anticancer Activity

Several studies have revealed that kaempferol possesses significant cytotoxic/pro-apoptotic, anti-proliferative, anti-angiogenic and anti-metastatic activity that can potentially contribute to chemoprevention of various tumours (Gacche *et al.*, 2011; Calderón-Montaño *et al.*, 2011). Since oxidative stress and inflammations underlie many forms of carcinogenesis, it is rather unsurprising that kaempferol with its prominent anti-oxidant and anti-inflammatory property will have cancer chemopreventive potential. However, published reports also suggest that the anti-cancer activity of kaempferol can also be attributed to its ability to modulate the activity or downregulate the expression levels of various proteins that are components of signalling pathways associated with cell growth and proliferation (ribosomal S6 kinase 2, MAP kinases, DNA topoisomerases, cyclin dependent kinase-1, Src kinase, phosphatidylinositol-3 kinase), apoptosis (p53, caspases 3, 7 and 9, pro-survival BCl-2, pro-survival XIAP and pro-apoptotic Bad and Bax, apoptosis inducing factor -1, TRAIL receptor 2, phosphatidylinositol-3 kinase, estrogen receptors) (Gacche *et al.*, 2011; Calderón-Montaño *et al.*, 2011; Gopalakrishnan *et al.*, 2006; Wang *et al.*, 2009; Luo *et al.*, 2011; Szliszka *et al.*, 2011); angiogenesis (hypoxia inducible factor-1, vascular endothelial growth factor) (Mylonis *et al.*, 2010; Luo *et al.*, 2012), invasion and metasis (hepatocyte growth factor, matrix metalloproteases) (Phromnoi *et al.*, 2009; Labbé *et al.*, 2009). Apart from these, kaempferol has also been reported to promote cellular differentiation, which could contribute to its anti cancer activity (Guo *et al.*, 2012; Nakamura *et al.*, 2005; Miyake *et al.*, 2003). Like kaempferol, astragalin (and its analogs) has also been reported to have anti cancer properties (Burmistrova *et al.*, 2011; Lee *et al.*, 1981).

Anti-microbial Activity

Numerous papers have reported that kaempferol, its glycosides, or plants containing kaempferol have significant antibacterial, antiviral, antifungal and antiprotozoal activities. For example, kaempferol and some of its glycosides have been reported to be active against pathogenic strains of *Helicobacter pylori, Klebsiella pneumonia, S.aureus, S. epidermidis, P.aerusinosa, Enterococcus faecalis, Escherichia coli, Mycobacterium tuberculosis, Bacillus subtilis, Vibrio cholera, Clostridium perfringens* etc. (Calderón-Montaño *et al.*, 2011). Kaempferol and its glycosides can also act synergistically with antibiotics (*e.g.* rifampicin, vancomycin, methicillin, erythromycin and clindamycin) against antibiotic-resistant bacteria and thus could be used in combination with these drugs in cases of resistance (Lim *et al.*, 2007; Süzgeç *et al.*, 2005; Kataoka *et al.*, 2001). One of the possible mechanisms that allow kaempferol to work against antibiotic-resistant bacteria includes its ability to inhibit bacterial efflux pump (Falcão-Silva *et al.*, 2009).

Apart from bacteria, kaempferol has shown anti-viral activity against several viruses, including herpes simplex virus, cytomegalovirus, influenza virus, and human immunodeficiency virus (HIV) possibly through its ability to inhibit some key viral enzymes such as reverse transcriptase and neuraminidase (Calderón-Montaño *et al.*, 2011).

Other Activities

Apart from the aforementioned activities of kaempferol, there are reports of various other activities of this compound (and some of its glycosides) that includes hypoglycemic activity, hypolipidemic activity, vasorelaxant activity, antiplatelet and anti-thrombotic activity, estrogenic as well as anti-estrogenic activities, neuroprotective activity, anxiolytic activity and immunosuppressive activity (Calderón-Montaño *et al.*, 2011).

Although apart from emodin, aloe-emodin and kaempferol, there are several other bioactive compounds belonging to the anthraquinone, flavonoid and terpenoid family present in *Cassia alata*, we refrained from presenting their bioactivity profiles comprehensively here because of their presence as minor constituents. The latter still can contribute to diverse biological activities of this plant. It is evident from the above that the anthraquinones and the flavonoids are endowed with various types of biological activity as if they possess some 'privileged structures', which in one hand should pave the way for future drug development against various forms of cancers, inflammations and pathogenic microorganisms. On the other hand, such versatility of action would place considerable challenge towards achieving desired level of target specificity as well as toxicity. Future study should aim for applying modern drug discovery and development tools such as computational modeling, chemogenomics, lead optimization, chemical biology and other *de novo* drug design methodologies with a view to modifying structural features of these bioactive chemical scaffolds en route to developing more specificity and least toxicity.

References

Abubacker, M.N., Ramanathan, R., and Kumar, T.S. (2008). *In vitro* antifungal activity of *Cassia alata* Linn. flower extract. *Natural Product Radiance*, **7**: 6-9.

Acevedo-Duncan, M., Russell, C., Patel, S., and Patel, R. (2004). Aloe-emodin modulates PKC isozymes, inhibits proliferation, and induces apoptosis in U-373MG glioma cells. *Internatinonal Immunopharmacology*, **4**: 1775-1784

Ademola, I.O., and Eloff, J.N. (2011). Ovicidal and larvicidal activity of *Cassia alata* leaf acetone extract and fractions on *Haemonchus contortus*: *in vitro* studies. *Pharmaceutical Biology*, **49**: 539-544.

Adjanohoum, J.E., Aboubacar, N., Dramane, K., Ebot, M.E., Ekpere, J.A., Enow-Orock, E.G., Focho, D., Gbile, Z.O., Kamanyi, A., Kamsu, K.J., Keita, A., Mbenku, T., Mbi, C.N., Mbiele, A.L., Mbome, I.L., Mubiru, N.K., Nancy, W.L., Kongmeneck, B., Satabie, B., Sofora, A., Tamze, V., and Wirmun, C.K. (1996). Traditional medicine and pharmacopoeia. Contribution to ethnobotanical and floristic studies in Cameroon. Organisation of African Unity Scientific and Research Commission. Centre National de Production de manuels Scolaire. Porto-novo, p. 227.

Agarkar, S., and Jadge, D.R. (1999). Phytochemical and pharmacological investigations of genus *Cassia*: a review. *Asian Journal of Chemistry*, **11**: 295-299.

Agarwal, S.K., Singh, S.S., Verma, S., and Kumar, S. (2000). Antifungal activity of anthraquinone derivatives from *Rheum emodi*. *Journal of Ethnopharmacology*, **72**: 43-46.

Agnaniet, H., Bikanga, R., Bessiere, J.M., and Menut, C. (2005). Essential oil constituents of *Cassia alata* (L.) from Gabon. *Journal of Essential Oil Research*, **17**: 410-412.

Ajaiyeoba, E.O., Ashidi, J.S., Okpako, L.C., Houghton, P.J., and Wright, C.W.(2008). Antiplasmodial compounds from *Cassia siamea* stem bark extract. *Phytotherapy Research*, **22**: 254-255

Ajose, F.O.A. (2007). Some Nigerian plants of dermatologic importance. *International Journal of Dermatology*, **46** Suppl: 148-155.

Alam, A., Mamedov, V.A., Gubaidullin, A.T., Kalita, D., and Tsuboi, S. (2003). Isolation and identification of 2,3,7-tri-O-methylellagic acid from *Cassia alata* leaves. *Natural Medicines (Tokyo, Japan)*, **57**: 73.

Alam, M.T., Karim, M.M., and Khan, S.N. (2009). Antibacterial activity of different organic extracts of *Achyranthes aspera* and *Cassia alata*. *Journal of Scientific Research (Rajshahi, Bangladesh)*, **1**: 393-398.

Ali-Emmanuel, N., Moudachirou, M., Akakpo, J.A., and Quetin-Leclercq, J. (2003). Treatment of bovine dermatophilosis with *Senna alata*, *Lantana camara* and *Mitracarpus scaber* leaf extracts. *Journal of Ethnopharmacology*, **86**: 167-171.

Alisi, A., Pastore, A., Ceccarelli, S., Panera, N., Gnani, D., Bruscalupi, G., Massimi, M, Tozzi, G, Piemonte, F., and Nobili, V. (2012). Emodin prevents intrahepatic fat accumulation, inflammation and redox status imbalance during diet-induced hepatosteatosis in rats. *International Journal of Molecular Science*, **13**: 2276-2289.

Alves, D.S., Pérez-Fons, L., Estepa, A., and Micol, V. (2004). Membrane-related effects underlying the biological activity of the anthraquinones emodin and barbaloin. *Biochemical Pharmacology*, **68**: 549-561.

Anand, S., Muthusamy, V.S., Sujatha, S., Sangeetha, K.N., Bharathi Raja, R., Sudhagar, S., Poornima Devi, N., and Lakshmi, B.S. (2010). Aloe emodin glycosides stimulates glucose transport and glycogen storage through PI3K dependent mechanism in L6 myotubes and inhibits adipocyte differentiation in 3T3L1 adipocytes. *FEBS Letters*, **584**: 3170-3188.

Anandan, R., Jayakar, B., Karar, B.S., Manavalan, R., and Kumar, R.S. (2008). Protective effect of leaves of *Cassia alata* Linn. in CCl_4 induced hepatotoxicity in rats. *Biosciences, Biotechnology Research Asia*, **5**: 487-490.

Aniya, Y., Koyama, T., Miyagi, C., Miyahira, M., Inomata, C., Kinoshita, S., and Ichiba, T. (2005). Free radical scavenging and hepatoprotective actions of the medicinal herb, *Crassocephalum crepidioides* from the Okinawa Islands. *Biological Pharmaceutical Bulletin*, **28**: 19-23.

Appiah-Opong, R., Commandeur, J.N.M., Axson, C., and Vermeulen, N.P.E. (2008). Interactions between cytochromes P450, glutathione S-transferases and Ghanaian medicinal plants. *Food and Chemical Toxicology*, **46**: 3598-3603.

Arosio, B., Gagliano, N., Fusaro, L.M., Parmeggiani, L., Tagliabue, J., Galetti, P., De Castri, D., Moscheni, C., and Annoni, G. (2000). Aloe-Emodin quinone pretreatment reduces acute liver injury induced by carbon tetrachloride. *Pharmacology and Toxicology*, **87**: 229-233

Awal, M.A., Nahar, A., Hossain, S., Bari, M.A., Rahman, M., and Haque, M.E. (2004). Brine shrimp toxicity of leaf and seed extracts of *Cassia alata* Linn., and their antibacterial potency. *Journal of Medical Sciences*, **4**: 188-193.

Babitha, S., Nguyen, D.H., Park, S.-J., Shin, J.-H., Reyes, G.A., Caburian, A., and Kim, E.-K. (2010) Potential of *Cassia alata* leaf extract in inducing differentiation and migration of mouse melanoblasts. *Biotechnology and Bioprocess Engineering*, **15**: 1071-1076.

Balboa, J.G., and Lim-Sylianco, C.Y. (1992). Antigenotoxic effects of drug preparations Akapulko and Ampalaya. *Philippine Journal of Science*, **121**: 399-411.

Ban, S.H., Kwon, Y.R., Pandit, S., Lee, Y.S., Yi, H.K., and Jeon, J.G. (2010). Effects of a bio-assay guided fraction from *Polygonum cuspidatum* root on the viability, acid production and glucosyltranferase of mutans streptococci. *Fitoterapia*, **81**: 30-34.

Bardey, V., Danoux, L., Pauly, G., Jeanmaire, C., and Henry, F. (2005). Cosmetic use of an active ingredient of *Cassia alata* that increases GADD45 gene expression. *European Patent Application*, EP 1607098 A1 20051221.

Burmistrova, O., Quintana, J., Díaz, J.G., and Estévez, F. (2011). Astragalin heptaacetate-induced cell death in human leukemia cells is dependent on caspases and activates the MAPK pathway. *Cancer Letters*, **309**: 71-77.

Calderón-Montaño, J.M., Burgos-Morón, E., Pérez-Guerrero, C., and López-Lázaro, M. A. (2011). A review on the dietary flavonoid kaempferol. *Mini Reviews in Medicinal Chemistry*, **11**: 298-344

Chang, Y.C., Lai, T.Y., Yu, C.S., Chen, H.Y., Yang, J.S., Chueh, F.S., Lu, C.C., Chiang, J.H., Huang, W.W., Ma, C.Y., and Chung, J.G. (2011) Emodin induces apoptotic death in murine myelomonocytic leukemia WEHI-3 cells *in vitro* and enhances phagocytosis in leukemia mice *in vivo*. *Evidence-Based Complementary and Alternative Medicine*, **2011**: 523596.

Chen, S.H., Lin, K.Y., Chang, C.C., Fang, C.L., and Lin, C.P. (2007). Aloe-emodin-induced apoptosis in human gastric carcinoma cells. *Food Chemistry and Toxicology*, **45**: 2296-2303.

Chen, X.H., Sun, R.S., Hu, J.M., Mo, Z.Y., Yang, Z.F., Jin, G.Y., Guan, W.D., and Zhong, N.S. (2009). Inhibitory effect of emodin on bleomycin-induced pulmonary fibrosis in mice. Clinical and Experimental Pharmacology and Physiology, **36**: 146-153.

Chen, Z., Zhang, L., Yi, J., Yang, Z., Zhang, Z., Li, Z. (2012). Promotion of adiponectin multimerization by emodin, a novel AMPK activator with PPARγ-agonist activity. *Journal of Cellular Biochemistry*, (in press).

Cho, C.H., Bae, J.S., and Kim, Y.U. (2010) 5α-Reductase inhibitory components as antiandrogens from herbal medicine. *Journal of Acupuncture and Meridian Studies*, 3: 116-8.

Choi, S.G., Kim, J., Sung, N.D., Son, K.H., Cheon, H.G., Kim, K.R., and Kwon, B.M. Anthraquinones, Cdc25B phosphatase inhibitors, isolated from the roots of *Polygonum multiflorum* Thunb. *Natural Products Research*, **21**: 487-93

Chomnawang, M.T., Surassmo, S., Nukoolkarn, V.S., and Gritsanapan, W. (2005). Antimicrobial effects of Thai medicinal plants against acne-inducing bacteria. *Journal of Ethnopharmacology*, **101**: 330-333.

Chu, X., Wei, M., Yang, X., Cao, Q., Xie, X., Guan, M., Wang, D., and Deng, X.(2012). Effects of an anthraquinone derivative from Rheum officinale Baill, emodin, on airway responses in a murine model of asthma. *Food Chemistry and Toxicology*, **50**: 2368-2375.

Crockett, C.O., Guede-Guina, F., Pugh, D., Vangah-Manda, M., Robinson, T.J., Olubadewo, J.O., and Ochillo, R.F. (1992). *Cassia alata* and the preclinical search for therapeutic agents for the treatment of opportunistic infections in AIDS patients. *Cellular and Molecular Biology*, **38**: 505-511.

Damodaran, S., and Venkataraman, S. (1994). A study on the therapeutic efficacy of *Cassia alata* Linn. leaf extract against *Pityriasis versicolor*. *Journal of Ethnopharmacology*, **42**: 19-23.

Ding, X., Yin, B., Qian, L., Zeng, Z., Yang, Z., Li, H., Lu, Y., and Zhou, S. (2011). Screening for novel quorum-sensing inhibitors to interfere with the formation of Pseudomonas aeruginosa biofilm. *Journal of Medical Microbiology*, **12**: 1827-1834.

Duvoix, A., Delhalle, S., Blasius, R., Schnekenburger, M., Morceau, F., Fougère, M., Henry, E., Galteau, M.M., Dicato, M., and Diederich, M. (2004). Effect of chemopreventive agents on glutathione S-transferase P1-1 gene expression mechanisms via activating protein 1 and nuclear factor κB inhibition. *Biochemical Pharmacology*, **68**: 1101-1111.

Eja, M.E., Arikpo, G.E., Enyi-Idoh, K.H., Etim, S.E., and Etta, H.E. (2009). Efficacy of local herbal therapy in the management of dermatophytosis among primary school children in Cross River State, South-south Nigeria. *African Journal of Medicine and Medical Sciences*, **38**: 135-141.

Elujoba, A.A., Ajulo, O.O., and Iweibo, G.O. (1989). Chemical and biological analyses of Nigerian *Cassia* species for laxative activity. *Journal of Pharmaceutical and Biomedical Analysis*, **7**: 1453-1457.

Falcão-Silva, V.S., Silva, D.A., Souza Mde, F, and Siqueira-Junior, J.P. (2009). Modulation of drug resistance in *Staphylococcus aureus* by a kaempferol glycoside from *Herissantia tiubae* (Malvaceae). *Phytotherapy Research*, **23**: 1367-1370.

Feitosa, C.M., Freitas, R.M., Luz, N.N.N., Bezerra, M.Z.B., and Trevisan, M.T.S. (2011). Acetylcholinesterase inhibition by some promising Brazilian medicinal plants. *Brazilian Journal of Biology (Revista Brasleira de Biologia)*, **71**: 783-789.

Feng, Y., Huang, S.L., Dou, W., Zhang, S., Chen, J.H., Shen, Y., Shen, J.H., and Leng, Y. (2010). Emodin, a natural product, selectively inhibits 11β-hydroxysteroid dehydrogenase type 1 and ameliorates metabolic disorder in diet-induced obese mice. *British Journal of Pharmacology*, **161**: 113-126.

Fernand, V.E., Dinh, D.T., Washington, S.J., Fakayode, S.O., Losso, J.N., van Ravenswaay, R.O., and Warner, I.M. (2008). Determination of pharmacologically active compounds in root extracts of *Cassia alata* L. by use of high performance liquid chromatography. *Talanta*, **74**: 896-902.

Gacche, R.N., Shegokar, H.D., Gond, D.S., Yang, Z., and Jadhav, A.D. (2011). Evaluation of selected flavonoids as antiangiogenic, anticancer, and radical scavenging agents: an experimental and in silico analysis. *Cellular Biochemistry and Biophysics*, **61**: 651-663

Ghani, A. (1998). Medicinal Plants of Bangladesh. 1st Edn. Dhaka, Asiatic Society of Bangladesh, pp. 120-122.

Ghosh, S., Das Sarma, M., Patra, A., and Hazra, B. (2010). Anti-inflammatory and anticancer compounds isolated from *Ventilago madraspatana* Gaertn., *Rubia cordifolia* Linn., and *Lantana camara* Linn. *Journal of Pharmacy and Pharmacology*, **62**: 1158-1166.

Giron, L.M., Freire, V., Alonzo, A., and Caceres, A. (1991) Ethnobotanical survey of the medicinal flora used by the Caribs of Guatemala. *Journal of Ethnopharmacology*, **34**: 173-187.

Gopalakrishnan, A., Xu, C.J., Nair, S.S., Chen, C., Hebbar, V., and Kong, A.N. (2006). Modulation of activator protein-1 (AP-1) and MAPK pathway by flavonoids in human prostate cancer PC3 cells. *Archives of Pharmacal Research*, **29**: 633-644.

Gong, J.H., Shin, D., Han, S.Y., Kim, J.L., and Kang, Y.H. (2012). Kaempferol suppresses eosionphil infiltration and airway inflammation in airway epithelial cells and in mice with allergic asthma. *Journal of Nutrition*, **142**: 47-56.

Grabher, P., Durieu, E., Kouloura, E., Halabalaki, M., Skaltsounis, L.A., Meijer, L., Hamburger, M., and Potterat, O. (2012). Library-based discovery of DYRK1A/ CLK1 inhibitors from natural product extracts. *Planta Medica*, **8**: 951-956.

Guo, H.C., Bu, H.Q., Luo, J., Wei, W.T., Liu, D.L., Chen, H., Tong, H.F., Wang, Z.H., Wu, H.Y., Li, H.H., Zuo, M.M., Li, W., Lin, S.Z. (2012). Emodin potentiates the antitumor effects of gemcitabine in PANC-1 pancreatic cancer xenograft model *in vivo* via inhibition of inhibitors of apoptosis. *International Journal of Oncology*, **40**: 1849-1857.

Guo, J.M., Xiao, B.X., Liu, Q., Zhang, S., Liu, D.H., and Gong, Z.H. (2007). Anticancer effect of aloe-emodin on cervical cancer cells involves G2/M arrest and induction of differentiation. *Acta Pharmacologica Sinica*, **28**: 1991-1995.

Guo, J., Xiao, B., Zhang, S., Liu, D., Liao, Y., and Sun, Q. (2007). Growth inhibitory effects of gastric cancer cells with an increase in S phase and alkaline phosphatase activity repression by aloe-emodin. *Cancer Biology and Therapeutics*, **6**: 85-88.

Guo, Q., Chen, Y., Zhang, B., Kang, M., Xie, Q., Wu, Y. (2009). Potentiation of the effect of gemcitabine by emodin in pancreatic cancer isassociated with survivin inhibition. *Biochemical Pharmacology*, **77**: 1674-1683.

Guo, A.J., Choi, R.C., Zheng, K.Y., Chen, V.P., Dong, T.T., Wang, Z.T., Vollmer, G., Lau, D.T., and Tsim, K.W. (2012). Kaempferol as a flavonoid induces osteoblastic differentiation via estrogen receptor signaling. *Chinese Medicine*, **7**: 10.

Gupta, D., and Singh, J. (1991). Flavonoid glycosides from *Cassia alata*. *Phytochemistry*, **30**: 2761-2763.

Ha, M.K., Song, Y.H., Jeong, S.J., Lee, H.J., Jung, J.H., Kim, B., Song, H.S., Huh, J.E., and Kim, S.H. (2011). Emodin inhibits proinflammatory responses and inactivates histone deacetylase 1 in hypoxic rheumatoid synoviocytes. *Biological Pharmaceutical Bulletin*, **34**: 1432-1437.

Hämäläinen, M., Nieminen, R., Vuorela, P., Heinonen, M., and Moilanen, E. (2007). Anti-inflammatory effects of flavonoids: genistein, kaempferol, quercetin, and daidzein inhibit STAT-1 and NF-κB activations, whereas flavone, isorhamnetin, naringenin, and pelargonidin inhibit only NF-κB activation along with their inhibitory effect on iNOS expression and NO production in activated macrophages. *Mediators of Inflammation*, **2007**: 45673.

Han, Y.M., Lee, S.K., Jeong, D.G., Ryu, S.E., Han, D.C., Kim, D.K., and Kwon, B.M. (2012). Emodin inhibits migration and invasion of DLD-1 (PRL-3) cells via inhibition of PRL-3 phosphatase activity. *Bioorganic and Medicinal Chemistry Letters*, **22**: 323-326.

Hazni, H., Ahmad, N., Hitotsuyanagi, Y., Takeya, K., and Choo, Chee-Yan (2008). Phytochemical constituents from *Cassia alata* with inhibition against methicillin-resistant *Staphylococcus aureus* (MRSA). *Planta Medica*, **74**: 1802-1805.

He, T.P., Yan, W.H., Mo, L.E., and Liang, N.C (2008). Inhibitory effect of aloe-emodin on metastasis potential in HO-8910PM cell line. *Journal of Asian Natural Products Research*.**10**: 383-390.

He, Z.H., He, M.F., Ma, S.C., and But, P.P. (2009). Anti-angiogenic effects of rhubarb and its anthraquinone derivatives. *Journal of Ethnopharmacology*, 121: 313-317.

Heijnen, C.G., Haenen, G.R., van Acker, F.A., van der Vijgh, W.J., and Bast, A. (2001). Flavonoids as peroxynitrite scavengers: the role of the hydroxyl groups. *Toxicology In Vitro*, **15**: 3-6.

Hemlata, K., and Suraj, B. (1993). Alatinone, an anthroquinone from *Cassia alata*. *Phytochemistry*, **32**: 1616-1617.

Hemlata, K., and Kalidhar, S.B. (1994a). Alatonal, an anthraquinone derivative from *Cassia alata*. *Indian Journal of Chemistry, Section B: Organic Chemistry Including Medicinal Chemistry*, **33B**: 92-93.

Hemlata, K., and Kalidhar, S.B. (1994b). Alarone, an anthrone from *Cassia alata*. *Proceedings of the Indian National Science Academy, Part A: Physical Sciences*, **60**: 765-767.

Hibatallah, J., Carduner, C., and Poelman, M.C. (1999). *In vivo* and *in vitro* assessment of the free-radical-scavenger activity of Ginkgo flavone glycosides at high concentration. *The Journal of Pharmacy and Pharmacology*, **51**: 1435-40.

Hiraoka, A., Koike, S., Sakaguchi, S., Masuda, Y., Terai, F., Miura, I., Kawasaki, T., Kawai, H., and Sakakibara, M. (1989). The sennoside constituents of rhei rhizoma and sennae folium as inhibitors of serum monoamine oxidase. *Chemical and Pharmaceutical Bulletin (Tokyo)*, **37**: 2744-2746.

Hirose, E., Matsushima, M., Takagi, K., Ota, Y., Ishigami, K., Hirayama, T., Hayashi, Y., Nakamura, T., Hashimoto, N., Imaizumi, K., Baba, K., Hasegawa, Y., and Kawabe, T. (2009). Involvement of heme oxygenase-1 in kaempferol-induced anti-allergic actions in RBL-2H3 cells. *Inflammation*, **32**: 99-108.

Hofilena, J.G., Ragasa, C.Y., and Rideout, J.A. (2000). An antimicrobial and antimutagenic anthraquinone from *Cassia alata*. *ACGC Chemical Research Communications*, **10**: 15-20.

Hong, Chang-Eui and Lyu, Su-Yun (2011). Genotoxicity detection of five medicinal plants in Nigeria. *The Journal of Toxicological Sciences*, **36**: 87-93.

Hsiang, C.Y., and Ho, T.Y. (2008). Emodin is a novel alkaline nuclease inhibitor that suppresses herpes simplex virus type 1 yields in cell cultures. *British Journal of Pharmacology*, **155**: 227-235.

Huang, Q., Shen, H.M., and Ong CN. (2004). Inhibitory effect of emodin on tumor invasion through suppression of activator protein-1 and nuclear factor-κB. *Biochemical Pharmacology*, **68**: 361-71.

Huang, H.C., Chang, J.H., Tung, S.F., Wu, R.T., Foegh, M.L., and Chu, S.H. (1992). Immunosuppressive effect of emodin, a free radical generator. *European Journal of Pharmacology*, **211**: 359-364.

Ibrahim, D., and Osman, H. (1995). Antimicrobial activity of *Cassia alata* from Malaysia. *Journal of Ethnopharmacology*, **45**: 151-156.

Idu, M., Oronsaye, F.E., Igeleke, C.L., Omonigho, S.E., Omogbeme, O.E., and Ayinde, B.A. (2006). Preliminary investigation on the phytochemistry and antimicrobial activity of *Senna alata* L. leaves. *Journal of Applied Sciences*, **6**: 2481-2485.

Idu, M., Omonigho, S.E., and Igeleke, C.L. (2007). Preliminary investigation on the phytochemistry and antimicrobial activity of *Senna alata* L. flower. *Pakistan Journal of Biological Sciences*, **10**: 806-809.

Jain, R., Chitale, G., Sharma, P., and Jain, S.C. (2010). Phytochemical and antimicrobial investigations of *Cassia alata* Linn. roots. *Journal of Medicinal and Aromatic Plant Sciences*, **32**: 13-15.

Jin, J.H., Ngoc, T.M., Bae, K., Kim, Y.S., and Kim, H/P. (2011). Inhibition of experimental atopic dermatitis by rhubarb (rhizomes of *Rheum tanguticum*) and 5-lipoxygenase inhibition of its major constituent, emodin. Phytotherapy Research, **25**: 755-759.

Kataoka, M., Hirata, K., Kunikata, T., Ushio, S., Iwaki, K., Ohashi, K., Iked,a M., and Kurimoto, M. (2001). Antibacterial action of tryptanthrin and kaempferol, isolated from the indigo plant (*Polygonum tinctorium* Lour.), against *Helicobacter pylori*-infected Mongolian gerbils. *Journal of Gastroenterology*, **36**: 5-9.

Kaneshiro, T., Morioka, T., Inamine, M., Kinjo, T., Arakaki, J., Chiba, I., Sunagawa, N., Suzui, M., and Yoshimi, N. (2006). Anthraquinone derivative emodin inhibits tumor-associated angiogenesis through inhibition of extracellular signal-regulated kinase 1/2 phosphorylation. *European Journal of Pharmacology*, **553**: 46-53.

Kayembe, J.S., Taba, K.M., Ntumba, K., Tshiongo, M.T.C., and Kazadi, T.K. (2010). *In vitro* anti-malarial activity of 20 quinones isolated from four plants used by traditional healers in the Democratic Republic of Congo. *Journal of Medicinal Plants Research*, **4**: 991-994.

Kempuraj, D., Madhappan, B., Christodoulou, S., Boucher, W., Cao, J., Papadopoulou, N., Cetrulo, C.L., and Theoharides, T.C. (2005). Flavonols inhibit proinflammatory mediator release, intracellular calcium ion levels and protein kinase C è phosphorylation in human mast cells. *British Journal of Pharmacology*, **145**: 934-944.

Khan, M.R., Kihara, M., and Omoloso, A.D. (2001). Antimicrobial activity of *Cassia alata*. *Fitoterapia*, **72**: 561-564.

Khare, C.P. (2007). Indian Medicinal Plants: An Illustrated Dictionary, Springer-Verlag, Berlin/Heidelberg.

Kim, J.M., Lee, E.K., Kim, D.H., Yu, B.P., and Chung, H.Y. (2010). Kaempferol modulates pro-inflammatory NF-κB activation by suppressing advanced glycation endproducts-induced NADPH oxidase. *Age (Dordrecht, Netherlands)*, **32**: 197-208.

Kim, M.S., Park, M.J., Kim, S.J., Lee, C.H., Yoo, H., Shin, S.H., Song, E.S., and Lee, S.H. (2005). Emodin suppresses hyaluronic acid-induced MMP-9 secretion and invasion of glioma cells. *International Journal of Oncology*, **27**: 839-46.

Kim, Y.M., Lee, C.H., Kim, H.G., and Lee, H.S. (2004). Anthraquinones isolated from *Cassia tora* (Leguminosae) seed show an antifungal property against phytopathogenic fungi. *Journal of Agricultural and Food Chemistry*, **52**: 6096-6100.

Kelly, T.R., Ma, Z., and Xu, W. (1994). Revision of the structure of alatinone to emodin. *Phytochemistry*, **36**: 253-254.

Kong, L.D., Cheng, C.H., and Tan, R.X. (2004). Inhibition of MAO A and B by some plant-derived alkaloids, phenols and anthraquinones. *Journal of Ethnopharmacology*, **91**: 351-355.

Kuo, P.L., Lin, T.C., and Lin, C.C. (2002). The antiproliferative activity of aloe-emodin is through p53-dependent and p21-dependent apoptotic pathway in human hepatoma cell lines. *Life Sciences*, **71**: 1879-1892.

Kundu, S., and Lyndem, L.M. (2012). *In vitro* screening for cestocidal activity of three species of Cassia plants against the tapeworm *Raillietina tetragona*. *Journal of Helminthology*, 1-6 (available on CJO 2012, doi: 10.1017/S0022 149X12000156).

Kundu, S., Roy, S., and Lyndem, M.L. (2012). *Cassia alata* L: potential role as anthelmintic agent against *Hymenolepis diminuta*. *Parasitology Research*, (in press).

Kurokawa, T., He, G., and Siddik, Z.H. (2010). Protein kinase inhibitors emodin and dichloro-ribofuranosylbenzimidazole modulate the cellular accumulation and cytotoxicity of cisplatin in a schedule-dependent manner. *Cancer Chemotherapy and Pharmacology*, **65**: 427-436.

Kwak, H.J., Park, M.J., Park, C.M,. Moon, S.I., Yoo, D.H., Lee, H.C., Lee, S.H., Kim, M.S., Lee, H.W., Shin, W.S., Park, I.C., Rhee, C.H., and Hong, S.I. (2006). Emodin inhibits vascular endothelial growth factor-A-induced angiogenesis by blocking receptor-2 (KDR/Flk-1) phosphorylation. *International Journal of Cancer*, **118**: 2711-2720.

Labbé, D., Provençal, M., Lamy, S., Boivin, D., Gingras, D., and Béliveau, R. (2009). The flavonols quercetin, kaempferol, and myricetin inhibit hepatocyte growth factor-induced medulloblastoma cell migration. *Journal of Nutrition*, **139**: 646-652.

Lans, C., Harper, T., Georges, K., and Bridgewater, E. (2000). Medicinal plants used for dogs in Trinidad and Tobago. *Preventive Veterinary Medicine*, **45**: 201-220.

Lee, K.H., Tagahara, K., Suzuki, H., Wu, R.Y., Haruna, M., Hall, I.H., Huang, H.C., Ito, K., Iida, T., and Lai, J.S. (1981). Antitumor agents. 49 tricin, kaempferol-3-O- β-D-glucopyranoside and (+)-nortrachelogenin, antileukemic principles from *Wikstroemia indica*. *Journal of Natural Products*, **44**: 530-535.

Lee, C.H., Kim, S.I., Lee, K.B., Yoo, Y.C., Ryu, S.Y., and Song, K.S. (2003). Neuraminidase inhibitors from *Reynoutria elliptica*. *Archives of Pharmacal Research*, **26**: 367-374.

Lee, H.Z., Lin, C.J., Yang, W.H., Leung, W.C., and Chang, S.P. (2006). Aloe-emodin induced DNA damage through generation of reactive oxygen species inhuman lung carcinoma cells. *Cancer Letters*, **239**: 55-63.

Lee, S.U., Shin, H.K., Min, Y.K., and Kim, S.H. (2008). Emodin accelerates osteoblast differentiation through phosphatidylinositol 3-kinase activation and bone morphogenetic protein-2 gene expression. *International Immunopharmacology*, **8**: 741-7.

Lee, S., Kim, Y.J., Kwon, S., Lee, Y., Choi, S.Y., Park, J., and Kwon, H.J. (2009). Inhibitory effects of flavonoids on TNF-α-induced IL-8 gene expression in HEK 293 cells. *BMB Reports*, **42**: 265-70.

Lee, S.J., Jeong, D., Park, W.K., Kong, J.Y., Choi, G., Kim, H., Kang, S., and Cho, H. (2010a). Screening of Kit inhibitors: suppression of Kit signaling and melanogenesis by emodin. *Phytotherapy Research*, **24**: 308-312.

Lee, Y.S., Kang, O.H., Choi, J.G., Oh, Y.C., Keum, J.H., Kim, S.B., Jeong, G.S., Kim, Y.C., Shin, D.W., and Kwon, D.Y. (2010b). Synergistic effect of emodin in combination with ampicillin or oxacillin against methicillin-resistant *Staphylococcus aureus*. *Pharmaceutical Biology*, **48**: 1285-1290.

Lee, B.H., Huang, Y.Y., Duh, P.D., and Wu, S.C. (2012). Hepatoprotection of emodin and Polygonum multiflorum against CCl_4-induced liver injury. *Pharmaceutical Biology*, **50**: 351-359.

Leu, Y.L., Hwang, T.L., Hu, J.W., and Fang, J.Y. (2008). Anthraquinones from *Polygonum cuspidatum* as tyrosinase inhibitors for dermal use. *Phytotherapy Research*, **22**: 552-556.

Levy, A., and Lewis, A. (2011). *Cassia alata* leaf extract induces cytotoxicity in A549 lung cancer cells via a mechanism that is caspase 8 dependent. *The West Indian Medical Journal*, **60**: 608-614.

Lewis, A., and Levy, A. (2011). Anti-inflammatory activities of *Cassia alata* leaf extract in complete Freund's adjuvant arthritis in rats. *The West Indian Medical Journal*, **60**: 615-621.

Liang, Y.C., Huang, Y.T., Tsai, S.H., Lin-Shiau, S.Y., Chen, C.F., and Lin, J.K. (1999). Suppression of inducible cyclooxygenase and inducible nitric oxide synthase by apigenin and related flavonoids in mouse macrophages. *Carcinogenesis*, **20**: 1945-52.

Lim, Y.H., Kim, I.H., and Seo, J.J. (2007). *In vitro* activity of kaempferol isolated from the *Impatiens balsamina* alone and in combination with erythromycin or clindamycin against *Propionibacterium acnes*. *Journal of Microbiology*, **45**: 473-477.

Lin, J.G., Chen, G.W., Li, T.M., Chouh, S.T., Tan, T.W., and Chung, J.G. (2006). Aloe-emodin induces apoptosis in T24 human bladder cancer cells through the p53 dependent apoptotic pathway. *Journal of Urology*, **175**: 343-347.

Li, Y., Luan, Y., Qi, X., Li, M., Gong, L., Xue, X., Wu, X., Wu, Y., Chen, M., Xing, G., Yao, J., and Ren, J. (2010). Emodin triggers DNA double-strand breaks by stabilizing topoisomerase II-DNA cleavage complexes and by inhibiting ATP hydrolysis of topoisomerase II. *Toxicological Sciences*, **118**: 435-443.

Li, Z.F., Xia, X.M., Huang, C., Zhang, S., Zhang, J., and Zhang, A.J. (2009). Emodin and baicalein inhibit pancreatic stromal derived factor-1 expression in rats with acute pancreatitis. *Hepatobiliary and Pancreatic Diseases International*, **8**: 201-8.

Limsong, J., Benjavongkulchai, E., and Kuvatanasuchati, J. (2004). Inhibitory effect of some herbal extracts on adherence of *Streptococcus mutans*. *Journal of Ethnopharmacology*, **92**: 281-289.

Lin, H.J., Lai, C.C., Lee Chao, P.D., Fan, S.S., Tsai, Y., Huang, S.Y, Wan, L., and Tsai, F.J. (2007). Aloe-emodin metabolites protected N-methyl-d-aspartate-treated retinal ganglion cells by Cu-Zn superoxide dismutase. *Journal of Ocular Pharmacology and Therapeutics*, **23**: 152-171.

Lin, C.W., Wu, C.F., Hsiao, N.W., Chang, C.Y., Li, S.W., Wan, L., Lin, Y.J., and Lin, W.Y. (2008). Aloe-emodin is an interferon-inducing agent with antiviral activity against Japanese encephalitis virus and enterovirus 71. *International Journal of Antimicrobial Agents*, **32**: 355-359.

Lin, M.L., Lu, Y.C., Chung, J.G., Wang, S.G., Lin, H.T., Kang, S.E., Tang, C.H., Ko, J.L., and Chen, S.S. (2010a). Down-regulation of MMP-2 through the p38 MAPK-NF-κB-dependent pathway by aloe-emodin leads to inhibition of nasopharyngeal carcinoma cell invasion. *Molecular Carcinogenesis*, **49**: 783-797.

Lin, S.Z., Chen, K.J., Tong, H.F., Jing, H., Li, H., and Zheng, S.S. (2010b). Emodin attenuates acute rejection of liver allografts by inhibiting hepatocellular apoptosis and modulating the Th1/Th2 balance in rats. *Clinical and Experimental Pharmacology and Physiology*, **37**: 790-794.

Lin, S.Z., Wei, W.T., Chen, H., Chen, K.J., Tong, H.F., Wang, Z.H., Ni, Z.L., Liu, H.B., Guo, H.C., and Liu, D.L. (2012). Antitumor activity of emodin against pancreatic cancer depends on its dual role: promotion of apoptosis and suppression of angiogenesis. *PLoS One*, **7**: e42146.

Liu, Y., Jia, L., Liu, Z.C., Zhang, H., Zhang, P.J., Wan, Q., and Wang, R. (2009). Emodin ameliorates high-glucose induced mesangial p38 over-activation and hypocontractility via activation of PPARγ. *Experimental and Molecular Medicine*, **41**: 648-655.

Liu, A., Chen, H., Wei, W., Ye, S., Liao, W., Gong, J., Jiang, Z., Wang, L., and Lin, S (2011). Antiproliferative and antimetastatic effects of emodin on human pancreatic cancer. *Oncology Report*. **26**: 81-89.

Liu, K., Park, C., Li, S., Lee, K.W., Liu, H., He, L., Soung, N.K, Ahn, J.S., Bode, A.M., Dong, Z., Kim, B.Y., and Dong, Z. (2012). Aloe-emodin suppresses prostate cancer by targeting the mTOR complex 2. *Carcinogenesis*. **33**: 1406-1411.

Lu, Y., Zhang, J., and Qian, J. (2008). The effect of emodin on VEGF receptors in human colon cancer cells. *Cancer Biotherapy and Radiopharmaceuticals*, **23**: 222-228.

Lu, Y., Yang, J.H., Li, X., Hwangbo, K., Hwang, S.L., Taketomi, Y., Murakami, M., Chang, Y.C., Kim, C.H., Son, J.K., and Chang, H.W. (2011). Emodin, a naturally occurring anthraquinone derivative, suppresses IgE-mediated anaphylactic reaction and mast cell activation. *Biochemical Pharmacology*, **82**: 1700-1708.

Luo, H., Rankin, G.O., Li, Z., Depriest, L., and Chen, Y.C. (2011). Kaempferol induces apoptosis in ovarian cancer cells through activating p53 in the intrinsic pathway. *Food Chemistry*, **128**: 513-519.

Luo, H., Rankin, G.O., Juliano, N., Jiang, B.H., and Chen, Y.C. (2012). Kaempferol inhibits VEGF expression and *in vitro* angiogenesis through a novel ERK-NFêB-cMyc-p21 pathway. *Food Chemistry*, **130**: 321-328.

Mahat, M.Y., Kulkarni, N.M., Vishwakarma, S.L., Khan, F.R., Thippeswamy, B.S., Hebballi, V., Adhyapak, A.A., Benade, V.S., Ashfaque, S.M., Tubachi, S., and Patil, B.M. (2010). Modulation of the cyclooxygenase pathway via inhibition of nitric oxide production contributes to the anti-inflammatory activity of kaempferol. *European Journal of Pharmacology*, **642**: 169-176.

Malterud, K.E., Farbrot, T.L., Huse, A.E., and Sund, R.B. (1993). Antioxidant and radical scavenging effects of anthraquinones and anthrones. *Pharmacology*, **47 Suppl 1**: 77-85.

Mizuno, M., Kawamura, H., Takei, N., and Nawa, H. (2008). The anthraquinone derivative Emodin ameliorates neurobehavioral deficits of a rodent model for schizophrenia. *Journal of Neural Transmission*, **115**: 521-530.

Mohideen, S., Sasikala, E., and Aruhaj, P. (2005). Pharmacognosy of *Cassia alata* Linn – leaves. *Ancient Science of Life*, **24**: 192-198.

Moriyama, H., Iizuka, T., and Nagai, M. (2001). A stabilized flavonoid glycoside in heat-treated *Cassia alata* leaves and its structural elucidation. *Yakugaku Zasshi: Journal of the Pharmaceutical Society of Japan*, **121**: 817-820.

Moriyama, H., Iizuka, T., Nagai, M., Miyataka, H., and Satoh, T. (2003a). Antiinflammatory activity of heat-treated *Cassia alata* leaf extract and its flavonoid glycoside. *Yakugaku Zasshi*, **123**: 607-611.

Moriyama, H., Iizuka, T., Nagai, M., and Murata, Y. (2003b). HPLC quantification of kaempferol-3-O-gentiobioside in *Cassia alata*. *Fitoterapia*, **74**: 425-430.

Moriyama, H., Iizuka, T., Nagai, M., and Hoshi, K. (2003c). Adenine, an inhibitor of platelet aggregation, from the leaves of *Cassia alata*. *Biological and Pharmaceutical Bulletin*, **26**: 1361-1364.

Mijatovic, S., Maksimovic-Ivanic, D., Radovic, J., Popadic, D., Momcilovic, M., Harhaji, L., Miljkovic, D., and Trajkovic, V. (2004). Aloe-emodin prevents cytokine-induced tumor cell death: the inhibition of auto-toxic nitric oxide release as a potential mechanism. *Cellular and Molecular Life Sciences*, **61**: 1805-1815.

Miyake, M., Arai, N., Ushio, S., Iwaki, K., Ikeda, M., and Kurimoto, M. (2003). Promoting effect of kaempferol on the differentiation and mineralization of murine pre-osteoblastic cell line MC3T3-E1. *Bioscience, Biotechnology, and Biochemistry*, **67**: 1199-1205.

Muto, A., Hori, M., Sasaki, Y., Saitoh, A., Yasuda, I., Maekawa, T., Uchida, T., Asakura, K., Nakazato, T., Kaneda, T., Kizaki, M., Ikeda, Y., and Yoshida, T. (2007). Emodin has a cytotoxic activity against human multiple myeloma as a Janus-activated kinase 2 inhibitor. *Molecular Cancer Therapeutics*, **6**: 987-994.

Mylonis, I., Lakka, A., Tsakalof, A., and Simos, G. (2010). The dietary flavonoid kaempferol effectively inhibits HIF-1 activity and hepatoma cancer cell viability under hypoxic conditions. *Biochemical and Biophysical Research Communications*, **398**: 74-78.

Nakamura, Y., Chang, C.C., Mori, T., Sato, K., Ohtsuki, K., Upham, B.L., and Trosko. J.E. (2005). Augmentation of differentiation and gap junction function by kaempferol in partially differentiated colon cancer cells. *Carcinogenesis*, **26**: 665-671.

Nagao, A., Seki, M., and Kobayashi, H. (1999). Inhibition of xanthine oxidase by flavonoids. *Bioscience, Biotechnology, and Biochemistry*, **63**: 1787-1790.

Neharkar, V.S., Gaikwad, K.G., and Ghongade, D.B. (2010). Hepatoprotective activity of *Cassia alata* Linn. in rifampicin induced liver injury in rats. *Advances in Pharmacology and Toxicology*, **11**: 27-36.

Ogunwande, I.A., Flamini, G., Cioni, P.L., Omikorede, O., Azeez, R.A., Ayodele, A.A., and Kamil, Y.O. (2010). Aromatic plants growing in Nigeria: essential oil constituents of *Cassia alata* (Linn.) Roxb., and *Helianthus annuus* L. *Records of Natural Products*, **4**: 211-217.

Okpuzor, J., Ogbunugafor, H., Kareem, G.K., and Igwo-Ezipke, M.N. (2009). *In vitro* investigation of antioxidant phenolic compounds in extracts of *Senna alata*. *Research Journal of Phytochemistry*, **3**: 68-76.

Okwu, D.E., and Nnamdi, F.U. (2011). Cannabinoid dronabinol alkaloid with antimicrobial activity from *Cassia alata* Linn. *Chemica Sinica*, **2**: 247-254.

Ozyurek, M., Bektasoglu, B., Guclu, K., and Apak, R. (2009). Measurement of xanthine oxidase inhibition activity of phenolics and flavonoids with a modified cupric reducing antioxidant capacity (CUPRAC) method. *Analytica Chimica Acta*, **636**: 42-50.

Palanichamy, S., Nagarajan, S., and Devasagayam, M. (1988). Effect of *Cassia alata* leaf extract on hyperglycemic rats. *Journal of Ethnopharmacology*, **22**: 81-90.

Palanichamy, S., and Nagarajan, S. (1990a). Analgesic activity of *Cassia alata* leaf extract and kaempferol 3-O-sophoroside. *Journal of Ethnopharmacology*, **29**: 73-78.

Palanichamy, S., and Nagarajan, S. (1990b). Anti-inflammatory activity of *Cassia alata* leaf extract and kaempferol 3-O-sophoroside. *Fitoterapia*, **61**: 44-47.

Panichayupakaranant, P., and Kaewsuwan, S. (2004). Bioassay-guided isolation of the antioxidant constituent from *Cassia alata* L. leaves. *Songklanakarin Journal of Science and Technology*, **26**: 103-107.

Park, H.H., Lee, S., Son, H.Y., Park, S.B., Kim, M.S., Choi, E.J., Singh, T.S., Ha, J.H., Lee, M.G., Kim, J.E., Hyun, M.C., Kwon, T.K., Kim, Y.H., and Kim, S.H. (2008). Flavonoids inhibit histamine release and expression of proinflammatory cytokines in mast cells. *Archives of Pharmacal Research*, **31**: 1303-1311.

Park, M.Y., Kwon, H.J., and Sung, M.K. (2009). Evaluation of aloin and aloe-emodin as anti-inflammatory agents in aloe by using murine macrophages. *Bioscience, Biotechnology, and Biochemistry*, **73**: 828-832.

Patrick-Iwuanyanwu, K.C., Onyeike, E.N., and Dar, A. (2011). Anti-inflammatory effect of crude methanolic extract and fractions of ring worm plant *Senna alata* (L. Roxb) leaves from Nigeria. *Pharmacia Sinica*, **2**: 9-16.

Phromnoi, K., Yodkeeree, S., Anuchapreeda, S., and Limtrakul, P. (2009). Inhibition of MMP-3 activity and invasion of the MDA-MB-231 human invasive breast carcinoma cell line by bioflavonoids. *Acta Pharmacologica Sinica*, **30**: 1169-1176.

Pickhardt, M., Gazova, Z., von Bergen, M., Khlistunova, I., Wang, Y., Hascher, A., Mandelkow, E.M., Biernat, J., and Mandelkow E. (2005). Anthraquinones inhibit tau aggregation and dissolve Alzheimer's paired helical filaments *in vitro* and in cells. *Journal of Biological Chemistry*, **280**: 3628-3635.

Pieme, C.A., Penlap, V.N., Ngogang, J., Kuete, V., Catros, V., and Moulinoux, J.Ph. (2009). *In vitro* effects of extract of *Senna alata* (Caesalpiniaceae) on the polyamines produced by leukemia cells (L1210). *Pharmacognosy Magazine*, **5**: 8-13.

Putalun, W., Pimmeuangkao, S., De-Eknamkul, W., Tanaka, H., and Shoyama, Y. (2006). Sennosides A and B production by hairy roots of *Senna alata* (L.) Roxb. *Zeitschrift fur Naturforschung. C. Journal of Biosciences*, **61**: 367-371.

Qian, Z.J., Zhang, C., Li, Y.X., Je, J.Y., Kim, S.K., and Jung, W.K. (2011). Protective effects of emodin and chrysophanol isolated from marine fungus *Aspergillus* sp. on ethanol-induced toxicity in HepG2/CYP2E1 cells. *Evidence-based Complementary and Alternative Medicine*, **2011**: 452621.

Radha, K.S., Madhyastha, H.K., Nakajima, Y., Omura, S., and Maruyama, M. (2008). Emodin upregulates urokinase plasminogen activator, plasminogen activator inhibitor-1 and promotes wound healing in human fibroblasts. *Vascular Pharmacology*, **48**: 184-190.

Rahman, M.S., Hasan, A.J.M.M., Ali, M.Y., and Ali, M.U. (2005a). Studies on lignin, carbohydrates and fatty acid compositions of *Cassia alata* leaves. *Bangladesh Journal of Scientific and Industrial Research*, **40**: 69-74.

Rahman, M.S., Hasan, A.J.M.M., Ali, M.Y., and Ali, M.U. (2005b). A new flavanone from the leaves of *Cassia alata*. *Bangladesh Journal of Scientific and Industrial Research*, **40**: 287-290.

Rahman, M.S., Hasan, A.J.M.M., Ali, M.Y., and Ali, M.U. (2005c). An isoflavone from the leaves of *Cassia alata*. *Bangladesh Journal of Scientific and Industrial Research*, **40**: 123-126.

Rahman, M.S., Hasan, A.J.M.M., Ali, M.Y., and Ali, M.U. (2006a). Studies on the isolation of parahydroxy benzoic acid from the leaves of *Cassia alata*. *Bangladesh Journal of Scientific and Industrial Research*, **41**: 89-92.

Rahman, M.S., Hasan, A.J.M.M., Ali, M.Y., and Ali, M.U. (2006b). A flavones from the leaves of *Cassia alata*. *Bangladesh Journal of Scientific and Industrial Research*, **41**: 93-96.

Rahman, M.S., Ali, M.Y., Ali, M.U., and Hasan, A.J.M.M. (2006c). Studies on the lipid and glyceride compositions of *Cassia alata* seed oil. *Bangladesh Journal of Scientific and Industrial Research*, **41**: 83-88.

Rai, P.P. (1978). Anthracene derivatives in leaves and fruits of *Cassia alata*. *Current Science*, **47**: 271-272.

Rai, K.N., and Prasad, S.N. (1994). Chemical examination of the stem of *Cassia alata* Linn. *Journal of the Indian Chemical Society*, **71**: 653-654.

Ranganathan, S., and Balajee, S.S. (2000). Anti-*Cryptococcus* activity of combination of extracts of *Cassia alata* and *Ocimum sanctum*. *Mycoses*, **43**: 299-301.

Rao, J.V.L.N.S., Sastry, P.S.R., Rao, R.V.K., and Vimaladevi, M. (1975). Occurrence of kaempferol and aloe-emodin in the leaves of *Cassia alata*. *Current Science*, **44**: 736-737.

Ravindran, R., Juliet, S., Sunil, R.A., Ajith Kumar, K.G., Nair, S.N., Amithamol, K.K., Bandyopadhyay, A., Rawat Ajay, K.S., and Ghosh, S. (2012). Acaricidal activity of *Cassia alata* against *Rhipicephalus* (*Boophilus*) *annulatus*. *Experimental and Applied Acaralogy*, **56**: 69-74.

Saito, T.S., Trentin, da Silva, D., Macedo, A.J., Pungartnik, C., Gosmann, G., Silveira, de Deos, J., Guecheva, T.N., Henriques, J.A.P., and Brendel, M. (2012a). Bioguided fractionation shows *Cassia alata* extract to inhibit *Staphylococcus epidermis* and *Pseudomonas aeruginosa* growth and biofilm formation. *Evidence-Based Complementary and Alternative Medicine*, **2012**: 2012867103.

Saito, S., Silva, G., Santos, R.X., Gosmann, G., Pungartnik, C., and Brendel, M. (2012b). Astragalin from *Cassia alata* induces DNA adducts *in vitro* and repairable DNA damage in the yeast *Saccharomyces cereviseae*. *International Journal of Molecular Sciences*, **13**: 2846-2862.

Sakunpak, A., Sirikatitham, A., and Panichayupakaranant, P. (2009). Preparation of anthraquinone high-yielding *Senna alata* extract and its stability. *Pharmaceutical Biology (London, United Kingdom)*, **47**: 236-241.

Sanz, M.J., Ferrandiz, M.L., Cejudo, M., Terencio, M.C., Gil, B., Bustos, G., Ubeda, A., Gunasegaran, R., and Alcaraz, M.J. (1994). Influence of a series of natural flavonoids on free radical generating systems and oxidative stress. *Xenobiotica*, **24**: 689-699.

Singh, K.B., and Trigun, S.K. (2011). Apoptosis of Dalton's lymphoma due to *in vivo* treatment with emodin is associated with modulations of hydrogen peroxide metabolizing antioxidant enzymes. *Cell Biochemistry and Biophysics*, (in press).

Singh, B., Nadkarni, J.R., Vishwakarma, R.A., Bharate, S.B., Nivsarkar, M., and Anandjiwala, S. (2012). The hydroalcoholic extract of *Cassia alata* (Linn.) leaves and its major compound rhein exhibits antiallergic activity via mast cell stabilization and lipoxygenase inhibition. *Journal of Ethnopharmacology*, **141**: 469-473.

Smith, R.M., and Ali, S. (1979). Anthraquinones from the leaves of *Cassia alata* from Fiji. *New Zealand Journal of Science*, **22**: 123-125.

Sodipo, O.A., Effraim, K.D., and Emmagun, E. (1998). Effect of aqueous leaf extract of *Cassia alata* (Linn.) on some hematological indices in albino rats. *Phytotherapy Research*, **12**: 431-433.

Somchit, M.N., Reezal, I., Nur, *I.E.*, and Mutalib, A.R. (2003). *In vitro* antimicrobial activity of ethanol and water extracts of *Cassia alata*. *Journal of Ethnopharmacology*, **84**: 1-4.

Sridhar, J., Liu, J., Foroozesh, M., and Klein Stevens, C.L. (2012). Inhibition of cytochrome p450 enzymes by quinones and anthraquinones. *Chemical Research in Toxicology*, **25**: 357-365.

Subash-Babu, P., and Alshatwi, A.A. (2012). Aloe-emodin inhibits adipocyte differentiation and maturation during *in vitro* human mesenchymal stem cell adipogenesis. *Journal of Biochemical and Molecular Toxicology*, (in press).

Suboj, P., Babykutty, S., Valiyaparambil Gopi, D.R., Nair, R.S., Srinivas, P., and Gopala, S. (2012). Aloe emodin inhibits colon cancer cell migration/angiogenesis by downregulating MMP-2/9, RhoB and VEGF via reduced DNA binding activity of NF-κB. *European Journal of Pharmaceutical Sciences*, **45**: 581-591.

Subramanian, D.P., and Venugopal, S. (2011). Phytochemical investigation of *Cassia alata* Linn. flowers through various *in vitro* antioxidant assays. *International Journal of Pharmacy and Technology*, **3**: 3521-3534.

Süzgeç, S., Meriçli, A.H., Houghton, P.J., and Cubukçu, B. (2005). Flavonoids of *Helichrysum compactum* and their antioxidant and antibacterial activity. *Fitoterapia*. **76**: 269-272.

Swanbeck G. (1966). Interaction between deoxyribonucleic acid and some anthracene and anthraquinone derivatives. *Biochimica et Biophysica Acta*, **123**: 630-633.

Szliszka, E., Zydowicz, G., Janoszka, B., Dobosz, C., Kowalczyk-Ziomek, G., and Krol, W. (2011). Ethanolic extract of Brazilian green propolis sensitizes prostate cancer cells to TRAIL-induced apoptosis. *International Journal of Oncology*, **38**: 941-953.

Tabolacci, C., Lentini, A., Mattioli, P., Provenzano, B., Oliverio, S., Carlomosti, F., and Beninati, S. (2010). Antitumor properties of aloe-emodin and induction of transglutaminase 2 activity in B16-F10 melanoma cells. *Life Sciences*, **87**: 316-324.

Tabolacci, C., Oliverio, S., Lentini, A., Rossi, S., Galbiati, A., Montesano, C., Mattioli, P., Provenzano, B., Facchiano, F., and Beninati, S. (2011). Aloe-emodin as antiproliferative and differentiating agent on human U937 monoblastic leukemia cells. *Life Sciences*, **89**: 812-820.

Tai, Z.F., Zhang, G.L., and Wang, F. (2012). Identification of small molecule activators of the janus kinase/signal transducer and activator of transcription pathway using a cell-based screen. *Biological and Pharmaceutical Bulletin*, **35**: 65-71.

Thamlikitkul, V., Bunyapraphatsara, N., Dechatiwongse, T., Theerapong, S., Chantrakul, C., Thanaveerasuwan, T., Nimitnon, S., Boonroj, P., Punkrut, W., and Gingsungneon, V. (1990). Randomized controlled trial of *Cassia alata* Linn. for constipation. *Journal of the Medical Association of Thailand*, **73**: 217-222.

Tiwari, R.D., and Yadava, O.P. (1971). Structural study of the quinine pigments from the roots of *Cassia alata*. *Planta Medica*, **19**: 299-305.

Traore, M., Diallo, A., Nikiema, J.B., Tinto, H., Dakuyo, Z.P., Ouedraogo, J.B., Guissou, I.P., and Guiguemde, T.R. (2008). *In vitro* and *in vivo* antiplasmodial activity of 'saye', an herbal remedy used in Burkina Faso traditional medicine. *Phytotherapy Research*, **22**: 550-551.

Upadhyay, R.K., Yadav, N., and Ahmad, S. (2011). Toxic effects of solvent and aqueous extracts of *Cassia alata* against bio-molecules and enzymatic parameters of *Callosobruchus chinensis* L. (Coleoptera: Bruchidae). *Advances in Applied Science Research*, **2**: 367-381.

Verma, A.R., Vijayakumar, M., Mathela, C.S., and Rao, C.V. (2009). *In vitro* and *in vivo* antioxidant properties of different fractions of *Moringa oleifera* leaves. *Food and Chemical Toxicology*, **47**: 2196-2201.

Villaroya, M.L.E., and Bernal-Santos, R. (1976). A chemical investigation of *Cassia alata* L. (fam. Leguminosae). *Asian Journal of Pharmacy*, **3**: 10-12, 17-24.

Villasenor, I.M., Canlas Arlyn, P., Pascua Marcy Paul, I., Sabando May, N., and Soliven Leen Aloha, P. (2002). Bioactivity studies on *Cassia alata* Linn. leaf extracts. *Phytotherapy Research*, **16 Suppl 1**: S93-S96.

Villasenor, I.M., and Sanchez, A.C. (2009). Cassiaindoline, a new analgesic and anti-inflammatory alkaloid from *Cassia alata*. *Zeitschrift fur Naturforschung. C, Journal of Biosciences*, **64**: 335-338.

Wang, L., Tu, Y.C., Lian, T.W., Hung, J.T., Yen, J.H., and Wu, M.J. (2006). Distinctive antioxidant and antiinflammatory effects of flavonols. *Journal of Agricultural and Food Chemistry*, **54**: 9798-9804.

Wang, C., Zhang, D., Ma, H., and Liu, J. (2007). Neuroprotective effects of emodin-8-O-β-D-glucoside *in vivo* and *in vitro*. *European Journal of Pharmacology*, **577**: 58-63.

Wang, C., Jiang, Z., Yao, J., Wu, X., Sun, L., Liu, C., Duan, W., Yan, M., Sun, L., Liu, J., and Zhang, L. (2008). Participation of cathepsin B in emodin-induced apoptosis in HK-2 Cells. *Toxicology Letters*, **181**: 196-204.

Wang, J., Fang, F., Huang, Z., Wang, Y., and Wong, C. (2009). Kaempferol is an estrogen-related receptor alpha and gamma inverse agonist. *FEBS Letters*, **583**: 643-7.

Wang, M., Zhao, R., Wang, W., Mao, X., and Yu, J. (2012). Lipid regulation effects of *Polygoni Multiflori* Radix, its processed products and its major substances on steatosis human liver cell line L02. *Journal of Ethnopharmacology*, **139**: 287-93.

Wei, W.T., Chen, H., Ni, Z.L., Liu, H.B., Tong, H.F., Fan, L., Liu, A., Qiu, M.X., Liu, D.L., Guo, H.C., Wang, Z.H., and Lin, S.Z. (2011). Antitumor and apoptosis-promoting properties of emodin, an anthraquinone derivative from *Rheum officinale* Baill, against pancreatic cancer in mice via inhibition of Akt activation. *International Journal of Oncology*, **39**: 1381-90.

Wuthi-udomlert, M., Kupittayanant, P., and Gritsanapan, W. (2010a). *In vitro* evaluation of antifungal activity of anthraquinone derivatives of *Senna alata*. *Journal of Health Research*, **24**: 117-122.

Xiong, H.R., Luo, J., Hou, W., Xiao, H., and Yang, Z.Q. (2011). The effect of emodin, an anthraquinone derivative extracted from the roots of *Rheum tanguticum*, against herpes simplex virus *in vitro* and *in vivo*. *Journal of Ethnopharmacology*, **133**: 718-23.

Yadav, S.K., and Kalidhar, S.B. (1994). Alquinone: an anthraquinone from *Cassia alata*. *Planta Medica*, **60**: 601.

Yagi, S.M., El Tigani, S., and Adam, S.E.I. (1998). Toxicity of *Senna obtusifolia* fresh and fermented leaves (kawal), Senna alata leaves and some products from *Senna alata* to rats. *Phytotherapy Research*, **12**: 324-330.

Yang, Y.C., Lim, M.Y., and Lee, H.S. (2003). Emodin isolated from *Cassia obtusifolia* (Leguminosae) seed shows larvicidal activity against three mosquito species. *Journal of Agricultural and Food Chemistry*, **51**: 7629-31.

Yang, Y., Shang, W., Zhou, L., Jiang, B., Jin, H., and Chen, M. (2007). Emodin with PPARgamma ligand-binding activity promotes adipocyte differentiation and increases glucose uptake in 3T3-Ll cells. *Biochemical and Biophysical Research Communications*, **353**: 225-30.

Yang, H., Xu, L.N., He, C.Y., Liu, X., Fang, R.Y., and Ma, T.H. (2011). CFTR chloride channel as a molecular target of anthraquinone compounds in herbal laxatives. *Acta Pharmacologica Sinica*, **32**: 834-9

Yerbanga, S.R., Lucantoni, L., Lupidi, G., Dori, U.G., Tepongning, R.N., Nikiema, B.J., Esposito, F., and Habluetzel, A. (2012). Antimalarial plant remedies from Burkina Faso: their potential for prophylactic use. *Journal of Ethnopharmacology*, **140**: 255-260.

Yim, H., Lee, Y.H., Lee, C.H., and Lee, S.K. (1999). Emodin, an anthraquinone derivative isolated from the rhizomes of *Rheum palmatum*, selectively inhibits the activity of casein kinase II as a competitive inhibitor. *Planta Medica*, **65**: 9-13.

Yon, J.M., Baek, I.J., Lee, B.J., Yun, Y.W., and Nam, S.Y. (2011). Emodin and [6]-gingerol lessen hypoxia-induced embryotoxicities in cultured mouse whole embryos via upregulation of hypoxia-inducible factor 1α and intracellular superoxide dismutases. *Reproductive Toxicology*, **31**: 513-8.

Yuan, Y.F., Hu, X.Y., He, Y., and Deng, J.G. (2012). Synthesis and anti-tumor activity evaluation of rhein-aloe emodin hybrid molecule. *Natural Product Communications*, **7**: 207-10.

Zhao, L.M., Zhang, L.M., Liu, J.J., Wan, L.J., Chen, Y.Q., Zhang, S.Q., Yan, Z.W., and Jiang, J.H. (2012). Synthesis and antitumor activity of conjugates of 5-Fluorouracil and emodin. *European Journal of Medicinal Chemistry*, **47**: 255-60.

Zhang, L., Lau, Y.K., Xia, W., Hortobagyi, G.N., and Hung, M.C. (1999). Tyrosine kinase inhibitor emodin suppresses growth of HER-2/neu-overexpressing breast cancer cells in athymic mice and sensitizes these cells to the inhibitory effect of paclitaxel. *Clinical Cancer Research*, **5**: 343-53.

Zhou, X., Song, B., Jin, L., Hu, D., Diao, C., Xu, G., Zou, Z., and Yang, S. (2006). Isolation and inhibitory activity against ERK phosphorylation of hydroxyanthraquinones from rhubarb. *Bioorganic and Medicinal Chemistry Letters*, **16**: 563-8.

Medicinal Plants: Phytochemistry, Pharmacology and Therapeutics Vol. 3 (2014) *Pages* **79–123**
Editor-in-Chief: **V.K. Gupta**
Published by: **DAYA PUBLISHING HOUSE, NEW DELHI**

3

GCMS Screening for Phytocontents in the Essential Oil of *Coleus forskohlii*

Ashok Kumar[1]

ABSTRACT

Coleus forskohlii, a plant of Lamiaceae family, is widely used in ayuerveda for treating inflammation, skin disease, digestive complaints, heart and lungs conditions, asthma, insomnia, muscles spasm convulsions. The aim of the present study was find out the composition of essential oil from Coleus forskohlii. The oil was extracted from fresh roots of Coleus forskohlii by hydro distillation and their physical character recorded. Chemical composition of oil was studied by GC-MS which detected total 125 compounds, out of bornyl acetate (23.52 per cent), 2-Decen-1-ol (19.32 per cent), cyclohexene,3-(1,5-di methyl-4-hexenyl)-6-methylene (6.05), a-pinene (5.39 per cent), Bicyclo [3,1,1] heptanes, 6, 6-dimethyl-2-methylene (5.05 per cent) shows the highest area per cent. The phytochemicals of experimental plant have various medicinal values and may be used for herbal preparation against human diseases.

Keywords: Bornyl acetate, *Coleus forskohlii*, Essential oil, GC-MS, Medicinal value, Phytochemicals.

Introduction

India is one of the twelve mega biodiversity centers in the world with a wealth of 8000 species of medicinal plants. The world of naturals is storming the globe with scientific rationale and trends that are fast emerging to support better health and life through plant and plant products. The demand for the products obtained from these

1 Department of Biotechnology, Himachal Institute of Life Sciences, Rampurghat Road, Paonta Sahib – 173 025, Himachal Pradesh, India

* *Corresponding author*: E-mail: asokumr@gmail.com

plants such as phytochemical, steroidal, biologically active compounds, alkaloids, etc. is increasing in the national and international market.

Coleus forskohlii (wild) Briq [Syn. *C. barbatus* (Andr.) Benth.] is a plant of Indian origin (Valdes, 1987) and belongs to the family Lamiaceae (previously Labiatae). *Coleus forskohlii* is a botanical that has been used since ancient times in Hindu and Ayurvedic traditional medicine. The root portion of the plant has been traditionally used for medicinal purposes and contains the active constituent, forskolin. Forskolin was named after the Finnish botanist Forskal. It is the most important species of genus *Coleus* popularly known as *Mainamool* or *Manganiberu* or *Makandi beru* in Karnataka and *garmar* in Maharashtra. It is distributed in sub-tropical Himalayas from Gharwal to Nepal up to an altitude of 2500 m above mean sea level including Pakistan and Sri Lanka. Traditionally, the roots have been used for preparation as condiments in pickles and preparation of pickles (Anonymous, 1950) and also for medicinal purposes by the Ayurvedic schools of medicines (Ammon and Muller, 1985). The roots also yield a viscous, dark brown colored essential oil containing a range of aroma constituents that collectively impart a pleasing delicate aroma with a spicy note (Misra *et al.*, 1994). Root juice is given to children suffering from constipation (Singh and Singh, 1994). Kothas, the native tribes of Trichigadi in Nilgiri, South India consider the decoction of tuberous roots as tonic (Abraham, 1981). Roots are eaten for curing cough in Kumaon Himalayas and one to three teaspoonful of root decoction is recommended for treatment of asthma in Maharashtra. Paste prepared from the roots is mixed with mustard oil and used in the treatment of skin infection by the natives of Kumaon Himalayas. Concurrent research by Hoescht India identified the same compound as forskolin (Bhat *et al.*, 1977). Various studies show it has number of compounds.

Materials and Methods

Collection of Plant

The plant *Coleus forskohlii* was collected from G. B. Pantnagar University Ranichauri, Chamba, Uttarakhand (INDIA). The root of the plant *Coleus forskohlii* was removed and washed in tape water finally dried.

Isolation of Essential Oil

The air dried and finally ground plant materials were subjected to hydro distillation for three hour using Clevenger apparatus (Anonymous, 1950).The distillation of essential oil was done and dried over anhydrous sodium sulphate. The filtrate stored at 4°C until analyzed (Saksena *et al.*, 1985).

Chromatographic (GC-MS) Analysis

Analysis was performed with Perkin Elmer Clarus 500 (Turbomass software Ver.5.0.0) data handling system with manual injection/injector with mass detector. The components of GC-MS are listed in Table 3.1.

Identification of Compounds

The identification of the components was based on comparison of their mass spectra with those of NIST man spectral library and as well as comparison of their retention indices either with those of authentic compound or with literature value (Vishwakarma *et al.*, 1988).

Table 3.1: Components of GCMS.

Sl.No.	*Components*	*Description*
1.	Column	Rtx[R]-5 capillary column (60 m x 0.32 mm ID X film thickness 0.25 μm) crossbond®5 per cent diphenyl siloxane.
2.	Carrier Gas	Helium gas flow rate 1 ml/min
3.	Injector Temp (Inlet)	210 °C
4.	Source Temp	200 °C
5.	Injection Volume	0.2 μl
6.	Split Ratio	1: 50

Results and Discussion

The essential oil was estimated form the experimental plant. The essential oil includes sesquiterpenes, sesquiterpene alcohols, monoterpenoids and diterpenoids. The gas chromatogram of the essential oil of *Coleus forskohlii* showed more than 125 peaks. The major chemical components identified (29 Nos) in the present study along with their respective chemical classes are given below.

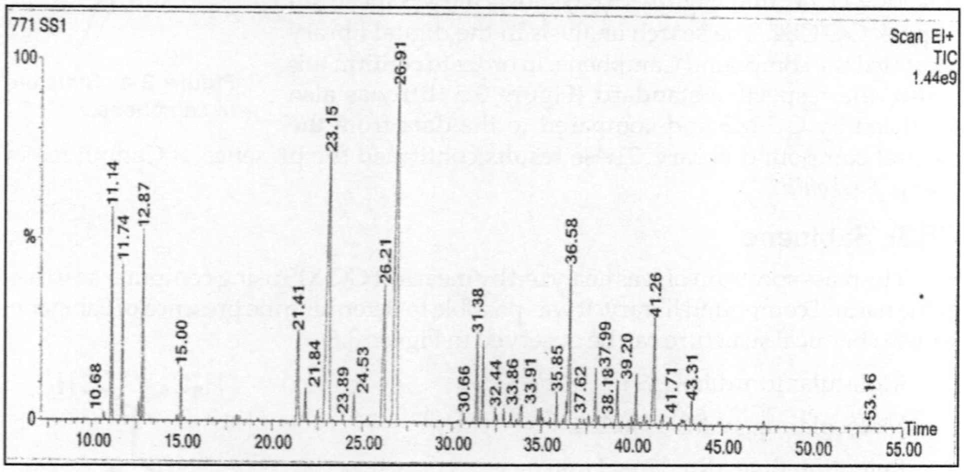

Figure 3.1: GC-MS of essential oil.

CB1: α-pinene

The mass spectrum of was analyzed by means of GC-MS using computer searches in the natural compound library. It was possible to determine the presence of α-pinene, whose chemical structure can be observed in Figure 3.2.

Molecular formula-$C_{10}H_{16}$

Molecular mass: M = 136.24

Retention time (Method GC-MS): tR = 11.14 min

Figure 3.1 exhibited a peak GC-CB1 with a retention time tR = 11.14 min. Figure 3.3 (A) shows the MS spectrum of peak GC-CB1. The search analysis in the digital library indicated the compound α-pinene. In order to confirm this result, the respective standard (Figure 3.3 (B)) was also analyzed by GC-MS and compared to the data from the natural compound library. These results confirmed the presence of α-pinene in *Coleus forskohlii*.

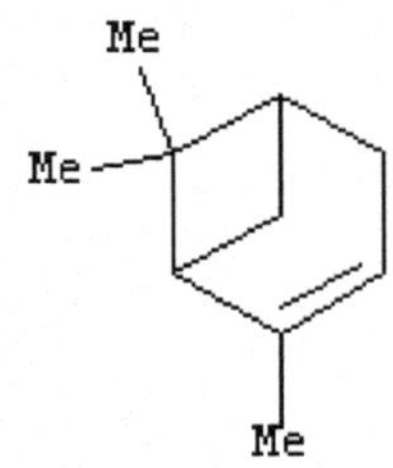

Figure 3.2: Structure of α-pinene.

CB2: Camphene

The mass spectrum of was analyzed by means of GC-MS using computer searches in the natural compound library. It was possible to determine the presence of Camphene, whose chemical structure can be observed in Figure 3.4.

Molecular formula-$C_{10}H_{16}$

Molecular mass: M = 136.234

Retention time: tR = 11.741 min

Figure 3.1 exhibited a peak GC-CB2 with a retention time tR = 11.741 min. Figure 3.5 (A) shows the MS spectrum of peak GC-CB2. The search analysis in the digital library indicated the compound Camphene. In order to confirm this result, the respective standard (Figure 3.5 (B)) was also analyzed by GC-MS and compared to the data from the natural compound library. These results confirmed the presence of Camphene in *Coleus forskohlii*.

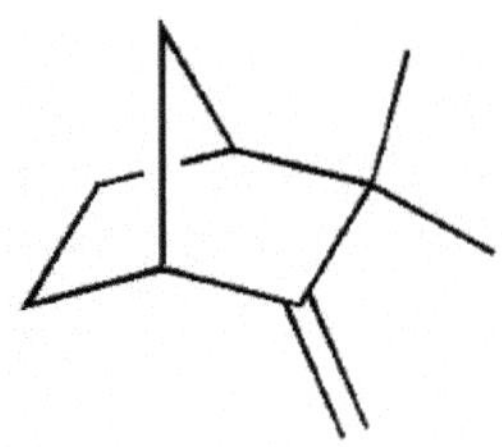

Figure 3.4: Structure of camphene.

CB3: Sabinene

The mass spectrum of was analyzed by means of GC-MS using computer searches in the natural compound library. It was possible to determine the presence of Sabinene, whose chemical structure can be observed in Figure 3.6.

Molecular formula-$C_{10}H_{16}$

Molecular mass: M = 136.234

Retention time: tR = 12.647 min

Figure 3.1 exhibited a peak GC-CB3 with a retention time tR = 12.647 min. Figure 3.7 (A) shows the MS spectrum of peak GC-CB3. The search analysis in the digital library indicated the compound Sabinene. In order to confirm this result, the respective standard (Figure 3.7 (B)) was also analyzed by GC-MS and compared to the data from the natural compound library. These results confirmed the presence of Sabinene in *Coleus forskohlii*.

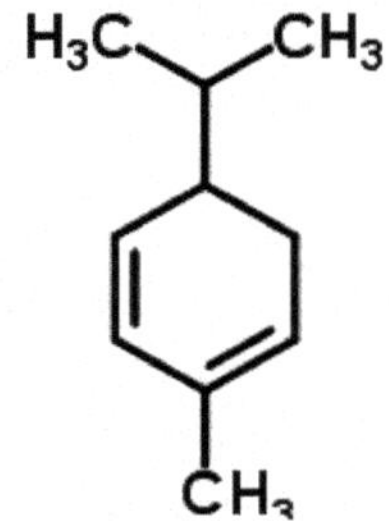

Figure 3.6: Structure of Sabinene.

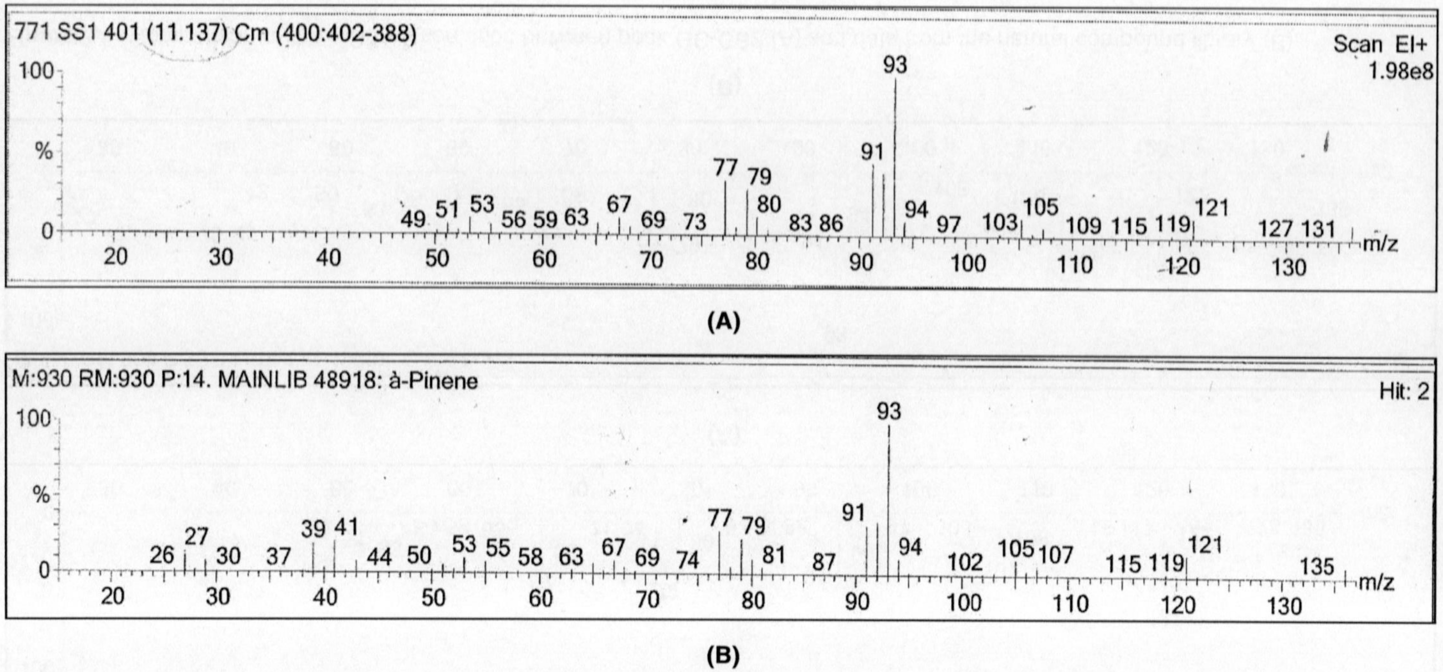

Figure 3.3: Comparison of the MS fragmentation between peak GC-CB1 (A) and data from the natural compound library (B).

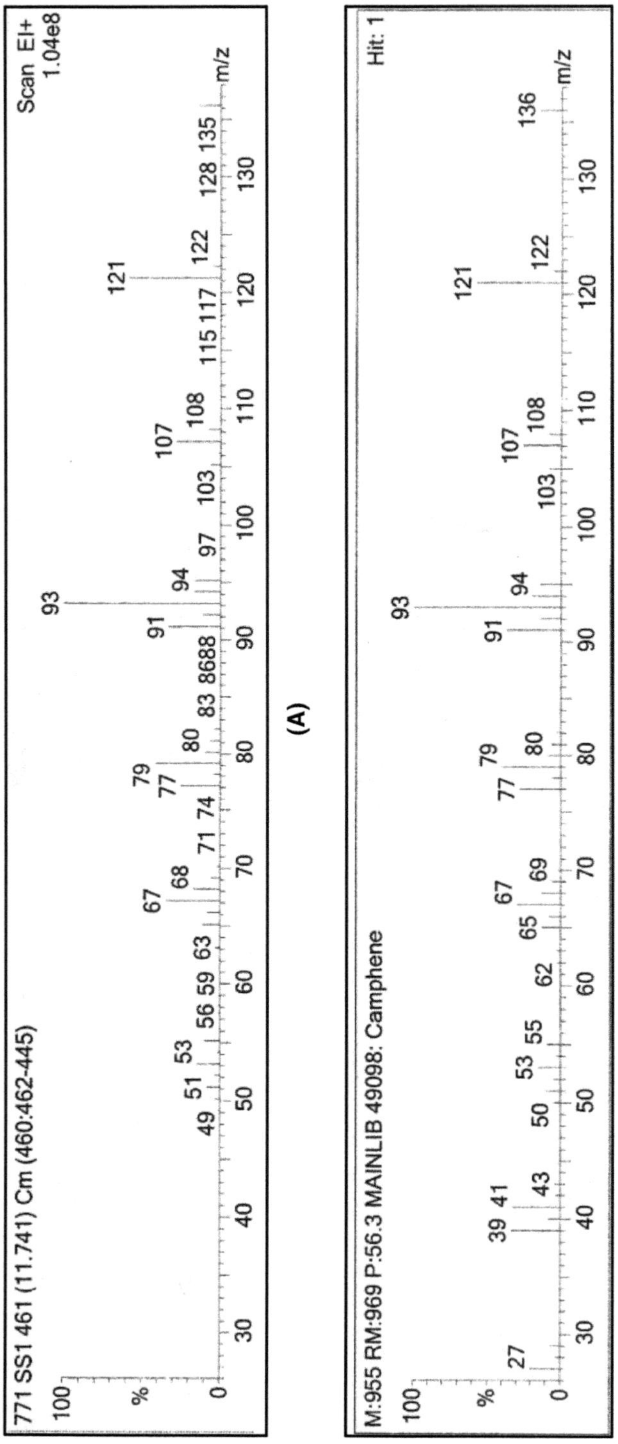

Figure 3.5: Comparison of the MS fragmentation between peak GC-CB2 (A) and data from the natural compound library (B).

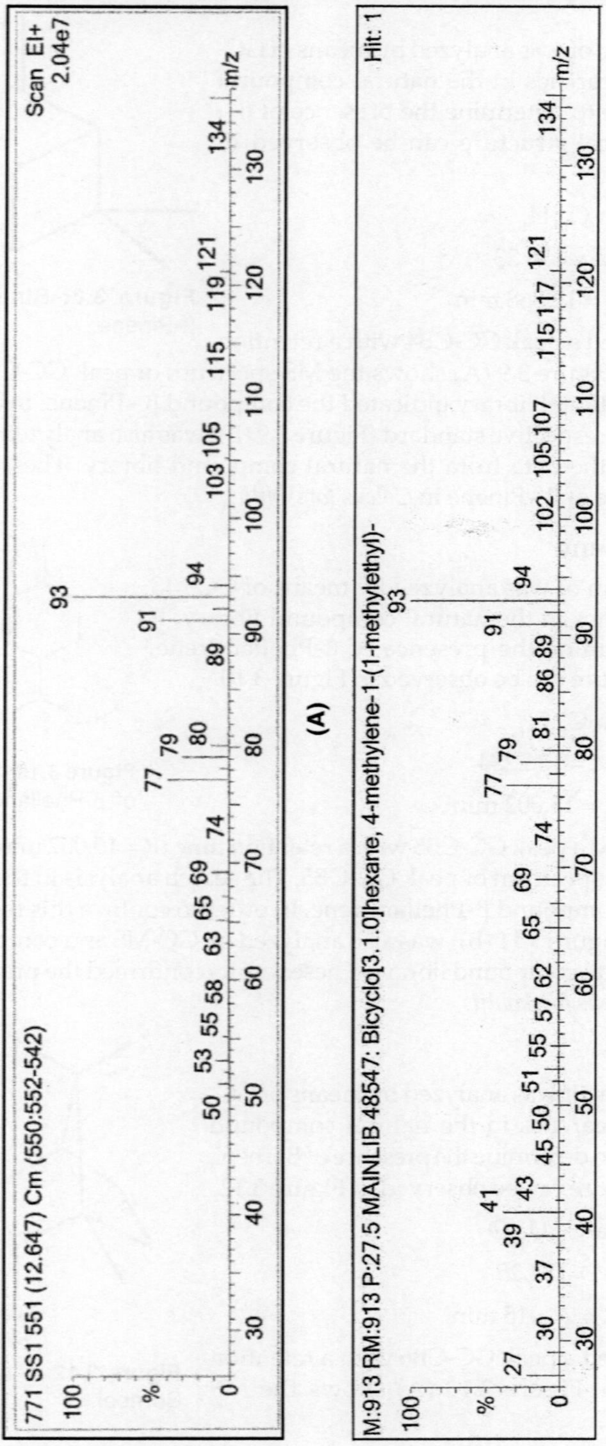

Figure 3.7: Comparison of the MS fragmentation between peak GC-CB3 (A) and data from the natural compound library (B).

CB4: β-Pinene

The mass spectrum of was analyzed by means of GC-MS using computer searches in the natural compound library. It was possible to determine the presence of β –Pinene, whose chemical structure can be observed in Figure 3.8.

Molecular formula-$C_{10}H_{16}$

Molecular mass: M = 136.23

Retention time: tR = 12.868 min

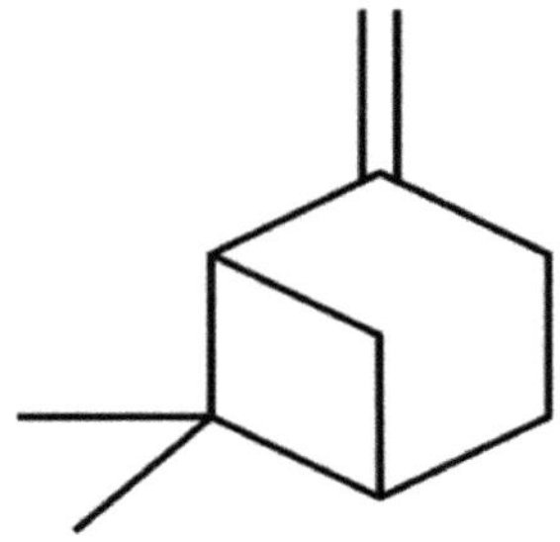

Figure 3.8: Structure of β-Pinene.

Figure 3.1 exhibited a peak GC-CB4 with a retention time tR = 12.868 min. Figure 3.9 (A) shows the MS spectrum of peak GC-CB4. The search analysis in the digital library indicated the compound β –Pinene. In order to confirm this result, the respective standard (Figure 3.9 (B)) was also analyzed by GC-MS and compared to the data from the natural compound library. These results confirmed the presence of β –Pinene in *Coleus forskohlii.*

CB5: β-Phellandrene

The mass spectrum of was analyzed by means of GC-MS using computer searches in the natural compound library. It was possible to determine the presence of β-Phellandrene, whose chemical structure can be observed in Figure 3.10.

Molecular formula-$C_{10}H_{16}$

Molecular mass: M = 136.234

Retention time: tR = 15.002 min

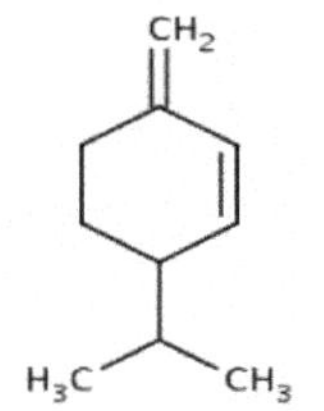

Figure 3.10: Structure of β-Phellandrene.

Figure 3.1 exhibited a peak GC-CB5 with a retention time tR = 15.002 min. Figure 3.11 (A) shows the MS spectrum of peak GC-CB5. The search analysis in the digital library indicated the compound β-Phellandrene. In order to confirm this result, the respective standard (Figure 3.11 (B)) was also analyzed by GC-MS and compared to the data from the natural compound library. These results confirmed the presence of β-Phellandrene in *Coleus forskohlii.*

CB6: Borneol

The mass spectrum of was analyzed by means of GC-MS using computer searches in the natural compound library. It was possible to determine the presence of Borneol, whose chemical structure can be observed in Figure 3.12.

Molecular formula-$C_{10}H_{18}O$

Molecular mass: M =154.25

Retention time: tR = 21.415 min

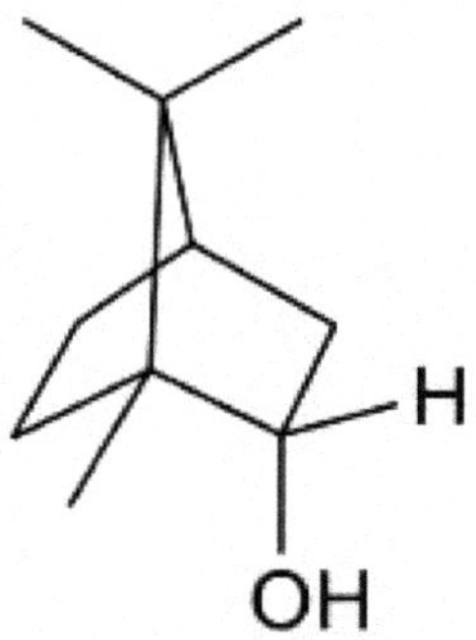

Figure 3.12: Structure of Borneol.

Figure 3.1 exhibited a peak GC-CB6 with a retention time tR = 21.415 min. Figure 3.13 (A) shows the MS

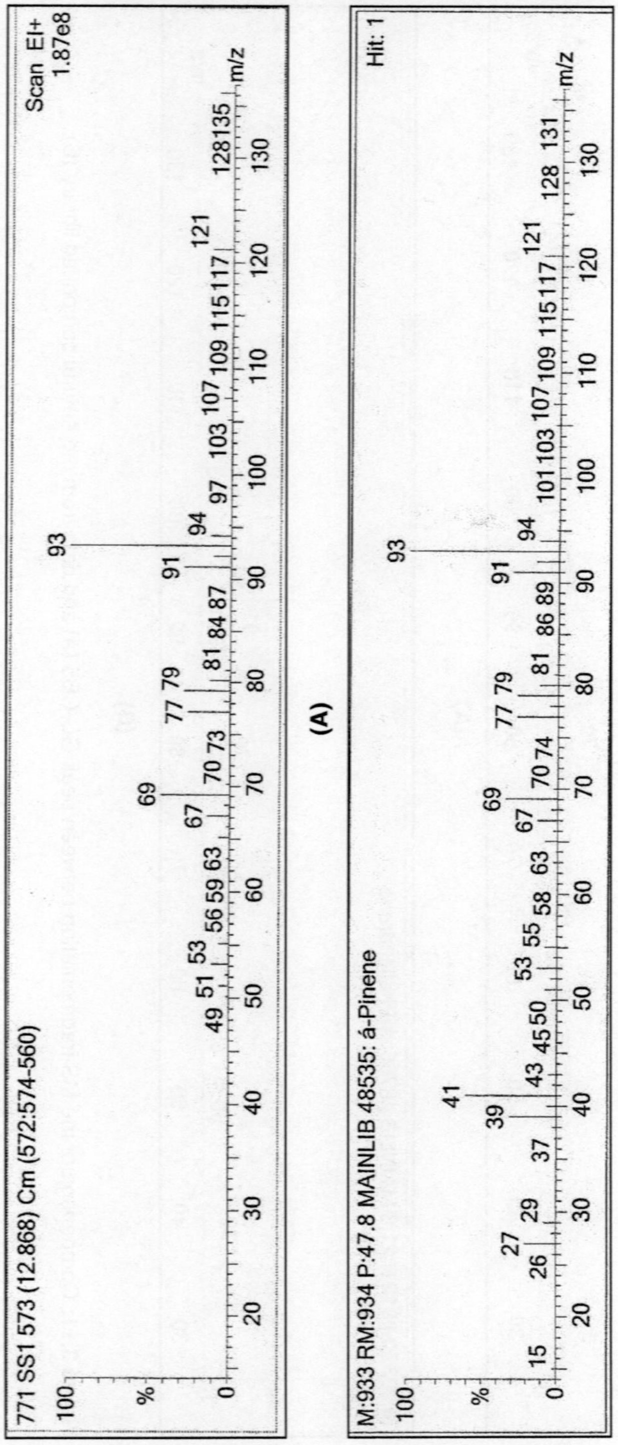

Figure 3.9: Comparison of the MS fragmentation between peak GC-CB4 (A) and data from the natural compound library (B).

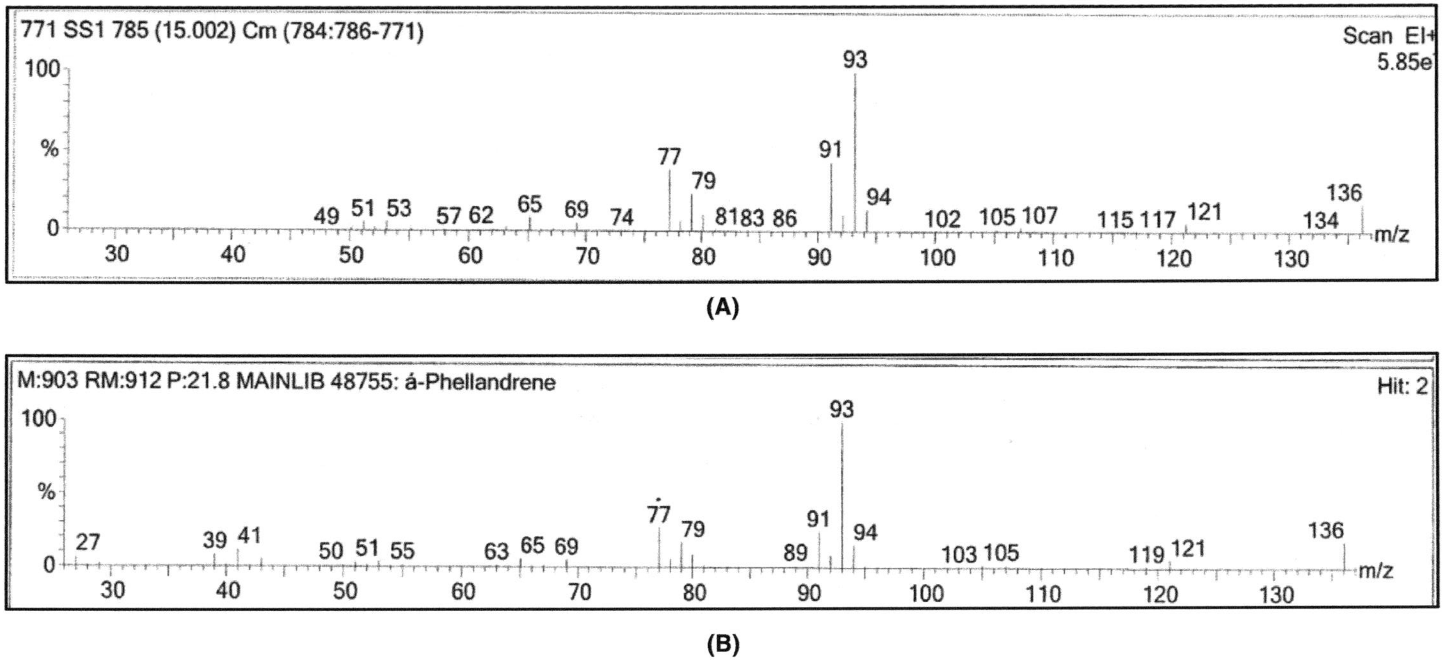

Figure 3.11: Comparison of the MS fragmentation between peak GC-CB5 (A) and data from the natural compound library (B).

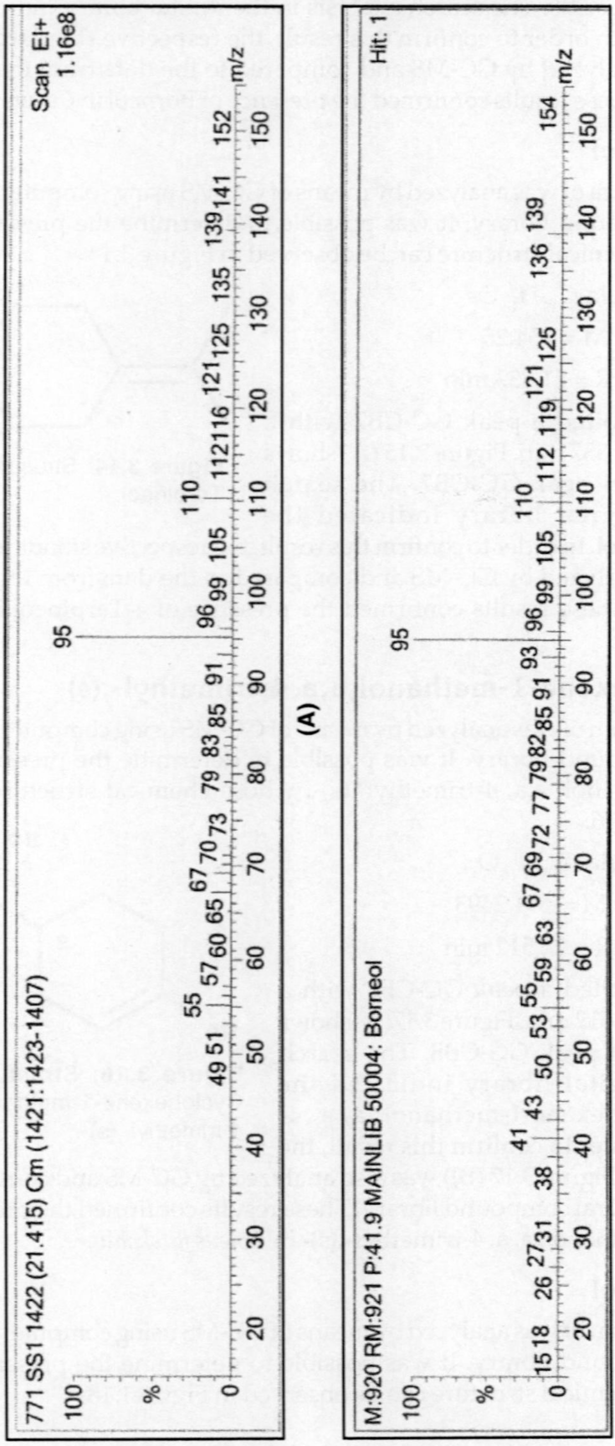

Figure 3.13: Comparison of the MS fragmentation between peak GC-CB6 (A) and data from the natural compound library (B).

spectrum of peak GC-CB6. The search analysis in the digital library indicated the compound Borneol. In order to confirm this result, the respective standard (Figure 3.13 (B)) was also analyzed by GC-MS and compared to the data from the natural compound library. These results confirmed the presence of Borneol in *Coleus forskohlii.*

CB7: 4-Terpineol

The mass spectrum of was analyzed by means of GC-MS using computer searches in the natural compound library. It was possible to determine the presence of 4-Terpineol, whose chemical structure can be observed in Figure 3.14.

Molecular formula-$C_{10}H_{18}$Os

Molecular mass: M = 154.25

Retention time: tR = 21.837 min

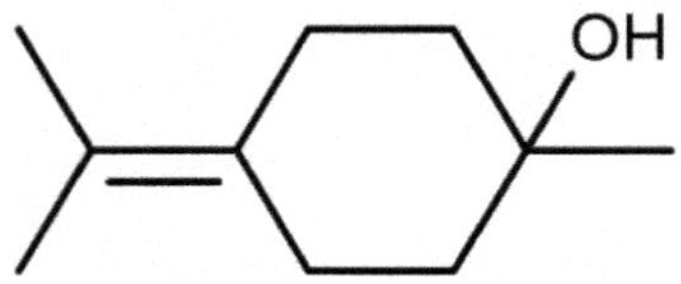

Figure 3.14: Structure of 4-Terpineol.

Figure 3.1 exhibited a peak GC-CB7 with a retention time tR = 21.837 min. Figure 3.15 (A) shows the MS spectrum of peak GC-CB7. The search analysis in the digital library indicated the compound 4-Terpineol. In order to confirm this result, the respective standard (Figure 3.15 (B)) was also analyzed by GC-MS and compared to the data from the natural compound library. These results confirmed the presence of 4-Terpineol in *Coleus forskohlii.*

CB8: 3-Cyclohexene-1-methanol,a,a,4-trimethyl-,(s)-

The mass spectrum of was analyzed by means of GC-MS using computer searches in the natural compound library. It was possible to determine the presence of 3-Cyclohexene-1-methanol, a,a, 4-trimethyl-,(s)-, whose chemical structure can be observed in Figure 3.16.

Molecular formula-$C_{10}H_{18}O$

Molecular mass: M = 154.2493

Retention time: tR = 22.512 min

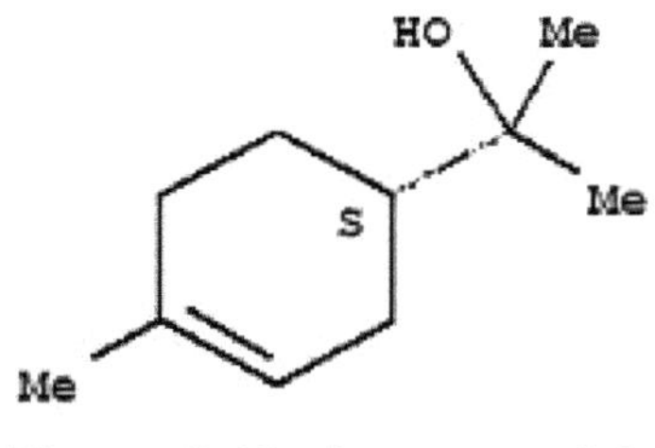

Figure 3.16: Structure of 3-Cyclohexene-1-methanol, a, a, 4-trimethyl-,(s)-.

Figure 3.1 exhibited a peak GC-CB8 with a retention time tR = 22.512 min. Figure 3.17(A) shows the MS spectrum of peak GC-CB8. The search analysis in the digital library indicated the compound 3-Cyclohexene-1-methanol, a, a, 4-trimethyl-, (s)-. In order to confirm this result, the respective standard (Figure 3.17 (B)) was also analyzed by GC-MS and compared to the data from the natural compound library. These results confirmed the presence of 3-Cyclohexene-1-methanol, a, a, 4-trimethyl-,(s)- in *Coleus forskohlii.*

CB9: 2-decen-1-ol

The mass spectrum of was analyzed by means of GC-MS using computer searches in the natural compound library. It was possible to determine the presence of 2-decen-1-ol, whose chemical structure can be observed in Figure1.18.

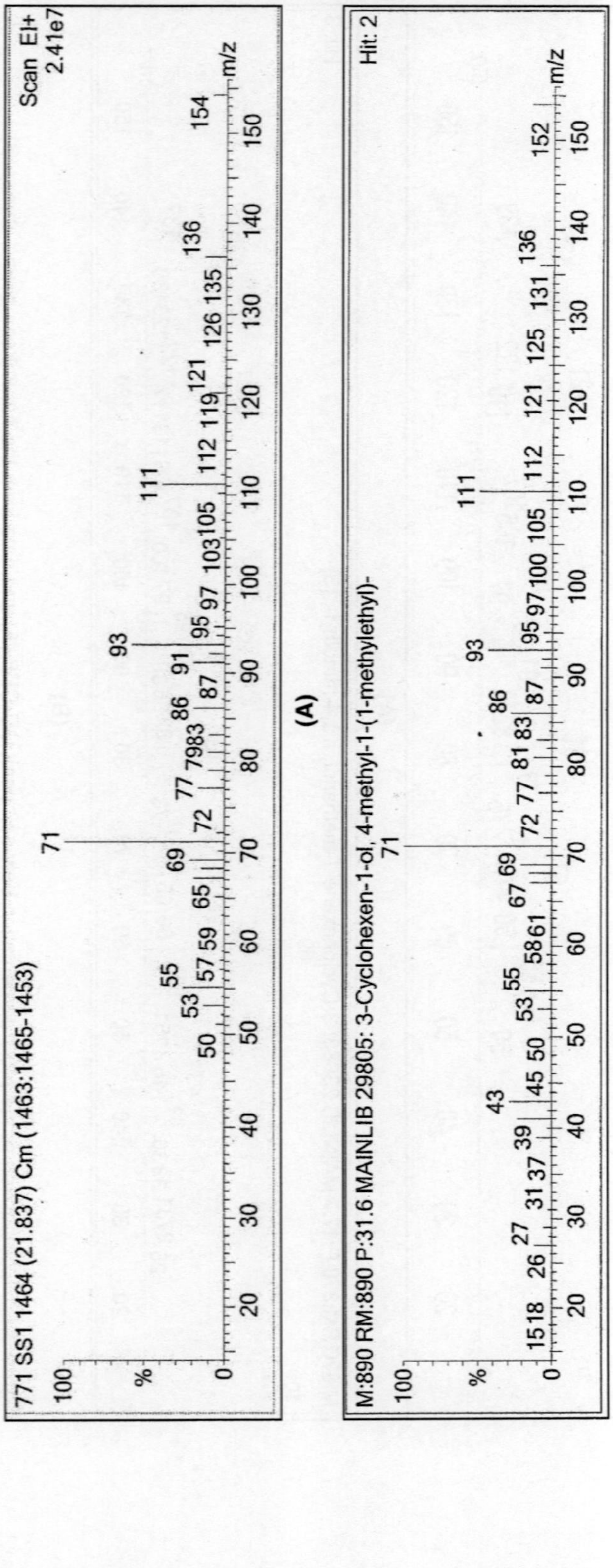

Figure 3.15: Comparison of the MS fragmentation between peak GC-CB7 (A) and data from the natural compound library (B).

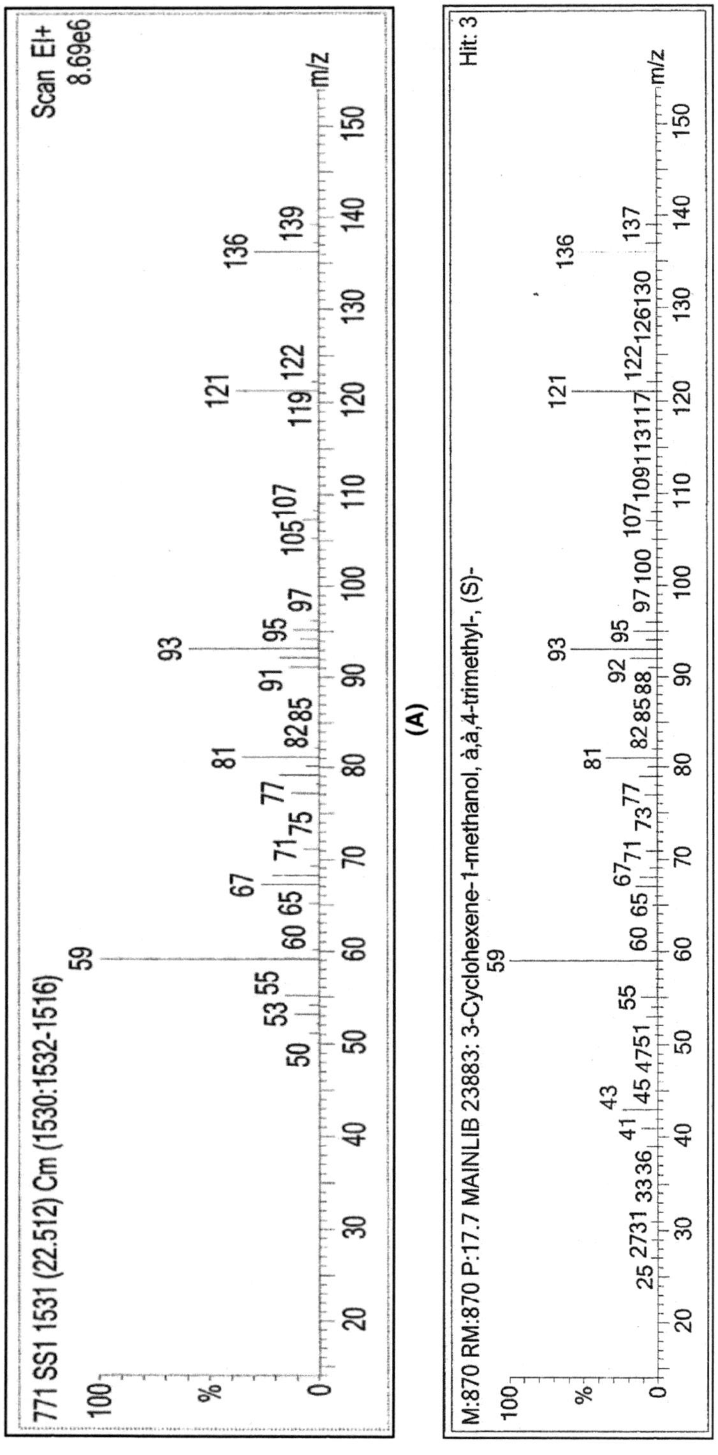

Figure 3.17: Comparison of the MS fragmentation between peak GC-CB8 (A) and data from the natural compound library (B).

Molecular formula-$C_{10}H_{20}O$

Molecular mass: M =156.265

Retention time : tR = 23.14 min

Figure 3.18: Structure of 2-decen-1-ol.

Figure 3.1 exhibited a peak GC-CB9 with a retention time tR = 23.14 min. Figure 3.19 (A) shows the MS spectrum of peak GC-CB9. The search analysis in the digital library indicated the compound 2-decen-1-ol. In order to confirm this result, the respective standard (Figure 3.19 (B)) was also analyzed by GC-MS and compared to the data from the natural compound library. These results confirmed the presence of 2-decen-1-ol in *Coleus forskohlii*.

CB10: Benzoic acid, 2,6-dimethyl-, methyl ester

The mass spectrum of was analyzed by means of GC-MS using computer searches in the natural compound library. It was possible to determine the presence of Benzoic acid, 2,6-dimethyl-, methyl ester, whose chemical structure can be observed in Figure 3.20.

Molecular formula- $C_{10}H_{12}O_2$

Molecular mass: M = 164.20

Retention time: tR = 24.525 min

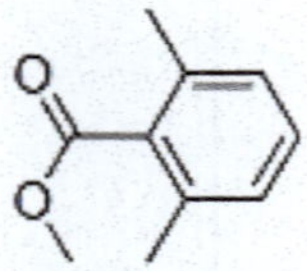

Figure 3.20: Structure of Benzoic acid, 2,6-dimethyl-, methyl ester.

Figure 3.1 exhibited a peak GC-CB10 with a retention time tR = 24.525 min. Figure 3.21 (A) shows the MS spectrum of peak GC-CB10. The search analysis in the digital library indicated the compound Benzoic acid, 2,6-dimethyl-, methyl ester. In order to confirm this result, the respective standard (Figure 3.21 (B)) was also analyzed by GC-MS and compared to the data from the natural compound library. These results confirmed the presence of Benzoic acid, 2,6-dimethyl-, methyl ester in *Coleus forskohlii*.

CB11: Cyclopropane, nonyl-

The mass spectrum of was analyzed by means of GC-MS using computer searches in the natural compound library. It was possible to determine the presence of Cyclopropane, nonyl: , whose chemical structure can be observed in Figure 3.22.

Molecular formula-$C_{12}H_{24}$

Molecular mass: M =168.319

Retention time: tR = 26.21 min

MS data (GC-MS): see in Figure 3.1

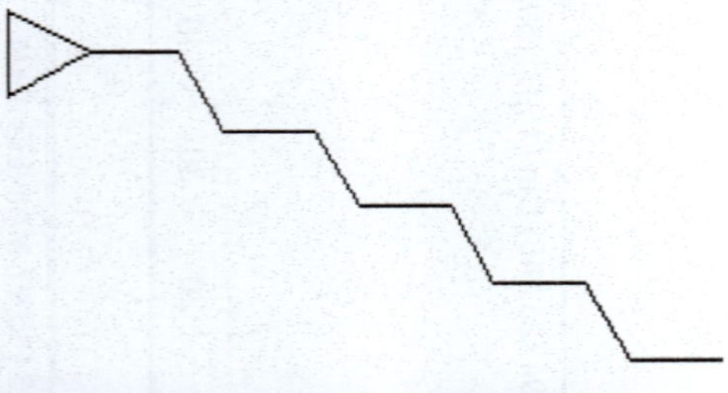

Figure 3.22: Structure of Cyclopropane, nonyl.

Figure 3.1 exhibited a peak GC-CB11 with a retention time tR = 26.21 min. Figure 3.23 (A) shows the MS spectrum of peak GC-CB11. The search analysis in the digital library indicated the compound Cyclopropane, nonyl. In order to confirm this result, the respective standard (Figure 3.23 (B)) was also analyzed by GC-MS and compared to the data

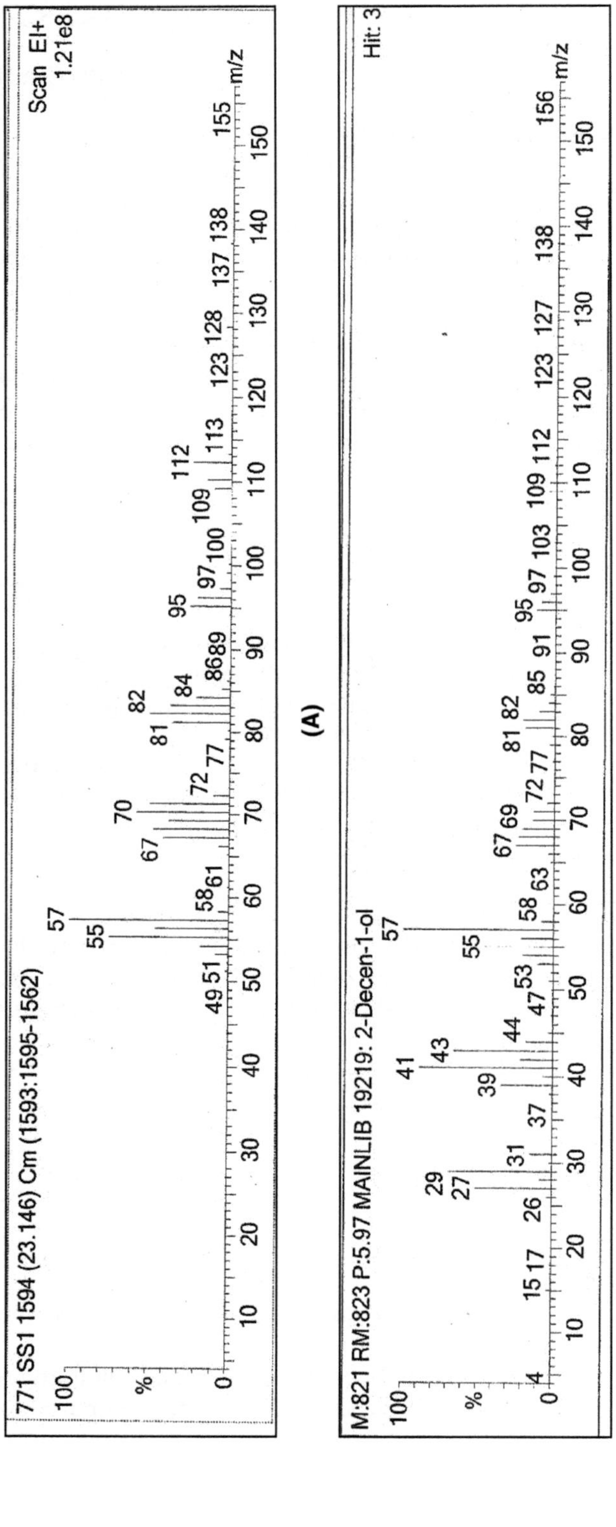

Figure 3.19: Comparison of the MS fragmentation between peak GC-CB9 (A) and data from the natural compound library (B).

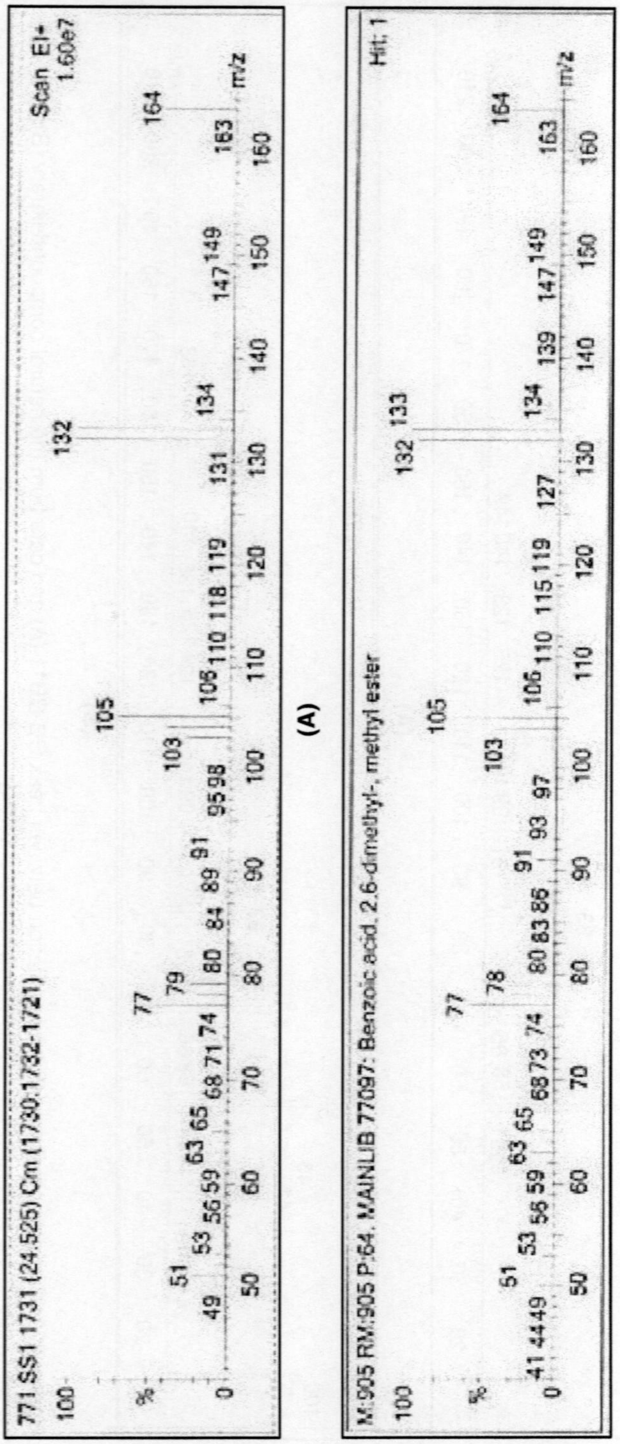

Figure 3.21: Comparison of the MS fragmentation between peak GC-CB10 (A) and data from the natural compound library (B).

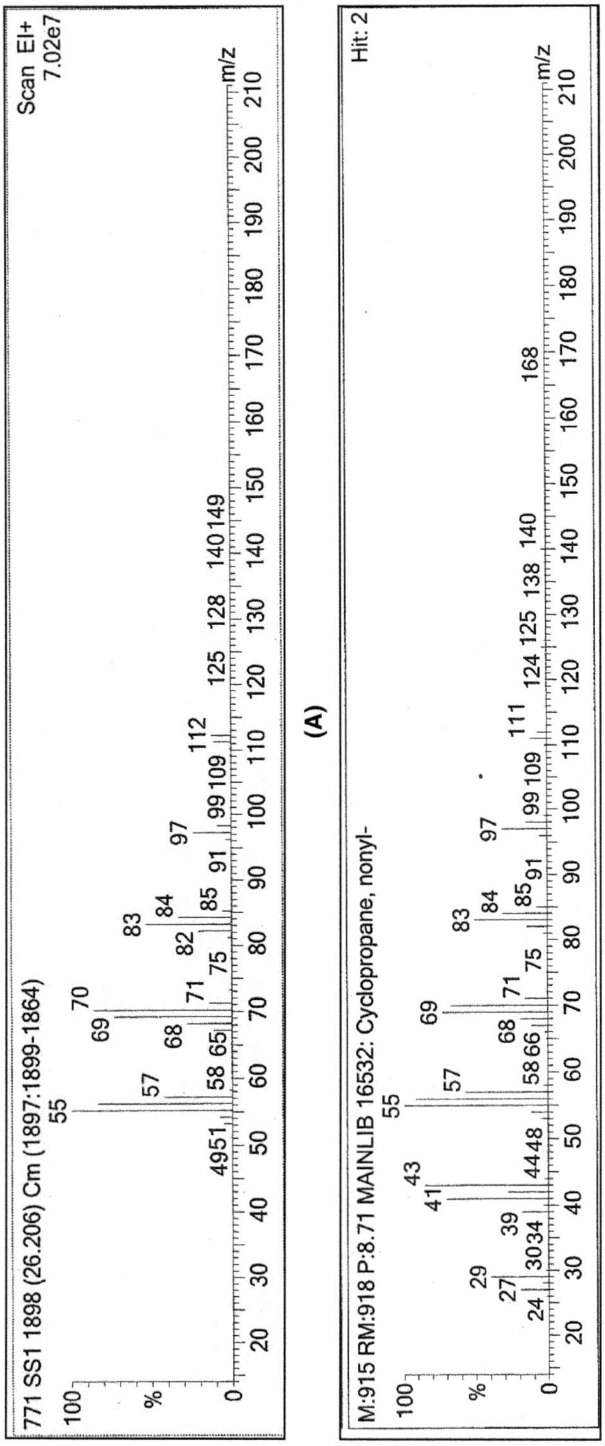

Figure 3.23: Comparison of the MS fragmentation between peak GC-CB11 (A) and data from the natural compound library (B).

from the natural compound library. These results confirmed the presence of Cyclopropane, nonyl in *Coleus forskohlii.*

CB12: Bornyl Acetate

The mass spectrum of was analyzed by means of GC-MS using computer searches in the natural compound library. It was possible to determine the presence of Bornyl acetate, whose chemical structure can be observed in Figure 3.24.

Molecular formula-$C_{12}H_{20}O_2$

Molecular mass: M =196.29

Retention time: tR = 26.91 min

Figure 3.1 exhibited a peak GC-CB12 with a retention time tR = 26.91 min. Figure 3.25 (A) shows the MS spectrum of peak GC-CB12. The search analysis in the digital library indicated the compound Bornyl acetate. In order to confirm this result, the respective standard (Figure 3.25 (B)) was also analyzed by GC-MS and compared to the data from the natural compound library. These results confirmed the presence of Bornyl acetate in *Coleus forskohlii.*

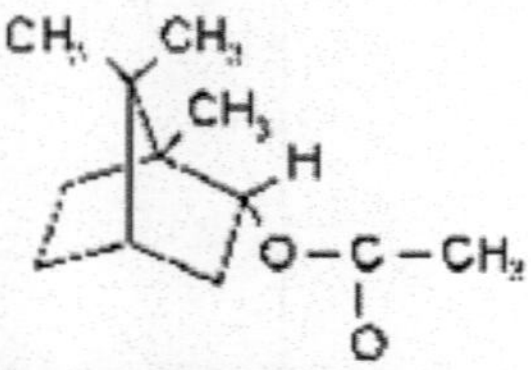

Figure 3.24: Structure of Bornyl acetate.

IUPAC- (1,7,7-trimethyl-6-bicyclo[2.2.1]heptanyl) acetate.

CB13: Azulene, 1,2,3,4,5,6,7,8,8a-octahydro-1,4-dimethyl-7-(1-methylethenyl)-

The mass spectrum of was analyzed by means of GC-MS using computer searches in the natural compound library. It was possible to determine the presence of alpha-bulnesene, whose chemical structure can be observed in Figure 3.26.

Molecular formula-$C_{15}H_{24}$

Molecular mass: M = 204.351

Retention time: tR = 31.381 min

Figure 3.1 exhibited a peak GC-CB13 with a retention time tR = 31.381 min. Figure 3.27 (A) shows the MS spectrum of peak GC-CB13. The search analysis in the digital library indicated the compound alpha-bulnesene. In order to confirm this result, the respective standard (Figure 3.27 (B)) was also analyzed by GC-MS and compared to the data from the natural compound library. These results confirmed the presence of alpha-bulnesene in *Coleus forskohlii.*

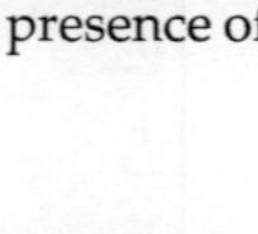

Figure 3.26: Structure of alpha-bulnesene.

CB14: Decyl Acetate

The mass spectrum of was analyzed by means of GC-MS using computer searches in the natural compound library. It was possible to determine the presence of Decyl Acetate, whose chemical structure can be observed in Figure 3.28.

Molecular formula-$C_{12}H_{24}O_2$

Molecular mass: M = 200.317

Retention time: tR = 31.723 min

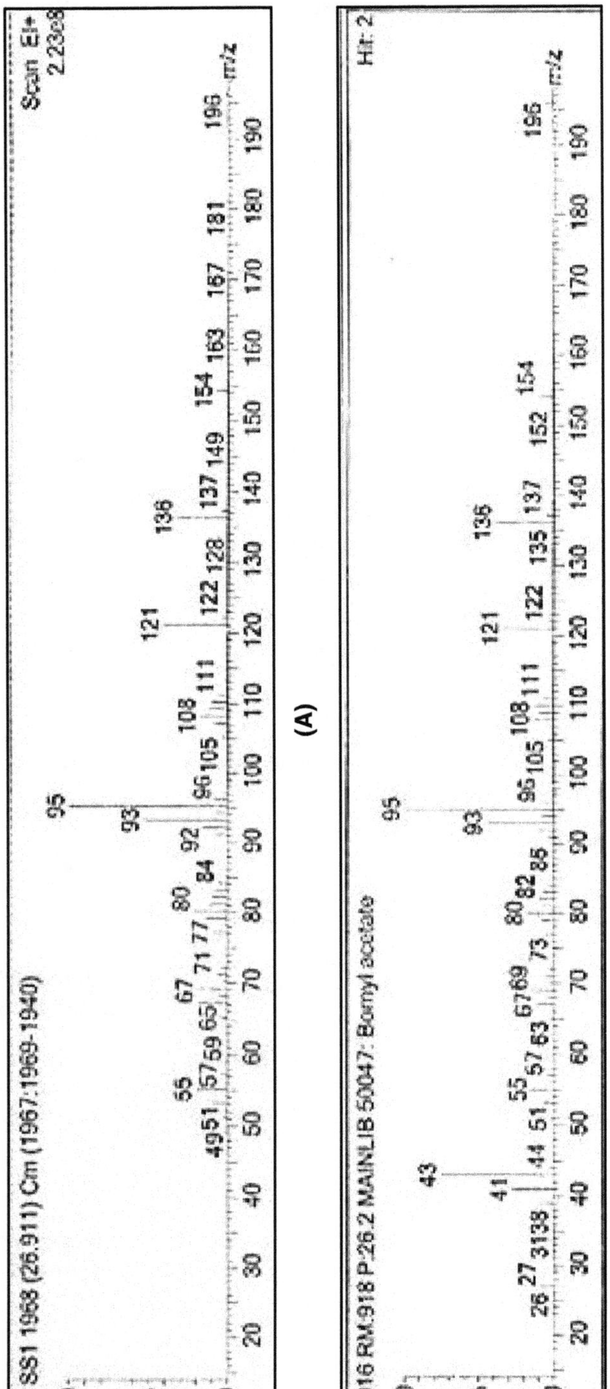

Figure 3.25: Comparison of the MS fragmentation between peak GC-CB12 (A) and data from the natural compound library (B).

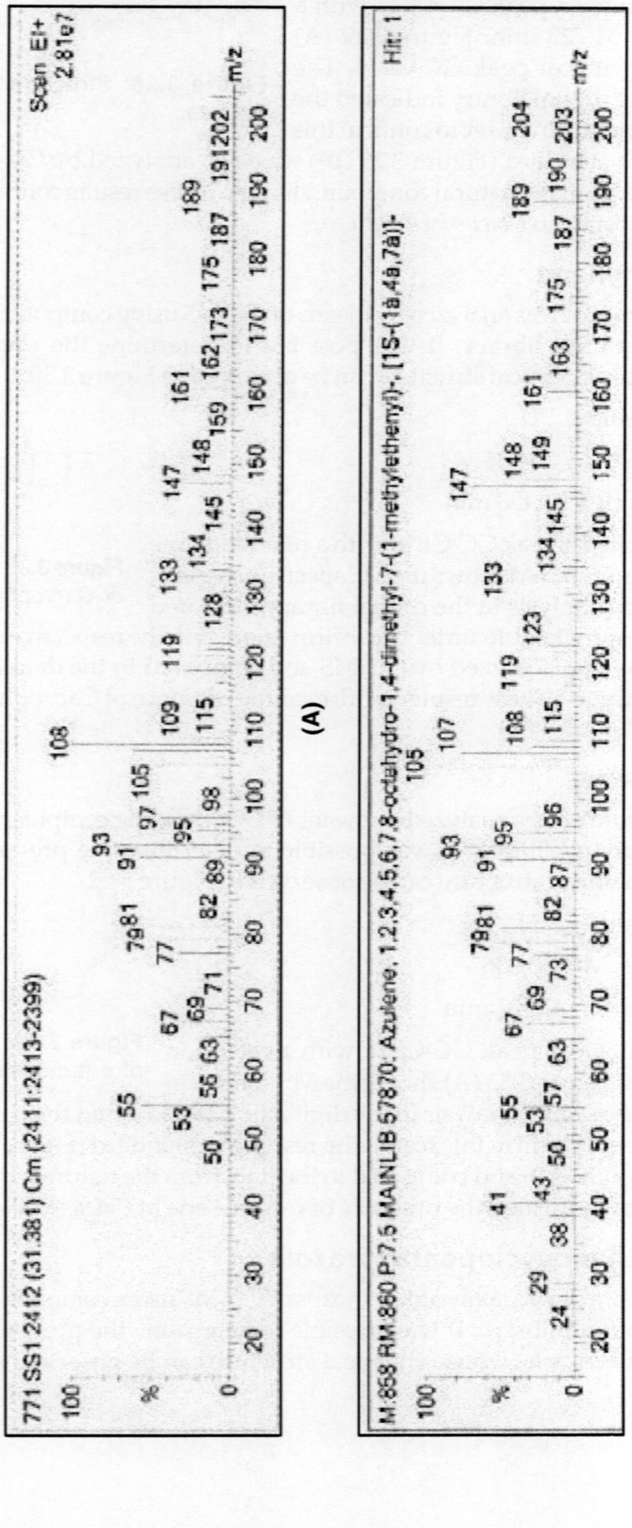

Figure 3.27: Comparison of the MS fragmentation between peak GC-CB13 (A) and data from the natural compound library (B).

Figure 3.1 exhibited a peak GC-CB14 with a retention time tR = 31.723 min. Figure 3.29 (A) shows the MS spectrum of peak GC-CB14. The search analysis in the digital library indicated the compound Decyl Acetate. In order to confirm this result, the respective standard (Figure 3.29 (B)) was also analyzed by GC-MS and compared to the data from the natural compound library. These results confirmed the presence of Decyl Acetate in *Coleus forskohlii*.

Figure 3.28: Structure of Decyl Acetate.

CB15: Caryophyllene

The mass spectrum of was analyzed by means of GC-MS using computer searches in the natural compound library. It was possible to determine the presence of Caryophyllene, whose chemical structure can be observed in Figure 3.30.

Molecular formula-$C_{15}H_{24}$

Molecular mass: M = 204.36

Retention time: tR = 32.438 min

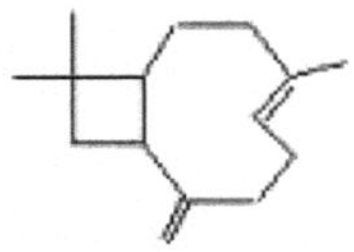

Figure 3.30: Structure of Caryophyllene.

Figure 3.1 exhibited a peak GC-CB15 with a retention time tR = 32.438 min. Figure 3.31 (A) shows the MS spectrum of peak GC-CB15. The search analysis in the digital library indicated the compound Caryophyllene. In order to confirm this result, the respective standard (Figure 3.31 (B)) was also analyzed by GC-MS and compared to the data from the natural compound library. These results confirmed the presence of Caryophyllene in *Coleus forskohlii*.

CB16: α-farnesene

The mass spectrum of was analyzed by means of GC-MS using computer searches in the natural compound library. It was possible to determine the presence of α-farnesene, whose chemical structure can be observed in Figure 3.32.

Molecular formula-$C_{15}H_{24}$

Molecular mass: M = 204.36

Retention time: tR = 32.921 min

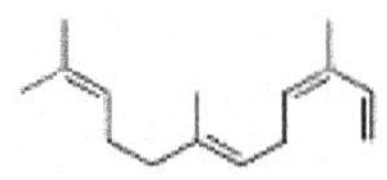

Figure 3.32: Structure of α-farnesene.

Figure 3.1 exhibited a peak GC-CB16 with a retention time tR = 32.921 min. Figure 3.33 (A) shows the MS spectrum of peak GC-CB16. The search analysis in the digital library indicated the compound α-farnesene. In order to confirm this result, the respective standard (Figure 3.33 (B)) was also analyzed by GC-MS and compared to the data from the natural compound library. These results confirmed the presence of α-farnesene in *Coleus forskohlii*.

CB17: 3,5-Methanocyclopentapyrazole

The mass spectrum of was analyzed by means of GC-MS using computer searches in the natural compound library. It was possible to determine the presence of 3,5-Methanocyclopentapyrazole, whose chemical structure can be observed in Figure 3.34.

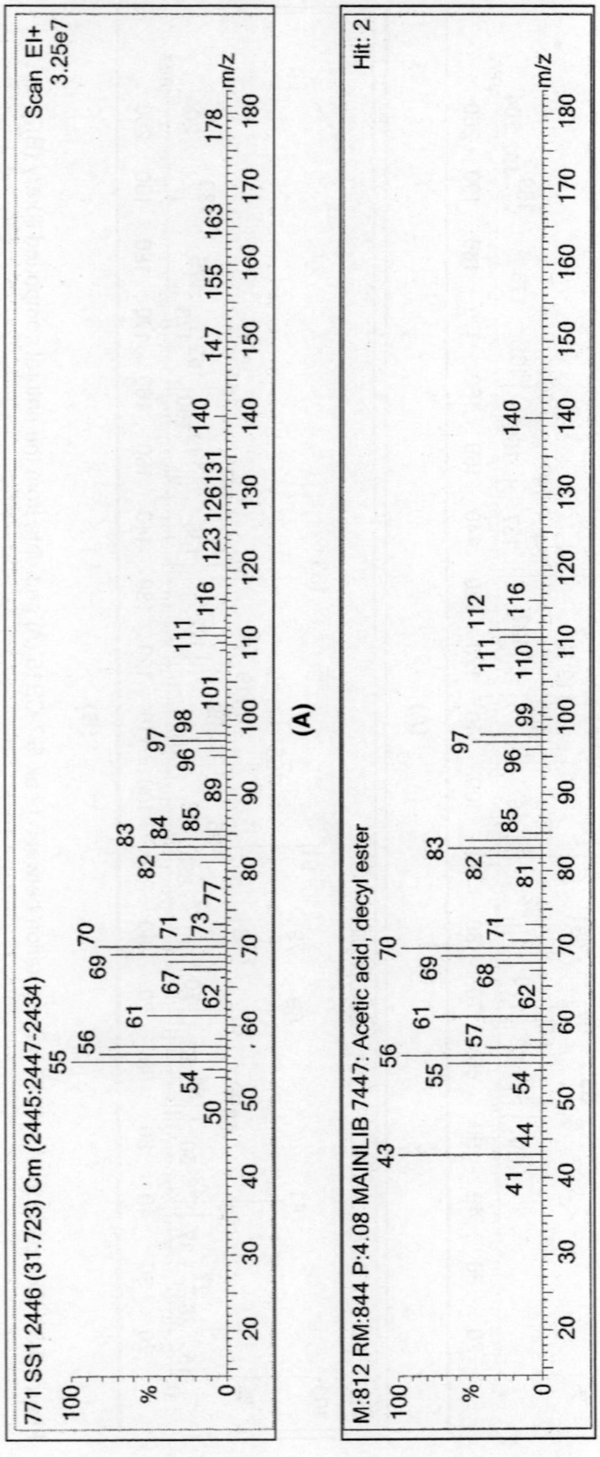

Figure 3.29: Comparison of the MS fragmentation between peak GC-CB14 (A) and data from the natural compound library (B).

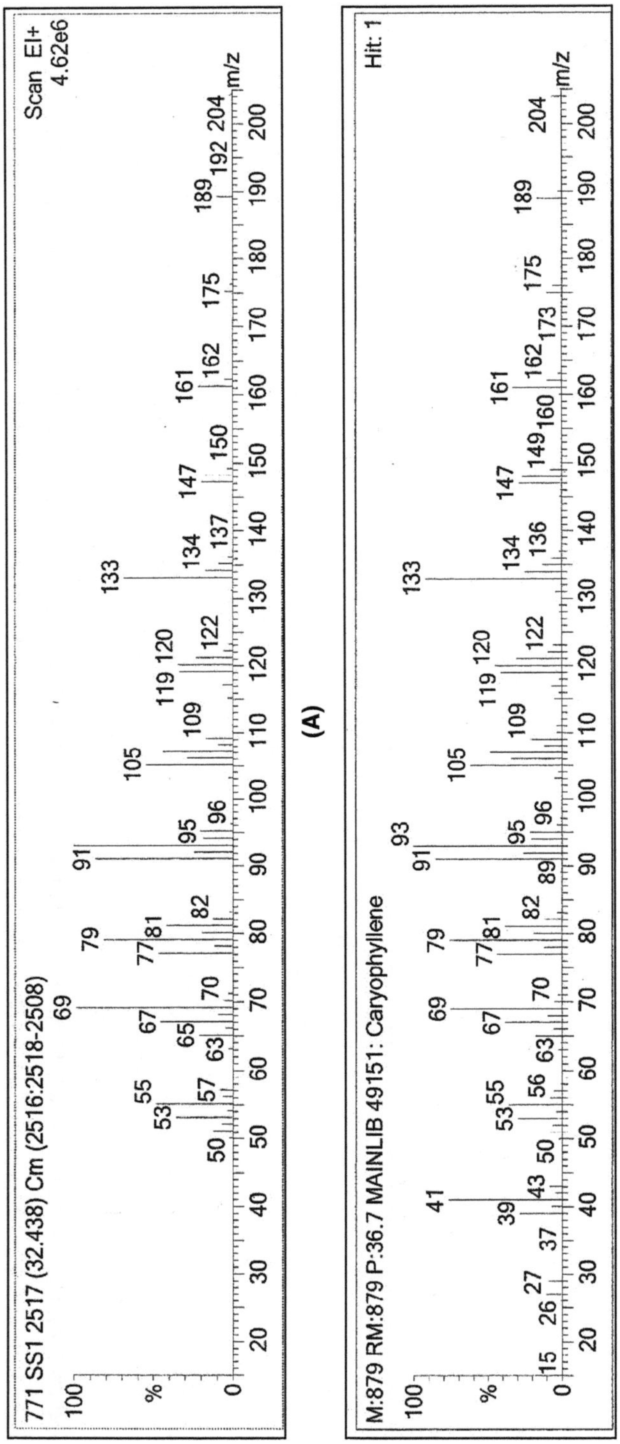

Figure 3.31: Comparison of the MS fragmentation between peak GC-CB15 (A) and data from the natural compound library (B).

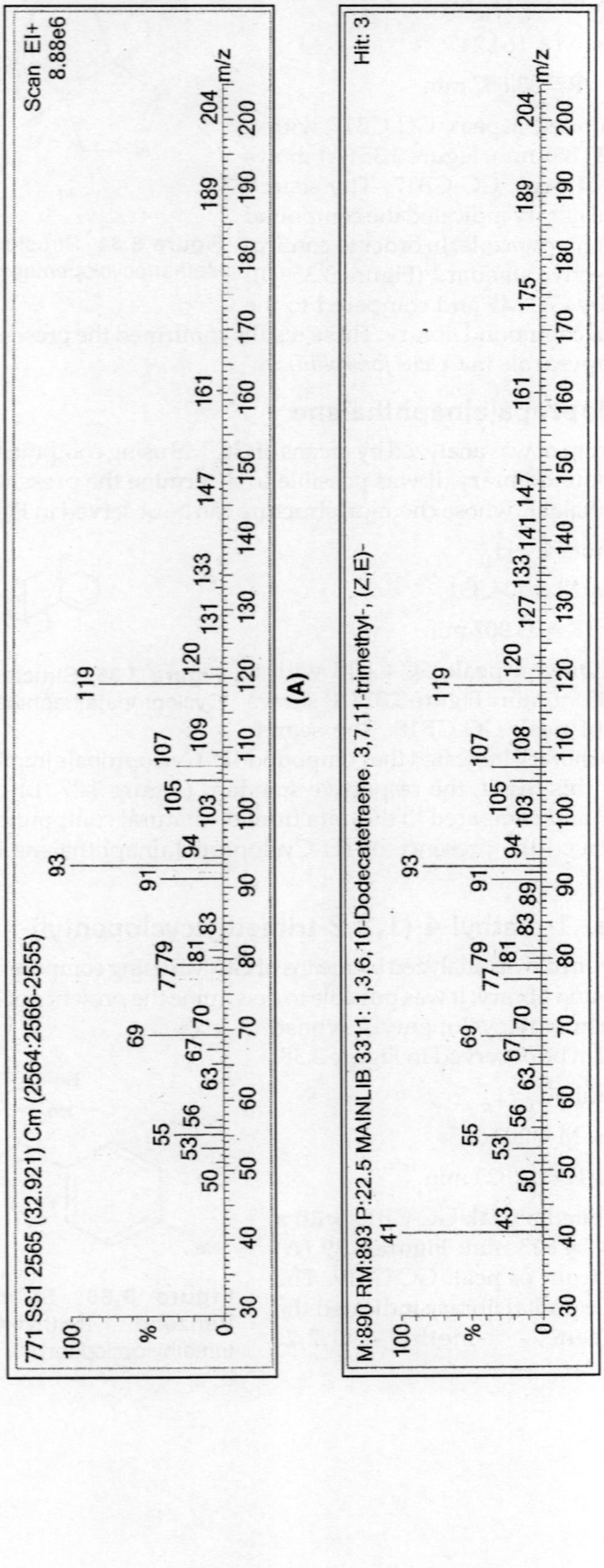

Figure 3.33: Comparison of the MS fragmentation between peak GC-CB16 (A) and data from the natural compound library (B).

Molecular formula-$C_{10}H_{16}N_2$

Molecular mass: M = 164.247

Retention time: tR = 33.857 min

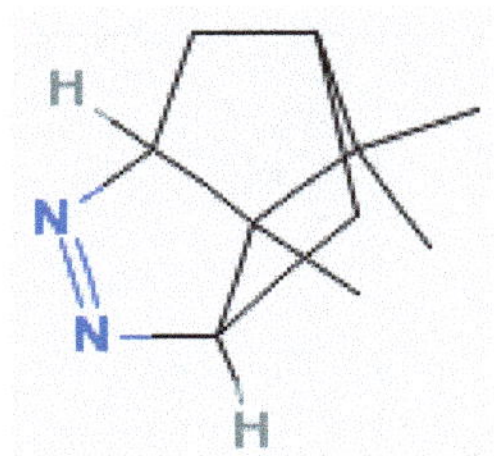

Figure 3.34: Structure of 3,5-Methanocyclopentapyrazole.

Figure 3.1 exhibited a peak GC-CB17 with a retention time tR = 33.857 min. Figure 3.35 (A) shows the MS spectrum of peak GC-CB17. The search analysis in the digital library indicated the compound 3, 5-Methanocyclopentapyrazole. In order to confirm this result, the respective standard (Figure 3.35 (B)) was also analyzed by GC-MS and compared to the data from the natural compound library. These results confirmed the presence of 3,5-Methanocyclopentapyrazole in *Coleus forskohlii.*

CB18: 1H-Cyclopropa[a]naphthalene

The mass spectrum of was analyzed by means of GC-MS using computer searches in the natural compound library. It was possible to determine the presence of 1H-Cyclopropa[a]naphthalene, whose chemical structure can be observed in Figure 3.36.

Molecular formula-$C_{15}H_{24}$

Molecular mass: M = 204.351

Retention time: tR = 33.907 min

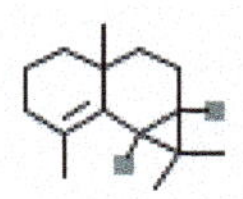

Figure 3.36: Structure of 1H-Cyclopropa[a]naphthalene.

Figure 3.1 exhibited a peak GC-CB18 with a retention time tR = 33.907 min. Figure 3.37 (A) shows the MS spectrum of peak GC-CB18. The search analysis in the digital library indicated the compound 1H-Cyclopropa[a]naphthalene. In order to confirm this result, the respective standard (Figure 3.37 (B)) was also analyzed by GC-MS and compared to the data from the natural compound library. These results confirmed the presence of 1H-Cyclopropa[a]naphthalene in *Coleus forskohlii*.

CB19: Benzene, 1-methyl-4-(1,2,2-trimethylcyclopentyl)-, (R)-

The mass spectrum of was analyzed by means of GC-MS using computer searches in the natural compound library. It was possible to determine the presence of Benzene, 1-methyl-4-(1,2,2-trimethylcyclopentyl), whose chemical structure can be observed in Figure 3.38.

Molecular formula-$C_{15}H_{22}$

Molecular mass: M = 202.335

Retention time: tR = 34.823 min

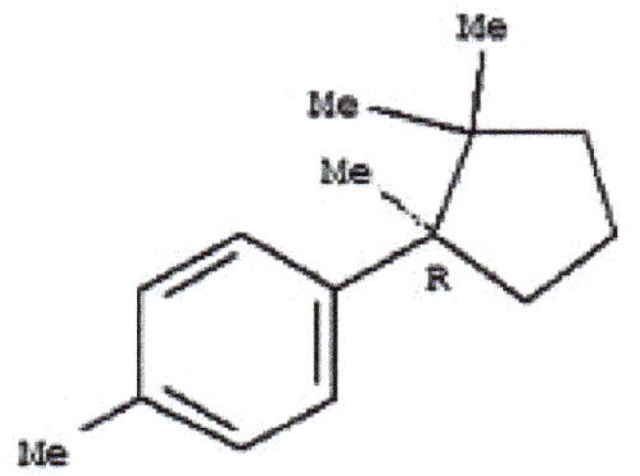

Figure 3.38: Structure of Benzene, 1-methyl-4-(1,2,2-trimethylcyclopentyl)-

Figure 3.1 exhibited a peak GC-CB19 with a retention time tR = 34.823 min. Figure 3.39 (A) shows the MS spectrum of peak GC-CB19. The search analysis in the digital library indicated the compound Benzene, 1-methyl-4-(1,2,2-

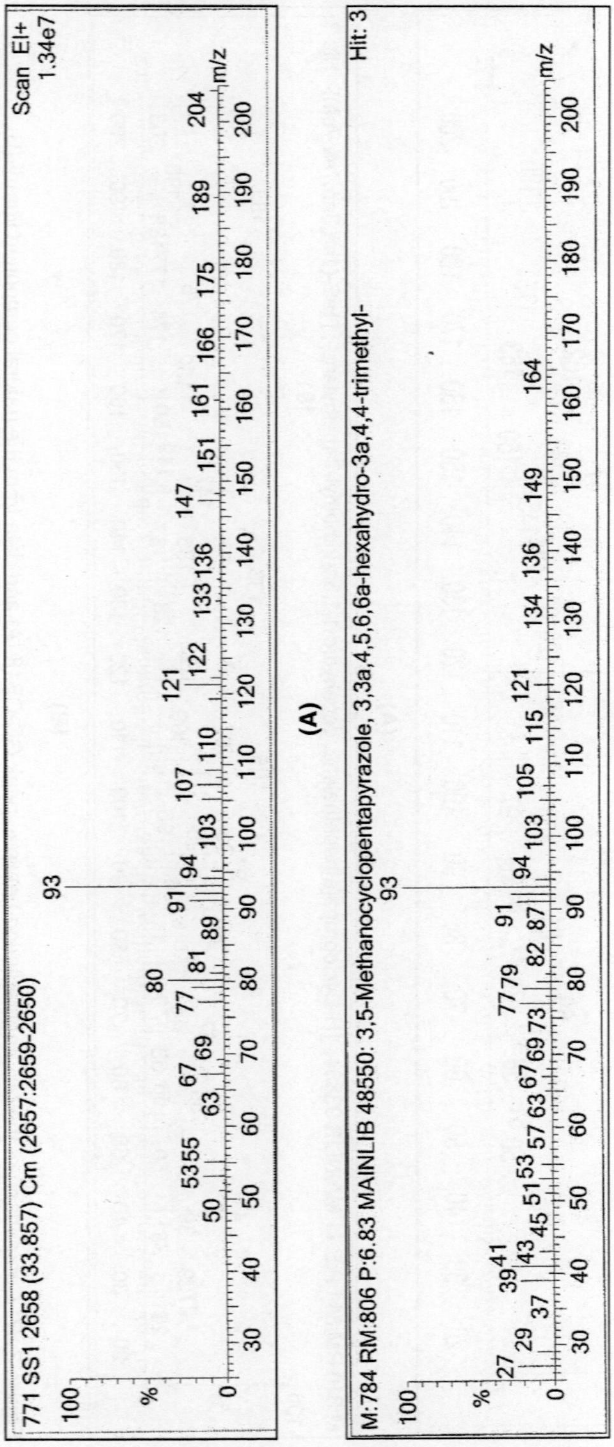

Figure 3.35: Comparison of the MS fragmentation between peak GC-CB17 (A) and data from the natural compound library (B).

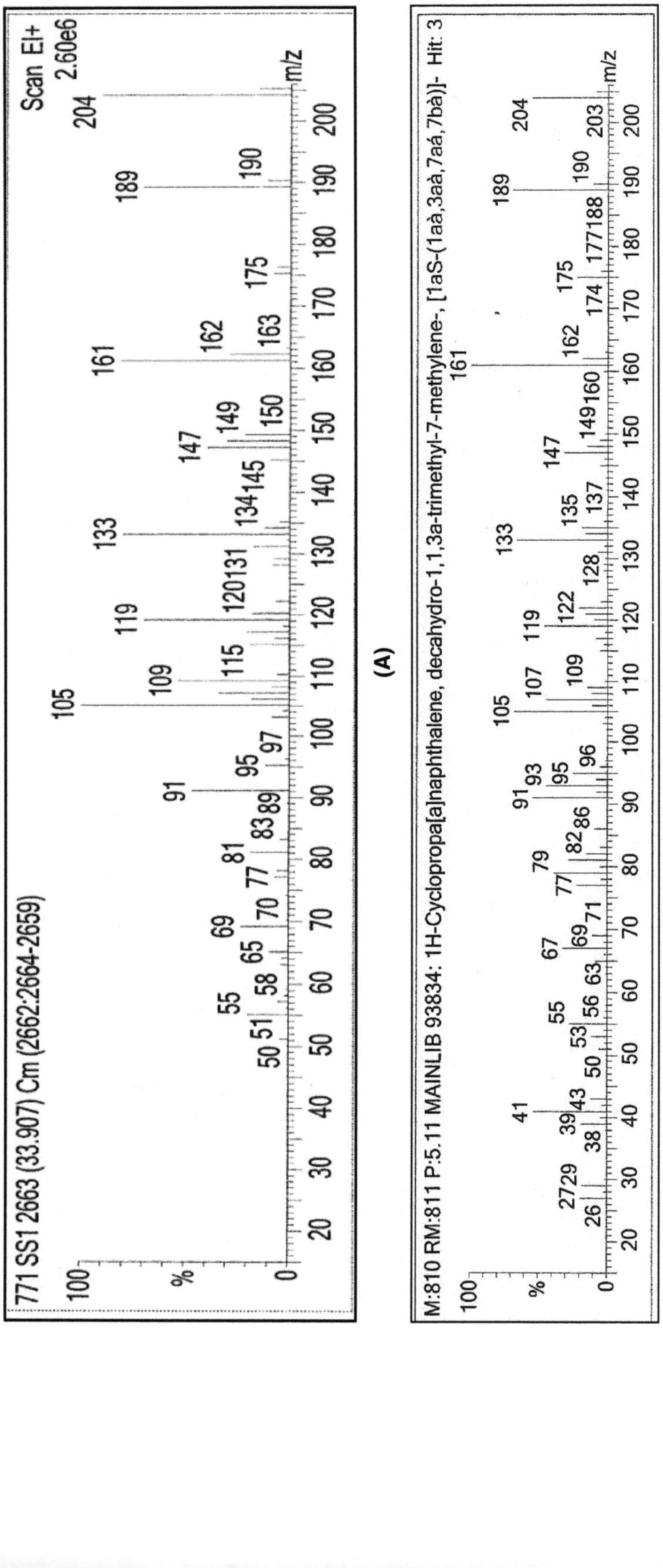

Figure 3.37: Comparison of the MS fragmentation between peak GC-CB18 (A) and data from the natural compound library (B).

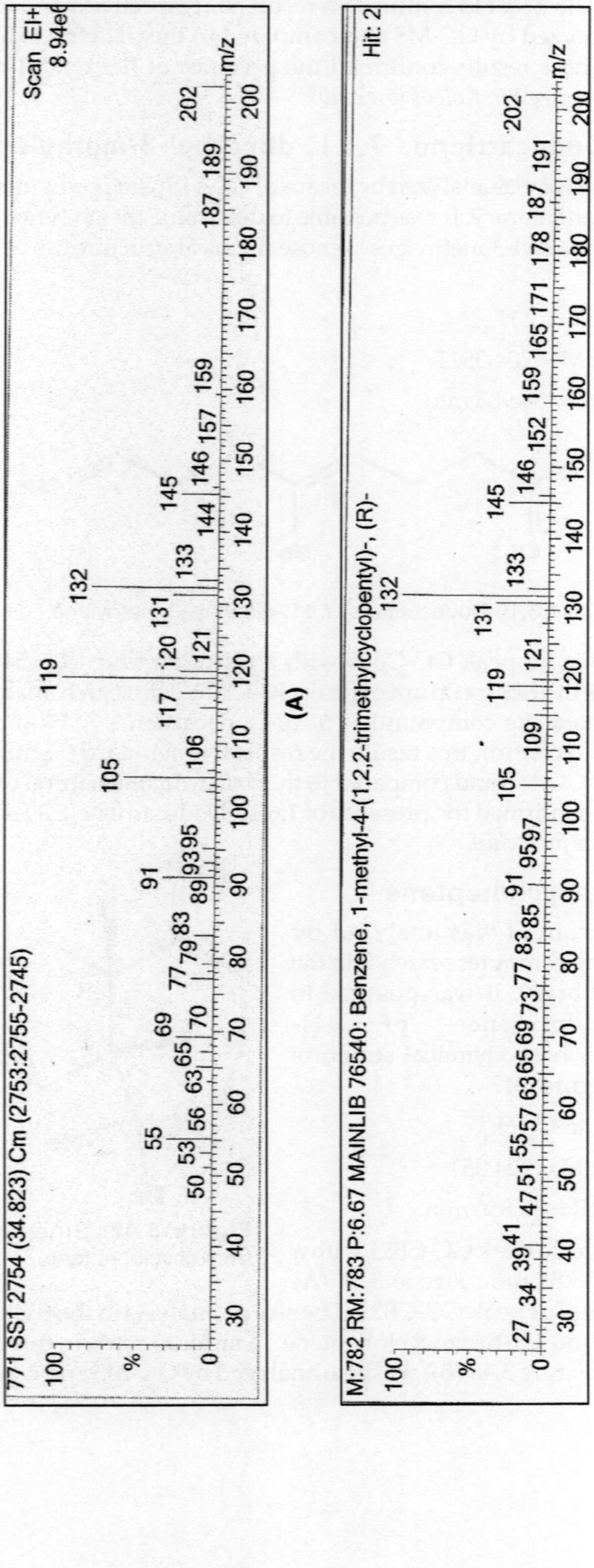

Figure 3.39: Comparison of the MS fragmentation between peak GC-CB19 (A) and data from the natural compound library (B).

trimethylcyclopentyl). In order to confirm this result, the respective standard (Figure 3.39 (B)) was also analyzed by GC-MS and compared to the data from the natural compound library. These results confirmed the presence of Benzene, 1-methyl-4-(1,2,2-trimethylcyclopentyl) in *Coleus forskohlii*.

CB20: 1,6,10-Dodecatriene, 7,11,-dimethyl-3-methylene-

The mass spectrum of was analyzed by means of GC-MS using computer searches in the natural compound library. It was possible to determine the presence of 1, 6,10-Dodecatriene, 7,11,-dimethyl-3-methylene-, whose chemical structure can be observed in Figure 3.40.

Molecular formula-$C_{15}H_{24}$

Molecular mass: M = 204.3511

Retention time: tR = 34.985 min

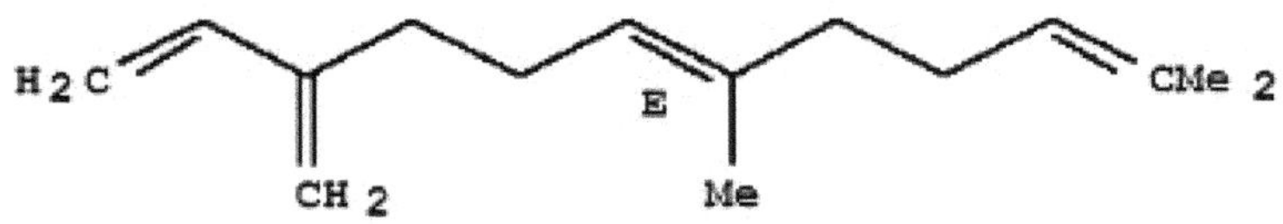

Figure 3.40: Structure of 1,6,10-Dodecatriene, 7,11,-dimethyl-3-methylene.

Figure 3.1 exhibited a peak GC-CB20 with a retention time tR = 34.985 min. Figure 3.41 (A) shows the MS spectrum of peak GC-CB20. The search analysis in the digital library indicated the compound 1, 6, 10-Dodecatriene, 7, 11,-dimethyl-3-methylene-. In order to confirm this result, the respective standard (Figure 3.41 (B)) was also analyzed by GC-MS and compared to the data from the natural compound library. These results confirmed the presence of 1,6,10-Dodecatriene, 7,11,-dimethyl-3-methylene- in *Coleus forskohlii*.

CB21: 1H-Benzocycloheptene

The mass spectrum of was analyzed by means of GC-MS using computer searches in the natural compound library. It was possible to determine the presence of 1H-Benzocycloheptene, whose chemical structure can be observed in Figure 3.42.

Molecular formula-$C_{15}H_{24}$

Molecular mass: M = 204.351

Retention time: tR = 35.850 min

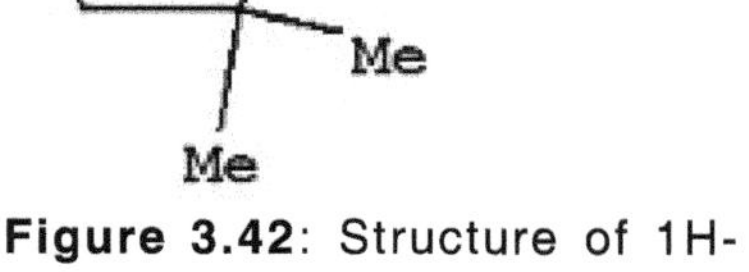

Figure 3.42: Structure of 1H-Benzocycloheptene.

Figure 3.1 exhibited a peak GC-CB21 with a retention time tR = 35.85 min. Figure 3.43 (A) shows the MS spectrum of peak GC-CB21. The search analysis in the digital library indicated the compound 1H-Benzocycloheptene. In order to confirm this result, the respective standard (Figure 3.43 (B)) was also analyzed by GC-MS and compared to

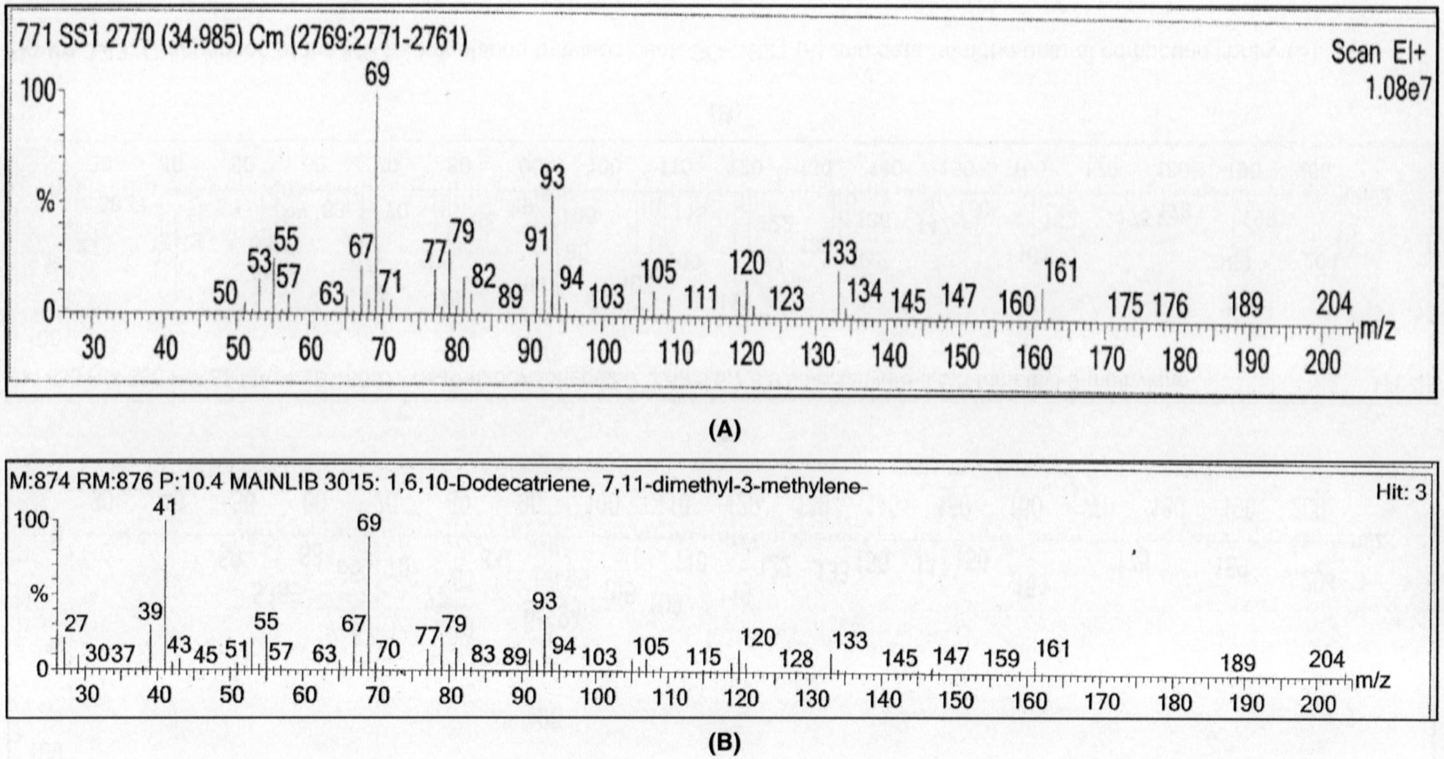

Figure 3.41: Comparison of the MS fragmentation between peak GC-CB20 (A) and data from the natural compound library (B).

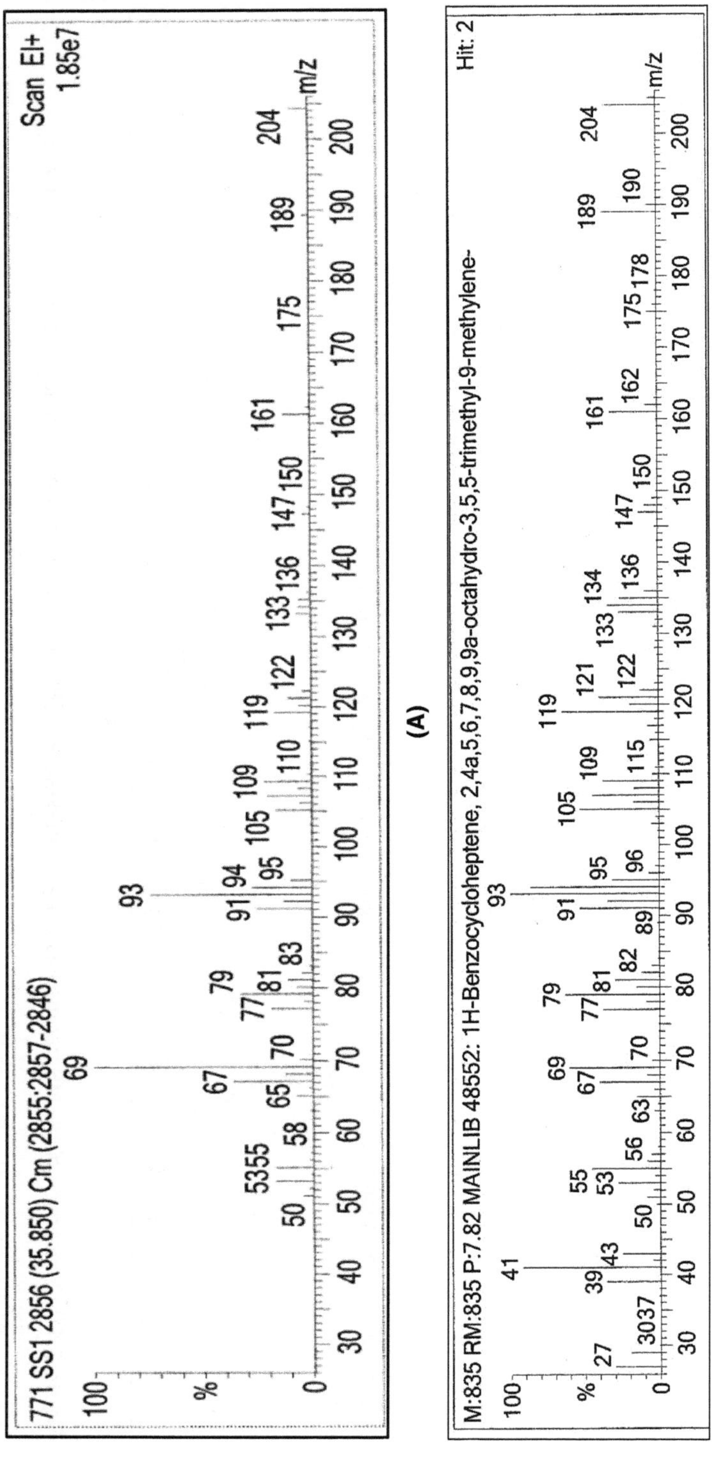

Figure 3.43: Comparison of the MS fragmentation between peak GC-CB21 (A) and data from the natural compound library (B).

the data from the natural compound library. These results confirmed the presence of 1H-Benzocycloheptene in *Coleus forskohlii*.

CB22: Azulene, 1,2,3,4,5,6,7,8-octahydro-1,4-dimethyl-7-(1-methylethenyl)-

The mass spectrum of was analyzed by means of GC-MS using computer searches in the natural compound library. It was possible to determine the presence of Azulene, 1,2,3,4,5,6,7,8-octahydro-1,4-dimethyl-7-(1-methylethenyl), whose chemical structure can be observed in Figure 3.44.

Molecular formula-$C_{15}H_{24}$

Molecular mass: M = 204.351

Retention time: tR = 36.364 min

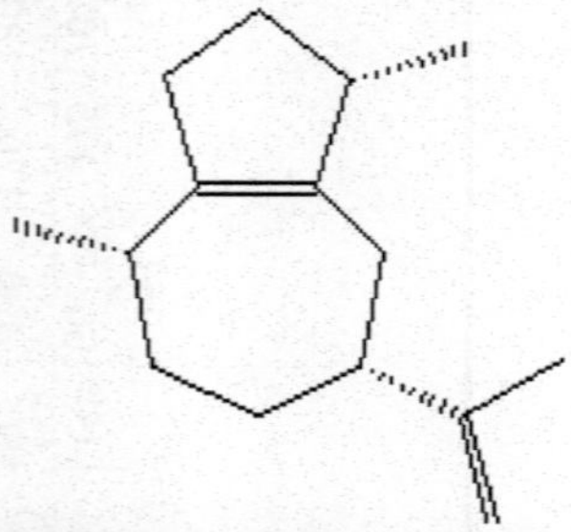

Figure 3.44: Structure of Azulene, 1,2,3,4,5,6,7,8-octahydro-1,4-dimethyl-7-(1-methylethenyl).

Figure 3.1 exhibited a peak GC-CB22 with a retention time tR = 36.394 min. Figure 3.45 (A) shows the MS spectrum of peak GC-CB22. The search analysis in the digital library indicated the compound Azulene, 1,2,3,4,5,6,7,8-octahydro-1,4-dimethyl-7-(1-methylethenyl). In order to confirm this result, the respective standard (Figure 3.45 (B)) was also analyzed by GC-MS and compared to the data from the natural compound library. These results confirmed the presence of 1 Azulene, 1,2,3,4,5,6,7,8-octahydro-1,4-dimethyl-7-(1-methylethenyl), in *Coleus forskohlii*.

CB23: β-farnesene

The mass spectrum of was analyzed by means of GC-M using computer searches in the natural compound library. It was possible to determine the presence of β-farnesene, whose chemical structure can be observed in Figure 3.46.

Molecular formula-$C_{15}H_{24}$

Molecular mass: M =204.35

Retention time: tR = 36.57 min

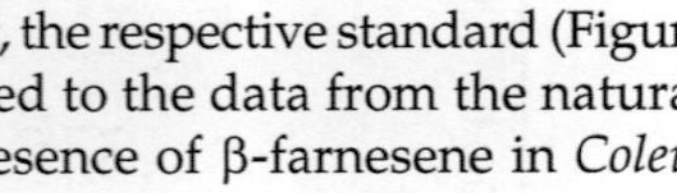

Figure 3.46: Structure of β-farnesene.

IUPAC- (E)-7,11-dimethyl-3-methylene-1,6,10-dodecatriene.

Figure 3.1 exhibited a peak GC-CB23 with a retention time tR = 36.57 min. Figure 3.47 (A) shows the MS spectrum of peak GC-CB23. The search analysis in the digital library indicated the compound β-farnesene. In order to confirm this result, the respective standard (Figure 3.47 (B)) was also analyzed by GC-MS and compared to the data from the natural compound library. These results confirmed the presence of β-farnesene in *Coleus forskohlii*.

CB24: Nerolidol

The mass spectrum of was analyzed by means of GC-MS using computer searches in the natural compound library. It was possible to determine the presence of Nerolidol, whose chemical structure can be observed in Figure 3.48.

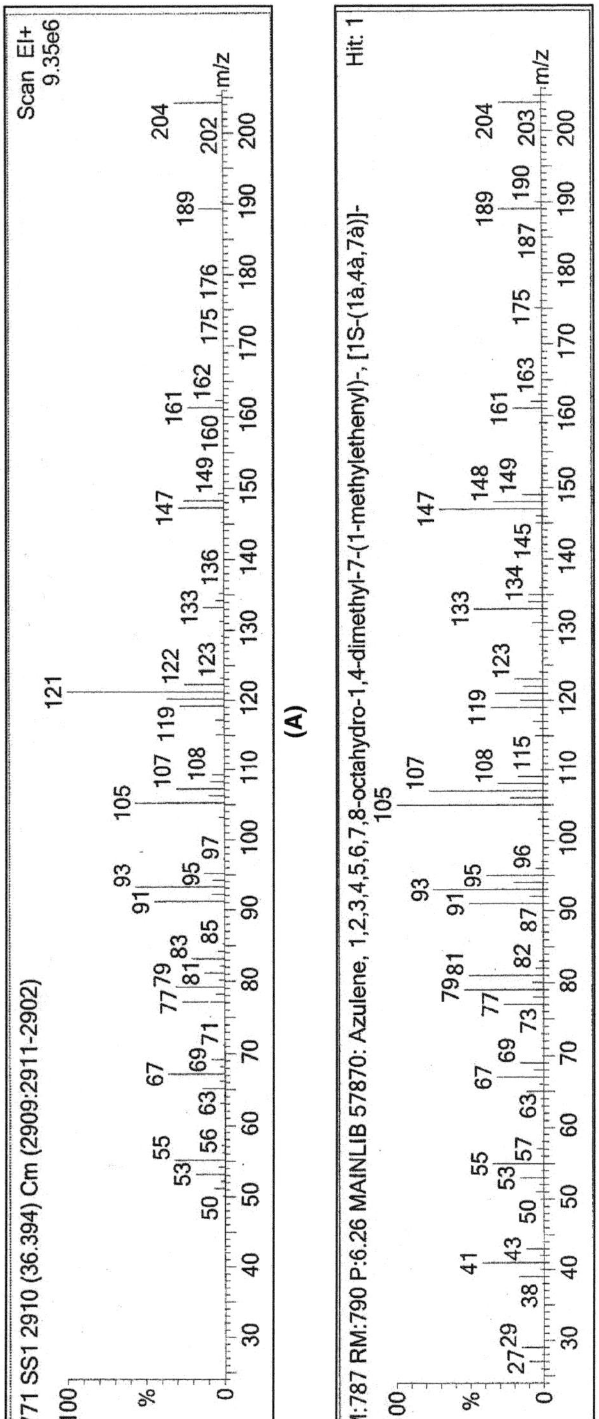

Figure 3.45: Comparison of the MS fragmentation between peak GC-CB22 (A) and data from the natural compound library (B).

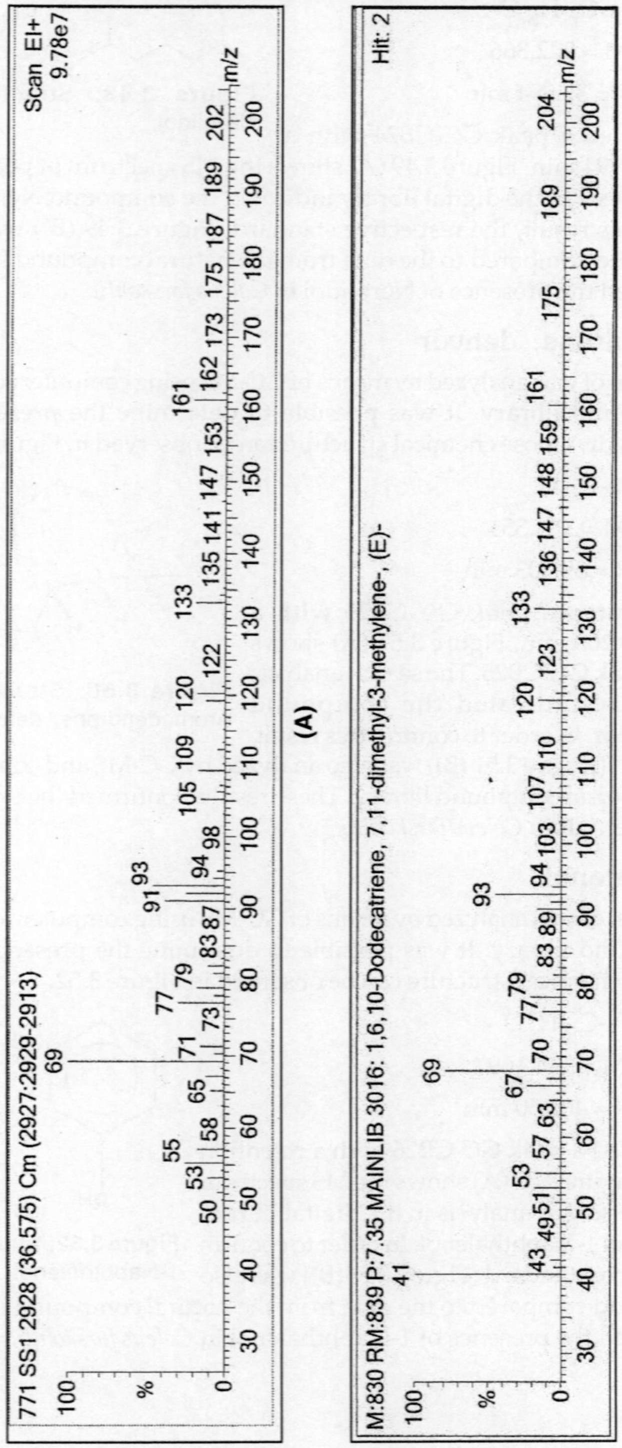

Figure 3.47: Comparison of the MS fragmentation between peak GC-CB23 (A) and data from the natural compound library (B).

Molecular formula-$C_{15}H_{26}O$

Molecular mass: M = 222.366

Retention time: tR = 37.994 min

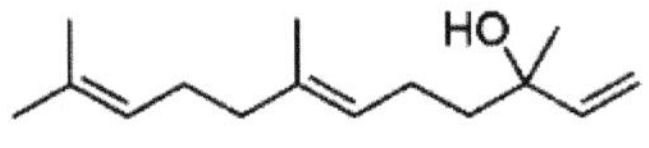

Figure 3.48: Structure of Nerolidol.

Figure 3.1 exhibited a peak GC-CB24 with a retention time tR = 37.994 min. Figure 3.49 (A) shows the MS spectrum of peak GC-CB24. The search analysis in the digital library indicated the compound Nerolidol. In order to confirm this result, the respective standard (Figure 3.49 (B)) was also analyzed by GC-MS and compared to the data from the natural compound library. These results confirmed the presence of Nerolidol in *Coleus forskohlii*.

CB25: Aromadendrene, dehydr

The mass spectrum of was analyzed by means of GC-MS using computer searches in the natural compound library. It was possible to determine the presence of Aromadendrene, dehydr, whose chemical structure can be observed in Figure 3.50.

Molecular formula-$C_{15}H_{24}$

Molecular mass: M = 204.351

Retention time: tR = 39.203 min

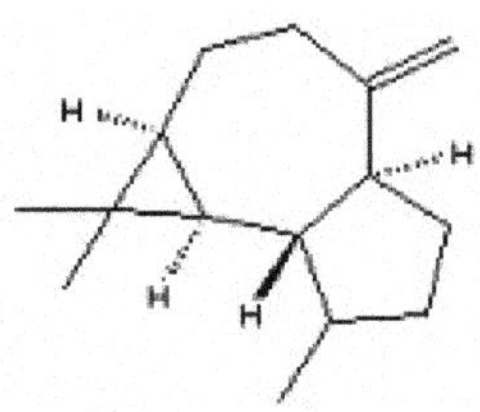

Figure 3.50: Structure of Aromadendrene, dehydr.

Figure 3.1 exhibited a peak GC-CB25 with a retention time tR = 39.203 min. Figure 3.51 (A) shows the MS spectrum of peak GC-CB25. The search analysis in the digital library indicated the compound Aromadendrene, dehydr. In order to confirm this result, the respective standard (Figure 3.51 (B)) was also analyzed by GC-MS and compared to the data from the natural compound library. These results confirmed the presence of Aromadendrene, dehydr in *Coleus forskohlii*.

CB26: 1-Naphthalenol

The mass spectrum of was analyzed by means of GC-MS using computer searches in the natural compound library. It was possible to determine the presence of 1-Naphthalenol, whose chemical structure can be observed in Figure 3.52.

Molecular formula-$C_{10}H_8O$

Molecular mass: M = 144.169

Retention time: tR = 40.250 min

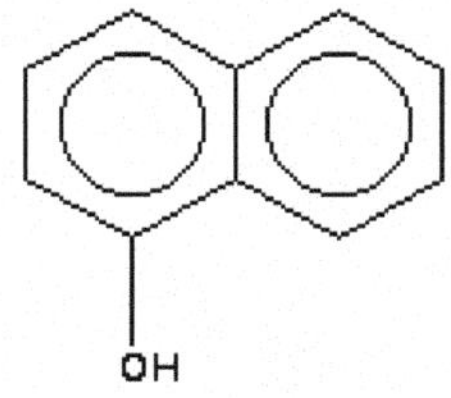

Figure 3.52: Structure of 1-Naphthalenol.

Figure 3.1 exhibited a peak GC-CB26 with a retention time tR = 40.250 min. Figure 3.53 (A) shows the MS spectrum of peak GC-CB26. The search analysis in the digital library indicated the compound 1-Naphthalenol. In order to confirm this result, the respective standard (Figure 3.53(B)) was also analyzed by GC-MS and compared to the data from the natural compound library. These results confirmed the presence of 1-Naphthalenol in *Coleus forskohlii*.

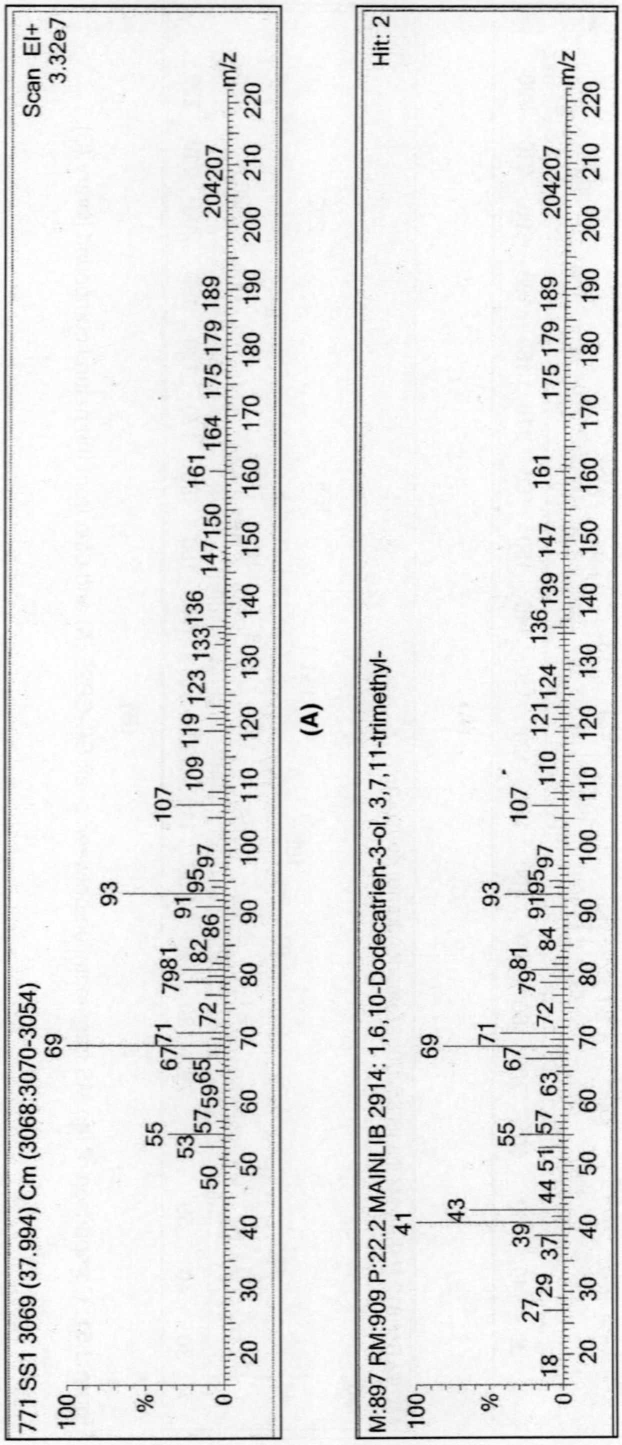

Figure 3.49: Comparison of the MS fragmentation between peak GC-CB24 (A) and data from the natural compound library (B).

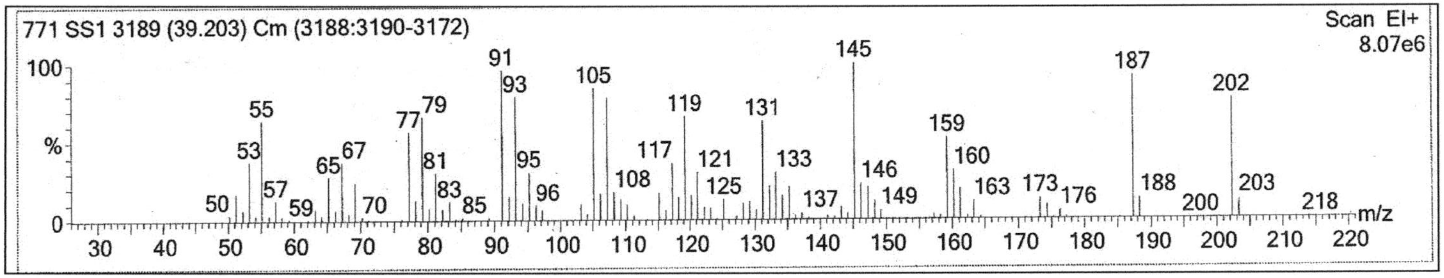

(A)

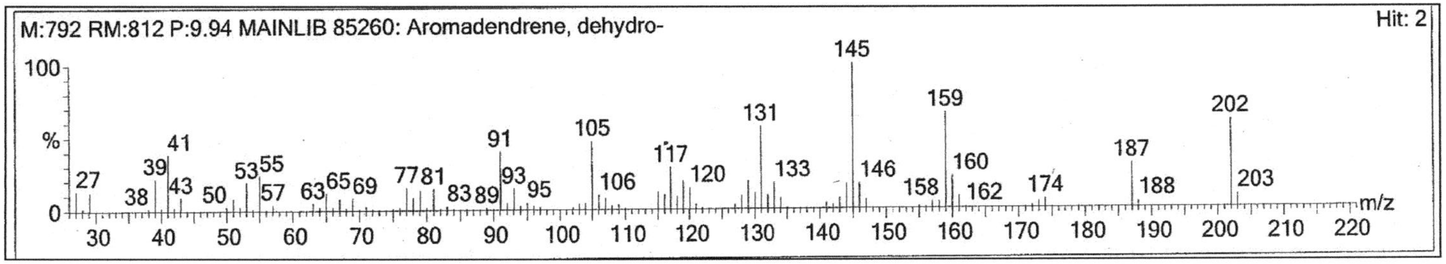

(B)

Figure 3.51: Comparison of the MS fragmentation between peak GC-CB25 (A) and data from the natural compound library (B).

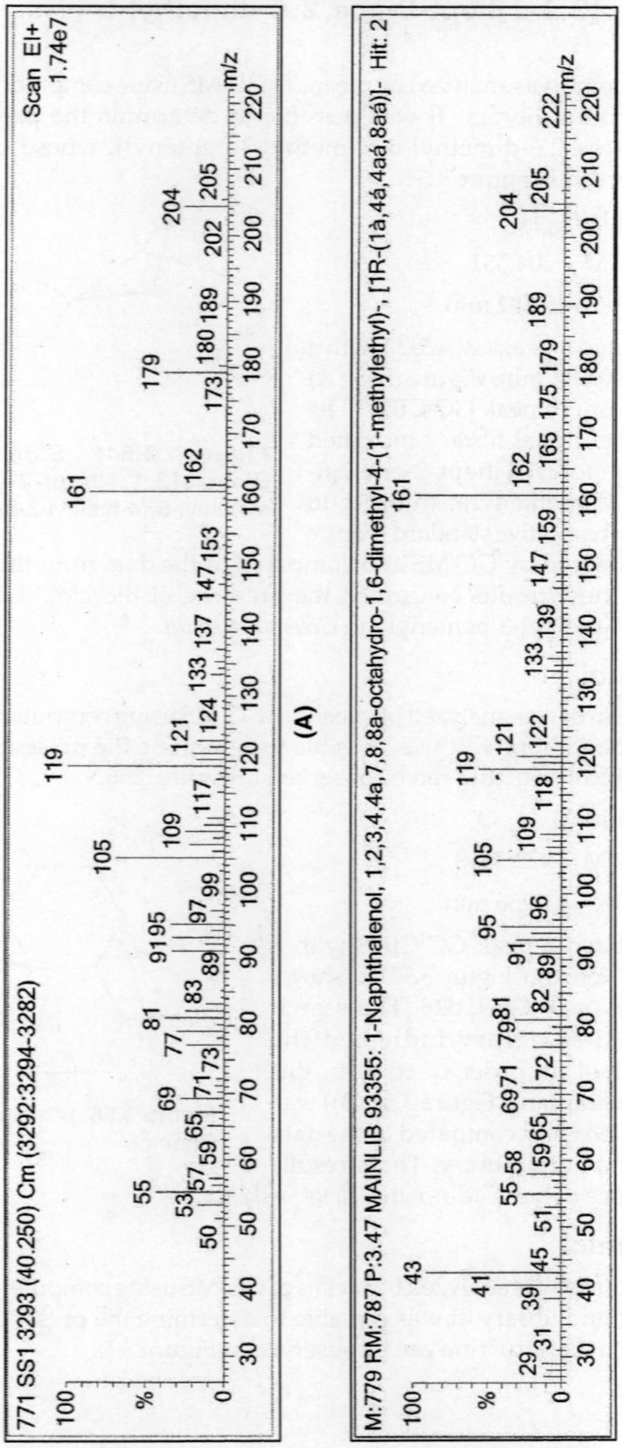

Figure 3.53: Comparison of the MS fragmentation between peak GC-CB26 (A) and data from the natural compound library (B)

CB27: Bicyclo[3,1,1]hept-2-ene,2,6-dimethyl-6-(4-methyl-3-pentenyl)-

The mass spectrum of was analyzed by means of GC-MS using computer searches in the natural compound library. It was possible to determine the presence of Bicyclo[3,1,1]hept-2-ene,2,6-dimethyl-6-(4-methyl-3-pentenyl), whose chemical structure can be observed in Figure 3.54.

Molecular formula-$C_{15}H_{24}$

Molecular mass: M = 204.351

Retention time: tR = 40.682 min

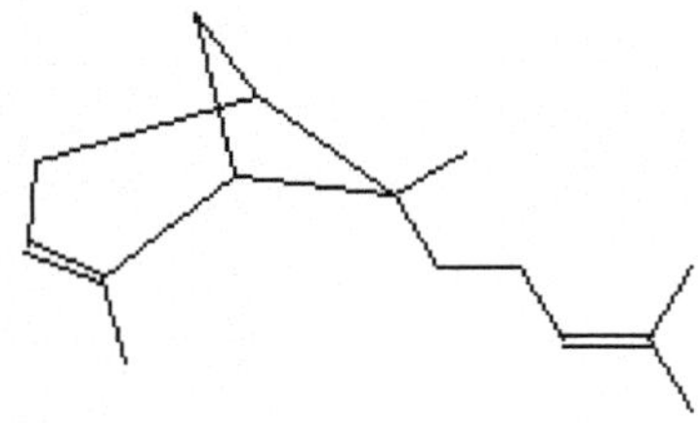

Figure 3.54: Structure of Bicyclo[3,1,1]hept-2-ene,2,6-dimethyl-6-(4-methyl-3-pentenyl).

Figure 3.1 exhibited a peak GC-CB27 with a retention time tR = 40.682 min. Figure 3.55 (A) shows the MS spectrum of peak GC-CB27. The search analysis in the digital library indicated the compound Bicyclo[3,1,1]hept-2-ene,2,6-dimethyl-6-(4-methyl-3-pentenyl)-. In order to confirm this result, the respective standard (Figure 3.55 (B)) was also analyzed by GC-MS and compared to the data from the natural compound library. These results confirmed the presence of Bicyclo[3,1,1]hept-2-ene,2,6-dimethyl-6-(4-methyl-3-pentenyl)- in *Coleus forskohlii.*

CB28: tau-Cadinol

The mass spectrum of was analyzed by means of GC-MS using computer searches in the natural compound library. It was possible to determine the presence of tau-Cadinol, whose chemical structure can be observed in Figure 3.56.

Molecular formula-$C_{15}H_{26}O$

Molecular mass: M = 222.3663

Retention time: tR = 41.256 min

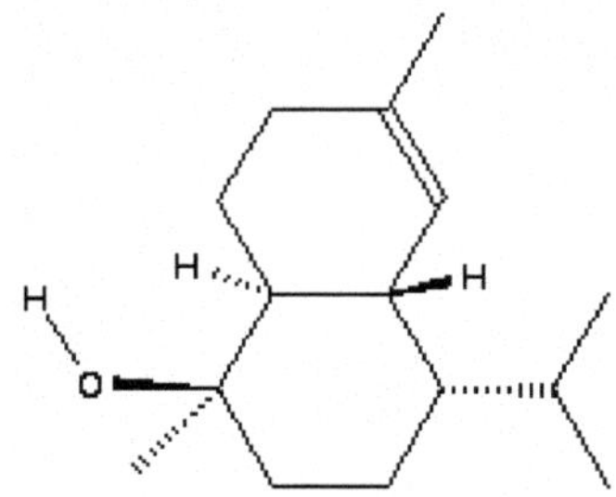

Figure 3.56: Structure of tau-Cadinol.

Figure 3.1 exhibited a peak GC-CB28 with a retention time tR = 41.256 min. Figure 3.57 (A) shows the MS spectrum of peak GC-CB28. The search analysis in the digital library indicated the compound tau-Cadinol. In order to confirm this result, the respective standard (Figure 3.57 (B)) was also analyzed by GC-MS and compared to the data from the natural compound library. These results confirmed the presence of tau-Cadinol in *Coleus forskohlii.*

CB29: tau-Muurolol

The mass spectrum of was analyzed by means of GC-MS using computer searches in the natural compound library. It was possible to determine the presence of tau-Muurolol, whose chemical structure can be observed in Figure 3.58.

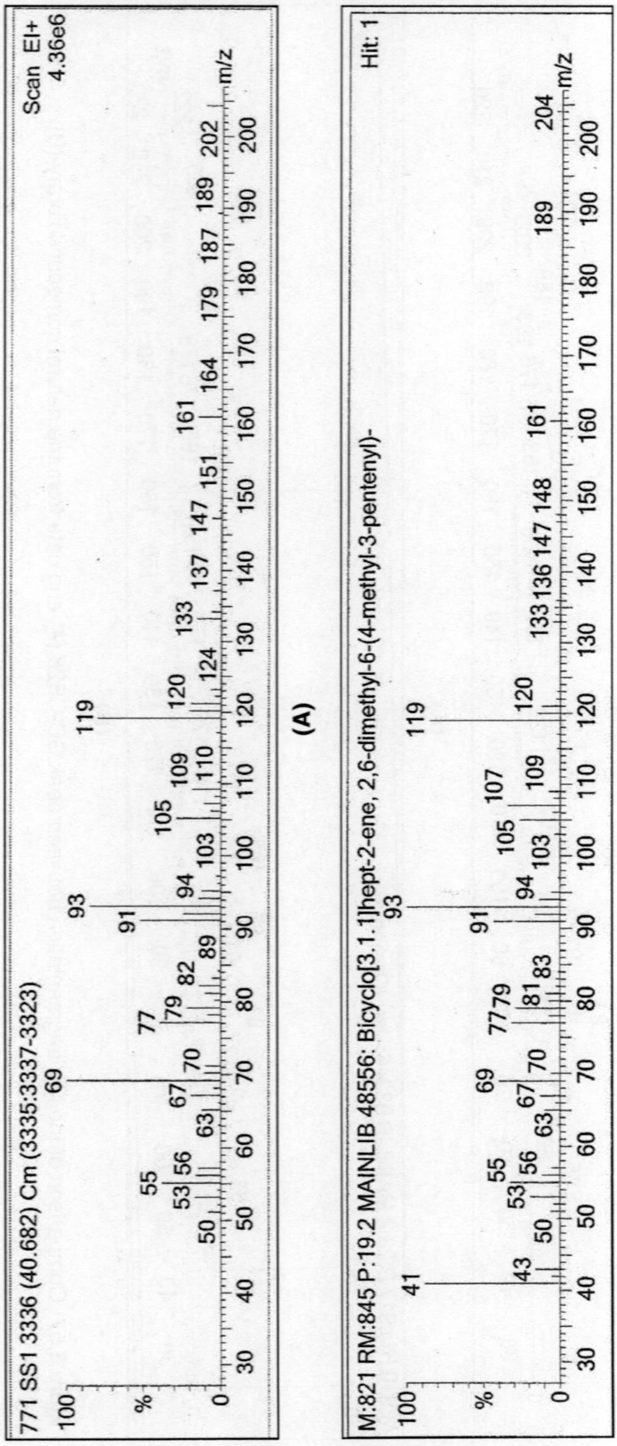

Figure 3.55: Comparison of the MS fragmentation between peak GC-CB27 (A) and data from the natural compound library (B).

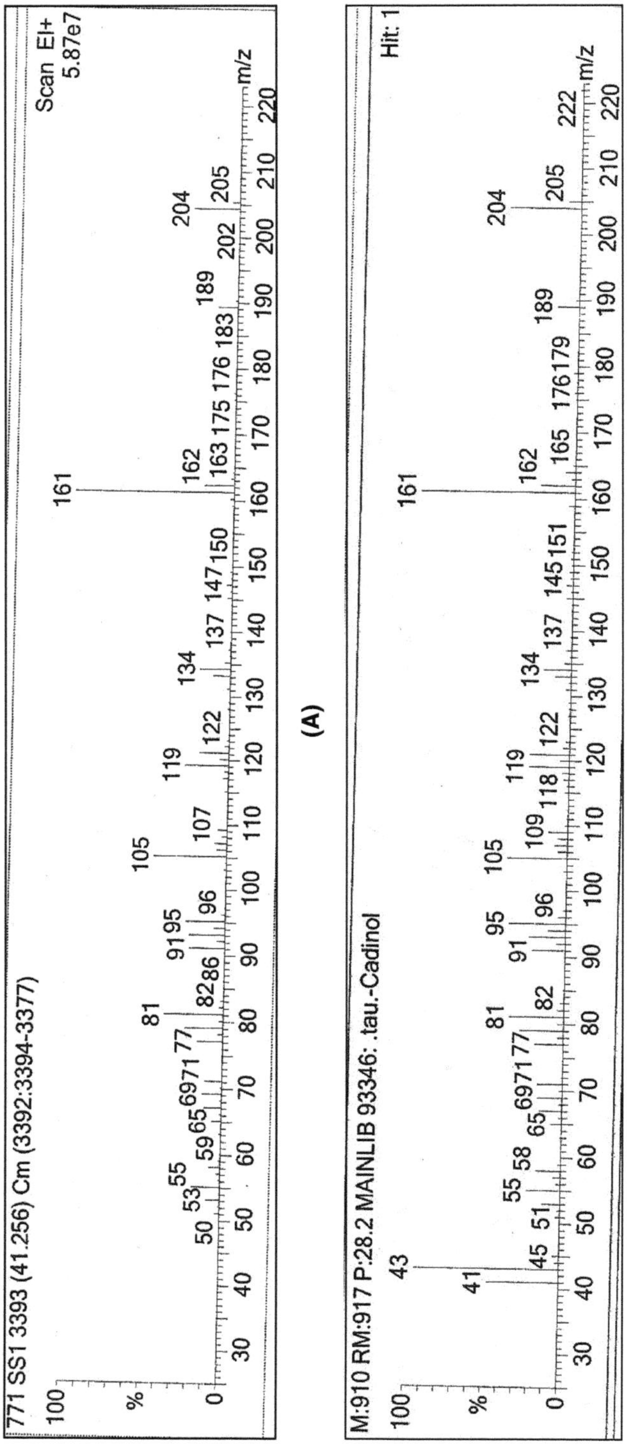

Figure 3.57: Comparison of the MS fragmentation between peak GC-CB28 (A) and data from the natural compound library (B).

Molecular formula-$C_{15}H_{26}O$

Molecular mass: M = 222.366

Retention time: tR = 41.709 min

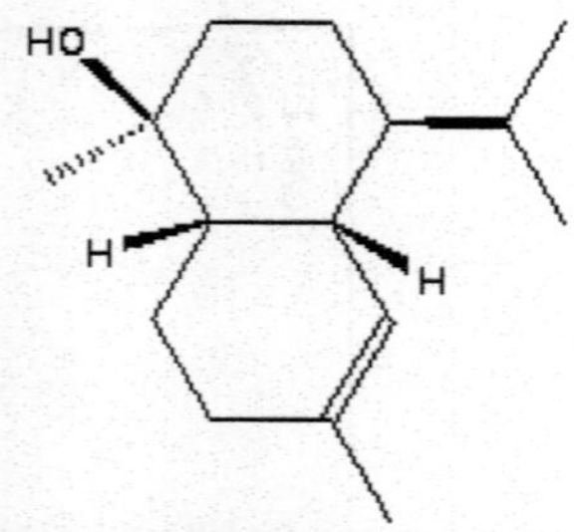

Figure 3.58: Structure of tau-Muurolol.

Figure 3.1 exhibited a peak GC-CB29 with a retention time tR = 41.709 min. Figure 3.59 (A) shows the MS spectrum of peak GC-CB29. The search analysis in the digital library indicated the compound tau-Muurolol. In order to confirm this result, the respective standard (Figure 3.59 (B)) was also analyzed by GC-MS and compared to the data from the natural compound library. These results confirmed the presence of tau-Muurolol in *Coleus forskohlii*.

In laboratory studies, (De Souza, 1991; Kavitha *et al.*, 2007) Coleus oil was found to more effectively inhibit the growth of skin pathogens (Patil *et al.*, 2001) including *Propionbacterium acnes* (associated with acne), *Staphylococcus aureus* a bacterial strain found in infected wounds and skin eruptions including acne (Prakash Krishnan, 1994), *Staphylococcus epidermidis* a bacterial strain occurring in a variety of opportunistic bacterial skin infections and in acne, than the well known tea tree oil. It was shown that forskolin can activate, in a unique manner, the enzyme that produces cyclic adenosine monophosphate (Seamon and Daly, 1981). The promising new drug suddenly generated immense interest as a research tool for the study of biochemical systems involving cyclic AMP. Dubey *et al.* (1974) reported coleonol, the component of *C. forskohli*, is hypotensive and spasmolytic.

Conclusion

Coleus forskohlii oil also represents a novel essential oil for use in oral and topical preparations for candidal infections. The essential oil extracted from the *Coleus forskohlii* had about 125 phytochemicals in it, out of 29 photochemical were identified in higher concentration than others. In the embodiments of this invention, the essential oil contains bornyl acetate (23.52 per cent), 2-Decen-1-ol (19.32 per cent), cyclohexene, 3-(1, 5-di methyl-4-hexenyl)-6 methylene (6.05), a-pinene (5.39 per cent), Bicyclo [3, 1, 1] heptanes, 6, 6-dimethyl-2 methylene (5.05 per cent) were identified in experimental studies on the oil. The Bornyl acetate (Acetic acid isobornyl ester; endo-1, 7, 7- Trimethylbicyclo [2.2.1] hept-2-yl acetate) contain highest area of percentage 23.52 per cent. Therefore, the oil of *Coleus forskohlii* is much effective for various human ailments.

Acknowledgements

The author is most grateful to G. B. Pant University of Agriculture and Technology, Pantnagar, Ranichauri Campus, Chamba, Uttarakhand (INDIA) for providing gift sample of *Coleus forskohlii*, Centre of Aromatic Plant (CAP) Dehradun for providing technical support and K.P. Rathoure (Mrs.) for her technical and valuable comments.

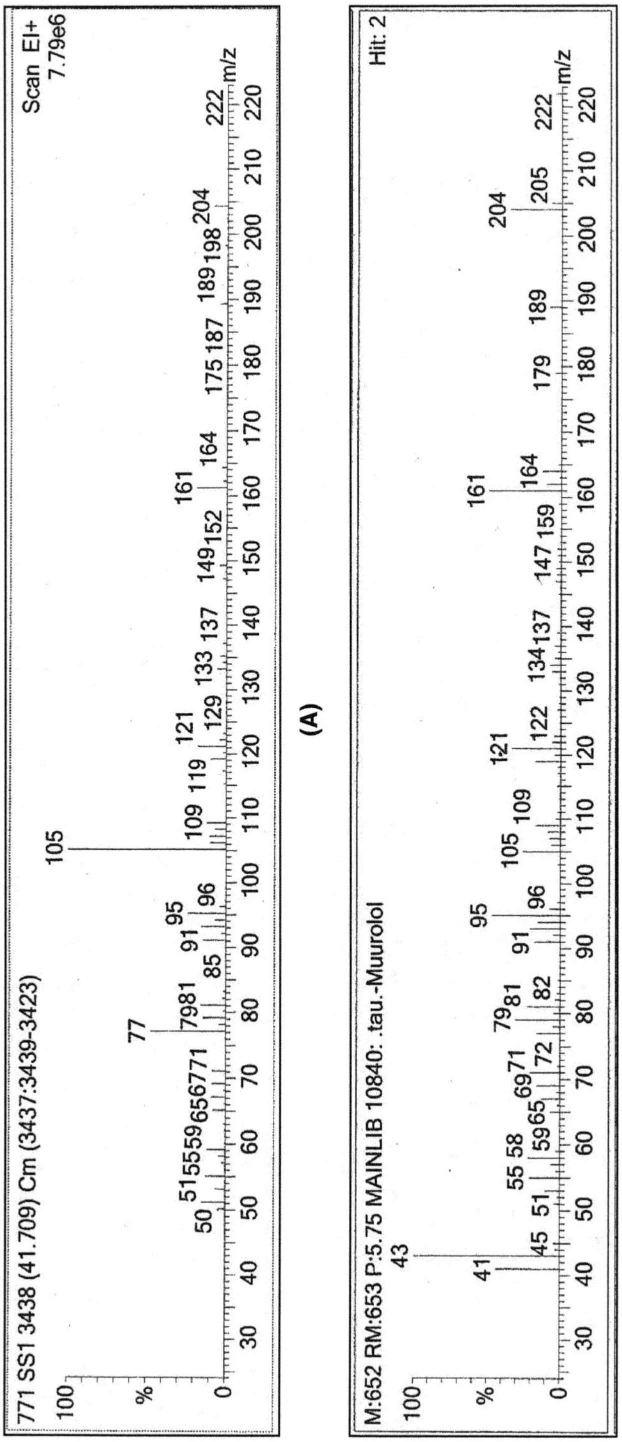

Figure 3.59: Comparison of the MS fragmentation between peak GC-CB29 (A) and data from the natural compound library (B).

References

Abraham, Z. (1981). Glimpses of Indian Ethanabotany (Eds. Jain, S.K.) Oxford and IBH Publishing Co., Bombay, p. 315-316.

Ammon, H.P., and Muller, A.B. (1985). Forskorin: From an ayurvedic remedy to a modern agent. *Planta Medica*, **46:** 473-477.

Anonymous (1950). *Coleus forskohlii*. In: Wealth of India-Raw materials, Vol. II. Central Scientific and Industrial Research, New Delhi; 308.

Bhat, S.V., Bajwa, B.S., Dornauer, H., and De Souza, N.J. (1977). Structure and stereochemistry of new labdane diterpenoid from *Coleus forskhlii* Briq. *Tetrahedron Lett.*, **19**: 1669.

De Souza, N.J. (1991); *Coleus forskohlii Briq*.- The Indian plant source for forskolin. Recent Advances in Medicinal, Aromatic and Spice crops, (ed: S.P. Raychaudhuri) Today and Tomorrow's Printers and Publishers, New Delhi, India, Vol I: 83-91.

Dubey, M.P., Srimal, R.C., Patnaik, G.K., and Dhawan, B.N. (1974). Blood pressure lowering and antispasmodic effects of *C. forskohlii* Briq. *Indian J. Pharm.*, **6**: 15.

Kavitha, C., Vadivel, E., Thangamani, C., and Rajamani, K. (2007). Analysis of variability for qualitative and quantitative traits in *Coleus forskohlii* Briq. *J. Hortic. Sci.*, **2**: 44-46.

Misra, L.N., Yagi, B.R., Ahmad, A., and Bahl, J.R. (1994). Variability of the chemical composition of the essential oil of *Coleus forskohlii* genotypes. *J. Essential Oil Res.*, **6:** 243-247

Patil, S., Hulamani, N.C., and Rokhade, A.K. (2001). Performance of genotypes of *Coleus forskohlii* Briq. for growth, yield and essential oil content. *Indian Perfumer*, **45**: 17-21.

Prakash Krishnan, R. (1994). Comparative performance of accessions and intervarietal hybrids in *Coleus forskohlii* Briq. *J. Root Crops.*, **20**: 70-73.

Saksena, A.K., Green, M.J., and Shue, H.J. (1985). Identity of coleonol with forskolin: structure revision of a base catalysed rearrangement product. *Tetrahedron Lett.*, **26**: 551-554.

Seamon, K.B., and Daly, J.W. (1981). Forskolin: a unique diterpene activator of cyclic AMP- generating systems. *J. Cyclic Nucleotide Res.*, **7**: 201.

Singh, A., and Singh, D. (1994). Biocontrol of *Sclerotium rolfsii* Sacc. causing collar rot of brinjal. *Journal of Biological Control*, **8**: 105-110.

Valdes, L.J., Mislankars, S.G., and Paul, A.G. (1987). *Coleus barabatus* (Lamiaceae) and the potential new drug forskolin (Colenol). *Eco. Bot.*, **41:** 474-483.

Vishwakarma, R.A., Tyagi, B.R., Ahmed, B., and Hussain, A. (1988). Variation in forskolin content in the roots of *Coleus forskohlii. Planta Medica*, **54**: 471-472.

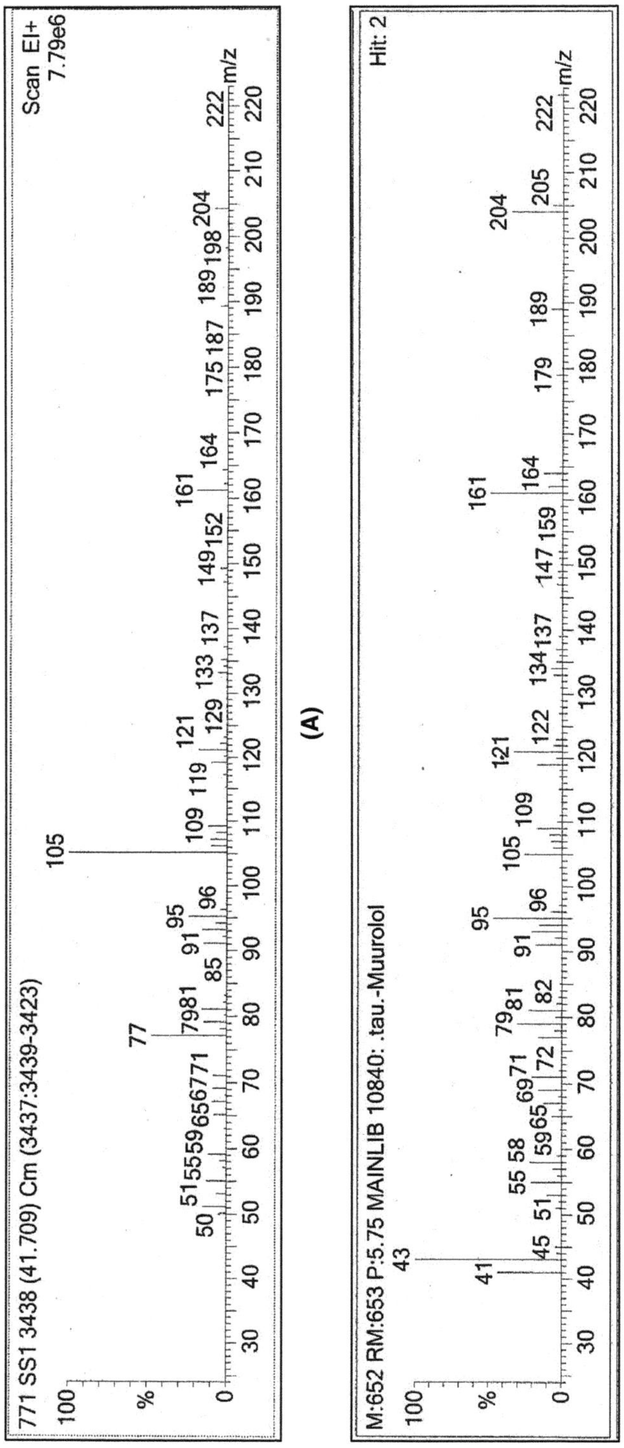

Figure 3.59: Comparison of the MS fragmentation between peak GC-CB29 (A) and data from the natural compound library (B).

Medicinal Plants: Phytochemistry, Pharmacology and Therapeutics Vol. 3 (2014) *Pages* **125–142**
Editor-in-Chief: **V.K. Gupta**
Published by: **DAYA PUBLISHING HOUSE, NEW DELHI**

4

Pharmacological Approaches to the Anti-inflammatory and Analgesic Effects of Methanol Extract of *Pluchea arguta* Boiss.

Yogeshkumar Vaghasiya[1], Jigna Parekh[2], Vinay Shukla[3] and Sumitra Chanda*[2]

ABSTRACT

Acute and chronic inflammatory diseases are still one of the most important health problems in the world. The management and treatment of pain is also one of the most common and yet difficult aspects of medicinal practice. Herbal drugs are being proved as effective as synthetic drugs with lesser side effects. So, there is a continuous search for indigenous drugs, which can provide relief from inflammation and pain. In the present study, methanol extract of Pluchea arguta Boiss. (PAM) was subjected to anti-inflammatory (Carrageenan, dextran, histamine and formalin induced paw edema model) and analgesic activity (acetic acid induced writhing test and formalin induced paw licking test) at the doses of 200 mg/kg b.w. (PAM-200) and 400 mg/kg b.w. (PAM-400). Extraction was done by soxhlet extraction method. Indomethacin was used as positive control. PAM-200 showed significant anti-inflammatory activity in different paw edema models like carrageenan ($P \leq 0.01$ at 1h, $P \leq 0.05$ at 3 h), dextran ($P \leq 0.05$ at 1 h and $P \leq 0.01$ at 3 h), histamine ($P \leq 0.05$ at 0.5 h, $P \leq 0.01$ at 1 h, $P \leq 0.05$ at 3 h) and formalin ($P \leq 0.05$ at 3 h, $P \leq 0.05$ at 24 h). PAM-400 showed significant anti-inflammatory activity in different paw edema models like carrageenan ($P \leq 0.001$ at 1h,

---

1 Department of Biochemistry, Saurashtra University, Rajkot- 360 005, Gujarat, India

2 Phytochemical, Pharmacological and Microbiological Laboratory, Department of Biosciences, Saurashtra University, Rajkot- 360 005, Gujarat, India

3 Pharmaceutical Chemistry Laboratory, Gujarat Ayurveda University, Jamnagar, Gujarat, India

* *Corresponding author*: E-mail: svchanda@gmail.com

$P \leq 0.05$ at 3 h), dextran ($P \leq 0.01$ at 1 h and $P \leq 0.01$ at 3 h), histamine ($P \leq 0.05$ at 0.5 h, $P \leq 0.01$ at 1 h, $P \leq 0.001$ at 3 h) and formalin ($P \leq 0.05$ at 3 h, $P \leq 0.05$ at 24 h). PAM-200 and PAM-400 showed significant analgesic activity in acetic acid induced writhing test ($P \leq 0.05$). PAM-400 showed significant analgesic activity in formalin test at 0-10 min ($P \leq 0.05$). Based on all these results it can be concluded that methanol extract of Pluchea arguta produced potent anti-inflammatory and analgesic activity.

Keywords: *Pluchea arguta,* Methanol extract, Anti-inflammatory, Analgesic effects.

Introduction

Inflammation is a pathophysiological response of living tissue to injuries that leads to the local accumulation of plasmatic fluid and blood cells (Speranza *et al.*, 2009; Gochman *et al.*, 2012). Inflammation manifests usually in form of painful swelling associated with some changes in skin covering the site (Jain and Bari, 2010). All inflammatory processes develop along a known sequence: locally increased blood supply, leakage of fluid, small molecules and proteins, and infiltration of cells (Roitt, 1997).

Inflammation can be classified as either acute or chronic. Acute inflammation is the initial response of the body to harmful stimuli and is achieved by the increased movement of plasma and leukocytes from the blood into the injured tissues (Latha *et al.*, 2012). This acute inflammatory response is believed to be a defense mechanism aimed at killing of bacteria, viruses and parasites while still facilitating wound repair (Iwalewa *et al.*, 2007). Inflammatory response is triggered by tissue mast cells and resident macrophages whose degranulation and activation sequentially release a battery of inflammatory mediators, including bioactive amines (Histamine and 5-HT), cytokines, chemokines as well as lipid mediators that collectively recruit and activate inflammatory cells as well as bring about oedema formation (Lawrence and Gilroy, 2007). Prolonged inflammation, known as *chronic inflammation*, leads to a progressive shift in the type of cells which are present at the site of inflammation and is characterized by simultaneous destruction and healing of the tissue from the inflammatory process.

There are several mediators involved in inflammation. Histamine, serotonine, bradykinin and prostaglandins (PG) are involved in the increased vascular permeability (Posadas *et al.*, 2004). Histamine, bradikinin and serotonin are released in the early phase (Di Rosa *et al.*, 1971; Hori *et al.*, 1988) while prostaglandins (PGs) are involved in both early and late phases of inflammation (Kikuchi *et al.*, 1996). Prostaglandins (PGs) elicit a variety of important biological responses. Among the properties of PGs are their abilities to induce pain, fever and symptoms associated with inflammatory responses. Most of nonsteroidal anti-inflammatory drugs (NSAIDs), which are clinically used to date, inhibit the production of PGs by inhibition of COX enzymes and are the main drugs used to reduce the untoward consequences of inflammation (Albert *et al.*, 2002). They are also involved in maintaining normal physiological processes (Moon *et al.*, 1999). There are two isoforms of COX: COX-1 and COX-2. COX-1 is detectable, but COX-2 is not detectable in most normal tissues, however, COX-2 can be induced by many factors such as pro-inflammatory cytokines, phlogistic factors, etc. Studies indicated that COX-2 plays an important role in inflammation (Subbaramaiah *et al.*, 1996). Thus, those agents that could suppress the activity or protein expression of COX-2 are likely to be valuable medicine for anti-inflammation and easing of pain (Shu *et al.*, 2006). Recent development of selective COX-2 inhibitors as anti-inflammatory agents is considered a major therapeutic breakthrough in the treatment of inflammatory diseases since the inhibition of COX-2-derived PGs is thought to be the major pharmacodynamic action responsible for their anti-inflammatory effects (Abad *et al.*, 2006).

Acute and chronic inflammatory diseases are still one of the most important health problems in the world. Various agents are adapted to treat inflammatory disorders; their prolonged use often leads to serious adverse reactions such as gastric intolerance, bone marrow depression, water and salt retention. Consequently, development of new anti-inflammatory drugs with low side-effects is still necessary (Xiao *et al.*, 2005; Yonathan *et al.*, 2006; Kupeli *et al.*, 2007). The management and treatment of pain is also one of the most common and yet difficult aspects of medicinal practice. Analgesic therapy is domain by two major classes of analgesic drugs; *viz.* opioids and non steroidal anti-inflammatory drugs (NSAIDs). Both classes of analgesic drugs produce serious side effects, such as gastrointestinal disturbance, renal damage reparatory depression etc. (Domaj *et al.*, 1999; Dahl and Reader, 2000; Jarald *et al.*, 2008; Nakajima *et al.*, 2012).

Herbal drugs are being proved as effective as synthetic drugs with lesser side effects. So, there is a continuous search for indigenous drugs, which can provide relief from inflammation (Balasuramanian *et al.*, 2005; Simpson *et al.*, 2010). Medicinal herbs have been used as a form of therapy for the relief of pain throughout history (Eisner, 1990; Almeida *et al.*, 2001). The search for plants with reported folkloric use as pain relievers, anti-inflammatory agents, should therefore be viewed as a fruitful and logical research strategy in the search for new anti-inflammatory drugs (Vaghasiya *et al.*, 2011; Loganayaki *et al.*, 2012).

Pluchea arguta Boiss is a member of family Asteraceae and about 1000 genera of this family are well known in the world (Jaffri, 1966). Different species of *Pluchea* are used in Ayurvedic system of medicine in various diseases. Pharmacological studies demonstrated anti-inflammatory and antioxidant activities of different *Pluchea* species (Barros *et al.*, 2006; Fernandez and Torres, 2006). In view of the therapeutic properties attributed by people to different species of this plant, anti-inflammatory and analgesic effects of methanol extract of *Pluchea arguta* Boiss in rats and mice was studied using various inflammatory and analgesic models.

Materials and Methods

Experimental Animals

Wistar albino rats (200-250 g) and Swiss albino mice (20-25 g) of either sex were used for anti-inflammatory and analgesic studies. The animals were obtained from Sarabhai Research Centre (SRC), Baroda. All the rats and mice were housed in standard plastic cages with stainless steel coverlids and wheat straw as bedding material at the animal house of Department of Biosciences, Saurashtra University, Rajkot, Gujarat, India. The animals were kept in group of 5-6 per cage and facilitated with standard environmental condition of photoperiod (12: 12 h dark: light cycle) and temperature (25 ± 2°C). They were provided with commercial rat and mice feed (Pranav Agro Industries Ltd., Amrut brand rat and mice pellet feed) and water given *ad labium*. The use of these animals and the study protocols were approved by CPCSEA recognised local ethical committee.

Plant Material

Fresh leaves and stem of *Pluchea arguta* Boiss. were collected from Junagadh Agriculture University, Junagadh, Gujarat, India. It was identified by Dr. P. S. Nagar, Department of Botany, M.S. University, Baroda, Gujarat, India (Voucher specimen no. PSN402). Fresh plant material was washed with tap water, air dried and then homogenized to fine powder and stored in airtight bottles.

Extraction

The dried crude powder of *Pluchea arguta* Boiss was extracted by soxhlet extraction method (Vaghasiya *et al.*, 2008). Plant material was defatted by petroleum ether. Defatted plant material was extracted in methanol. The methanol was evaporated from the extract, and the dried extract was collected, weighed and stored in air-tight bottles at 4°C for further experiments.

Drug Dose

The doses considered for the experiment on rat and mice were obtained from conversion of human dose of *Pluchea arguta* (3 g kg^{-1}). The conversion factor of human dose to rat (Per 200 g body weight) dose is 0.018 (Paget and Barnes, 1964) and the conversion factor of human dose to mice (Per 20g body weight) dose is 0.0026. Hence the calculated dose for the rat is 270 mg/kg and for the mouse is 390 mg/kg. From this calculation, all the anti-inflammatory and analgesic experiments were carried out in two different doses of 200 and 400 mg/kg body weight.

Preparation of Test Drug and Standard Drug

Test drug (Methanol extract of *Pluchea arguta*) and standard drug (Indomethacin) were prepared as a suspension in distilled water using mortal and pestle.

Animal Grouping

The animals (Rats and mice) were divided into four groups for anti-inflammatory and analgesic studies. Each group consisted of six animals of either sex. The groups were:

Group I: Negative control (Distilled water)

Group II: Test drug- Methanol extract of *Pluchea arguta*- 200 mg/kg body weight (PAM-200)

Group III: Test drug- Methanol extract of *Pluchea arguta*- 400 mg/kg body weight (PAM-400)

Group IV: Positive control- Standard drug indomethacin- 2.5 mg/kg body weight (Indo-2.5)

Acute Anti-inflammatory Studies

Carrageenan Induced Rat Paw Edema (Winter *et al.*, 1962)

The initial paw volume of each animal was measured using a plethysmograph (Bhatt *et al.*, 1977). Thereafter, methanol extract of *Pluchea arguta* (200, 400 mg/kg) or distilled water or indomethacin (2.5 mg/kg) was orally administered to the respective group of the animals. Acute inflammation was produced by the subplantar administration of 0.1 ml of 1 per cent (w/v) carrageenan in the right paw of the rat after 1 h of drug administration. The paw volume was measured at 1h, 2h, and 3h intervals after the administration of the carrageenan. Increase in the paw volume (per cent) was calculated by following formula:

Increase in paw volume (per cent) = [(Final paw volume – Initial paw volume)/ Initial paw volume] X 100

Dextran Induced Rat Paw Edema (Winter and Porter, 1957)

The animals were treated in a manner similar to that of carrageenan induced paw edema model. 0.1 ml 1 per cent dextran (60,000-90,000 m.w., HiMedia, India) was used for the study. Paw volume was measured as mentioned in carrageenan induced paw edema model at 1 h, 2 h and 3 h.

Histamine Induced Rat Paw Edema (Suleyman *et al.*, 1991)

In this model paw edema of a rat was induced by subplantar injection of 0.1 ml of 1 per cent freshly prepared histamine in normal saline and the paw oedema was measured as mentioned in carrageenan induced paw edema model. The paw volume was measured at 0.5 h, 1 h, 2 h and 3 h.

Formaldehyde Induced Paw Edema (Brownlee *et al.*, 1950)

Methanol extract of *Pluchea arguta* (200, 400 mg/kg) or distilled water or standard drug was administered once daily (p.o.) for seven consecutive days to all the groups. On seventh day, initial paw volume was measured before drug administration. After 1 h of drug administration, paw edema of the rat was induced by subplantar injection of 0.1 ml of 3 per cent formaldehyde solution in normal saline. Paw volume was measured at 3 h, 24 h and 48 h after formaldehyde injection as described earlier in carrageenan model.

Analgesic Studies

Analgesic study was determined in two different models:

Acetic Acid Induced Writhing in Mice (Hernandez-Perez and Rabanal, 2002)

Swiss albino mice were used for acetic acid induced writhing test. Methanol extract of *Pluchea arguta* (200, 400 mg/kg) or distilled water or standard drug was administered orally, 1 h prior to the injection of acetic acid. Writhing was induced by administrating of 3 per cent aqueous solution of acetic acid (10 ml/kg body weight) intraperitoneally. Immediately after the acetic acid injection, each animal was placed in a transparent observation cage and the number of writhes per mouse was counted for 30 min. Writhing movement was accepted as contraction of the abdominal muscles accompanied by stretching of hind limbs. The percentage inhibition was calculated using the following ratio:

[(Control mean–treated mean)/control mean] × 100

Formaldehyde Induced Paw Licking Response in Rats

The effect of methanol extract of *Pluchea arguta* Boiss upon formaldehyde induced paw licking response was evaluated by adopting the procedure explained by Magali *et al.* (2000). The procedure is the part of formaldehyde induced hind paw edema, which is mentioned earlier. After the injection of formaldehyde, the animals were kept under observation for half an hour. The amount of time spent licking the injected paw was noted, and was considered to be indicative of pain. The time taken for the onset of paw licking was initially measured. The first of the nociceptive responses normally peaked 5 min after formalin injection and the second phase 15-30 min after formalin injection, representing the neurogenic and inflammatory pain (Hunskaar and Hole K., 1987). Therefore the frequency of paw licking was measured in five intervals at 0-5 min., 6-10 min., 11-15 min., 16-20 min. and 21-30 min.

Results and Discussion

Anti-inflammatory Activities

Carrageenan Induced Paw Edema

Carrageenan-induced edema has been commonly used as an experimental animal model for acute inflammation and is believed to be biphasic. The early phase (1-2 h) of the carrageenan model is mainly mediated by histamine, serotonin and increased synthesis of prostaglandins in the damaged tissue surroundings. The late phase (3 h) is associated with neutrophil originated free radicals, such as hydrogen peroxide, superoxide and hydroxyl radicals, as well as prostaglandin release (Mazzon *et al.*, 2001; Ogonowski *et al.*, 1997; Vineger *et al.*, 1969; Brito and Antonio,1998; Arawwawala *et al.*, 2012).

The results of Anti-inflammatory activity of methanol extract of *Pluchea arguta* in carrageenan induced paw edema is shown in Table 4.1. PAM-200 group showed potent anti-inflammatory activity at 1 h (71.47 per cent, $P \leq 0.001$), 2 h (45.25 per cent, $P \leq 0.001$) and 3 h (39.06 per cent, $P \leq 0.05$). PAM-400 group showed significant decrease in paw volume at 1 h (72.97 per cent, $P \leq 0.001$), 2 h (68.45 per cent, $P \leq 0.001$) and 3 h (49.08 per cent, $P \leq 0.05$). Indomethacin group showed significant decrease in paw volume at 1 h (67.45 per cent, $P \leq 0.001$), 2 h (66.53 per cent, $P \leq 0.001$) and 3 h (72.61 per cent, $P \leq 0.01$). PAM-200 showed more anti-inflammatory activity than reference drug indomethacin at 1 h, while PAM-400 group showed more anti-inflammatory activity than reference drug indomethacin at 1 h and 2 h. Thus it can be concluded that methanol extract of *Pluchea arguta* has potent anti-inflammatory activity in carrageenan induced rat paw edema in early phase (1 h) and also in later phase (3 h).

Dextran Induced Paw Edema

It is well established that carrageenan and dextran induce rat paw edema by different mechanisms. Dextran is a polysaccharide of high molecular weight that induces anaphylactic reaction after injection in rat's extremities, which is characterized by extravasation and oedema formation, as a consequence of liberation of histamine and serotonin from mast cells (Rowley and Benditt, 1956; Van Wauve and Goosens, 1989; Leal *et al.*, 2011).

Table 4.1: Anti-inflammatory activity of methanol extract of *Pluchea arguta* in carrageenan induced paw edema.

Groups	*After 1 h*		*After 2 h*		*After 3 h*	
	Increase in Paw Volume (Per cent)	*Per cent Change*	*Increase in Paw Volume (Per cent)*	*Per cent Change*	*Increase in Paw Volume (Per cent)*	*Per cent Change*
Control	28.45 ± 2.88	–	38.98 ± 2.45	–	38.94 ± 5.33	–
PAM-200	8.12 ± 2.65**	71.47↓	19.90 ± 2.20**	45.25↓	23.73 ± 2.05*	39.06↓
PAM-400	7.69 ± 1.64***	72.97↓	12.30 ± 1.87***	68.45↓	19.83 ± 1.55*	49.08↓
Indo-2.5	9.26 ± 1.25***	67.45↓	13.05 ± 3.17***	66.53↓	10.67 ± 3.39**	72.61↓

Data are expressed as mean ± SEM.

P≤ (0.05*, 0.01**, 0.001***); F ≤ 0.05.

PAM: Methanol extract of *Pluchea arguta*

The results of anti-inflammatory activity of methanol extract of *Pluchea arguta* in dextran induced paw edema is shown in Table 4.2. PAM-200 group and PAM-400 group showed significant anti-inflammatory activity in dextran induced paw edema model when compared with the control group. PAM-200 group showed significant decrease in paw volume at 1 h (31.97 per cent, P≤0.05), 2 h (53.70 per cent, P≤0.01) and 3 h (72.04 per cent, P≤0.01). PAM-400 group showed significant decrease in paw volume at 1 h (32.26 per cent, P≤0.01), 2 h (55.64 per cent, P≤0.001) and 3 h (79.31 per cent, P≤0.01). Indomethacin group showed significant decrease in paw volume at 1 h (49.39 per cent, P≤0.001), 2 h (38.85 per cent, P≤0.05) and 3 h (65.61 per cent, P≤0.01). Methanol extract of *Pluchea arguta* showed anti-inflammatory activity in dose dependant manner. PAM-200 and PAM-400 groups showed almost similar anti-inflammatory activity at 1 h and 2 h, while at PAM-400 group showed higher anti-inflammatory activity than PAM-200 group at 3 h. PAM-200 and PAM-400 groups showed higher anti-inflammatory activity than reference drug indomethacin group at 2 h and 3 h, while it was lower at 1 h. The results tend to suggest that the anti-inflammatory activity of the methanol extract of *Pluchea arguta* is possibly backed by its anti-histamine or anti-serotonin activity.

Histamine Induced Paw Edema

The histamine is a basic amine related with inflammatory and allergic process causing, among several effects, both vasodilatation and increase of vascular permeability. Histamine is one of the important inflammation mediators and it is a potent vasodilator substance and increases the vascular permeability (Cuman *et al.*, 2001; Rang *et al.*, 2001; Linardi *et al.*, 2002; Muhammad *et al.*, 2012). The results of anti-inflammatory activity of methanol extract of *Pluchea arguta* in histamine induced paw edema is shown in Table 4.3. Anti-inflammatory activity of PAM-200 and PAM-400 groups was statistically significant at 0.5 h, 1 h, 2 h and 3 h as compared with the control group. PAM-200 group showed significant decrease in paw volume at 0.5 h (24.39 per cent, $P \leq 0.05$), 1 h (25.99 per cent, $P \leq 0.01$), 2 h (36.40 per cent, $P \leq 0.05$) and 3 h (39.27 per cent, $P \leq 0.05$). PAM-400 group showed significant decrease in paw volume at 0.5 h (26.10 per cent, $P \leq 0.05$), 1 h (26.16 per cent, $P \leq 0.01$) and highly significant at 2 h (42.72 per cent, $P \leq 0.001$) and 3 h (46.42 per cent, $P \leq 0.001$). Reference drug indomethacin group showed highly significant decrease in paw volume at 0.5 h (51.15 per cent, $P \leq 0.001$), 1 h (57.08 per cent, $P \leq 0.001$), 2 h (66.45 per cent, $P \leq 0.001$) and 3 h (66.34 per cent, $P \leq 0.001$) as compared with the control group. Thus PAM-200 and PAM-400 groups did not show higher anti-inflammatory activity than reference drug indomethacin.

Table 4.2: Anti-inflammatory activity of methanol extract of *Pluchea arguta* in dextran induced paw edema.

Groups	*After 1 h*		*After 2 h*		*After 3 h*	
	Increase in Paw Volume (Per cent)	*Per cent Change*	*Increase in Paw Volume (Per cent)*	*Per cent Change*	*Increase in Paw Volume (Per cent)*	*Per cent Change*
Control	67.73 ± 2.87	–	51.19 ± 4.28	–	42.59 ± 5.69	–
PAM-200	46.08 ± 5.85*	31.97↓	23.70 ± 3.30**	53.70↓	11.91 ± 2.79**	72.04↓
PAM-400	45.88 ± 4.96**	32.26↓	22.71 ± 1.50***	55.64↓	8.81 ± 2.61**	79.31↓
Indo-2.5	34.28 ± 3.53***	49.39↓	31.30 ± 3.41*	38.85↓	14.65 ± 2.11**	65.61↓

Data are expressed as mean ± SEM.

P≤ (0.05*, 0.01**, 0.001***); F ≤ 0.05.

PAM: Methanol extract of *Pluchea arguta*

Formaldehyde Induced Paw Edema

The results of Anti-inflammatory activity of methanol extract of *Pluchea arguta* in formaldehyde induced paw edema is shown in Table 4.4. It is well known that inhibition of formaldehyde induced paw edema in rats is one of the most suitable test procedures to screen anti-arthritic and anti-inflammatory agents as it closely resembles human arthritis (Greenwald, 1991). Thus formaldehyde-induced paw edema is a model used for the evaluation of an agent with anti-proliferative activity (Banerjee *et al.*, 2000; Hajhashemi *et al.*, 2012). Injection of formaldehyde subcutaneously into hind paw of rats produces localized inflammation. PAM-200, PAM-400 and indomethacin administered daily for 7 days successfully inhibited edema induced by formaldehyde (Table 4.4). PAM-200 group showed decrease in paw volume at 3 h (40.79 per cent, $P \leq 0.05$), 24 h (34.98 per cent, $P \leq 0.05$) and 48 h (35.96 per cent). PAM-400 group showed decrease in paw volume at 3 h (42.96 per cent, $P \leq 0.05$), 24 h (36.50 per cent, $P \leq 0.05$) and at 48 h (39.67 per cent). Indomethacin group showed decrease

Table 4.3: Anti-inflammatory activity of methanol extract of *Pluchea arguta* in histamine induced paw edema

Groups	*After 0.5 h*		*After 1 h*		*After 2 h*		*After 3 h*	
	Increase in Paw Volume	*Per cent Change*	*Increase in Paw Volume*	*Per cent Change*	*Increase in Paw Volume*	*Per cent Change*	*Increase in Paw Volume*	*Per cent Change*
Control	60.65 ± 3.33	–	58.42 ± 3.15	–	45.23 ± 2.76	–	35.98 ± 2.52	–
PAM-200	45.85 ± 4.15*	24.39↓	43.24 ± 2.48**	25.99↓	28.77 ± 3.53*	36.40↓	21.85 ± 3.30*	39.27↓
PAM-400	44.82 ± 3.36*	26.10↓	43.14 ± 2.46**	26.16↓	25.91 ± 1.64***	42.72↓	19.28 ± 1.11***	46.42↓
Indo-2.5	29.63 ± 3.42***	51.15↓	25.07 ± 2.72***	57.08↓	15.18 ± 2.80***	66.45↓	12.11 ± 2.20***	66.34↓

Data are expressed as mean ± SEM.

P≤ (0.05*, 0.01**, 0.001***); F ≤ 0.05.

PAM: Methanol extract of *Pluchea arguta*

in paw volume at 3 h (30.98 per cent, P≤0.05), 24 h (42.64 per cent, P≤0.01) and 48 h (21.67 per cent). PAM-200 and PAM-400 groups showed almost similar anti-inflammatory activity at 3 h, 24 h and 48 h. PAM-200 and PAM-400 groups showed higher anti-inflammatory activity than reference drug indomethacin at 3 h and 48 h. Thus, from the results, it can be concluded, that methanol extract of *P. arguta* has higher anti-inflammatory activity in formaldehyde induced paw edema test. PAM showed significant decrease in paw volume till 48 h with both doses, which suggests its long duration of action.

Table 4.4: Anti-inflammatory activity of methanol extract of *Pluchea arguta* in formaldehyde induced paw edema.

Groups	*Per cent Increase in Paw Volume*					
	After 3 h		*After 24 h*		*After 48 h*	
	Increase in Paw Volume	*Per cent Change*	*Increase in Paw Volume*	*Per cent Change*	*Increase in Paw Volume*	*Per cent Change*
Control	53.03 ± 4.98	–	54.03 ± 4.97	–	38.57 ± 4.23	–
PAM-200	31.40 ± 3.76*	40.79↓	35.13 ± 2.12*	34.98↓	24.70 ± 4.31	35.96↓
PAM-400	30.25 ± 4.76*	42.96↓	34.31 ± 3.45*	36.50↓	23.27 ± 4.71	39.67↓
Indo-2.5	36.60 ± 3.73*	30.98↓	30.99 ± 2.52**	42.64↓	30.21 ± 3.89	21.67↓

Data are expressed as mean ± SEM.

P≤ (0.05*, 0.01**); F ≤ 0.05.

PAM: Methanol extract of *Pluchea arguta*.

Analgesic Activities

Acetic Acid Writhing Test

Methanol extract of *Pluchea arguta* was used to distinguish between the central and peripheral analgesic action by acetic acid induced writhing response in mice. This method is not only simple and reliable but also affords rapid evaluation of peripheral type of analgesic action (Gene *et al.*, 1998). In this test the animals react with characteristic stretching behavior, which is called writhing (Ganesh *et al.*, 2008; Mbiantcha *et al.*, 2011). The writhing response of the mouse to an intraperitoneal injection of noxious chemical is used to screen for both peripherally and centrally acting analgesic activities (Trongsakul *et al.*, 2003).

Intraperitoneal injection of diluted solutions of acetic acid is a well established animal model for tonic visceral pain in rodents (Martinez *et al.*, 1999). Intraperitoneal injection of acetic acid induces a stereotypic response pattern in the form of abdominal contractions (lengthwise stretches of the torso with a concomitant concave arching of the back) that may persist beyond 6 h after the administration, although most contractions occur within 30 min of the application of the irritant (Koster *et al.*, 1959; Mogil *et al.*, 1996).

The results of analgesic activity of methanol extract of *Pluchea arguta* in acetic acid induced writhing test is shown in Table 4.5. PAM-200 and PAM-400 groups showed 31 per cent and 35 per cent inhibition significantly in abdominal writhes produced by acetic acid respectively as compared with the control group. Thus, PAM-200 and PAM-400 groups produced a significant ($P \leq 0.05$) and dose depend inhibition of analgesic effect produced by acetic acid. Indomethacin group showed significant inhibition with 46 per cent ($P \leq 0.01$) as compared with the control group. This probably means that the extract is able to reduce the receptor sensitivity to the acetic acid induced pain in a dose dependent manner. The abdominal constriction is related to the sensitization of nociceptive receptors to prostaglandins. This therefore shows anti-nociceptive activity (Nwinyi *et al.*, 2006).

Table 4.5: Analgesic activity of methanol extract of *Pluchea arguta* in acetic acid writhing test in Swiss albino mice.

Groups	*Onset Time (Sec.)*	*Per cent Change*	*Frequency*	*Per cent Change*
Control (D/W)	29 ± 5	–	84 ± 7	–
PAM-200	226 ± 49**	675↓	58 ± 5*	31↓
PAM-400	246 ± 52**	743↓	54 ± 5*	35↓
Indo-2.5	495 ± 51***	1594↓	45 ± 8**	46↓

Data are expressed as mean ± SEM.

P≤ (0.05*, 0.01**, 0.001***).

PAM: Methanol extract of *Pluchea arguta*.

Formalin Induced Paw Licking Test

The formalin test is a valid and reliable model of nociception, and it is sensitive to various classes of analgesic drugs (Karabay-Yavasoglu *et al.*, 2007). The Formalin test may be more useful as model of clinical pain in which the first phase seems to be due to direct chemical activation on nociceptive afferent fibers, whereas the second phase is dependent of peripheral inflammation and changes in central processing (Tjolsen *et al.*, 1992; Sayyah *et al.*, 2004; Liao *et al.*, 2012).

The results of analgesic activity of methanol extract of *Pluchea arguta* in formalin induced paw licking are shown in Table 4.6. The onset time of paw licking was measured after formalin injection. PAM-200 and PAM-400 showed 7.29 per cent and 21.53 per cent increase in onset time respectively, while indomethacin group showed 25 per cent increase in onset time as compared with the control group. Onset time of paw licking response of PAM-400 group was almost the same as that of the indomethacin group. PAM-400 and indomethacin groups showed statistically significant increase in paw licking response ($P \leq 0.05$).

After formalin injection, the frequency of paw licking was measured between 0-5 min, 6-10 min, 11-15 min, 16-20 min and 21-30 min. As compared with the control group, PAM- 200 group showed 14.04 per cent decrease in frequency between 0-5 min, 38.34 per cent between 6-10 min, 62.24 per cent between 11-15 min, 2.74 per cent

Table 4.6: Analgesic activity of methanol extract of *Pluchea arguta* in formalin induced paw licking test in rats.

Groups	*Onset Time (Sec)*	*Per cent Change*	*0-5 min.*		*6-10 min.*		*11-15 min.*		*16-20 min.*		*21-30 min.*	
			Frequency	*Per cent Change*	*Frequency*	*Per cent Change*	*Frequency*	*Per cent Change*	*Frequency*	*Per cent Change*	*Frequency*	*Per cent Change*
Control	48 ± 3.13	–	16.67 ± 1.05	–	4.33 ± 0.56	–	6.17 ± 1.58	–	5.83 ± 0.91	–	19.17 ± 2.52	–
PAM-200	51.50 ± 4.19	7.29↓	14.33 ± 1.96	14.04↓	2.67 ± 0.61	38.34↓	2.33 ± 0.95	62.24↓	5.67 ± 1.43	2.74↓	16.67 ± 2.03	13.04↓
PAM-400	58.33 ± 1.56*	21.53↓	10.33 ± 2.23*	38↓	0.83 ± 0.40**	80.83↓	2.83 ± 1.14	54.13↓	5 ± 0.68	14.24↓	16.50 ± 1.09	13.93↓
Indo-2.5	60 ± 2.14*	25↓	12 ± 1.34*	28.01↓	1.17 ± 0.98*	72.98↓	3.17 ± 0.91	48.62↓	5.50 ± 1.45	5.66↓	17 ± 1.86	11.32↓

Data are expressed as mean ± SEM.

P≤ (0.05*, 0.01**).

PAM: Methanol extract of *Pluchea arguta.*

between 16-20 min and 13.04 per cent between 21-30 min. PAM-400 group showed 38 per cent decrease in frequency between 0-5 min, 80.83 per cent between 6-10 min, 54.13 per cent between 11-15 min, 14.24 per cent between 16-20 min and 13.93 per cent between 21-30 min. Indomethacin group showed 28.01 per cent decrease in frequency between 0-5 min, 72.98 per cent between 6-10 min, 48.62 per cent between 11-15 min, 5.66 per cent between 16-20 min and 11.32 per cent between 21-30 min. Statistically significant decrease in frequency between 0-5 min was seen in PAM-400 group ($P \leq 0.05$) and indomethacin group ($P \leq 0.05$). Statistically significant decrease in frequency between 6-10 min was seen in PAM-400 group ($P \leq 0.01$) and indomethacin group ($P \leq 0.05$). Higher level of frequency in paw licking was observed between 0-5 min and 21-30 min.

Conclusion

The results of the present study demonstrated that methanol extract of *Pluchea arguta* produced dose related acute anti-inflammatory activity (Carrageenan, dextran, histamine and formalin) and analgesic activity (Acetic acid and formalin). These studies have shown that the methanol extract of *Pluchea arguta* contains some active ingredients with the potential of being good anti-inflammatory and analgesics. There is a need for detailed investigation of the mechanism of action of methanol extract of *Pluchea arguta* which based on a possible therapy can be visualized.

References

Abad, M.J., Bessa, A.L., Ballarin, B., Aragon, O., Gonzales, E., and Bermejo, P. (2006). Anti-infammatory activity of four Bolivian *Baccharis* species (Compositae). *Journal of Ethnopharmacology*, **103:** 338-344.

Albert, D., Zundorf, I., Dingermann, T., Muller, W.E., Steinhilber, D., and Werz, O. (2002). Hyperforin is a dual inhibitor of cyclooxygenase-1 and 5- lipoxygenase. *Biochemical Pharmacology*, **64**: 1767-1775.

Almeida, R.N., Navarro, D.S., and Barbosa-Filho, J.M. (2001). Plants with central analgesic activity. *Phytomedicine*, **8**: 310-322.

Aggarwal, L.D., Arambewela, L.S., and Ratnasooriya, W.D. (2012). *Alpinia calcarata* Roscoe: a potent anti-inflammatory agent. *Journal of Ethnopharmacology*, **139**: 889-92.

Balasuramanian, A., Ramalingam, K., Krishnan, S., and Chisitina, A.J.M. (2005). Anti-inflammatory activity of *Morus indica* Linn. *Iranian Journal of Pharmacology and Therapeutics*, **4**: 13-15.

Banerjee, S., Sur, T.K., Mandal, S., Das, P.C., and Sikdar, S. (2000). Assessment of the anti-inflammatory effects of *Swertia chirata* in acute and chronic experimental models in male albino rats. *Indian Journal of Pharmacology*, **32**: 21-24.

Barros, I.M.C., Lopes, L.D.G., Borges, M.O.R., Borges, A.C.R., Ribeiro, M.N.S., and Freire, S.M.F. (2006). Anti-inflammatory and anti-nociceptive activities of *Pluchea quitoc* (DC.) ethanolic extract. *Journal of Ethnopharmacology*, **106**: 317-320.

Bhatt, K.R., Mehta, R.K., and Shrivastava, P.N. (1977). A simple method for recording anti-inflammatory effects on rat paw edema. *Indian Journal of Physiology and Pharmacology*, **21**: 399-400.

Brito, A.R.M.S., and Antonio, M.A. (1998). Oral anti-inflammatory and antiulcerogenic activities of a hydroalcoholic extract and partitioned fractions of *Turnera ulmifolia* (Turneraceae). *Journal of Ethnopharmacology*, **61**: 215-228.

Brownlee, G. (1950). Effect of deoxycotone and ascorbic acid in formaldehyde-induced arthritis in normal and adrenalectomized rats. *The Lancet*, **1**: 157-159.

Cuman, R.K.N., Bersani-Amadio, C.A., and Fortes, Z.B. (2001). Influence of type 2 diabetes on the inflammatory response in rat. *Inflammation Research*, **50**: 460-465.

Dahl, V., and Reader, J.C. (2000). Non-opioid postoperative analgesia. *Acta Anaesthesiologica Scandinavica*, **44**: 1191-1203.

Di Rosa, M., Giround, J.P., and Willoughby, D.A. (1971). Studies of the acute inflammatory response induced in rats in different sites by carrageenan and turpentine. *Journal of Pathology*, **104**: 15-29.

Domaj, M.I., Glassco, W., Aceto, M.D., and Martin, B.R. (1999). Antinociceptive and pharmacological effects of metanicotina, a selective nicotine agonist. *Journal of Pharmacology and Experimental Therapy*, **291**: 390-398.

Eisner, T. (1990). Chemical prospecting. A call for action. In: Ecology, Economic and Ethics (Eds.) Borman FH, Kellert SR. The Broken Circle, Yale University Press, New Haven.

Farnsworth, N.R. (1989). Screening Plants for New Medicines. In: Biodiversity, Part II (Ed.) Wilson EO. National Academy Press, Washington pp. 83-97.

Fernandez, F., and Torres, M. (2006). Evaluation of *Pluchea carolinensis* extracts as antioxidants by the epinephrine oxidation method. *Fitoterapia*, **77**: 221-226.

Ganesh, M., Vasudevan, M., Kamalakannan, K., Saravana Kumar, A., Vinoba, M., Ganguly, S., and Sivakumar, T. (2008). Anti-inflammatory and analgesic effects of *Pongamia glabra* leaf gall extract. *Pharmacologyonline*, **1**: 497-512.

Gene, R.M., Segura, L., Adzet, T., Marin, E., and Inglesias, J. (1998). *Heterotheca inuloides*: anti-inflammatory and analgesic effects. *Journal of Ethnopharmacology*, **60:** 157-162.

Gochman, E., Mahajna, J., Shenzer, P., Dahan, A., Blatt, A., Elyakim, R., and Reznick, A. Z. (2012). The expression of iNOS and nitrotyrosine in colitis and colon cancer in humans. *Acta Histochemica* (In press).

Greenwald, R.A. (1991). Animal models for evolution of arthritic drug. *Methods and Findings in Experimental and Clinical Pharmacology*, **13**: 75-83.

Hajhashemi, V., Zolfaghari, B., and Yousefi, A. (2012). Antinociceptive and anti-inflammatory activities of *Satureja hortensis* seed essential oil, hydroalcoholic and polyphenolic extracts in animal models. *Medicinal Principles and Practice*, **21**, 178-82.

Hernandez-Perez, M., and Rabanal, R.M. (2002). Evaluation of the anti-inflammatory and analgesic activity of *Sideritis canariensis* var. *pannosa* in mice. *Journal of Ethnopharmacology*, **81**: 43-47.

Hori, Y., Jyoyama, H., Yamada, K., Takagi, M., Hirose, K., and Katori, M. (1988). Time course analyses of kinins and other mediators in plasma exudation of rat kaolin-induced pleurisy. *European Journal of Pharmacology*, **152**: 235- 245.

Hunskaar, S., and Hole, K. (1987). The formalin test in mice: dissociation between inflammatory and non-inflammatory pain. *Pain*, **30**: 103-114.

Iwalewa, E.O., McGaw, L.J., Naidoo, V., and Eloff, J.N. (2007). Inflammation: the foundation of diseases and disorders. A review of phytomedicines of South African origin used to treat pain and inflammatory conditions. *African Journal of Biotechnology*, **6**: 2868-2885.

Jaffri, S.M.H. (1966). Flora of Karachi, Pakistan. pp.335.

Jarald, E.E., Edwin, S., Saini, V., Deb, L., Gupta, V.B., Wate, S.P., and Busari, K.P. (2008). Anti-inflammatory and anthelmintic activities of *Solanum khasianum* Clarke. *Natural Product Research*, **22**: 269-274.

Karabay-Yavasoglu, N.U., Karamenderes, C., Baykan, S., and Apaydin, S. (2007). Antinociceptive and anti-inflammatory activities and acute toxicity of *Achillea nobilis* subsp. *neilreichii* extract in mice and rats. *Pharmaceutical Biology*, **45**: 162-168.

Kikuchi, M., Tsuzurahara, K., Suzuki, T., Yato, N., and Naito, K. (1996). Involvement of leukotrienes in allergic pleurisy in actively sensitized rats: inhibition by the lipoxygenase inhibitor T-0757 of the increase in vascular permeability and leukotriene E4 production. *Inflammation Research*, **45**: 192-197.

Koster, R., Anderson, M., and De Beer, E.J. (1959). Acetic acid for analgesic screening. *Federation Proceeding*, **18**: 418-420

Kupeli E., Orhan I., and Yesilada E. (2007). Evaluation of some plants used in Turkish folk medicine for their anti-inflammatory and antinociceptive activity. *Pharmaceutical Biology*, **45:** 547-555.

Lawrence T., and Gilroy, D.W. (2007). Chronic inflammation: a failure of resolution?. *International Journal of Experimental Pathology*, **88**: 85-94.

Leal, L. K., Pierdona, T. M., Goes, J. G., Fonseca, K. S., Canuto, K. M., Silveira, E. R., Bezerra, A. M., and Viana, G. S. (2011). A comparative chemical and pharmacological study of standardized extracts and vanillic acid from wild and cultivated *Amburana cearensis* A.C. Smith. *Phytomedicine*, **18**, 230-3.

Liao, C. R., Kao, C. P., Peng, W. H., Chang, Y. S., Lai, S. C., and Ho, Y. L. (2012). Analgesic and anti-inflammatory activities of methanol extract of *Ficus pumila* L. in mice. *Evidence Based Complementary and Alternative Medicine*, doi: 10.1155/2012/340141.

Linardi, A., Costa, S.K.P., De Silva, G.R., and Antunes, E. (2002). Involvement of kinins, mast cells and sensory neurons in the plasma exudation and paw edema induced by styphylococcal entrotoxin B in the mouse. *European Journal of Pharmacology*, **399**: 235–242.

Loganayaki, N., Siddhuraju, P., and Manian, S. (2012). Antioxidant, anti-inflammatory and anti-nociceptive effects of *Ammannia baccifera* L. (Lythracceae), a folklore medicinal plant. *Journal of Ethnopharmacology*, **140**, 230-3.

Latha, M.S., Latha, K.P., Vagdevi, H.M., and Virupaxappa, S.B. (2012). Anti inflammatory activity of *Mangifera indica* L. Var. Rasapuri root extracts. *Journal of Chemical and Pharmaceutical Research*, **4**: 333-336.

Magali, B., DeSonza, M.M., Yunus, R.A., Lento, R., Monache, F.D., and Fillo, V.C. (2000). Antinociceptive activity of 13, 118 Binaringenin, a biflavonoid present in plants of the Guttiferae. *Planta Medica*, **66**: 84-86.

Martýnez, V., Thakura, S., Mogilb, J.S., Tache,Y., and Mayera, E.A. (1999). Differential effects of chemical and mechanical colonic irritation on behavioral pain response to intraperitoneal acetic acid in mice. *Pain*, **81**: 179-186.

Mazzon, E., Serraino, I., Li, J.H., Dugo, L., Caputi, A.P., Zhang, J., and Cuzzocrea, S. (2001). GPI 6150, a poly (ADP-ribose) polymerase inhibitor, exhibits an anti-inflammatory effect in rat models of inflammation. *European Journal of Pharmacology*, **415**: 85-94.

Mbiantcha, M., Kamanyi, A., Teponno, R., Tapondjou, A., Watcho, P., and Nguelefack, T. (2011). Analgesic and anti-inflammatory properties of extracts from the bulbils of *Dioscoreabulbi fera* L. var *sativa* (Dioscoreaceae) in mice and rats. *Evidence Based Complementary and Alternative Medicine*, doi: 10.1155/2011/912935.

Mogil, J.S., Kest, B., Sadowski, B., and Belknap, J.K. (1996). Differential genetic mediation of sensitivity to morphine in genetic models of opiate antinociception: influence of nociceptive assay. *Journal of Pharmacology and Experimental Therapeutics*, **276**: 532-544.

Moon,T.C., Murakami, M., Kudo, I., Son, K.H., Kim, H.P., Kang, S.S., and Chang, H.W. (1999). A new class of COX-2 inhibitor, rutaecarpine from *Evodia rutaecarpa*. *Inflammation Research*, **48**: 621-625.

Muhammad, N., Saeed, M., and Khan, H. (2012). Antipyretic, analgesic and anti-inflammatory activity of *Viola betonicifolia* whole plant. *BMC Complementary and Alternative Medicine*,**12**, 59.

Nakajima, A., Fukui, T., Takahashi, Y., Kishimoto, M., Yamashina, M., Nakayama, S., Sakaguchi, Y., Yoshida, K., Uchida, K., Nishio, A., Yodoi, J., and Okazaki, K. (2012). Attenuation of indomethacin-induced gastric mucosal injury by prophylactic administration of sake yeast-derived thioredoxin. *Journal of Gastroenterology*, DOI 10.1007/s00535-012-0564-5.

Nwinyi, F.C., Ajoku, G.A., Aniagu, S.O., Kubmarawa, D., Enwerem, N., Dzarma, S., and Inyang, U.S. (2006). Pharmacological justification for the ethnomedicinal use of *Amblygonocarpus andongensis* stem bark in pain relief. *African Journal of Biotechnology*, **5**: 1566-1571.

Ogonowski, A.A., May, S.W., Moore, A.B., Barret, L.T., O'Bryant, C.L., and Pollock, S.H. (1997), Anti-inflammatory and analgesic activity of an inhibitor of neuropeptide amidation. *Journal of Pharmacology and Experimental Therapeutics*, **280**: 846–853.

Jain, P.S., and Bari, S.B. (2010). Anti-inflammatory activity of *Abelmoschus manihot* extracts. *International Journal of Pharmacology*, **6**: 505-509.

Paget, G.E., and Barnes, J.M. (1964). In: Evaluation of drug activities. Pharmacometrics, Lawrence DR, Bachrach AL (Eds.) Vol. 1, Academic press, New York.

Posadas, I., Bucci, M., Roviezzo, F., Rossi, A., Parente, L., Sautebin, L., and Cirino, G. (2004). Carrageenan-induced mouse paw oedema is biphasic, age-weight dependent and displays differential nitric oxide cyclooxygenase-2 expression. *British Journal of Pharmacology*, **142**: 331-338.

Rang, H.P., Dale, M.M., and Ritter, J.M. (2001). Pharmacology, 4th ed. Guanabara-Koogan, Rio de Janeiro pp. 174-188, 691.

Roitt, I.M. (1997). Essential immunology, 9th Ed., Oxford: Blackwell Scientific: Oxford.

Rowley, D.A., and Benditt, E.P. (1956). 5-Hydroxytryptamine and histamine as mediators of the vascular impurity produced by agents which damage mast cells in rats. *Journal of Experimental Medicine*, **103**: 399-415.

Sayyah, M., Hadidi, N., and Kamalinejad, M. (2004). Analgesic and anti-inflammatory activity of *Lactuca sativa* seed extract in rats. *Journal of Ethnopharmacology*, **92**: 325-329.

Shu, X.S., Gao, Z.H., and Yang, X.L. (2006). Anti-inflammatory and antinociceptive activities of *Smilax china* L. aqueous extract. *Journal of Ethnopharmacology*, **103**: 327-332.

Simpson, B., Claudie, D., Smith, N., Wang, J., McKinnon, R., and Semple, S. (2010). Evaluation of the anti-inflammatory properties of *Dodonaea polyandra*, a Kaanju traditional medicine. *Journal of Ethnopharmacology*, **132**: 340-343.

Speranza, L., Franceschelli, S., Pesce, M., Menghini, L., Patruno, A., Vinciguerra, I., De Lutiis, M. A., Felaco, M., Felaco, P., and Grilli, A. (2009). Anti-inflammatory properties of the plant *Verbascum mallophorum*. *Journal of Biological Regulators and Homeostatic Agents*, **23**, 189-95.

Subbaramaiah, K., Telang, N., Ramonetti, J.T., Araki, R., Devito, B., Weksker, B.B., and Dannenberg, A.J. (1996). Transcription of cyclooxygenase-2 is enhanced in transformed mammary epithelial cells. *Cancer Research*, **56**: 4424-4429.

Suleyman, H., Demirezer, L.O., Kuruuzum, A., Banoglu, Z.N., Gocer, F., Ozbakir, G., and Gepdiremen, A. (1991). Anti-inflammatory effect of the aqueous extract from *Rumex patientia* L. roots. *Journal of Ethnopharmacology*, **65**: 141-148.

Tjolsen, A., Berge, O.G., Hunskaar, S., Rosland, J.H., and Hole, K. (1992). The formalin test: an evaluation of the method. *Pain*, **51**: 5-17.

Trongsakul, S., Panthong, A., Kanjanapothi, D., and Taesotikul, T. (2003). The analgesic, antipyretic and anti-inflammatory activity of *Diospyros variegata* Kruz. *Journal of Ethnopharmacology*, **85**: 221-225.

Vaghasiya, Y., Nair, R., and Chanda, S. (2008). Antibacterial and preliminary phytochemical and physico-chemical analysis of *Eucalyptus citriodora* Hk leaf. *Natural Product Research*, **22**: 754-762.

Vaghasiya, Y., Parekh, J., Shukla, V., and Chanda, S. (2011). Antimicrobial and anti-inflammatory screening of four Indian medicinal plants. *Latin American Journal of Pharmacy*, **30**: 661-662.

Van Wauve, J.P., and Goosens, J.G. (1989). Arabinolactan and dextran induced ear inflammation in mice: differential inhibition by H1-antihistamines, 5- HT-serotonin antagonists and lipoxygenase blockers. *Agents Action*, **28**: 78–82.

Vineger, R., Schreiber, W., and Hugo, R. (1969). Biphasic development of carrageenan edema in rats. *Journal of Pharmacology and Experimental Therapeutics*, **166**: 96-103.

Winter, C.A., and Porter, C.C. (1957). Effect of alteration in side chains upon anti-inflammatory and liver glycogen activities in hydrocortisone ester. *Journal of the American Pharmaceutical Association*, **46**: 515-519.

Winter, C.A., Risley, E.A., and Nuss, G.V. (1962). Carrageenan-induced edema in hind paws of the rat as an assay for anti inflammatory drugs. *Proceedings of the Society for Experimental Biology and Medicine*, **111**: 544-547.

Xiao, J., Jiang, X., and Chen, X. (2005). Antibacterial, anti-inflammatory and diuretic effect of flavonoids from *Marchantia convolute. African Journal of Traditional, Complementary and Alternative Medicines*, **2**: 244-252.

Yonathan, M., Asres, K., Assefa, A., and Bucar, F. (2006). *In vivo* anti-inflammatory and antinociceptive activities of *Cheilanthes farinose. Journal of Ethnopharmacology*, **108**: 462-470.

Medicinal Plants: Phytochemistry, Pharmacology and Therapeutics Vol. 3 (2014) *Pages* **143–152**
Editor-in-Chief: **V.K. Gupta**
Published by: **DAYA PUBLISHING HOUSE, NEW DELHI**

5

Assessing the Beneficial Effects of a Magnolia Extract Supplemented with Phosphatidylserine: A Randomised, Double-blind, Placebo-controlled Clinical Trial

Elizabeth A. Nelson[1,2], Simon R. Knowles[2], Snehal R. Jadhav[1], Aoiffe Walsh[3] and Enzo A. Palombo[1*]

ABSTRACT

Stress is common aspect of everyday life, often associated with reduced well-being. There is increasing interest in the use of natural herbal remedies for the relief of stress. The aim of the current study was to investigate the biological and psychological mechanisms underlying the beneficial effects of a propriety blend of magnolia extract and phosphatidylserine in a double-blind, placebo-controlled, clinical trial comprising 34 healthy adults (26.09 ± 6.09 years; range 19-45). Participation took place over a consecutive 4-week period. This included a baseline period where participants did not take the herbal supplement, two experimental periods where participants were instructed to take the recommended dose of herbal supplement (two capsules immediately upon awakening), and a follow-up period where participants ceased taking the supplement. Throughout the study period, participants were required to complete a series of questionnaires assessing concentration, mood, perceived stress, anxiety,

1 Environment and Biotechnology Centre, Faculty of Life and Social Sciences, Swinburne University of Technology, John Street Hawthorn, VIC 3122, Australia.

2 Brain and Psychological Sciences Research Centre, Faculty of Life and Social Sciences, Swinburne University of Technology, John Street Hawthorn, VIC 3122, Australia

3 Canopus Corporation, Baybush, Straffan, Co. Kildare, Ireland

* *Corresponding author*: E-mail: epalombo@swin.edu.au

and depression. Participants also provided saliva samples on three select days during each week, to measure the biological markers of stress, cortisol and alpha-amylase. Waking cortisol levels decreased throughout the study period in the treatment group, while there was no significant change in levels for the control group. For the treatment group, the cortisol awakening response also tended to decrease throughout the study period. Paralleling this, there was a significant decrease in the daily measure of stress and the weekly measure of depression in the treatment group, with similar trends seen for the other measures of psychological well-being. There were no significant changes in well-being measures over the study period for the control group. Our study showed that a dietary supplement containing proprietary extracts of magnolia and phosphatidylserine may have beneficial anxiolytic effects through targeting the hypothalamic-pituitary-adrenal axis regulator of the stress response, with the supplement having significant beneficial effects on waking cortisol, daily stress and weekly depression.

Keywords: Magnolia extract, Phosphatidylserine, Salivary cortisol, Salivary alpha-amylase, HPA-axis, Psychological well-being.

Introduction

Stress is a familiar aspect of everyday life encountered by everyone in one form or another. Nearly a century of stress research has suggested that stress is associated with reduced well-being and has been linked to increased disease states (Drossman, 1998; Gabry *et al.*, 2002) and increased susceptibility to common viral and bacterial infections (Segerstrom and Miller, 2004). There is also evidence that stress adversely affects various aspects of psychological well-being (Talbott, 2000; Lozovaya and Miller, 2003). Recently, there has been an increasing interest in the use of natural herbal remedies for the relief of stress.

An ancient traditional Chinese herb, magnolia extract, is known for its anxiolytic properties and is commonly used for decreasing symptoms and signs of stress and depression. Magnolia extract is a standardised herb extract made from the bark of the *Magnolia officinalis* tree. The bark contains two active phytochemicals: honokiol and magnolol, with honokiol exerting an anti-stress effect, and magnolol exerting an anti-depressant effect. Xu *et al.* (2008) found that elevated levels of the stress hormone corticosterone induced by a chronic mild stress model where ameliorated following administration of this mixture. These findings provide evidence that the mixture of honokiol and magnolol are effective at normalizing hypothalamic-pituitary-adrenal (HPA) dysfunction in rats (the neuroendocrine regulator of the stress response). Phosphatidylserine is a phospholipid (including amino and fatty acids) that is vital to brain cell function (Kennedy *et al.*, 2007). It is important in learning and memory and has been shown to have a direct effect on improvements in stress and depression (Castilho *et al.*, 2004; Hellhammer *et al.*, 2004).

There has been limited research investigating the anxiolytic activity of magnolia extract amongst humans. Kalman *et al.* (2008) measured the effects of a proprietary blend of magnolia and phellodendron extracts on anxiety, stress and sleep in healthy premenopausal women. The 16 patients in the treatment group received 750 mg of the mixture, daily for six weeks. They found the blend was not effective in reducing long standing feelings of anxiety or depression. Other assessments conducted in this study, including salivary cortisol and amylase levels, were not significantly changed in comparison to placebo.

Although it is widely accepted that magnolia extract exerts an anti-stress effect, the underlying mechanisms involved in this process are poorly understood. There have been limited studies to date that have specifically explored the role of magnolia extract on the stress response in humans. One of the main responses to stress is the activation of the HPA axis, with the primary consequence being the secretion of cortisol (Lozovaya and Miller, 2003). Several studies have demonstrated a relationship between perceived stress and cortisol levels, with findings providing evidence that higher cortisol secretion is related to higher levels of stress (Ockenfels *et al.*, 1995; Smyth *et al.*, 1998; Vedhara *et al.*, 1999). The transition from sleep to wakefulness provides a measure of HPA-axis activity, characterised by a marked increase in cortisol levels immediately after awakening. This process is referred to as the cortisol awakening response (CAR). There is a general consensus that psychological variables are involved in the dysregulation of the HPA-axis, as evident by effects on the CAR (Pruessner *et al.*, 2003). Several studies have also demonstrated that alpha-amylase is a sensitive biomarker of the stress response (Nater *et al.*, 2009). Salivary alpha-amylase reflects the activity of the sympathetic nervous system, with studies showing increases in alpha-amylase during times of stress (Nater *et al.*, 2009). The current research involved a placebo-controlled clinical trial to determine the possible psychological and biological mechanisms underlying the beneficial effects of the proprietary blend of magnolia extract and phosphatidylserine, Total Calm and Focus® (TC&F®).

Materials and Methods

Participant Selection and Study Design

The protocol was approved by the Swinburne University of Technology Human Research Ethics Committee and informed consent was obtained from all the subjects who agreed to participate in the protocol. To be eligible to participate in the study, participants had to be aged 18 years or over; have no history of significant psychological or physiological conditions; no allergies to magnolia extract or phosphatidylserine, or other ingredients in the supplement; not taking any stimulant medication including MAO-I, anxiolytics, psychotropics, or steroid hormones; not taking any dietary supplement purported to alter stress levels; not pregnant, breast feeding, or planning to become pregnant during the study.

Recruitment advertisements were posted around the Hawthorn campus of Swinburne University of Technology (Victoria, Australia). Students were also notified about the study via the online teaching resource (Blackboard), and the study was also advertised in the University Staff Bulletin.

Participants were randomly allocated to either the treatment group or the placebo-control group. Participants in the treatment group were administered the proprietary supplement, TC&F®. Participants in the placebo control group were administered an inactive, matched capsule containing olive oil, identical in size, shape and colour.

Participation in the study was over a period of four consecutive weeks. This included a baseline period where a 'normal' or baseline measure of well-being before treatment was obtained (week 1); an experimental period (weeks 2 and 3) where possible effects of treatment on well-being were assessed. Participants were asked to take two capsules immediately upon awakening each day throughout this period; and a follow-up period (week 4) where possible continuing effects of treatment on well-being were assessed after treatment had ceased.

Questionnaire Measures

Data were collected using three questionnaires; the demographic questionnaire, the daily questionnaire, and the weekly questionnaire. The demographic questionnaire assessed demographic information (*e.g.*, age and sex) and was completed at the onset of the study. The daily diary questionnaire consisted of a series of scales assessing daily incidences of health-related symptoms, and daily mood, stress, depression, anxiety, and alertness, measured on a 10cm visual analog scale. This was to be completed in the evening of any three selected days during each week of testing. The weekly questionnaire assessed weekly stress, depression, and anxiety (DASS21, Lovibond and Lovibond, 1995) and was completed at the end of each week of testing.

Saliva Sampling and Analysis

Participants were asked to provide three timed saliva samples on any three selected days during each week of testing. For each day of sampling, a saliva sample was to be provided immediately upon awakening, 30 minutes post-awakening, and at 9.00 pm. A written protocol instructing participants how to collect saliva samples was provided. Participants were asked to provide as uncontaminated a sample as possible, and refrain from eating or drinking 1-hour prior to the saliva collection. Timed 3-min samples of whole unstimulated saliva were collected by placing the cotton dental swab of a Salivette saliva sampling device (Sarstedt, Nümbrecht, Germany) under the tongue for an accurately timed 3-min period. Participants were instructed not to swallow during this sampling period. After sampling, the swab was returned to the Salivette tube and immediately placed in the freezer. Upon receipt at the laboratory, saliva samples were stored at -20°C, until assayed.

Before analysis, saliva samples were thawed and centrifuged (15 min at 3,000 *g*). Salivary alpha-amylase was assayed with a Salivary Alpha-Amylase Assay kit (Salimetrics, Biocore Pty. Ltd.). A 1: 20 dilution was made with Assay Buffer. A competitive enzyme immunoassay kit (Salimetrics, Biocore Pty. Ltd.) was used for the determination of cortisol. Analyses were carried out according to the manufacturer's specifications. All samples from a participant were analysed in one assay. Absorbance measurements were carried out with an Emax micro-plate reader (Molecular Devices, Victoria, Australia) at 405 nm. The concentration of samples was calculated using an immunoassay software package (SoftMAX Pro), utilising a four parameter logistic curve fitting programme.

Statistical Methods

Descriptive data are presented as mean ± *SD* when applicable. Due to the small sample size, all tests were performed using non-parametric techniques. Differences were tested for significance using a non-parametric Friedman one-way within subjects ANOVA. Statistical significance was defined at a p value of less than 0.05. Data are presented as means. All statistical analyses were performed using Statistical Package for Social Sciences (SPSS) for Windows Statistical Package Version 18.0.

Results

Study Population

The intent-to-treat population consisted of 40 participants; however 34 participants aged between 19-45 years (M=26.09, SD=6.09) completed the study. The sample had an equal number of females and males. Based on random assignment, two groups were established, the treatment group and the control group. Eighteen in the treatment group and 16 in the control group completed the study. Six subjects (two in the treatment group and four in the control group) terminated the study early for various reasons (*i.e.*, due to date constraints related to the study protocol). Baseline characteristics were similar in both groups. The only notable difference between groups was that the control group showed lower waking cortisol levels at baseline, but this was not considered of clinical importance as all values were within the normal range of human salivary cortisol levels. Participants with missing observations were removed from relevant analyses.

Laboratory Measures

Salivary Cortisol

A non-parametric Friedman within subjects ANOVA was conducted to compare the change in the CAR across the study period for the treatment group and control group. Statistical analysis indicated there was no significant change in the CAR over the study period for the control group, $\chi^2 (3, N = 14) = 1.63, p = 0.653$. However, in the treatment group, statistical analyses indicated a trend towards significance, $\chi^2 (3, N = 16) = 4.21, p = 0.240$. Further examination of the CAR showed that it tended to decrease following baseline week for the treatment group, with this drop in CAR maintained throughout the study.

Closer examination of cortisol levels from saliva samples provided immediately upon waking was performed, as the process of awakening results in HPA-axis activity, often characterised by a marked increase in cortisol levels, therefore providing an accurate measure of the stress response. Mean waking cortisol levels over the study period for the treatment group and control group are shown in Figure 5.1.

A non-parametric Friedman one-way within subjects ANOVA found that there was no significant difference in waking cortisol levels across the study period for the control group $\chi^2 (3, N = 14) = 0.86, p= 0.836$. However, there was a significant effect for the study sampling period for waking cortisol levels for the treatment group $\chi^2 (3, N = 17) = 8.08, p =0.044$. Pairwise comparisons with non-parametric t-tests (Wilcoxon Signed Rank Test) indicated that the amount of waking cortisol detected in the baseline

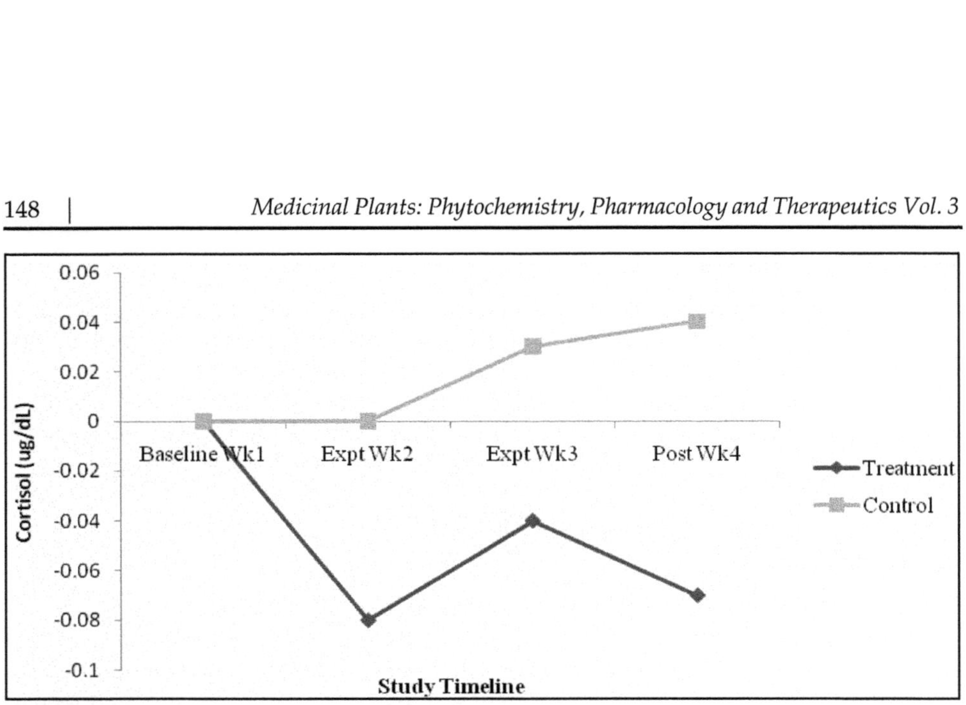

Figure 5.1: Pattern of waking cortisol over the study period for treatment group and control (placebo) group.

condition was significantly greater than the amount detected in experimental week 1 (Z= -2.33, p= 0.02) and follow-up week 4 (Z= -2.20, p= 0.028) of the treatment group.

Salivary Alpha-amylase

A non-parametric Friedman one-way within subjects ANOVA was conducted to compare the levels of alpha-amylase detected from evening saliva samples over the study period for the treatment and control groups. For both groups, there were no significant differences in salivary alpha-amylase levels over the study period, χ^2 (3, N = 18) = 6.02, p= 0.111 and χ^2 (3, N = 15) = 4.36, p= 0.225, respectively.

Questionnaire Measures of Psychological Well-being

Daily Diary Questionnaire

A non-parametric Friedman within subjects ANOVA was conducted to compare the changes in the daily diary questionnaire measures of psychological well-being over the study. For the control group, statistical analysis found there was no significant change in the daily measures of well-being over the study period. However, in the treatment group, statistical analyses found a significant difference in the daily measure of stress (χ^2 (3, N = 18) = 11.01, p = 0.012). There was no significant difference in the daily measures of depression, anxiety and alert-and-focus in the treatment group. However, further examination of these daily measures (Figure 5.2) showed an interesting pattern for the ill-being measures of stress, depression, and anxiety, with levels continuing to decrease up until experimental week 3, followed by a rise in levels in the follow-up week of the treatment group. However, the well-being measure of alert-and-focus remained relatively stable over the study period.

Weekly Questionnaire

A non-parametric Friedman within subjects ANOVA was conducted to compare the changes in the weekly questionnaire measures of psychological well-being over

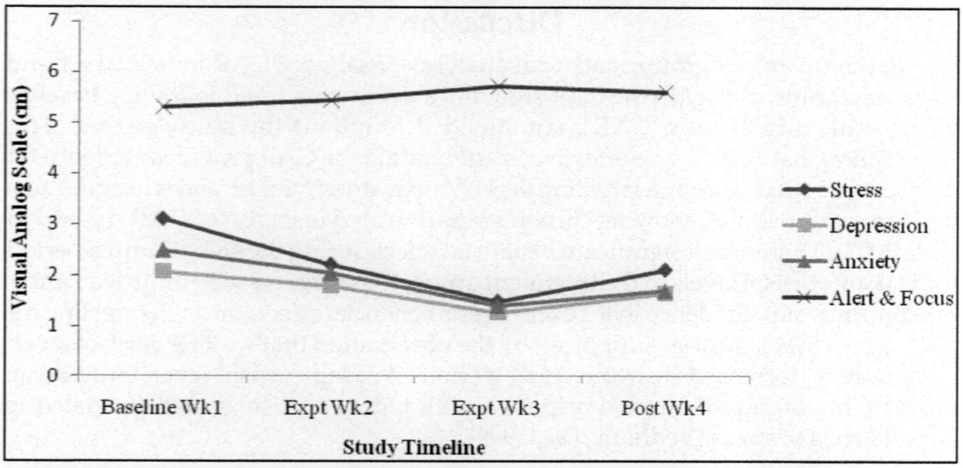

Figure 5.2: Pattern of daily diary psychological measures over the study period for the treatment group.

the study. For the control group, statistical analysis found there was no significant change in the daily measures of stress, depression, and anxiety. However, in the treatment group, statistical analyses found a significant difference in the weekly measure of depression (χ^2 (3, N = 18) = 8.79, p = 0.032). There was no significant difference in the weekly measures of stress and anxiety in the treatment group (χ^2 (3, N = 18) = 4.83, p = 0.185, χ^2 (3, N = 18) = 0.773, p = 0.856, respectively). However, further examination of these daily measures (Figure 5.3) showed that all weekly measures of ill-being showed a similar pattern, with levels continuing to decrease up until experimental week 3 in the treatment group.

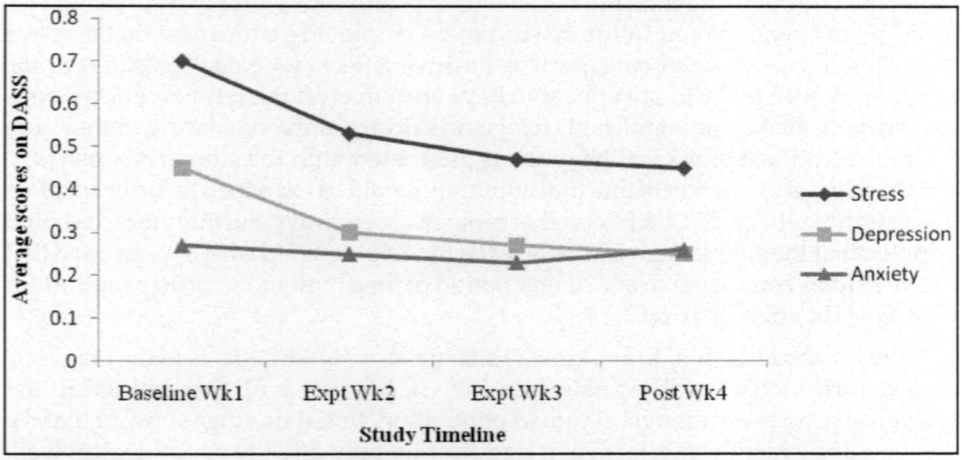

Figure 5.3: Pattern of weekly stress, depression, and anxiety over the study period for the treatment group.

Discussion

In the treatment group, statistical analyses of salivary CAR indicated a trend towards significance, with the CAR showing a decreasing trend following baseline week, with this drop in CAR maintained throughout the study period. This observation that TC&F® was effective in attenuating the CAR provides evidence for its anxiolytic effects through targeting the HPA-axis stress marker, and is in agreement with previous research showing chronic stress is related to increased CAR (Pruessner *et al.*, 2003). There was a significant beneficial effect during the study sample period for waking cortisol levels in the treatment group; there was no such improvement in the control group. Evidence that TC&F® exerts beneficial effects through targeting the HPA-stress axis is further supported by the observation that waking cortisol levels significantly decreased over the study period. Waking cortisol levels provide an accurate measure of the stress response, with higher cortisol secretion related to higher levels of stress (Vedhara *et al.*, 1999).

The beneficial biological effects of TC&F® on the stress response through targeting the HPA-axis is further supported by the findings from the questionnaire measures of well-being. Data from the daily diary questionnaire found a significant reduction in daily stress in the treatment group but not in the control group. The dietary supplement was also shown to significantly reduce depression measured in the weekly questionnaire. The other psychological measures of well-being also showed similar patterns, with levels tending to decrease during the study period. These trends suggest that future research employing a larger sample size is likely to result in significant reductions in these measures.

However, TC&F® did not appear to have any effect on the psychological measure of alert-and-focus, with daily measures remaining relatively stable over the study period. This may indicate that TC&F® demonstrates beneficial effects through improvements on adverse measures of well-being (*i.e.*, stress, anxiety and depression), rather than through targeting improvements in positive measures of well-being. This could be further explored in future studies by employing a measure that assesses both dimensions of well-being, such as Positive Affect and Negative Affect Scale. Furthermore, TC&F® did not appear to have any effect on the sympathetic nervous system, indicated by no significant changes in salivary alpha-amylase over the study period. Additional biological parameters associated with the stress response (*e.g.*, secretory IgA, a biomarker of mucosal immunity) could be examined to further explore the anxiolytic effects of TC&F® on other biological pathways. Furthermore, it should be noted that the supplement was only taken for a short period of two weeks, and that future studies conducted over a longer period of time may show further support for the anxiolytic effect of TC&F®.

This study is the first to explore the beneficial anxiolytic effects of the magnolia extract formulated in phosphatidylserine, TC&F®, on both psychological and biological processes amongst a human population. Initial findings showed that the dietary supplement was effective in significantly reducing the psychological well-being measures of daily stress and weekly depression, and significantly reducing the biological stress marker, waking salivary cortisol.

Acknowledgements

The present study was supported by Canopus Corporation, who gave advice for the study design and assisted in preparing the final manuscript for publication.

References

Castilho, J.C., Perry, J.C., Andreatini, R., and Vital, M. (2004). Phosphatidylserine: an antidepressive or a cognitive enhancer? *Progress in Neuro-Psychopharmacology and Biological Psychiatry*, **28**: 731-738.

Drossman, D.A. (1998). Gastrointestinal illness and the biopsychosocial model. *Psychosomatic Medicine*, **60**: 258-267.

Gabry, K.E., Chrousos, G.P., Rice, K.C., Mostafa, R.M., Sternberg, E., Negrao, A.B., Webster, E.L., McCann, S.M., and Gold, P.W. (2002). Marked suppression of gastric ulcerogenesis and intestinal responses to stress by a novel class of drugs. *Molecular Psychiatry*, **7**: 474-483.

Hellhammer, J., Fries, E., Buss, C., Engert, V., Tuch, A., Rutenberg, D., and Hellhammer, D. (2004). Effects of soy lecithin phosphatidic acid and phosphatidylserine complex (PAS) on the endocrine and psychological responses to mental stress. *Stress*, **7**: 119-26.

Kalman, D.S., Feldman, S., Feldman, R., Schwartz, H.I., Krieger, D. R., and Garrison, R. (2008). Effect of a proprietary *Magnolia* and *Phellodendron* extract on stress levels in healthy women: a pilot, double-blind, placebo-controlled clinical trial. *Nutrition Journal*, **7**: 1-6.

Kennedy, D.O., Haskell, C.F., Mauri, P.L., and Scholey, A.B. (2007). Acute cognitive effects of standardised *Ginkgo biloba* extract complexed with phosphatidylserine. *Human psychopharmacology*, **22**: 199-210.

Lovibond, P. F., and Lovibond, S. H. (1995). The structure of negative emotional states: comparison of the depression anxiety stress scales (DASS) with the Beck depression and anxiety inventories. *Behavioral Research Therapy*, **33**: 335-343.

Lozovaya, N., and Miller, A.D. (2003). Chemical neuroimmunology: health in a nutshell. Bidirectional communication between immune and stress (limbichypothalamic-pituitary adrenal) systems. *ChemBioChem.*, **4**: 466-484.

Nater, U.M., and Rohleder, N. (2009). Salivary alpha-amylase as a non-invasive biomarker for the sympathetic nervous system: Current state of research. *Psychoneuroendocrinology*, **34**: 486-496.

Ockenfels, M. C., Porter, L., Smyth, J., Kirschbaum, C., Hellhammer, D. H., and Stone, A.A. (1995). Effect of chronic stress associated with unemployment on salivary cortisol: overall cortisol levels, diurnal rhythm, and acute stress reactivity. *Psychosomatic Medicine*, **57**: 460-467.

Pruessner, M., Boekestyn, L., Béchard-Evans, L., Abadi, S., Vracotas, N., Joober, R., Pruessner, J.C., and Malla, A.K. (2008). Sex differences in the cortisol response to awakening in recent onset psychosis. *Psychoneuroendocrinology*, **33**: 1151-1154.

Segerstrom, C., and Miller, G.E. (2004). Psychological stress and the human immune system: a meta-analytic study of 30 years of inquiry. *Psychological Bulletin*, **130**: 601–630.

Smyth, J., Ockenfels, M. C., Porter, L., Kirschbaum, C., Hellhammer, D. H., and Stone, A.A. (1998). Stressors and mood measured on a momentary basis are associated with salivary cortisol secretion. *Psychoneuroendocrinology*, **23**: 353-370.

Talbott, S. (2002). The Cortisol Connection: Why Stress Makes You Fat and Ruins Your Health- and What You Can Do About It. Hunter House Publishers, USA.

Vedhara, K., Cox, K. M., Wilcock, G. K., Perks, P., Hunt, M., Anderson, S., Lightman, S. L., and Shanks, N. M. (1999). Chronic stress in elderly careers of dementia patients and antibody response to influenza vaccination. *Lancet*, **353**: 627-631.

Xu, Q., Yi L.T., Pan Y., Wang X., Li Y.C., Li J.M., Wang C.P., and Kong L.D. (2008). Antidepressant-like effects of the mixture of honokiol and magnolol from the barks of *Magnolia officinalis* in stressed rodents. *Progress in Neuro Psychopharmacology and Biological Psychiatry*, **32:** 715-725.

Medicinal Plants: Phytochemistry, Pharmacology and Therapeutics Vol. 3 (2014) *Pages* **153–177**
Editor-in-Chief: **V.K. Gupta**
Published by: **DAYA PUBLISHING HOUSE, NEW DELHI**

6

Naringin: A Citrus Flavonoid Protects Mice Against the Benzo(a)pyrene-Induced Forestomach Carcinoma in Mice

Ganesh Chandra Jagetia[1]*

ABSTRACT

The chemically-induced forestomach carcinogenesis inhibitory activity of naringin, a citrus flavonoid was investigated by administering 1, 2.5, 5, 10, 25, 50, 100 or 250 mg/kg b. wt. of narinign orally or intraperitoneally before (pre-treatment), after (post-treatment) or pre-post-treatment to the female mice receiving benzo(a)pyrene for the induction of forestomach carcinoma. The naringin treatment reduced tumor incidence and tumor multiplicity in a dose dependent manner. The 10 mg/kg of naringin alleviated the tumor multiplicity by 78 per cent and tumor incidence by 20 per cent. The oral administration was more effective than the intraperitoneal treatment. A maximum inhibition in tumorigenesis by naringin was observed when it was given before, during and after (pre-post-treatment) benzo(a)pyrene treatment followed by when naringin was given either before or after chemically induced carcinogenesis. These observations were further confirmed by PCNA labeling where naringin pretreatment reduced the PCNA labeling index by four folds. The protection in benzo(a)pyrene-induced DNA damage was studied by micronuclei analysis in splenocytes, where benzo(a)pyrene treatment increased the frequency of micronuclei bearing splenocytes (MNBNC) by six folds, whereas. naringin treatment during and after benzo(a)pyrene administration reduced the frequency of MNBNC in a dose dependent manner up to 10 mg/kg naringin, where a highest reduction in MNBNC was observed, thereafter the frequency of MNBNC showed a marginal increase up to a dose of 250 mg/kg naringin. Despite this increase, the MNBNCs were two folds lower in naringin treated group than benzo(a)pyrene treatment alone. A similar effect on

1 Department of Zoology, Mizoram University, Aizawl – 796 004, India
* *Corresponding author*: E-mail: gc.jagetia@gmail.com

MNBNC decline was observed for pre or post-naringin treated group however, the degree of decline in MNBNC was lesser when compared to pre-post-naringin treatment.

The naringin treatment before-during and after benzo(a)pyrene administration inhibited the chemical carcinogenesis and the chemopreventive activity of naringin may be due to reduction in benzo(a)pyrene-induced DNA damage and cell proliferation.

Keywords: Naringin, Citrus Flavonoid, Benzo(a)pyrene, Forestomach carcinoma.

Introduction

The development of malignant tumor involves complex interactions between several factors, both exogenous (environmental) and endogenous (genetic, hormonal and immunological). Many evidences indicate that 50-80 per cent of human cancers are potentially preventable, because its causation *i.e.,* the factors that determine the incidence, are largely exogenous. Exogenous causative agents or factors that have been identified in humans include: cigarette smoking, occupational and environmental chemicals, radiation and dietary factors, and specific viruses, bacteria and parasites (Sporn, 1999). Environmental pollutants are one of the main risk factors for the induction of cancer and are major public health concern and also a leading cause of death in both developing and developed countries. These pollutants include benzo(a)pyrene (BaP), which is a polycyclic aromatic hydrocarbon (PAH) formed by the pyrolytic process during smoking of cigarettes and other tobacco products as well as in combusted organic matter including automobile exhaust. BaP is frequently used as a representative indicator of total PAH levels. It is a complete carcinogen, as it produces initiation and promotion of carcinogenesis (Halliwell and Gutteridge, 1989). BaP, a PAH, is released into the atmosphere from incomplete combustion processes. Humans ingest BaP and other PAHs from charbroiled meat, cigarette smoke, and automobile emissions (Collins *et al.,* 1991). CYP-mediated metabolic activation of BaP is necessary for its conversion to form the reactive metabolite, BaP 7,8-diol-9,10-epoxide, which binds to DNA and initiates tumor formation (Levin *et al.,* 1982). The cytochrome P450 (CYP)-dependent monooxygenase system consists of CYP, NADPH-CYP reductase, and phospholipids. This monooxygenase plays a major role in the detoxification and bioactivation of xenobiotics including drugs, environmental pollutants, and chemical carcinogens (Guengerich, 1988; Nebert, 1989). The importance of the metabolic activation of BaP has led to the suggestion that inhibitors of BaP hydroxylation (AHH) activity could diminish the mutagenicity and carcinogenicity of BaP (Cassady *et al.,* 1990; Castonguay, 1992). Many studies on rat hepatic CYP enzymes have suggested that CYPs including CYP1A, CYP2C, and CYP3A were involved in the oxidation of BaP (Wood *et al.,* 1976). CYP1A has the highest AHH activity and CYP3A is the most abundant CYP in mammalian livers (Wood *et al.,* 1976; Bauer *et al.,* 1995; Guengerich, 1995). CYP1A1 and CYP1A2 are two major members of the CYP1A subfamily. CYP1A2 is the main hepatic CYP1A member and CYP1A1 is essentially an extrahepatic CYP1A in rodents and humans (Guengerich, 1995).

Despite of the great advances made in the clinical management of neoplastic disorders the stated goal of NCI to reduce the cancer related mortality by 50 per cent by the year 2000 has not been met. This has led to the thinking that instead of curing the cancer, its occurrence shall be prevented. This can be achieved by pharmacological intervention with certain chemical compounds, phytoceuticals or natural products. Cancer chemoprevention is recognized as the prevention, inhibition or reversal of carcinogenesis by intervention with pharamcological agents, which can be administered either as individual drugs or as naturally occurring dietary constituents. Human beings consume several natural products in their daily diet. Many dietary ingredients have been found to alter the development of tumors in experimental animals (Slaga, 1980). Dietary habits have also been linked to various types of cancers (Weisburger, 1991; Reddy *et al.*, 1992).

Fruits and certain vegetables have been found to reduce the risk of cancer (Singletary, 2000). Grapefruit is part of human diet in most countries, where it is usually consumed as a juice. The chemical responsible for the characteristic sour odor of the fruit is naringin, a flavanone that is rapidly transformed into naringenin by the action of enzymes alpha-rhamnosidase and beta-glucosidase (Kim *et al.*, 1998). Naringin (glycone of naringenin), a grapefruit bioflavonoid, has been demonstrated as an antioxidant, anti-cancer, free radical scavenging and potent cholesterol-lowering agent (Kroyer, 1986; Francis *et al.*, 1989; Haenen, 1997; Shin *et al.*, 1999; Jagetia *et al.*, 2003). Naringin has been reported to alleviate radiation-induced chromosome damage a as well as bleomycin-induced DNA damage *in vivo* and *in vitro* (Jagetia and Reddy, 2002; Jagetia *et al.*, 2003; Jagetia *et al.*, 2007). It has also been reported to reduce iron-induced oxidative stress (Jagetia *et al.*, 2004; Jagetia and Reddy, 2011). The genotoxicity study of naringin in Ames/Salmonella system (strains TA98 and TA100), showed a 36 per cent reduction in BaP induced DNA damage and it was found to be non-mutagenic (Calomme *et al.*, 1996). It has also been reported to protect rat hepatocytes against the damage induced by 2-amine-1-methyl-6-phenylimidazo (4,5-b) pyridine (Lake *et al.*, 1999). naringin is a potent inhibitor of cytochrome P450 1A2 protein levels *in vivo*, which is a major cytochrome P4501A protein present in the mouse and human liver (Ueng *et al.*, 1999). The main interest of the present study was to know whether naringin, a dietary bioflavonoid would be able to reduce the benzo(a)pyrene-induced forestomach carcinogenesis in mice or not? If so what is the optimum dose and best route of administration. Therefore, present study was conceptualized to obtain an insight into the chemopreventive activity of naringin in benzo(a)pyrene-induced forestomach carcinoma in female Swiss albino mice.

Materials and Methods

Chemicals

Benzo[*a*]pyrene [BaP purity > 98 per cent Cat no. B 1760] was procured from Sigma Chemical Company, St. Louis, MO, USA. Naringin was procured from Acros Organics Ltd., Geel, Belgium. Proliferating Cell Nuclear Antigen (PCNA) staining kit was requisitioned from Zymed Laboratories Inc. South San Francisco, CA, USA.

Animal Care and Handling

The animal care and handling were done according to the guidelines issued by the World Health Organization, Geneva, Switzerland and the INSA (Indian National Science Academy, New Delhi, India). Eight to ten weeks old female Swiss albino mice weighing 22 to 24 g were selected from an inbred colony maintained under the controlled conditions of temperature (23 ± 2°C), humidity (50 ± 5 per cent) and light (12 h of light and dark, respectively). The animals had free access to the sterile food and water. Four animals were housed in a polypropylene cage containing sterile paddy husk (procured locally) as bedding throughout the experiment. The study was approved by the Institutional Animal Ethical Committee.

Preparation of Drug and Mode of Administration

NIN was freshly dissolved in sterile double distilled water (DDW) immediately before use, whereas the benzo(a)pyrene was dissolved in sesame oil. The animals were administered with NIN either orally (*p.o.*), or intraperitoneally as the case may be, whereas benzo(a)pyrene was given orally through an oral gavage.

Determination of Acute Toxicity of NIN

The acute toxicity of NIN was determined according to Prieur *et al.* (1973) and Ghosh (1984). Briefly, the animals were allowed to fast by withdrawing food and water for 18 h. The fasted animals were divided into several groups of 10 each. Each group of animals was administered with 0, 5, 10, 50, 100, 250, 500, 1000, 2500, or 5000 mg/kg b. wt. of freshly dissolved NIN through an oral gavage (*p.o*). Another set of animals was administered intraperitoneally with 0, 1, 2.5, 5, 10, 25, 50, 100, 250, 500, 1000 or 1500 mg/kg b. wt. of NIN. The animals were provided with food and water immediately after the drug administration. Mortality of the animals was observed up to 14 days post drug treatment. A minimum of 10 animals was used for each drug dose, and a total of 220 animals were used for this experiment.

Chemopreventive Effect

Induction of BaP Induced Forestomach Tumorigenesis

Forestomach is a dilated lower esophagus (Leininger and Jokinen, 1994), and the forestomach carcinoma was induced in the female Swiss albino mice as described by Wattenberg *et al.* (1980) with minor modifications. Briefly, the animals were administered with 100 µl of sesame oil or 1 mg of BaP in 100 µl sesame oil *p.o.* twice a week for four weeks (a total of eight administrations). The animals were divided into the following groups:

DDW Group

The animals of this group received sterile double distilled water (DDW) as a sole source of drinking throughout the study period.

Oil Group

The animals of this group received 100 µl of sesame oil by *p.o.* through an oral gavage twice a week for four consecutive weeks.

BaP Group

The animals of this group received 1 mg of BaP in 100 µl sesame oil *p.o.* twice a week for four weeks.

NIN Group

The animals of this group received various doses of NIN either orally (1, 2.5, 5, 10, 25, 50, 100 or 250 mg/kg b. wt.) or intraperitoneally (1, 2, 4, 6, 8, 10, 25 or 50 mg/kg b. wt.).

NIN + BaP Group (Pre-post treatment)

The animals of this group received various doses of NIN and BaP, depending on the experimental design.

Effect of Oral Administration

A separate experiment was conducted to study the effect of oral administration of naringin on benzo(a)pyrene-induced forestomach carcinogenesis, where the grouping and other conditions were essentially similar except that the animals were divided into several groups of 10 each. Each group of animals was administered orally with 1, 2.5, 5, 10, 25, 50, 100 or 250 mg/kg b. wt. NIN for five days a week, for 2 weeks before treatment with BaP, followed by the concomitant treatment with NIN and BaP or DDW for 4 weeks during and 2 weeks after the last dose of BaP.

Effect of Intraperitoneal Administration

In another experiment the animals were divided into several groups of 10 each. Each group of animals was administered intraperitoneally with 1, 2, 4, 6, 8, 10, 25 or 50 mg/kg b. wt. of NIN for five days a weeks for 2 weeks before treatment with BaP, followed by the concomitant treatment with NIN and BaP or DDW for 4 weeks during and 2 weeks after the last dose of BaP.

Effect of Altered Administration Schedule

A separate experiment was conducted to evaluate the effect of alteration in the NIN administration schedule. Since oral administration was most effective further experiments were carried out using this model of treatment. The animals were divided into following two additional groups in addition to the groups described above.

NIN + BaP Group (Pre-treatment)

The animals of this group received 5, 10 and 25 mg/kg b. wt. p.o. of NIN for 5 days a week, for 2 weeks. The last treatment of NIN was given forty-eight hours before the first administration of BaP or sesame oil.

BaP + NIN Group (Post-treatment)

The animals of this group were provided with normal drinking water and food during the carcinogen treatment for 4 weeks. Forty-eight hours after the last administration of BaP or sesame oil the animals were administered with 5, 10 and 25 mg/kg b. wt. NIN *p.o.* for 8 weeks as described above.

The body weights of the animals from each group were recorded at the beginning and at the end of the experiments. The mice were sacrificed 14 weeks after the last dose of BaP and buffered formalin was immediately injected into the stomach of each mouse, so as to distend the stomach and fix the tumors. Thereafter, forestomach was cut open longitudinally and kept in 10 per cent buffered formalin for 24 h for fixation. The stomach papillomas that were 1 mm or larger in diameter were counted under a stereozoom microscope (Leitz, Ernst, Wetzlar, Germany). A total of 440 (220 each for oral and IP administration) animals were used for this experiment. The chemopreventive activity of NIN was assessed on the basis of tumor incidence, mean and multiplicity of tumors (Jagetia *et al.*, 2003). The tumor incidence and multiplicity were compared between the animals fed with BaP or NIN or BaP and NIN.

Tumor incidence, mean and multiplicity of tumors were determined as follows (Jagetia *et al.*, 2003):

Tumor incidence = Number of animals presenting tumors

Tumor Multiplicity= Total number of tumors scored/Total number of animals with tumors

Tumor Mean = Lowest number of tumors + Highest number of tumors/2

Histological Examination of Tumors

Tumors observed by visual examinations were further confirmed histologically. The stomach samples were excised, fixed in 10 per cent buffered formalin, embedded in paraffin wax and 5 μm thick sections were cut using a rotary microtome. The slides containing sections were stained with hematoxylin and eosin for the histopathological examination. The slides were scored blindly by a pathologist and the tumors were classified according to well-established criteria.

Proliferating Cell Nuclear Antigen (PCNA) Staining

The sections were dewaxed, in xylene, passed through graded alcohol series and rehydrated in distilled water. The hydrated sections were immersed for 30 min in 3 per cent hydrogen peroxidase in methanol to block endogenous peroxidase activity and subsequently washed in PBS. Thereafter, the sections were immersed in citrate buffer and microwaved and allowed to cool for another 20 min, while still being immersed in cirtrate buffer. The PCNA staining was carried out according to the manufacturer's instructions. The scorer was unaware of the group to which these specimens belonged.

The PCNA-positive index number (PPIN) was calculated as follows:

$$\text{PPIN} = \frac{\text{Number of PCNA positive nuclei}}{\text{Total number of nuclei scored}} \times 100$$

Splenocyte Culture and Micronuclei Analysis

The splenocytes were cultured as described by Jagetia *et al.* (2001). Briefly, four animals from each group were killed by cervical dislocation at the end of the experiment. The animals were thoroughly wiped with 70 per cent ethanol and their abdominal cavities were cut open with the help of sterile scissors and forceps in an aseptic environment. The spleens of the animals were removed aseptically and washed with sterile phosphate buffered saline (PBS). The splenocytes were isolated and cultured in RPMI-1640 medium containing 10 per cent fetal calf serum (FCS) and concanavalin-A as mitogen. The splenocytes were incubated at 37°C and allowed to grow for 72 h. Duplicate cultures were set up for each animal for each drug dose. The micronuclei (MN) were prepared according to the method of Fenech and Morley (1985) with minor modifications. Briefly, 40 h after the initiation of splenocyte culture, 3 µg/ml of cytochalasin-B was added to each culture. The cells were allowed to grow for another 32 h. The splenocytes were centrifuged, subjected to mild hypotonic treatment (0.7 per cent ammonium oxalate) so as to retain the cytoplasm and fixed in Carnoy's fixative (3: 1 methanol, acetic acid). The cells were centrifuged again and resuspended in a small volume of fixative and dropped on to precleaned coded slides to avoid observer's bias. Cells were stained with 0.125 per cent acridine orange (BDH, England, Gurr Cat. No. 34001 9704640E) in Sorensen's buffer (pH 6.8). The slides were washed twice in Sorensen's buffer and observed under a fluorescent microscope, equipped with 450-490 nm BP filter set with excitation at 453 nm (Carl Zeiss Photomicroscope III, Germany), using a 40X Neofluar objective. A minimum of 2000 (1000 per culture) binucleate cells (BNC) with well-preserved cytoplasm was scored from each animal for each NIN dose for the presence of MN and the frequency of micronucleated binucleate splenocytes (MNBNC) were determined. The MN identification was carried out according to the criteria of Countryman and Heddle (1976) and Krisch-Volders *et al.* (2000).

Statistical Analysis

The statistical significance in the alteration in the body weights after various treatments was determined using Student's t-test. The tumor incidence and multiplicity have been expressed as the percentage of animals presenting tumors and the analysis was carried out using χ^2 –test. The statistical significance for micronuclei test was carried out using one-way analysis of variance ANOVA with the application of Bonferroni's Post-hoc test.

Results

Acute Toxicity

Oral Administration

The acute toxicity of NIN was determined by administering mice with 5, 10, 50, 100, 250, 500, 1000, 2500, or 5000 mg/kg b. wt. of NIN. It was found to be non-toxic up to a dose of 1000 mg/kg b. wt., where no drug induced mortality was found. A further increase in the drug dose to 2500 mg and 5000 mg resulted in 10 and 20 per cent mortality, respectively (Tables 6.1–6.2). The LD_{50} dose after oral administration could not be determined owing to the problem in drug administration and making of drug solution.

Intraperitoneal Administration

The administration of 1, 2.5, 5, 10, 25, 50, or 100 mg/kg b. wt. NIN to mice did not induce drug related toxicity in the animals. The administration of 250 and 500 mg/kg resulted in 10 per cent mortality. A further increase in the drug dose to 1000 and

Table 6.1: Effect of different doses of naringin on the acute toxicity in mice.

	Naringin mg/kg	*Mortality on Different Post-Drug Treatment Days*															
		1	*2*	*3*	*4*	*5*	*6*	*7*	*8*	*9*	*10*	*11*	*12*	*13*	*14*	*Survivors (Per cent)*	*Total No. of Animals Used*
Oral administration	0	–	–	–	–	–	–	–	–	–	–	–	–	–	–	10 (100)	10
	5	–	–	–	–	–	–	–	–	–	–	–	–	–	–	10 (100)	10
	10	–	–	–	–	–	–	–	–	–	–	–	–	–	–	10 (100)	10
	50	–	–	–	–	–	–	–	–	–	–	–	–	–	–	10 (100)	10
	100	–	–	–	–	–	–	–	–	–	–	–	–	–	–	10 (100)	10
	250	–	–	–	–		–	–	–	–	–	–	–	–	–	10 (100)	10
	500	–	–	–	–	–	–	–	–	–	–	–	–	–	–	10 (100)	10
	1000	–	–	–	–	–	–	–	–	–	–	–	–	–	–	10 (100)	10
	2500	–	–	–	–	–	–	–	1	–	–	–	–	–	–	9 (90)	10
	5000	–	–	1	–	–	–	–	–	–	–	–	–	–	–	9 (90)	10
Intraperitoneal administration	0	–	–	–	–	–	–	–	–	–	–	–	–	–	–	10 (100)	10
	1	–	–	–	–	–	–	–	–	–	–	–	–	–	–	10 (100)	10
	2.5	–	–	–	–	–	–	–	–	–	–	–	–	–	–	10 (100)	10
	5.0	–	–	–	–	–	–	–	–	–	–	–	–	–	–	10 (100)	10
	10	–	–	–	–	–	–	–	–	–	–	–	–	–	–	10 (100)	10
	25	–	–	–	–		–	–	–	–	–	–	–	–	–	10 (100)	10
	50	–	–	–	–	–	–	–	–	–	–	–	–	–	–	10 (100)	10
	100	–	–	–	–	–	–	–	–	–	–	–	–	–	–	10 (100)	10
	250	–	–	–	–	–	–	–	1	–	–	–	–	–	–	9 (90)	10
	500	–	–	1	–	–	–	–	–	–	–	–	–	–	–	9 (90)	10
	1000	–	–	–	–	–	–	–	–	–	1	1	–	–	–	8(80)	10
	1500	–	1	–	1	–	1	–	1	–	–	–	–	–	–	6(60)	10

Table 6.2: Effect of BaP, naringin or naringin and BaP treatment on the survival of mice during chemical carcinogenesis.

Treatment			Mortality on Different Post-Drug Treatment Days																			
			1	2	3	4	1	2	3	4	5	6	7	8	9	10	11	12	13	14	Survivors (Per cent)	Total No. of Animals Used
Intraperitoneal administration	Pre	Control	–	–	–	–	–	–	–	–	–	–	–	–	–	–	–	–	–	–	10(10)	0(0/10)
	and	Bap	–	–	–	–	–	–	–	–	–	–	–	–	–	–	–	–	1	1	8(80)	49(8/8)
	post	NIN1	–	–	–	–	–	–	–	–	–	–	–	–	–	–	–	–	–	–	10(100)	0(0/10)
		NIN2	–	–	–	–	–	–	–	–	–	–	–	–	–	–	–	–	–	–	10(100)	0(0/10)
		NIN4	–	–	–	–	–	–	–	–	–	–	–	–	–	–	–	–	–	–	10(100)	0(0/10)
		NIN6	–	–	–	–	–	–	–	–	–	–	–	–	–	–	–	–	–	–	10(100)	0(0/10)
		NIN8	–	–	–	–	–	–	–	–	–	–	–	–	–	–	–	–	–	–	10(100)	0(0/10)
		NIN10	–	–	–	–	–	–	–	–	–	–	–	–	–	–	–	–	–	–	10(100)	0(0/10)
		NIN25	–	–	–	–	–	–	–	–	–	–	–	–	–	–	–	–	–	–	10(100)	0(0/10)
		NIN50	–	–	–	–	–	–	–	–	–	–	–	–	–	–	–	–	–	–	10(100)	0(0/10)
		NIN1+ Bap	–	–	–	–	–	–	–	–	–	–	–	–	–	–	–	–	–	–	10(100)	34(10/10)
		NIN2+ Bap	–	–	–	–	–	–	–	–	–	–	–	–	–	–	–	–	–	–	10(100)	33(10/10)
		NIN4+ Bap	–	–	–	–	–	–	–	–	–	–	–	–	–	–	–	–	–	–	10(100)	25(10/10)
		NIN6+ Bap	–	–	–	–	–	–	–	–	–	–	–	–	–	–	–	–	–	–	10(100)	21(10/10)
		NIN8+ Bap	–	–	–	–	–	–	–	–	–	–	–	–	–	–	–	–	–	–	10(100)	20(10/10)
		NIN10+ Bap	–	–	–	–	–	–	–	–	–	–	–	–	–	–	–	–	–	–	10(100)	24(10/10)
		NIN25+ Bap	–	–	–	–	–	–	–	–	–	–	–	–	–	–	–	–	–	–	10(100)	28(10/10)
		NIN50+ Bap	–	–	–	–	–	–	–	–	–	–	–	–	–	–	–	–	–	–	10(100)	32(10/10)

Contd...

Table 6.2–*Contd...*

	Treatment		*Mortality on Different Post-Drug Treatment Days*																			
			1	*2*	*3*	*4*	*1*	*2*	*3*	*4*	*5*	*6*	*7*	*8*	*9*	*10*	*11*	*12*	*13*	*14*	*Survivors (Per cent)*	*Total No. of Animals Used*
	Pre	Bap	–	–	–	–	–	–	–	–	–	–	–	–	–	–	–	–	1	1	8(80)	49(8/8)
	and	NIN1	–		–	–	–		–	–	–		–	–	–		–	–	–		10(100)	0(0/10)
	post	NIN2.5	–	–	–	–	–	–	–	–	–	–	–	–	–	–	–	–	–	–	10(100)	0(0/10)
		NIN5	–	–	–	–	–	–	–	–	–	–	–	–	–	–	–	–	–	–	10(100)	0(0/10)
		NIN10	–	–	–	–	–	–	–	–	–	–	–	–	–	–	–	–	–	–	10(100)	0(0/10)
		NIN25	–		–	–	–		–	–	–		–	–	–		–	–	–		10(100)	0(0/10)
		NIN50	–	–	–	–	–	–	–	–	–	–	–	–	–	–	–	–	–	–	10(100)	0(0/10)
		NIN100	–	–	–	–	–	–	–	–	–	–	–	–	–	–	–	–	–	–	10(100)	0(0/10)
		NIN250	–	–	–	–	–	–	–	–	–	–	–	–	–	–	–	–	–	–	10(100)	0(0/10)
		NIN1+Bap	–		–	–	–		–	–	–		–	–	–		–	–	–		10(100)	32(10/10)
Oral administration		NIN2.5+ Bap	–	–	–	–	–	–	–	–	–	–	–	–	–	–	–	–	–	–	10(100)	24(10/10)
		NIN+5Bap	–	–	–	–	–	–	–	–	–	–	–	–	–	–	–	–	–	–	10(100)	19(10/10)
		NIN10+Bap	–	–	–	–	–	–	–	–	–	–	–	–	–	–	–	–	–	–	10(100)	12(9/10)
		NIN25+Bap	–		–	–	–		–	–	–		–	–	–		–	–	–		10(100)	16(9/10)
		NIN50+Bap	–	–	–	–	–	–	–	–	–	–	–	–	–	–	–	–	–	–	10(100)	19(10/10)
		NIN100+Bap	–	–	–	–	–	–	–	–	–	–	–	–	–	–	–	–	–	–	10(100)	21(10/10)
		NIN250+Bap	–	–	–	–	–	–	–	–	–	–	–	–	–	–	–	–	–	–	10(100)	22(10/10)
		NIN5+BaP	–	–	–	–	–	–	–	–	–	–	–	–	–	–	–	–	–	–	10(100)	38(10/10)
		NIN10+BaP	–	–	–	–	–	–	–	–	–	–	–	–	–	–	–	–	–	–	10(100)	37(10/10)
		NIN25+BaP	–	–	–	–	–	–	–	–	–	–	–	–	–	–	–	–	–	–	10(100)	35(10/10)
	Post	Bap+NIN5	–		–	–	–		–	–	–		–	–	–		–	–	–		10(100)	40(9/10)
		Bap+NIN10	–	–	–	–	–	–	–	–	–	–	–	–	–	–	–	–	–	–	10(100)	39(9/10)
		Bap+NIN25	–	–	–	–	–	–	–	–	–	–	–	–	–	–	–	–	–	–	10(100)	36(10/10)

1500 mg resulted in 20 and 40 per cent mortality, respectively (Tables 6.1–6.2). The administration of higher doses was not possible owing to drug solubility and administration problems. Therefore, LD_{50} could not be determined.

Selection of Optimum Dose and Route of Administration

The optimum drug dose and route of administration was selected by treating the animals with various doses of NIN before or after carcinogen (BaP) treatment. The chemopreventive effect of NIN on BaP induced forestomach papillomas in mice is presented in Table 6.3. The administration of 1 mg Bap by oral gavage twice a week for 4 weeks resulted in 100 per cent incidence of forestomach tumors with a mean of 4.5 tumors per mouse. The administration of BaP alone resulted in 20 per cent mortality during the course of experiment. The mice treated with DDW, oil or NIN without BaP did not show any tumor-induction throughout the study period.

Oral Administration

The treatment of mice with oil or NIN alone did not induce any tumors. Treatment of mice with 1 mg/kg b. wt. NIN *p.o.* caused an approximate 22 per cent reduction in the mean number of tumors, whereas the tumor multiplicity was reduced by 48 per cent (Table 6.3). An increase in the dose of NIN reduced the mean number of tumors and tumor multiplicity further and a highest reduction of 78 per cent in tumor multiplicity and 77 per cent reduction in mean number of tumors was observed in animals that received 10 mg/kg NIN. An increase in NIN dose did not reduce the mean number of tumors and tumor multiplicity significantly (Table 6.3). However, these were always lower than the BaP treated group that did not receive NIN. Treatment of mice with 1, 2.5, 5, 10, 25, 50, 100 and 250 mg NIN treatment brought about 1.9, 2.5, 3.2, 4.6, 3.4, 3.2, 2.9 and 2.8 folds reduction in the tumor multiplicity, respectively. The tumor incidence also declined for 10 and 25 mg/kg NIN (Table 6.3). The oral administration was superior to intraperitoneal administration in reducing the tumor incidence and tumor multiplicity in female mice.

Intraperitoneal Administration

The treatment of mice with DDW or NIN alone did not induce any tumors. The intraperitoneal administration of different doses of NIN significantly reduced the mean number of tumors when compared with the DDW + BaP group. Treatment of mice with 1 mg/kg b. wt. NIN caused an approximate 22 per cent reduction in the mean number of tumors, whereas the tumor multiplicity was reduced by 36 per cent (Table 6.3). A further increase in NIN dose reduced the mean tumor number (45 per cent and 33 per cent) and tumor multiplicity and a highest reduction of 60 per cent and 62 per cent in tumor multiplicity was observed in animals that received 6 and 8 mg/kg NIN respectively. An increase in NIN dose further did not reduce the mean tumor number and tumor multiplicity (Table 6.3). However, these were always lower than the BaP treated group that did not receive NIN. Treatment of mice with 1, 2, 4, 6, 8, 10, 25 or 50 mg NIN treatment brought about 1.5, 1.6, 2, 2.5, 2.6, 2.2, 1.9 and 1.6 folds reduction in the tumor multiplicity, respectively. Treatment of mice with various doses of NIN intraperitoneally did not reduce the tumor incidence.

Table 6.3: Effect of various doses of naringin on BaP-induced carcinogenesis.

Treatment			*Tumor Incidence*	*Mean Number*	*Tumor Multiplicity*	*Initial Weight*	*Final Weight*
Oral administration	Pre and post	DDW	0 (0/10)	0	0	22.6 ± 0.26	31.5 ± 0.34
		BaP	100 (8/8)	4.5 (100)	6.125(100)	21.8 ± 0.24	31.7 ± 0.51
		NIN1	0 (0/10)	0	0	21.4 ± 0.22	31.9 ± 0.5
		NIN2.5	0 (0/10)	0	0	21.7 ± 0.3	32.3 ± 0.49
		NIN5	0 (0/10)	0	0	22.0 ± 0.25	32.3 ± 0.21
		NIN10	0 (0/10)	0	0	21.3 ± 0.33	32.0 ± 0.29
		NIN25	0 (0/10)	0	0	21.1 ± 0.34	32.4 ± 0.16
		NIN50	0 (0/10)	0	0	21.3 ± 0.3	32.8 ± 0.53
		NIN100	0 (0/10)	0	0	21.2 ± 0.35	30.8 ± 0.592
		NIN250	0 (0/10)	0	0	21.4 ± 0.49	30.6 ± 0.76
		Bap+NIN1	100 (10/10)	3.5(77.77)	3.2(52.28)	21.6 ± 0.6	30.3 ± 0.63
		Bap+NIN2.5	100 (10/10)	3(66.66)	2.4(39.18	21.1 ± 0.34	30.7 ± 0.80
		Bap+NIN5	100 (10/10)	2(44.44)	1.9(31.02)	21.3 ± 0.3	30.6 ± 0.85
		Bap+NIN10	80 (8/10)	1(22.22)	1.33(21.71)	21.5 ± 0.37	30.2 ± 0.7
		Bap+NIN25	80 (8/10)	1.5(33.33)	1.77(28.89)	21.6 ± 0.3	29.5 ± 0.68
		Bap+NIN50	100 (10/10)	2.5(55.55)	1.9(31.02)	21.5 ± 0.5	31.0 ± 0.64
		Bap+NIN100	100 (10/10)	2(44.44)	2.1(34.08)	21.5 ± 0.34	30.6 ± 0.30
		Bap+NIN250	100 (10/10)	2(44.44)	2.2(35.91)	21.5 ± 0.5	30.3 ± 0.78
	Pre	NIN5+BaP	100 (10/10)	3.5(77.77)	3.8(62.04)	22.3 ± 1.22	31.2 ± 0.54
		NIN10+BaP	100 (10/10)	3.5(77.77)	3.7(60.40)	21.4 ± 0.98	32.5 ± 0.55
		NIN25+BaP	100 (10/10)	4(88.88)	3.5(57.14)	23.5 ± 0.54	31.5 ± 0.64

Contd...

Table 6.3–*Contd...*

Treatment			*Tumor Incidence*	*Mean Number*	*Tumor Multiplicity*	*Initial Weight*	*Final Weight*
	Post	Bap+NIN5	100 (10/10)	3.5(77.77)	4.0(65.30)	22.1 ± 0.34	30.8 ± 0.66
		Bap+NIN10	100 (10/10)	3(66.66)	3.9(63.67)	23.4 ± 0.5	34.21 ± 0.3
		Bap+NIN25	100 (10/10)	3(66.66)	3.6(58.77)	22.4 ± 0.35	31.7 ± 0.19
Intraperitoneal administration	Pre and post	NIN1	0 (0/10)	0	0	22.18 ± 2.11	34.56 ± 2.33
		NIN2	0 (0/10)	0	0	20.98 ± 2.04	34.57 ± 3.45
		NIN4	0 (0/10)	0	0	23.14 ± 0.97	35.62 ± 2.51
		NIN6	0 (0/10)	0	0	24.11 ± 1.87	35.78 ± 1.58
		NIN8	0 (0/10)	0	0	22.54 ± 2.11	34.55 ± 2.44
		NIN10	0 (0/10)	0	0	21.04 ± 1.11	33.65 ± 3.14
		NIN25	0 (0/10)	0	0	23.14 ± 2.44	36.14 ± 2.58
		NIN50	0 (0/10)	0	0	22.12 ± 2.21	34.28 ± 2.11
		NIN1+ Bap	100 (10/10)	3.5 (77.77)	3.4(63.78)	23.11 ± 1.11	30.56 ± 1.98
		NIN2+ Bap	100 (10/10)	3 (66.66)	3.3(61.91)	21.11 ± 2.35	29.88 ± 1.44
		NIN4+ Bap	100 (10/10)	3 (66.66)	2.5(46.9)	22.45 ± 3.22	31.25 ± 1.22
		NIN6+ Bap	100 (10/10)	2.5 (55.55)	2.1(39.39)	24.11 ± 1.21	31.78 ± 1.48
		NIN8+ Bap	100 (10/10)	3 (66.66)	2.0(37.52)	22.55 ± 1.01	29.87 ± 1.45
		NIN10+ Bap	100 (10/10)	3.5 (77.77)	2.4(45.02)	21.12 ± 1.54	31.01 ± 1.49
		NIN25+ Bap	100 (10/10)	3.5 (77.77)	2.8(52.53)	22.44 ± 1.87	30.11 ± 1.78
		NIN50+ Bap	100 (10/10)	3 (66.66)	3.2(60.03)	23.11 ± 1.25	29.87 ± 1.57

Altered Administration Schedule

The effect of alteration in drug administration was studied for 5, 10 and 25 mg/kg NIN. Treatment of mice with 5, 10 or 25 mg/kg NIN before carcinogen administration resulted in a reduction in mean tumor number and multiplicity when compared with BaP treatment alone. The effect of 5 and 10 mg NIN was similar, as both doses caused 22 per cent reduction in mean tumor number, whereas tumor multiplicity was also reduced by 40 per cent for these doses (Table 6.3). Altered administration schedule did not reduce the incidence of tumors. Similarly, mean tumor number and tumor multiplicity also declined when 5, 10 or 25 mg/kg NIN was given to the animals after BaP treatment (Post-treatment). The effect was almost similar for 10 and 25 mg/kg NIN, where 33 per cent reduction in mean tumor number was observed, while tumor multiplicity reduced to 36 and 41 per cent respectively for 10 and 25 mg/kg NIN, respectively (Table 6.3).

Micronucleus Assay

Treatment of mice with oil or various doses of NIN did not significantly alter the frequency of micronuclei in the splenocytes of mice (Table 6.4). Treatment of mice with 8 mg/kg b. wt. BaP for the induction of forestomach carcinoma resulted in an approximate six fold increase in MNBNC frequency ($p<0.0001$). Treatment of mice with various doses of NIN before and after carcinogen treatment resulted in a dose dependent but drastic decline in the MNBNCs and the lowest frequency of MNBNC was observed for 10 mg/kg NIN ($p<0.0001$), the dose that gave best chemopreventive effect (Table 6.4). BaP treatment induced not only cells with one MN but also cells bearing two or more than two MN. Treatment of mice with various doses of NIN before and after BaP treatment significantly reduced the occurrence of MNBNCs bearing one, two or more than two micronuclei and a greatest reduction was observed for 10 mg/kg NIN (Table 6.4). This reduction was almost 2 fold for 10 and 25 mg/kg NIN, respectively. Similarly, the animals that were treated with different doses of NIN before or after treatment of carcinogen exhibited a dose dependent decline in the frequency of MNBNC when compared with the BaP treated animals. NIN treatment before or after BaP administration also reduced the frequency of cells bearing one, two or more than two MN significantly. However, this reduction in MNBNC was not to that extent as was observed for the mice that received NIN before and after (Pre-post) the carcinogen treatment (Table 6.4).

Histopathological Analysis

The majority of papillomas scored in BaP and BaP+NIN group were histologically found to be squamous cell carcinomas (SCC). Forestomach squamous cell carcinomas, often accompanied by inflammation and the presence of abundant PCNA-positive nuclei were observed in BaP treated animals. The PCNA labeling indices in the forestomach tumors are presented in Table 6.5. Histopathological examination of sections from BaP treated group showed an approximate 3.7 fold elevation in the PCNA labeling index, whereas treatment of mice with different doses of NIN *p.o.* before and after BaP treatment resulted in an almost 4 fold decline in PCNA labeling index, when compared to BaP treatment alone (Table 6.5). Similarly, treatment of mice with 5, 10 or 25 mg/kg NIN before or after treatment with BaP also

Table 6.4: Alteration in the BaP-induced micronuclei in the splenocytes of mice treated with different doses of naringin orally during chemical carcinogenesis.

		One MN		Two MN		Multiple MN		Total MN	
		NIN+oil	*NIN+BaP*	*NIN+oil*	*NIN+BaP*	*NIN+oil*	*NIN+BaP*	*NIN+oil*	*NIN+BaP*
Pre and post treatment	Control	21.9 ± 0.67	134.7 ± 2.4	7.1 ± 0.52	40.8 ± 1.29	0 ± 0	1 ± 0.2582	29.5 ± 1.3	176.5 ± 4.4
	NIN1	21.7 ± 0.7	86.1 ± 0.95[a]	5.5 ± 0.56	30 ± 0.73[b]	0 ± 0	0.7 ± 0.15	26.9 ± 1.0	117.2 ± 1.0[a]
	NIN2.5	21.6 ± 0.52	81 ± 1.37[a]	4.9 ± 0.50	30.2 ± 0.5[b]	0 ± 0	0.5 ± 0.16	26.9 ± 0.67	111.9 ± 1.1[a]
	NIN5	21.8 ± 0.59	65.6 ± 1.2[a]	4.9 ± 0.58	22.8 ± 0.7[a]	0 ± 0	0.3 ± 0.15	26.6 ± 1.03	88.3 ± 1.41[a]
	NIN10	23.2 ± 0.67	57.4 ± 1.49[a]	5.7 ± 0.36	19 ± 0.91[a]	0 ± 0	0.3 ± 0.15	29 ± 0.77	76.6 ± 1.6[a]
	NIN25	22.2 ± 0.48	60.5 ± 1.19[a]	5.4 ± 0.22	19 ± 0.93[a]	0 ± 0	0.5 ± 0.16	27.6 ± 0.4	80.4 ± 1.3[a]
	NIN50	22.5 ± 0.40	60.1 ± 1.39[a]	4.8 ± 0.29	21.2 ± 0.6[a]	0 ± 0	0.5 ± 0.16	27.7 ± 0.4	81.7 ± 1.0[a]
	NIN100	21.8 ± 0.59	60.5 ± 1.02[a]	5 ± 0.365	21.5 ± 1.01[a]	0 ± 0	0.5 ± 0.16	26.8 ± 0.7	82.1 ± 1.4[a]
	NIN250	22.4 ± 0.42	61.7 ± 1.23[a]	5.9 ± 0.5	22.2 ± 0.71[a]	0 ± 0	0.6 ± 0.16	29.10 ± 2.4	84 ± 1.67[a]
Pre	NIN5	22.31 ± 1.21	100 ± 4.210[c]	4.19 ± 0.46	25.14 ± 2.8	0 ± 0	0.45 ± 0.02	27.50 ± 1.54	126.2 ± 4.1[b]
	NIN10	23.47 ± 1.95	102.4 ± 5.1[c]	5.87 ± 0.78	20.45 ± 1.5[c]	0 ± 0	0.74 ± 0.09	29.210 ± 2.4	124.0 ± 3.6[b]
	NIN25	21.55 ± 0.98	97.5 ± 3.94[c]	4.45 ± 1.07	21.87 ± 2.1[c]	0 ± 0	0.54 ± 0.08	26.140 ± 2.6	119.1 ± 4.6[b]
Post	NIN5	22.54 ± 1.44	129.5 ± 4.17	5.24 ± 0.97	31.24 ± 3.4	0 ± 0	1.14 ± 0.24	27.90 ± 1.41	162.2 ± 5.68
	NIN10	22.41 ± 1.11	130.9 ± 5.64	4.14 ± 0.87	32.56 ± 2.8	0 ± 0	0.98 ± 0.31	26.540 ± 1.1	163.92 ± 4.1
	NIN25	23.17 ± 2.10	131.2 ± 4.55	5.89 ± 1.01	29.87 ± 3.5	0 ± 0	1.22 ± 0.12	29.510 ± 2.1	162.54 ± 5.8

No symbol – non significant, a-p<0.001, b-p<0.01 and c-p<0.05. n=4.

reduced PCNA labeling index (Figure 6.1). However, this reduction in PCNA labeling index was lesser when compared with the group of animals that received NIN before, during as well as after the BaP treatment.

Table 6.5: PCNA labeling index in the forestomach carcinoma of mice treated with various doses of naringin during BaP- induced carcinogenesis.

	PCNA Positive Index	
NIN mg/kg b.wt.	*NIN*	*NIN+BaP*
NINO	132 ± 2.87	487 ± 4.12
NIN1	120 ± 3.22	272 ± 2.19[a]
NIN2.5	129 ± 3.01	265 ± 3.84[a]
NIN5	135 ± 2.97	269 ± 4.12[a]
NIN10	141 ± 5.11	271 ± 4.22[a]
NIN25	121 ± 3.4	275 ± 5.31[a]
NIN50	129 ± 3.78	274 ± 3.56[a]
NIN100	100 ± 5.11	300 ± 3.55[a]
NIN250	105 ± 3.1	305 ± 3.61[a]

No symbol – non significant, a-p<0.001, b-p<0.01 and c-p<0.05. n=4.

Discussion

The changes in the environment, lifestyle, dietary habits and increased exposure to mutagenic agents have been responsible for increasing number of cancers worldwide. In spite of the great advances being made to cure cancer, the treatment of advanced invasive disease, such as carcinoma of the lung, colon, breast, prostate and pancreas remains elusive. This indicates that new approaches to control cancer are critically needed. In general the old adage that "prevention is better than cure" may be the best approach to control cancer. Chemoprevention, therefore, is a mean to control cancer, where the occurrence of disease, as a consequence of exposure to carcinogens, can be slowed, blocked, or reversed by the administration of one or more naturally occurring dietary and or other pharmacological agents (Wattenberg, 1990; Morse and Stoner, 1993; Kelloff *et al.*, 1994; Stoner and Mukhtar, 1995). The present study has been undertaken to evaluate the effect of naringin a grapefruit flavanone on the BaP-induced forestomach carcinoma in mice.

Acute toxicity studies indicate that NIN is non-toxic up to 1000 mg/kg when administered orally and 500 mg/kg when administered intraperitoneally. However, earlier report has found NIN non-toxic up to 5000 mg/kg in mice after oral administration (Alvarez-Gonzalez *et al.*, 2001). This variation may be owing to the species differences. Similarly, NIN has also been found to be non-toxic up to 2000 mg in human receiving oral dose (Guengerich and Kim, 1990).

Treatment of mice with BaP caused 100 per cent incidence of forestomach tumors, whereas the oil treatment did not induce any tumors in the recipient animals. The tumor multiplicity in the BaP group was also high. A similar effect of BaP induced

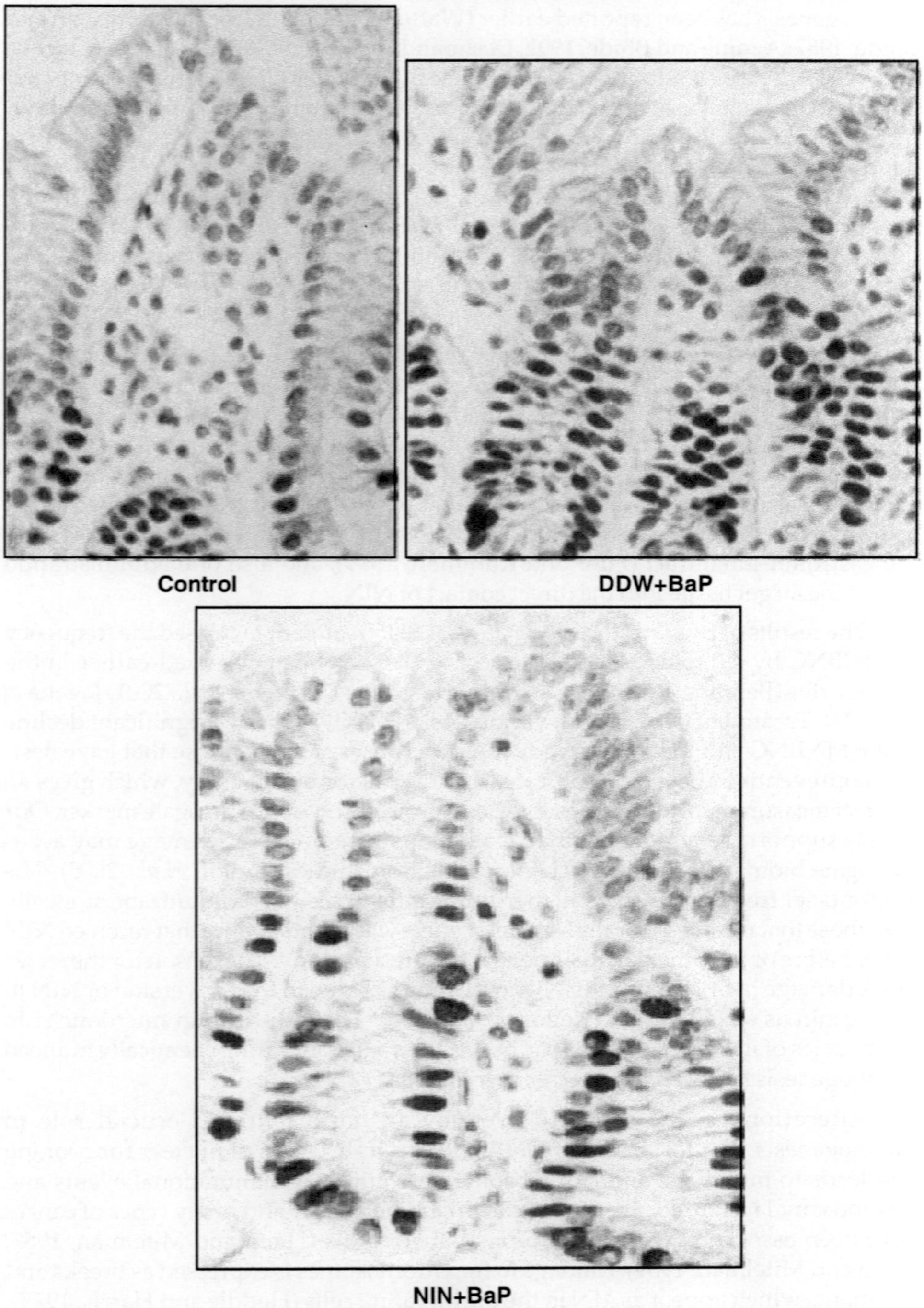

Figure 6.1: Effect of various doses of NIN on the BaP- induced PCNA in the forestomach carcinoma of mice.

carcinogenesis has been reported earlier (Wattenberg *et al.*, 1980; Nagabhushan and Bhide, 1987; Azuine and Bhide, 1992; Deshpande *et al.*, 1997; Agha *et al.*, 2001; Jagetia *et al.*, 2003). The oral administration of dietary ingredient, NIN before-during and after BaP treatment reduced the mean tumor number, tumor multiplicity and tumor incidence in mice. Phytochemicals like *Xanthophylls, Astaxanthin* and *Canthaxanthin* have been reported to inhibit chemically induced oral carcinogenesis in rats (Tanaka *et al.*, 1995). Several other synthetic or phytochemicals have been found to inhibit chemical carcinogenesis (Ito and Hirose, 1989; Balansky *et al.*, 1996). Several dietary ingredients like β–carotene, ascorbic acid and tochopherols have been found to be chemopreventive in human studies (Omenn *et al.*, 1996; De Flora *et al.*, 2001; Lee *et al.*, 2003).

NIN treatment also inhibited tumors when it was given either before or after the BaP treatment. However, chemopreventive effect of NIN was lesser when compared with the continuous treatment (pre-post). A similar effect has been observed for turmeric and *Alstonia scholaris* earlier (Deshpande *et al.*, 1997; Jagetia *et al.*, 2003). The intraperitoneal administration of NIN although inhibited BaP-induced carcinogenesis, its effect was lesser than that of oral administration. The reason of this difference lies in the conversion of NIN to naringenin by bacterial hydrolysis in the gastrointestinal tract (Fuhr and Kummert, 1995), and also oral administration brings the target tissue into the direct contact of NIN.

The results of present study indicate that BaP treatment increased the frequency of MNBNC by six folds. A similar effect of BaP has been observed earlier in the splenocytes (Benning *et al.*, 1994; Winker *et al.*, 1995; Dertinger *et al.*, 2001; Jagetia *et al.*, 2003). Treatment of mice with various doses of NIN caused a significant decline in the MNBNC and the highest reduction was observed at the dose that gave best-chemopreventive effect. This indicates that the micronucleus assay, which gives an indirect measurement of chromosome damage, may serve as a surrogate marker. Our results support the hypothesis that evaluation of chromosome damage may act as surrogate biomarker in cancer chemoprevention studies (Kelloff *et al.*, 2000). The micronuclei frequencies were higher in mice that received NIN intraperitoneally than those that received it orally. A similar effect was found in mice that received NIN either before or after the BaP treatment in the present study. The reason for the lesser DNA damage in mice receiving NIN orally may be due to the conversion of NIN to naringenin as stated earlier (Kelloff *et al.*, 2000). The reduction in micronuclei in splenocytes of mice receiving extract of *Alostonia scholaris* during chemically induced carcinogenesis has been reported earlier (Jagetia *et al.*, 2003).

Alterations in the native DNA play an important and crucial role in carcinogenesis. The formation of DNA adducts inflicts loss of nuclear functioning that leads to mutations and chromosome aberrations. The mutational events and chromosomal instability are key events in carcinogenesis and many types of cancer have been associated with chromosomal aberrations (Heim and Mitelman, 1987; Heim and Mitelman, 1989). Damage to the chromosomes is expressed as breaks and fragments which appear as MN in the proliferating cells (Heddle and Harris, 1975). The chromatid breaks and exchanges that lead to chromosome anomalies must occur during carcinogenesis to attain the neoplastic property. The tumor cell population is composed of one or few clones of cells with particular chromosome anomalies and this phenomenon is detectible by an increased frequency of micronuclei in the proliferating cells (Heim and Mitelman, 1989; Kelloff *et al.*, 1994). The cell proliferation also allows endogenous adducts to be converted into mutations and makes more sensitive single-stranded DNA target for carcinogenesis. Therefore the evaluation of micronuclei gives an indication on the quantitative aspect on the insult to the cellular DNA. In chronic administration studies it is always desirable to study the MN formation in the spleen, because the irreparable damage to the cellular DNA gets accumulated, which becomes manifest when the splenocytes are stimulated to divide. BaP has been reported to be immunotoxic and has profound inhibitory effect on the spleen (Ginsberg *et al.*, 1989; Zhao, 1990). BaP treatment has been reported to increase the BaP-DNA adducts formation and frequency of mutation in the spleen cells (Benning *et al.*, 1994; Winker *et al.*, 1995; Dertinger *et al.*, 2001). BaP alone has increased the frequency of micronuclei indicating that it has caused DNA damage and subsequently tumor development in the recipient mice. The treatment of mice with NIN reduced the DNA damage as evident by a significant reduction in the frequency of micronuclei. Therefore, protection against the DNA damage by NIN during tumorigenesis may be one of the reasons for lesser frequency of tumors in this group. In this study we have observed the best chemoreventive effect when the NIN was administered before, during and after (pre-post) BaP treatment.

There may be several possible mechanisms by which NIN may have inhibited BaP-induced forestomach carcinogenesis. It is well established that BaP requires metabolic activation, mediated by cytochrome P450 (CYP) dependent monooxygenase, for carcinogenesis (Singh *et al.*, 1998; Ueng *et al.*, 1999). Covalent interaction of carcinogen with nucleophilic sites in DNA is a critical event in BaP induced tumerogenesis (Singh *et al.*, 1998). NIN has been reported to inhibit benzo[*a*]pyrene hydroxylase (AHH), cytochrome P4501A2, CYP450 3A4 and expression of CYP4501A2 proteins (Guengerich and Shimada, 1991; Ueng *et al.*, 1999; Alvarez-Gonzalez *et al.*, 2001). BaP-induces carcinogenesis by its metabolic activation, mediated by cytochrome-450 (CYP) dependent monooxygenase (Singh *et al.*, 1998; Ueng *et al.*, 1999). Cytochrome P4503A4 also appears to play a more dominant role in the conversion of 7,8-dihydroxy-7,8 dihydrobenzo(a)pyrene to 9, 10-epoxide during carcinogenesis (Guengerich and Shimada, 1991). The consumption of grapefruit juice led to a marked reduction of CYP3A4 activity in the gastrointestinal tract (Ho *et al.*, 2001). Therefore, it is clear that chemopreventive activity of NIN is because of inhibition of AHH and CYP4503A4. The crucial importance of DNA damage in carcinogenesis and inhibition of BaP-induced DNA damage by NIN may also be one of the important mechanisms of chemoprevention in this study. This is evident from the reduction in the frequency of micronuclei in the NIN+BaP treated groups when compared with BaP treatment alone. Recently, naringin has been reported to inhibit the the molecular DNA damage in V79 cell exposed to bleomycin (Jagetia *et al.*, 2007). BaP has been reported to activate NF-κB, which plays an important role in carcinogenesis by making mutated cells immortal (Bolotina *et al.*, 2007). The suppression of NF-κB transactivation by naringin may be responsible for reduced number of tumors in naringin treated animals. Matrix metalloproteinases (MMPs) and in particular MMP9 are associated with tumor progression and metastasis (Wu *et al.*, 2008; Chou *et al.*, 2012). Naringin has been reported to inhibit MMP9 (Lee *et al.*, 2009) indicating that it can inhibit tumor promotion.

The chemopreventive action of NIN may also be due to inhibition of enzyme lipooxygenase and cyclooxygenase. Since these enzymes are known to cause hyperproliferative responses (Frstenberger and Marks, 1980). NIN has been reported to inhibit lipooxygenase and cyclooxygenase (Corvazier and Maclouf, 1985). NIN may have also inhibited cell proliferation thereby reducing the tumor multiplicity. This is supported by PCNA labeling studies, where the PCNA labeling index has been found to be significantly lower in the animals receiving NIN and BaP than those receiving BaP only. Last but not least NIN may have scavenged free radicals and raised the antioxidant status, resulting in the reduction of tumors. NIN has been reported to scavenge free radicals and increase antioxidant status *in vitro* earlier (Jagetia *et al.*, 2003). NIN has been found to be an antioxidant and anti-inflammatory agent and protect cells from ethanol, okadaic acid, ifosfamide, aflotoxin B_1 and H_2O_2 induced oxidative stress (Guengerich and Kim, 1990; Martin *et al.*, 1994; Gordon *et al.*, 1995; Wilcox *et al.*, 2001; Alvarez-Gonzalez *et al.*, 2001; Kanno *et al.*, 2003).

NIN treatment has reduced the BaP induced forestomach carcinoma in receipent mice and the chemoreventive effect of NIN may be due to inhibition of BaP metabolism, suppression of NFkB activation, inhibition of MMP9, and BaP-induced DNA damage and scavenging of free radicals as well as elevation in the antioxidant status of mouse receiving NIN during chemically induced carcinogenesis.

References

Agha, A. M., El-Fattah, A. A., Al-Zuhair, H. H., and Al-Rikabi, A. C. (2001). Chemopreventive effect of *Ginkgo biloba* extract against benzo[*a*]pyrene -induced forestomach carcinogenesis in mice: amelioration of doxorubicin cardiotoxicity. *Journal of Experimental and Clinical Cancer Research*, **20:** 39-50.

Alvarez-Gonzalez, I., Madrigal-Bujaidar, E., Dorado, V., and Espinosa-Aguirre, J. J. (2001). Inhibitory effect of naringin on the micronuclei induced by ifosfamide in mouse, and evaluation of its modulatory effect on the Cyp3a subfamily. *Mutation Research*, **1:** 480-481.

Azuine, M. A., and Bhide, S. V. (1992). Chemopreventive effect of turmeric against stomach and skin tumors induced by chemical carcinogens in Swiss mice. *Nutrition and Cancer*, **17:** 77-83.

Balansky, R., Izzotti, A., Scatolini, L., D'Agostini, F., and De Flora, S. (1996). Induction by carcinogens and chemoprevention by N-acetylcysteine of adducts to mitochondrial DNA in rat organs. *Cancer Research*, **56:** 1642-1647.

Bauer, E., Guo, Z., Ueng, Y. F., Bell, L. C., Zeldin, D., and Guengerich, F. P. (1995). Oxidation of benzo[*a*]pyrene by recombinant human cytochrome P450 enzymes. *Chemical Research in Toxicology*, **8:** 136-142.

Benning, V., Brault, D., Duvinage, C., Thybaud, V., and Melcion, C. (1994). Validation of the *in vivo* CD1 mouse splenocyte micronucleus test. *Mutagenesis*, **9:** 199-204.

Calomme, M., Pieters, L., Vlietinck, A., and Vanden, D. (1996). Inhibition of bacterial mutagenesis by citrus flavonoids. *Planta Medica*, **62**: 222–226.

Cassady, J. M., Baird, W. M., and Chang, C. J. (1990). Natural products as a source of potential cancer chemotherapeutic and chemopreventive agents. *Journal of Natural Products*, **53**: 23-41.

Castonguay, A. (1992). Methods and strategies in lung cancer control. *Cancer Research*, **52**: 2641s-2651s.

Chou, C.H., Teng, C.M., Tzen, K.Y., Chang, Y.C., Chen, J.H., and Cheng, J.C. (2012). MMP-9 from sublethally irradiated tumor promotes Lewis lung carcinoma cell invasiveness and pulmonary metastasis. *Oncogene*, **31(4):** 458-468.

Collins, J. F., Brown, J. P., Dawson, S. V., and Marty, M. A. (1991). Risk assessment for benzo[*a*]pyrene. *Regulatory Toxicology and Pharmacology*, **13**: 170-184.

Corvazier, E., and Maclouf, J. (1985). Interference of some flavonoids and non-steroidal anti-inflammatory drugs with oxidative metabolism of arachidonic acid by human platelets and neutrophils. *Biochimca Biophysica Acta*, **9**: 315-21.

Countryman, P. I., and Heddle, J. A. (1976). The production of micronuclei from chromosome aberrations in irradiated cultures of human lymphocytes. *Mutation Research*, **41**: 321-332.

De Flora, S., Izzotti, A., D'Agostini, F., Balansky, R. M., Noonan, D., and Albini, A. (2001). Multiple points of intervention in the prevention of cancer and other mutation-related diseases. *Mutation Research*, **1**: 480-481.

Dertinger, S. D., Nazarenko, D. A., Silverstone, A. E., and Gasiewicz, T. A. (2001). Aryl hydrocarbon receptor signaling plays a significant role in mediating benzo[*a*]pyrene- and cigarette smoke condensate-induced cytogenetic damage *in vivo*. *Carcinogenesis*, **22:** 171-177.

Deshpande, S. S., Ingle, A. D., and Maru, G. B. (1997). Inhibitory effects of curcumin-free aqueous turmeric extract on benzo[a]pyrene-induced forestomach papillomas in mice. *Cancer Letters*, **16**: 79-85.

Fenech, M., and Morley, A. A. (1985). Measurement of micronuclei in lymphocytes. *Mutation Research*, **147:** 29-36.

Francis, A. R., Shetty, T. K., and Bhattacharya, R. K. (1989). Modulating effect of plant flavonoids on the mutagenicity of N-methyl-N-nitro-N-nitrosoguanidine. *Carcinogenesis*, **10**: 1953–1955.

Frstenberger, G., and Marks, F. (1980). Early prostaglandin E synthesis is an obligatory event in the induction of cell proliferation in mouse epidermis *in vivo* by the phorbol ester TPA. *Biochemical Biophysical Research Communications*, **92**: 749-756.

Fuhr, U., and Kummert, A. L. (1995). The fate of naringin in humans: a key to grapefruit juice-drug interactions. *Clinical Pharmacology and Therapeutics*, **58**: 365-73.

Ghosh, M. N. (1984). Toxicity studies. In: Fundamentals of Experimental Pharmacology. Ghosh MN (ed). Scientific Book Agency, Calcutta, pp. 153–158.

Ginsberg, G. L., Atherholt, T. B., and Butler, G. H. (1989). Benzo[a]pyrene-induced immunotoxicity: comparison to DNA adduct formation *in vivo*, in cultured splenocytes and in microsomal systems. *Journal of Toxicology and Environmental Health*, **28**: 205-220.

Gordon, F. B., and Ingunn, H., Pr, O. (1995). Seglen protection by naringin and some other flavonoids of hepatocytic autophagy and endocytosis against inhibition by okadaic acid. *Journal of Biological Chemistry*, **270**: 5830–5838.

Guengerich, F. P. (1988). Roles of cytochrome P-450 enzymes in chemical carcinogenesis and cancer chemotherapy. *Cancer Research*, **48**: 2946-2954.

Guengerich, F. P. (1995). Human cytochrome P450 enzymes. In "Cytochrome P450: Structure, Mechanism and Biochemistry". Ortz de Montellano P R (ed). Plenum press, New York, pp. 473-535.

Guengerich, F. P., Kim. (1990). *In vitro* inhibition of dihydropyridine oxidation and aflotoxin B_1 activation in human liver microsomes by naringenin and other flavonoids. *Carcinogenesis*, **11**: 2275-2279.

Guengerich, F. P., and Shimada, T. (1991). Oxidation of toxic and carcinogenic chemicals by human cytochrome P-450 enzymes. *Chemical Research in Toxicology*, **4**: 391-407.

Haenen, G. R. M. M., Paquay, J. B. G., Korthouwer, R. E. M., and Bast, A., 1997. Peroxynitrite scavenging by flavonoids. *Biochemicaland Biophysical Research Communication*, **236**: 591–593.

Halliwell, B., and Gutteridge, J. M. (1989). Free Radicals in Biology and Medicine. Second Ed. New York Oxford University.

Heddle, J. A., and Harris, J. W. (1975). Letter: rapid screening of radioprotective drugs *in vivo*. *Radiation Research*, **61**: 350-353.

Heim, S., and Mitelman, F. (1987). Ninteen of 26 cellular oncogenes precisely localized in the human genome map to one of the 83 bands involved in primary cancer specific rearrangements. *Human Genetics*, **75**: 70-72.

Heim, S., and Mitelman, F. (1989). Primary chromosome abnormalities in human neoplasia. *Advances in Cancer Research*, **52**: 1 -43.

Ho, P. C., Saville, D. J., and Wanwimolruk, S. (2001). Inhibition of human CYP3A4 activity by grapefruit flavonoids, furanocoumarins and related compounds. *Journal of Pharmacy and Pharmaceutical Sciences*, **4**: 217-27.

Ito, N., and Hirose, M. (1989). Antioxidants—carcinogenic and chemopreventive properties. *Advances in Cancer Research*, **53**: 247-302.

Jagetia, A., Jagetia, G.C., and Jha, S. (2007). Naringin, a grapefruit flavanone, protects V79 cells against the bleomycin-induced genotoxicity and decline in survival. *Journal of Applied Toxicology*, **27(2)**: 122-32.

Jagetia, G. C., Baliga, M. S., and Venkatesh, P. (2003). Effect of Sapthaparna (*Alstonia scholaris* Linn) in modulating the benzo(*a*)pyrene-induced forestomach carcinogenesis in mice. *Toxicology Letters*, **144:** 183-193.

Jagetia, G. C., and Reddy, T. K. (2002). The grapefruit flavanone naringin protects against the radiation-induced genomic instability in the mice bone marrow: a micronucleus study. *Mutation Research*, **26**: 37-48.

Jagetia, G. C., Venkatesha, V. A., and Reddy, T. K. (2003). Naringin, a citrus flavonone, protects against radiation-induced chromosome damage in mouse bone marrow. *Mutagenesis*, **18**: 337-343.

Jagetia, G. C., Menon, K. S., and Jain, V. (2001). Genotoxic effect of hydroquinone on the cultured mouse spleenocytes. *Toxicology Letters*, **8:** 15-20.

Jagetia, G.C., and Reddy T.K. (2011). Alleviation of iron induced oxidative stress by the grape fruit flavanone narinign *in vitro. Chemico-Biological Interactions,* **190:** 121–128.

Jagetia, G.C., and Reddy, T.K., Venkatesha, V. A., and Kedlaya, R. (2004). Influence of naringin on ferric iron induced oxidative damage *in vitro. Clinica Chimica Acta*, **347**: 189-197.

Kanno, S., Shouji, A., Asou, K., and Ishikawa, M. (2003). Effects of naringin on hydrogen peroxide-induced cytotoxicity and apoptosis in P388 cells. *Journal of Pharmacological Sciences,* **92:** 166-170.

Kelloff, G.J., Boone, C. W., Steele, V. E., Crowell, J. A., Lubet, R., and Sigman, C. C. (1994). Progress in cancer chemoprevention: perspectives on agent selection and short-term clinical intervention trials. *Cancer Research*, **1**: 2015s-2024s.

Kelloff, G. J., Sigman, C. C., Johnson, K. M., Boone, C. W., Greenwald, P., Crowell, J. A., Hawk, E. T., and Doody, L. A. (2000). Perspectives on surrogate end points in the development of drugs that reduce the risk of cancer. *Cancer Epidemiology Biomarkers and Prevention,* **9**: 127-37.

Kim, D. H., Jung, E. A., Shong, I. S., Han, J. A. (1998). Intestinal bacterial metabolism of flavonoids and its relation to some biological activities. *Archives of Pharmacological Research*, **21**: 17–23.

Krisch-Volders, M., Souni, T., Aardema, M., Albertini, S., Eastmond, D., Fenech, M., Ishidate, M., Lorge, E., Norppa, H., Surralles, J., Von Dur Hude, W., and Wakata, A. (2000). Report from the *in vitro* micronucleus assay-working group. *Environmental and Molecular Mutagenesis*, **35**: 167-172.

Kroyer, G. (1986). The antioxidant activity of citrus fruit peels. *Zeitschrift fur Ernahrungswissenschaft*, **25**: 63–69.

Lake, B. G., Beamand, J. A., Tredger, J. M., and Barton, P. T. (1999). Inhibition of xenobiotic-induced genotoxicity in cultured precision-cut human and rat liver slices. *Mutation Research*, **440**: 91–100.

Lee, K. W., Lee, H. J., Surh, Y. J., and Lee, C. Y. (2003). Vitamin C and cancer chemoprevention: reappraisal. *American Journal of Clinical Nutrition*, **78**: 1074-1078.

Leininger, J. R., and Jokinen, M. P. (1994). Tumours of the oral cavity, pharynx, oesophagus, and stomach. Tumours of the Mouse. Turusov VS, Mohr U (eds). International Agency for Research on Cancer. Lyon, France, pp. 2 167-193.

Levin, W., Wood, A., Chang, R., Ryan, D., and Thomas, P. (1982). Oxidative metabolism of polycyclic aromatic hydrocarbons to ultimate carcinogens. *Drug Metabolism Reviews*, **13**: 555-580.

Martin, M. J., Marhuenda, E., Perez-Guerrero, C., and Franco, J. M. (1994). Antiulcer effect of naringin on gastric lesions induced by ethanol in rats. *Pharmacology*, **49:** 144-50.

Morse, M. A., and Stoner, G. D. (1993). Cancer chemoprevention: principles and prospects. *Carcinogenesis*, **14**: 1737-1746.

Nagabhushan, M., and Bhide, S. V. (1987). Antimutagenicity and anticarcinogenicity of turmeric (*Curcuma longa*). *Journal of Nutrition Growth and Cancer*, **4**: 83-89.

Nebert, D. W. (1989). The Ah locus; genetic differences in toxicity, cancer, mutation, and birth defects. *Critical Reviews in Toxicology*, **20**: 153-174.

Nguyen, C. F., Porter, L., Ren, W. W., Roberts, R. F., Roche, A. B., Subbaramaiah, N. S., and Sporn, M. B. (1999). Inhibitory effects of phenolic compounds on benzo(a)pyrene-induced neoplasia. *Cancer Research*, **40**: 2820-2823.

Omenn, G. S., Goodman, G. E., Thornquist, M. D., Balmes, J., Cullen, M. R., Glass, A., Keogh, J. P., Meyskens, F. L Jr., Valanis, B., Williams, J. H Jr., Barnhart, S., Cherniack, M. G., Brodkin, C. A., and Hammar, S. (1996). Risk factors for lung cancer and for intervention effects in caret, the beta-carotene and retinol efficacy trial. *Journal of National Cancer Institute*, **6**: 1550-1559.

Prieur, D. J., Young, D. M., Davis, R. D., Cooney, D. A., Homan, E. R., Dixon, R. L., and Guarino, A. M. (1973). Procedures for preclinical toxicologic evaluation of cancer chemotherapeutic agents, protocols of the laboratory of toxicology. *Cancer Chemotherapy Reports*, **4**: 1–28.

Reddy, B. S., Tokumo, K., Kulkarni, N., Aligia, C., and Kelloff, G. (1992). Inhibition of colon carcinogenesis by prostaglandin synthesis inhibitors and related compounds. *Carcinogenesis*, **13**: 1019-1023.

Shin, Y. W., Bok, S. H., Jeong, T. S., Bae, K. H., Jeoung, N. H., Choi, M. S., Lee, S. H., and Park, Y. B. (1999). Hypocholesterolemic effect of naringin associated with hepatic cholesterol regulating enzyme changes in rats. *International Journal of Vitamin Nutrition Research*, **69**: 341–347.

Singh, S. V., Hu, X., Srivastava, S. K., Singh, M., Xia, H., Orchard, J. L., and Zaren, H. A. (1998). Mechanism of inhibition of benzo[a]pyrene-induced forestomach cancer in mice by dietary curcumin. *Carcinogenesis*, 19: 1357-60.

Singletary, K. (2000). Diet, natural products and cancer chemoprevention. *Journal of Nutrition*, **130:** 465S-466S.

Slaga, T. J. (1980). Cancer: etiology, mechanisms, and prevention—a summary. *Carcinogenesis; a Comprehensive Survey*, **5:** 243-262.

Sporn, B. M. (1999). Prevention of cancer in the next millennium: Report of the chemoprevention working group to the American association of cancer research. *Cancer Research*, **59:** 4743-4758.

Stoner, G. D., and Mukhtar, H. (1995). Polyphenols as cancer chemopreventive agents. *Journal Cellular Biochemistry Suppl.*, **22**: 169-180.

Tanaka, T., Makita, H., Ohnishi, M., Mori, H., Satoh, K., and Hara, A. (1995). Chemoprevention of rat oral carcinogenesis by naturally occurring xanthophylls, astaxanthin and canthaxanthin. *Cancer Research,* **15**: 4059-4064.

Ueng, Y. F., Chang, Y. L., Oda, Y., Park, S.S., Liao, J. F., Lin, M. F., and Chen, C. F. (1999). *In vitro* and *in vivo* effects of naringin on cytochrome P450-dependent monooxygenase in mouse liver. *Life Science,* **65**: 591-602.

Wattenberg, L. W. (1990). Inhibition of carcinogenesis by minor anutrient constituents of the diet. *Proceedings of Nutrition Society,* **49**: 173-183.

Wattenberg, L. W., Coccia, J. B., and Lam, L. K. (1980). Inhibitory effects of phenolic compounds on benzo(*a*)pyrene-induced neoplasia. *Cancer Research,* **40**: 2820-2823.

Weisburger, J. H. (1991). Nutritional approach to cancer prevention with emphasis on vitamins, antioxidants, and carotenoids. *American Journal of Clinical Nutrition,* **53:** 226S-237S.

Wilcox, L. J., Borradaile, N. M., de Dreu, L. E., and Huff, M. W. (2001). Secretion of hepatocyte apoB is inhibited by the flavonoids, naringenin and hesperetin, via reduced activity and expression of ACAT2 and MTP. *Journal Lipid Research,* **42**: 725-734.

Winker, N., Weniger, P., Klein, W., Ott, E., Kocsis, F., Schoket, B., and Korpert, K. (1995).Detection of polycyclic aromatic hydrocarbon exposure damage using different methods in laboratory animals. *Journal of Applied Toxicology,* **15**: 59 - 62.

Wood, A. W., Levin, W., Lu, A. .H., Yagi, H., Hernandez, O., Jerina, D. W., and Conney, A. H. (1976). Metabolism of benzo[*a*]pyrene and benzo(a)pyrene derivatives to mutagenic products by highly purified hepatic microsomal enzymes. *Journal of Biological Chemistry,* **251**: 4882- 4890.

Wu, Z.S., Wu, Q., Yang, J.H., Wang, H.Q., Ding, X.D., Yang, F., and Xu, X.C. (2008). Prognostic significance of MMP-9 and TIMP-1 serum and tissue expression in breast cancer. *International Journal of Cancer,* **122**: 2050–2056.

Zhao, X. L. (1990). Effects of benzo[*a*]pyrene on the humoral immunity of mice exposed by single intraperitoneal injection. *Zhonghua,* **24:** 220-222.

Medicinal Plants: Phytochemistry, Pharmacology and Therapeutics Vol. 3 (2014) *Pages* **179–211**
Editor-in-Chief: **V.K. Gupta**
Published by: **DAYA PUBLISHING HOUSE, NEW DELHI**

7

Pharmacological and Therapeutical Profile of Terpenoids: An Avenue for the Development of Lead Molecules for New Generation Phytopharmaceuticals

Shrinivas B. Bumrela[1] and Suresh R. Naik[1]*

ABSTRACT

Medicinal plants have been used for therapeutic purposes from the time of civilization. In the recent past, there has been a renewed interest in natural product's research due to the failure of drug discovery methods to deliver lead compounds in key therapeutic areas. Among natural products, terpenoid constituents are abundantly distributed in wide variety of plants/herbs. Terpenoids are hydrocarbons of plant origin of the general formula $(C_5H_8)_n$ *as well as their oxygenated, hydrogenated and dehydrogenated derivatives. Terpenoids are volatile substances responsible for fragrance in plants and flowers. They form an important group of traditional plant derived medicine which exhibits (antibacterial, antineoplastic, anthelmintic, diuretic, local anaesthetic etc.) activities. Some of the established and therapeutically used terpenoid preparations are citral, menthol, cannabinoids, camphor, lupeol, carotenoids and many more. Diterpenoids are the active constituent of number of reported medicinal plants, which are of current interest for exploring their pharmacological and therapeutic activity and subsequently developing as drug candidate. Abietic acid, pimaric acid and their isomers are the integral components of diterpenoid. Abietic acid is known to elicit wide spectrum of pharmacological activity such as antimicrobial, antiulcer and cardiovascular. Gentian, a diterpene contains seco-iridoid gentiopicroside, used as a bitter tonic for inflammatory disorders. Diterpene lactone (ginkgolides A, B, C, J, M) were characterized and showed antagonist activity to platelet activating factor. Sesqueterpenoids are biogenetically derived*

1 Department of Pharmacology, Sinhgad Institute of Pharmaceutical Sciences, Kusgaon (Bk), Lonavala, Pune – 410 401, Mh., India

* *Corresponding author*: E-mail: srnaik5@rediffmail.com

farnesyl pyrophosphate and exert antitumor, anti-leukemic, cytotoxic and antimicrobial activity. Sesqueterpene lactones isolated from arnica flowers showed anti-rheumatic, antiarthritic and hypolipidemic activity. Quassia contains amaroidterpenoid compound (quassin) which is widely used as a bitter tonic for vermicidal activity. Carotenoids another important phytoconstituent belongs to tetraterpenoid exhibit antioxidant, immune system booster and precursor for Vitamin-A. The review describes the different classes of terpenoids, isolated or derived phytochemicals including some important chemical structures, pharmacological and therapeutic profile including underlying mechanism(s) of action. The scope and prospectus of terpenoids in the development of phytopharamceuticals are also discussed. The review will be useful for teaching professionals, researchers in the field of natural products, as well as organizations pursuing active research in the quest of new drug development at large.

Keywords: Terpenoids, Antiinflammatory-antiarthritic, Anticancer, Antidiabetic, Antioxidant, Hepato and Cardioprotective, Immunomodulatory, Antimicrobial.

Introduction

All plants produce chemical compounds as a part of their normal metabolic activities. They are mainly categorised as primary and secondary metabolites. Primary metabolites (sugars, fats, amino acids etc.) are required for normal cellular processes of the plant whereas secondary metabolites (alkaloids, phenolic compounds, glycosides, terpenes etc.) are produced in response to stress such as the case when acting as deterrent against herbivorous (Keeling, 2006). These secondary metabolites have been subsequently exploited by humans for their beneficial role in health and disease and referred as natural products. Normally occurring three broad categories of secondary metabolites (Figure 7.1) and reported in natural products are: terpenes and terpenoids (~25,000 types), alkaloids (~12000 types) and phenolic compounds (~8000 types) (Croteau *et al.*, 2000).

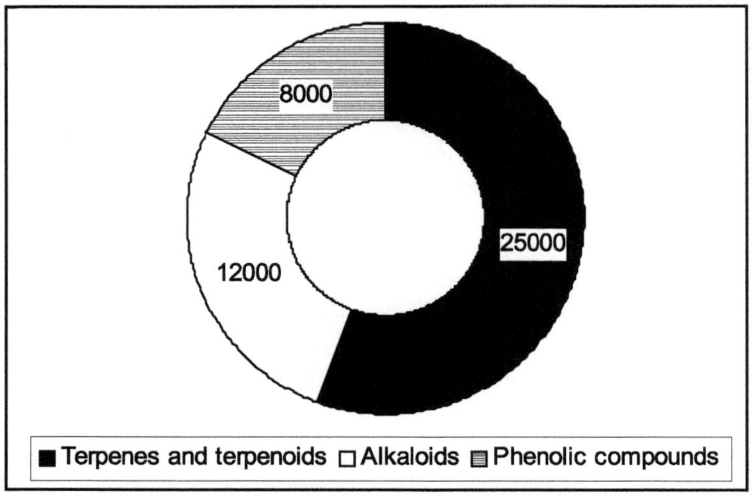

Figure 7.1: Distribution of major secondary metabolites in plants.

Apart from terpenes synthesized by plants in response to stress factors, it was also observed that flowers can emit terpenoids to attract pollinating insects. Interestingly, terpenoids also attract beneficial mites, which feed on herbivores insects. This is scent chemistry signaling mechanism of terpenes (pheromones) to attract predatory species, beetles and wasps for pollinating purpose. Terpenes play important role in ecology by acting as chemical messenger largely involved in plant defense mechanism. They are also of great importance from earth's atmosphere protection point of view because it is estimated that annual global emission of isoprene is 500 teragrams and thus, tempted to speculate their interaction with solar radiations to reduce its associated harmful effect (Zwenger and Basu, 2008).

Terpenoids are diverse group of compounds that are found abundantly in plant kingdom. Terpenoids can be defined as natural products whose structure contains several isoprene (C_5H_8) units. It is also referred as isoprenoids. Isoprene units are biogenetically derived from acetate via mevalonic acid. Terpenoids usually have such isoprene units joined together in a head to tail manner (Kar, 2006).

In the plant kingdom; friedelin, ursane and oleanane triterpenoid skeletons are common and have functional importance (Dev *et al.*, 1989). Triterpenoids exhibits diverse pharmacological activities *viz.* gastroprotective, antiprurutic, hepatoprotective activity, anti-inflammatory, anti-tumor and antimicrobial (Oliveira *et al.*, 2004; Oliveira *et al.*, 2005). Few therapeutically important terpenoids are: artemisinin (anti-malarial), parthenolide (serotonin antagonist), bisabolane (anti-inflammtory), forskolin (anti-hypertension), ginkgolides (useful as bronchocodialator, hypotensive agent, anti-diabetes, vasodilator, anti-inflammatory), taxol (anticancer), quassin (antifertility), andrographolide (anti-inflammatory, immunomodulatory and anti-diabetic) and many more (Kar, 2006).

Terpenoids constitute large group of secondary metabolites present in nature. Approximately 20000 different terpenes and terpenoids have been isolated, identified and known to play major role in growth and development of living systems (Devarenne, 2009). Plant derived terpenoids are used for food flavouring and clinically active therapeutic agents in cancer, diabetes, inflammation etc.

Classification

Terpenoids are classified on the basis of number of isoprene units $(C_5H_8)n$ incorporated into unsaturated hydrocarbon terpenoids molecule. Few examples are described below (Kar, 2006).

Monoterpenoids

Monoterpenes prevent the carcinogenesis process at both initiation and progression stages. Monoterpenes are effective at early and advanced stages of cancers. Limonene and perillylalcohol are found to be effective in preventing mammary, liver, lung and other cancers (Wattenberg, 1983; Elson *et al.*, 1988) mainly by inhibiting the isoprenylation of certain proteins (Crowell *et al.*, 1991), the synthesis of ubiquinone (Co-Q), and the conversion of lathesterol to cholesterol (Ren and Gould, 1994). The latter two cellular events/activities have the potential to modify both cell structure and cell energy production.

Table 7.1: Classification of terpenoids.

Number of Carbon Atoms	*Value of n (Number of isoprene units)*	*Class*	*Molecular Formula*	*Chemical Structure*
10	2	Monoterpenoid	$C_{10}H_{16}$	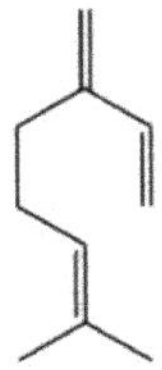 Myrcene
15	3	Sesquiterpenoid	$C_{15}H_{24}$	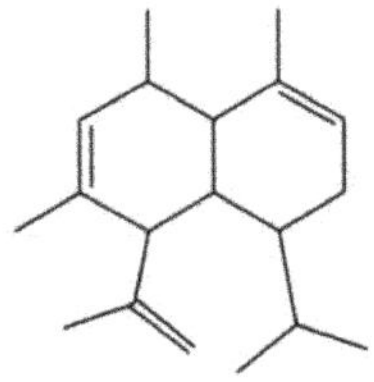 Cadinene
20	4	Diterpenoid	$C_{20}H_{32}$	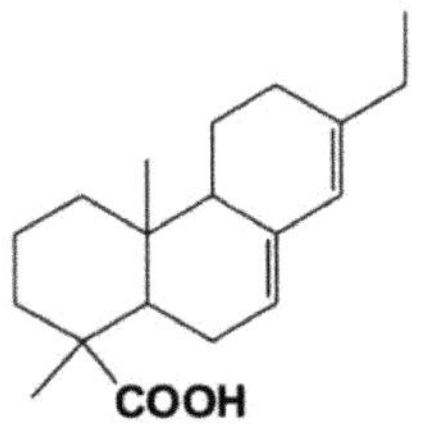 Abietic acid
30	6	Triterpenoid	$C_{30}H_{48}$	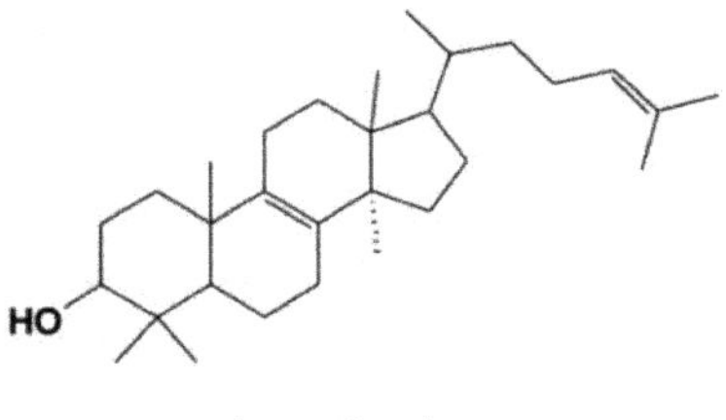 Lanosterol
40	8	Tetraterpenoid	$C_{40}H_{64}$	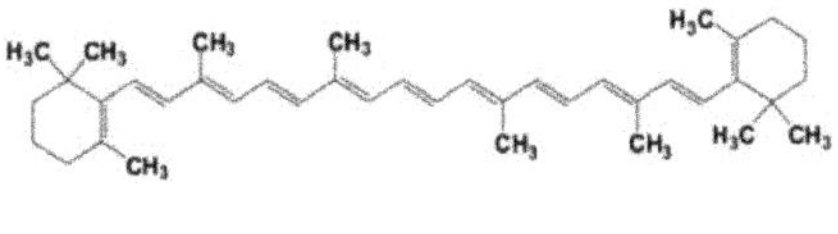 β-carotene
>40	>8	Polyterpenoid	$(C_5H_8)_n$	Resins

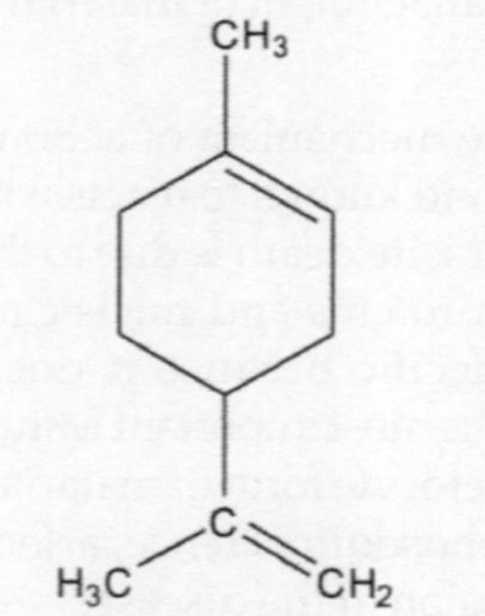

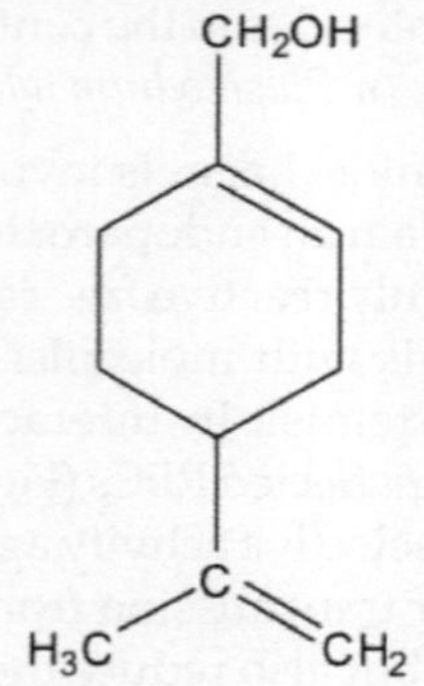

d-limonene **Perillyl alcohol**

A significant antimicrobial activity of three monoterpenes, linalyl acetate, (+)menthol, and thymol, against gram-positive and gram-negative bacteria (*S. aureus* and *E. coli*, respectively) was observed. Each of the monoterpenes inhibited the growth of both the microbial strains, although the gram-positive (*S. aureus*) appears more sensitive than the gram-negative (*E. coli*) toward all the three compounds. The mechanism of action of monoterpenes was further studied using water-soluble fluorescent marker carboxyfluorescein from large unilamellar vesicles with different lipid compositions. The antimicrobial activity of monoterpenes was found to be largely due to the perturbation of the lipid fraction of the plasma membrane (Domenico *et al.*, 2005).

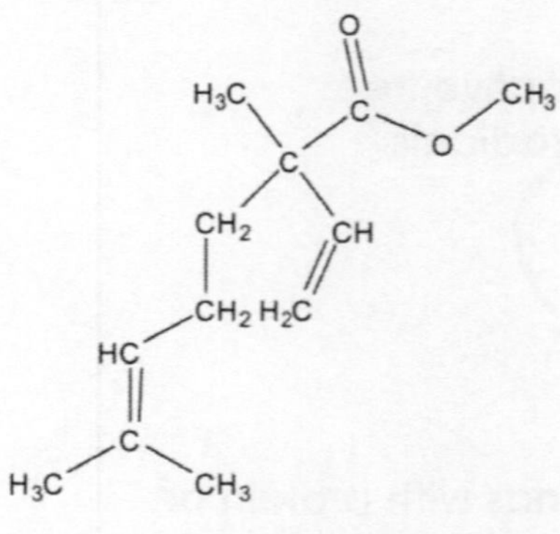

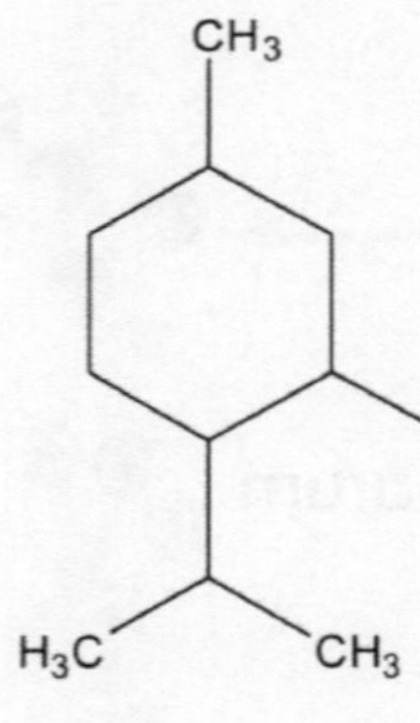

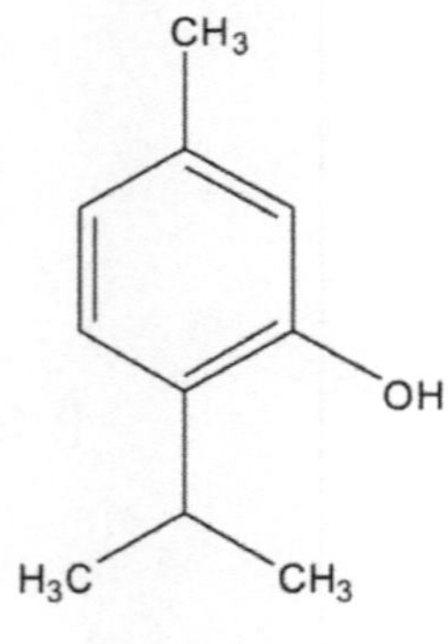

Linalyl acetate **(+)menthol** **Thymol**

Sesquiterpenoids

Artemisia annuna Linn is a herb belonging to family Asteraceae. It grows abundantly in China. In Chinese system of medicine this plant has been used to cure fever including malaria for more than hundreds of years. In India it is being grown in Gujarat, Uttar Pradesh, Himachal Pradesh, Jammu and Kashmir and Karnataka for research purpose (Gupta *et al.*, 2005). The leaves and flowers contain bioactive compound, artemisinin a sesquiterpene lactone. In the body, artemisinin is converted into dihydroartemisinin which is more active than parent compound. The artemisinin is used as schizontocides in the control and management of malerial fever caused by *Plasmodium vivax* or *Plasmodium falciparum*.

As with quinine, haem is involved in the mechanism of action of artemisinin. Artemisinin contain an endoperoxide group and known to reacts with iron in haem, giving rise to highly reactive free radicals. Parasite death is due to the interaction of these free radicals with molecules, such as proteins and nucleic acid of parasite. Furthermore, artemisinin interaction is specific because it doesn't react with haemoglobin of unaffected RBCs (Figure 7.2). The most important feature of artemisinin derivatives is its selective activity against gametocyte form of malarial parasite which is responsible for transmission from man to mosquito. Hence, artemisinin not only cure the malaria but also reduce the transmission of the disease.

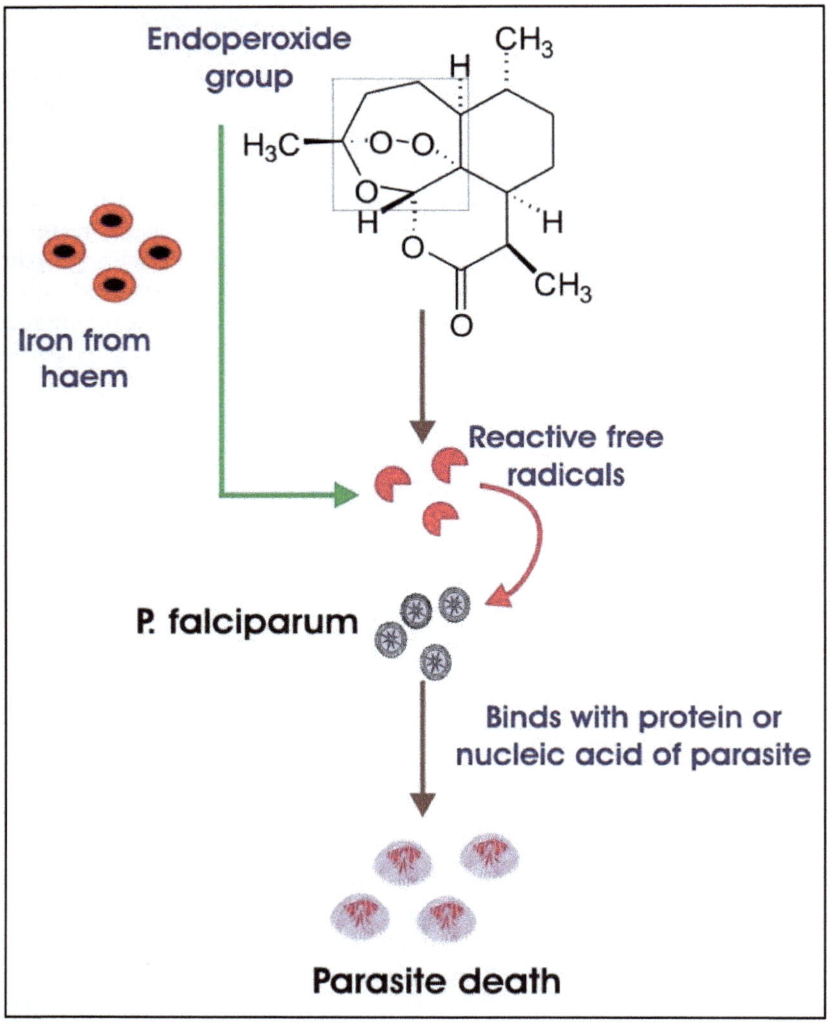

Figure 7.2: Mechanism of action of artimesinin.

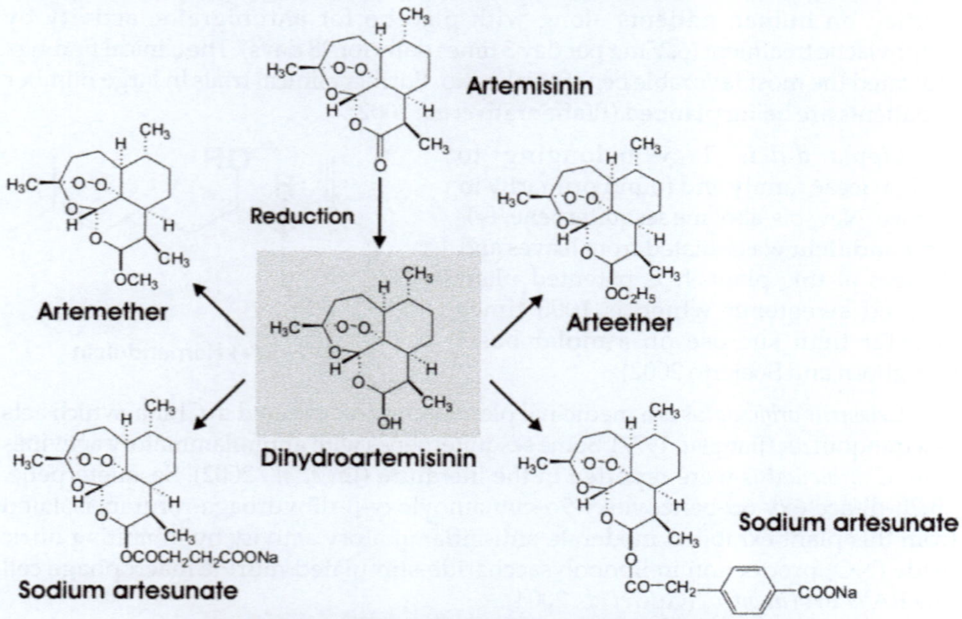

Artemisinin and its dihydro, ether and ester derivatives

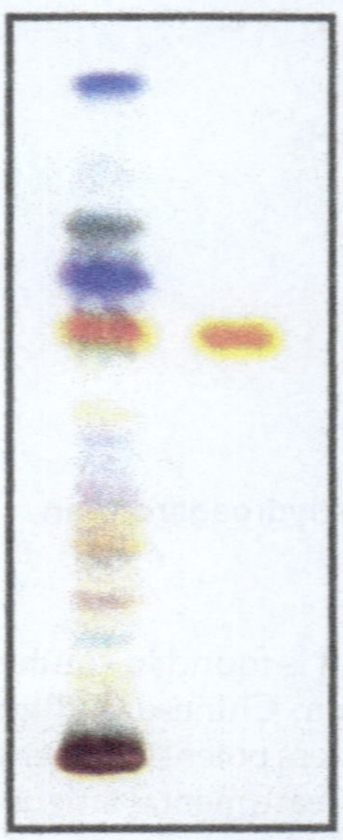

Figure 7.3: TLC of methanolic extract of entire plant of *Artemisia annuna* Linn.

Track 1: Test solution; Track 2: Artemisinin standard.

Artemisinin TLC identity test was performed on silica gel precoated plates as stationary phase and toluene: ethyl acetate: formic acid (7: 3: 1 v/v) as mobile phase. Artemisinin appear as yellowish pink spot (R_f= 0.59) after derivatization with anisaldehyde-sulphuric acid reagent (Figure 7.3) (Gupta *et al.*, 2005).

Tanacetum parthenium Linn is a herb belonging to family Asteraceae. It is commonly known as *feverfew* and is used for centuries as antipyretic. Parthenolide is a sesquiterpenoid lactone which acts as serotonin antagonist thereby causing inhibition of release of serotonin from blood platelets (Kar, 2006). The stable extract was prepared by supercritical CO_2 method and

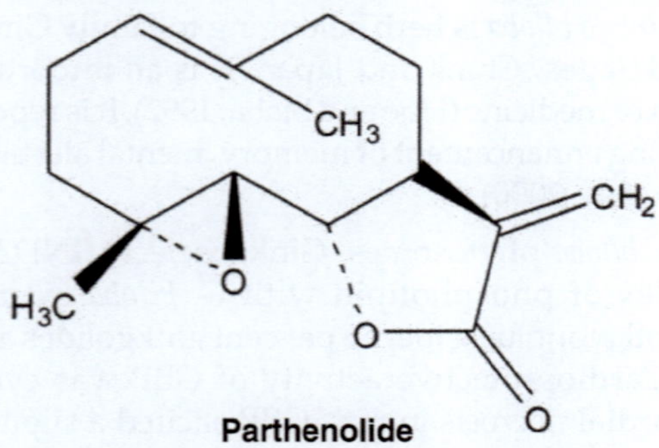

Parthenolide

studied on human patients along with placebo for antimigraine activity by prophylactic treatment (6.25mg per day 3 times daily for 28 days). The clinical findings indicated the most favorable benefit risk ratio. Further clinical trials in large number of patients are being planned (Pfaffenrath *et al.*, 2002).

Lippia dulcis Trev belonging to Verbenaceae family and found primarily in Mexico. New bis-abolane sesquiterpene, **(+)-Hernandulcin** was isolated from leaves and flowers of this plant. It is patented plant derived sweetener which is 1000 times sweeter than sucrose on a molar basis (Kinghorn and Soejarto 2002).

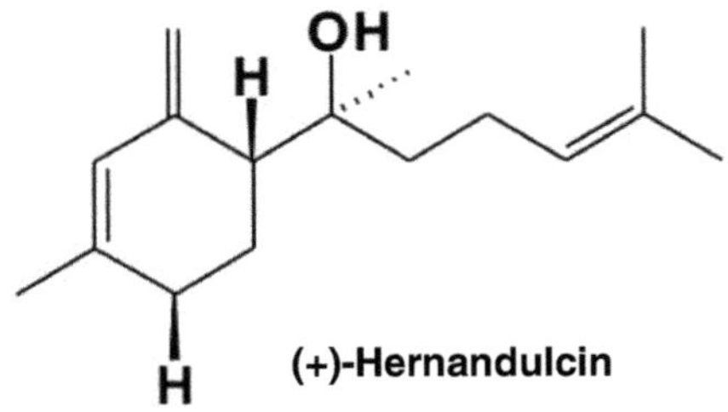

(+)-Hernandulcin

Celastrus orbiculatus is a medicinal plant widely distributed in China, which acts as a tranquilizer (Jiangsu, 1977). Some sesquiterpenes with antiinflammatory activities from *C. orbiculatus* were reported in the literature (Jin *et al.*, 2002). Sesqueterpene, 1β,2β-di-acetoxy-6α-benzoyloxy-9α-cinnamoyloxy-β-dihydroaga-rofuran, isolated from this plant exhibited moderate anti-inflammatory activity by inhibiting nitric oxide (NO) production in lipopolysaccharide-stimulated murine macrophage cell line RAW264 (*in vitro*) (Guo *et al.*, 2004).

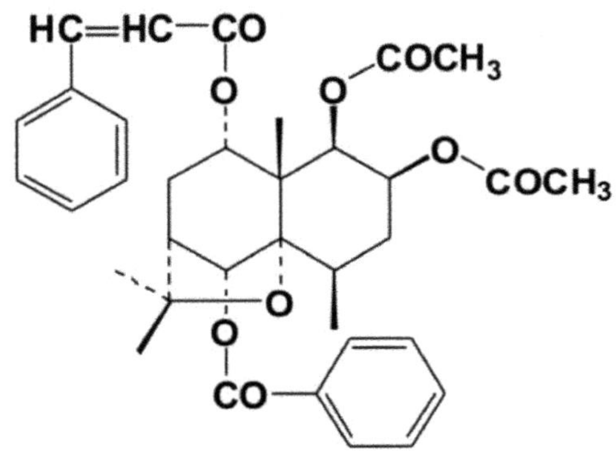

1β,2β-di-acetoxy-6α-benzoyloxy-9α-cinnamoyloxy-β-dihydroagarofuran

Diterpenoids

Ginkgo biloba is herb belonging to family Ginkgoaceae. It is found in south east United States, China and Japan. It is an important herb from Chinese traditional system of medicine (Hue and Staba, 1992). It is reported to possess potent CNS activity including enhancement of memory, mental alertness and decrease mental fatigue (Di Renzo *et al.*, 2000).

G. biloba phytosomes, Ginkgoselect® (INDANA, Italy) is 3 : 1 stoichiometric complex of phospholipid with *G. biloba* extract which contains ≥24 per cent ginkgoflavonglucoside, ≥6 per cent ginkgolides and bilobalide and ≤5ppm ginkgoic acid. Cardioprotective activity of GBP was evaluated in isoproterenol-induced myocardial necrosis in rats. GBP elicited a significant cardioprotective activity by lowering the elevated levels of serum marker enzymes *viz.* aspartate transaminase (AST), lactate dehydrogenase (LDH) and creatinine kinase (CPK) and lipid peroxidation and elevated the depleted antioxidants like glutathione (GSH), superoxide dismutase (SOD), catalase (CAT), glutathione peroxidase (GPx) and glutathione reductase (GR) in heart (Panda and Naik, 2008).

Antioxidant activity of GBP in rat brain was evaluated in sodium nitrite induced cerebral hypoxia. GBP treatment was found to increase SOD, CAT, GPx and GR activities in discrete nuclei of the brain *viz.*, cerebral cortex, cerebellum, hippocampus and striatum. These finding indicate its potential therapeutic use in neurodegenerative disorders such as Alzheimer's disease and Parkinsonism (Naik *et al.*, 2006). In addition, *G. biloba* phytosomes showed mild to moderate CNS stimulant activity. Phytosomes also exhibited antidepressant and antiamnestic/nootropic activity (Naik *et al.*, 2006a).

Standardized extract of dried leaves equivalent to 6 per cent terpenoids and 24 per cent flavones glycoside is used to enhance blood fluidity and circulation. Mixture of Ginkgolide A,B,C and M showed inhibition of the platelet activating factor (PAF) thus causing bronchocodilation, hypotension and inhibit the release of inflammatory mediators. The extract of the leaves exhibited cardioprotective, antiasthematic, antidiabetic and potent CNS activities (Pietri *et al.*, 1997; Pinakini *et al.*, 2005, Tighilet and Lacour, 1995). Constituents of *G. biloba* are the potent scavenger of free radicals as well as inhibit the lipid peroxidation thereby able to maintain the integrity and permeability of cell membranes. Antioxidant and hepato protective activities of *G. biloba* phytosomes in carbon tetrachloride induced liver injury in rodents have been reported by Naik and Panda (2007).

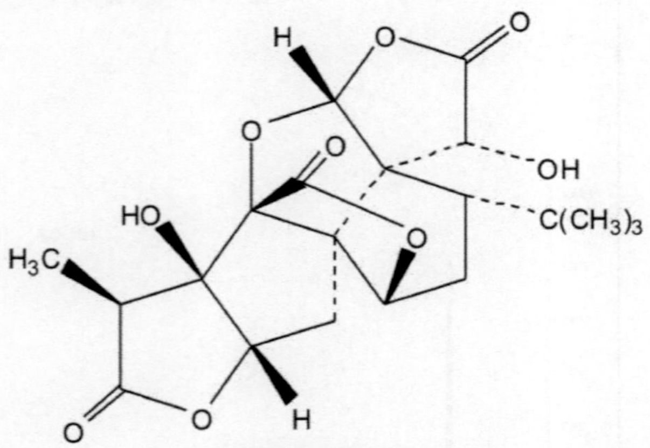

Ginkgolide B

Andrographis paniculata Nees is an erect herb of Acanthaceae family widely distributed in tropical Asian countries (South India, Malaysia, Indonesia and southern part of China, Sri Lanka). In traditional system of medicine it is used for diabetes, dysentery, cholera, influenza, bronchitis, and gonorrhoea and as a "blood purifier"(*Gupta et al.*, 1990). The major constituents are diterpene lactones (free and in glycosidic forms) including andrographolide, deoxyandrographolide, andrographiside, deoxyandrographiside and andropanoside (WHO, 2002; Sharma *et al.*, 1992).

The diterpenoids andrographolide, 14-deox yandrographolide, and 14-deoxy-11,12-didehydroandrographolide were extracted from *A. paniculata* and analyzed by HPLC (Hewlett-Packard). High resolution and sharp symmetrical peaks of the diterpenoids were obtained using a LiChrospher RP-18 re-versed-phase column (125×4 mm i.d.) eluted isocratically with 50-55 per cent methanol in water at a f low rate of 1.2 ml/min. The retention times of andrographolide, 14-deoxyandrographolide, and 14-deoxy-11,12-didehydroandrographolide were 3.2, 10.5, and 11.6 min, respectively (Naik and Hule, 2009) (Figure 7.4).

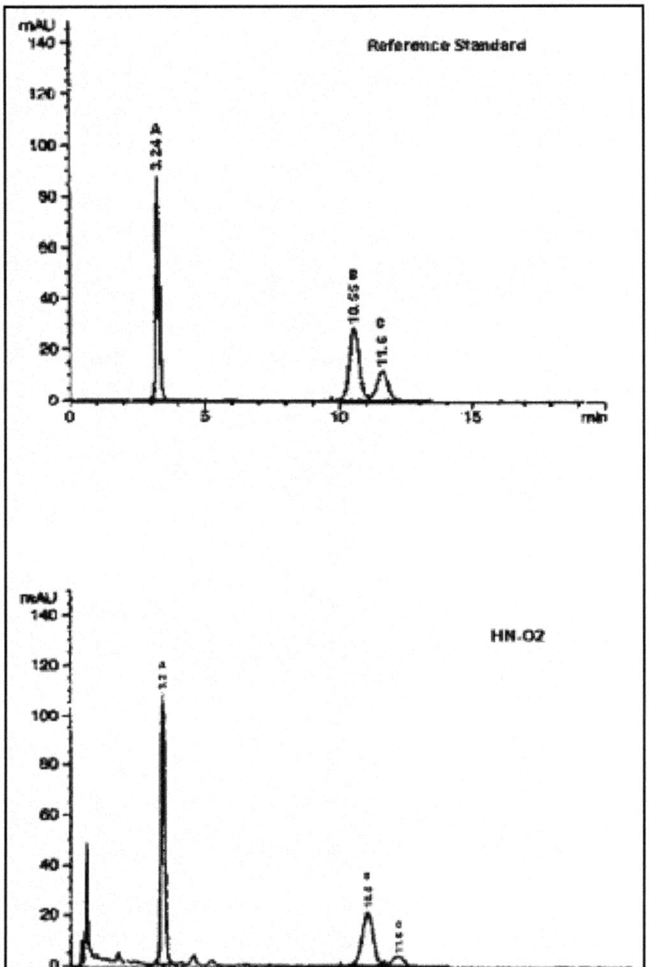

Figure 7.4: HPLC of reference standard and isolated andrographolides.

A: Andrographolide, B: 14-deoxyandrographolide, C: 14-deoxy-11,12-didehydro-andrographolide.

Isolation of andrographolides

The air-dried and powdered whole plant of *A. paniculata* was extracted with methanol under reflux and the combined extracts were evaporated in vacuo to yield dark brown residue. This was fractionated by its solubility in chloroform, dimethyl ether and methanol respectively. The chloroform fraction was evaporated to dryness in vacuo and defatted with n-hexane and then separated by silica gel column chromatography using n-hexane-ethyl acetate step gradient as eluents to get andrographolide, 14,deoxyandrographolide and 14,deoxy-11,12-didehydroandrographolide (Rao *et al.*, 2004).

Immunomodulatory activity was evaluated using both *in vitro and in vivo* methods for standardized extract (AP extract) containing andrographolide (88 ± 5 per cent) and 14,deoxyandrographolide + 14,deoxy-11,12-didehydroandrographolide (12 ± 5 per cent). Mice treated with cyclophosphamide and different doses of standardized extract of *A. paniculata* showed a significant dose related suppression of paw edema induced by sheep RBC challenge in mice (delayed hypersensitive reaction), rise in hemagglutnin antibody titre, and decreased antibody secreting cells in mouse spleen (Plaque Forming Colony). The phagocytic activity of reticulo-endothelial system (RES) was demonstrated by enhanced carbon clearance in mice treated with AP extract (Naik and Hule,2009). Ethanolic extract of aerial parts (25mg/kg, p.o.) or purified andrographolide (1mg/kg, p.o.) to mice stimulated antibody production and delayed type hypersensitivity reactions to sheep red blood cells. The extract also stimulated non-specific immune response in mice, measured by macrophage migration index and proliferation of splenic lymphocytes (Puri *et al.*, 1993).

Intragastric administration of deoxyandrographolide, andrographolide, neo-andrographolide and 11,12-didehydrodeoxyandrographolide to mice inhibited xylene induced increased cutaneous capillary permeability and reduced acute exudation in Selye's granuloma pouch induced by croton oil. 14-dexoy-11,12-didehydrodeoxyandrographolide found to be the most potent anti-inflammatory agent (Chang and But, 1986).

A hot aqueous extract parts of *A. paniculata* reduced the percentage of HIV antigen-positive H9 cells. Dehydroandrographolide inhibited HIV-1 infection of H9 cells and human lymphocytes. Methanolic extract of the leaves suppressed syncytia formation in co-cultured uninfected and HIV-1 infected MOLT cells (Chang and Yeung, 1988; Otake *et al.*, 1995). A 50 per cent ethanolic extract of aerial parts inhibited the growth of *Plasmodium berghei*- both in *in vitro* and *in vivo* experiments (Misra *et al.*, 1991). Andrographolide and neoandrographolide were also showed antimalerial activity at 5 and 2.5mg/kg respectively (Misra *et al.*, 1992). Ethanol extract of *A. paniculata* exhibited inhibitory effect on human hepatic cytochrome enzyme (CYP_{2C19}) involved in the metabolism of many drugs (Yan *et al.*, 2011).

Taxus brevifolia Nutt is evergreen small tree belonging to family Taxaceae. It is native to northwest United States. The other similar species, *Taxus wallichiana* is found in Himalayan region of India (Gupta *et al.*, 2005a). Paclitaxel (taxane-type diterpenes) is the most clinically effective anticancer drug. Paclitaxel was isolated from this species and widely used in the management of metastatic carcinoma of the

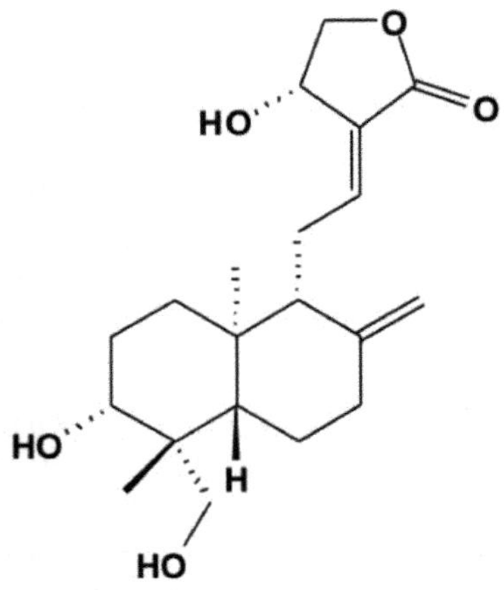

Andrographolide

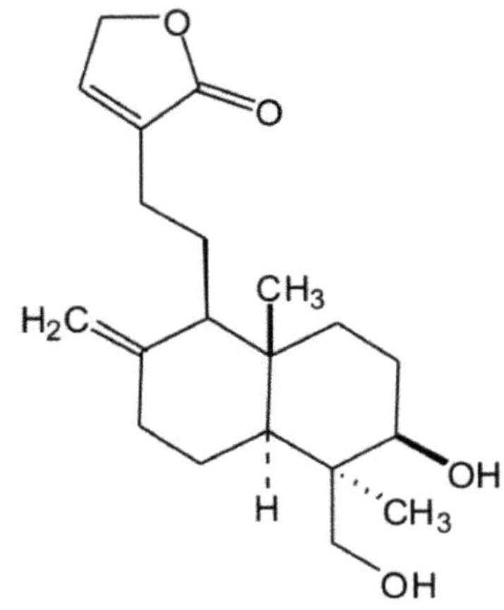

14-deoxyandrographolid

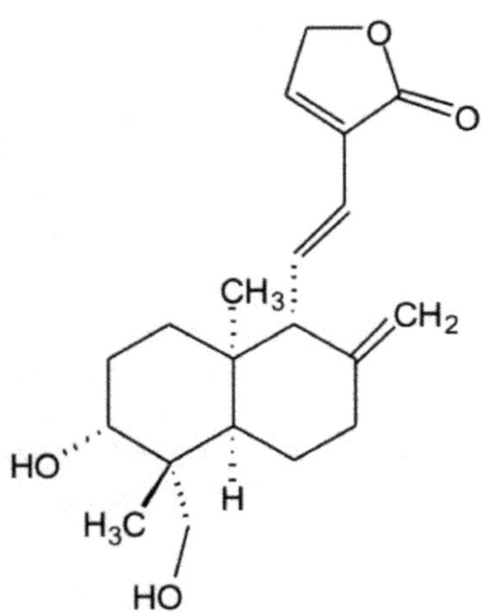

11,12-didehydrodeoxyandrographolide

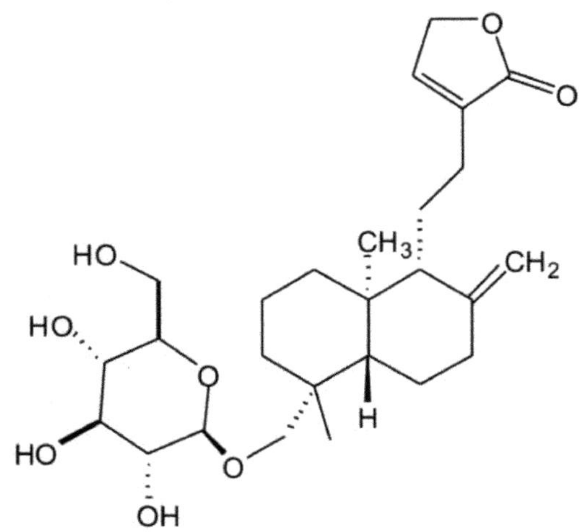

Neoandrographolide

ovarian glands and breast cancer. During the past two decades, extensive phytochemical studies have been undertaken in search of more effective derivatives of Paclitaxel. As a result, more than 500 taxane-type diterpenes have been isolated from various *Taxus* species (Baloglu and Kingston, 1999).

Paclitaxel TLC identity test can be performed on silica gel precoated plates as stationary phase and chloroform: acetone: methanol (8: 1.5: 0.5 v/v) as mobile phase. Paclitaxel appear as blue spot (R_f= 0.46) (Figure 7.5) after derivatization with anisaldehyde-sulphuric acid reagent (Gupta *et al.*, 2005a).

Hedychium spicatum is rhizomatous plant belonging to Zingiberaceae family and

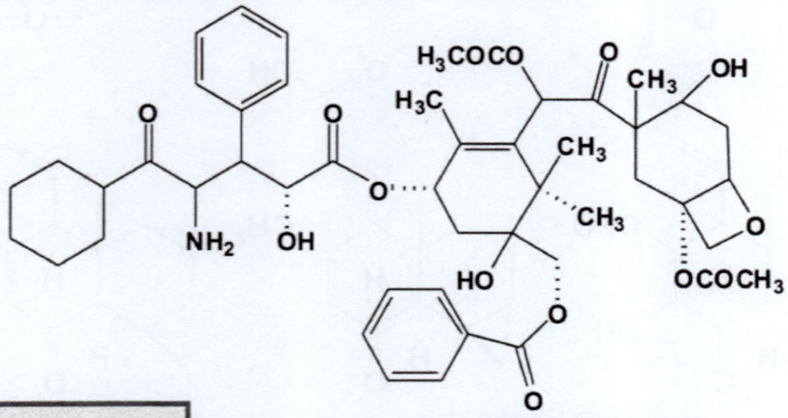

Taxol (Paclitaxel)

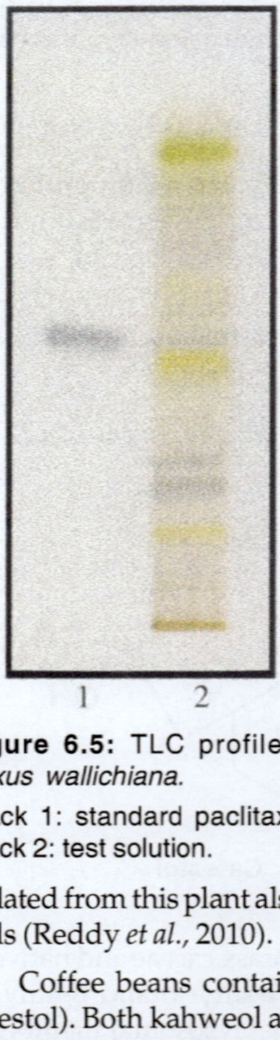

Figure 6.5: TLC profile of *Taxus wallichiana.*

Track 1: standard paclitaxel; Track 2: test solution.

is widely distributed in central Himalaya at 1100–2500meter altitude, East India, and hills of South India. Because of its camphor-like smell and a strong aromatic odour, it is locally called as 'kapurakachri'. The rhizomes are popular as an insect repellants and having tobacco smell (WOI, 1959). Essential oils of the rhizomes are used for the treatment of skin diseases, stomach ailments (Chopra *et al.*, 1956), analgesic (Tandon *et al.*, 1997), anti-emetic, blood purifier, anti-inflammatory (Itokawa *et al.*, 1988), antimicrobial (Srimal *et al.*, 1984), antidote for snake bite, pediculicidal (*in vitro*) (Jigna *et al.*, 2006), mild sedative activity of short duration (Jadhav *et al.*, 2007), anthelmintic activity (Dixit *et al.*, 1979) and cytotoxic activity (Akhtar *et al.*, 2000). Two new labdane diterpenoids, 7-hydroxy hydichinal and spicatanoic acid has been isolated and showed promising cytotoxic activity against the THP-1 (human acute monocytic leukemia), HL-60 (human promyelocytic leukemia), A-375 (human malignant melanoma) and A-549 (human lung carcinoma) cancerous cell lines (Reddy *et al.*, 2009). In addition, hedychinone, a labdane diterpenoid isolated from this plant also showed potent *in vitro* cytotoxic activity against cancerous cells (Reddy *et al.*, 2010).

Coffee beans contain considerable amount of two diterpenes (kahweol and cafestol). Both kahweol and cafestol increases the blood cholesterol level in humans and animals (De Roos *et al.*, 1999). However, animal studies have demonstrated protective activity of kahweol and cafestol against well-documented carcinogens in animal models (Huber *et al.*, 1997; Cavin *et al.*, 2001). The epidemiological evidence

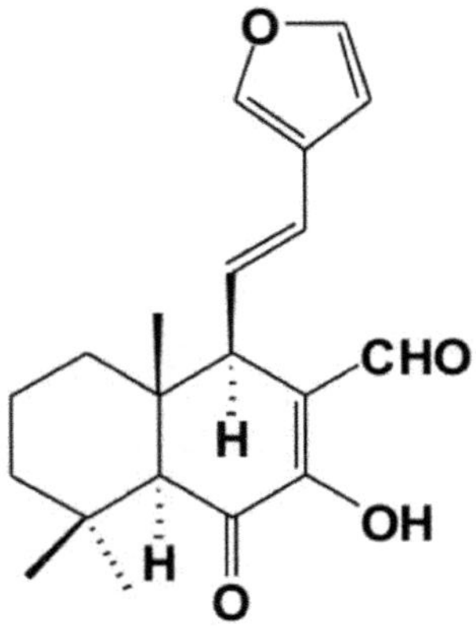

7-hydroxy

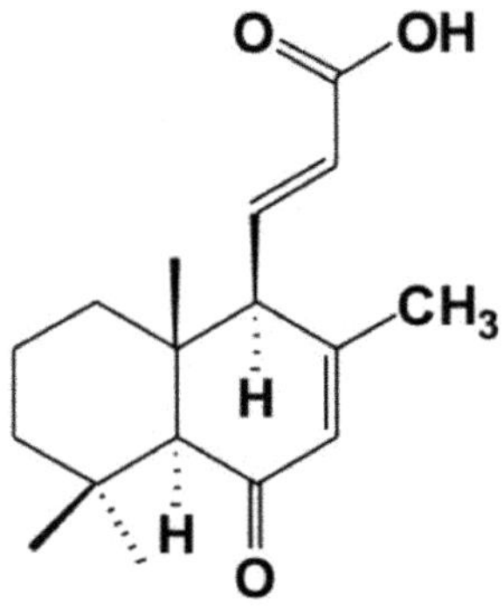

Hydichinal

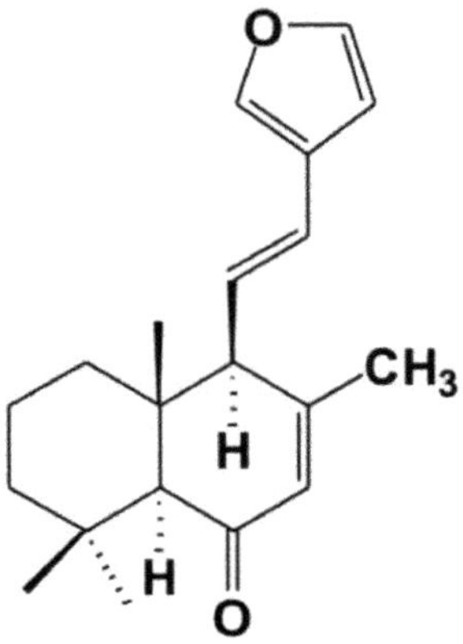

Spicatanoicacidhedychinone

suggest the consumption of coffee with a high amount of kahweol and cafestol minimizes the occurrence of colon cancer in humans, which is one of the most common cancers found in the western world (Giovannucci *et al.,* 1998). The chemoprotective activity of kahweol and cafestolis related to the beneficial modifications of the xenobiotic metabolism (Cavin *et al.,* 2001, Huber *et al.,* 2002). Both kahweol and cafestol have been demonstrated experimentally a potent hepato-protective agent in carbon tetrachloride induced hepatic damage in mice. The hepatoprotective activity of kahweol and cafestol attributed due to their ability to prevent the biodegradation of CCl_4 by inhibiting microsomal enzyme (CYP_{2E1}) activity resulting decreased formation of trichloromethyl radicals. Both have exhibited antioxidant activity along with free radicals scavenging and prevention of lipid peroxidation (Lee *et al.,* 2007).

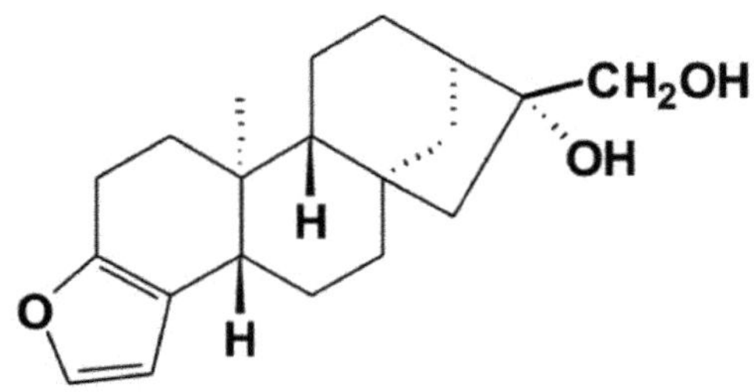

Kahweol

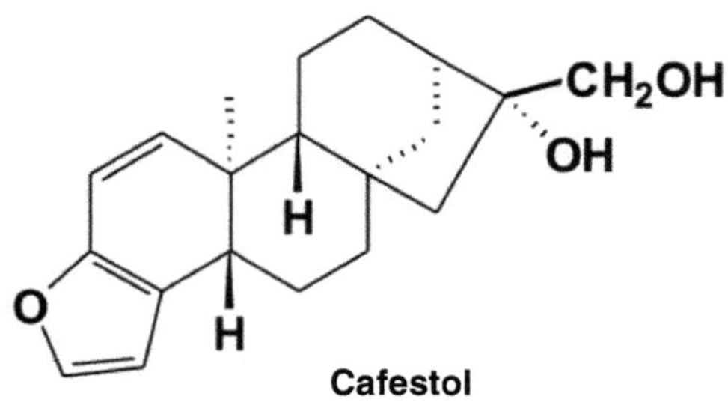

Cafestol

Triterpenoids

Brassica oleracea L. (Wild cabbage) belonging to family Brassicaceae and native of coastal southern and western Europe, contains Lupeol, α-amyrin and β-amyrin. (Martelanc *et al.,* 2007). Lupeol and its esters protect against lysosomal membrane damage provoked by cyclophosphamide, which is widely used as an alkylating agent in cancer chemotherapy (Sudharsan *et al.,* 2006). Lupeol itself exhibits antitumor

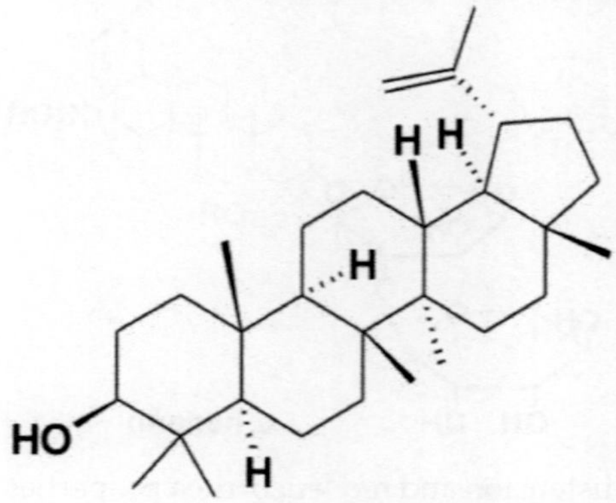

Lupeol

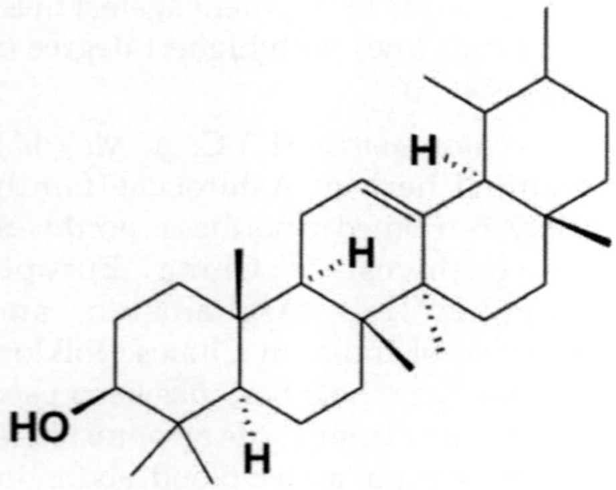

α-amyrin

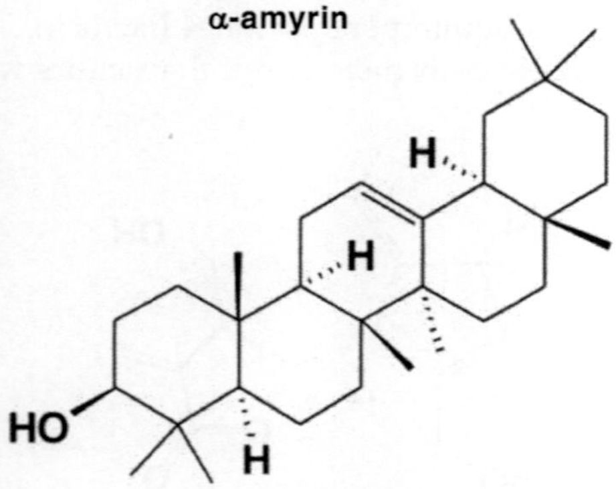

β-amyrin

activity in murine skin and may inhibit tumor cellular processes (Saleem *et al.*, 2001). Mixtures of α- and β-amyrins exhibited gastroprotective (Oliveira *et al.*, 2004), antipruritic (Oliveira *et al.*, 2004a) and hepatoprotective activity against acetoaminophen-induced hepatotoxicity (Oliveira *et al.*, 2005). Alpha-amyrin also possesses anti-inflammatory properties (Otuki *et al.*, 2005). The chemical composition is largely responsible for antioxidant activity hence can be used in neutraceutical preparations as well as in human daily diet (Pdsedek *et al.*, 2007).

Polygala japonica Houtt belonging to family Polygalaceae is widely distributed in Asia, especially in Eastern China. In Chinese traditional medicine it is used for the treatment of various inflammatory disorders, such as acute tonsillitis, pharyngitis, myelitis and nephritis (Liu and Li, 1998; Teng and Teng, 1994; Wan *et al.*, 1997). Triterpene saponins and flavones were isolated from *Polygala japonica* (Liu and Li, 1989; Keun *et al.*, 1993). New triterpene glycoside isolated from *Polygala japonica* showed anti-inflammatory activity in carrageenan induced rat paw edema by inhibiting the release of histamine and prostaglandin (Wang *et al.*, 2008).

Clematis ganpiniana, is rhizomatous plant belonging to family Ranunculaceae. The roots and rhizomes are used in Chinese medicine as an analgesic, abirritative, antibacterial, antiphlogistic, anticancer and diuretic agent (Chinese Pharmacopocia, 2005). Four triterpene glycosides (3β-[(α-L-arabinopyranosyl)-oxy]olean-12-en-28-oic acid, hederagenin-3β-*O*-α-L-arabinopyranoside, 3 β-*O*-α-L-rhamnopyranosyl-(1→2)-α-L-arabinopyranosyloleanolic acid and α-hederin) were isolated. These isolates showed significant inhibitory activities against cancer cell lines (MCF-7, MDA-MB-231) with IC_{50} (0.7–16.5 µg/ml), and significant apoptosis was observed in these cell lines (Ding *et al.*, 2009). Amongst these four triterpene glycosides, α-hederin found to be the most potent against these cancer cells lines with highest degree of apoptosis.

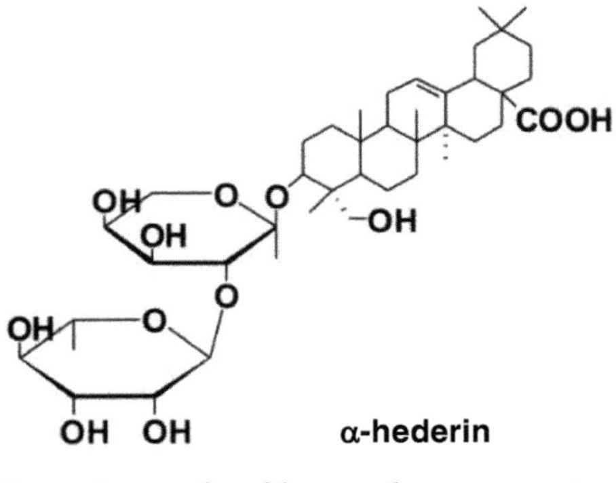

α-hederin

Lactuca tatarica (L.) C. A. Mey is a perennial herb of Asteraceae family mainly distributed in northeast, northwest and southwest of China, Europe, Mongolia, Iran, Afghanistan, and northwest of India. In Chinese folklore medicine the whole herb has been used for the treatment of appendicitis, erysipelas, extravasated blood, abdominal distension and red leucorrhea properties (Wu *et al.*, 1990). The chemical studies on this plant, lead to isolation of triterpenoids and sesquiterpene lactones (lactucin, 11β,13-dihydrolactucin, lactucopicrin) and subsequently their chemical structures were elucidated. Furthermore, lanostane type

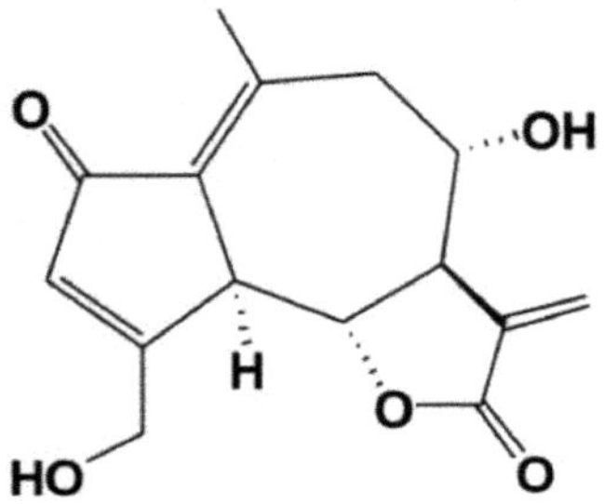

Lactucin

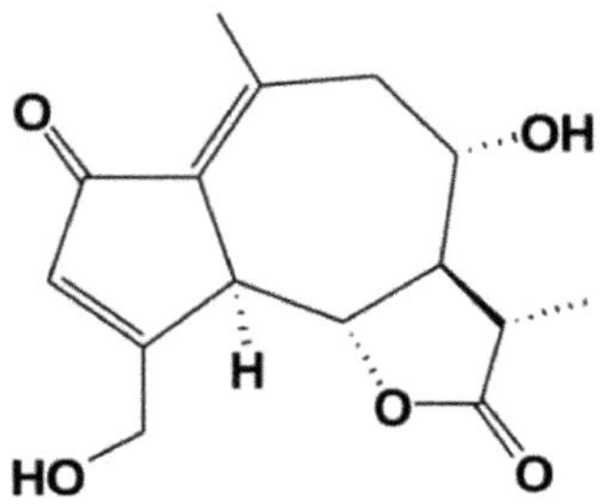

11β,13-dihydrolactucin

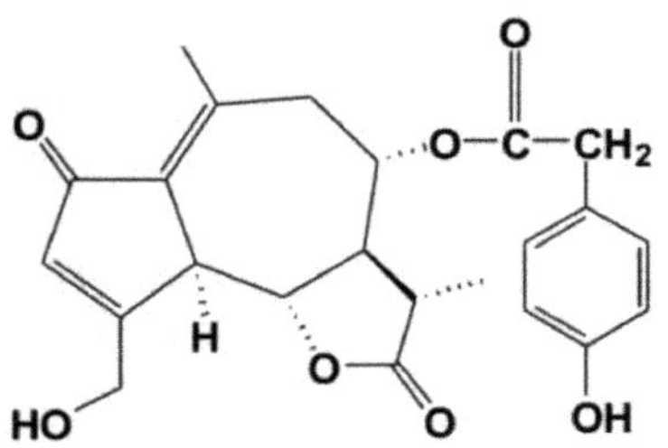

Lactucopicrin

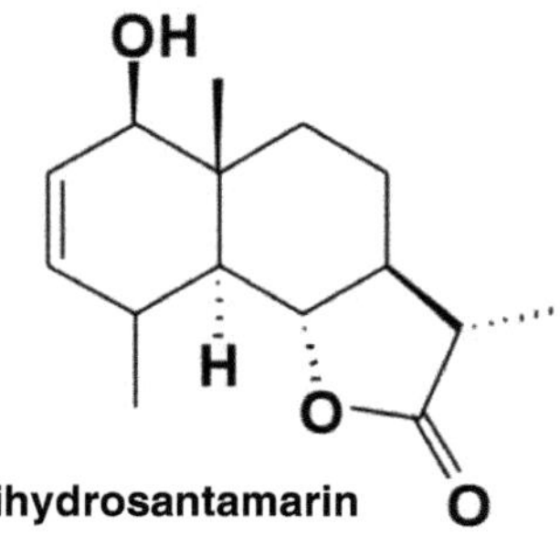

Dihydrosantamarin

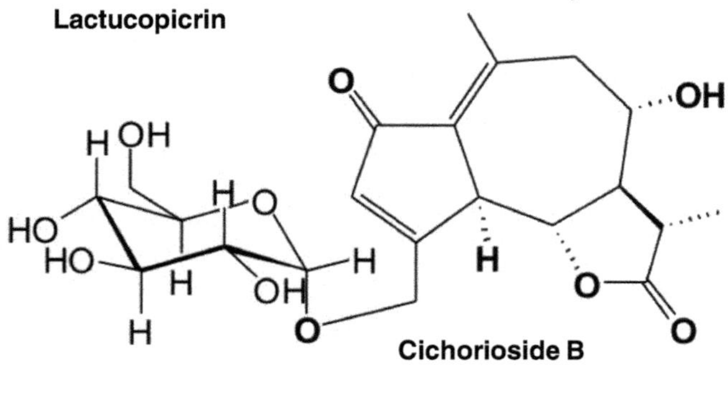

Cichorioside B

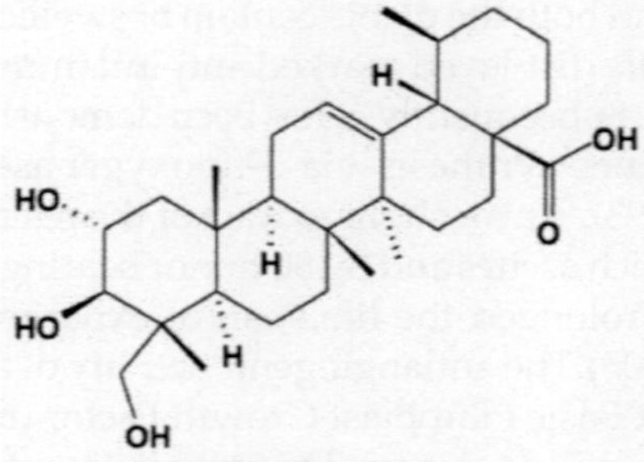

Asiatic acid

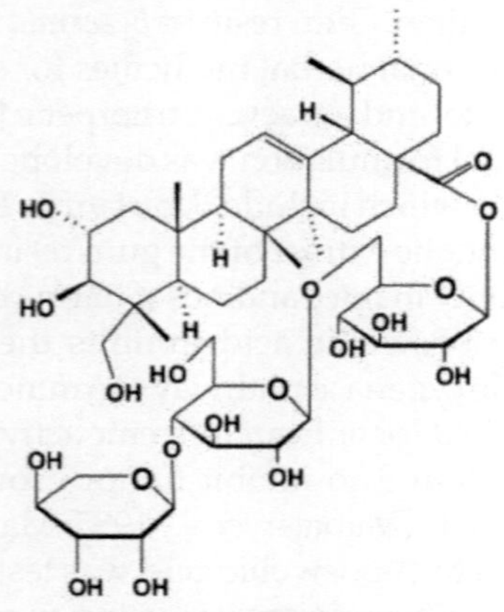

Asiaticoside

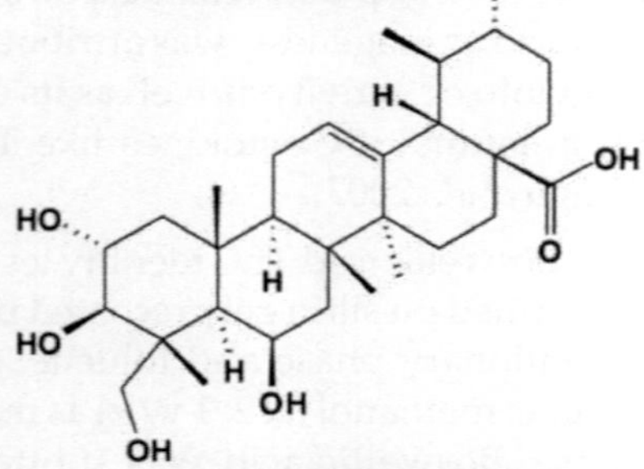

Madecassic acid

triterpenoids elicited antibacterial activity whereas the guaiane type sesquiterpenoids elicited significant cytotoxic activity (Wang *et al.*, 2006).

Centella asiatica Linn. Urban, is a small herbaceous annual plant belonging to the family Apiaceae. It is native to India, Sri Lanka, Northern Australia, Indonesia and many parts of Asia (Miyako *et al.*, 2005). In Indian popular medicine herb *C. asiatica* has long been used for the treatment of wound healing, mild diuresis, boosting memory, improving the ability of degree of concentration, alertness, and also anti-anxiety and anti-stress activity (Rao *et al.*, 1973). The medicinal properties of *C. asiatica* have been ascribed due to three triterpenoids: asiatic acid, asiaticoside and madecassic acid (Inamdar *et al.*, 1996). Dichloromethane extract of *C. asiatica* demonstrated moderate to marked inhibition on human cytochrome enzyme (CYP_{2C19}). Therefore a proper care should be taken when *C. asiatica* extracts is co-administered with CYP_{2C19} substrates such as omeprazole, barbiturates, citalopram, diazepam etc (Yan *et al.*, 2011).

Terminalia pallid Brandis (TP) is a small evergreen endemic tree belongs to family Combretaceae (Chadha *et al.*, 1976). TP is found in South Eastern Ghats of India (Tirumala hills, Rayalaseema region, Andhra Pradesh India). Tribal people use TP fruit for the treatment of diabetes and decoction to cure diarrhea (Nagaraju and Rao, 1989; Nagaraju *et al.*, 1992; Rai *et al.*, 1995). The phytochemical screening of TP fruit revealed the presence of flavonoids, phenolic acids, sterols/triterpenoids, alkaloids, tannins and anthocyanins. The oral administration of ethanolic extract exhibited a significant antihyperglycemic activity in alloxan diabetic rats, via stimulation of remnant β-islets (Rao *et al.*, 2003).

Boswellia serrata (Burseraceae) is medium or large tree found in India in Vindhyan and Satpura hills of deccan, Bihar, Odisha, North Gujarat and Aravali hills of

Rajasthan. Gum resin of *B. serrata* is regularly used to treat inflammatory disorders in traditional herbal medicines for several years. The plant contain diterpene alcohol serratol and tetracyclic triterpene β-boswellic acid (Gupta *et al.,* 2005b). The traditional herbal formulation was developed to treat inflammatory diseases both in China and India which include *B. carterii* or *B. serrata* and both the plants contain boswellic acid. Alcoholic extract of the gum resin of *B. serrata* displayed marked anti-inflammatory activity in mice and rats (Sharma *et al.,* 1989). Subsequently, it has been demonstrated that boswellic acid inhibits the leukotrienes synthesis via 5-lipoxygenase and cycloxygenase pathways (Ammon *et al.,* 1993). An alcoholic extract of the resin was studied for anticarcinogenic activity on Ehrlich ascites and S-180 tumor bearing mice and found to inhibit tumor growth and prolonged the life span of experimental animals (Mahajan *et al.,* 1995; Marc *et al.,* 2005).The antiangiogenic activity of acetyl 11-keto-β-boswellic acid was tested against Basic Fibroblast Growth Factor (bFGF) induced angiogenesis using *in vivo* Matrigel Plug Assay. The inhibitory effect of acetyl 11-keto-β-boswellic acid on bFGF-induced angiogenesis was attributed to its cytotoxic activity as well as its effect on inflammatory cytokines like TNFα (Singh *et al.,* 2007).

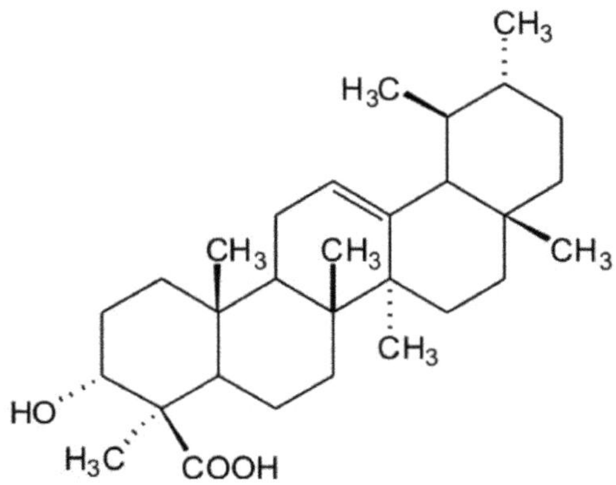

Boswellic acid

Boswellic acid TLC identity test was performed on silica gel precoated plates as stationary phase and toluene: ethyl acetate: methanol (8: 2: 1 v/v) as mobile phase. Boswellic acid was spotted as yellowish pink (R_f= 0.53) (Figure 7.6) after derivatization with anisaldehyde-sulphuric acid reagent (Gupta *et al.,* 2005b).

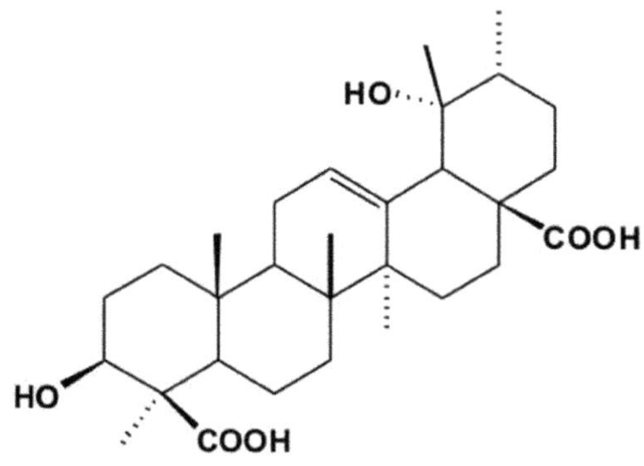

Ilexgenin A

Ilex hainanensis, Ilex stewardii and *Ilex pubescens* is regularly used as herbal tea in China. Two triterpenoids, ilexgenin A and ilexsaponin A were isolated from these *Ilex* species. The major components from ethanolic extract of this plant were tested (*in vitro*) against oral pathogens, *Streptococcus mutans* and *Actinomyces viscosu.* Of these two, Ilexgenin A was found to be most effective with MIC (7.8 and ≤3.9 μg/ml) against *Streptococcus mutans* and actinomyces respectively. Antibacterial activity of Ilexgenin A, may be useful to reduce dental plaque biofilm formation. Dental plaque biofilm plays an important role in the progression of dental disorders. Formation of dental plaque biofilm is a biological process associated with the attachment, detachment and proliferation of oral bacteria on the surface of tooth (Chen *et al.,* 2011).

Saussurea petrovii is a perennial herbaceous plant of Compositae family. It grows mainly in North China. Its rhizome has been used as a traditional Chinese medicine for the treatment of rheumatism and bleeding since ancient times (Ma *et al.*, 1982). Its phytochemical or pharmacological properties have not been systematically studied till date. In the acetone extract of *S. petrovii*, two new triterpenoids [taraxast-20-ene-3β,30-diol **(1)** and 20α,21α-epoxy-taraxastane- 3β,22α-diol**(2)**], as well as four known triterpenes[taraxast-20(30)ene-3β,21α-diol, taraxast-20(30)-ene-3β-ol,taraxast-20-ene-3β-oland taraxastane-3β,20α-diol**(3)**were reported (Su *et al.*, 2000). The result of the evaluation of antitumor and antibacterial activities of these triterpenoid and triterpenes indicated that compound **(1)** and **(2)** possesses marked cytotoxicity against mouse melanoma (B16) cells and compound **(1)** and **(3)** elicited significant antibacterial activity against *Bacillus subtilis, Escherichia coli* and *Staphylococcus aureus (in vitro)* (Dai, 2001).

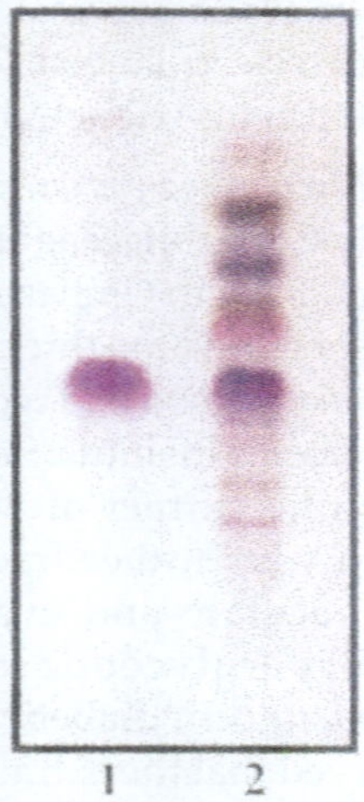

Figure 7.6: TLC details of methanolic extract of oleo gum resin of *B. serrata*.

Track 1: standard boswellic acid; Track 2: test sample.

Ganoderma lucidum (Fr.) P. Karst belongs to the family of Ganodermataceae (Ziegenbein *et al.*, 2006), and its clinical use can be traced back to 100 BC (Leung and Lin, 2002) to treat and cure many types of diseases, and it was considered as an elixir that could revive the dead. It was reported that *G. lucidum* possesses antimicrobial (Yoon *et al.*, 1994; Wang and Ng, 2006), antiviral especially (anti-HIV activity) (Min *et al.*, 1998), anti-tumor (Liu *et al.*, 2009; Müller *et al.*, 2006), and anti-aging activity (Shieh *et al.*, 2001).The isolated triterpenoids showed promising cytotoxicity against

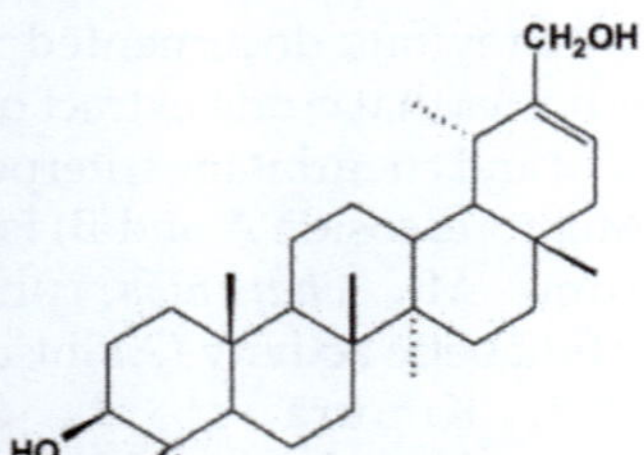

1

taraxastane-3β,20α-diol

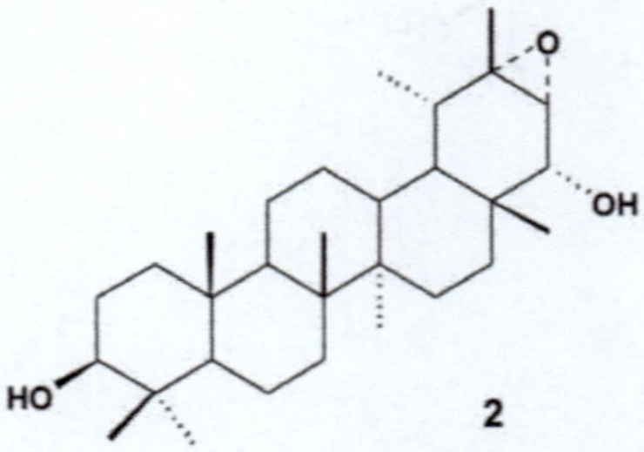

2

20α,21α-epoxy-taraxastane-3β,22α-diol

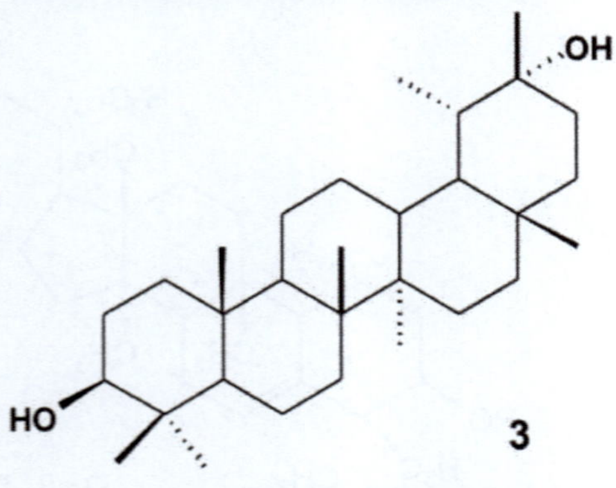

3

Taraxast-20-ene-3β,30-diol

human HeLa cervical cancer cells. Among these triterpenoids studied, 15α,26-dihydroxy-5α-lanosta-7,9,24(*E*)-tri-en-3-one, lucidadiol and ganoderiol F showed marked cytotoxic activities (IC_{50} 1, 5 and 8 μM, respectively) (Cheng *et al.*, 2010).

Momordica charantia Lin. is widely cultivated as a vegetable crop belonging to family Cucurbitaceae and grow in many tropical and subtropical countries. Its fruit called *kugua* in Chinese or *bitter melon* in English and *karela* in Hindi, is herbaceous climber cultivated throughout India. *M. charantia L.* (Cucurbitaceae) has been used as a functional food and recommended for diabetes mellitus. Standardized ethanolic extract of *M. charantia* which was further fractionated in ethyl acetate and evaluated for antihyperglycemic activity in streptozotocin diabetic rats. It was observed that the extract decreased the blood glucose, increased glucose uptake by muscles and enhanced utilization of glucose by liver tissue via glycogenesis and gluconeogenesis (Fernandes *et al.*, 2007; Naik *et al.*, 2009a).

Previous documented report indicates that crude extract of fruit, stem and cucurbitane triterpenoids (Momordicoside A and B) isolated from *M. charantia* showed antidiabetic activity (Akihisa *et al.*, 2007; Kimura *et al.*, 2005). Cucurbitanetriterpenoids (Kuguacin-C and Kuguacin-E) isolated from *M. charantia* was also

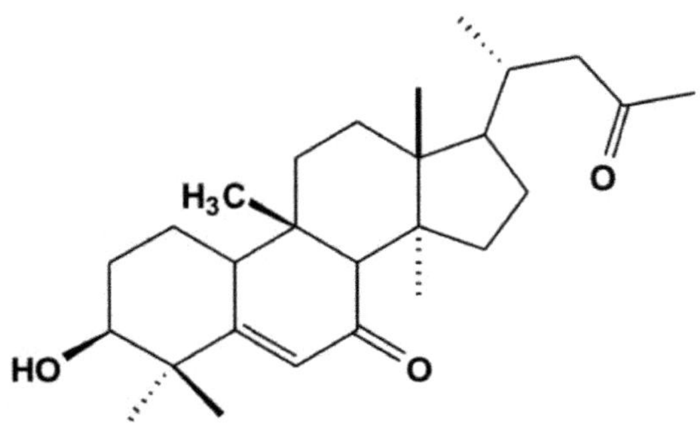

Kuguacin C

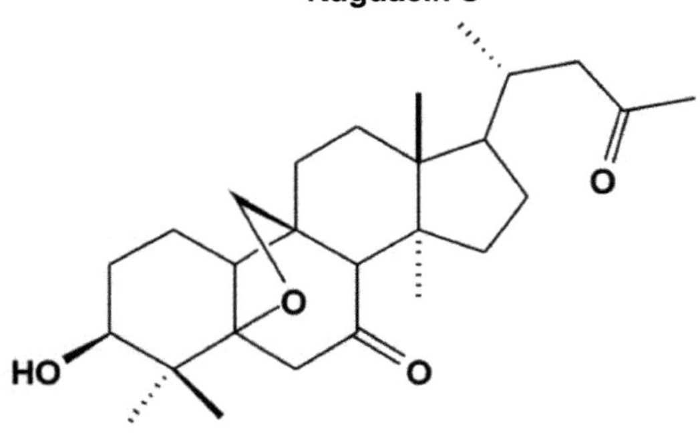

Kuguacin E

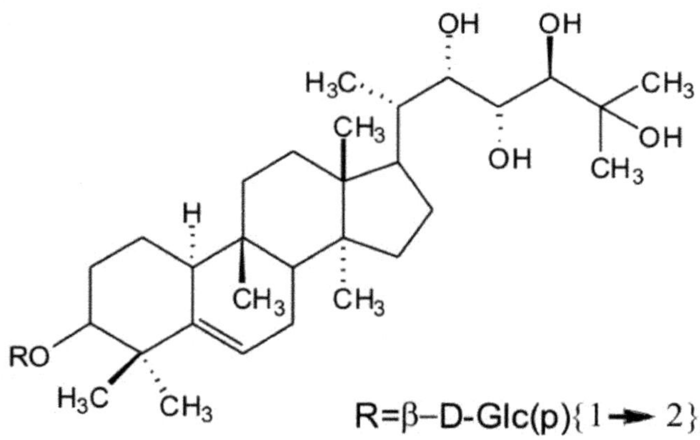

R=β–D-Glc(p){1→2}β–D-Glc(p)

Momordicoside A

studied and demonstrated moderate anti-HIV-1 activity with EC_{50} (8.45 and 25.62 μg/ml) and exert minimal cytotoxicity against C8166 cells (IC_{50}> 200μg/ml), with a selectivity index more than 23.68 and 7.81, respectively (Chen *et al.*, 2008).

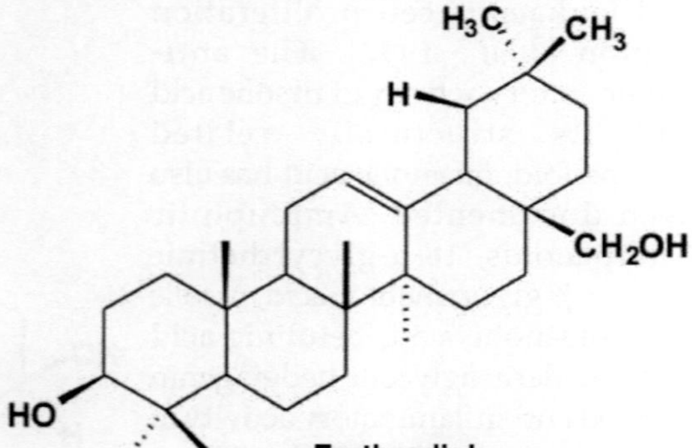

Erythrodiol

Natural triterpenoids betulin, betulinic acid, erythrodiol and ursolic acid were tested on 12-*O*-tetradecanoylphorbol-13-acetate multiple-dose model of skin chronic inflammation in Swiss albino mice. Among the tested triterpenoids, erythrodiol and ursolic acid were found to be more active than others. This mice model is reported to be a specific model for the evaluation of topical agents for proliferative inflammatory conditions such as psoriasis (Stanley *et al.*, 1991).

The acidic triterpenoids represent a particular subset of triterpenoid compounds and reported to elicit variety of pharmacological activities. The reported pharmacological activities of such acidic triterpenoids include cytotoxic, anti-tumour and anti-inflammatory (Ying *et al.*, 1991; Najiid *et al.*, 1992; Stormer *et al.*, 1993) and anti-viral activity (Hudson *et al.*, 1990). In particular, the acidic triterpenoid, ursolic acid inhibited lipoxygenase and cyclo-oxygenase *in vivo* (Najiid *et al.*, 1992), histamine release from mast cells (Tsuruga *et al.*, 1991), mutagenicity (Niikawa *et al.*, 1993)

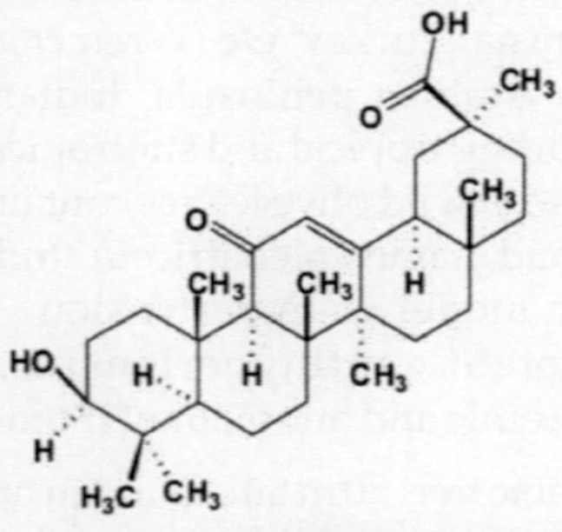

18-α-Glycyrrhetinic acid

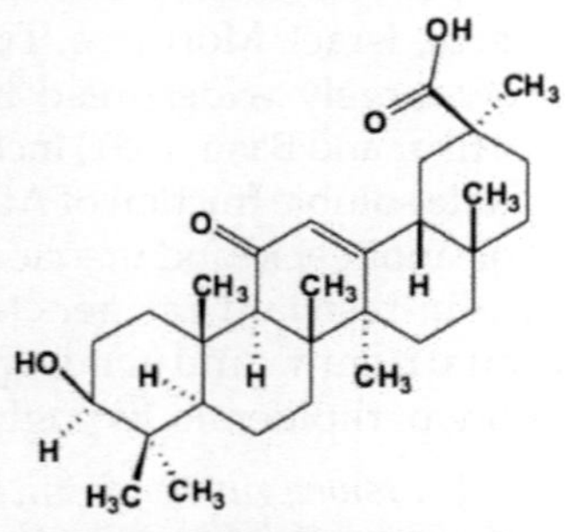

18-β-Glycyrrhetinic acid

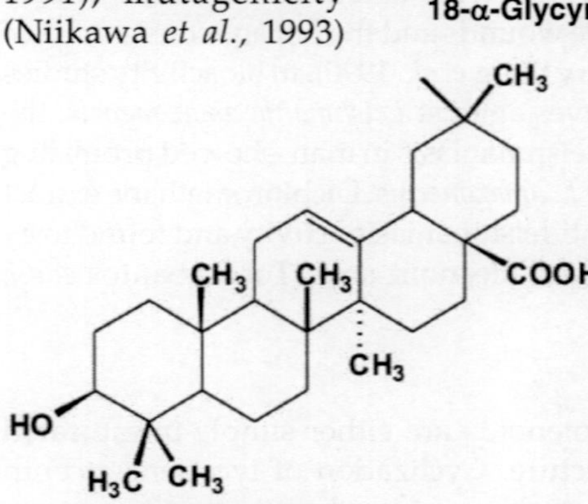

Oleanolic acid

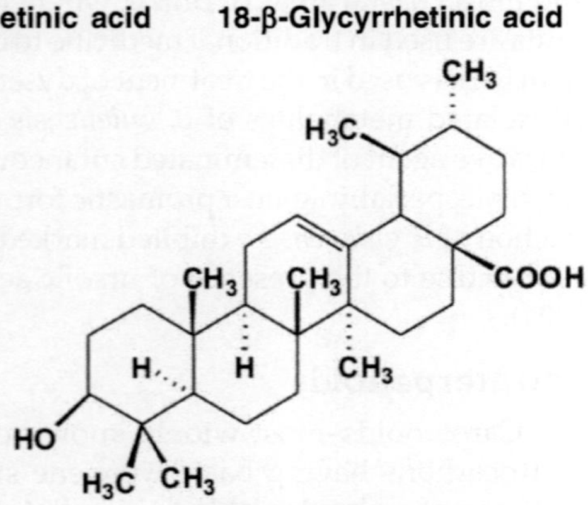

Ursolic acid

and leukaemic cell proliferation (Simon *et al.*, 1992). The anti-inflammatory activity of ursolic acid and its structurally related triterpenoid, oleanolic acid has also been documented. Amphiphilic triterpenoids, 18-α-glycyrrhetinic acid, 18-β-glycyrrhetinic acid, ursolic acid, oleanolic acid, betulinic acid and α-hederin aglycone hederagenin showed anti-inflammatory activity in different animal models. These triterpenoids found to inhibit leukocyte elastase which may be one of the key mechanism responsible for anti-inflammatory activity (Ying *et al.*, 1991). However, other mechanisms responsible for anti-inflammatory activity may be inhibition of histamine and other mediators (Tsuruga *et al.*, 1991) including prostaglandin oxidation (Lewis *et al.*, 1992). Moreover, ursolic acid and oleanolic acid known to inhibit selectively cyclic AMP-dependent protein kinases (Wang *et al.*, 1996).

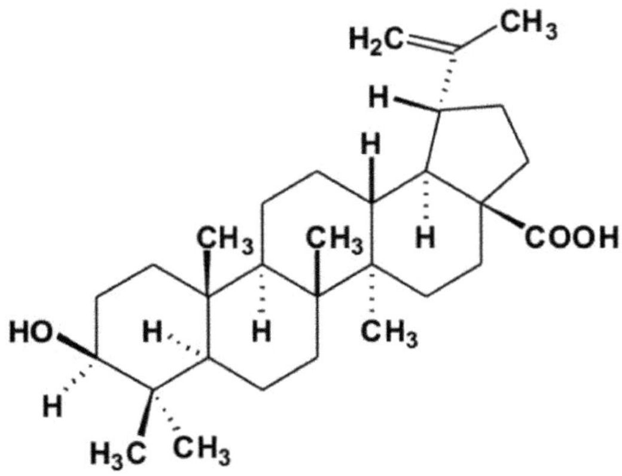

Betulinic acid

Olea europaea Lin. of Oleaceae family has been used widely in folk medicine in European Mediterranean islands and other countries such as Spain, Italy, France, Greece, Israel, Morrocco, Tunisia, Turkey, etc (Wren *et al.*, 1994). The plant is also grow largely widespread in Arabian peninsula, Indian sub-continent and Asia (Kirtikar and Basu, 1991) including tropical and subtropical parts of the world. Ethyl acetate soluble fraction of African wild olive leaves contain 0.27 per cent 1: 1 mixture of oleanolic acid and ursolic acid, named oleuafricein studied on Dahl salt-sensitive, insulin-resistant rat genetic model of hypertension. The results showed that oleuafricein induced potent antihypertensive, diuretic/natriuretic, antihyperlipidemic, hypoglycemic and antioxidant (Somova *et al.*, 2003).

Pourouma guianensis is a dioic tree, 30 m tall, found in northern and South America and most coastal areas of Brazil, where it is known as "embaúba da mata". Unripe fruits are used in traditional medicine to cure wounds and the hot aqueous extract of stem bark is used in the treatment of dysentery (Berg *et al.*, 1990). In bioactivity studies on isolated metabolites of *P. guianensis* leaves against *Leishmania amazonensis*, the causative agent of disseminated cutaneous leishmaniasis in man, showed promising activity especially against promastic form of *L. amazonensis*. Dichloromethane extract fraction of *P. guianensis* exhibited marked anti-leishmaniasis activity and found to be mainly due to the presence of ursolic acid and oleanolic acid (Torres-Santos *et al.*, 2004).

Tetraterpenoids

Carotenoids–most widely known carotenoids are either simple unsaturated hydrocarbons having basic lycopene structure. Cyclization of lycopene at both terminals of molecule yielded a bicyclic hydrocarbon commonly known as β-carotene, which occurs most abundantly in nature. Most of the carotenoids are potent natural antioxidant. Biochemically, carotene and precursor of the vitamin-A substances like β-carotenoids undergo cessation by the presence of β-carotene oxygenase in mucosal cells of intestine to give rise to retinal and get reduced by NADH to retinol. It is essential for proper vision, growth and tissue differentiation. It is responsible for the synthesis of specific glycoproteins (Kar, 2007).

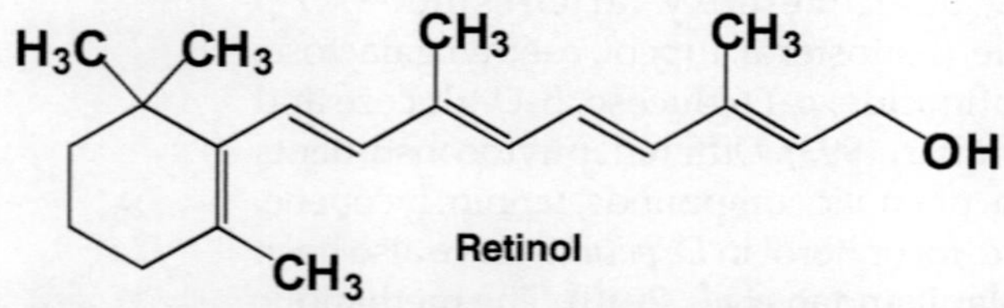

Retinol

Dipteracanthus is an under-shrub genus distributed in tropical and subtropical

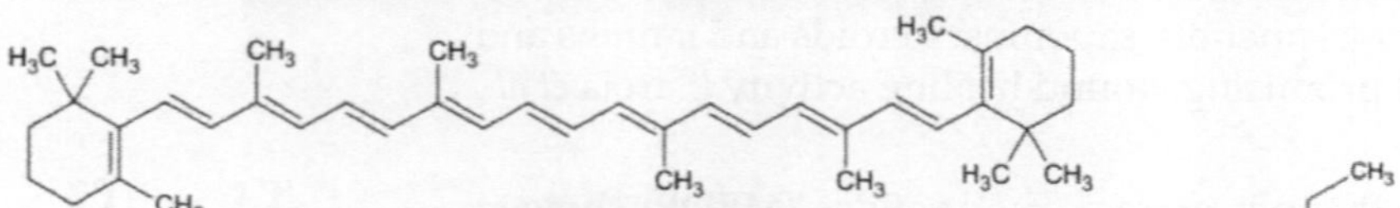

β-carotene

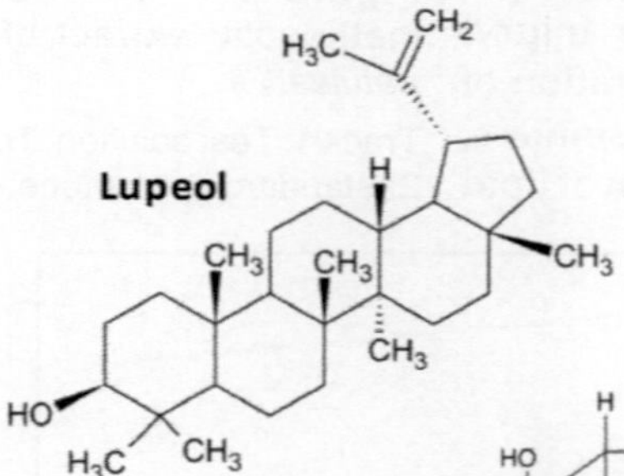

Lupeol

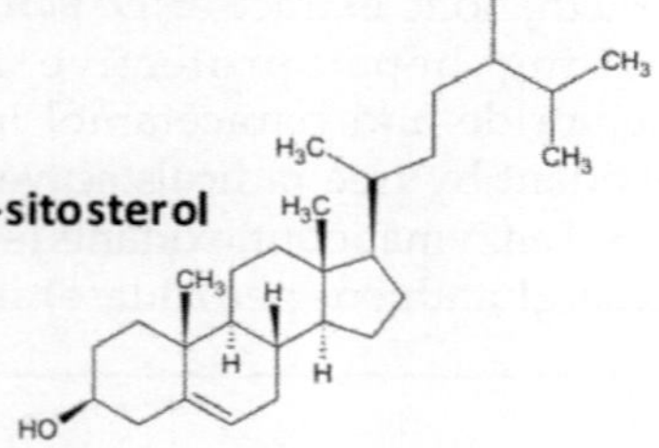

β-sitosterol

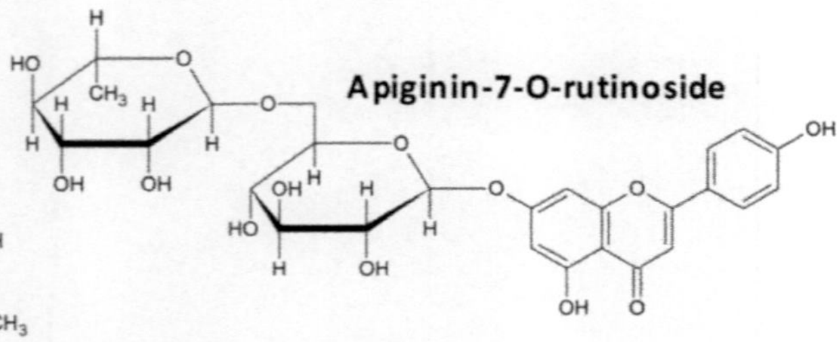

Apiginin-7-O-rutinoside

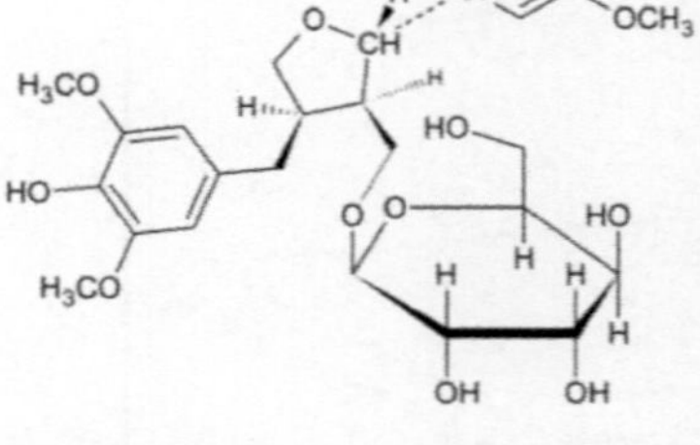

5,5'-dimethoxy lariciresinol-9-O-β-D-glucopyranoside

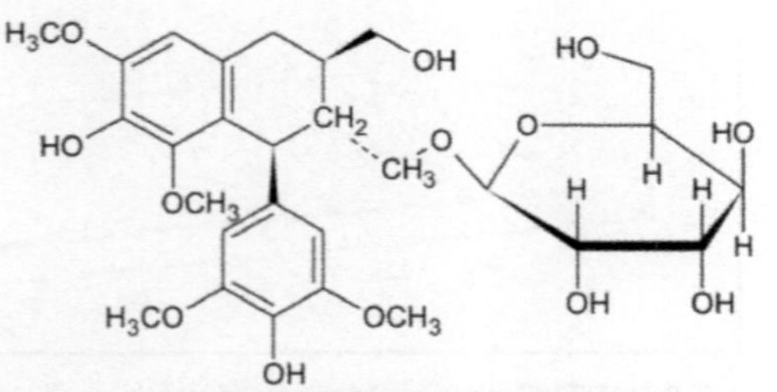

Lyoniresinol-9'-O- β-glucoside

regions and about 7 dipteracanthus species are found in India. This genus was earlier included under *Ruellia* (The WOI, 2003). The *Dipteracanthus patulus* (Jacq) nees is belonging to the family acantheaceae. *Ruellia patula* Jacq is synonym of *D. patulus* and found abundantly in Africa, Arabia and South West India, Pakistan (Jaffri, 1966).

D. patulus showed the presence of lyoniresinol-9′-O-β-D-glucoside, 5,5-dimethoxy-lariciresinol-9-O-β glucopyranoside, β-sitosterol, lupeol, α-ethyl galactose, apiginin-7-O-rutinoside, α-D-glucose, β-D-glucose and β-D-fructose (Akthar, 1993). Different phytoconstituents viz ascorbic acid, phenolic compounds, tannin, lycopene, carotenoid and α-tocopherol in *D. patulus* have also been documented (Manikandan *et al.*, 2010). The methanolic extract of *D. patulus* leaves showed the presence of flavonoids, phenols, saponins, steroids and tannins and elicited promising wound healing activity (Saroja *et al.*, 2009).

Methanolic extract of *D. patulus* (MEDP) showed promising hepatoprotective activity in carbon tetrachloride and paracetamol induced liver injury, antioxidant by free radicals scavenging, restoration of depleted enzymatic antioxidants (superoxide dismutase, catalase, glutathione peroxidase) and prevention of lipid

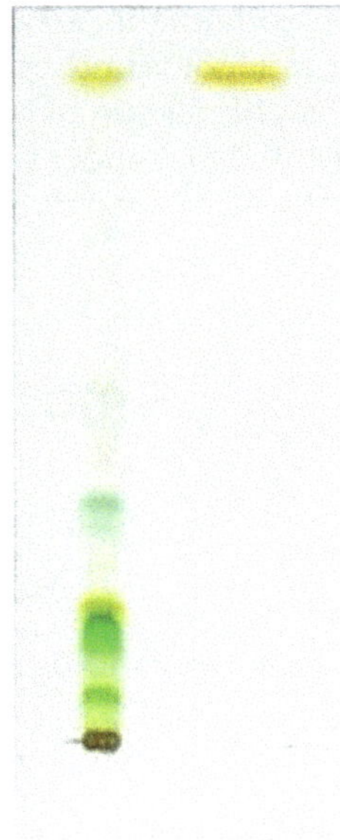

Figure 7.7: HPTLC of methanolic extract of *D. patulus*.

Track 1: Test solution, Track 2: standard β-carotene.

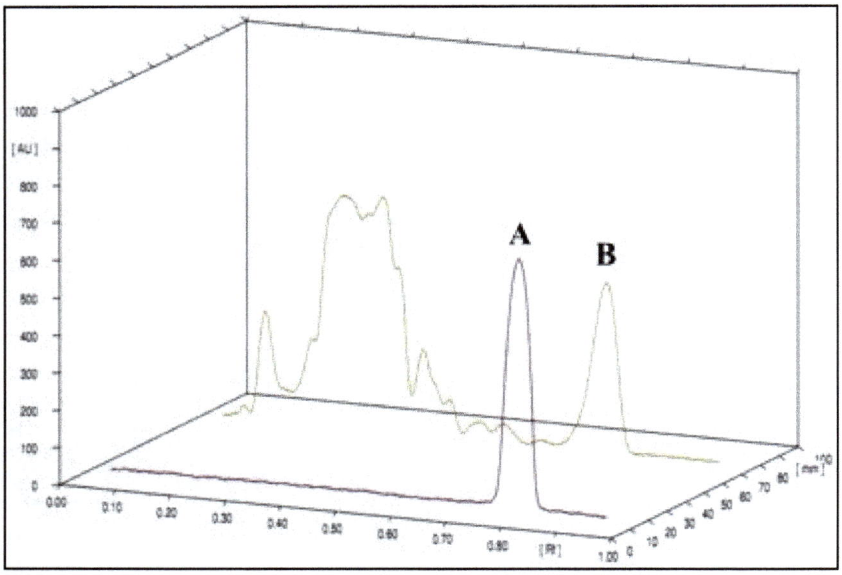

Figure 7.8: HPTLC chromatogram of methanolic extract of *D. patulus*.
(A) Standard β-carotene and (B) Test solution.

peroxidation. MEDP exhibited antimicrobial activity against gram positive (*B. subtilis, B. cereus, S. aureus, S. pyogens*) and gram negative (*E. coli, P aerogenosa,* K.) bacteria (Bumrela and Naik, 2011, 2012).

Isolation of β-carotene was carried out by column chromatography technique using alumina as stationary phase. The first yellow layer was separated with n-hexane and collected in volumetric flask. Carotene absorbs blue-green and blue light and appears yellow. Yellow fraction thus obtained was further evaporated to dryness to obtain β-carotene powder. HPTLC fingerprinting of β-carotene was carried out using CAMAG HPTLC system, silica gel aluminium TLC plates $60F_{254}$ (E-Merck, Germany) and n-hexane: acetone (70: 30 v/v) as mobile phase. At R_f 0.71, β-carotene was appeared as yellow spot in visible light (Bumrela and Naik, 2011) (Figures 7.7 and 7.8).

Conclusions

Terpenoids, secondary metabolites abundantly distributed in wide variety of plants. The present review on terpenoid cover the use and practice of plants containing terpenoids in many traditional medicines for the treatment of large number of diseases, and such traditional knowledge has been explored scientifically for the isolation and development of natural products: as functional food and potential therapeutic agents (artemisinin, forskolin, ginkolides, taxol, quassin, andrographolides and many more). Some of the terpenoids have also been employed as a tool to study and understand the role of physiological and biochemical function(s) in the etiology of diseases. The studies on terpenoids have opened up new vistas as well as avenues for the development of novel breed of phytopharmaceuticals. Such impetus on natural product research would certainly encourage and facilitate the development of new drugs in cancer and antibacterial chemotherapy, metabolic and immune disorders.

The present review will be immensely useful and stimulate the researchers, postgraduates and teachers in their endeavor as well as organizations pursuing drug research in the area of natural products.

References

Akhtar, M.S., Iqbal, Z., Khan, M.N., and Lateef, M. (2000). Anthelmintic activity of medicinal plants with particular reference to their use in animals in the Indo–Pakistan subcontinent. *Small Ruminant Res.*, **38**: 99-107.

Akihisa, T., Higo, N., Tokuda, H., Ukiya, M., Akazawa, H., Tochigi, Y., Kimura, Y., Suzuki, T., and Nishino, H. (2007).Cucurbitane-type triterpenoids from the fruits of *Momordica charantia* and their cancer chemopreventive effects.*J. Nat. Prod.*, **70:** 1233–1239.

Akthar, M.F. (1993). Chemical and biological investigations of medicinal herbs *phyla nodiflora*, *ruellia patula* and *ruellia brittoniana*. Dissertation, University of Karachi, Department of Pharmacognosy, Faculty of Pharmacy, Pakistan, pp.59.

Ammon, H., Safayhi, H., Mack, T., and Sabieraj, J. (1993). Mechanism of anti-inflammatory actions of curcumine and boswellic acids. *J Ethnopharmacol.*, **38:** 105-112.

Baloglu, E., and Kingston, D.G.I. (1999).The taxane diterpenoids [J]. *J Nat Prod.*, **62**: 1448-1472.

Bumrela, S.B., and Naik S.R. (2012). Hepatoprotective activity of methanolic extract of *Dipteracanthus patulus (Jacq) nees:* Possible involvement of antioxidant and membrane stabilization property. *Int J PharmPharmSci.*, **4(2)**: 685-690.

Bumrela, S.B., and Naik, S.R. (2011).Identification of β-carotene and β-sitosterol in methanolic extract of *Dipteracanthus patulus (Jacq) nees* and their role in antimicrobial and antioxidant activity. *Int JPhytomed.*, **3**: 204-215.

Cavin, C., Mace, K., Offord, E.A., and Schilter, B. (2001). Protective effects of coffee diterpenes against aflatoxin B1-induced genotoxicity: Mechanisms in rat and human cells. *Food Chem. Toxicol.*, **39**: 549-556.

Chadha, Y.R. (1976). The Wealth of India.Vol. 10.pp.185.

Chang, H.M., and But, P.P.H. (eds). (1986). Pharmacology and applications of Chinese material medica.Vol. 1.World Scientific, Singapore, pp.918-928.

Chang, P.C., and Wu, C.C. (1979). Herbs of Taiwan in Color, Vol. 1, Nan Cun (South Villa) Trade Co., Ltd., Taipei, ROC, pp. 99.

Chang, R.S., and Yeung, H.W. (1988). Inhibition of growth of human immunodeficiency virus *in vitro* by crude extracts of Chinese medicinal herbs. *Antiviral Res.*, **9**: 163-157.

Chen J., Renrong, T., Minghua, Q., Lu, L., Yongtang, Z., and Zhongquan, Z. (2008). Trinorcucurbitane and cucurbitane triterpenoids from the roots of *Momordica charantia*. *Phytochemistry*,**69**: 1043-1048.

Chen, X.Q., Zan, K., Yang, J., Liu, X.X., Mao, Q., Zhang, Li., Lai, M., and Wang, Q. (2011) Quantitative analysis of triterpenoids in different parts of *Ilex hainanensis*, *Ilex stewardii* and *Ilex pubescens* using HPLC–ELSD and HPLC–MS^n and antibacterial activity. *Food Chem.*, **126:** 1454-1459.

Cheng, C., Qing-Xi, Y., Zhi-Yuan, W., Xiao-Yi, S., Si-Jia, T., Xiao-Hui, W., Ping-Ping, X., Xuan, L., Shu-Hong, G., and De-An, Guo. Cytotoxic triterpenoids from *Ganodermalucidum*.*Phytochemistry*, **71:** 1579-1585.

Chinese Pharmacopoeia Comittee (Ed.). (2005). Chinese Pharmacopoeia. Chemical Industry Press, Beijing.

Chopra, R.N., Nayar, S. L., and Chopra, I.C. (1956).Glossary of Indian Medicinal Plant. C.S.I.R., New Delhi.

Croteau, R., Kutchan, T.M., and Lewis, N.G. (2000). Natural products (secondary metabolites). In Buchanan, B., Gruissem, W., Jones, R. (Eds.), Biochemistry and molecular biology of plants. American Society of Plant Physiologist, Rockville, MD, pp. 1250-1318.

Crowell, P.L., Chang, R.R., Ren, Z., Elson, C.E., and Gould, M.N. (1991).Selective inhibition of isoprenylation of 21-26 kDa proteins by the anticarcinogen *d*-limonene and its metabolites. *J Biol Chem.*, **266**: 17679-17685.

Dai, J., Zhao, C., Zhang, Q., Liu, Z., Zheng, R., and Yang L. (2001). Taraxastane-type triterpenoids from *Saussurea petrovii*. *Phytochem.*, **58:** 1107–1111.

De Roos, B., Sawyer, J.K., Katan, M.B., and Rudel, L.L. (1999).Validity of animal models for the cholesterol-raising effects of coffee diterpenes in human subjects. *Proc. Nutr. Soc.*, **58**: 551–557.

Dev, S., Gupta, A.S., and Patwardhan, S.A. (Eds.). (1989). Pentacyclic and HexacyclicTriterpenoids, vol. II, CRC Press, Boca Raton, FL, pp. 17

Devarenne, T.P. (2009). Terpenoids: Higher. In: Encyclopedia of life sciences. New Jersey: John Wiley and Sons., Ltd. pp. 1-11.

Di Renzo, G. (2000). *Ginkgo biloba* and the central nervous system. *Fitoterapia*,**71**(Suppl 1): S43-S47.

Ding Q., Yang, L., Yang, H., Jiang, C., Wang, Y., and Wang, S. (2009). Cytotoxic and antibacterial triterpenoids derivatives from *Clematis ganpiniana*. *J Ethnopharmacol.*, **126**: 382–385.

Dixit, V.K., and Varma, K.C. (1979).Effect of essential oils of rhizomes of *Hedychium coronarium* and *Hedychium spicatum* on central nervous system. *Indian J. Pharmacol.*, **11**: 147-150.

Domenico, T., Castelli, F., Sarpietro, M.G., Venuti, V., Cristani, M., Daniele, C., Saija, A., Mazzanti, G., and Bisignano, G. (2005). Mechanisms of antibacterial action of three monoterpenes. *Antimicrob. Agents Chemother.*, **49**: 2474.

Elson, C.E., Maltzman, T.H., Boston, J.L., Tanner, M.A., and Gould, M.N. (1988).Anti-carcinogenic activity of *d*-limonene during the initiation and promotion/ progression of DMBA-induced rat mammary carcinogenesis. *Carcinogenesis*, **9:** 331-332.

Fang, Z.P., and Yin, G.J. (1989).Studies on the structures of two new triterpenesaponins C and D from *Polygala japonica Houtt*. *Acta Bot. Sin.*, **31:** 708–712.

Fernandes, N.P.C., Lagishetty, C.V. Panda, V.S., and Naik, S.R. (2007). An experimental evaluation of the antidiabetic and antilipidemic properties of a standardized *Momordica charantia* fruit extract. *BMC Complem Altern Med.*, **7**: 29

Giovannucci, E. (1998). Meta-analysis of coffee consumption and risk of colorectal cancer. *Am. J. Epidemiol.*, **147**: 1043-1052.

Guo, Y.Q., Xian, Li., Jing, Xu., Ning, Li., Da, Li., and Jin, H.W. (2004). Sesquiterpene Esters from the Fruits of *Celastrus orbiculatus*. *Chem. Pharm. Bull.*, **52:** 1134-1136.

Gupta, A.K., Tandon, N., and Sharma M. (2005). Quality standards of Indian Medicinal Plants. Indian Council of Medical Research, New Delhi, Vol.1, pp.21-26

Gupta, A.K., Tandon, N., and Sharma M. (2005a).Quality standards of Indian Medicinal Plants. Indian Council of Medical Research, New Delhi, Vol.2, pp. 235-242.

Gupta, A.K., Tandon, N., and Sharma M. (2005b).Quality standards of Indian Medicinal Plants. Indian Council of Medical Research, New Delhi, Vol.2, pp. 19-24.

Gupta, S., Choudhry, M.A., Yadava, J.N.S., Srivastava, V., Tandon, J.S., and Eichelbaum, M. (1990).Antidiarrheal activity of diterpenes of *Andrographis paniculata* (Kalmegh) against Escherichia coli enterotoxin in *in vivo* models. *Int J Crude Drug Res.*, **28:** 273–283.

Huber, W.W., McDaniel, L.P., Kaderlik, K.R., Teitel, C.H., Lang, N.P., and Kadlubar, F.F. (1997).Chemoprotection against the formation of colon DNA adducts from the food-borne carcinogen 2amino-1-methyl-6-phenylimidazo[4,5-b]-pyridine (PhIP) in the rat. *Mutat. Res.*, **376**: 115–122.

Huber, W.W., Scharf, G., Rossmanith, W., Prustomersky, S., Gra-slKraupp, B., Peter, B., Turesky, R.J., and Schulte-Hermann, R. (2002). The coffee components kahweol and cafestol induced glutamylcysteine synthetase, the rate limiting enzyme of chemoprotective glutathione synthesis, in several organs of the rat. *Arch. Toxicol.*, **75**: 685–694.

Hudson, J.B. (1990). Antiviral compounds from Plants. CRC Press, Boston.

Hue, H., and Staba, E.J. (1992). Botany and chemistry of *Ginkgo biloba* L. *J Herbs Spices Med Plants.*, **1**: 92-124.

Itokawa, H., Morita, H., Katou, I., Takeya, K., Cavalheiro, A.J., Oliveira, R.C.B., Ishige, M., and Moti dome, M. (1988). Cytotoxic diterpenes from the rhizomes of *Hedychium coronarium*. *Planta Med.*, **54:** 311.

Jadhav, V., Kore, A., and Kadam, V. (2007). *In vitro* pediculicidal activity of *Hedychium spicatum* essential oil. *Fitoterapia*,**78**: 470-473.

Jaffri, S.M.H. (1966).The flora of Karachi. Karachi: The Book Corporation,pp. 312-313.

Jiangsu New Medicinal College. (1977). "Dictionary of Chinese Herbal Medi-cine," Vol. 2, Shanghai People,s Publishing House, Shanghai, pp.1563.

Jigna, P., Nehal, K., and Sumitra, C.J. (2006). Screening of some traditionally used medicinal plants for potential antibacterial activity. *Indian J. Pharm. Sci.*, **68**: 832-834.

Jin, H.Z., Hwang, B.Y., Kim, H.S., Lee, J.H., Kim, Y.H., and Lee, J.J. (2002). Antiinflammatory constituents of *Celastrus orbiculatus* Inhibit the NF-κB Activation and NO Production. *J. Nat. Prod.*, **65:** 89-91.

Kar, A. (2006). Terpenoids. In: Pharmacognosy and Pharmacobiotechnology. New Delhi: New Age International (P) Limited., pp. 235-359.

Keun, Y.J., Jae, C.D., and Kun, H.S. (1993). Kaempferol 3-*O*-[63-*O*-(3-hydroxy-3-methyl glutaroyl) glucoside] from leaves of *Polygala japonica*. *Phytochemisty*,**34:** 1196–1197.

Kimura, Y., Akihisa, T., Yuasa, N., Ukiya, M., Suzuki, T., Toriyama, M., Motohashi, S., and Tokuda, H. (2005). Cucurbitane-type triterpenoids from the fruit of *Momordica charantia*. *J. Nat. Prod.*, **68**: 807–809.

Kinghorn A.D., and Soejarto, D.D. (2002). Discovery of terpenoid and phenolic sweeteners from plants. *Pure Appl. Chem.*, **74**: 1169–1179.

Lee K.J., Choi J.H., and Jeong H.G. (2007). Hepatoprotective and antioxidant effects of the coffee diterpeneskahweol and cafestol on carbon tetrachloride-induced liver damage in mice. *Food Chem Toxicol.*, **45:** 2118–2125.

Leung, S.W.S., and Lin, Z.B., and Ling, Z. (2002). (Ganoderma) research-the past, present and future perspectives. In: *Ganoderma*: Genetics, Chemistry, Pharmacology and Therapeutics, Proceedings of International Symposium on *Ganoderma* Research, Shanghai, pp.1-9.

Lewis, D. A. (1992) *Chemistry in Britain* February, 141.

Liu, Y.L., and Li, Y.F. (1998). Clinic effects of *Polygala japonica* on acute pharyngitis. *Acta Chin. Med. Pharmacol.*, **5**: 12.

Liu, Y.W., Gao, J.L., Guan, J., Qian, Z.M., Feng, K., and Li, S.P. (2009). Evaluation of antiproliferative activities and action mechanisms of extracts from two species of *Ganoderma* on tumor cell lines. *J Agri Food Chem.*, **57:** 3087–3093.

Ma, Y.C., Fu, H.C., and Chen, S. (1982). Flora Intramongolica 2nd ed., Vol. 6. Intramongolica People's Press, Huhhot.

Mahajan, B., Taneja, S., Sethi, V., and Dhar, K. (1995).Two triterpinoids from *Boswellia serrata* gum resin. *Phytochemistry*, **39:** 453-455.

Manikandan, A., and Doss, D.V.A. (2010). Evaluation of biochemical contents, nutritional value, trace elements, SDS-PAGE and HPTLC profiling in the leaves of *Ruellia tuberose* L. and *Dipteracanthus patulus* (Jacq.). *J Chem Pharm Res.*, **2(3)**: 295-303.

Marc, R. Chevrier, Abigail, R., David, Y., Lee, M., Zhang, W., and Charles, S. (2005). *Boswellia carterii* extract inhibits TH1 cytokines and promotes TH2 cytokines *in vitro*. *Clin. Diagn. Lab. Immunol.*, 12: 575–580.

Martelanc, M., Vovk, I., and Simonovska, B. (2007).Determination of three major triterpenoids in epicuticular wax of cabbage (*Brassica oleracea* L.) by high-performance liquid chromatography with UV and mass spectrometric detection. *J Chromatogr A.*, **1164**: 145–152.

Min, B.S., Nakamura, N., Miyashiro, H., Bae, K.W., and Hattori, M. (1998).Triterpenes from the spores of *Ganoderma lucidumand* their inhibitory against HIV-1 protease. *Chem Pharm Bull.*, **46:** 1607–1612.

Misra, P., Pal, N.L., Guru, P.Y., Katiyar, J.C. Srivastava, V., and Tandon, J.S. (1992). Antimalarial activity of *Andrographis paniculata* (kalmegh) against *Plasmodium berghei* NK65 in *Mastomysnatalensis*. *Int J Pharmacog.*, **30:** 263-274.

Misra, P., Pal, N.L., Guru, P.Y., Katiyar, J.C., and Tandon, J.S. (1991) Antimalarial activity of traditional plants against erythrocytic stages of *Plasmodium berghei. Int J Pharmacog.*, **29:** 19-23.

Morikawa, H., Ryoji, K., Hideaki, O., Eiji, H., Takakazu, S., Mitsunori, A., and Yoshio, T. (2004).Terpenic and phenolic glycosides from leaves of *Breynia officinalis* HEMSL. *Chem. Pharm. Bull.*, **52:** 1086-1090.

Müller, C.I., Kumagai, T., Okelly, J., Seeram, N.P., Heber, D., and Koeffler, H.P. (2006). *Ganoderma lucidum* causes apoptosis in leukemia, lymphoma and multiple myeloma cells. *Leukemia Res.*, **30:** 841–848.

Nagaraju, N. (1992). Biochemical Studies on Some Medicinal Plants of Rayalaseema Region.Ph.D. Thesis, S.V. University, Tirupati, India.

Nagaraju, N., and Rao, K.N. (1989). Folk–medicine for diabetes from Rayalaseema of Andhra Pradesh. *Ancient Sci Life.*, **9:** 31–35.

Naik, S.R., Pilgaonkar, V.W., and Panda, V.S. (2006). Evaluation of antioxidant activity of *Ginkgo biloba* phytosomes in rat brain. *Phytother Res.*, **20:** 1013-1016.

Naik, S. R., Pilgaonkar, V.W., and Panda, V. S. (2006a). Neuropharmacological evaluation of *Ginkgo biloba* phytosomes in rodents. *Phytother Res.*, **20:** 901-905.

Naik, S.R., and Panda, V.S. (2007). Antioxidant and hepatoprotective effects of *Ginkgo biloba* phytosomes in carbon tetrachloride-induced liver injury in rodents. *Liver Int.*, **27(3)**: 393-399.

Naik, S.R., and Hule, A. (2009).Evaluation of immunomodulatory activity of extract of andrographolides from *Andrographis paniculata.Planta Med.*, **75:** 785-791.

Naik, S.R., Bumrela, S. B., Lagishetty, C.V., and Mandlik, R.V. (2009a). Probable mechanism of the antihyperglycemic effect of standardised extract from *Momordica charantia* in streptozotocin induced diabetic rats. *J Complementary Integrative Med.*, **6:** 1-16.

Najiid, A., Simon, A., Cook, J., Chable-Rabinovitch, H., Delage, C., Chulia, A. J., and Rigaud, M. (1992). Characterization of ursolic acid as a lipoxygenase and cyclooxygenase inhibitor using macrophages, platelets and differentiated HL60 leukemic cells. *FEBS Lett.*, **299:** 213.

Oliveira, F.A., Lima, R.C.P., Cordeiro, W.M., Vieira-Junior, G.M., Chaves, M.H., Almeida, F.R.C., Santos, F.A., and Rao, V.S.N. (2004). Pentacyclictriterpenoids, α,β-amyrins, suppress the scratching behaviour in a mouse model of pruritus. *Pharm. Biochem. Behav.*, **78:** 719.

Oliveira F.A., Chaves, M.H., Almeida, F.R.C., Lima, R.C.P., Silva, R.M., Maia, J.L., Brito, G.A.A.C., Santos, F.A., and Rao, V.S. (2005). Protective effect of α- and β-amyrin, a triterpene mixture from *Protium heptaphyllum* (Aubl.) March. trunk wood resin, against acetaminophen-induced liver injury in mice. *J. Ethnopharmacol.*, **98:** 103.

Otake, T., Mori, H., Morimoto, M., Ueba, N., Sutardjo, S., Kusumoto, IT., Hattori, M., and Namba, T. (1995). Screening of Indonesian plant extracts for anti-human immunodeficiency virus type-1 (HIV-1) activity. *PhytotherRes.*, **9:** 6-10.

Otuki, M.F., Vieira-Lima, F., Malheiros, A., Yunes, R.A., and Calixto, J.B. (2005) Topical antiinflammatory effects of the ether extract from Protiumkleinii and α-amyrin pentacyclictriterpene. *Eur. J. Pharmacol.*, **507:** 253.

Panda, V.S., and Naik, S.R. (2008) Cardioprotective activity of *Ginkgo biloba* Phytosomes in isoproterenol-induced myocardial necrosis in rats: a biochemical and histoarchitectural evaluation. *Exp Toxicol Pathol.*, **60(4-5):** 397-404.

Pdsedek A. (2007).Natural antioxidants and antioxidant capacity of *Brassica* vegetables: A review. *LWT-Food Sci. Technol.*, **40**: 1-11.

Pfaffenrath, V., Diener, H.C., Fischer, M., Friede, M., and Henneicke-von Zepelin, H.H. (2002). The efficacy and safety of *Tanacetum parthenium* (feverfew) in migraine prophylaxis—a double-blind, multicentre, randomized placebo-controlled dose-response study. *Cephalalgia*, **22(7)**: 523-532.

Pietri, S., Maurelli, E., Drieu, K., and Culcasi, M. (1997). Cardioprotective and anti-oxidant effects of the terpenoid constituents of *Ginkgo biloba* extract (EGb 761). *J Mol Cell Cardiol.*, **29:** 733-742.

Pinakini, K.S., Vasanth, K., and Rao N. (2005). Evaluation of antidiabetic activity of *Ginkgo biloba* in streptozotocin induced diabetic rats. *Iran J Pharmacol Ther.*, **4:** 16-19.

Puri, A., Saxena, R., Saxena, R.P., Saxena, K.C., Srivastava, V., and Tandon, J.S. (1993). Immunostimulant agents from *Andrographis paniculata. J Nat Prod.*, **56:** 995-999.

Rai, M.K. (1995). A review on some antidiabetic plants of India. *Ancient Sci Life.*, **14**: 43–54.

Rao B.K., Renuka, S., Rajasekhar, M.D., Nagaraju, N., and Rao, C.A. (2003). Antidiabetic activity of *Terminalia pallida* fruit in alloxan induced diabetic rats. *J Ethnopharmacol.*, **85**: 169–172.

Rao, K.Y., Vimalamma, G., Rao, C.V., and Tzeng, Y. (2004). Flavonoids and andrographolides from andrographispaniculata. *Phytochemistry*, **65**: 2317-2321.

Reddy P.P., Lavekar, A.G., Suresh Babu, K., Rao, R.R., Shashidhar, J., Shashikiran, G., and Madhusudana Rao, J. (2010). Synthesis, cytotoxic activity and structure–activity relationships of hedychenone analogues. *Bioorg Medicinal Chem Lett.*, **20**: 2525–2528.

Reddy, P.P., Rao, R.R., Shashidhar, J., Sastry, B.S., Rao, J.M., and Suresh Babu, K. (2009). Phytochemical investigation of labdane diterpenes from the rhizomes of *Hedychium spicatum* and their cytotoxic activity. *Bioorg Medicinal Chem Lett.* (**19**): 6078–6081.

Ren, Z., and Gould, M.N. (1994). Inhibition of ubiquinone and cholesterol synthesis by the monoterpeneperillyl alcohol.*Cancer Lett.*, **76**: 185-190.

Saleem, M., Alam A., Arifin, S., Shah, M.S., Ahmed, B., and Sultana, S. (2001). Lupeol, a triterpene, inhibits early responses of tumor promotion induced by benzoyl peroxide in murine skin. *Pharmacol. Res.*, **43**: 127-134.

Salvador, M.M., Recio, C., Giner, R.M., and Rios, J. (1997). Effect of selected triterpenoids on chronic dermal inflammation.*European J Pharmacol.*, **334:** 103–105.

Saroja, K., Elizabeth, J.D., and Gopalkrishnan, S. (2009). Wound healing activity of leaves of *Dipteracanthus patulus (Jacq.). Pharmacologyonline*; **2**: 462-469.

Sharma, A., Krishan, L., and Handa, S.S. (1992). Standardization of the Indian crude drug Kalmegh by high pressure liquid chromatographic determination of andrographolide. *Phytochem Anal.*, **3**: 129–131.

Sharma, M., Bani, S., and Singh, G. (1989). Anti-arthritic activity of boswellic acids in bovine serum albumin (BSA)-induced arthritis.*Int J Immunopharmacol.,* **11:** 647-652.

Shieh, Y.H., Liu, C.F., Huang, Y.K., Yang, J.Y., Wu, I.L., Lin, C.H., and Li, S.C. (2001). Evaluation of the hepatic and renal-protective effects of *Ganoderma lucidum* in mice. *The Am J Chin Med.,* **29:** 501–507.

Singh, S.K., Bhusari, S., Singh, R., Saxena, A., Mondhe, D., and Qazi, G.N. (2007). Effect of acetyl 11-keto β-boswellic acid on metastatic growth factor responsible for angiogenesis.*Vascular Pharmacol.,* **46**: 333 – 337.

Srimal, R.C., Sharma, S.C., and Tandon, J.S. (1984). Anti-inflammatory and other pharmacological effects of *Hedychium spicatum* (Buch-Hem).*Indian J. Pharmacol.,* **16**: 143-147.

Stormer, F.C., Reistad, R., and Alexander, J. (1993). Glycyrrhizic acid in liquorice-evolution of health hazard. *Food Chem. Toxicol.,* **31:** 303.

Su, B.N., Yang, L., Gao, K., and Jia, Z.J. (2000). Aquaticol, a bis-sesquiterpene and iridoidglucosides from *Yeronicaanagallis-aquatica. Planta Med.,* **66(3):** 281-283.

Sudharsan, P.T., Mythili, Y., Selvakumar, E., and Varalakshmi, P. (2006).Lupeol and its ester ameliorate the cyclophosphamide provoked cardiac lysosomal damage studied in rat. *Mol. Cell. Biochem.,* **282**: 23-29.

Tandon, S. K., Chandra, S., Gupta, S., and Lal, J.J. (1997). Analgesic and anti-inflammatory effects of *Hedychium spicatum. Indian J. Pharm. Sci.,* **59**: 148-149.

Teng, Y.H., and Teng, H.L. (1994). Intensified therapy on acute osteomyelitis by *Polygala japonica. J. Railway Med.* 1, 35.

The Wealth of India. (1959). C.S.I.R, New Delhi.Vol. V, pp.13.

The wealth of India. (2003). A dictionary of Indian raw materials and industrial products.Raw materials.Vol-III (D-E). CSIR, New Delhi, pp.90

Tighilet, B., and Lacour, M. (1995). Pharmacological activity of the *Ginkgo biloba* extract (EGb 761) on equilibrium function recovery in the unilateral vestibular neurectomized cat. *J Vestib Res.,* **5(3):** 187-200.

Tsuruga, T., Chun, Y.-T., Ebizuka, Y., and Sankawa, V. (1991). Biologically active constituents of *Melaleuca leucadendron*: inhibitors of induced histamine release from rat mast cells. *Chem. Pharm. and Bull.,* **39:** 3276-3278.

Wan, J.Z., Li, S.F., and Pao, S.J. (1997). Pharmacodynamic study on Shenyan capsule in the treatment of chronic nephritis. *Chin. J. New Drugs.,* **6**: 450–452

Wang, B.H., and Polya G.M. (1996). Selective inhibition of cyclic AMP-dependent protein kinase by amphiphilictriterpenoids and related compounds. *Biol Chem Hoppe Seyler.,* **377(6):** 373-84.

Wang, H., and Ng, T.B. (2006).Ganodermin, an antifungal protein from fruiting bodies of the medicinal mushroom *Ganoderma lucidum.Peptides***27:** 27–30.

Wang, H., Gao, J., Kou, J., Zhu, D., and Yu, B. (2008). Anti-inflammatory activities of triterpenoidsaponins from *Polygala japonica*. *Phytomedicine*,**15:** 321 – 326.

Wang, X.X., Lin, C.J., and Jia, Z.J. (2006). Triterpenoids and Sesquiterpenes from *Mulgedium tataricum*. *Planta Med.*, **72**: 764-767.

Wattenberg, L.W. (1983). Inhibition of neoplasia by minor dietary constituents.*Cancer Res.*, **43**: 2448-2453.

WHO monographs on selected medicinal plants. (2002). Vol. 2. World Health Organization, Geneva, pp. 15.

Wu, Z.Y. (1990). Xinhua Bencao Gangyao. Science Technical Publishing Press, Shanghai, pp. 435.

Yan, P., Badrul, A.A., Zakiah, I., Rusli, I., Joon, W.M., Peter, C.K., Pook, H.M., and Chin, E.O. (2011). *In vitro* modulatory effects of *Andrographis paniculata*, *Centella asiatica* and *Orthosiphon stamineus* on cytochrome P450 2C19 (CYP2C19). *J Ethnopharmacol.*, **133:** 881–887.

Ying, Q.L., Rinehart, A.R., Simon, S.R., and Cheronis, J.C. (1991). Inhibition of human leucocyte elastase by ursolic acid. Evidence for a binding site for pentacyclic triterpenes. *Biochem J.*, **277:** 521-526.

Yoon, S.Y., Eo, S.K., Kim, Y.S., Lee, C.K., and Han, S.S. (1994). Antimicrobial activity of *Ganoderma lucidum* extract alone in combination with some antibiotics. *Arch. Pharmacal Res*.**,** **17:** 438–442.

Ziegenbein, F.C., Hanssen, H.P., and König, W.A. (2006). Secondary metabolites from *Ganoderm alucidum* and *Spongiporus leucomallellus*. *Phytochemistry*, **67:** 202–211.

Zwenger, S., and Basu C. (2007). *In silico* analysis of terpene synthase genes in *Arabidopsis thaliana. EXCLI Journal*, **6:** 203-211.

Medicinal Plants: Phytochemistry, Pharmacology and Therapeutics Vol. 3 (2014) *Pages* **213–234**
Editor-in-Chief: **V.K. Gupta**
Published by: **DAYA PUBLISHING HOUSE, NEW DELHI**

8

Biological Activities, Phytochemistry and Industrial Potentials of *Khaya senegalensis* (Meliaceae) A. Juss

Mohammed Auwal Ibrahim[1,2] and Md. Shahidul Islam[1*]

ABSTRACT

Khaya senegalensis A. Juss (African mahogany) is a large tree growing mainly in the sub-Saharan Africa and highly reputed for numerous medicinal effects. Scientific investigations have provided evidences that the various part of the plant contained antitrypanosomal, antiplasmodial, antihelminthic, antioxidant, antifungal, antibacterial and anticancer agents. Other biological activities attributed to the plant are antiinflammatory, larvicidal and immunomodulatory properties. Phytochemical analysis using different chromatographic and spectroscopic tools revealed that the different parts of the plant contain mainly limonoids and more than 45 have been identified so far. However, khayanolides and seneganolides are the most commonly found limonoids among other forms and are linked to the numerous pharmacological effects of the plant. Furthermore, phenolics such as gallic acid, rutin, procyanidins, catechin and quarcetin have been detected in different parts of the plant. On the other hand, the gum polymer produced by the plant had high binding (mechanical and disintegrational) properties in tablet formulation and could serve as a binder in the pharmaceutical industries. This review will thoroughly discuss the possible implications of the above-mentioned findings to the pharmaceutical industries of developing countries.

Keywords: *Khaya senegalensis*, African mahogany, Khayanolides, Seneganolides.

1 Department of Biochemistry, School of Life Sciences, University of KwaZulu-Natal (Westville Campus), Durban, 4000, South Africa

2 Department of Biochemistry, Ahmadu Bello University, Zaria, Nigeria

* *Corresponding author*: E-mail: islamd@ukzn.ac.za

Introduction

Khaya senegalensis (Desr.) A. Juss, also known as African mahogany, is a member of the Meliaceae family, within the sub tribe Swietenioideae (Styles, 1972), which contains many of world's most highly valued timber species (Sexton *et al.*, 2010). The species occurs naturally in the tropics of 19 central African countries discontinuously from Senegal and neighboring countries in the west (mostly between about 8°N and 15°N but extending to about 6°N in Nigeria) through to Sudan and Uganda in the east where it extends southward to about 2°45′N. This distribution covers a wide range of climatic, altitudinal, ecological and edaphic conditions from sea level to 1800 m, encompassing mean annual rainfalls from <700 mm (under which it is often associated with seasonal watercourses) to about 1750 mm, with dry season lengths of 2–8 months. Also, the species occurs on a wide variety of soil types including those within gallery forest (where it attains heights of around 35 m with a bole diameter of

Figure 8.1: A picture of *Khaya senegalensis*.

1.5m and length of 10 m) and drier open savanna with lateritic soils and on rocky places where it mostly occurs as solitary trees attaining heights of 15–20 m, often with a sinuous bole (Nikles *et al.*, 2008).

Throughout its natural range, *K. senegalensis* has been exploited for timber, which is used for fine furniture, construction, and fuel. It was one of the first west African forest tree species to be imported by Europeans due to its excellent timber and veneer. Furthermore, the African mahogany is now widely used in United States furniture sector as a substitute for American mahogany. Indigenous American companies have been keen to replace increasingly costly and scarce American mahogany with more competitive timbers whilst also reacting to domestic and international pressures associated with use of American mahogany from natural stands (Arnold, 2004). Due to such factors, the United States' imports of African mahogany (as logs and sawn timber) increased dramatically during the 1990s – from 4,100 m^3 in 1991 to more than 20,000 m^3 in 1998 (Traffic, 2000). Also due to the increasing cost and scarcity of American mahogany, there has been increasing interest in African mahogany from timber importers and furniture manufacturers in China (Arnold, 2004).

On the other hand, medicinal compounds, extracted from bark, leaf, and seed, are used to treat a multitude of tropical diseases, and its foliage is used to sustain livestock during the dry period of the year. The dynamics of such demands coupled with the timber demands have already led to local extinctions (especially in the north of its range), and the species is now classified as vulnerable under the Red List of Threatened Species (IUCN, 2010). Luckily, the demonstrated wide adaptability, substantial yield potential and proven timber quality of African mahogany led to various domestication and plantation programmes in some parts of Asia and Australia (Nikles *et al.*, 2008) which may consequently assist in the conservation of this highly valued species.

Apart from the wood and timber uses of this plant, more relevant to this paper, is the medicinal applications of this species in the treatment of various tropical diseases. Almost all parts of the plant are claimed to cure many diseases prevalent in the areas where it grows and as such, it is considered as a primary health care facility to millions of Africans. However, this is not surprising when considering an array of phytochemicals isolated from the different parts of the plant. In spite of the widely use of this plant in the traditional circles and the scientifically documented evidences on some of its pharmacological actions, there is no document that could provide an overall picture of scientifically validated bioactivities and industrial potentials of this plant. This chapter will examine the documented ethnobotanical uses of this plant and systematically discuss the biological and pharmacological activities attributed to the different parts and extracts of the plant as well as the phytochemicals isolated from the plant. This is with a view to provide an overall medicinal and industrial potential of this highly respected plant for possible exploitation by relevant sectors to solve a number of agricultural, health and economic problems.

Ethnobotanical Uses of *K. senegalensis*

Khaya senegalensis is well known as an important medicine within the traditional healers of West Africa for the treatment of malaria, anemia, diabetes, gastrointestinal diseases among others. In an ethnobotanical survey, Atawodi *et al.* (2002) reported that among all other plants, *K. senegalensis* was the most commonly used plant for the treatment of trypanosomiasis in Kaduna state of Nigeria where the disease is prevalent. In fact, >60 per cent of respondents in the study prescribed recipes containing *K. senegalensis.* Another study from north western Nigeria also indicated that *K. senegalensis* is highly ranked in the list of medicinal plants used for the traditional management of diabetes mellitus (Etuk *et al.*, 2010). Similarly, in a survey of Ghanian herbal market, *K. senegalensis* was among the medicinal plants that had highest market demand. Indeed, it is the fourth most commonly purchased medicinal plant in the Ghanian market with an average daily market stock of 788 kg where the bark is used as blood tonic, aphrodisiac and for treating fever (van Andel *et al.*, 2012). Reports from northern Cote-d 'ivoire also indicate that the stem bark of *K. senegalensis* is among the highly utilized medicinal plants for the treatment of wounds, diarrhoea and dysentery (Kone *et al.*, 2004). Furthermore, a study on herbal remedies among the Tem tribe of Togo also suggests that *K. senegalensis* was among the list of plants used to cure several diseases but unfortunately, it also tops the list of medicinal plants with highest adverse effects (Tchacondo *et al.*, 2011). Some of the adverse effects linked to this plant are diarrhoea, polyuria, hunger, abdominal pains and general weakness (Tchacondo *et al.*, 2011). However, the reported adverse effects may be associated to the difference in geographical location which is known to affect the phytochemistry and hence, the biological activities of a plant material.

These reported ethnobotanical uses of the plant has attracted the attention of scientists from different parts of the world to evaluate the plant for a variety of pharmacological and biological activities with a view to find alternative therapies to a number of ailments. Below, we provide a rundown of available pharmacological activities investigated in this plant and briefly discuss all available scientific literatures associated with the activity under study.

Antimalarial Activity of *K. senegalensis*

Malaria is one of the tropical diseases traditionally treated with the parts of *K. senegalensis* and unfortunately only few studies have been carried out to validate the claim. According to a study by Ndjonka *et al.* (2012), ethanolic extract of *K. senegalensis* stem bark inhibited *in vitro* proliferation of *Plasmodium falciparum* 3D7 with an IC_{50} and IC_{90} values of 40.1 ± 4.3 µg/ml and 214.7 ± 0.3 µg/ml respectively. It was also found to synergistically interact with an extract of *Anogeissus leiocarpus* (IC_{50} of the combination was 12.5 ± 2.7 µg/ml) to inhibit the *in vitro* proliferation of the parasites which support the combination of *K. senegalensis* with other plants in the traditional treatment of malaria. In order to identify the biological activity of some limonoids isolated from *K. senegalensis*, other authors reported that fassinolide was active *in vitro* against chloroquine sensitive strains of *P. falciparum* (Khalid *et al.*, 1998). In another study from Burkina Faso, ethanolic extract of the bark and leaves of *K. senegalensis* inhibited *in vitro* maturation of fresh clinical isolates of *P. falciparum* with IC_{50} values of 82.17 µg/ml and 58.48 µg/ml respectively (Karou *et al.*, 2003). Although these authors did not appreciate the observed antiplasmodial activities of *K. senegalensis* extracts when considering other plants used in the study but postulated that the traditional use of *K. senegalensis* parts for treating malaria could be linked to its tendency to treat malarial symptoms like fever and does not actually mediate the killing of the parasites. Thus, the contrasting scientific reports on the actual antimalarial activity of *K. senegalensis* extracts needs further verification, especially in an *in vivo* animal model.

Antihelminthic Action of *K. senegalensis*

Some investigators have studied the efficacy of *K. senegalensis* extracts against different species of gastrointestinal nematode parasites since the plant is also used in the traditional circles for the treatment of gastrointestinal disorders. Using sheep with naturally acquired infection of gastrointestinal nematodes, Ademola *et al.* (2004) observed that administration of 500 mg/kg body weight (bw) of ethanolic extract of *K. senegalensis* stem bark caused the reduction of fecal egg count (FEC) after 2 days and that the percentage reduction of FEC of sheep parasitized with *Oesophagostomum columbianum*, *Stronglyloides* sp. and *Trichuris* sp. was 100 per cent by the third day of administration while the percentage reduction of FEC of *Haemonchus contortus* and *Trichostrongylus colubriformis* was 68.2 and 66.1 per cent, respectively, by the twelfth day of administration. The study further reveals the potentials of the plant extract to mediate the killing of both intestinal and abdominal parasite species in an *in vivo* system. In another study, these authors used bioassay guided fractionation of the *K. senegalensis* ethanolic extract in order to identify the possible group of bioactive antihelmenthic agents using *H. contortus* as a model organism (Ademola *et al.*, 2009). The results indicate that the antihelminthic activity of the plant extract is due to some synergistic relationships between saponins, terpenoids, flavonoids and condensed tannins. Using *Onchocerca ochengi* and *Caenorhabditis elegans* as model organisms, other workers from Cameroon and Germany reported that the stem bark *K. senegalensis* ethanolic extract displayed the highest antihelminthic activity among all other tested plants in the study (Ndjonka *et al.*, 2011). Although limited in number, but these scientific investigations tend to lend credence to the widely use of *K. senegalensis* stem bark in the traditional treatment of gastrointestinal disorders by herdsmen. Further studies on the subject could lead to the development of novel antihelminthic agent(s).

Antitrypanosomal Activity of *K. senegalensis* Parts

Antitrypanosomal effects is perhaps the most widely scientifically validated pharmacological property of this plant extracts. In an *in vitro* study with *Trypanosoma brucei brucei* and *T. congolense*, methanol extract from the bark of *K. senegalensis* was found to contain the most potent trypanocidal agents among twenty three plants harvested from the savannah belt region of Nigeria (Atawodi *et al.*, 2003). Similarly, Wurochekke and Nok (2004) reported that aqueous and methanol extract of the plant had the highest *in vitro* activity against *T. brucei brucei* than other thirteen medicinal plants used for the traditional management of trypanosomiasis in northern Nigeria. Subsequently, an excellent *in vivo* activity of stem bark aqueous extract of the plant against *T. brucei* was demonstrated (Ibrahim *et al.*, 2008) where 100 mg/kg bw of the extract completely eliminated the parasites from infected animals within three days but unfortunately could not reverse the trypanosome-associated liver damage. These findings support, though indirectly, another investigation from Germany where dichloromethane extract of *K. senegalensis* stem bark demonstrated very high trypanocidal activity but was also toxic to mammalian cells which makes its selectivity index very low (Aderbauer *et al.*, 2008). Other workers from south western Nigeria also demonstrated that *K. senegalensis* extracts contain the most effective trypanocidal agents than other ten medicinal plants from the region (Antia *et al.*, 2009). In another study, Adeiza *et al.* (2010) compared the *in vitro* anti-*T. evansi* activity of *K. senegalensis* parts and found that the methanol extract of the stem bark had the highest activity. Furthermore, ethanolic extract of the stem bark was found to inhibit the multiplication of *T. evansi* in an *in vivo* system but could not ameliorate the parasite induced anemia (Umar *et al.*, 2010). It seems therefore that *K. senegalensis* stem bark extract contain highly trypanocidal constituents but are not effective in ameliorating some of the disease associated pathological changes. Identification of the active trypanocidal agents and possible toxic components from the stem bark would certainly provide a clue on the potential application of this plant part or its bioactive agent in the treatment of trypanosomiasis. It is hoped that an on-going study in our lab would provide the much needed clue on the subject.

Antibacterial Activity of *K. senegalensis*

Although different parts of *K. senegalensis* are used in the traditional treatment of several bacterial infections but there is a dearth of scientific research on the subject. Perhaps the first study on the antibacterial action of *K. senegalensis* parts was reported by Kone *et al.* (2004) which demonstrated that ethanolic extract of *K. senegalensis* stem bark was among the ten most active plant extracts (out of fifty) against some antibiotic resistant bacteria strains. Subsequently, a relatively more elaborate investigation was reported by Kubmarawa *et al.* (2008). These authors prepared aqueous and ethanolic extracts from the root, stem bark and leaf of the plant and tested their antibacterial activity using both gram positive and gram negative bacteria. Their findings revealed that the aqueous and ethanolic extracts of the stem bark as well as the ethanolic extract of the roots contained more potent antibacterial agents because of their ability to inhibit all the bacteria species tested in a disc diffusion assay method. Furthermore, a most recent study indicated that aqueous and ethanolic extracts of the root of *K. senegalensis* inhibited the growth of four different species of bacteria in a punch hole method for antibacterial testing (Ide and Igeleke, 2012). Summarily, reports on the *in vitro* antibacterial activity of *K. senegalensis* looks promising but this might not be a reflection of its efficacy in physiological systems and therefore more work is needed in that respect.

Anti-inflammatory Effect of *K. senegalensis*

So far, most of the available scientific investigations on the anti-inflammatory activity of *K. senegalensis* emanate from Senegal. Freeze dried aqueous extract of *K. senegalensis* have been shown to reduce carrageenan induced oedema and was linked to the presence of methylangolensate and methyl hydroxyangolensate (Thioune *et al.*, 1999). Furthermore, in a subsequent study, ointments made from a hydro-alcoholic extract was found to inhibit ear oedema in croton oil induced ear oedema method which demonstrate the possibility to maintain the anti-inflammatory activity of *K. senegalensis* in a gel form (Thioune *et al.*, 2000; 2003). In order to explain the mechanism

Figure 8.2: Some of the isolated bioactive compounds from different parts of *K. senegalensis* (Khalid *etal.*, 1998; Abdulgaleil *et al.*, 2001; Kayser and Abreu, 2001; Nakatani *et al.*, 2001; Nakatani *et al.*, 2002; Abdulgaleil *et al.*, 2004; El-Aswad *et al.*, 2004; Zhang *et al.*, 2007).

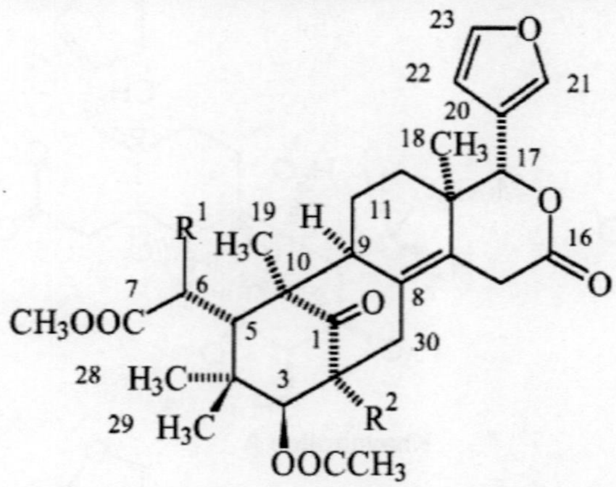

2,6-dihydroxyfissinolide: R1= R2= OH
Fissinolide: R1= R2= H

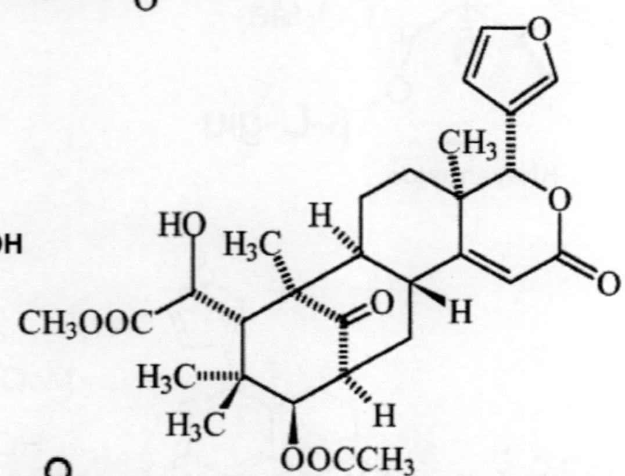

methyl 3b-acetoxy-6-hydroxy-1-oxomeliac-14-enoate

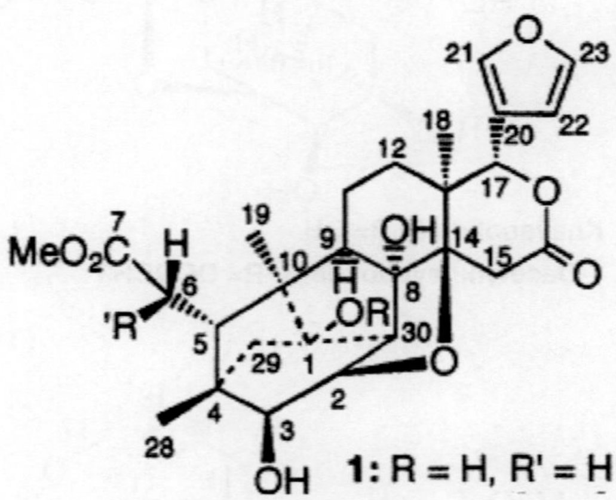

1: R = H, R' = H

Khayanolide D

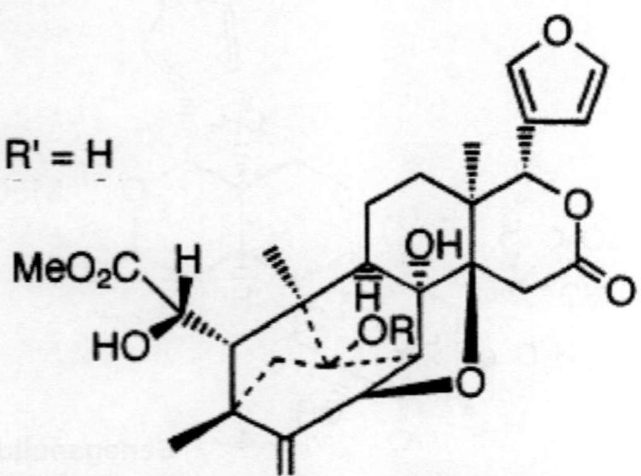

Khayanolide E **2: R = Ac**

Contd...

Figure 8.2–*Contd...*

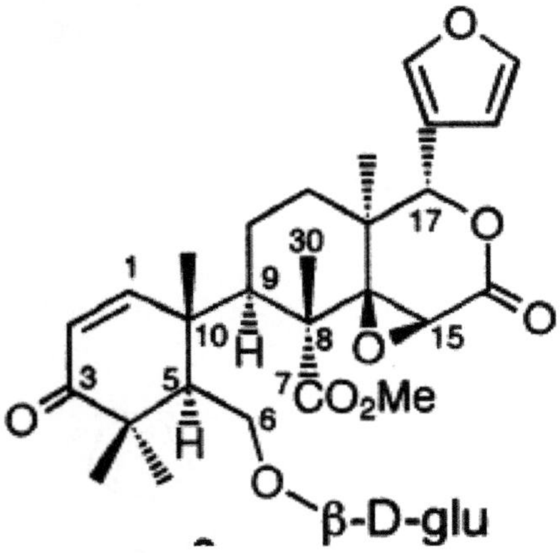

Khayanoside

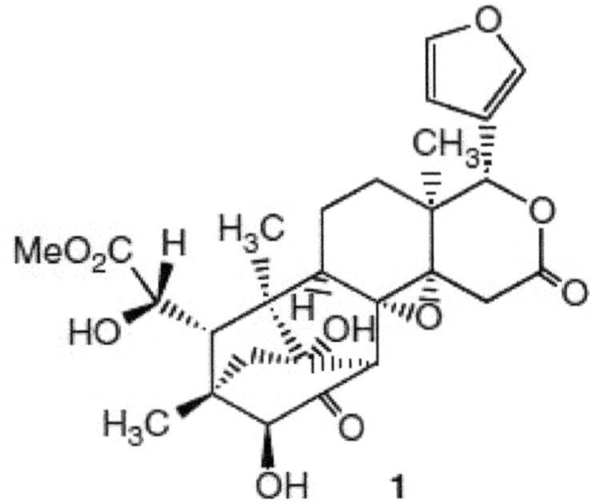

1

Khayanolide A

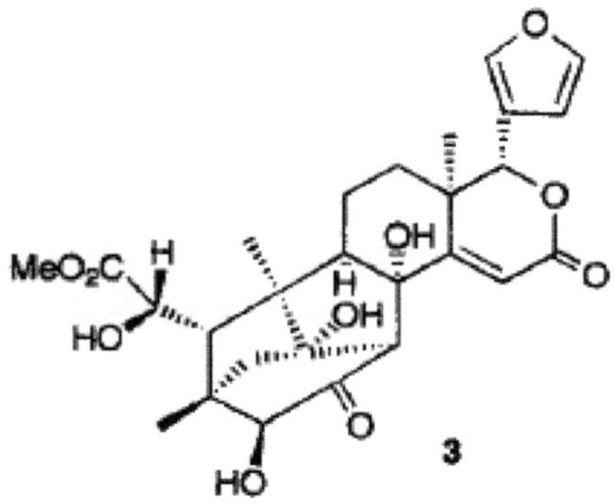

3

Khayanolide C

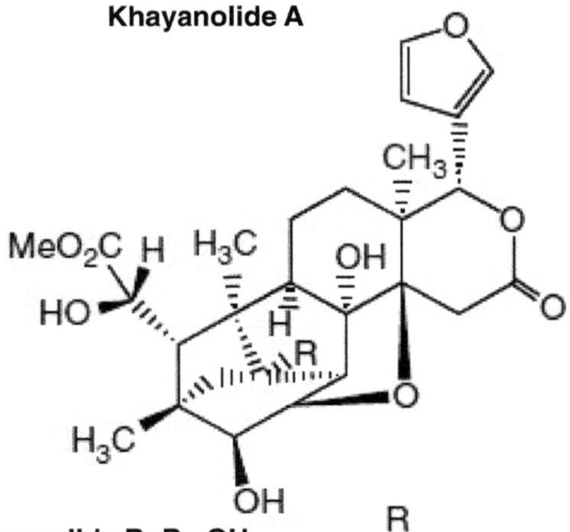

Khayanolide B: R= OH

1-Oacetylkhayanolide B: R= OCOCH3

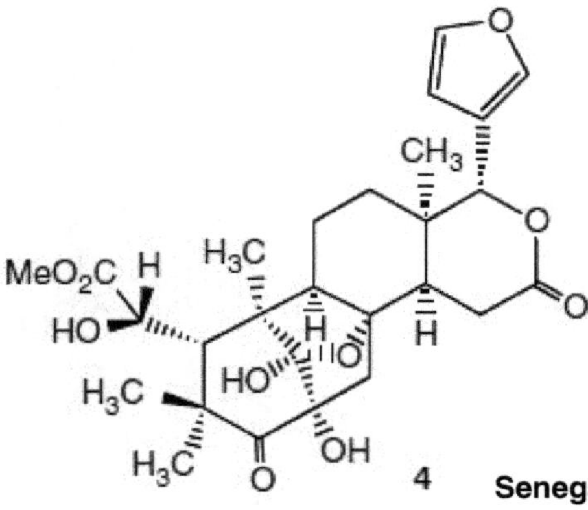

4

Khayalactol

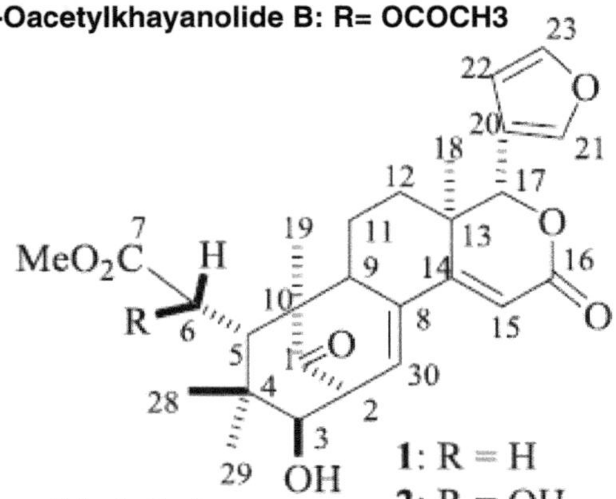

Seneganolide A: R=1

2-hydroxyseneganolide A: R=OH

2-acetoxyseneganolide A: R=OAc

Contd...

Figure 8.2–*Contd...*

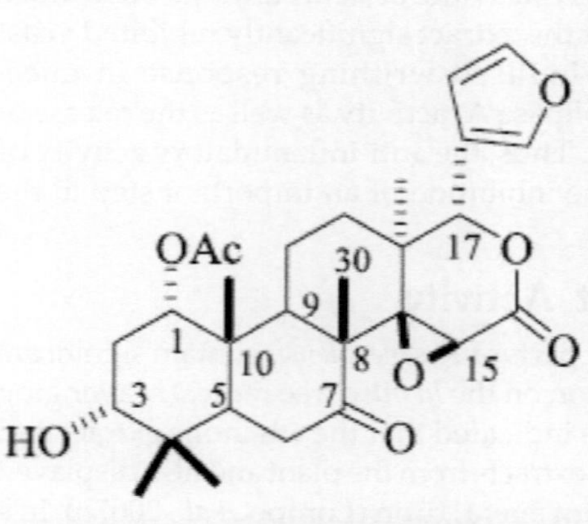

3-deacetyl-7-deacetoxy-7-oxokhivorin

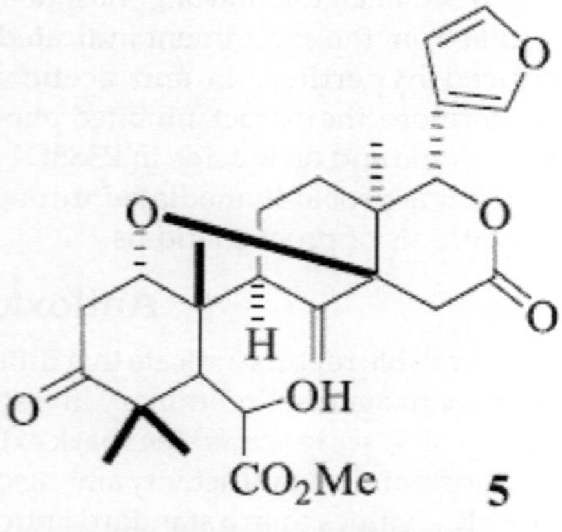

5

methyl 6-hydroxyangolensate

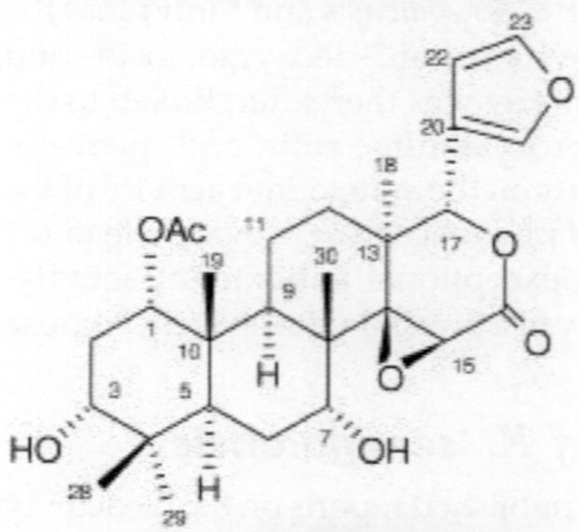

3α,7α–dideacetylkhivorin

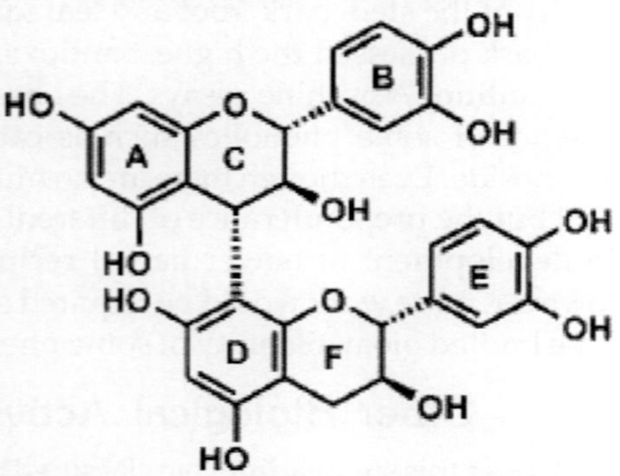

Catechin–(4α,8)-catechin (Proanthocyanidin B3)

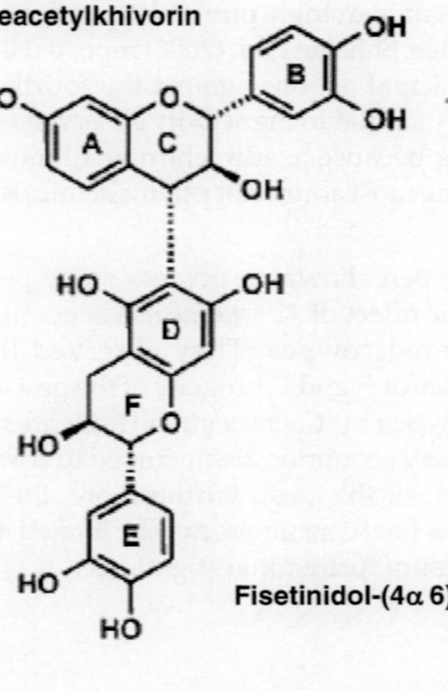

Fisetinidol-(4α 6)-catechin

of anti-inflammatory activity of *K. senegalensis*, Lompo *et al.* (2007a) investigated the antipyretic, analgesic and antiphospholipase A_2 activities of stem bark aqueous extract. Results from the experiment indicated that the extract significantly inhibited yeast induced hyperthermia and acetic acid induced writhing response in mice. Furthermore, the extract inhibited phospholipase A_2 activity as well as the release of arachidonic and oleic acids in P388D1 cells. Thus, the anti-inflammatory activity of the plant is probably mediated through the inhibition of an important step in the biosynthesis of prostaglandins.

Antioxidant Activity

Available reports indicate that different parts of *K. senegalensis* contain significant antioxidant agents. Preliminary investigation on the *in vitro* free radical scavenging activity of *K. senegalensis* stem bark extracts indicated that the ethanolic extract had the highest antioxidant activity among other extracts from the plant and also displayed lower IC_{50} values than a standard antioxidant agent, rutin (Lompo *et al.*, 2007b). In a relatively more comprehensive study, Atawodi *et al.* (2009) investigated the antioxidant activity of the stem bark, root and leaf samples of *K. senegalensis* and found that the stem bark possessed the highest antioxidant activity using 2-deoxyguanosine and hypoxanthine/xanthine assays. The observed activity was thereafter, linked to the presence of some phenolics such as catechin, procyanidins, rutin and quercetin rhamnoside. Even though there are no much reports on the antioxidant activity of the plant but the preponderance of different kinds of phenolics may provide a lead for the development of newer herbal recipes with exceptional antioxidant activity. However, more work would be required especially in a physiological system because of the limited bioavailability of some phenolics.

Other Biological Activities of *K. senegalensis*

Under this subheading, we classified all other published reports on the bioactivity of *K. senegalensis* extracts that appear in a single publication with no complementary information elsewhere. For instance, Shaalan *et al.* (2006) reported that the stem bark extracts displayed excellent larvicidal actions against the fourth instars of *Culex annulirostris* which was statistically similar to the activity observed with azadirachtin. This finding could be interesting because azadirachtin is a limonoid and several reports have demonstrated that the most abundant phytochemicals of *K. senegalensis* are limonoids.

Khaya senegalensis has also been shown to possess some pesticidal activity. Bamaiyi *et al.* (2006) evaluated the effect of *K. senegalensis* seed oil in the control of *Callosobruchus maculatus* on stored cowpea. They observed that the seed oil significantly reduced the emergence of F_1 and F_2 progeny of the pest and consequently reduced the damage made to cowpea by *C. maculatus.* These area warrant serious consideration considering the heavy economic loss incurred to cowpea farmers and traders during storage as a result of this pest. Furthermore, the use of synthetic chemicals which may be toxic to the final consumers may be limited if more promising results are obtained over the course of further investigations.

Investigation on the anticoccidal activity of *K. senegalensis* stem bark aqueous extract showed that the extract had no *in vitro* antisporulation effects (Nwosu *et al.*, 2011). However, oral treatment of pullet chicks infected with sporulated *Eimeria* oocysts with the extract significantly reduced fecal oocyst output and improved packed cell volume and live weight of the birds. This study demonstrate that the plant contain some bioactive compounds that could potentially be exploited in the treatment of avian coccidiosis, a disease that wreak havoc in poultry industry.

A research group from USA and Guinea investigated the effects of *K. senegalensis* bark extract on colorectal cancer (Androulakis *et al.*, 2006) by measuring cell growth inhibition, cyclooxegenase (COX) and prostaglandin E_2 suppression as well as apoptosis by western blot analysis and ELISA. Data obtained from the work unequivocally revealed that the bark extract possessed anti-proliferative and pro-apoptotic effects on a range of cell lines which further suggested that the extract has potent antitumor effects by causing both cell cycle arrest and apoptosis via COX 2-dependent pathway. These observations support a hypothesis, we earlier made whilst discussing the anti-inflammatory action of the plant.

Other investigators reported that the simultaneous administration of aqueous extract of *K. senegalensis* stem bark with phenylhydrazine was able to protect rats from the phenyl hydrazine induced anemia (Sanni *et al.*, 2008). However, the same extract did not show any significant curative effect after the induction of anemia. Although the stem bark is used in the traditional treatment of anemic conditions and as blood tonic, but results from the above mentioned work suggest that the extract simply block or reduce the contact of the anemia agents with red blood cells rather than a direct effect on the replenishment of red blood cells by the bone marrow.

In order to study the immunomodulatory activity of some Sudanese medicinal plants, Koko *et al.* (2008) investigated the inhibitory effects of plants on the whole blood oxidative burst activity using luminal/lucigenin-based chemiluminescence assay and *K. senegalansis* extracts top the list. Further investigation was conducted to monitor the effects of 14 plants on oxidative burst of isolated polymorphonuclear and mononuclear cells by using two different phagocytosis activators and also the plant extracts displayed the best inhibitory actions. It appears therefore that *K. senegalensis* extracts contain some immunomodulatory agents but their efficacy needs to be confirm in an *in vivo* study. Perhaps some of the previously discussed *in vivo* pharmacological activities are simply the result of immunomodulatory action of the *K. senegalensis*.

The traditional use of *K. senegalensis* in diarrhoeal management prompted a research group to validate the claim. Administration of an aqueous and methanolic extract of the leaf samples of the plant delayed the onset of diarrhoea and significantly reduced gastrointestinal motility in rats (Nwosu *et al.*, 2012).

Even though *K. senegalensis* is commonly used by traditional healers to treat diabetes mellitus but the only available scientific investigation on the subject was recently published. Aqueous stem bark extract of *K. senegalensis* was found to exhibit a dose dependent reduction in the fasting blood glucose levels of alloxan induced diabetes model of rats (Kolawole *et al.*, 2012) and the result was comparable to the effect of glibenclamide. No much information can be obtained on the actual antidiabetic effects of this plant because of some limitations and preliminary nature of the experiment and therefore an all encompassing work that measures most of the diabetes related parameters especially in a type 2 diabetes model would provide a more convincing evidence on the subject. Our lab is currently working on this important area of research.

Toxicological Reports on *K. senegalensis*

Traditionally, this plant was reported to have the highest adverse effects by the Tem tribe of Togo and some of the adverse effects reported were diarrhea, polyuria, hunger, abdominal pains and general weakness (Tchacondo *et al.*, 2011). This observation clearly points the need for careful scientific assessments in order to identify the safe dosage and the possible adverse effect of using this plant.

Various toxicological studies have been carried out *in vivo* on different subjects to confirm the toxicity of *K. senegalensis.* Adebayo *et al.* (2003) are the first to report on the toxicity aspect of *K. senegalensis.* Their findings revealed that daily oral administration of 2 mg/kg bw of ethanolic extract of *K. senegalensis* bark for 18 days exerted deleterious effect on the functioning of the kidney by modulating the activity of the kidney alkaline phosphatase and increasing serum sodium ion concentration. In another study, daily oral administration of aqueous extract of *K. senegalensis* bark for 21 days increased the plasma levels of aspartate and alanine transaminases, alkaline phosphatase, urea, creatinine, total proteins, sodium and potassium ions (Kolawole *et al.*, 2011). On the other hand, the extract caused a reduction in the red blood cells, packed cell volume and heamoglobin level in rats. These observations connote that the *K. senegalensis* extract adversely affect the functioning of vital organs of the body especially liver and kidney.

On the contrary, both single intraperitoneal administration and prolonged daily oral administration of aqueous extract of the bark of the plant to four-week old pullet chicks did not produce clinical manifestations of toxicity and/or mortality of the chicken indicating that the extract is relatively safe by both parenteral and oral routes (Nwosu *et al.*, 2011). Other investigators evaluated the toxicological effects of aqueous leaf extract and found that sub chronic administration of the extract for 28 days resulted in nonsignificant changes in liver and renal functions as well as histopathology of the organs (Nwosu *et al.*, 2012). Based on these studies, it is possible to surmise that the stem bark extract of the plant is toxic to mammals, especially after prolonged exposure, but not to birds. However, the leaves are relatively safer compared to the bark. Thus, it will be expedient to fully characterize the toxic components of the bark so that they can be excluded from the list of possible therapeutic candidates to be exploited from this plant.

Identified Phytochemical Contents of *K. senegalensis* Extracts

Using a preliminary reagent-based phytochemical analysis of extracts from the stem bark, root and leaves of *K. senegalensis*, Kubramawa *et al.* (2008) showed that aqueous and ethanolic extracts from all the parts contained saponins, tannins and phenolics but did not contain flavonoids, glycosides and resin. On the contrary, the presence of glycosides has been reported in the stem bark aqueous extract (Sanni *et al.*, 2005) whereas flavonoids were detected in the leaf extract (Ndjonka *et al.*, 2011). Thus the contrasting findings could be related to the geography and/or collection time of the plant material. On the other hand, quantitative phytochemical analysis of root sample of the plant showed a high quantity of tannins (7.12 mg/100g), phytate (4.75 mg/100g) and alkaloids (3.36 mg/100g) (Idu and Igeleke, 2012). Also, the presence of mineral elements which include calcium, potassium, sodium, magnesium, iron, copper and zinc has been demonstrated in the stem bark and root samples (Sanni *et al.*, 2005; Idu and Igeleke, 2012). Even though not much information can be derived from these preliminary informations but the overall picture of the plant's phytochemistry can be snapped.

Isolated Bioactive Agents from *K. senegalensis* and their Biological Activity

Extensive literature search reveals that the major bioactive compounds isolated from different parts and extracts of *K. senegalensis* are limonoids. The nomenclature limonoid is derived from the compound limonin, one of the primary bitter components in citrus fruits. The structural determination of limonin in 1960 marked the beginning of limonoid (tetranortriterpenoid) chemistry (Arigoni *et al.*, 1960). Today, limonoids are regarded as group of structurally diverse and highly oxygenated tetracyclic triterpene derivatives (tetranortriterpenoids), with fascinating structural features and are commonly found in plants of the families Meliaceae, Rutaceae, Cneoraceae and Simaroubaceae of the order Rutales. They are derived from a precursor with a 4,4,8-trimethyl-17- furanylsteroid skeleton derived from euphane (H-20β) or tirucallane (H-20α) triterpenoids, and, in general, have intensely bitter taste. Approximately 1300 limonoids, exhibiting more than 35 different carbon frameworks created through ring fission, re-cyclization, reopening, re-closure, and skeletal rearrangements, have been observed over the past five decades, with new structural types continuing to appear (Zhang *et al.*, 2009). So far, more than forty five limonoids have been isolated from different parts of *K. senegalensis* but in this chapter, we will discuss only those with established chemical structure and reported biological activity.

The first limonoid isolated from *K. senegalensis* in 1996 was khayanolide limonoid which results from a cleavage between C-1, C-2 and a linkage between C-1, C-30 of phragmalin (Olmo *et al.*, 1996; Zhang *et al.*, 2009). Subsequently, Khalid *et al.* (1998) reported the isolation of a novel limonoid, 2,6-dihydroxyfissinolide and two known limonoids, fissinolide and methyl 3b-acetoxy-6-hydroxy-1-oxomeliac-14-enoate from the bark of *K. senegalensis*. Fissinolide was found to be active *in vitro* against chloroquine sensitive strain of *P. falciparum* and *Leishmania major* whereas 2,6-dihydroxy fissinolide was slightly active against the two protozoans but methyl 3b-acetoxy-6-hydroxy-1-oxomeliac-14-enoate had no antiprotozoal activity.

In 2001, another research group identified the presence of two mexicanolide-type limonoids named khayanone and 2-hydroxyseneganolide as well as one rearranged phragmalin limonoid 1-O-acetylkhayanolide A (Nakatani *et al.*, 2001). All the compounds were tested for antifeedant activity against the third-instar larvae of *Spodoptera littoralis* (Boisduval) as well as antiviral activity against virus-induced cytopathicity in MT-4 cells. Results from the experiment suggest that only 1-O-acetylkhayanolide A showed appreciable antifeedant activity whereas all the compounds did not show antiviral activity. Furthermore, this research group isolated three new modified limonoids from the acetone extract of *K. senegalensis* bark and were named khayanolide D, khayanolide E and khayanoside (Nakatani *et al.*, 2002). Unfortunately, however, only khayanolide E showed good antifeedant activity against the third-instar larvae of *S. littoralis* and none of three isolated compound showed antiviral activity against HIV-1 replication. Eventhough these compounds did not show remarkable antifeedant and antiviral activity but in our opinion, these findings could not be used to disregard this compounds from the list of bioactive phytochemicals from this plant because they may possess exceptional activity in other bioassays.

In other studies by the Nakatani group, the isolation, characterization, antifeedant and growth inhibitory properties of another three new rearranged phragmalin-type limonoids named khayanolides A, B and C as well as seneganolide, methyl angolensate with its 6-hydroxy and 6-acetoxy derivatives, 1-O-acetylkhayanolide B, and a mexicanolide-type limonoid, khayalactol (Abdelgaleil *et al.*, 2001; El Aswad *et al.*, 2003). Although all the four compounds displayed remarkable antifeedant and growth inhibitory activity against *S. littoralis* larval growth but khayanolide B was the most potent in the two studies with an EC_{50} of 2.19 mg/kg and 6.96 mg/kg bw respectively. This study demonstrate the great possibility of developing new antifeedant and/or larvicidal agent(s) from these *K. senegalensis* derived phytochemicals when pursues by pharmaceutical industries.

Although El Aswad *et al.* (2003) described the isolation of 1-O-acetylkhayanolide B from the acetone extract but Zhang *et al.* (2007) reported the isolation and characterization of this compound from the methanol extract of the stem bark in addition to another limonoid named 3α,7α-dideacetylkhivorin. Both the two compounds were subjected to extensive anticancer studies and results obtained suggested that 3α,7α-dideacetylkhivorin is a potent anticancer and antitumor agent and the activity was linked to14,15β epoxide moiety of the compound. However, 1-O-acetylkhayanolide B showed a poor anticancer activity. It seems therefore that 1-O-acetylkhayanolide B, though abundant and repeatedly isolated from the stem bark, did not possess much bioactivity based on the existing data but an array of pharmacological assays need to be conducted on the compound before a definite statement can be made.

The fruits of *K. senegalensis*, though not commonly used in traditional medicine, were also investigated for the presence of limonoids. Three new mexicanolide limonoids containing a rare conjugated diene lactone system named seneganolide A, 2-hydroxyseneganolide A and 2-acetoxyseneganolide A as well as two known limonoids, 3- deacetyl-7-deacetoxy-7-oxokhivorin and methyl 6-hydroxyangolensate, were isolated and characterized from the fruits (Abdelgaleil *et al.*, 2004). Radial growth technique was then used to evaluate the antifungal activity of seneganolide A, 2-acetoxyseneganolide A and methyl 6-hydroxyangolensate on the fungus

Botrytis cinerea and 2-acetoxyseneganolide A displayed the best inhibitory activity on the mycelial growth of the fungus among other phytochemicals. These compounds are also worth pursuing for other pharmacological properties.

Another research group from China is worth mentioning because of their extensive studies on limonoids and elucidation of another type of limonoids from *K. senegalensis* named khayalenoids (not khayanolides) but in most cases failed to subject the isolated limonoids to a biological screening. They described the isolation of khayalenoids A and B with unprecedented 8-oxa-tricyclo[4.3.2.02,7]undecane motif (Yuan *et al.*, 2009) and seven novel limonoids with different arrangements, named khayalenoids C – I (Yuan *et al.*, 2010). These khayalenoids need to be subjected to extensive bioassays in order to understand the likely pharmacological effects they exert.

Apart from limonoids, another group of phytochemicals isolated from *K. senegalensis* are dimeric proanthocyanidins fisetinidol-(4α,6)-catechin (proanthocyanidin B3) and catechin-(4α,8)-catechin (Kayser and Abreu, 2001). These compounds were tested against the promastigotes form of *L. donovani*, *L. major*, *L. infantum* but activity was not detected in the experiment. Interestingly, authors observed significant activity when the compounds were tested against the amastigote forms, the stage after invading the *Leishmania* parasite in the macrophages. These observations were explained by a subsequent experiment which revealed that there was an auxiliary stimulation of macrophages mediated defense mechanisms by the compounds. Also from this study, it is possible to deduce that not only limonoids and related terpenoid compounds are responsible for the claimed pharmacological activities of this plant and as such the roles of other bioactive agents need to be elucidated.

In an attempt to identify the phenolics content in the different parts of *K. senegalensis*, Atawodi *et al.* (2009) used analytical HPLC, GC-MS and LC-MS analyses to identify phenolics in the leaf extract which include catechin, rutin and quercetin rhamnoside whilst catechin and procyanidins were identified in the methanol extract of the stem bark. Procyanidins of unknown identity were detected in the root extract.

Potential Application of *Khaya senegalensis* Gum in Pharmaceutical Formulations

Another important product of *K. senegalensis* is the gum usually deposited at the bark. It is a pale greenish yellow to golden yellow gum. The *K. senegalensis* gum is a polysaccharide with a galactan in which the 1-3 linked β-D galactopyranosyl residues are concentrated in the inner chain. It has also been reported to contain both D-glucoronic and D-galactoronic acids (Olayemi *et al.*, 2011). Furthermore, analysis of the gum showed that approximately 5 per cent was in the free acid form and the remainder was largely a calcium salt (Aslam *et al.*, 2006). Also, it is slightly soluble in water and this is a peculiar characteristic of the meliaceae gum producers. It swells to about ten times its original weight in water and the swelling of a linear polymer without dissolution has been said to be an indication that it is crossed linked. Thus, the unique properties of the *K. senegalensis* gum have attracted scientific attention to establish its suitability in some pharmaceutical formulations and consequently be exploited by the pharmaceutical industries.

Table 8.1: Some scientifically validated pharmacological activities of *K. senegalensis* extracts.

Pharmacological Activity	*Parts Used*	*Type of Extract*	*References*
Antimalarial	Stem bark	Ethanolic	Ndjonka *et al.*, 2012
	Stem bark and leaf	Ethanolic	Karou *et al.*, 2003
Antihelminthic	Stem bark	Ethanolic	Ademola *et al.*, 2004; 2009
	Stem bark	Ethanolic	Ndjonka *et al.*, 2011
Antitrypanosomal	Stem bark	Methanolic	Atawodi *et al.*, 2003
	Stem bark	Aqueous and methanolic	Wurochekke and Nok 2004
	Stem bark	Aqueous	Ibrahim *et al.*, 2008
	Stem bark	Dichloromethane	Aderbauer *et al.*, 2008
	Stem bark	Ethanolic	Umar *et al.*, 2010
Antibacterial	Stem bark	Ethanolic	Kone *et al.*, 2004
	Stem bark and root	Aqueous and ethanolic	Kubmarawa *et al.*, 2008
	Root	Aqueous and ethanolic	Ide and Igeleke, 2012
Antiinflammatory	Stem bark	Aqueous	Thioune *et al.*, 2000; 2003
Analgesic	Stem bark	Aqueous	Lompo *et al.*, 2007a
Antioxidant	Stem bark	Ethanolic	Lompo *et al.*, 2007b
	Stem bark, root and leaf	Methanol	Atawodi *et al.*, 2009
Larvicidal	Stem bark	Methanol, acetone, ethanol and hexane	Shaalan *et al.*, 2006
Pesticidal	Seed	Oil	Bamaiyi *et al.*, 2006
Anticoccidal	Stem bark	Aqueous	Nwosu *et al.*, 2011
Anticancer	Stem bark	Methanol	Androulakis *et al.*, 2006
Antianaemic	Stem bark	Aqueous	Sanni *et al.*, 2008
Immunomodulatory	Stem bark	Ethanolic	Koko *et al.*, 2008
Antidiarrhoea	Leaf	Aqueous and methanolic	Nwosu *et al.*, 2012
Antidiabetic	Stem bark	Aqueous	Kolawole *et al.*, 2012

Adenuga *et al.* (2008) evaluated the binding properties of the *K. senegalensis* gum in a paracetamol tablet formulation and found that the gum produced tablets with strong mechanical properties, less tendency to laminate as well as long disintegration and dissolution times which suggest that it will be an appropriate commercial binding agent especially when higher mechanical strength and slower release profiles of tablets are desired. Other authors investigated the physicochemical properties of the gum and reported that the gum had a swelling ability that may provide potentials for its use as a disintegrant in tablet formulation and as a hydro gel in modified release dosage forms (Mahmud *et al.*, 2008). Furthermore, the rheological flow properties may also provide potentials for its use as suspending and emulsifying agents owing to its pseudo plastic and thixotropic flow patterns. In a follow up study, these authors also investigated the suspending properties of the gum in paracetamol suspensions (Mahmud *et al.*, 2009). Results from the experiment revealed that the *K. senegalensis* gum at 5 per cent w/v concentration produced a flocculated, re-dispersible and crystal free paracetamol suspension which was stable throughout the storage period of 8 weeks. The results further demonstrated that *K. senegalensis* gum has potential to be used as suspending agent in paracetamol suspensions and may provide substitute excipients in the liquid formulation of paracetamol and perhaps curb the problem of paediatric mortality associated with the use of alternative organic chemicals such as di ethylene glycol.

On the other hand, the swelling ability of the *K. senegalensis* gum attracted some investigators to study the release property of the gum in a chloroquine phosphate tablet formulation (Olayemi *et al.*, 2011). Batches of tablets containing the gum were observed to be softer, disintegrated faster and had greater dissolution than the batches of tablets with microcrystalline cellulose. An increase in the concentration of the gum led to slower disintegration of the tablets. Furthermore, an increase in the concentration of the gum led to a retardation in the release of the drug and the effective concentration for fast disintegration and quick drug release was found to be 5 per cent w/w.

Although more studies could be needed in this subject but results from these experiments have provided clear and unequivocal evidence that the gum of *K. senegalensis* could potentially be exploited in pharmaceutical formulations, both tablets and syrups, so as to reduce over dependence on synthetic agents. These observations also have far reaching economic consequences to the pharmaceutical industries (especially in developing countries) and drug end users because huge amount of resources are currently used in the importation of pharmaceutical excipients which usually increase the production cost and market prices of the products.

Conclusions

Khaya senegalensis possesses multiple bioactivities that justify its wide usage for the treatment of an array of tropical diseases in different countries of Africa. The plant also presents numerous opportunities that could be exploited by developing countries for health management and economic development. However, the level of toxicity is a serious concern that warrants further experimental and clinical investigations before the full utilization of this plant by humans. A number of compounds present in this plant may be involved in the toxicological effects when most other compounds may have beneficial effects on health. So the isolation of more pure compounds from this plant and investigating the individual pharmacological and toxicological effects can reveal the more benefits and potentials of this plant. A number of studies on this plant are currently underway from our lab which may answer many of these unanswered questions.

Acknowledgements

This study was supported by a competitive research grant from the Research office, University of KwaZulu-Natal (UKZN), Durban; an incentive grant for rated researchers and a grant support for women and young researchers from the National Research Foundation (NRF), Pretoria, South Africa.

References

Abdelgaleil, S.A.M., Iwagawa, T., Doe, M., and Nakatani, M. (2004). Antifungal limonoids from the fruits of *Khaya senegalensis*. *Fitoterapia*,**75**: 566–572.

Abdelgaleil, S.A.M., Okamura, H., Iwagawa, T., Sato, A., Miyahara, I., Doe, M., and Nakatani, M. (2001). Khayanolides, rearranged phragmalin limonoids antifeedants from *Khaya senegalensis*. *Tetrahedron*, **57**: 119–126.

Adebayo, J. O., Yakubu, M. T., Egwim, E. C., Owoyele, V. B., and Enaibe, B. U. (2003). Effect of ethanolic extract of *Khaya senegalensis* on some biochemical parameters of rat kidney. *Journal of Ethnopharmacology*, **88**: 69-72.

Adeiza, A. A., Makeri, H. K., and Mohammed, M. (2010). Comparative *in vitro* trypanocidal activities of water and methanol extracts of three parts of *Khaya senegalensis* on *Trypanosoma evansi*. *Journal of Medicinal Plant Research*, **4(14)**: 1365-1369.

Ademola, I.O., Fagbemi, B.O., and Idowu, S.O. (2004). Evaluation of the anthelmintic activity of *Khaya senegalensis* extract against gastrointestinal nematodes of sheep: *in vitro* and *in vivo* studies. *Veterinary Parasitology*, **122**: 151–164.

Ademola, I. O., Fagbemi, B. O., and Idowu, S. O. (2009). Bioseparation and activity of *Khaya senegalensis* fractions against infective larvae of *Haemonchus contortus*. *Veterinary Parasitology*, **165:** 170-174

Adenuga, Y. A., Odeku, O. A., Adegboye, T. A., and Itiola, O. A. (2008). Comparative evaluation of the binding properties of two species of Khaya gum polymer in a paracetamol tablet formulation. *Pharmacy Development and Technology*,**13(6):** 473-480.

Aderbauer, B., Clausen, P., Kershawc, O., and Melzig, M. F. (2008). *In vitro* and *in vivo* trypanocidal effect of lipophilic extracts of medicinal plants from Mali and Burkina Faso. *Journal of Ethnopharmacology*, **119**: 225-231.

Antia, R. E., Olayemi, J. O., Aina, O. O., and E. O. Ajaiyeoba (2009). *In vitro* and *in vivo* animal model antitrypanosomal evaluation of ten medicinal plant extracts from south west Nigeria. *African Journal of Biotechnology*, **8 (7)**: 1437-1440.

Androulakis, X. M., Muga, S. J., Chen, F., Koita, Y., Toure, B., and Wargovich, M. J. (2006). Chemopreventive effects of *Khaya senegalensis* bark extract on human colorectal cancer. *Anticancer Research*, **26**: 2397-2406.

Arigoni, D., Barton, D.H.R., Corey, E.J., Jeger, O., Caglioti, L., Dev, S., Ferrini, P.G., Glazier, E.R., Melera, A., Pradhan, S.K., Schaffner, K., Sternhell, S., Templeton, J.F., and Tobinaga, S. (1960). The constitution of limonin. *Experientia*, **16:** 41–49.

Arnold, R. (2004). *Khaya senegalensis*: current use from its natural range and its potential in Sri Lanka and elsewhere in Asia. In: Bevege *et al*. (2004). 9 pp.

Aslam, M., Geoffrey, P., and Glyn, O. (2006). Properties of *Khaya grandifoliola* Gum. *Journal of the Science of Food and Agriculture*, **29(6):** 563-568.

Atawodi, S.E., Ameh, D.A., Ibrahim, S., Andrew, J.N., Nzelibe, H.C., Onyike, E., Anigo, K.M., Abu, E.A., James, B.D., Njoku, G.C., and Sallau, A.B. (2002). Indigenous knowledge system for treatment of trypanosomiasis in Kaduna State of Nigeria. *Journal of Ethnopharmacology*, **79:** 279-282.

Atawodi, S.E., Atawodi, J.C., Pala, Y., and Idakwo, P. (2009). Assessment of polyphenol profile and antioxidant properties of leaves, stem and root barks of *Khaya senegalensis* A Juss. *Electronic Journal of Biology*, **5:** 80-84.

Atawodi, S.E., Bulus, T., Ibrahim, S., Ameh, D.A., Nok, A.J., Mamman, M., and Galadima, M. (2003). *In vitro* trypanocidal effects of methanolic extracts of some Nigeria savannah plants. *African Journal of Biotechnology*, **2:** 317-321.

Bamaiyi, L.J., Ndams, I.S., Toro, W.A., and Odekina, S. (2006). Effect of mahogany *Khaya senegalensis* seed oil in the control of *Callosobruchus maculatus* on stored cowpea. *Plant Protection Science*, **42:** 130–134.

El-Aswad, A.F., Abdelgaleil, S.A.M., and Nakatani, M. (2004). Feeding deterrent and growth inhibitory properties of limonoids from *Khaya senegalensis* against the cotton leaf worm, *Spodoptera littoralis*. *Pest Management Science*, **60:** 199–203.

Etuk, E.U., Bello, S.O., Isezuo, S.A., and Mohammed, B.J. (2010). Ethnobotanical survey of medicinal plants used for the treatment of diabetes mellitus in the North Western region of Nigeria. *Asian Journal of Experimental Biological Sciences*, **1:** 55-59

Ibrahim, M.A., Njoku, G.C., and Sallau, A.B. (2008). *In vivo* activity of stembark aqueous extract of *Khaya senegalensis* against *Trypanosoma brucei*. *African Journal of Biotechnology*, **7:** 661-663.

Idu, M. C., and Igeleke, C. L. (2012). Antimicrobial Activity and Phytochemistry of *Khaya senegalensis* Roots. *International Journal of Ayurvedic and Herbal Medicine*, **2(3):** 416-422.

IUCN (2010). IUCN Red List of Threatened Species. Version 2010.3.

Karou, D., Dicko, M. H., Sanon, S., Simpore, J., and Traore, A. S. (2003). Antimalarial activity of *Sida acuta* Burm. f. (Malvaceae) and *Pterocarpus erinaceus* Poir. (Fabaceae). *Journal of Ethnopharmacology*, **89:** 291–294.

Kayser, O., and Abreu, P.M. (2001). Antileishmania and immunostimulating activities of two dimeric proanthocyanidins from *Khaya senegalensis*. *Pharmaceutical Biology*, **39(4):** 284-288.

Khalid, S.M., Friedrichsen, G.M., Kharazmi, A., Theander, T.G., Olsenk, C.E., and Christensen, S.B. (1998). Limonoids from *Khaya senegalensis*. *Phytochemistry*, **49:** 1769-1772.

Koko, W. S., Mesaik, M. A., Yousaf, S., Galala, M., and Choudhary, M. I. (2008). *In vitro* immunomodulating properties of selected Sudanese medicinal plants. *Journal of Ethnopharmacology*, **118:** 26–34.

Kolawole, S. O., Kolawole, A. T., and Akanji, M. A. (2011). Effects of aqueous extract of *Khaya senegalensis* stem bark on biochemical and haematological parameters in rats. *Journal of Pharmacology and Toxicology*, **6(6):** 602-607.

Kolawole, O. T., Kolawole, S. O., Ayankunle, A. A., and Olaniran, O. I. (2012). Anti-hyperglycemic effect of *Khaya senegalensis* stem bark aqueous extract in Wistar rats. *European Journal of Medicinal Plants*, **2(1):** 66-73.

Koné, W.M., Atindehou, K.K., Terreaux, C., Hostettmann, K., Traoré, D., and Dosso, M. (2004). Traditional medicine in North Côte-d'Ivoire: screening of 50 medicinal plants for antibacterial activity. *Journal of Ethnopharmacology*, **93**: 43–49.

Kubmarawa, D., Khan, M. E., Punah, A. M., and Hassan, M.(2008).Phytochemical screening and antimicrobial efficacy of extracts from *Khaya senegalensis* against human pathogenic bacteria. *African Journal of Biotechnology*, **7(24):** 4563-4566.

Lompo, M., Dubious, J., and Guissou, I P. (2007b). *In vitro* preliminary study of free radical scavenging activity of extracts from *Khaya senegalensis*. A. Juss. *Journal of Biological Sciences*, **7(4):** 677-680.

Lompo, M., Guissou, I. P., Dubious, J., Dehaye, J. P., Ouedraogo, S., Traore, A., and Some, N. (2007a). Mechanism of anti-inflammatory activity of *Khaya senegalensi* A. Juss. *International Journal of Pharmacology*, **3(2):** 137-142.

Mahmud, H. S., Oyi, A.R., and Allagh, T.S. (2008). Studies on some physico-chemical properties of *Khaya senegalensis* gum. *Nigerian Journal of Pharmaceutical Sciences*, **7(1):** 147-153.

Mahmud, H. S., Oyi, A. R., and Allagh, T.S. (2009). Evaluation of the suspending properties of *Khaya senegalensis* gum in paracetamol suspensions.*Nigerian Journal of Pharmaceutical Sciences*, **8(1):** 128-134.

Nakatani, M., Abdelgaleil, S.A.M., Kassem, S.M.I., Takezaki, K., Okamura, H., Iwagawa, T., and Doe, M. (2001). Antifeedant rings B and D opened limonoids from *Khaya senegalensis*. *Journal of Natural Product*, **64:** 1261–1265.

Nakatani, M., Abdelgaleil, S.A.M., Kassem, S.M.I., Takezaki, K., Okamura, H., Iwagawa, T., and Doe, M. (2002). Three new modified limonoids from *Khaya senegalensis*. *Journal of Natural Product*, **65:** 1219–1221.

Ndjonka, D., Agyare. C., Lüersen, K., Djafsia, B., Achukwi, D., Nukenine, E.N., Hensel, A., and Liebau, E. (2011). *In vitro* activity of Cameroonian and Ghanaian medicinal plants on parasitic (*Onchocerca ochengi*) and free-living (*Caenorhabditis elegans*) nematodes. *Journal of Helminthology*, **85**: 304–312.

Ndjonka, D., Bergmann, B., Agyare, C., Zimbres, F. M., Lüersen, K., Hensel, A., Wrenger, C., and Liebau, E. (2012). *In vitro* activity of extracts and isolated polyphenols from West African medicinal plants against *Plasmodium falciparum*. *Parasitology Research*. DOI 10.1007/s00436-012-2905-y.

Nikles, D. G., Bevege, D. I., Dickinson, G. R., Griffiths, D. W., Reilly, D. F., and Lee, D. J. (2008). Developing African mahogany (*Khaya senegalensis*) germplasm and its management for a sustainable forest plantation industry in northern Australia: progress and needs. *Australian Forestry*, **71(1):** 33-47.

Nwosu, C.O., Yahayah, K., and Igbokwe, I. O. (2011). Toxicity and anticoccidial efficacy of some plants used in the traditional treatment of avian coccidiosis in semi-arid northeastern Nigeria. *Research Journal of Parasitology*, **6(1)**: 18-30.

Nwosu, C. U., Hassan, S. W., Abubakar, M. G., and Ebbo, A. A. (2012). Anti-diarrhoeal and toxicological studies of leaf extracts of *Khaya senegalensis*. *Journal of Pharmacology and Toxicology*, **7(1)**: 1-10.

Olayemi, O., Abolude, O., and Mahmud, H. S. (2011). Effect of concentration on the release property of *Khaya senegalensis* gum in chloroquine phosphate tablet formulation. *International Journal of Pharmaceutical Research and Development*, **2(11):** 8.17.

Olmo, L.R.V., Silva, M.F., Fo, E.R., Vieira, P.C., Fernandes, J.B., Marsioli, A.J., Pinheiro, A.L., and Vilela, E.F. (1996). Rearraged limonoids from *Khaya senegalensis*. *Phytochemistry*, **42**: 831–837.

Sanni, F.S., Ibrahim, S., Esievo, K.A.N., and Sanni, S. (2005). Effect of oral administration of aqueous extract of *Khaya senegalensis* stem bark on phenylhydrazine-induced anaemia in rats. *Pakistan Journal of Biological Science*, **8(2)**: 255-258.

Sexton G. J., Frere, C. H., Dieters, M. J., Godwin, I. D., and Prentis, P. J. (2010). Development and characterization of microsatellite loci for *Khaya senegalensis* (meliaceae). *American Journal of Botany*, 111–113.

Shaalan, E. A., Canyon, D.V., Younes, M.W., Abdel-Wahab, H., and Mansour, A.H. (2006). Efficacy of eight larvicidal botanical extracts from *Khaya senegalensis* and *Daucus carota* against *Culex annulirostris*. *Journal of American Mosquito Control Association*, **22(3):** 433-436.

Styles, B. T. (1972). The flower biology of the Meliaceae and its bearing on tree breeding. *Silvae Genetica*, **21** : 175 – 182.

Tchacondo, T. Karou, S. D., Batawila, K., Agban, A., Ouro-Bang'na, K., Anani, K. T., Gbeassor, M., and de Souza C. (2011).Herbal remedies and their adverse effects in Tem tribe traditional medicine in Togo. *African Journal of Traditional,Complementary and Alternative Medicine*, **8(1):** 45-60.

Thioune, O., Ahodikpe, D., Dieng, M., Diop, A., Ngom, S., and Lo, I.(2000). Inflammatory ointment from shea butter and hydro-alcoholic extract of *Khaya senegalensis* barks (Cailcederat). *Dakar Medical*, **45:** 113–116.

Thioune, O., Diallo, A. K., Diop, A.B., Ngom, S., and Lo, I. (2003).Fractionation of the total extract of *Khaya senegalensis* barks: Value of the formulation of an anti-inflammatory ointment. *Dakar Medical*, **48(1):** 68-71.

Thioune, O., Pousset, J.L., and Lo, I. (1999). Anti-inflammatory activity of the bark of *Khaya senegalensis* (A Juss). Preliminary research of structure/activity relationship. *Dakar Medical*, **44(1)**: 12-15.

Traffic (2000). Mahogany Matters: The U.S. market for big-leafed mahogany and its implications for the conservation of the species. Report by Traffic, North America.

Umar, I. A., Ibrahim, M.A., Fari, N.A., Isah, S., and Balogun, D.A. (2010). *In vitro* and *in vivo* anti-Trypanosoma evansi activities of extracts from various parts of *Khaya senegalensis*. *Journal of Cell and Animal Biology*, 4: 91-95.

van Andel, T., Myren, B., and van Onselen, S. (2012). Ghana's herbal market. *Journal of Ethnopharmacology*, **140**: 368-378

Wurochekke, A.U., and Nok, A. J. (2004). *In vitro* antitrypanosomal activity of some medicinal plants used in the treatment of trypanosomiasis in Northern Nigeria. *African Journal of Biotechnology*, **3**: 481-483.

Yuan, T., Yang, S., Zhang, C., Zhang, S., and Yue, J. (2009). Two Limonoids, Khayalenoids A and B with an Unprecedented 8-Oxa-tricyclo[4.3.2.02,7]undecane Motif, from *Khaya senegalensis*. *Organic letters*, **11**: 617-620.

Yuan, T., Zhang, C., Yang, S., and Yue, J. (2010). Limonoids and Triterpenoids from *Khaya senegalensis*. *Journal of Natural Product*, **73**: 669-674.

Zhang, H., Tan, J., VanDerveer, D., Wang, X., Wargovich, M. J., and Chen F. (2009). Khayanolides from African mahogany *Khaya senegalensis* (Meliaceae): A revision. *Phytochemistry*, **70**: 294-299.

Zhang, H., Wang, X., Chen, F., Androulakis, X.M., and Wargovich, M. J. (2007). Anticancer Activity of Limonoid from *Khaya senegalensis*. *Phytotherapy Research*, **21**: 731–734.

Medicinal Plants: Phytochemistry, Pharmacology and Therapeutics Vol. 3 (2014) *Pages* **235–252**
Editor-in-Chief: **V.K. Gupta**
Published by: **DAYA PUBLISHING HOUSE, NEW DELHI**

9

Phytochemistry and Pharmacology of *Portulaca olearacea*

Neeraj Upmanyu[1*], Surya Prakash Gupta [2], Madhu Gupta[2], Abhinav Kumar[1], Manish Gupta[1], Amol Chandekar[1], Durgesh Nandini[3] and N.S. Chauhan[4]

ABSTRACT

Portulaca oleracea (Portulacaceae) is an important medicinal plant that has been used in traditional medicine. Purslane is beneficial in urinary and digestive problems. It has antifungal and antimicrobial effects. It is high in vitamins, with vitamins A, B, C, and E contained in the plant and minerals and has been used as a food source. Purslane is high in carotenoid content, including beta-carotene. Calcium, magnesium, potassium, folate, and lithium are also present. It possesses marked antioxidant activity. These potential uses have not been verified by clinical studies. Purslane is a rich source of omega-3 fatty acids, which are beneficial in congenital heart disease (CHD) and certain cancers. This review is an attempt to compile the pharmacological and phytochemical literature on Portulaca oleracea, to highlight and critically assess the pharmaceutical potential of this plant.

Keywords: Phytochemisty, Pharmacology, *Portulaca olearacea.*

Introduction

Portulaca oleracea L. (Portulaceae) is a warm climate annual commonly known as purslane in the U.S The plant is succulent, herbaceous, erect or decumbent growing

1 RKDF College of Pharmacy, Bhopal – 462 047, M.P., India
2 Rajiv Gandhi Institute of Pharmacy, Satna – 485 001, M.P., India
3 Sagar Institute of Pharmaceutical Sciences, Sagar, M.P., India
4 Drugs Testing Laboratory and Research Centre, Raipur, C.G., India
* *Corresponding author*: E-mail: neerajupmanyu@rediffmail.com

up to 30 cm height with cylindrical stem of 2-3 mm in diameter. It can be found growing in almost any unshaded area, including flower beds, corn fields, and waste places. Purslane can be found growing in cold climate areas (*e.g.*, Canada) as well as warm areas (*e.g.*, the Caribbean) (Okwuasaba *et al.*, 1987). It has been used in salads and as a medicinal plant (for people) for hundreds of years.

Vernacular Names

Purslane (U.S.), Pusley (U.S.), (**Sanskrit) Lonika, Brihat loni, Khotika**, (Hindi) Lunia, (Manipuri) Leibak kundo, (Tamil) Paruppu keerai, (Malayalam) Koluppa, (Kannada) Dudagorai, (Bengali) Nunia sag.

Taxonomic Hierarchy

Kingdom: Plantae

Subkingdom: Tracheobionta

Superdivision: Spermatophyta

Division: Magnoliophyta

Class: Magnoliopsida

Subclass: Caryophyllidae

Order: Caryophyllales

Family: Portulacaceae

Genus: *Portulaca* L.

Species: *Portulaca oleracea* L.

Geographic Distribution

Portulaca oleracea is a cosmopolitan weed occurring especially in warm areas. It occurs throughout tropical Africa. The origin of its cultivation is uncertain, possibly western Asia or India. It is one of the oldest leafy vegetables, used from Europe to Japan, Australia and the Americas. It is eaten in many African countries, *e.g.*, Côte d'Ivoire, Benin, Cameroon, Kenya, Uganda, Angola and South Africa. It is especially popular in Sudan and Egypt, where it is known in Arabic as 'rigla'. It is distributed all over India, up to 170 m in the Himalaya and in all warm countries. It can be found growing in almost any unshaded area, including flower beds, corn fields, and waste places. Purslane is found all over the world, in the temperate countries of Europe, America, Canada, New Zealand, Australia, abundant in India (The Wealth of India, 2003).

Botanical Description

Purslane is an annual, herbaceous plant, with branched, decumbent on fairly ascending stems of up to 50 cm, 2-3 mm in diameter and which are reddish, fleshy and glabrous (Figure 9.1). The leaves measure 0.5 to 3.3 x 0.2 to 1.5 cm, are obovate, entire and fairly papillose. The flowers are yellow and solitary or in axillary groups of two or three. The fruit is in a capsule (pyxidium) of up to 7 mm. The seeds measure 0.6 to 1 mm; they are reniform, black, and maintain their germinating capacity for

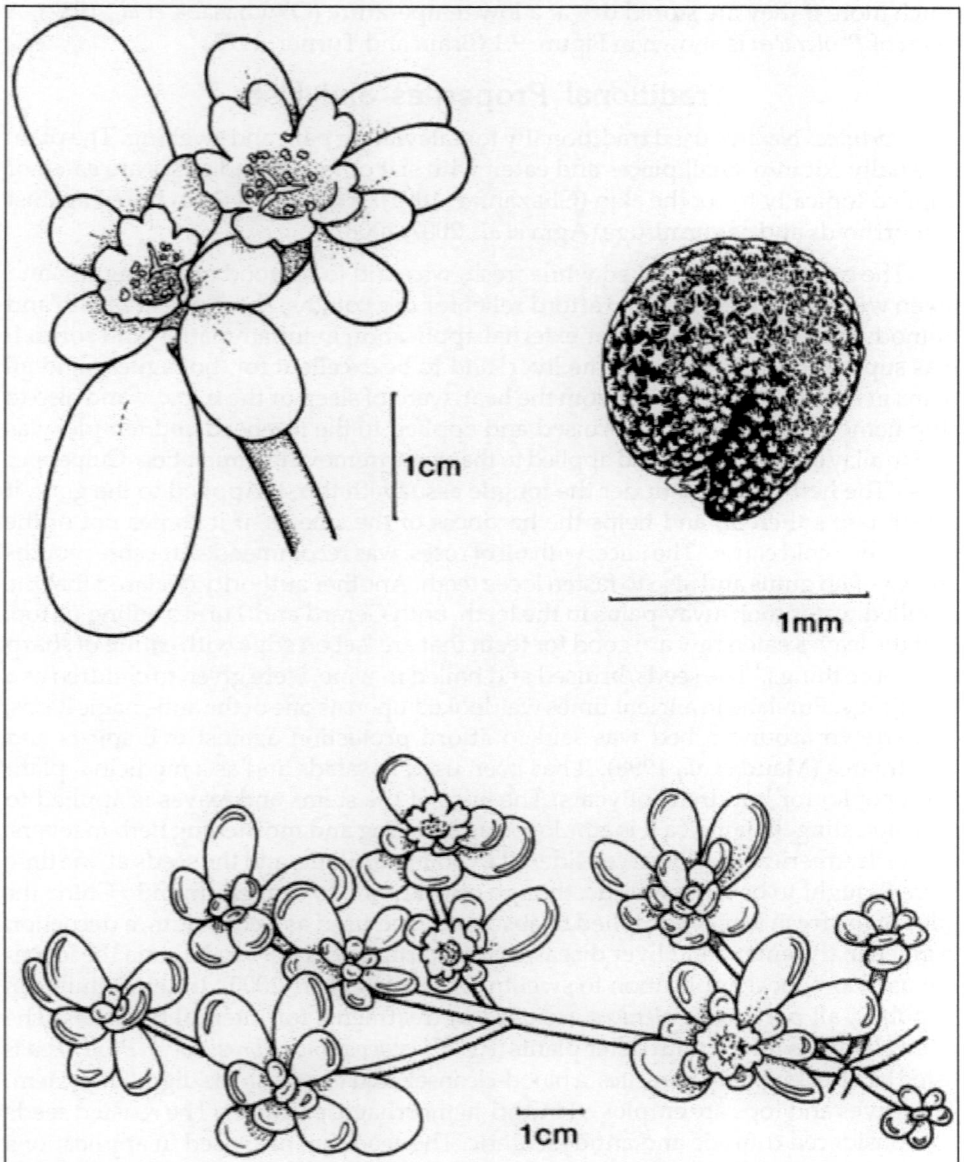

Figure 9.1: Plant of *Portulaca oleracea.*

eight to ten years. Of orthodox behaviour in germination, their viability is maintained much more if they are stored dry at a low temperature (Okwuasaba *et al.*, 1987). A plant of *P. oleracea* is shown in Figure 9.1 (Brain and Turner, 1975)

Traditional Properties and Uses

Portulaca oleracea used traditionally for alleviating pain and swelling. The plant is usually cut into small pieces and eaten with slat or leaves and seeds are eaten or applied topically to soothe skin (Ghazanfar, 1994). It is employed in Brazil against hemorrhoids and as vermifuge (Agra *et al.*, 2007, 2008).

The expressed juice, taken while fresh, was said to be good for strangury, and taken with sugar and honey to afford relief for dry coughs, shortness of breath and immoderate thirst, as well as for external application in inflammation and sores. It was supposed to cool 'heat in the liver' and to be excellent for 'hot agues,' and all pains in the head 'proceeding from the heat, want of sleep or the frenzy,' and also to stop hemorrhages. The herb bruised and applied to the forehead and temple, was said to allay excessive heat, and applied to the eyes to remove inflammation. Culpepper says: 'The herb if placed under the tongue assuayeth thirst. Applied to the gout, it easeth pains thereof, and helps the hardness of the sinews, if it comes not of the cramp, or a cold cause.' The juice, with oil of roses, was recommended for sore mouths and swollen gums and also to fasten loose teeth. Another authority declared that the distilled water took away pains in the teeth, both Gerard and Turner telling us too, that the leaves eaten raw are good for teeth that are 'set on edge with eating of sharp and soure things.' The seeds, bruised and boiled in wine, were given to children as a vermifuge. Purslane in ancient times was looked upon as one of the anti-magic herbs, and strewn around a bed was said to afford protection against evil spirits and nightmares (Maud *et al.*, 1998). It has been used in salads and as a medicinal plant (for people) for hundreds of years. The juice of the stems and leaves is applied to scorpion sting. In Jamaica it is employed as a cooling and moistening herb in fevers. In North America it has been considered a cooling diuretic, and the seeds at one time were thought to be anthelmintic, though now known to be inert. In Indo China the juice of the fresh leaves is applied to abscesses, and used as a collyrium, a decoction is given in dysentery and liver diseases (Nadkarni *et al.*, 1999). In Nigeria the leaves are used as a local application to swellings (Kirtikar *et al.*, 2000). In the Dominican Republic, all parts of *P. oleracea* are used in treatments for internal parasites. The plant always is mixed with other plants (*e.g.*, *Chenopodium ambrosioides*). *P. oleracea* is listed as a treatment for parasites, a blood-cleanser, and to refresh the digestive system. The leaves and tops are employed in anti-hemorrhagic poultices. The roasted seeds are considered diuretic and antidysenteric. The seeds are also used in applications for burns and scalds. In general, their medicinal uses are similar to those mentioned for the herb (Maheshwari *et al.*, 1984).

Traditional Formulations and their Uses

Portulaca Oleracea (Ma Chi Xian) is a product which is used for indications like Anti cancer effect, promote normalization of the biological function of epithelium and healing of ulcers, Anti-bacteria effect, Anti-fungus effect, Anti-blood cell coagulation effect, Anti-hypertensive effect (Oh *et al.*, 2000).

Compound Paste

Paste made with Gokhru, Kakdibij and Javakhar is used in gonorrhea, scanty urine, dose of 2 to 3 ounces.

Juice of Stem

Juice of the stems is applied for the relief to prickly heat as well as to the hands and feet when a burning sensation is felt.

Patents

Use of Purslane to Treat Facial Wrinkles

The invention relates to topical compositions containing components of plants from the Purslane family and methods for improving the aesthetic appearance of skin, particularly, preventing, ameliorating, treating and/or reducing fine lines and/or wrinkles. More particularly, the present invention relates to the use of topical compositions containing extract from the *Portulaca oleracea* plant to treat signs of dermatological aging, especially facial lines and deep wrinkles, and/or improve the aesthetic appearance of the skin. Preferably, the composition is applied once daily to the skin.

Patent number: US7060303

Inventor: Brian C. Jones

Original Assignee: Avon Products, Inc.

Current Assignee: Search USPTO Assignment Database

Primary Examiner: Michael Meller

Attorneys: Anthony M. Santini, Charles J. Zeller, Joan M. McGillycuddy

Portulaca oleracea and Tumor Cell Growth

This invention is directed to the novel use of *P. oleracea* for the treatment of cancer. More specifically it is disclosed that *P. oleracea* has a specific and distinct effect on the inhibition and/or suppression of gastric tumor cell growth *in vitro* and *in vivo*. An aqueous extract of *P. oleracea* showed a tumoricidal activity against KATO III (human gastric carcinoma cell line) and COLO 320 HSR cells (human colon adenoma cell line) in a dose-dependent and time-dependent manner, but not against the non-tumorous cell lines, L929 (murine lung connective tissue) and W138 (human lung diploid cell) cells. Subcutaneous injection of six week old CD1 nude mice with COLO 320 HSR cells and subsequently *Portulaca oleracea* extract showed a clear inhibition of tumor growth as compared to the control nude mice which received only COLO 320 HSR cells.

Patent number: US 5869060

Inventors: Ji-WonYoon, SeungShiHam, HeeSookJun

Original Assignee: Eastwood Biomedical Research, Inc.

Current Assignee: Search USPTO Assignment Database

Phyto-constituents

Leaves and Stems

P. oleracea contains many biologically active compounds and is a source of many nutrients. Some of the biologically active (and, in some case, potentially toxic compounds) include free oxalic acids, alkaloids, omega-3 fatty acids, coumarins, flavonoids, cardiac glycosides, and anthraquinone glycosides. It has high contents of Omega-3 fatty acids and protein (compared to other vegetables) (Ezekwe *et al.*, 1999). The quantity of these compounds in *P. oleracea* varies with the growing conditions (*e.g.*, planting date, soil quality, fertilization) and the age of the plant.

Analysis of edible leaves and stems demonstrated the presence of protein, carbohydrates, mineral matter, calcium, magnesium, oxalic acid, thiamine, riboflavin, nicotinic acid, and vitamin C (29 mg/100 g); carotene (as vitamin A, 3820 I.U/100 g). Vitamin C is highest in the green leaves of young plants and decreases after flowering. A sample from North India contained 16 mg/100 g of total carotenoids, of which about one-third were active in terms of -carotene (vitamin A, 7500 I.U/100 g). The oxalic acid content is by far in excess and cannot be compensated by the fairly high amount of calcium present. Purslane is also rich in sodium and potassium (Sakai *et al.*, 1996). Purslane is rich source of vitamin E, fatty acids and other nutrients, which make it a prime vegetable crop. It is also rich source of omega-3-fatty acid called –linolenic acid. It is suggested that purslane would be a possible crop in arid areas of the Southwestern United States due to its adaptability to both dry conditions and salty soils (Norman, 1992). Leaf wax of *P. oleracea* was studied by Tulloch, 1974 (Tulloch AP, 1974). Total lipids and omega-3-fatty acids in *P. oleracea* were determined in leaves, stems and whole plants at three ages. Significant differences existed in levels of total lipids among ages and between leaves and stems, but no relationship of age to plant part was found (Omara *et al.*, 1991). From the fresh leaves of *P. oleracea* various hydrocarbons have been characterized and their relative distribution determined through GLC studies. The considerable occurrence of branched chain hydrocarbons was considered an indication of the characteristics of lower plant based on taxonomy (Laskar *et al.*, 2002).

From the methanol extract of *P. oleracea*, a monoterpene glucoside, portuloside A, has been isolated was confirmed to be (3S)-3-(3,7-dimethyl-octa-1,7-dien- 6-onyl)–D-glucopyranoside by synthesis from linalool (Sakai *et al.*, 1996). Fresh aerial parts of *P. oleracea* growing in Jordan were found to contain -sitosterol, -sitosterol glucoside, *N, N′*-dicyclohexylurea and allantoin (Asia *et al.*, 2004). Two clerodane-type diterpenes, porwenins A and B have been reported from *P. okinawensis* from biosynthetic and chemosystematic points of view and the structures were elucidated by spectroscopic data (Ohsaki *et al.*, 2001). Oleracins I and II (5-*O*–cellobiosides of betanidin and isobetanidin, respectively) and acylated betacyanins have been reported along with mucilage from leaves. Its acidic fraction consisted of galacturonic acid residues joined by -(14) linkages and neutral fraction composed of arabinose (Su *et al.*, 2005) and galactose with traces of rhamnose (Wealth of India, 2003). *Portulaca oleracea* leaves were found to contain 0.42 per cent of a mucilage mixture.

The mucilage was fractionated into an acidic and a neutral fraction. The acidic fraction consists of galacturonic acid residues joined by á-(1>4)–linkages; 60 per cent of these residues are present as the calcium salt, and esterified galacturonic acid residues are absent. The neutral fraction is composed of 41 per cent of arabinose and 43 per cent of galactose residues, besides traces of rhamnose residues (Sayed *et al.*, 1977). Five flavonoids (kaempferol, apigenin, myricetin, quercetin and luteolin) in different parts of *P. oleracea* L. were identified by a method based on capillary electrophoresis with electrochemical detection (CE-ED) (Xu and Chen, 2006).

Xiang and his coworkers isolated five alkaloids (oleraceins A, B, C, D and E) from *Portulaca oleracea* L., and their structures determined by spectroscopic methods as 5-hydroxy-1-p-coumaric acyl-2,3-dihydro-1H-indole-2-carboxylic acid-6-O-beta-D-glucopyranoside, 5-hydroxy-1-ferulic acyl-2,3-dihydro-1Hindole- 2-carboxylic acid-6-O-beta-D-glucopyranoside, 5-hydroxy-1-(p coumaric acyl-7'-O-beta-D-glucopyranose)-2,3-dihydro-1H-indole-2- carboxylic acid-6-O-beta-D-glucopyranoside, 5-hydroxy-1-(ferulic acyl-7'-Obeta-D-glucopyranose)-2,3-dihydro-1H-indole-2-carboxylic acid-6-O-beta-D glucopyranoside and8,9-dihydroxy-1,5,6,10b-tetrahydro-2H-pyrrolo[2,1-a]isoquinolin-3-one, respectively (Xiang *et al.*, 2005). Besides isolating oleracein A, oleracein B and oleracein E Yang and his coworkers isolated hesperidin and caffeic acid for the first time by various column chromatography methods from *Portulaca oleracea* (Yang *et al.*, 2007). Chromatographic fractionation of the chloroform extract of *Portulaca oleracea* L. growing in Egypt afforded a new clerodene diterpene portulene, in addition to the known compounds lupeol, beta-sitosterol and daucosterol, which were reported for the first time from the title plant (Elkhayat *et al.*, 2008).

Seed

The seeds on extraction with petroleum ether give light green oil (17.4 per cent) with the following constituent fatty acids: palmitic, 10.9; stearic, 3.7; behenic, 1.3; oleic, 28.7; linoleic, 38.9; and linolenic, 9.9 per cent; unsaponifiable fraction yields – sitosterol (Kamil *et al.*, 1998). A crude protein-free extract gave a strong pressor response when injected intravenously into anaesthetized dogs; it was found to contain biologically active 1-noradrenaline, dopamine [4-(2-aminoethyl)-pyrocatechol], and dopa [3-(3,4-dihydroxyphenyl)-alanine], besides an unidentified catechol. The concentration of l-noradrenaline in the fresh plant (2.5 mg/g in one sample) is likely to be greater than that extractable from the suprarenal glands of the mammals. The herb probably contains the bioflavonoid liquiritin. Macerated herb exhibited carbonic anhydrase activity (*The Wealth of India*, 2003). The fatty acid profile and -carotene content of a number of Australian varieties of purslane (*P. oleracea*) were determined by GC and HPLC. These results indicate that Australian purslane varieties are a rich source of alpha linolenic acid and beta carotene (Liu *et al.*, 1992).

Physico-chemical Analysis

Physico-chemical values such as the percentage of ash values (Table 9.1) and extractive values (Table 9.2) were performed according to official methods prescribed (Indian Pharmacopoeia, 1996) and the Who guidelines on quality control methods for medicinal plant materials (WHO/QCMMPM guidelines, 1992).

Table 9.1: Ash values of *P. oleracea*

Sl.No.	*Parameters*	*Values per cent (w/w)*
1.	Total ash	7.37
2.	Acid insoluble ash	0.95
3.	Water soluble ash	1.04
4.	Sulphated ash	11.35

Table 9.2: Extractive values of *P. oleracea*

Sl.No.	*Parameters*	*Values per cent (w/w)*
1	Alcohol soluble extractive	1.91
2	Water soluble extractive	5.16

Preliminary Phytochemical Screening

Preliminary phytochemical screening (Table 9.3) was carried out by using standard procedure described by Kokate (1986) and Harborne (1998).

Table 9.3: Preliminary phytochemical screening of *P. oleracea*

Sl.No.	*Test*	*Methanol extract*
1.	Alkaloids	+
2.	Carbohydrates	+
3.	Glycosides	–
4.	Tannins and phenolic compounds	+
5.	Flavonoids	+
6.	Fixed oil	–
7.	Saponins	+
8.	Proteins and amino acids	+
9.	Steroids	+

Pharmacological Properties

According to the literature *Portulaca oleracea* has been reported to possess hepatoprotective, analgesic and anti inflammatory, antioxidant, anticancer, wound healing, bronchodilator, neuroprotective, hypochloresterolemic and many other biological activities. Further pharmacological studies regarding these activities have been undertaken by various workers which are given below:

Antitussive Activity

The antitussive effect of boiled extract of *Portulaca oleracea* was evaluated using a standard method applied previously by several investigators (Karlsson *et al.*, 1990). Data showed relatively a potent antitussive effect for boiled extract of the plant that was even greater than the effect of codeine (Mohammad *et al.*, 2004).

Hepatoprotective Activity

Treatment of CCl_4 hepatic injured rats with 70 per cent alcohol extract of *P. oleracea* significantly restored the hepatic marker enzymes and total bilirubin to near normal values demonstrating hepatoprotective activity (Elkhayat *et al.*, 2008).

Anti-hyperglycemic Activity

The oral administration of the homogenates of *P. oleracea* reduced the blood sugar level of alloxan-diabetic rabbits to normal (Akhtar *et al.*, 1985).

Hypochloresterolemic Effects

The investigation of hydroalcoholic extract of *P.oleracea* leaves on serum lipids of rabbits fed with hyperchloresteromic diet was performed. Different groups of animals were fed with diet enriched in cholesterol (0.5 per cent). *P. oleracea* extract (200, 400, 800 mg/kg body weight) orally for 12 weeks was administered to hyperchloresterolemic animals. It was found out that serum total cholesterol and atherogenic index decreased in all groups treated with *P. oleracea* extract with respect to positive control group thus indicating that plant may be useful for treatment of hypercholesterolemia (Movahedian *et al.*, 2007).

Neuroprotective Effect

Neuroprotective effects of purslane herb aqueous extracts (PHAS) at doses of 2.5, 5 and 10 mg/(kg day) on SD mice injected daily with D-gal (50 mg/(kg day) by behavioral tests. PHAS-fed mice showed higher activity upon induction by new environmental stimuli, lower anxiety and higher novelty-seeking behavior in the open field tasks, and significantly improved learning and memory ability in step-through compared with D-gal-treated mice. The mechanisms involved in neuroprotective effects of PHAS on mouse brain was significantly increased superoxide dismutase (SOD) activity and decreased the malondialdehyde (MDA) level (Hongxing *et al.*, 2007).

Anti-Nephrotoxic Activity

Aqueous and ethanolic extracts of *P. oleracea* were studied on cisplatin-induced renal toxicity and changes in renal function. Doses (0.2, 0.4, 0.8 g/kg, i.p.) of aqueous and doses (0.5, 1, 2 g/kg, i.p.) of ethanolic extracts, were injected 6 hr or 12 hr before cisplatin and the mentioned doses of aqueous and ethanolic extracts, were injected 6 hr or 12 hr after cisplatin. Functional nephrotoxicity indicators such as BUN (blood urea nitrogen) and Scr (serum creatinine) were elevated in cisplatin-treated rats compared with control. Treatment with aqueous and ethanolic extracts in the highest dose (0.8 and 2 g/kg), 6 and 12 hr before cisplatin injection reduced BUN and Scr. Tubular necrotic damage was not observed either the protective effect of aqueous and ethanolic extracts before cisplatin injection were relatively similar and these effects were dose dependent. Rats treated with aqueous and ethanolic extract, 6 and 12 hr after cisplatin injection had BUN and Scr levels significantly lower than those receiving cisplatin alone but mild to moderate cell injury was observed thus indicating that treatment with aqueous and ethanolic extracts of *P. oleracea* in the highest dose

(0.8 and 2 g/kg), 6 and 12 hr before cisplatin injection reduced BUN and Scr. Tubular necrotic damage was not observed either (Gholamreza *et al.*, 2010).

Intestinal Parasitical Activity

Bioactive compounds appear to be present in all of these plants commonly used to treat intestinal worms. These were *Ambrosia hispida* (Asteraceae), *Aristolochia trilobata* (Aristlochiaceae), *Chenopodium ambrosioides* (Chenopodiaceae), *Portulaca oleracea* (Portulacaceae), and *Artemisia absinthium* (Asteraceae). The cognitive salience of these plant remedies coupled with evidence of biochemical properties suggest that they provide efficacious treatments for controlling intestinal parasite loads (Quinlan *et al.*, 2002).

Antioxidant Activity

P. oleracea was studied for its ability to reduce oxidative stress induced by vitamin A deficiency. Vitamin A-deficient male Wistar rats were divided into four groups which were treated for 30 days with different diets: AIN-93G vitamin A-deficient diet (DD), DD supplemented with pure beta-carotene (beta-D) and DD supplemented with malanga (*Xanthosoma sagittifolium*) (MD) or purslane (*Portulaca oleracea*) (PD) leaves as the only source of vitamin A. The thiobarbituric acid-reactive substances (TBARS), reduced (GSH) and oxidized (GSSG) glutathione, and antioxidant enzyme activities were determined in the heart and liver. The rats fed beta-D, MD and PD showed liver and heart TBARS concentrations lower than did DD rats. The liver GSH concentration of beta-D, MD and PD rats was lower compared to DD rats. Thus suggesting that the ingestion of purslane or malanga leaves may have a protective effect against oxidative stress caused by vitamin A deficiency (Arruda *et al.*, 2004). Three phenolic alkaloids, *i.e.*, oleracein A (OA), oleracein B (OB) and oleracein E (OE), isolated from *Portulaca oleracea* were studied for antioxidant activity, based on scavenging activity against 1,1-diphenyl- 2-picryl-hydrazyl (DPPH) radical and inhibitory effect on hydrogen peroxide-induced lipid peroxidation in rat brain homogenates. The DPPH radical scavenging activities of these phenolic alkaloids were lower than caffeic acid but higher than ascorbic acid and α-tocopherol, being in the following order: OB > OA > OE. OE was most potent in preventing formation of malondialdehyde (MDA) with an EC50 value of 73.13 μM, close to that of caffeic acid (72.09 μM). It was demonstrated that phenolic alkaloids served as a new class of antioxidant agents in this plant (Zijuan *et al.*, 2009).

Anti-fertility Effect

The antifertility effects of alcoholic extract of *P. oleracea* seeds were observed on the reproductive organs of male albino mice after s.c. administrations of 15, 20 and 30 doses (1 dose=50 mg/mouse per alternate day). The treatment produced mass atrophy of spermatogenic elements. Epididymal lumina were devoid of spermatozoa or contained debris. Treatment led to significant decrease in absolute weights of testes, epididymides, vas deferens and seminal vesicles. Administration of 30 doses produced a significant decrease in protein content and sialic acid of testes, epididymides and seminal vesicles remained unaltered after 30 doses while it was drastically reduced in testes. The administration (s.c.) of alcoholic extract of *P. oleracea* seed induced an effective impairment of spermatogenesis (Verma *et al.*, 1982).

Gastric Antiulcerogenic Activity

Aqueous and ethanolic extracts of *P. oleracea* extracts showed a dose-dependent reduction in severity of ulcers induced by HCl or absolute ethanol. The highest dose of extracts exerted similar activity to sucralfate. The oral and intraperitoneal administration of extracts reduced the gastric acidity in pylorus-ligated mice. These results suggested that *P. oleracea* has gastroprotective action and validates its use in folk medicine for gastrointestinal diseases (Karimi *et al.*, 2004).

Antihypertensive Activity

An aqueous extract of the stems and leaves of *P. oleracea* abolished the twitch contraction of the directly stimulated rat hemidiaphragm preparation. The effects of the extract mimic qualitatively the action of potassium oxalate-a known constituent of *P. oleracea* on the diaphragm. Removal of K + ions from the methanol extract by passing it through a cation exchange resin reduced the inhibitory effect of the extract. There was a positive correlation between the concentration of K+ ions in the extract and the effects of potassium chloride of similar molarity. It was concluded that the K+ ion content of *P. oleracea* is at least partly responsible for the relaxant effect observed on the isolated rat diaphragm (Parry *et al.*, 1993). An aqueous extract of *P. oleracea* leaves and stems produced a dose dependent relaxation of guinea pig fundus, taenia coli and rabbit jejunum and a dose dependent contraction of the rabbit aorta. On spontaneously beating rabbit right atria and electrically paced left atria, the extract produced a dose dependent negative inotropic and chronotropic effects. On rat blood pressure, the extract produced dose dependent pressor responses (Parry *et al.*, 1987).

Anti-inflammatory and Analgesic Effect

The significant anti-inflammatory and analgesic effect of ethanolic extract of the aerial parts (dried leaves and stem) of *P. oleracea* ssp *sativa*) showed when compared with the synthetic drug, diclofenac sodium as the active control (Chan *et al.*, 2000).

Wound Healing Activity

The preliminary wound healing activity of *P. oleracea* was studied using *Mus musculus* JVI1. For this purpose fresh homogenized crude aerial parts of *P.oleracea* were applied topically on the excision wound surface as single and two doses in different amounts. Wound contraction and tensile strength measurements were used to evaluate the effect of the plant on wound healing. The results obtained indicated that *P. oleracea* accelerates the wound healing process by decreasing the surface area of the wound and increasing the tensile strength. The greatest contraction was obtained at a single dose of 50mg and the second greatest by two doses of 25mg. Measurements of tensile strength and healed area were in agreement (Rasheed *et al.*, 2003).

Antimicrobial Effect

Aqueous and ether extracts of the herb showed activity against gram-negative bacteria. The antifungal activity of *P. oleracea* extracts against hyphal growth of various fungi was evaluated in real time using an automatic single-cell bioassay system. The antifungal activity of each fraction of *P. oleracea* was evaluated based on the dynamic hyphal growth response curves of test fungi *Aspergillus* and *Trichophyton* and the yeast *Candida*. A crude sample obtained by ethylacetate extract showed a specific and marked activity against dermatophytes of the genera *Trichophyton* (Oh *et al.*, 1998).Whole plant of *P. oleracea* extracted in ethanol was found inhibitory to *Bacillus subtilis* and those extracted in chloroform, ethanol and hexane to *Rhizobium leguminosarum*. The species failed to prove antagonistic to *E. coli* (Banerjee *et al.*, 2003). Fungitoxicity of aqueous and organic solvent (*e.g.*, hexane, ethanol and chloroform) extracts were tested against *Aspergillus niger*, *Rhizopus artocarpi* and *Fusarium* sp. by agar cup assay and filter disc methods. Hexane and aqueous extracts showed antifungal activity against *Fusarium* sp., while ethanol and chloroform extracts of the same herb inhibited the growth of *Rhizopus artocarpi* (Banerjee *et al.*, 2002).

Neuropharmacological Effect

Ethanolic extract of *P. oleracea* var. *sativa*, on intraperitoneal administration, showed a significant reduction in the locomotor activity in mice, antinociceptive activity in rats using tail flick method, an increase in the onset time of pentylenetetrazole-induced convulsions in mice and muscle relaxant activity in *in vitro* (rat hemidiaphragm) and *in vivo* (grip strength) experiments. The anti-nociceptive activity of the extract in rats was attenuated by naloxone pre-treatment indicating the involvement of opioid receptors in its anti-nociceptive effects. It indicated that *P. oleracea* var. *sativa* possesses varied effects on both the central and peripheral nervous system (Radhakrishnan *et al.*, 2001).

Bronchodilatory Effect

The bronchodilatory effect of the boiled extract of *P. oleracea* in the airway of asthmatic patients was examined and the results showed that the boiled extract of *P. oleracea* caused significant increases in all measured pulmonary function tests (PFTs), ($P < 0.05$ to $P < 0.01$). There was no significant difference between the maximum increase in measured PFTs due to the boiled extract and theophylline. However, maximum increase in PEF and MEF(25- 75) due to the boiled extract were significantly lower than those of salbutamol ($P < 0.05$ for both cases). The onset of brochodilatory effect of extract was similar to that of theophylline beginning 60 min, but the effect of extract decline after 120 min after administration. The results of the study showed that *P. oleracea* has a relatively potent but transient bronchodilatory effect on asthmatic airways (Malek *et al.*, 2004).

Urinary Problems

The study was based on ethnobotanical interviews conducted from 1996-2000 in Trinidad and Tobago with thirty male and female respondents. A nonexperimental validation was conducted on the different plants used for urinary problems and diabetes mellitus. Thus establishing that the plants used are safe or effective, to help direct clinical trials, and to inform Caribbean physicians of the plants' known properties to avoid counter-prescribing. *P. oleraceae* was one of the plants having sufficient evidence to support their traditional use for urinary problems (Lans *et al.*, 2006).

Hypoxia Tolerance Activity

Portulaca oleracea extracts had protective effects on hypoxic nerve tissue. After being orally administrated with the PO extracts or distilled water for seven days, adult male BALB/c mice were adapted to a normobaric low oxygen environment (10 per cent oxygen and 90 per cent nitrogen) for different time and then were sacrificed. The results showed that the PO extracts enhanced the EPO mRNA and protein expression in the mouse cortices. Compared to the control group, the mouse in the group treated with the PO extracts by 1 g/day had significantly higher activities of PF, PFK, LDH and higher levels of ATP in the cortices, especially under the hypoxic environment for 24 hours. Histological analysis indicated that the extracts lessened the inflammation damage of the mouse brain. MTT assay results showed the PO extracts raised the viability of the cells under the tested hypoxic conditions and decreased the degree of LDH in the culture medium in a dose-dependent manner demonstrating that the PO extracts had protective effects on hypoxic nerve tissue (Wang *et al.*, 2007).

Anti-phenolic Endocrine Disruptors

Portulaca oleracea showed the ability to efficiently remove from water bisphenol A (BPA), which is well known as an endocrine disrupting compound (EDC) having estrogenic properties. In water culture, 50 muM BPA was almost completely removed within 24 h when the ratio of whole plant weight to the water volume was set up at 1 g to 25 ml. The estrogenic activity of the water decreased in parallel with the elimination of BPA. This plant also rapidly removed other EDCs having a phenol group including octylphenol (OP), nonylphenol (NP), 2,4-dichlorophenol (2,4-DCP) and 17beta-estradiol and, thereby, removed the endocrine disrupting activities. In addition, the ability of *P. oleracea* to remove BPA was not affected by BPA concentration (up to 250 microM), by cultivation in the dark, by temperatures ranging from 15 degrees Celcius to 30 degrees celcius, or by pH ranging from 4 to 7. Moreover, the ability of *P. oleracea* to individually remove BPA, NP, and OP was the same as when they were all present thus suggesting that *P. oleracea* is a promising material for practical phytoremediation of landfill leachates and industrial wastewater contaminated with the tested EDCs (Imai *et al.*, 2007).

TNF-α and IL-6 Inhibitory Activity

The effects of drug-carried serum of the different parts of *Portulace oleracea* on cytokine TNF-α and IL-6 secreted by adipose cell *in vitro* method was explored. The cell viability of each group was tested by Methy thiazolyl tetrazolium (MTT) assay. The levels of TNF-α and IL-6 in the supernatant of cultured adipose cell were assayed by RIA. MTT assay results showed that the drug-carried serum of *P. oleracea* and its different parts act on adipose cell damaged by the high lipid serum, significantly increase the cell viability in thegroups in 40 per cent and 20 per cent concentration, and improve the disorder of lipid in different degree by lowering the levels of TNF-α and IL-6 that adipose cell secreted *in vitro* (Xiao *et al.*, 2005).

Skeletal Muscle Relaxant Property

The skeletal muscle relaxant properties of an aqueous extract of *Portulaca oleracea* were examined on the twitch and tetanus tension evoked by electrical stimulation using the rat phrenic nerve-hemidiaphragm and frog sciatic nervesartorius muscle preparations and on contractures induced by nicotinic agonijsts using the rat *rectus abdominus* muscle preparation. Observations indicated that the aqueous extract possesses unique skeletal muscle relaxant properties which do not appear to involve interference with cholinoceptor mechanism(s) and that the mechanism of action of the extract may involve interference with Ca2+ mobilization in skeletal muscle (Okwuasaba *et al.*, 1986). The aqueous extract of *P. oleracea* produced skeletal muscle relaxation in rats following i.p. or oral administration, as assessed by the prolongation of pullup time (Parry *et al.*, 1987). The i.p. route of administration was more effective. When compared with chlordiazepoxide (20 mg/kg, i.p.), diazepam (40 mg/kg, i.p.) and dantrolene sodium (30 mg/kg, oral), the extract (200-1000 mg/kg, i.p.) proved a more effective skeletal muscle relaxant. With 1000 mg/kg i.p., 80 per cent lethality was seen. The LD50 in an acute toxicity test in mice was 1040 mg/kg i.p (Okwuasaba *et al.*, 1987). The effects of aqueous (AEE), dialysable (DIF) and methanol (MEE) extracts of *P. oleracea* stems and leaves were compared with those of dantrolene sodium and methoxyverapamil (D-600) with respect to inhibition of twitch tension on the rat phrenic nerve-hemidiaphragm and with respect to contracture induced by nicotinic agonists on the frog rectus abdominis preparations. The extracts, dantrolene and D-600 inhibited twitch tension due to indirect electrical stimulation via the phrenic nerve (NS) on hemidiaphragm muscle, whereas the extracts and dantrolene inhibited, in addition, twitch amplitude due to direct muscle stimulation (MS). The extracts and D-600 proved more effective in attenuating nicotinic agonist (acetylcholine, carbachol and nicotine)-induced contractures on the rectus abdominis muscle than dantrolene. From these observations, it appears that the *P. oleracea* extracts mimic, in part, the effect of D-600 and dantrolene on the rat hemidiaphragm and frog rectus abdominis muscles; therefore, the muscle relaxant properties of the extracts may be due, in part, inhibition of trans-membrane Ca influx, interference with the Ca-induced Ca release process and/or inhibition of the release of intracellular Ca from stores in the sarcoplasmic reticulum (Okwuasaba *et al.*, 1987).

Toxicity Studies

Dried powder of *P. oleracea* were extracted by the dipping method with methanol, ethanol, acetone, ethyl acetate, ether, trichloromethane, dichloromethane, benzene and petroleum ether, and bioactivity of the extracts against *Aphis gossypii* Glover including contact toxicity and antifeeding toxicity were approached. The results indicated that the methanol extract showed the highest contact toxicity among the 9 different extracts and the dichloromethane extract had the highest antifeeding toxicity. No data on the toxicity of *P. oleracea* could be found in the literature. However, the plant does contain cardiac glycosides and oxalic acids, which can be toxic (Su *et al.*, 2005).

Conclusion

The herbals occupied a distinct place in the life right from the primitive period till date and provided information on the use of plants or plant products and products as medicine. The plant appears to have broad spectrum of activity on several ailments. It is seen from the literature that *Portulaca oleracea* is a very important plant for its large number of medicinal properties as well as medicinally important chemicals like oxalic acids, alkaloids, omega-3 fatty acids, coumarins, flavonoids, cardiac glycosides, anthraquinone glycosides, alpha linolenic acid, beta carotene and protein. The plant shows many pharmacological activities like anti hyperglycemic activity, skeletal muscle relaxant property, anti nephrotoxic activity, bronchodilatory effect, antioxidant activity, wound healing activity, tumoricidal activity and anti wrinkle activity. The presence of a wide range of chemical compounds indicates that the plant could lead the way for the development of novel agents having good biological activity.

Keeping in mind the preliminary data accumulated from various animal studies, more studies are warranted to unearth the therapeutic potential of the plant. Thus, *Portulaca oleracea* is quite promising as a multipurpose medicinal agent so further clinical trials should be performed to prove its efficacy.

References

Arruda, S.F., Siqueira, E.M., and Souza, E.M. (2004). Malanga (*Xanthosoma sagittifolium*) and purslane (*Portulaca oleracea*) leaves reduce oxidative stress in vitamin A-deficient rats. *Annals of Nutrition and Metabolism*, **48(4):** 288-295.

Akhtar, M.S., Khan, Q.M., and Khaliq, T. (1985). Effects of *Portulaca oleracea* (*Kulfa*) and *Taraxacum officinale* (*Dhudhal*) in normoglycaemic and alloxan-treated hyperglycaemic rabbits. *Journal of Pakistan Medical Association*, **35(7):** 207-210.

Agra, M.F., França, P.F., and Barbosa-Filho, J.M. (2007). Synopsis of the plants known as medicinal and poisonous in Northeast of Brazil. *Rev Bras Farmacogn.*, **17:** 114-140.

Asia, N.R., Fatma, U.A., Mayadeh, S., and Mustaqeem, O.T. (2004). Investigation of active constituents of *Portulaca oleracea* L. (Portulacaceae) grown in Jordan. *Pakistan Journal of Pharmaceutical Sciences*, **17(1):** 37-46.

Brain, K.R., and Turner, T.D. (1975). In: *The Practical Evaluation of Phytopharmaceuticals*, Wright-scienctechnica, Bristol, 4-9.

Banerjee, G., and Mukherjee, A. (2003). Antibacterial activity of a common weed, *Portulaca oleracea* L. *Geobios*, **30(2-3):** 143-144.

Banerjee, G., and Mukherjee, A. (2002). Biological activity of a common weed: *Portulaca oleracea* L.-II. Antifungal activity. *Acta Botanica Hungarica*, **4(3-4):** 205-208.

Chan, K., Islam, M.W., Kamil, M., Radhakrishnan, R., Zakaria, M.N.M., and Habibullah, M. (2000). The analgesic and anti-inflammatory effects of *Portulaca oleracea* L. subs. *Sativa* (Haw.) Celak. *Journal of Ethnopharmacology*, **73(3):** 445-51.

Elkhayat, E.S., Ibrahim, S.R., and Aziz, M.A. (2008). Portulene, a new diterpene from *Portulaca oleracea* L. *Journal of Asian Natural Product Research*, **10(11):** 1039-1043.

E.l. Sayed Amin and Sanaa M. El-Deeb. (1977). Isolation of *portulaca oleracea* (regla) mucilage and identification of its structure. *Carbohydrate Research*, **56(1):** 123-28.

Elkhayat, E.S., Ibrahim, S.R., and Aziz, M.A. (2008). Portulene, a new diterpene from *Portulaca oleracea* L. *Journal of Asian Natural Product Research*, **10(11):** 1039-1043.

Ezekwe Michael, O., Omara Alwala., Thomas, R., and Membrahtu, Tadesse. (1999). Nutritive characterization of purslane accessions as influenced by planting date. *Plant Foods for Human Nutrition (Dordrecht)*, **54(3):** 183-191.

Ghazanfar, S.A. (1994). In: *Handbook of Arabian Medicinal Plants*, CRC press, Boca Raton, FL, pp. 176.

Gholamreza, K., Alireza, K., Abbas, O., and Mahmudreza, K. (2010). Protective effect of aqueous and ethanolic extracts of *Portulaca oleracea* against cisplatin induced nephrotoxicity. *Iranian Journal of Basic Medical Sciences*, **13(2):** 33-35.

Hongxing, Z., Nancai, Y., Guofu, H., Jianbo, S., Yanxia, W., and Hanju, H. (2007). Neuroprotective effects of purslane herb aquenous extracts against D-galactose induced neurotoxicity. *Chemical and Biological Interactions*, **170(3):** 145-152.

Imai, S., Shiraishi, A., Gamo, K., Watanabe, I., and Okuhata, H. (2007). Removal of phenolic endocrine disruptors by *Portulaca oleracea. Journal of Bioscience and Bioengineering*, **103(5):** 420-426.

Kamil, M., Jayaraj, A.F., Ahmad, F., and Gunasekhar, C. (1998). Chemical standardization of *Portulaca oleracea* v. *sativa. Journal of Pharmacy and Pharmacology*, **50**(Suppl.): 259-263.

Karlsson, J.A., Lanner, A.S., and Persson, G.A. (1990). Airway opioid receptors mediate inhibition of cough and reflex bronchoconstriction in guinea pigs. *J. Pharm.Exp. Ther.*, **252:** 863-868.

Karimi, G., Hosseinzadeh, H., and Ettehad, N. (2004). Evaluation of the gastric antiulcerogenic effects of *Portulaca oleracea* L. extracts in mice. *Phytotherapy Research*, **18(6):** 484-487.

Kirtikar, K.R., and Basu, B.D. (2000). Illustrated Indian Medicinal Plants Eds. Mahaskar K.S., Blatter E., Caius J.F. Sri Satguru Publications, Delhi, India, 329-336.

Liu, L., He P., Zhou, Y.F., Xu, Z.Q., Hocart, C., and Zhang, R. (1992). Fatty acids and beta carotene in Australian purslane (*Portulaca oleracea*) varieties. *Journal of Chromatography A*, **893(1):** 207-213.

Laskar, S., Banerjee, G., and Mukherjee, A. (2002). Surface hydrocarbons from the leaves of *Portulaca oleracea* Linn. *Asian Journal of Chemistry*, **14(2):** 1114-1116.

Lans, C.A. (2006). Ethnomedicines used in Trinidad and Tobago for urinary problems and diabetes mellitus. *Journal of Ethnobiology and Ethnomedicine*, **13:** 45-49.

Mohammad, T.B., Mohammad, H.B., and Farhad, M. (2004). Antitussive effect of *Portulaca oleracea* L. in guinea pigs. *Iranian Journal of Pharmaceutical Research*, **3:** 187-190.

Malek, F., Boskabady, M.H., Borushaki, M.T., and Tohidi, M. (2004). Bronchodilatory effect of *Portulaca oleracea* in airways of asthmatic patients. *Journal of Ethnopharmacology*, **93(1):** 57-62.

Maud, G. A. (1998). Modern Herbal – the medicinal, culinary, cosmetic and economic properties,cultivation and folklore of herbs, grasses, fungi, shrubs and trees with all their modern scientific uses. Tiger Books International, London, 600-620.

Movahedian, A., Ghannadi, A., and Vashirnia, M. (2007). Hypochloresterolemic effects of pursalane extract on serum lipids in rabbits fed with high cholesterol levels. *International Journal of Pharmacology*, **3(3):** 285-289.

Maheshwari, J.K., and Singh, J.P. (1984). Contribution to the ethnobotany of Bhoxa tribe of Bijnor and Pauri Garhwal districts Uttar Pradesh India. *Journal of Economic and Taxonomic Botany*, **5(2):** 253-260.

Nadkarni, K.M., and Nadkarni, A.K. (1999). Indian Materia Medica – with Ayurvedic, Unani-Tibbi, Siddha, Allopathic, Homeopathic, Naturopathic and Home remedies. Revised ed., Popular Prakashan, Private Ltd.

Norman, A. (1992). Purslane eyed as rich food source. *Agriculture Research*,**40(2):** 20-21.

Okwuasaba, F., Ejike, C., and Parry, O. (1987). Skeletal muscle relaxant properties of the aqueous extract of *Portulaca oleracea. J Ethnopharmacol* **18:** 103-107.

Oh, K. (2000). Detection of antifungal activity in *Portulaca oleracea* by a single-cell bioassay system. *Phytother Res.*, **14:** 329-332.

Omara, A.T.R., Mebrahtu, T., Prior, D.E., and Ezekwe, M.O. (1991). Omega-3-fatty acids in Purslane *Portulaca oleracea* tissues. *Journal of American Oil Chemists' Society*, **68(3):** 198-199.

Ohsaki, A., Ogawa, M., Imai, Y., Shinzato, T., Shigemori, H., Kobayashi, J., and Porwenins, A. (2001). New clerodane diterpenoids from *Portulaca okinawensis. Journal of Natural Products*, **64(6):** 804-05.

Oh, K.B., Chang, I.M., Hwang, K.J., and Mar, W. (1998). Detection of antifungal activity in *Portulaca oleracea* by a single-cell bioassay system. *Phytotherapy Research*, **14(5):** 329-332.

Okwuasaba, F., Ejike, C., and Parry, O. (1986). Skeletal muscle relaxant properties of the aqueous extract of *Portulaca oleracea. Journal of Ethnopharmacology*, **17(2):** 139-160.

Okwuasaba, F., Ejike, C., and Parry, O. (1987). Comparison of the skeletal muscle relaxant properties of *Portulaca oleracea* extracts with dantrolene sodium and methoxyverapamil. *Journal of Ethnopharmacology*, **20(2):** 85-106.

Okwuasaba, F., Ejike, C., and Parry, O. (1987). Effects of extracts of *Portulaca oleracea* on skeletal muscle *in vitro. Journal of Ethnopharmacology*, **21(1):** 55-63.

Parry, O., Okwuasaba, F., and Ejike, C. (1987). Skeletal muscle relaxant action of an aqueous extract of *Portulaca oleracea* in the rat. *Journal of Ethnopharmacology*, **19(3):** 247-253.

Parry, O., Marks, J.A., and Okwuasaba, F.K. (1993). The skeletal muscle relaxant action of *Portulaca oleracea*: Role of potassium ions. *Journal of Ethnopharmacology*, **40(3):** 187-194.

Parry, O., Okwuasaba, F., and Ejike, C. (1987). Skeletal muscle relaxant action of an aqueous extract of *Portulaca oleracea* in the rat. *Journal of Ethnopharmacology*, **19(3):** 247-254.

Quinlan, M.B., Quinlan, R.J., and Nolan, J.M. (2002). Ethnophysiology and herbal treatments of intestinal worms in Dominica, West Indies. *Journal of Ethnopharmacology*, **80(1):** 75-83.

Rasheed, A.N., Afifi, F.U., and Disi, A.M. (2003). Simple evaluation of the wound healing activity of a crude extract of *Portulaca oleracea* L. (growing in Jordan) in *Mus musculus* JVI-1. *Journal of Ethnopharmacology*, **88(2-3):** 131-136.

Radhakrishnan, R., Zakaria, M.N.M., Islam, M.W., Chen, H.B., Kamil, M., and Al-Attas, A. (2001). Neuropharmacological actions of *Portulaca oleracea* L v. *sativa* (Hawk). *Journal of Ethnopharmacology*, **76(2):** 171-76.

Sakai, N., Inada, K., Okamoto, M., Shizuri, Y., Fukuyama, Y., and Portuloside, A. (1996). A monoterpene glucoside, from *Portulaca oleracea*. *Phytochemistry*, **42(6):** 1625-1628.

Su, M.W., Luo, W.C., Zhang, J.Y., and Chen, Z.L. (2005). Bioactivity of extracts from dried powder of *Portulaca oleracea* L. against *Aphis gossypii*. *Journal of Plant Resources and Environment*, **14(2):** 10-14.

The Wealth of India (2003). A dictionary of Raw materials and Industrial Products, Raw Materials, CSIR, PID, New Delhi, VIII, 2003, 219-220.

Tulloch, A.P. (1974). Leaf wax of *Portulaca oleracea*. *Lipids*, **9(9):** 664-668.

Verma, O.P., Kumar, S., and Chatterjee, S.N. (1982). Anti-fertility effects of common edible *Portulaca oleracea* on the reproductive organs of male albino mice. *Indian Journal of Medical Research*, **75:** 301-310.

Wang, W., Gu, L., Dong, L., Wang, X., Ling, C., and Li, M. (2007). Protective effect of *Portulaca oleracea* extracts on hypoxic nerve tissue and its mechanism. *Asia Pacific Journal of Clinical Nutrition*, **16(1):** 227-233.

Xiao, F.Y., Lu, F.E., and Xu, L.J. (2005). Effect of different parts of *Portulace oleracea* on the levels of TNF-alpha and IL-6 in the supernatant of cultured adipose cell. *Zhongguo Zhong Yao Za Zhi.*, **30(22):** 1763-1766.

Xiang, L., Xing, D., Wang, W., Wang, R., Ding, Y., and Du, L. (2005). Alkaloids from *Portulaca oleracea* L. *Phytochemistry*, **66(21)**: 2595-601.

Zijuan, Y., Cejia, L., Lan, X., and Yinan, Z. (2009). Phenolic alkaloids as a new class of antioxidants in *Portulaca oleracea*. *Phytotherapy Research*, **23(7):** 1032–1035.

Medicinal Plants: Phytochemistry, Pharmacology and Therapeutics Vol. 3 (2014) *Pages* **253–284**
Editor-in-Chief: **V.K. Gupta**
Published by: **DAYA PUBLISHING HOUSE, NEW DELHI**

10

Mango Pulp: A Potential Source of Natural Antioxidant and Antimicrobial Agent

Kalpna D. Rakholiya[1], Mital J. Kaneria[1] and Sumitra V. Chanda[1*]

ABSTRACT

Fruits are an important component of diet. Mangifera indica L. (Anacardiaceae) is one of the important tropical fruits in the world and India. Mango is considered as a king of fruits in Indian delicacy. Present study was taken up to determine the influence of solvents and extraction methods on antioxidant and antimicrobial properties of unripe pulp of M. indica. Extraction was done by individual cold percolation method (petroleum ether, acetone, methanol and water) and decoction extraction method. Quantitative phytochemical analysis was done for total phenol, flavonoid and alkaloid contents. Antioxidant activity was evaluated using six different in vitro antioxidant assays (DPPH free radical scavenging assay, hydroxyl radical scavenging assay, superoxide anion radical scavenging assay, ABTS radical cation scavenging assay, ferric-reducing antioxidant power and reducing capacity assessment). Antimicrobial activity was done by agar well diffusion method against ten Gram positive bacteria, ten Gram negative bacteria and five fungi. The minimum inhibitory concentration (MIC) and minimum bactericidal concentration (MBC) were determined against selected organisms. Various phytochemical (alkaloids, flavonoids, tannins, phlobatanins, triterpenes, steroids, saponin and cardiac glycosides) and physicochemical parameters (total ash value, loss on drying, petroleum ether, methanol, acetone and water soluble extractive values) were also evaluated. Individual cold percolation extraction method and acetone extract proved to be the best extraction method and solvent respectively, to evaluate antioxidant and antimicrobial properties of M. indica pulp. Above results suggest that extraction method and appropriate solvent significantly influences antioxidant or antimicrobial activity of any plant. The extracts

---

1 Phytochemical, Pharmacological and Microbiological Laboratory, Department of Biosciences, Saurashtra University, Rajkot, Gujarat, India

* *Corresponding author*: E-mail: svchanda@gmail.com

showed better antibacterial activity than antifungal activity; Gram negative bacteria were more susceptible than Gram positive bacteria. The acetone extract showed better antioxidant as well as antimicrobial activity. Present study demonstrated that mango pulp, can serve as a potential source of natural antioxidants and antimicrobics for the use in preparation of dietary supplements or nutraceuticals, food ingredients, pharmaceutical and cosmetic products.

Keywords: ABTS, Antimicrobial activity, Antioxidant activity, DPPH, Extraction method, FRAP, Mango, Physicochemical parameters.

GRAPHICAL ABSTRACT

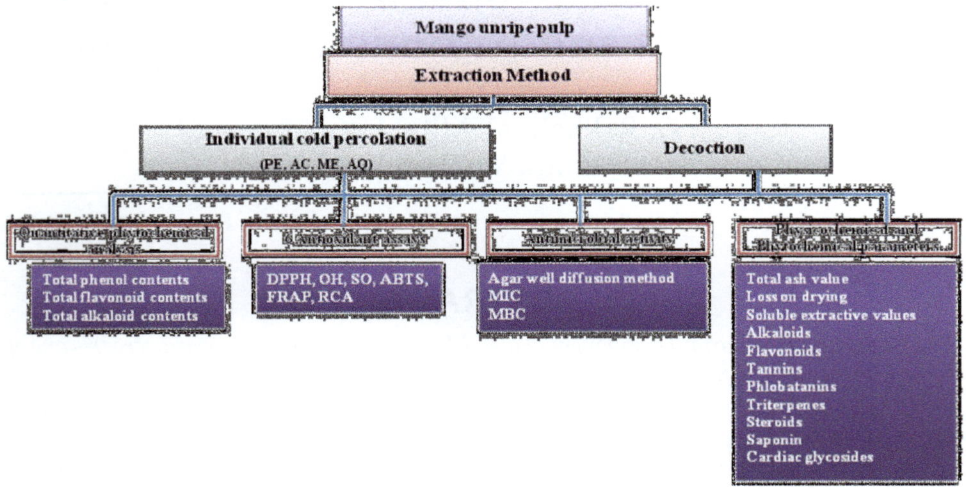

Introduction

Oxidative stress depict the existence of products called free radicals and reactive oxygen species (ROS), which are formed under normal physiological conditions but become deleterious when not being eliminated by the endogenous systems. The free radicals, which are physiologically produced, can exert diverse functions like signaling roles and provide defense against infections (Hancock *et al.*, 2001; Valko *et al.*, 2007) but excessive amount of free radicals can result in cellular damage and initiate several diseases and disorders. The reactive oxygen species (ROS) and the reactive nitrogen species (RNS) including diverse reactive entities namely superoxide ($O2^{\bullet -}$), hydroxyl ($OH^{\bullet}$), peroxyl ($ROO^{\bullet}$), peroxinitrite ($^{\bullet}ONOO^{-}$) and nitric oxide ($NO^{\bullet}$) radicals, as well as non free radical species such as hydrogen peroxide (H_2O_2), nitrous acid (HNO_2) and hypochlorous acid (HOCl) exert oxidative stress in the cells of human body rendering each cell to face about 10,000 oxidative hits per second (Mohammad *et al.*, 2007). When generation of free radicals overtakes the antioxidant defense system of the cells, the free radicals start destroying many major biological macromolecules (proteins, lipids, nucleic acid and carbohydrate) and are implicated in causation of several diseases such as diabetes (Eliza *et al.*, 2010), Parkinson's disease (Samoylenko *et al.*, 2010), inflammation (Banerjee *et al.*, 2012), Alzheimer's disease (Ustun *et al.*, 2012), aging process (Thiyagarajan *et al.*, 2012), cancer (Kim *et al.*, 2012), coronary heart ailment (Zhou *et al.*, 2008), atherosclerosis (Mutluay *et al.*, 2012), cataracts (Anitha *et al.*, 2011), etc. They may even result in cell death (Dean *et al.*, 1993). Moreover, the deterioration of some foods has been identified due to the oxidation of lipids or organoleptic rancidity and the formation of undesirable secondary lipid peroxidation products. Lipid oxidation by ROS such as superoxide anion, hydroxyl radicals and H_2O_2 also causes degrading effects like discoloration, vitamin destruction, nutritional losses, and polymerization and affect their safety and appearance (Ngo *et al.*, 2011).

Cells are equipped with different kinds of mechanisms to fight against ROS and to maintain the redox homeostasis of cell. They protect themselves from oxidative damage through several defense systems including antioxidant enzymes such as superoxide dismutase (SOD), catalase (CAT) and glutathione peroxidase (GPx) convert ROS into less toxic substances by scavenging the free radicals and preventing cell injury and non enzyme defenses like molecules such as glutathione, vitamin C and E inhibit lipid peroxidation in cell and also play an important role scavenging oxidants and preventing cell injury. When the mechanism of antioxidant protection becomes unbalanced in human body, antioxidant supplement may be used to help reduce oxidative damage.

Thus, antioxidants may help the body to protect itself against various types of oxidative damage caused by reactive oxygen species, which are linked to a variety of diseases by quenching reactive oxygen species, inhibiting lipid peroxidation or stimulating cellular antioxidant defenses (Maganha *et al.*, 2010). Antioxidants also provide protection against the hazardous effects of ultraviolet radiation (Tepe *et al.*, 2011). In addition, to their potential beneficial health effects, antioxidant compounds can increase shelf-life by delaying oxidative deterioration of unsaturated lipids which can lead to the development of rancidity (Rhee *et al.*, 1996) and is considered one of the main causes of deterioration of food products during processing and storage (Hayes *et al.*, 2011).

Therefore, antioxidants are considered to be important nutraceuticals with many health benefits and much attention is directed towards naturally occurring antioxidants and their protective mechanisms. In fact, antioxidant consumption will be helpful to prevent oxidative stress in the body. Synthetic antioxidants such as butylated hydroxyanisole (BHA) and butylated hydroxyltoluene (BHT), bisphenol (BP), octyl gallate (OG) and tert-butylhydroquinone (TBHQ) are used in food industry as preservatives which prevents food deterioration (Niciforovic *et al.*, 2010) in cosmetic industry (Guan *et al.*, 2005) and in pharmaceutical industry (Rayner *et al.*, 2006) but their use is being restricted because but they are reported to be toxic at high a high dose (Qi *et al.*, 2005; Labrador *et al.*, 2007). Therefore it is essential to develop effective, non toxic and natural antioxidants that can protect the human body from free radicals and retard the progress of many chronic diseases and also preserve food substances from deteriorating. Therefore the research has shifted towards naturally occurring antioxidants. Plant-derived antioxidants could function as singlet and triplet oxygen quenchers, peroxide decomposers, enzyme inhibitors or synergists (Larson, 1988).

Natural antioxidants occur in all parts of the plant. For *e.g.* aerial part (Juan-Badaturuge *et al.*, 2011), veins (Razali *et al.*, 2012), whole plant (Ooi *et al.*, 2011), flowers (Jang *et al.*, 2008), bulb (Luo *et al.*, 2012), seeds (Algamdi *et al.*, 2011), fruit (Srivastava *et al.*, 2012), pulp (Wang *et al.*, 2010), leaves (Wu *et al.*, 2011; Chanda and Kaneria, 2012; Kaneria and Chanda, 2012); stem (Chanda *et al.*, 2012), rhizomes (Ivanov *et al.*, 2011), nuts (John and Shahidi, 2010), wood (Saito *et al.*, 2008), bark (Badole *et al.*, 2011), pods (Khan *et al.*, 2012), roots (Saito *et al.*, 2008), Fruit rind (Chanda *et al.*, 2010b), etc. Turmerin from *Curcuma longa* (Srinivas *et al.*, 1992), PNP from *Phyllanthus niruri* (Sarkar, *et al.*, 2009) and APC from *Murraya koenigii* (Ningappa and Srinivas, 2008) are few well known antioxidative proteins from plants.

Plants offer a wide range of natural antioxidants due to the structural diversities of their secondary metabolites. Natural products from the plant sources are relatively safe and have immense potential to display many biological activities which depend upon their nature, structure and interaction with other molecules in the assortment (Djeridane *et al.*, 2006).

Ethno pharmacologists, microbiologists, botanists and natural-product chemists are searching compounds which could be developed for the treatment of infectious diseases especially in light of the emergence of multi drug resistant microorganisms and as well as food spoiling microbes. Microorganisms play a major role in food spoilage and life threatening diseases in human being (Negi, 2012). Food borne illness has been a vital concern to public health for years and the control of pathogens may reduce these food borne disease outbreaks which may facilitate the supply of consumer-safe and nutritious food. The food industries at present are using synthetic chemical preservatives to prevent the growth of pathogens, but these synthetic chemicals, by increasing the activity of microsomal enzymes, convert the ingested materials into toxins and carcinogens. This microbial contamination still poses important public health and economic concerns for the human society. Microbiological safety by inhibiting the growth of pathogenic microorganisms is one of the most important issues in the food and pharmaceutical industries. There has been an increasing consumer demand for food free from synthetic preservatives or drugs because it could be toxic to humans. Concomitantly, consumers have demand for wholesome and safe food with long shelf lives. Because of this situation, naturally occurring antimicrobial compounds derived from plant sources has great potential, as alternative to classically approved preservatives (Chanda and Kaneria, 2011). However, candidates for natural antimicrobial compounds need further research regarding aspects such as application type, target microorganism and toxicological information (Vegara *et al.*, 2011).

Generally, antimicrobial agents exert an inhibition effect by changing the synthesis and function of the sub-structure of microbial cells, which includes the cell wall, the membrane, protein synthesis and key enzyme metabolism. There is evidence that antimicrobial agents interfere with the formation of cell walls, damage the permeability of cell membranes, disturb the energy metabolism system and inhibit the formation of protein and nucleic acids (Yao *et al.*, 2012). Excessive use of antibiotics in the treatment of infections has resulted in the development of resistant microbial strains which increase the chance of treatment failure, thus necessitating the search for novel and potent therapeutic agents to treat bacterial and fungal infections (Chanda *et al.*, 2011a).

Antimicrobial agents, including food preservatives, have been used to inhibit food borne bacteria and extend the shelf life of processed food as well as human diseases caused by pathogenic microorganisms. Many naturally occurring extracts from plants, herbs and spices have been shown to possess antimicrobial functions and could serve as a source of antimicrobial agents against food spoilage and multidrug resistant pathogens (Chanda and Rakholiya, 2011; Gutierrez-Larrainzar *et al.*, 2012). The use of plant-based medicine in treating infectious diseases has been in existence for thousands of years and will continue to provide mankind with new remedies. Consumers are becoming more conscious of the nutritional value and safety of their food ingredients. The preference for natural foods and food ingredients that are believed to be safer, healthier and less subject to hazards is increasing compared to their synthetic counterparts (Sevil *et al.*, 2010; Chanda *et al.*, 2011b; Gutierrez-Larrainzar *et al.*, 2012).

Phytochemicals, especially phenolics in fruits are the major bioactive compounds known for their health benefits. *Mangifera indica* L. (Anacardiaceae) is one of the important tropical fruits in the world and India contributes a major part of the world production. Mango is considered as the king of fruits in India. There are many traditional medicinal uses of the different parts of *M. indica* throughout the world. The mango pulp has many therapeutic uses; ripe pulp used for rheological properties (Dak *et al.*, 2007), unripe pulp used as antibacterial agent against food borne bacteria (Gupta *et al.*, 2008). Mango pulp contains vitamins, organic acids, carbohydrates, amino acids, polyphenols and certain volatile compounds (Pino *et al.*, 2005).

In the present work, unripe pulp of *M. indica* was extracted by two different extraction methods: decoction method and individual cold percolation method and were evaluated for their antioxidant and antimicrobial potential.

Material and Methods

Chemicals and Reagents

2, 2-diphenyl-1-picrylhydrazyl (DPPH), nitroblue tetrazolium (NBT), phenazine methosulphate (PMS), nicotinamide adenine dinucleotide reduced (NADH), gallic acid, ascorbic acid, ferrous sulphate ($FeSO_4$), quercetin, Folin-Ciocalteu's reagent, aluminium chloride ($AlCl_3$), potassium acetate, ferric chloride ($FeCl_3$), 2,4,6- tripyridyl-5-triazine (TPTZ), Tris-HCl, sodium acetate, hydrochloric acid (HCl), potassium ferricyanide ($K_3Fe(CN)_6$), sodium carbonate, 2-deoxy-D-ribose, di-Potassium hydrogen phosphate (K_2HPO_4), Potassium biphosphate (KH_2PO_4), Potassium hydroxide (KOH), Ethylenediamine tetraacetic acid (EDTA), 2, 2'-Azino-bis-(3-ethyl)benzothiazoline)-6-sulfonic acid diammonium salt (ABTS), potassium persulfate, thiobarbituric acid (TBA), trichloroacetic acid (TCA), hydrogen peroxide (H_2O_2), petroleum ether, acetone, methanol, glacial acetic acid were obtained from Hi-media, Merck or Sigma. All reagents used were of analytical grade

Collection of the Plant Material

Mango fruits were purchased from local market of Rajkot, Gujarat, India in the month of May 2010. The fruits were washed thoroughly with tap water and unripe pulp (URP) was separated from the fruits, shade dried, homogenized to fine powder and stored in an air tight bottle.

Extraction

Extraction was done by two different methods as described below.

Individual Cold Percolation Extraction Method

The dried powder was individually extracted by cold percolation method (Parekh and Chanda, 2007a) using different organic solvents like petroleum ether, acetone, methanol and water. Ten grams of dried powder was taken in 100 ml of petroleum ether in a conical flask, plugged with cotton wool, and then kept on a rotary shaker at 120 rpm for 24 h. After 24 h, it was filtered through eight layers of muslin cloth, centrifuged at 5,000 rpm in a centrifuge (Remi Centrifuge, India) for 15 min and the supernatant was collected and the solvent was evaporated using a rotary vacuum evaporator (Equitron, India) to dryness. Petroleum ether was evaporated from the powder. This dry powder was then taken individually in 100 ml of each solvent (acetone, methanol and water) and was kept on a rotary shaker at 120 rpm for 24 h. The procedure followed in each case was the same, and the residues were weighed to obtain the extractive yield of all the extracts and were stored in air-tight bottles at 4°C.

Decoction Extraction Method

For the decoction method (Li *et al.*, 2007; Kaneria *et al.*, 2012a), 5 grams of dried powder was extracted with 100 ml of deionized water at 100°C for 30 min in a water bath. It was filtered through eight layers of muslin cloth and centrifuged at 5,000 rpm in centrifuge (Remi Centrifuge, India) for 15 min. The supernatant was collected and water was evaporated using rotary vacuum evaporator (Equitron, India) to dryness. The residue was weighed to obtain the extractive yield and it was stored in air-tight bottle at 4°C.

Quantitative Phytochemical Analysis

Determination of Total Phenol Content (TPC)

The amount of total phenol content (TPC), in different solvent extracts was determined by Folin-Ciocalteu's reagent method (Mc Donald *et al.*, 2001). The extract (0.5 ml) and 0.1 ml Folin-Ciocalteu's reagent (0.5 N) were mixed, and the mixture was incubated at room temperature for 15 min. Then, 2.5 ml saturated sodium carbonate solution was added and further incubated for 30 min at room temperature, and the absorbance was measured at 760 nm using a UV–VIS Spectrophotometer (Shimadzu, Japan) against a blank sample. Total phenol content is expressed in terms of gallic acid equivalent (mg g^{-1} of extracted compounds).

Determination of Flavonoid Content (TFC)

The amount of total flavonoid content (TFC) in different solvent extracts was determination by aluminium chloride colorimetric method (Chang *et al.*, 2002). The reaction mixture (3.0 ml), consisted of 1.0 ml sample (1 mg ml^{-1}), 1.0 ml methanol, 0.5 ml $AlCl_3$ (1.2 per cent) and 0.5 ml potassium acetate (120 mM) and was incubated at room temperature for 30 min. The absorbance of all samples was measured at 415 nm using a UV–VIS Spectrophotometer (Shimadzu, Japan), against a blank sample. The flavonoid content is expressed in terms of quercetin equivalent (mg g^{-1} of extracted compounds).

Antioxidant Assays

DPPH Free Radical Scavenging Assay (DPPH)

The free radical scavenging activity of different solvent extracts was measured by using DPPH by the modified method of McCune and Johns (2002). The reaction mixture (3.0 ml), consisted of 1.0 ml DPPH in methanol (0.3 mM), 1.0 ml methanol and 1.0 ml of different concentrations (200 to 1000 µg ml^{-1}) of different solvent extracts and fractions diluted by methanol, was incubated for 10 min, in dark, after which the absorbance was measured at 517 nm using a UV–VIS Spectrophotometer (Shimadzu, Japan), against a blank sample. Ascorbic acid (2 to 16 µg ml^{-1}) was used as positive control (Liu *et al.*, 2011). Percentage of inhibition was calculated using the following formula:

Per cent Inhibition = $[1- (A/B)] \times 100$

where,

B: Absorbance of blank (DPPH plus methanol)

A: Absorbance of sample (DPPH, methanol plus sample)

Hydroxyl Radical Scavenging Assay (OH)

The hydroxyl radical scavenging activity was measured by studying the competition between deoxyribose and test compound for hydroxyl radicals generated by Fe^{+3}-Ascorbic acid-EDTA-H_2O_2 system (Fenton reaction) according to the method of Kunchandy and Rao (1990). The reaction mixture (1.0 ml) consisted of 100 µl 2-deoxy-D-ribose (28 mM in 20 mM KH_2PO_4-KOH buffer, pH 7.4), 500 µl of different concentrations (40 to 1000 µg ml^{-1}) of different solvent extracts and fractions diluted by distilled water, 200 µl EDTA (1.04 mM) and 200 µM $FeCl_3$ (1: 1 v v^{-1}), 100 µl H_2O_2 (1.0 mM) and 100 µl ascorbic acid (1.0 mM), and was incubated at 37°C for 1 h. 1.0 ml TBA (1 per cent) and 1.0 ml TCA (2.8 per cent) was added, and incubated at 100°C for 20 min. After cooling, the absorbance of pink colour was measured at 532 nm using a UV–VIS Spectrophotometer (Shimadzu, Japan), against a blank sample. Gallic acid (20 to 200 µg ml^{-1}) was used as a positive control (Kunchandy and Rao, 1990). Percentage of inhibition was calculated using the formula described above.

ABTS Radical Cation Scavenging Assay (ABTS)

The ABTS radical cation scavenging activity of different solvent extracts was measured by the method as described by Re *et al.* (1999). ABTS radical cations are produced by reacting ABTS (7 mM) and potassium persulfate (2.45 mM) and incubating the mixture at room temperature in the dark for 16 h. The ABTS working solution obtained was further diluted with methanol to give an absorbance of

0.85 ± 0.20 at 734 nm. 1.0 ml of different concentrations (100 to 1000 µg ml^{-1}) of different solvent extracts and fractions diluted by methanol was added to 3.0 ml of ABTS working solution. The reaction mixture was incubated at room temperature for 4 min, and then the absorbance was measured at 734 nm using a UV–VIS Spectrophotometer (Shimadzu, Japan) against a blank sample. Ascorbic acid (1 to 10 µg ml^{-1}) was used as a positive control (Zhou *et al.*, 2011). Percentage of inhibition was calculated using the formula described above.

Superoxide Anion Radical Scavenging Assay (SO)

The superoxide anion radical scavenging activity of different solvent extracts was measured by the method as described by Robak and Gryglewski (1988). Superoxide radicals are generated by oxidation of NADH and assayed by the reduction of NBT. The reaction mixture (3.0 ml) consisted of 1.0 ml of different concentrations (100 to 1000 µg ml^{-1}) of different solvent extracts and fractions diluted by distilled water, 0.5 ml Tris-HCl buffer (16 mM, pH 8), 0.5 ml NBT (0.3 mM), 0.5 ml NADH (0.936 mM) and 0.5 ml PMS (0.12 mM). The superoxide radical generating reaction was started by the addition of PMS solution to the reaction mixture. The reaction mixture was incubated at 25°C for 5 min, and then the absorbance was measured at 560 nm using a UV–VIS Spectrophotometer (Shimadzu, Japan), against a blank sample. Gallic acid (50 to 225 µg ml^{-1}) was used as a positive control (Robak and Gryglewski, 1988). Percentage of inhibition was calculated using the formula described above.

Reducing Capacity Assessment (RCA)

The reducing capacity of different solvent extracts was determined using the modified method of Athukorala *et al.* (2006). 1.0 ml of different concentrations (20 to 180 µg ml^{-1}) of different solvent extracts and fractions diluted by distilled water, was mixed with 2.5 ml phosphate buffer (200 mM, pH 6.6) and 2.5 ml $K_3Fe(CN)_6$ (30 mM). The mixture was then incubated at 50°C for 20 min. Thereafter, 2.5 ml of TCA (600 mM) was added to the reaction mixture, and then centrifuged for 10 min at 3,000 rpm. The upper layer of solution (2.5 ml) was mixed with 2.5 ml distilled water and 0.5 ml $FeCl_3$ (6 mM), and the absorbance was measured at 700 nm using a UV–VIS Spectrophotometer (Shimadzu, Japan) against a blank sample. Ascorbic acid (20 to 180 µg ml^{-1}) was used as positive control (Kumar *et al.*, 2011).

Ferric Reducing Antioxidant Power (FRAP)

The reducing ability of different solvent extracts was determined by ferric reducing antioxidant power (FRAP) assay of Benzie and Strain (1996). FRAP assay is based on the ability of antioxidants to reduce Fe^{3+} to Fe^{2+} in the presence of TPTZ, forming an intense blue Fe^{2+}-TPTZ complex with an absorption maximum at 593 nm. This reaction is pH-dependent (optimum pH 3.6). 100 µl extract was added to 3.0 ml FRAP reagent (10 parts 300 mM sodium acetate buffer at pH 3.6, 1 part 10 mM TPTZ in 40 mM HCl, and 1 part 20 mM $FeCl_3$), and the reaction mixture was incubated at 37°C for 10 min. Then the absorbance was measured at 593 nm using a UV–VIS Spectrophotometer (Shimadzu, Japan), against a blank sample. The antioxidant capacity based on the ability to reduce ferric ions of sample was calculated from the linear calibration curve and expressed as mM $FeSO_4$ equivalents per gram of extracted compounds.

Antimicrobial Activity

Microorganisms Tested

The microorganisms studied were obtained from National Chemical Laboratory (NCL), Pune, India. The microorganisms were maintained at 4°C. The Gram positive bacteria studied were *Staphylococcus aureus* ATCC25923 (*SA1*), *Staphylococcus aureus* ATCC29737 (*SA2*), *Staphylococcus epidermidis* ATCC12228 (*SE*), *Staphylococcus albus* NCIM 2178 (*SAL*), *Bacillus megaterium* ATCC9885 (*BM*), *Bacillus subtilis* ATCC6633 (*BS*), *Bacillus cereus* ATCC11778 (*BC*), *Corynebacterium rubrum* ATCC14898 (*CR*), *Listeria monocytogenes* ATCC19112 (*LM*), *Micrococcus flavus* ATCC10240 (*MF*), Gram negative bacteria studied were *Proteus morganii* NCIM2040 (*PMO*), *Proteus vulgaris* NCIM2857 (*PV*), *Enterobacter aerogenes* ATCC13048 (*EA*), *Klebsiella pneumoniae* NCIM2719 (*KP*), *Proteus mirabilis* NCIM2241 (*PM*), *Pseudomonas aeruginosa* ATCC27853 (*PA*), *Salmonella typhimurium* ATCC23564 (*ST*), *Citrobacter frundii* NCIM2489 (*CF*), *Klebsiella aerogenes* NCIM2098 (*KA*), *Escherichia coli* NCIM2931 (*EC*) and Fungi were *Candida albicans* ATCC2091 (*CA*), *Candida neoformans* NCIM3542 (*CN*), *Candida glabrata* NCIM3448 (*CG*), *Candida epicola* NCIM3367 (*CE*), *Trichosporon beigelii* NCIM3404 (*TB*).

Antimicrobial Susceptibility Assay (Agar Well Diffusion Assay)

In vitro antimicrobial activity of the different solvent extracts was studied against pathogenic microbial strains by the agar well diffusion method (Perez *et al.*, 1990; Chanda *et al.*, 2010a). Muller Hinton No. 2/Sabouraud dextrose agar (Hi-media) was used for the antibacterial and antifungal susceptibility test respectively. The different solvent extracts were diluted in 100 per cent DMSO to give a concentration of 20 mg ml^{-1}. The Muller Hinton agar/Sabouraud dextrose agar was melted and cooled to 48-50°C and a standardized inoculum (1.5×10^8 CFU ml^{-1}, 0.5 McFarland) was then added aseptically to the molten agar and poured into sterile Petri dishes; wells (8.5 mm) were prepared in the seeded agar plates. The test compound (100 µl) was introduced into the well. The plates were incubated over night at 37°C and 28°C for 24 h and 48 h respectively, for bacteria and fungi. DMSO was used as negative control. The microbial growth was determined by measuring the diameter of the zone of inhibition and the mean values are presented with ± S.E.M.

Preparation of Bacterial Inocula for MIC and MBC Study

The inocula of the test organisms were prepared using the colony suspension method (EUCAST, 2003). Colonies picked from 24 h old cultures grown on nutrient agar were used to make suspension of the test organisms in saline solution to give an optical density of approximately 0.1 at 600 nm. The suspension was then diluted 1: 100 by transfer of 0.1 ml of the bacterial suspension to 9.9 ml of sterile nutrient broth before use to yield 6×10^5 CFU ml^{-1}.

Preparation of the Extracts and/or Antibiotic for MIC and MBC Study

Two-fold serial dilutions using 100 per cent DMSO were carried out from the 2500 µg ml^{-1} stock plant extract to make 6 test concentrations ranging from 78 to 2500 μg ml^{-1} for each solvent extract. Ceftazidime (1 to 32 μg ml^{-1}) was used as a positive control.

Determination of Minimum Inhibitory Concentration (MIC)

The MICs were determined only for the test organisms that had shown > 15 mm zone of inhibition of the crude extracts. Micro broth dilution method, with sterile flat bottom 96 well micro test plates (Tarsons Products Pvt. Ltd.), was performed to evaluate MIC of the plant extracts (Sampaio *et al.*, 2009). 150 μl volume of Muller-Hinton broth was introduced into all the 96 wells and 20 μl of varying concentrations of the extract was added in decreasing order along with 30 μl of the test organism suspension. A final volume of 200 μl was achieved in each well (150 μl Muller-Hinton broth, 30 μl of the test organism suspension, and 20 μl plant extract/antibiotic). Three control wells were maintained for each test batch. The positive control (antibiotic, Muller-Hinton broth and test organism) and sterility control (Muller-Hinton broth and DMSO) and organism control (Muller-Hinton broth, test organism and DMSO). Plates were then incubated at 37°C for 24 h overnight. Experiments were carried out in duplicate. After incubation, 40 μl of 2-(4-Iodo phenyl)-3-(4-nitro phenyl)5-phenyltetrazolium chloride (I.N.T.) (Himedia, India) solution (0.2 mg ml^{-1}) dissolved in sterile distilled water was added to each well (Meyers and Meyers, 2011). The plates were incubated for further 30 min, and estimated visually for any change in color to pink indicating reduction of the dye due to bacterial growth. The highest dilution (lowest concentration) that remained clear corresponded to the MIC.

Determination of Minimum Bactericidal Concentration (MBC)

MBC was determined from all wells showing no growth as well as from the lowest concentration showing growth in the MIC assay for all the samples. Bacterial cells from the MIC test plate were sub-cultured on freshly prepared solid nutrient agar by making streaks on the surface of the agar. The plates were incubated at 37°C for 24 h overnight. Plates that did not show growth were considered to be the MBC for the extract or drug used (Akinyemi *et al.*, 2005).The experiment was carried out in duplicate.

Phytochemical Analysis

The crude powder of unripe pulp was subjected to qualitative phytochemical analysis (Harborne, 1973; Parekh and Chanda, 2007b).

Physicochemical Parameters

The physicochemical parameters like total ash value, loss on drying, petroleum ether, methanol, acetone and water soluble extractive values were determined as per WHO guidelines (WHO, 2002).

Statistical Analysis

Each sample was analyzed individually in triplicate and the results are expressed as the mean value (n = 3) ± Standard Error of Mean (S.E.M.).

Results

Extractive Yield

The extraction yield depends on solvents, time and temperature of extraction, as well as the chemical nature of the sample. The dry powder of unripe pulp was extracted by individual cold percolation and decoction extraction methods. The extractive yield of unripe pulp in different solvent is given in Figure 10.1a. The extractive yield was higher in decoction method than in individual cold percolation method. The extractive yield varied among different solvents. In individual cold percolation method, different extractive yields can be ranked from high to low in the following order: Aqueous extract > Methanol extract > Acetone extract > Petroleum ether extract.

Total Phenol and Flavonoid Content (TPC and TFC)

Total phenol and flavonoid content of different solvent extracts is shown in Fig. 1b. The TPC was more than the TFC in both, individual cold percolation and decoction extraction methods. In both extraction methods, the TPC ranged from 5.99 to 14.57 mg g^{-1}, while TFC ranged from 0.21 to 3.13 mg g^{-1} (Figure 10.1b). The acetone extract showed highest TPC and TFC as compared to other solvent extracts.

Antioxidant Property

Antioxidant methods and modifications have been proposed to evaluate antioxidant characteristics and to explain how antioxidants function. It is essential to use more than one method to evaluate antioxidant capacity of plant materials

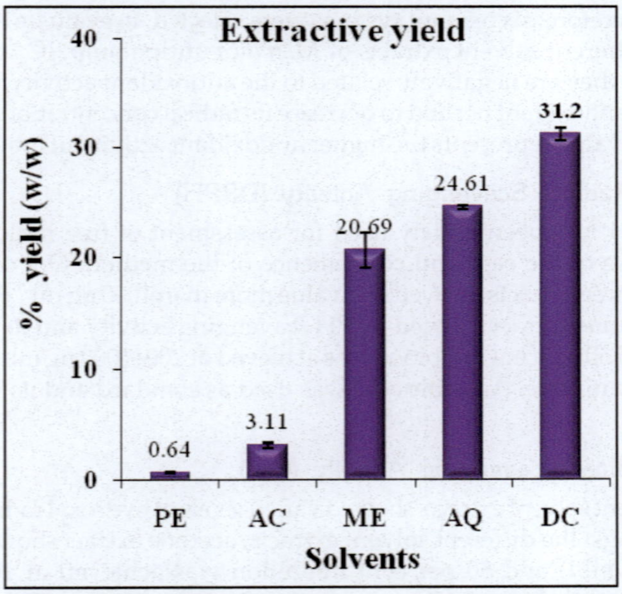

Figure 10.1a: Extractive yield of different solvent extracts of unripe pulp of *M. indica* (individual cold percolation: Petroleum ether (PE), Acetone (AC), Methanol (ME), Aqueous (AQ); Decoction extract (DC).

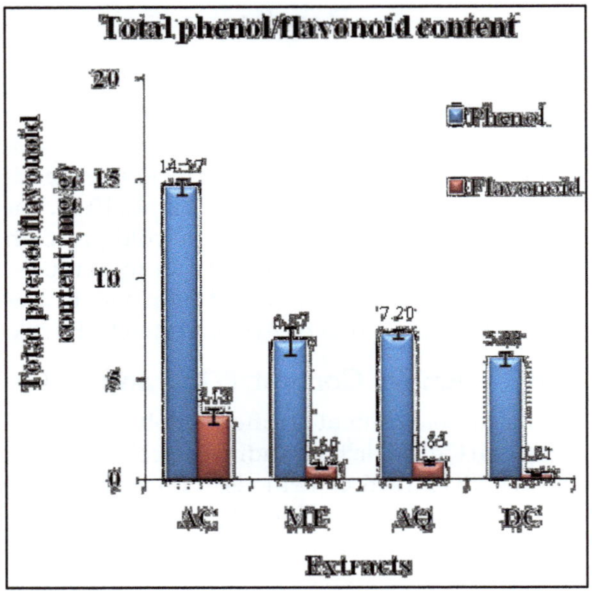

Figure 10.1b: Total phenol and flavonoid content of different solvent extracts of unripe pulp of *M. indica* (individual cold percolation: Acetone, Methanol, Aqueous; Decoction extract).

because of the complex nature of phytochemicals (Perez-Jimenez *et al.*, 2008; Rakholiya *et al.*, 2011). Therefore, six oxidant systems were selected, to evaluate the antioxidant potential of different solvent extracts of *M. indica* unripe pulp. IC_{50} values are also calculated and they are negatively related to the antioxidant activity, as it expresses the amount of antioxidant needed to decrease its radical concentration by 50 per cent. The lower IC_{50} value represents the higher antioxidant activity of the sample tested.

DPPH Free Radical Scavenging Activity (DPPH)

The DPPH• has been widely used for assessment of free radical scavenging activity because of the ease and convenience of the method. Out of four extracts investigated, three extracts showed IC_{50} value more than 1000 µg ml^{-1} (Table 10.1), the remaining acetone extract showed DPPH scavenging activity and its IC_{50} value was 720 µg ml^{-1} and 50 per cent inhibition was achieved at 200–1000 µg ml^{-1} concentration in the reaction mixture. Ascorbic acid was used as standard and its IC_{50} value was 11.4 µg ml^{-1}.

Hydroxyl Radical Scavenging Activity (OH)

The different solvent extracts showed varied levels of hydroxyl radical scavenging activity. Amongst the different solvent extracts, acetone extract showed lowest IC_{50} value (220 µg ml^{-1}) and 50 per cent inhibition was achieved at 40–280 µg ml^{-1} concentration, while highest IC_{50} value was in decoction extract (Table 10.1). Gallic acid was used as standard (140 µg ml^{-1}).

ABTS Radical Cation Scavenging Activity (ABTS)

ABTS activity was quantified in terms of percentage inhibition of the $ABTS^{\bullet+}$ radical cation by antioxidants in each sample. There was a significant variation in the percentage inhibition of the extracts obtained by both the extraction methods (Table 10.1). Individual extraction method displayed a greater $ABTS^{\bullet+}$ scavenging capacity than the decoction extraction method (IC_{50} = >1000 µg ml^{-1}). Among all solvents used, acetone extract showed highest ABTS radical cation scavenging activity and its IC_{50} value was 530 µg ml^{-1} (Table 10.1).

Superoxide Anion Radical Scavenging Activity (SO)

All the different solvent extracts showed poor superoxide anion radical scavenging activity. All extracts showed activity >1000 µg ml^{-1} (Table 10.1). Gallic acid was used as a standard and its IC_{50} value was 185 µg ml^{-1}.

Ferric Reducing Antioxidant Power (FRAP)

FRAP values as a measure of the antioxidant activity is shown in Figure 10.2a. The decoction extract possessed highest FRAP value (1310 mM g^{-1}) as compared to the different solvent extracts of individual extraction method. In individual extraction method, acetone extract showed more FRAP content as compared to the other extracts (Figure 10.2a).

Reducing Capacity Assessment (RCA)

The reducing power (RP) of compounds serves as a significant indicator of its antioxidant activity. The reduction of the Fe^{3+}/ferricyanide complex to the ferrous form (Fe^{2+}) through the donation of an electron was monitored at 700 nm (Sousa *et al.*,

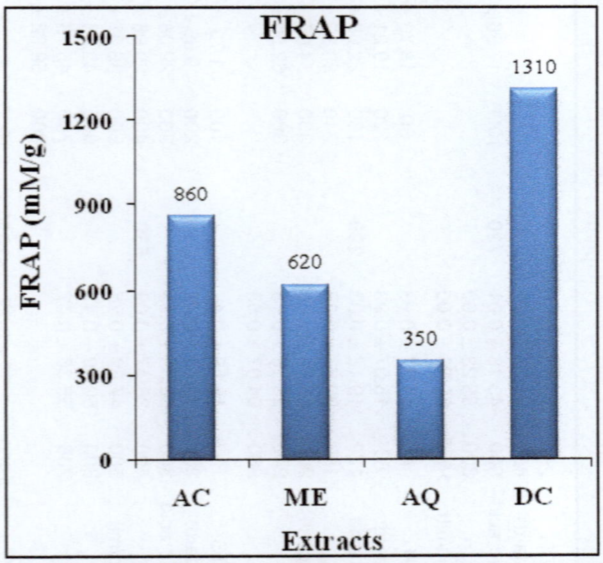

Figure 10.2a: FRAP content of different solvent extracts of unripe pulp of *M. indica* (individual cold percolation: Acetone, Methanol, Aqueous; Decoction extract).

Table 10.1: Concentration, per cent inhibition and IC_{50} values of DPPH free radical, hydroxyl radical scavenging activity, ABTS radical cation and superoxide anion radical, scavenging activity of different solvent extracts of *M. indica* unripe pulp.

Assay	*Acetone Extract*			*Methanol Extract*			*Aqueous Extract*			*Decoction Extract*		
	Conc. (µg/ml)	*Per cent Inhibition**	*IC_{50} Value µg/ml*	*Conc. (µg/ml)*	*Per cent Inhibition**	*IC_{50} Value µg/ml*	*Conc. (µg/ml)*	*Per cent Inhibition**	*IC_{50} Value µg/ml*	*Conc. (µg/ml)*	*Per cent Inhibition**	*IC_{50} Value µg/ml*
DPPH	200	19.19 ± 0.00										
Standard:	400	31.91 ± 0.01										
Ascorbic acid	600	43.18 ± 0.01	**720**	1000	17.56 ± 0.02	>1000	1000	4.32 ± 0.08	>1000	1000	0.86 ± 0.20	>1000
(IC_{50}=	800	53.23 ± 0.00										
11.4 µg/ml)	1000	60.26 ± 0.00										
OH	40	11.37 ± 0.93		60	18.73 ± 0.50		100	14.50 ± 1.23		200	12.59 ± 0.28	
Standard:	80	16.07 ± 0.03		120	19.21 ± 0.23		200	21.34 ± 0.20		400	39.46 ± 0.20	
Gallic acid	120	19.12 ± 0.03	**220**	180	22.08 ± 0.18	330	300	36.32 ± 0.05	405	600	75.39 ± 0.13	520
(IC_{50}=	160	33.67 ± 0.08		240	30.62 ± 0.08		400	53.79 ± 0.03		800	82.01 ± 0.13	
140 µg/ml)	200	45.56 ± 0.00		300	54.05 ± 0.08		500	61.28 ± 0.48		1000	86.41 ± 0.10	
	240	53.18 ± 0.13		360	62.50 ± 0.08							
	280	64.07 ± 0.63										
ABTS	100	13.56 ± 0.42		100	1.79 ± 0.56		200	18.98 ± 0.09				
Standard:	200	25.41 ± 0.42		200	9.66 ± 0.42		400	30.71 ± 0.98				
Ascorbic acid	300	32.39 ± 0.33		300	22.08 ± 0.37		600	32.10 ± 0.09				
(IC_{50}=	400	38.47 ± 0.09	**530**	400	29.06 ± 0.00	700	800	40.97 ± 0.04	990	1000	27.60 ± 0.00	>1000
6.5 µg/ml)	500	47.89 ± 0.75		500	35.31 ± 0.05		1000	50.77 ± 0.36				
	600	50.49 ± 0.47		600	42.13 ± 0.61							
	700	55.36 ± 0.47		700	45.78 ± 0.37							
				800	59.82 ± 0.52							
SO	1000	1.53 ± 0.29	**>1000**	1000	7.83 ± 0.01	>1000	1000	9.02 ± 0.48	>1000	1000	8.86 ± 0.00	>1000
Standard:												
Gallic acid												
(IC_{50}= 185 µg/ml)												

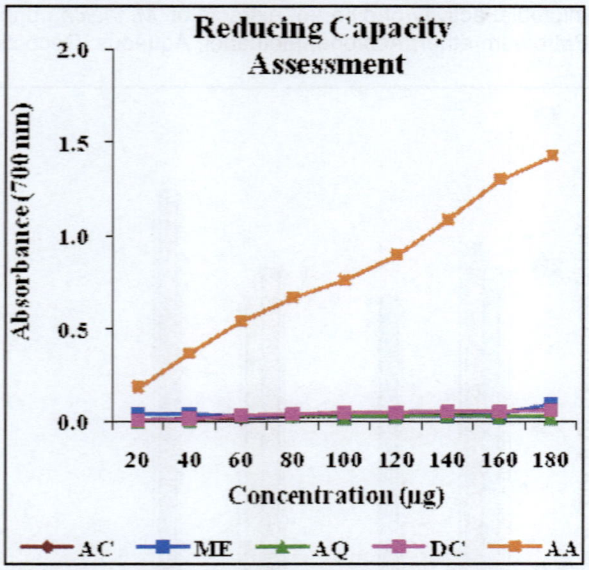

Figure 10.2b: Reducing capacity assessment of different solvent extracts of unripe pulp of *M. indica* (individual cold percolation: Acetone, Methanol, Aqueous; Decoction extract).

2008). In Figure 10.2b, all the extracts showed poor and similar degree of electron-donating capacity in a linear concentration dependent manner.

Antimicrobial Activity

The antimicrobial effect of the extracts, obtained by both methods is shown in Figure 10.3. In Gram positive bacteria petroleum ether extract, acetone and methanol extracts inhibited 40 per cent, 50 per cent and 20 per cent bacteria respectively. The aqueous and decoction extracts inhibited 10 per cent Gram positive bacterial strains. *L. monocytogens* (Figure 10.3a) and *M. flavus* (Figure 10.3b) were resistant towards all the different solvent extracts. The decoction extract showed highest activity against *S. aureus* 2 (Figure 10.3b). In Gram negative bacteria petroleum ether and aqueous extracts inhibited 50 per cent bacteria. The acetone, methanol and decoction extracts inhibited 80 per cent, 70 per cent and 20 per cent Gram negative bacterial strains respectively. *E. aerogenes* was resistant to all the different solvent extracts (Figure 10.3c), while *K. pneumoniae* and *S. typhimurium* were susceptible to all the extracts (Figures 10.3c-d). The highest activity was in acetone and methanol extracts against *S. typhimurium* and *K. pneumoniae* respectively. Petroleum ether and decoction extracts did not show any activity against fungi. The acetone, methanol and aqueous extract inhibited 60 per cent, 20 per cent and 40 per cent fungi respectively. *C. epicola* was resistant to all the extracts and the highest activity was in acetone extract (Figure 10.3e).

Amongst the different solvent extracts, acetone extract showed maximum zone of inhibition against tested microorganisms. The extracts of *M. indica* unripe pulp showed better antibacterial activity than antifungal activity; Gram negative bacteria were more susceptible than Gram positive bacteria.

Figure 10.3: Antimicrobial activity of different extracts of *M. indica* unripe pulp (individual cold percolation: Petroleum ether, Acetone, Methanol, Aqueous; Decoction extract).

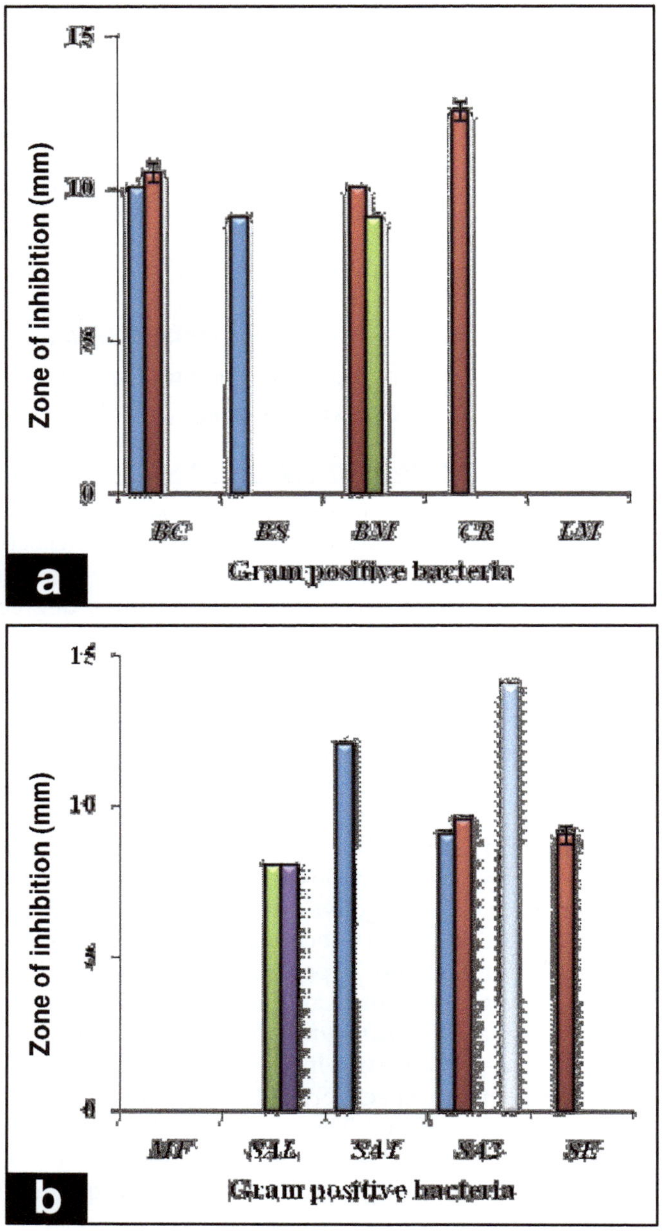

Contd...

Figure 10.3–*Contd...*

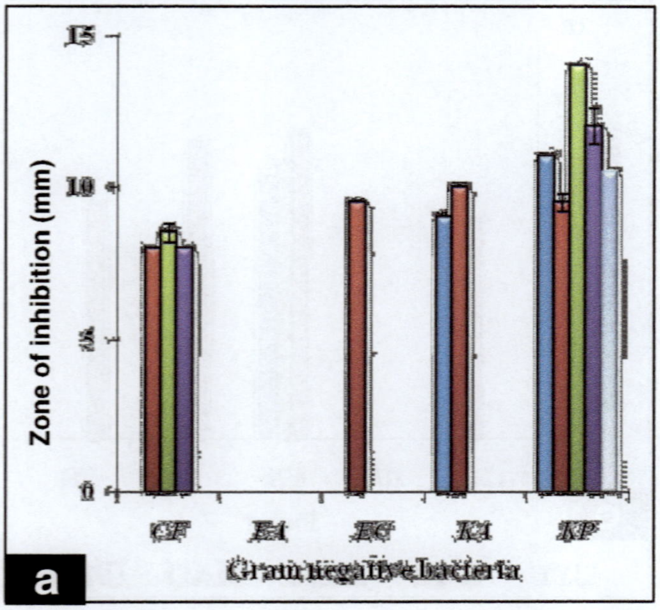

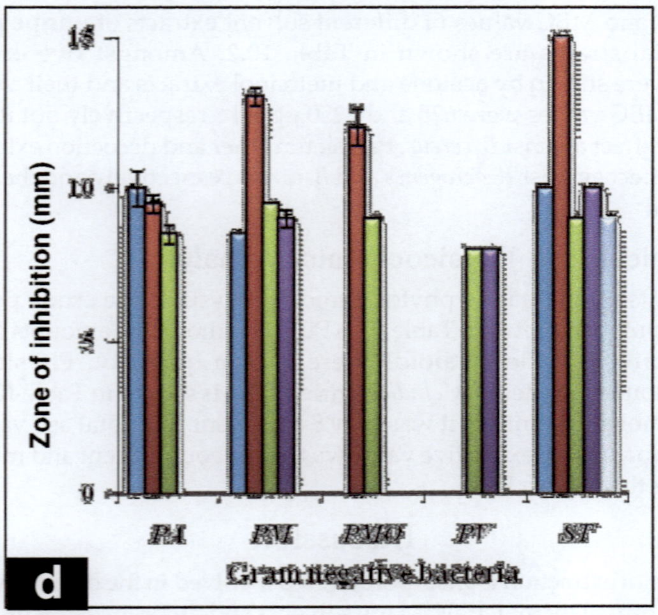

Contd...

Figure 10.3–*Contd...*

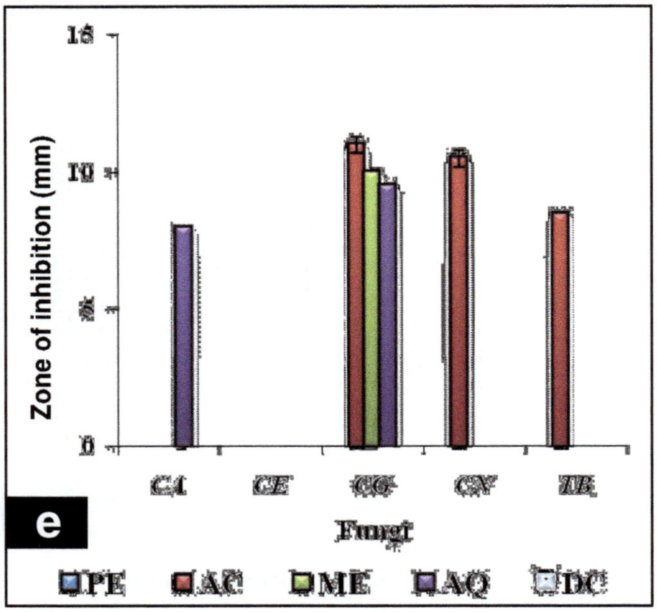

The MIC and MBC values of different solvent extracts of unripe pulp against some bacterial strains are shown in Table 10.2. Amongst five extracts, least MIC values were shown by acetone and methanol extracts and their values was 78 μg ml^{-1} and MBC values were 625 and 1250 μg ml^{-1} respectively but it showed the bacteriostatic effect against *B. cereus*. Petroleum ether and decoction extracts showed bactericidal effect against *K. aerogenes* and *B. cereus* respectively and their MIC value was 312 μg ml^{-1}.

Phytochemical and Physicochemical Analysis

The results of qualitative phytochemical analysis of the crude powder of *M. indica* unripe pulp are shown in Table 10.3. Pulp had moderate amounts of flavonoids and phlobatanins while alkaloids were less in quantity. Physico-chemical characterization of powder of *M. indica* unripe pulp is shown in Table 10.4. The pulp showed less moisture content; it was only 8.5 per cent. The total ash value was 2.17 per cent. The maximum extractive value was in aqueous solvent and minimum was in petroleum ether (Table 10.4).

Discussion

Efficiency of extraction is an important step involved in the discovery of bioactive components from plant materials. Many solvents and different extracting techniques have been used for the extraction of bioactive compounds from plant materials (Wang and Weller, 2006). The technique of extraction, as well as the extracting solvent, significantly affected the extractive yield, total phenol content and biological activities

Table 10.2: Minimim Inhibitory concentration (MIC) and Minimum Bactericidal Concentration (MBC) values of different solvent extracts unripe pulp of *M. indica* and Ceftazidime against some bacterial strains.

Bacteria		*Petroleum Ether Extract*		*Acetone Extract*		*Methanol Extract*		*Aqueous Extract*		*Decoction Extract*		*Ceftazidime*	
		*MIC**	*MBC**	*MIC**	*MBC**	*MIC**	*MBC**	*MIC**	*MBC**	*MIC**	*MBC**	*MIC**	*MBC**
Gram positive	SA2	1250	>2500	1250	>2500	**312**(2)	**2500**(2)	2500	>2500	2500	>2500	4	>32
	MF	1250	>2500	1250	>2500	2500	>2500	2500	>2500	2500	>2500	16	32
	BC	625	>2500	**78**(2)	**625**(2)	**78**(2)	**1250**(2)	1250	>2500	**312**(1)	**1250**(1)	32	>32
	SE	2500	>2500	1250	>2500	>2500	>2500	>2500	>2500	2500	>2500	16	>32
	SAL	1250	>2500	1250	>2500	>2500	>2500	**312**(2)	**2500**(2)	>2500	>2500	2	16
	CR	625	>2500	1250	>2500	>2500	>2500	>2500	>2500	2500	>2500	16	32
	LM	625	>2500	1250	>2500	>2500	>2500	>2500	>2500	2500	>2500	>32	>32
Gram negative	CF	1250	>2500	2500	>2500	>2500	>2500	1250	>2500	1250	>2500	4	>32
	EC	1250	>2500	1250	>2500	>2500	>2500	2500	>2500	>2500	>2500	4	>32
	PMO	625	>2500	1250	>2500	>2500	>2500	>2500	>2500	>2500	>2500	16	>32
	ST	>2500	>2500	2500	>2500	>2500	>2500	>2500	>2500	1250	>2500	<1	8
	KA	**312**(1)	**625**(1)	1250	>2500	>2500	>2500	>2500	>2500	>2500	>2500	>32	>32
	EA	>2500	>2500	>2500	>2500	>2500	>2500	>2500	>2500	>2500	>2500	4	>32
	PA	625	>2500	2500	>2500	>2500	>2500	>2500	>2500	>2500	>2500	>32	>32

*: Values are expressed in μg ml^{-1}; (1): Bactericidal effect; (2): Bacteriostatic effect.

(antioxidant and antimicrobial) of different extracts of *M. indica* unripe pulp. In the present study, the most efficient solvent for highest extractive yield was decoction extract followed by individual cold percolation method aqueous and methanol extracts. Regardless of the method used, the decoction extraction method resulted in higher yield as compared to individual cold percolation method. It appears from these results that in *M. indica* unripe pulp polar compounds are more than non polar compounds; similar results were reported earlier by Yang *et al.* (2007) in acetone and methanol extract of *Nelumbo nuficera*; Baravalia *et al.* (2009) in methanol extract of *Diospyros ebenum* and Kaneria *et al.* (2009) in methanol extract of *Manilkara zapota* Rakholiya *et al.* (2011) in acetone extract of *Mangifera indica*.

Table 10.3: Qualitative phytochemical analysis of unripe pulp of *M. indica.*

Phytochemicals	*Test*	*Result*
Alkaloids	Dragendroff's test	–
	Mayer's test	–
	Wagner's test	+
Flavonoids	Alkaline reagent	++
Tannins	$FeCl_3$ test	–
Phlobatanins	HCl test	++
Triterpenes	H_2SO_4 test	–
Steroids	Liebermann-Burchard test	–
Saponins	Frothing test	–
Cardiac glycosides	Keller-kilianni test	–

–: No presence; +: Less presence; ++: Moderate presence.

Table 10.4: Physicochemical parameters of unripe pulp of *M. indica.*

Sl.No.	*Parameters*	*Values*
1.	Loss on drying	8.50 ± 0.76
2.	Total ash	2.17 ± 0.33
3.	Petroleum ether soluble extractive value	0.64 ± 0.03
4.	Acetone soluble extractive value	3.11 ± 0.20
5.	Methanol soluble extractive value	20.69 ± 1.60
6.	Aqueous soluble extractive value	24.61 ± 6.84

Phenolic compounds are bioactive substances widely distributed in plants and are important constituents of the human diet because they provide positive effects due to their antioxidant properties (Di Majo *et al.*, 2008) and food technologists are keen to harness this nutritional benefit and also make use of their food preservative, antimicrobial, anti-mutagenic, therapeutic and pharmaceutical properties. Phenolic compounds such as flavonoids, phenolic acids, diterpenes and tannins have received attention for their high antioxidant and antimicrobial activity (Upadhyay *et al.*, 2010).

In all the extracts, total phenol content was considerably more than the total flavonoid content. Total phenol and flavonoid content varied in different solvent extracts though highest content was in acetone extract. Similar results were also reported by Chanda and Kaneria (2012) in *S. cumini*. It was observed that there was a correlation between phenol and flavonoid content in all the solvent extracts *i.e.* the extract which had more phenol content had more flavonoid content and vice versa as also reported by Serbetci *et al*. (2012).

Many methods have been proposed to evaluate the antioxidant potential of natural products. The antioxidant capacity cannot be fully assessed by using a single method because antioxidant capacity is influenced by many factors such as the complexity of the oxidation-antioxidation processes, the complex nature of phyto chemicals present in them, the diverse mechanisms of protective effects, solvent used for extraction and lastly the mechanism of different antioxidant assays is different. There is no single method capable of providing a comprehensive picture of the antioxidant profile of a studied sample (Chanda and Dave, 2009); it is also essential to use more than one organic solvent (Chanda and Nagani, 2010). Accordingly, we have used six chemical assays for evaluating antioxidant properties of *M. indica* pulp extracts.

According to the mode of action, the antioxidants can be classified as free radical terminators, chelators of metal ions capable of catalyzing lipid oxidations or as oxygen scavengers that react with oxygen in closed systems. Among the various methods to evaluate the radical scavenging activity of natural compounds, 2, 2-diphenyl-1-picryl hydrazyl DPPH method received more attention due to its fast and reliable results. DPPH radical is scavenged by antioxidants through the donation of a proton forming the reduced DPPH. The color changes from purple to yellow after reduction, which can be quantified by its decrease of absorbance at wavelength 517 nm. Radical scavenging activity increases with increasing percentage of the free radical inhibition. In the present study, acetone extract was the most effective antioxidant for scavenging DPPH free radicals than other solvent extracts; though it was significantly lower than that of standard ascorbic acid. *In situ,* there are various free radicals, such as polyaromatic hydrocarbon cations, which are associated with carcinogenesis. Thus, compounds that will scavenge DPPH *in vitro* may also scavenge radicals like polyaromatic hydrocarbon cations *in vivo*. The DPPH free radical scavenging activity and the total phenolic content showed a good positive correlation. The interaction of phenolic compounds with DPPH depends upon their structural conformation. Certain compounds can react very rapidly with the DPPH radicals and reduce a large number of DPPH molecules corresponding to the number of available hydroxyl groups. The free radical scavenging activity of the phenolic compounds depends on different structural features such as O–H bound dissociation energy, resonance delocalization of the antioxidant and steric-hindrance derived from bulky groups substituting hydrogen in the antioxidant compound.

Hydroxyl radicals are generated by a mixture of ascorbic acid, H_2O_2 and $FeCl_3$-EDTA. Ascorbic acid increases the rate of $OH^{\bullet}$ generation by reducing iron and maintaining a supply of Fe^{2+}-EDTA. The $OH^{\bullet}$ radical generated can degrade deoxyribose into a series of fragments that react with TBA upon heating and at low pH. This reaction results in a pink colour that absorbs at 532 nm light. When added to the reaction mixture, the test compounds compete with deoxyribose for $OH^{\bullet}$ radicals and inhibit deoxyribose degradation. If an antioxidant is capable of scavenging an $OH^{\bullet}$ present in the system (*e.g.* polyphenols), it will compete with deoxyribose for $OH^{\bullet}$ radicals and will inhibit the deoxyribose degradation (Rivas-Arreola *et al.*, 2010). All the extracts in the present study showed hydroxyl radical scavenging activity and highest activity was in acetone extract. There was a positive correlation between hydroxyl radical scavenging activity and the total phenolic content.

ABTS radical scavenging assay is better for evaluating the total antioxidant activity of extracts because it can be used in both organic and aqueous solvent systems as compared to other antioxidant assays. $ABTS^{+\bullet}$ is a blue chromophore produced by the reaction between ABTS and potassium persulfate in the presence of the plant extract, preformed cation radical gets reduced and employs a specific absorbance at 734 nm, a wavelength remote from the visible region, and it requires a short reaction time. The unripe pulp acetone extracts had more antioxidant activity than other solvent extracts. Similar results were also reported by Lamien-Meda *et al.* (2008). Like DPPH and hydroxyl radical scavenging activity, ABTS also showed a positive correlation with total phenolic content.

Superoxide anion radicals are precursor to active free radicals that have potential of reacting with biological macromolecules and thereby inducing tissue damage. Superoxide anion radical (O_2^{-}) is a weak oxidant but it gives rise to the generation of powerful and dangerous hydroxyl radicals as well as singlet oxygen, both of which contribute to the oxidative stress and lead to the genesis of several chronic diseases in human beings. In PMS-NADH-NBT system, superoxide anion radicals are derived from dissolved oxygen by PMS-NADH coupling reaction reduces NBT. Antioxidants are able to inhibit the blue NBT formation. The decrease of absorbance at 560 nm with antioxidants thus indicates the consumption of superoxide anion radical in the reaction mixture. All the extracts showed IC_{50} values >1000 µg ml^{-1} which indicates poor superoxide anion radical scavenging activity. There was no correlation between super oxide anion radical scavenging activity and total phenol content. (Locatelli *et al.*, 2010). These results were entirely different from that of other antioxidant activities shown by the same solvent extracts. This difference in the antioxidant activities may be due to different mechanisms of action. This also suggests that compounds other than phenols also contribute to antioxidant activity.

FRAP method based on reduction of a ferric analogue, measures the ability of a compound to reduce Fe^{3+} to Fe^{2+} that depends on the reduction of a ferric tripyridyltriazine (Fe^{3+}-TPTZ) complex to the intensely blue-colored ferrous tripyridyltriazine (Fe^{2+}-TPTZ) in the presence of any antioxidant at acidic pH (Wong *et al.*, 2006; Fan *et al.*, 2012). A significant increase in reducing antioxidant power was found in decoction extract (1310 mM g^{-1}) followed by acetone extract (Figure 10.2a). It was demonstrated that these extracts showed a mechanism of action as electron donors and they terminated the oxidation chain reaction by reducing the oxidized intermediates into more stable forms (Rivas-Arreola *et al.*, 2010; Kaneria *et al.*, 2012b). Like superoxide radical scavenging activity, there was no correlation between total phenolic and FRAP content.

The reducing properties are generally associated with the presence of electron-donating groups or hydrogen atoms. Accordingly, extracts may contain reductones, which could react with free radicals to stabilize and block radical chain reactions (Chanda *et al.*, 2011b; Wang *et al.*, 2012). In the present study, all the extracts show similar but very small amount of a degree of electron donating capacity in a linear concentration dependent manner as compared to standard ascorbic acid.

The differential response of the extracts in various antioxidant tests may be explained by the fact that transfer of electrons/hydrogen from antioxidants occur at different redox potential in various systems and the transfer also depends on the structure of antioxidants (Loo *et al.*, 2008).

Researchers for many decades have been trying to develop new broad-spectrum antibiotics for treating the infectious diseases caused by pathogenic microorganisms. Prolonged usage of these broad-spectrum antibiotics has led to the emergence of drug resistance. There is a tremendous need for novel antimicrobial agents from different sources. Another problem is that spoilage of food, due to the presence of bacterial and fungal infection, has been a major concern for decades, and causes a considerable economic loss, worldwide.

The unripe pulp extracts of *M. indica* showed better antibacterial activity than antifungal activity; Gram negative bacteria were more susceptible than a Gram positive bacterium which contradicts the previous reports that plant extracts are more active against Gram positive bacteria than Gram negative bacteria (Chan *et al.*, 2007; Kaneria *et al.*, 2009; Chanda *et al.*, 2011c). This difference may be due structural differences in cell walls of these bacteria. The cell wall of Gram negative bacteria is complex and multilayered; it has an outer phospholipid membrane carrying the structural lipopolysaccharide components, which forms a barrier to many environmental substances including synthetic and natural antibiotics. The Gram positive bacteria contain a single outer peptidoglycan layer, which is not an effective permeability barrier (Chanda *et al.*, 2010b; Yagi *et al.*, 2012). These results are very promising because extracts showed better activity towards Gram negative bacteria and there is always a strong search for a natural extracts which show activity towards Gram negative bacterial. Further, these results suggest that unripe pulp possesses compounds that have antimicrobial properties and can be used as antimicrobial agents in new drugs for the treatment of infectious diseases caused by pathogenic microorganisms.

The MIC index (MBC/MIC) was calculated for each extract and positive control drug to determine whether an extract had bactericidal (MBC/MIC$\leq$4) or bacteriostatic (>4 MBC/MIC<32) effect on the growth of bacteria (Stefanovic and Comic, 2011). The MIC value of methanol extract was 312 $\mu g\ ml^{-1}$ against *S. aureus* while Rakotoniriana *et al.* (2010) reported that methanol extract of *Evonymopsis longipes* had MIC value 8000 $\mu g\ ml^{-1}$. This is 25 times less MIC value in pulp extract.

The physical constant evaluation of the drugs is an important parameter in detecting adulteration or improper handling of drugs. The extractive values are useful to evaluate the chemical constituents present in the crude drug and also help in the estimation of specific constituents soluble in a particular solvent (Dave *et al.*, 2010; Kaneria and Chanda, 2011; Nagani *et al.*, 2011).

In conclusion, further studies are needed for understanding the chemical structure of the components present in such extracts responsible for antioxidant and antimicrobial activity. The results of this work indicate that unripe pulp with proper solvent and extraction method can serve as a source of natural antioxidant or nutraceutical. Further studies are warranted for the isolation and identification of individual phenolic compounds and also *in vivo* studies are needed for a better understanding of their mechanism of action as antioxidant and antimicrobial agent. *M. indica* unripe pulp is a novel natural source of antioxidant and antimicrobial agent with numerous health benefits that should be considered in the fields of functional foods and nutraceuticals.

Acknowledgements

The authors thank Prof. S.P. Singh, Head, Department of Biosciences, Saurashtra University, Rajkot, Gujarat, India for providing excellent research facilities; and Mr. V. Chanda for editing and correcting the English language. The authors Ms. Kalpna Rakholiya and Mr. Mital Kaneria are thankful to University Grants Commission, New Delhi, India for providing financial support.

References

Akinyemi, K.O., Oladapo, O., Okwara, C.E., Ibe, C.C., and Fasure, K.A. (2005). Screening of crude extracts of six medicinal plants used in South-West Nigerian unorthodox medicine for anti-methicillin resistant *Staphylococcus aureus* activity. *BMC Complementary and Alternative Medicine,* **5:** 6-12.

Algamdi, N., Mullen, W., and Crozier, A. (2011). Tea prepared from *Anastatica hirerochuntica* seeds contains a diversity of antioxidant flavonoids, chlorogenic acids and phenolic compounds. *Phytochemistry,* **72:** 248–254.

Anitha, T.S., Annadurai, T., Thomas, P.A., and Geraldine, P. (2011). Prevention of selenite-induced cataractogenesis by an ethanolic extract of *Cineraria maritima*: an experimental evaluation of the traditional eye medication. *Biological Trace Element Research,* **143:** 425–436.

Athukorala, Y., Kim, K.N., and Jeon, Y.J. (2006). Antiproliferative and antioxidant properties of an enzymatic hydrolysate from brown alga, *Ecklonia cava*. *Food and Chemical Toxicology,* **44:** 1065-1074.

Badole, S.L., Zanwar, A.A., Khopade, A.N., and Bodhankar, S.L. (2011). *In vitro* antioxidant and antimicrobial activity cycloart-23-ene-3 B, 25-diol (B2) isolated from *Pongamia pinnata* (L. Pierre). *Asian Pacific Journal of Tropical Medicine,* **4:** 910-916.

Banerjee, A., Mondal, N. K., Das, D., and Ray, M.R. (2012). Neutrophilic inammatory response and oxidative stress in premenopausal women chronically exposed to indoor air pollution from biomass burning. *Inammation,* **35:** 671-683.

Baravalia, Y., Kaneria, M., Vaghasiya, Y., Parekh, J., and Chanda, S. (2009). Antioxidant and antibacterial activity of *Diospyros ebenum* Roxb. leaf extracts. *Turkish Journal of Biology,* **33:** 159-164.

Benzie, I.F., and Strain, J.J. (1996). The ferric reducing ability of plasma (FRAP) as a measure of "antioxidant power": the FRAP assay. *Analytical Biochemistry*, **239:** 70-76.

Chan, E.W.C., Lim, Y.Y., and Omar, M. (2007). Antioxidant and antibacterial activity of leaves of *Etlingera* species (Zingiberaceae) in Peninsular Malaysia. *Food Chemistry*, **104:** 1586–1593.

Chanda, S., and Dave, R. (2009). *In vitro* models for antioxidant activity evaluation and some medicinal plants possessing antioxidant properties: An overview. *African Journal of Microbiology Research*, **3:** 981-996.

Chanda, S., and Kaneria, M. (2011). Indian nutraceutical plant leaves as a potential source of natural antimicrobial agents. In: A. Mendez-Vilas, Ed., Science against microbial pathogens: communicating current research and technological advances. Formatex Re-search Center, Spain, **2:** 1251-1259.

Chanda, S., and Rakholiya, K. (2011). Indian Combination therapy: Synergism between natural plant extracts and antibiotics against infectious diseases. In: A. Mendez-Vilas, Ed., Science against microbial pathogens: communicating current research and technological advances, Formatex Re-search Center, Spain, **1:** 520-529.

Chanda, S., Baravalia, Y., Kaneria, M., and Rakholiya, K. (2010a). Fruit and vegetable peels – strong natural source of antimicrobics. *In:* Current Research, Technology and Education Topics in Applied Microbiology and Microbial Biotechnology, 2nd Series, : Ed. By Mendez – Vilas, A. Formatex, Spain, pp. 444-450.

Chanda, S., Dave, R., and Kaneria, M. (2011b). *In vitro* antioxidant property of some Indian medicinal plant. *Research Journal of Medicinal Plant*, **5:** 169-179.

Chanda, S., Dudhatra, S., and Kaneria, M. (2010b). Antioxidative and antibacterial effects of seeds and fruit rind of nutraceutical plants belonging to the family Fabaceae. *Food and Functiion*, **1:** 308-315.

Chanda, S., Kaneria, M., and Baravalia, Y. (2012). Antioxidant and antimicrobial properties of various polar solvent extracts of stem and leaves of four *Cassia* species. *African Journal of Biotechnology*, **11:** 2490-2503.

Chanda, S., Kaneria, M., and Nair, R. (2011c). Antibacterial activity of *Psoralea corylifolia* L. seed and aerial part with various extraction methods. *Research Journal of Microbiology*, **6:** 124-131.

Chanda, S., Rakholiya, K., and Nair, R. (2011a). Antimicrobial activity of *Terminalia catappa* L. leaf extracts against some clinically important pathogenic microbial strains. *Chinese Medicine* **2:** 171-177.

Chanda, S.V., and Kaneria, M.J. (2012). Optimization of conditions for the extraction of antioxidants from leaves of *Syzygium cumini* L. using different solvents. *Food Analytical Methods*, **5:** 332-338.

Chanda, S.V., and Nagani, K.V. (2010). Antioxidant capacity of *Manilkara zapota* L. leaves extracts evaluated by four *in vitro* methods. *Nature and Science*, **8:** 260-266.

Chang, C.C., Yang, M.H., Wen, H.M., and Chern, J.C. (2002). Estimation of total flavonoid content in propolis by two complementary colorimetric methods. *Journal of Food and Drug Analysis,* **10:** 178-182.

Dak, M., Verma, R.C., and Jaaffrey, S.N.A. (2007). Effect of temperature and concentration on Rheological properties of "Kesar" mango juice. *Journal of Food Engineering,* **80:** 1011-1015.

Dave, R., Nagani, K., and Chanda, S. (2010). Pharmacognostic studies and physicochemical properties of the *Polyalthia longifolia* var. *pendula* leaf. *Pharmacognosy Journal,* **2:** 572-576.

Dean, R. T., Gieseg, R., and Davies, M. J. (1993). Reactive species and their accumulation on radical damaged proteins. *Trends in Biochemical Science,* **18:** 437–441.

Di Majo, D., la Guardia, M., Giammanco, S., la Neve, L., and Giammanco, M. (2008). The antioxidant capacity of red wine in relationship with its polyphenolic constituents. *Food Chemistry,* **111:** 45-49.

Djeridane, A., Yousfi, M., Nadjemi, B., Boutassouna, D., Stocker, P., and Vidal, N. (2006). Antioxidant activity of some Algerian medicinal plants extracts containing phenolic compounds. *Food Chemistry,* **97:** 654–660.

Eliza, J., Daisy, P., and Ignacimuthu, S. (2010). Antioxidant activity of costunolide and eremanthin isolated from *Costus speciosus* (Koen ex. Retz) Sm. *Chemico-Biological Interactions,* **188:** 467-472.

European Committee for Antimicrobial Susceptibility Testing (EUCAST) (2003). Determination of minimum inhibitory concentrations (MICs) of antibacterial agents by broth dilution. *Clinical Journal of Microbiology and Infection,* **9:** 1-7.

Fan, L., Li, J., Deng, K., and Lianzhong, A. (2012). Effects of drying methods on the antioxidant activities of polysaccharides extracted from *Ganoderma lucidum. Carbohydyderate Polymers,* **87:** 1849-1854.

Guan, Y.Q., Chu, Q.C., Fu, L.A., and Ye, J.N. (2005). Determination of antioxidants in cosmetics by micellar electrokinetic capillary chromatography with electrochemical detection. *Journal of Chromatography A,* **1074:** 201–204.

Gupta, C., Garg, A.P., and Uniyal, R.C. (2008). Antibacterial activity of Amchur (Dried Pulp of Unripe *Mangifera indica*) extracts on some food borne bacteria. *Journal of Pharmacy Research,* **1:** 54-57.

Gutierrez-Larrainzar, M., Rua, J., Caro, I., Castro, C., Arriaga, D., Garcia-Armesto, M.R., and Valle, P. (2012). Evaluation of antimicrobial and antioxidant activities of natural phenolic compounds against foodborne pathogens and spoilage bacteria. *Food Control,* **26:** 555-563.

Hancock, J.T., Desikan, R., and Neill, S.J. (2001). Role of reactive oxygen species in cell signaling pathways. *Biochemical Society Transactions,* **29:** 345–350.

Harborne, J.B. (1973). Phytochemical methods. 2nd Ed. Chapman and Hall. London.

Hayes, J.E., Allen, P., Brunton, N., O'Grady, M.N., and Kerry, J.P. (2011). Phenolic composition and *in vitro* antioxidant capacity of four commercial phytochemical products: Olive leaf extract (*Olea europaea* L.), lutein, sesamol and ellagic acid. *Food Chemistry,* **126:** 948-955.

Ivanov, S.A., Nomura, K., Malfanov, I.L., Sklyar, I.V., and Ptitsyn, L.R. (2010). Isolation of a novel catechin from *Bergenia* rhizomes that has pronounced lipase-inhibiting and antioxidative properties. *Fitoterapia,* **82:** 212-218.

Jang, I.C., Park, J.H., Park, E., Park, H.R., and Lee, S.C. (2008). Antioxidative and antigenotoxic activity of extracts from Cosmos (*Cosmos bipinnatus*) flowers. *Plant Foods For Human Nutrition,* **63:** 205-210.

John, J.A., and Shahidi, F. (2010). Phenolic compounds and antioxidant activity of Brazil nut (*Bertholletia excelsa*). *Journal of Functional Foods,* **2:** 196-209.

Juan-Badaturuge, M., Habtemariam, S., and Thomas, M.J.K. (2011). Antioxidant compounds from a South Asian beverage and medicinal plant, *Cassia auriculata. Food Chemistry,* **125:** 221-225.

Kaneria, M., and Chanda, S. (2011). Phytochemical and pharmacognostic evaluation of leaves of *Psidium guajava* L. (Myrtaceae). *Pharmacognosy Journal,* **3:** 41-45.

Kaneria, M., and Chanda, S. (2012). The Effect of sequential fractionation technique on the various efficacies of pomegranate (*Punica granatum* L.). *Food Analytical Methods,* DOI 10.1007/s12161-012-9412-8

Kaneria, M., Baravalia, Y., Vaghasiya, Y., and Chanda, S. (2009). Determination of antibacterial and antioxidant potential of some medicinal plants from Saurashtra region, India. *Indian Journal of Pharmaceutical Sciences,* **71:** 406-412.

Kaneria, M., Kanani, B., and Chanda, S. (2012a). Assessment of effect of hydroalcoholic and decoction methods on extraction of antioxidants from selected Indian medicinal plants. *Asian Pacific Journal of Tropical Biomedicine,* **2:** 195-202.

Kaneria, M.J., Bapodara, M.B., and Chanda, S.V. (2012b). Effect of extraction techniques and solvents on antioxidant activity of pomegranate (*Punica granatum* L.) leaf and stem. *Food Analytical Methods,* **5:** 396-404.

Khan, A.B., Akhtar, N., Rasul, A., Mahmood, T., Shoaib Khan, H.M., Shahiq-Uz-Zaman, Iqbal, M., and Murtaza, G. (2012). Investigation of the effects of extraction solvent/technique on the antioxidant activity of *Cassia fistula* L. *Journal of Medicinal Plants Research,* **6:** 500–503.

Kim, J., Jayaprakasha, G.K., Uckoo, R.M., and Patil, B.S. (2012). Evaluation of chemopreventive and cytotoxic effect of lemon seed extracts on human breast cancer (MCF-7) cells. *Food and Chemical Toxicology,* **50:** 423–430.

Kumar, M.S.Y., Dutta, R., Prasad, D., and Misra, K. (2011). Subcritical water extraction of antioxidant compounds from Sea buckthorn (*Hippophae rhamnoides*) leaves for the comparative evaluation of antioxidant activity. *Food Chemistry,* **127:** 1309-1316.

Kunchandy, E., and Rao, M.N.A. (1990). Oxygen radical scavenging activity of curcumin. *International Journal of Pharmaceutics*, **58:** 237-240.

Labrador, V., Fernandez Freire, P., Perez Martin, J.M., and Hazen, M.J. (2007). Cytotoxicity of butylated hydroxyanisole *in Vero* cells. *Cell Biology and Toxicology*, **23:** 189–199.

Lamien-Meda, A., Lamien, C.E., . Compaore, M.M.Y., Meda, R.N.T., Kiendrebeogo, M., Zeba, B., Millogo, J.F., and Nacoulma, O.G. (2008). Polyphenol content and antioxidant activity of fourteen wild edible fruits from Burkina Faso. *Molecules*, **13:** 581-594.

Larson, R.A. (1988). The antioxidants of higher plants. *Phytochemistry*, **4:** 969–978.

Li, H.B., Jiang, Y., Wong, C.C., Cheng, K.W., and Chen, F. (2007). Evaluation of two methods for the extraction of antioxidants from medicinal plants. *Analytical and Bioanalytical Chemistry*, **388:** 483-488.

Liu, J., Wang, C., Wang, Z., Zhang, C., Lu, S., and Liu, J. (2011). The antioxidant and free-radical scavenging activities of extract and fractions from corn silk (*Zea mays* L.) and related flavone glycosides. *Food Chemistry*, **126:** 261-269.

Locatelli, M., Travaglia, F., Coisson, J.D., Martelli, A., Stevigny, C., and Arlorio, M. (2010). Total antioxidant activity of hazelnut skin (Nocciola Piemonte PGI): Impact of different roasting conditions. *Food Chemistry*, **119:** 1647-1655.

Loo, A.Y., Jain, K., and Darah, I. (2008). Antioxidant activity of compounds isolated from the pyroligneous acid, *Rhizophora apiculata*. *Food Chemistry*, **107:** 1151-1160.

Luo, J., Li, L., and Kong, L. (2012). Preparative separation of phenylpropenoid glycerides from the bulbs of *Lilium lancifolium* by high-speed counter-current chromatography and evaluation of their antioxidant activities. *Food Chemistry*, **131:** 1056–1062.

Maganha, *E.G.*, Halmenschlager, R.C., Rosa, R.M., Henriques, J.A.P., Ramos, A.L., and Saffi, J. (2010). Pharmacological evidences for the extracts and secondary metabolites from plants of the genus *Hibiscus*. *Food Chemistry*, **118:** 1-10.

Mc Donald, S., Prenzler, P.D., Autolovich, M., and Robards, K. (2001). Phenolic content and antioxidant activity of olive extracts. *Food Chemistry*, **73:** 73-84.

McCune, L.M., and Johns, T. (2002). Antioxidant activity in medicinal plants associated with the symptoms of diabetes mellitus used by the indigenous peoples of the North American boreal forest. *Journal of Ethnopharmacology*, **82:** 197-205.

Meyers, F.M., and Meyers, R. (2011). Antibacterial activity of traditional medicinal plants used by Haudenosaunee peoples of New York State. *BMC Complementary and Alternative Medicine* **10:** 64.

Mohammad, A.M., Koji, Y., Toshiki, M., Yumi, N., Katsumi, S., and Hiroaki, S., and Yoshifumi, T. (2007). Superoxide anion radical scavenging activities of herbs and pastures in Northern Japan determined using electron spin resonance spectrometry. *International Journal of Biological Sciences*, **6:** 349355.

Mutluay, R., Dege, S.M., Bahadir, E., Durmaz, A.O., Citil, R., and Sindel, S. (2012). Uric acid is an important predictor for hypertensive early atherosclerosis. *Advances in Therapy*, **29:** 276-286.

Nagani, K.V., Kevalia, J., and Chanda, S.V. (2011). Pharmacognostical and phytochemical evaluation of stem of *Cissus quadrangularis* L. *International Journal of Pharmaceutical Science and Research*, **2:** 2856-2862.

Negi, P.S. (2012). Plant extracts for the control of bacterial growth: Efficacy, stability and safety issues for food application. *International Journal of Food Microbiology*, doi: 10.1016/j.ijfoodmicro.2012.03.006

Ngo, D., Wijesekara, I., Vo, T., Van, Q., and Kim S. (2011) Marine food-derived functional ingredients as potential antioxidants in the food industry: An overview. *Food Research International*, **44:** 523-529.

Niciforovic, N., Mihailovic, V., Maskovic, P., Solujic, S., Stojkovic, A., and Muratspahic, P.D. (2010). Antioxidant activity of selected plant species; potential new sources of natural antioxidants. *Food and Chemical Toxicology*, **48:** 3125-3130.

Ningappa, M.B., and Srinivas, L. (2008). Purification and characterization of 35 kDa antioxidant protein from curry leaves (*Murraya koenigii* L.). *Toxicology in vitro*, **22:** 699–709.

Ooi, K.L., Muhammad, T.S.T., Tanb, M.L., and Sulaimana, S.F. (2011). Cytotoxic, apoptotic and anti-α-glucosidase activities of 3,4-di-O-caffeoyl quinic acid, an antioxidant isolated from the polyphenolic-rich extract of *Elephantopus mollis* Kunth. *Journal of Ethnopharmacology*, **135:** 685-695.

Parekh, J., and Chanda, S. (2007a). *In vitro* antibacterial activity of the crude methanol extract of *Woodfordia fruticosa* kurz. flower (Lythraceae). *Brazilian Journal of Microbiology*, **38:** 204-207.

Parekh, J., and Chanda, S. (2007b). Antibacterial and phytochemical studies on twelve species of Indian medicinal plants. *African Journal of Biomedicine Research*, **10:** 175-181.

Perez, C., Paul, M., and Bazerque, P. (1990). An antibiotic assay by the agar well diffusion method. *Acta Biologiae et Medicine Experimentalis*, **15:** 113-115.

Perez-Jimenez, J., Arranz, S., Tabernero, M., Diaz-Rubio, M.E., Serrano, J., and Goni, I. (2008). Updated methodology to determine antioxidant capacity in plant foods, oils and beverages: Extraction, measurement and expression of results. *Food Research International*, **41:** 274-285.

Pino, J.A., Mesa, J., Munoz, Y., Marti, M.P., and Marbot, R. (2005). Volatile components from mango (*Mangifera indica* L.) cultivars. *Journal of Agriculture and Food Chemistry*, **53:** 2213- 2223.

Qi, H., Zhang, Q., Zhao, T., Chen, R., Zhang, H., Niu, X., and Li, Z. (2005). Antioxidant activity of different sulfate content derivatives of polysaccharide extracted from

Ulva pertusa (Chlorophyta) *in vitro*. *International Journal of Biological Macromolecules,* **37:** 195–199.

Rakholiya, K., Kaneria, M., and Chanda, S. (2011). Vegetable and fruit peels as a novel source of antioxidants. *Journal of Medicinal Plants Research,* **5:** 63-71.

Rakotoniriana, E.F., Rajaonarison, J.F., Raoelison, E.G., Rajaonarivelo, J.P., Manga, N., Solofoniaina, M., Rakotonirina, B., Randriamampionona, D., Rabemanantsoa, C., Cheuk, K., UrvegRatsimamanga, S., and Leclercq, J.Q. (2010). Antimicrobial activity of 23 endemic plants in Madagascar. *Tropical Journal of Pharmaceutical Research,* **9:** 165-171.

Rayner, B.S., Duong, T.T.H., Myers, S.J., and Witting, P.K. (2006). Protective effect of a synthetic anti-oxidant on neuronal cell apoptosis resulting from experimental hypoxia re-oxygenation injury. *Journal of Neurochemistry,* **97:** 211–221.

Razali, N., Mat-Junit, S., Abdul-Muthalib, A.F., Subramaniam, S., and Abdul-Aziz, A. (2012). Effects of various solvents on the extraction of antioxidant phenolics from the leaves, seeds, veins and skins of *Tamarindus indica* L. *Food Chemistry,* **131:** 441–448.

Re, R., Pellegrini, N., Proteggentle, A., Pannala, A., Yang, M., and Rice-Evans, C. (1999). Antioxidant activity applying an improved ABTS radical cation decolourisation assay. *Free Radical Biology and Medicine,* **26:** 1231-1237.

Rhee, K.S., Anderson, L.M., and Sams, A.R. (1996). Lipid oxidation of beef, chicken and pork. *Journal of Food Science,* **61:** 8–12.

Rivas-Arreola, M.J., Rocha-Guzman, N.E., Gallegos-Infante, J.A., Gonzalez-Laredo, R.F., and Rosales-Castro, M., Bacon, J.R., Cao, R., Proulx, A., and Intriago-Ortega, P. (2010). Antioxidant activity of oak (*Quercus*) leaves infusions against free radicals and their cardioprotective potential. *Pakistan Journal of Biological Sciences,* **13:** 537-545.

Robak, J., and Gryglewski, R.J. (1988). Flavonoids are scavengers of superoxide anions. *Biochemical Pharmacology,* **37:** 837-841.

Saito, K., Kohno, M., Yoshizaki, F., and Niwano, Y. (2008). Extensive screening for edible herbal extracts with potent scavenging activity against superoxide anions. *Plant Foods for Human Nutrition,* **63:** 65–70.

Samoylenkoa, V., Rahmana, M., Tekwania, B.L., Tripathia, L.M., Wanga, Y.H., Khana, S.I., Khana, I.A., Millerd, L.S., Joshi, V.C., and Muhammada, I. (2010). *Banisteriopsis caapi*, a unique combination of MAO inhibitory and antioxidative constituents for the activities relevant to neurodegenerative disorders and Parkinson's disease. *Journal of Ethnopharmacology*, **127:** 357-367.

Sampaio, F.C., Pereira, S., Dias, C.S., Costa, V.C., Conde, N.C., and Buzalaf, M.A. (2009). *In vitro* antimicrobial activity of *Caesalpinia ferrea* martius fruits against oral pathogens. *Journal of Ethnopharmacology,* **124:** 289-294.

Sarkar, M.K., Kinter, M., Mazumdar, B., and Sil, P.C. (2009). Purification and characterisation of a novel antioxidant protein molecule from *Phyllanthus niruri*. *Food Chemistry,* **114:** 1405–1412.

Serbetci, T., Ozsoy, N., Demirci, B., Can, A., Kültür, S., and Baser, K.H.C. (2012). Chemical composition of the essential oil and antioxidant activity of methanolic extracts from fruits and flowers of *Hypericum lydium* Boiss. *Industrial Crops and Products*, **36:** 599-606.

Sevil, A., Ahmet, A., Osman, S., and Ergin, H. (2010). Compositions, antioxidant and antimicrobial activities of *Helichrysum* (Asteraceae) species collected from Turkey. *Food Chemistry*, **119:** 114-122.

Sousa, A., Ferreira, I.C.F.R., Barros, L., Bento, A., and Pereira, J.A. (2008). Antioxidant potential of traditional stoned table olives "Alcaparras": influence of the solvent and temperature extraction conditions. *LWT-Food Science and Technology*, **41:** 739-745.

Srinivas, L., Shalini, V.K., and Shylaja, M. (1992). Turmerin: A water-soluble antioxidant peptide from turmeric (*Curcuma longa*). *Archives of Biochemistry and Biophysics*, 292: 617–623.

Srivastava, P., Raut, H.N., Wagh, R.S., Puntambekar, H.M., and Kulkarni, M.J. (2012). Purification and characterization of an antioxidant protein (16 kDa) from *Terminalia chebula* fruit. *Food Chemistry*, **131:** 141-148.

Stefanovic, O., and Comic, L. (2011). Inhibitory effect of *Cytisus nigricans* L., and *Cytisus capitatus* scop on growth of bacteria. *African Journal of Microbiology Research*, **5:** 4725-4730.

Tepe, B., Degerli, S., Arslan, S., Malatyali, E., and Sarikurkcu, C. (2011). Determination of chemical profile, antioxidant, DNA damage protection and antiamoebic activities of *Teucrium polium* and *Stachys iberica*. *Fitoterapia*, **82:** 237-246.

Thiyagarajan, R., Sung-Won, K., Jong-Hwan, S., Seock-Yeon, H., Sang-Hyon, S., Sung-Kwang, Y., and Si-Kwan, K. (2012). Effect of fermented *Panax ginseng* extract (GINST) on oxidative stress and antioxidant activities in major organs of aged rats. *Experimental Gerontology*, **47:** 77–84.

Upadhyay, N.K., Kumar, M.S.Y., and Gupta, A. (2010). Antioxidant, cytoprotective and antibacterial effects of Sea buckthorn (*Hippophae rhamnoides* L.) leaves. *Food and Chemical Toxicology*, **48:** 3443-3448.

Ustuna, O., Senola, F.S., Kurkcuoglub, M., Orhana, *I.E.*, Kartald, M., and Baser, K.H.C. (2012). Investigation on chemical composition, anticholinesterase and antioxidant activities of extracts and essential oils of Turkish *Pinus* species and pycnogenol. *Industrial Crops and Products*, **38:** 115-123.

Valko, M., Leibfritz, D., Moncol, J., Cronin, M.T.D., and Mazur, M., Telser, J. (2007). Free radicals and antioxidants in normal physiological functions and human disease. *International Journal of Biochemistry and Cell Biology*, **39:** 44–84.

Vegara, S., Funes, L., Marti, N., Saura, D., Micol, V., and Valero, M. (2011). Bactericidal activities against pathogenic bacteria by selected constituents of plant extracts in carrot broth. *Food Chemistry*, **128:** 872–877.

Wang, H., Gan, D., Zhang, X., and Pan, Y. (2010). Antioxidant capacity of the extracts from pulp of *Osmanthus fragrans* and its components. *LWT Food Science and Technology*, **43:** 319–325.

Wang, L.J., and Weller, C.L. (2006). Recent advances in extraction of nutraceuticals from plants. *Trends in Food Science and Technology*, **17:** 300–312.

Wang, Y., Yang, Z., and Wei, X. (2012). Antioxidant activities potential of tea polysaccharide fractions obtained by ultra filtration. *International Journal of Biological Macromolecules* **50:** 558– 564.

WHO (2002). Quality Control Methods for Medicinal Plant Materials. (An authorized publication of World health organization, Geneva). A.I.T.B.S. Publishers and Distributors, New Delhi.

Wong, C.C., Li, H.B., Cheng, K.W., and Chen, F. (2006). A systematic survey of antioxidant activity of 30 Chinese medicinal plants using the ferric reducing antioxidant power assay. *Food Chemistry,* **97:** 705-711.

Wu, C.R., Lin, W.H., Hseu, Y.C., Lien, J.C., Lin, Y.T., Kuo, T.P., and Ching, H. (2011). Evaluation of the antioxidant activity of five endemic *Ligustrum* species leaves from Taiwan flora *in vitro*. *Food Chemistry,* **127:** 564–571.

Yagi, S., Chretien, F., Duval, R.E., Fontanay, S., Maldini, M., Piacente, S., Henry, M., Chapleur, Y., and Laurain-Mattar, D. (2012). Antibacterial activity, cytotoxicity and chemical constituents of *Hydnora johannis* roots. *South African Journal of Botany,* **78:** 228-234.

Yang, D., Wang, Q., Ke, L., Jiang, J., and Ying, T. (2007). Antioxidant activities of various extracts of lotus (*Nelumbo nuficera* Gaertn) rhizome. *Asian Pacific Journal of Clinical Nutrition,* **16:** 158-163.

Yao, X., Zhu, X., Pan, S., Fang, Y., Jiang, F., Phillips, G.O., and Xu, X. (2012). Antimicrobial activity of nobiletin and tangeretin against *Pseudomonas*. *Food Chemistry,* **132:** 1883–1890.

Zhou, H.C., Lin, Y.M., Li, Y.Y., Li, M., Wei, S.D., Chai, W.M., and Tam, N.F. (2011). Antioxidant properties of polymeric proanthocyanidins from fruit stones and pericarps of *Litchi chinensis* Sonn. *Food Research International,* **44:** 613-620.

Zhou, S., Wang, P., Pi, R., Gao, J., Fu, J., Fang, J., Qin, J., Zhang, H., Li, R., Chen, S., Tang, F., and Liu, P (2008). Reduced expression of GSTM2 and increased oxidative stress in spontaneously hypertensive rat. *Molecular Cell Biochemistry,* **309:** 99–107.

Medicinal Plants: Phytochemistry, Pharmacology and Therapeutics Vol. 3 (2014) *Pages* **285–313**
Editor-in-Chief: **V.K. Gupta**
Published by: **DAYA PUBLISHING HOUSE, NEW DELHI**

11

Antineoplastic Action of Rohituka (*Aphanamixis polystachya*) in Mice Transplanted with Ehrlich Ascites Carcinoma

Ganesh Chandra Jagetia[1*]

ABSTRACT

Anticancer activity of various doses of alcoholic extract of rohituka, Aphanamixis polystachya (APE) was studied in mice transplanted with Ehrlich ascites carcinoma (EAC). Administration of 0, 0.125, 0.25, 0.5, 0.75, 1, 1.25 or 1.5 g/kg body weight APE once daily for consecutive 9 days resulted in a dose dependent regression in the tumor mass and increase in tumor-free survival. However, the greatest anticancer activity was observed for 1 g/kg APE as evident by a maximum number of tumor-free survivors by 120 days post APE administration. The effect of spatial delivery of 1 g/kg APE was evaluated by splitting it into two equal fractions of 0.5 g/kg twice daily for consecutive 9 days. Administrations of 0.5 g/kg APE twice daily for nine consecutive days was superior to the single administration of 1 g concomitantly, or 1 mg/kg doxorubicin (positive control) that gave best anticancer activity when compared with the other doses. The stage specific evaluation revealed that APE treatment was effective in regressing the tumors at all the stages. However, it was most effective up to stage III and the effect lessened when the APE administration was given at late stages of tumor development. Biochemical estimation revealed that APE administration increased the lipid peroxidation accompanied by a decline in glutathione contents and reduction in the catalase, superoxide dismutase, glutathione peroxidase and glutathione-s-transferase activities. APE

1 Department of Zoology, Mizoram University, Aizawl – 796 004, Mizoram, India
* *Corresponding author*: E-mail: gc.jagetia@gmail.com

treatment caused a dose dependent retardation in the tumor mass and it also regressed tumors in the late stages of tumor development. The regression in tumors by APE may be due to increased lipid peroxidation and reduction in the antioxidant enzymes.

Keywords: *Aphanamixis polystachya,* Ehrlich ascites carcinoma, Mice, Survival, Antioxidant enzymes.

Introduction

Cancer is the second largest killer disease in the world and one in four deaths in United State is due to cancer. Despite several advances made in the treatment of cancer, the cancer mortality rates have not come down and remains at the level it was fifty years ago specifically for solid tumors (Jemal *et al.*, 2011). This indicates that new approaches are required to treat cancer. Natural products including plants have been used by mankind for health care since the advent of human history and several traditional medicinal systems solely depend on natural products for medicament. Plants in particular have been used for more than 3,500 years in the treatment of 'cancer,' it was only since the late 1950s that the evaluation of crude plant extracts for their antiproliferative potential was initiated (Mann, 2002; Cragg and Newman, 2005). In fact, several anticancer lead molecules including vincristine, vinblastine, vinorelbine, etoposide, teniposide, paclitaxel, irinotecan, topotecan and camptothecin have been initially isolated from plants (Cragg and Newman, 2005; Kinghorn *et al.*, 2009) before their chemical synthesis. All these agents have been used successfully to treat various neoplastic diseases in humans. Medicinal use of these active molecules has been associated with various undesired side effects and induction of second malignancies in the survivors (Swerdlow *et al.*, 2011). Owing to their biologic origin the plant based drugs may be more compatible, less toxic and better tolerated by humans than synthetic drugs (Jagetia, 2007). The herbal products therefore, may provide an alternate strategy to control cancer. Moreover, 80 per cent of world population as such relies on alternative and complimentary medicines for their health care needs and cancer is no exception. Therefore, screening of plants could provide newer paradigms for the treatment of neoplastic disorders with lesser toxic side effects.

Aphanamixis polystachya Wall.Parker (*Amoorarohituka* (Roxb.)Wight and Arn.) known as rohituka in Sanskrit is distributed throughout India in evergreen forests and is a member of the family Meliaceae. It has been reported to possess several medicinal properties. The stem bark is acrid, astringent, bitter, vulnerary, digestive, anthelmintic, depurative, urinary astringent, ophthalmic and refrigerant. It is useful in the treatment of splenomegaly, liver disorders, tumors, ulcers, dyspepsia, intestinal worms, skin diseases, leprosy, diabetes, ophthalmopathy, jaundice, haemorrhoids, burning sensation rheumatoid arthritis and leucorrhoea (Chopra *et al.*, 1956). The seeds are useful in ulcers, ophthalmopathy, otopathy, myalgia, skin diseases, intestinal worms, burning sensation and vitiated conditions of vata (Warrier, 1994). *Aphanamixis polystachya* has been reported to be an insect repellant, insecticidal, antifeedant, toxicant, and protectant against many species of insects and fungus (Talukder and Howse, 1995). The stem bark has been found to be useful in certain tumors including liver, spleen and leukemia in mice (Dhar *et al.*, 1968; Srivastava and Agnihotri, 1985; Rabi and Gupta, 1995). The leaf extract of *A. polystachya* has been found to be hepatoprotective against carbon tetrachloride-induced hepatotoxicity (Gole and Dasgupta, 2002). It has been reported to protect against radiation-induced micronuclei formation *in vivo* (Jagetia and Venkatesha, 2006).

The fruits and leaves of *Aphanamixis polystachya* has been reported to contain labdanediterpenoid, aphanamixol (Chandrasekharan and Chakraborty, 1968). The seeds and stem bark contain an alkaloid rohitukin (Harmon *et al.*, 1979; Kundu *et al.*, 1985) and a range of complex limonoids like, polystachin, prieurianin and hispidin C (Connolly *et al.*, 1979; Mulholland and Naidoo, 1999) and flavonone glycosides *viz.*, anthraquinone and naringenin glycosides and 8-C-methyl-quercetin-3-*O*-b-D-xylopyranoside (Srivastava and Agnihotri, 1985; Jain and Srivastava, 1985). Limonoids have been reported to possess anticancer activity in various human cancer cell lines (Ahn, 1994). It is noteworthy that most of limonoids are structurally so complex that they would never have emerged from a synthetic program alone or from a combinatorial approach to new drug discovery (Gunatilaka, 1998). Several activities attributed to *Aphanamixis polystachya* and its frequent use in India stimulated us to conceptualize and investigate the anticancer activity of alcoholic extract of the stem bark of rohituka (*Aphanamixis polystachya*) in mice bearing Ehrlich ascites carcinoma.

Materials and Methods

Chemicals

5,5'-dithiobis-2-nitrobenzoic acid (DTNB), reduced nicotinamide adenine dinucleotide (NADPH), epinephrine, 1-chloro-2, and 4-dinitrobenzene (CDNB) were purchased from Sigma Chemical Company, St. Louis, USA; Trichloroacetic acid (TCA) was purchased from Acros Organics, Geel, Belgium, while all other routine chemicals were procured from Ranbaxy Fine Chemicals, Mumbai, India.

Preparation of the Extract

Rohituka, *Aphanamixis polystachya* R. Br. (family-Meliaceae) was identified by Dr. G. K. Bhat (a well-known taxonomist) Department of Botany, Poorna Prajna College, Udupi, India and the herbarium specimen has been stored with us. The bark of the tree was carefully peeled off, shade dried, and coarsely powdered in a ball mill. The powdered material was extracted with petroleum ether in a Soxhlet apparatus at 40°C for thirty cycles and dried at 40°C overnight. The petroleum ether free powdered material was further extracted with 95 per cent ethanol in a Soxhlet apparatus at 60°C extensively for 3 days. The cooled liquid extract was concentrated by evaporating its liquid contents under reduced pressure at room temperature. The dried extract was subjected to freeze drying to obtain a fine powder of the extract. Henceforth the extract of *Aphanamixis polystachya* will be called as APE.

Experimental

Determination of Heavy Metal Contamination

The rohituka extract was subjected to atomic absorption spectrometry (AAS) for presence of cadmium and lead, the two most prevalent heavy metals according to

WHO guidelines (WHO, 1994) before undertaking the detailed investigation of its antineoplastic activity.

Animal Care and Handling

The animal care and handling were done according to the guidelines issued by the World Health Organization, Geneva, Switzerland and the INSA (Indian National Science Academy, New Delhi, India). Ten to twelve weeks old female Swiss albino mice weighing 30 to 36 g were selected from an inbred colony maintained under the controlled conditions of temperature (23 ± 2°C), humidity (50 ± 5 per cent) and light (10 and 14 h of light and dark, respectively). The animals had free access to sterile food and water. Four animals were housed in a polypropylene cage containing sterile paddy husk (procured locally) as bedding throughout the experiment. The study was approved by the Institutional Animal Ethical Committee (IAEC) of the Kasturba Medical College, Manipal, India, where the study was performed.

Tumor Model

Ehrlich ascites carcinoma (EAC) procured from the Cancer Research Institute (ACTREC), Mumbai, India, was used throughout the study.

Preparation of Drug and Mode of Administration

The APEwas dissolved freshly in 0.5 per cent CMC (carboxy methylcellulose) before use. The APE was administered intraperitoneally in the tumor bearing mice.

Acute Drug Toxicity Studies (OECD guidelines 420-425)

The acute toxicity (Prieur *et al.*, 1972; Ghosh, 1984) of 0, 0.05, 0.1, 0.25, 0.5, 1, 2 and 3 g/kg body weight of APE was determined according to OECD guidelines 420-425. Various behavioural and neurological responses in the mice fed with APE were recorded as per the standard criteria (Ghosh, 1984). The responses were scored with a scale of 0 to 8 and scoring was done at the time of the peak effect.

Influence of APE on Sleeping Time in Mice

The influence of APE on sleeping time in mice was assessed (Lovell, 1986) to test for the proper functioning of central and peripheral nervous system in the form of sleep induction as follows.

Female mice (25-35 g) fasted for 18 h were divided into three groups of eight each. Animals of group I were injected intraperitoneally with 40 mg/kg pentobarbitone, and the animals of group II were pretreated with APE 1g/kg body weight intraperitoneally one hour before the injection of pentobarbitone, whereas the animals of group III were pretreated with APE 1g/kg body weight orally one hour before the injection of pentobarbitone. A total of 24 animals were used for this study.

The period between the loss of righting reflex and regaining of righting reflex was taken as the sleeping time.

Antineoplastic Activity

The optimum dose of APE for its antineoplastic activity was selected following the standard protocol recommended by Cancer Chemotherapy National Service

Center (CCNSC) (Geran *et al.*, 1972). Ten animals were administered with 0, 0.125, 0.25, 0.5, 0.75, 1, 1.25 or 1.5 g/kg body weight APE once daily, consecutively for 9 days, 24 h after tumor inoculation. A group of ten animals concomitantly received doxorubicin (DOX) 1 mg/kg body weight intraperitoneally and served as a positive control. A total of 90 animals were used for this experiment.

Altered Administration Schedule

A separate experiment was designed to evaluate the effect of alterations in the drug administration schedule on the antineoplastic activity of APE in the tumor bearing mice. A group of ten EAC mice were treated with 1 g/kg body weight of APE once daily for 9 consecutive days or once every alternate day or once every alternate two days intraperitoneally. A still another group of ten EAC mice were treated with 0.5 g/kg body weight APE (1 g/kg APE was split into two parts) twice daily for nine consecutive days with a gap of twelve hours between two fractions. A group of ten matching saline treated animals served as control. A total of fifty animals were utilized for this experiment.

Stage Specific Evaluation of the Anticancer Activity

The tumors were inoculated and allowed to grow for 1, 3, 6, 9, 12 or 15 days before APE treatment and for reasons of clarity these days have been designated as stage I, II, III, IV, V and VI, respectively. A group of ten tumor bearing animals were treated with 1 g/kg APE (optimum dose, split into 0.5 mg/kg) or 1 mg/kg doxorubicin intraperitoneally at 1, 3, 6, 9, 12 and 15 days after tumor inoculation (stage I, II, III, IV, V and VI, respectively), consecutively for nine days. A total of 130 animals were used for this experiment.

The animals of all experiments, except acute toxicity studies were monitored regularly for body weight changes, signs of toxicity and mortality. The weight of animals was recorded every third day up to 120 days or until the animal survival after tumor inoculation in all the groups. A 33 per cent of drug related deaths or a weight loss of 5 g per mouse was considered as an index of toxicity. The animal survival was monitored daily up to 120 days, since the survival of animals up to 120 days is roughly equivalent to 5 years survival in man (Nias, 1990). The tumor response was assessed on the basis of median survival time (MST) and the average survival time (AST). The MST and AST were calculated from the animals dying within 120 days and those surviving 120 days were excluded from it. The increase in median life span (per cent IMLS) and the increase in average life span (per cent IALS) were calculated using the formulae:

$$\text{MST} = \frac{(X + Y)}{2}$$

where,

X is the earliest day when the cumulative number of deaths is $\geq N/2$, and Y is the earliest day when the cumulative number of deaths is $\geq (N/2) + 1$. $N = 10$ (total number of animals including survivors).

$$\text{AST} = \frac{\text{Sum of animal death on different days}}{\text{Number of animals}}$$

$$\text{IMLS (\%)} = \frac{\text{MST of treated mice} - \text{MST of control}}{\text{MST of control}} \times 100$$

$$\text{IALS (\%)} = \frac{\text{AST of treated mice} - \text{MST of control}}{\text{AST of control}} \times 100$$

Biochemical Estimations

A separate experiment was conducted to estimate the antioxidant enzyme and lipid peroxidation in the EAC cells after treatment with APE or DOX. The tumor bearing animals were administered with an intraperitoneal single injection of 1 g/kg body weight of the freshly prepared APE in sterile saline containing 1 per cent CMC or saline or 1 mg/kg doxorubicin once daily for 9 consecutive days. Four animals from each group were sacrificed at 0, 1.5, 3, 6, 9, 12, 18 or 24 h post last drug administration (PLDA). A total of 96 animals were used for this study.

The tumor cells were aspirated in an aseptic manner and were washed with SPS thrice. The cells were counted under the inverted microscope (Leica Microsystems, Wetzlar, Germany) and 1 x 10^6 cells from each EAC mice were suspended in 1 ml of 0.15 M Tris–HCl (pH 7.4) with 50 µl of Triton X 100 and sonicated on ice for 20 sec using a microprobe sonicator (Sonics Vibra-cell, CT, USA). The sonicated cell suspension was centrifuged at 13,000 rpm for 15 min. The supernatant (S) thus obtained was immediately used for the estimation of lipid peroxidation (LOO) (Ohkawa *et al.*, 1979), glutathione (GSH) (Moron *et al.*, 1979), catalase (CAT) (Abei, 1974), superoxide dismutase (SOD) (Misra and Fridovich, 1972), glutathione peroxidase (GSHPx) (Ho, 1994), glutathione-S-transferase (GST) (Habig and Pabst, 1974).

Statistical Analyses

The statistical significance between the treatments for survival studies was determined using the "Z" test (Abramowitz and Stegun, 1972), while the student's '*t*'test was used for biochemical estimations.

Results

The results are presented in Tables 11.1–11.7. All the data are expressed as mean ± SEM (standard error of the mean).

Determination of Heavy Metal Contamination

Atomic absorption spectrometry did not show any detectable amounts of lead and cadmium in the APE sample as well as corresponding blank. Therefore, the APE was considered free from heavy metal contamination.

Determination of Acute Toxicity

The acute toxicity of APE was determined by administering a single dose of 0, 0.05, 0.1, 0.25, 0.5, 1, 2 and 3 g/kg body weight of APE. The APE was found to be non-

Table 11.1(a): Effect of different doses of rohituka stem extract on the behavioural responses in the normal mice.

APE (g/kg) Body Weight	*Behavioural Responses*												
	Awareness				*Mood*					*Motor Activity*			
	Alertness	*Visual Placing*	*Passivity*	*Stereo-typy*	*Groo-ming*	*Vocali-zation*	*Restless-ness*	*Irrit-ability*	*Fearful-ness*	*Reac-tivity*	*Sponta-neous Activity*	*Touch Res-ponse*	*Pain Res-ponse*
0.0	4	4	0	0	4	0	0	0	0	4	4	4	4
0.05	4	4	0	0	4	0	0	0	0	4	4	4	4
0.1	4	4	0	0	4	0	0	0	0	4	4	4	4
0.25	4	4	0	0	4	0	0	0	0	4	4	4	4
0.5	4	4	0	0	4	0	0	0	0	4	4	4	4
1	4	4	0	0	4	0	0	0	0	4	4	4	4
2	4	4	0	0	4	0	0	0	1	5	4	4	5
3	5	4	0	0	4	0	1	0	0	4	5	4	4

Table 11.1(b): Effect of different doses of APE on the neurological responses in the normal mice.

APE (g/kg) Body Weight	*Neurological Profile*															
	CNS Excitation				*Posture*		*Motor Incoordination*			*Muscle Tone*				*Reflexes*		
	Startle Response	*Straubs Response*	*Tremors*	*Convulsions*	*Body Posture*	*Limb Posture*	*Staggering Gait*	*Abnormal Gait*	*Righting Reflex*	*Limb Tone*	*Grip Strength*	*Body Tone*	*Abdominal Tone*	*Pinna*	*Corneal*	*ISR*
0.0	0	0	0	0	4	4	0	0	0	4	4	4	4	4	4	4
0.05	0	0	0	0	4	4	0	0	0	4	4	4	4	4	4	4
0.1	0	0	0	0	4	4	0	0	0	4	4	4	4	4	4	4
0.25	0	0	0	0	4	4	0	0	0	4	4	4	4	4	4	4
0.5	0	0	0	0	4	4	0	0	0	4	4	4	4	4	4	4
1	0	0	0	0	4	4	0	0	0	4	4	4	4	4	4	4
2	0	0	0	0	5	4	0	0	0	4	4	4	4	4	4	4
3	0	0	0	0	5	4	0	0	0	5	4	4	4	4	5	4

ISR: Ispilateral reflex.

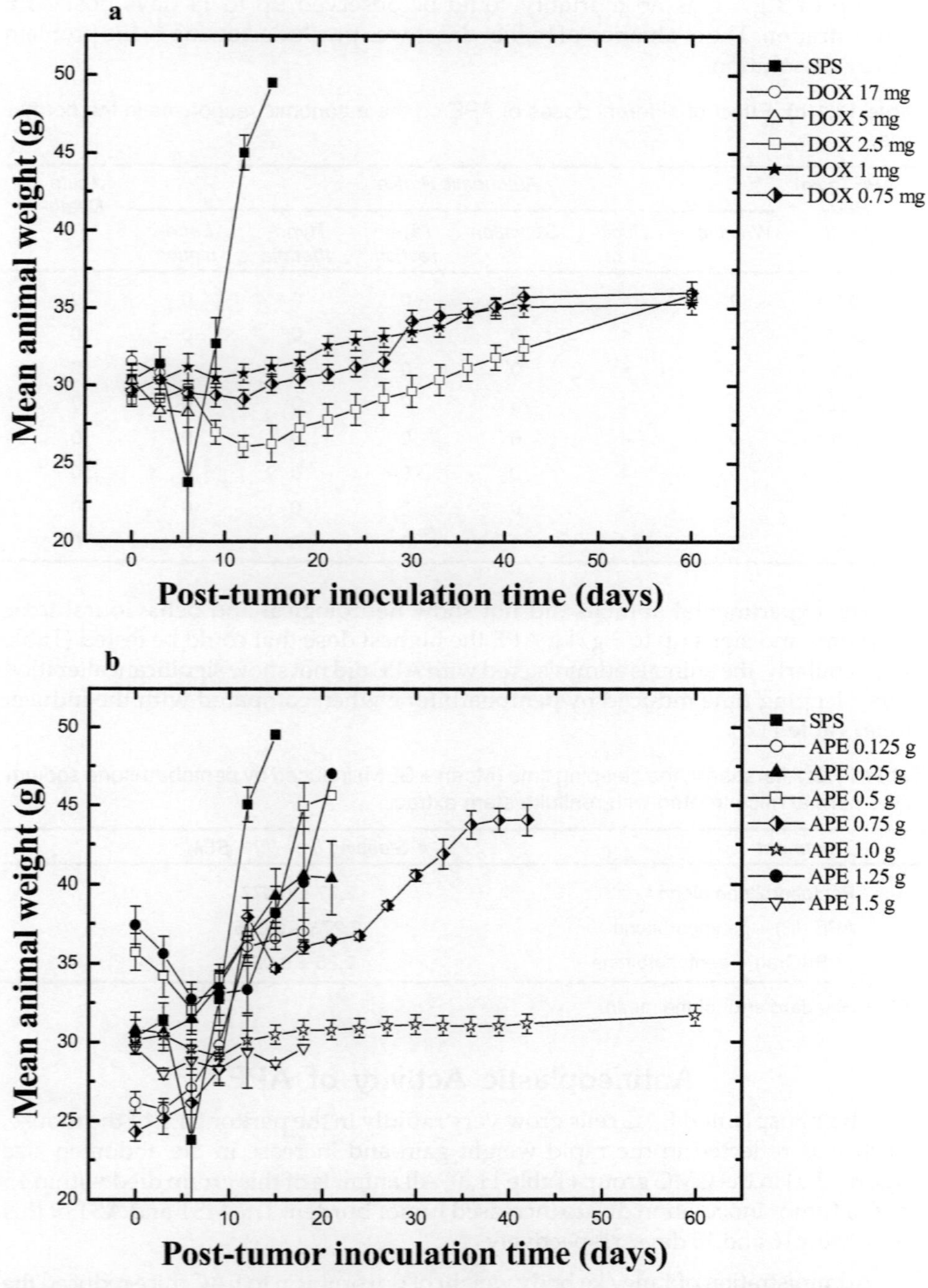

Figure 11.1: Alteration in the mean body weight of EAC mice treated with various doses of APE or DOX. a, Doxorubicin and b, APE.

toxic up to 3 g/kg as no mortality could be observed up to 14 days post APE administration. The evaluation of higher doses was precluded owing to the problem in drug dissolution.

Table 11.1(c): Effect of different doses of APE on the autonomic responses in the normal mice.

APE (g/kg) Body Weight	*Autonomic Profile*						*Acute Death*
	Writhing	*Pupil Size*	*Salivation*	*Piloe-rection*	*Hypo-thermia*	*Lacri-mation*	
0.0	0	4	0	0	0	0	0
0.05	0	4	0	0	0	0	0
0.1	0	4	0	0	0	0	0
0.25	0	4	0	0	0	0	0
0.5	0	4	0	0	0	0	0
1	0	4	0	0	0	0	0
2	0	4	0	0	0	0	0
3	0	4	1	0	0	0	0

The experimental animals did not show neurological and behavioural toxic symptoms and signs up to 3 g/kg APE the highest dose that could be tested (Table 11.1). Similarly, the animals administered with APE did not show significant alteration in the sleeping time induced by pentobarbitone when compared with the inducer alone (Table 11.2).

Table 11.2: Alteration in the sleeping time (Mean ± SEM) induced by pentobarbitone sodium in the normal mice treated with rohituka stem extract.

Treatment	*Sleeping time (h) ± SEM*
Pentobarbitone alone	2.37 ± 0.077
APE (IP) + pentobarbitone	2.275 ± 0.065
APE (Oral) + pentobarbitone	2.25 ± 0.059

SEM = Standard error of the mean.

Antineoplastic Activity of APE

The transplanted EAC cells grow very rapidly in the peritoneum of the mouse, which was reflected in the rapid weight gain and increase in the abdomen size (Figure 11.1) in the CMC group (Table 11.3). All animals of this group died within 15 days of tumor inoculation due to increased tumor burden. The MST and AST of this group were16 and 15 days, respectively.

Administration of 1 mg/kg body weight of doxorubicin to EAC mice reduced the weight gain (Figure 11.2) caused a marked increase in the survival time, where 50 per cent of the EAC animals survived up to day 40 and 10 per cent survived beyond 120 days post-tumor transplantation (Table 11.4). The MST and AST were 29.5 and 29

Table 11.3(a): Body weight changes (mean ± SEM) of tumorized mice treated with different concentrations of rohituka stem extract 24 h post-tumorization.

Days	*APE Dose mg/kg Body Weight*							
	0	*125*	*250*	*500*	*750*	*1000*	*1250*	*1500*
0	30.33 ± 1.09	26.17 ± 0.68	30.80 ± 1.09	24.31 ± 0.58	35.71 ± 1.16	29.35 ± 0.47	37.41 ± 1.18	29.64 ± 0.35
3	31.43 ± 1.06	25.74 ± 0.67	30.57 ± 1.09	25.17 ± 1.00	34.21 ± 1.35	30.03 ± 0.51	35.60 ± 1.11	27.99 ± 0.34
6	27.77 ± 6.04	27.12 ± 0.81	31.48 ± 0.97	26.15 ± 0.75	32.05 ± 1.30	29.47 ± 0.44	32.76 ± 1.06	28.84 ± 0.54
9	32.71 ± 1.66	29.82 ± 0.90	33.63 ± 0.94	28.28 ± 0.86	34.05 ± 0.88	29.37 ± 0.75	33.10 ± 1.40	28.31 ± 1.04
12	45.01 ± 1.12	36.04 ± 0.67	36.71 ± 1.65	37.93 ± 1.19	36.70 ± 1.16	31.19 ± 0.83	33.34 ± 1.49	29.39 ± 2.30
15	49.47 ± 0.96	36.55 ± 0.67	39.29 ± 1.69	34.68 ± 0.22	38.18 ± 0.96	32.88 ± 0.78	38.17 ± 1.71	28.64 ± 0.21
18	–	37.04 ± 0.69	40.54 ± 1.72	36.06 ± 0.44	44.94 ± 1.49	33.40 ± 0.70	40.12 ± 4.05	29.63 ± 0.00
21	–	–	40.37 ± 2.30	36.47 ± 0.17	45.62 ± 0.00	34.17 ± 0.70	47.00 ± 0.00	–
24	–	–	–	–	–	34.52 ± 0.75	–	–
27	–	–	–	–	–	34.96 ± 0.73	–	–
30	–	–	–	–	–	35.50 ± 0.72	–	–
33	–	–	–	–	–	35.61 ± 0.74	–	–
36	–	–	–	–	–	35.83 ± 0.71	–	–
39	–	–	–	–	–	36.07 ± 0.70	–	–
42	–	–	–	–	–	36.35 ± 0.73	–	–
60	–	–	–	–	–	41.01 ± 0.55	–	–
120	–	–	–	–	–	47.05 ± 0.68	–	–

Table 11.3(b): Body weight changes (mean ± SEM) of tumorized mice treated with different concentrations of doxorubicin 24 h post-tumorization.

Days	*Doxorubicin mg/kg Body Weight*					
	0.75	*1*	*1.5*	*2*	*2.5*	*5*
0	32.14 ± 0.51	34.18 ± 1.47	33.13 ± 0.80	31.65 ± 0.61	29.05 ± 0.43	30.32 ± 0.99
3	32.09 ± 0.69	33.63 ± 0.86	34.81 ± 1.37	32.83 ± 0.62	29.17 ± 0.54	28.40 ± 0.74
6	32.62 ± 0.58	34.59 ± 1.70	32.51 ± 0.73	31.19 ± 0.40	29.78 ± 0.49	28.23 ± 0.92
9	31.64 ± 0.94	34.36 ± 1.90	30.37 ± 1.23	30.63 ± 0.45	26.99 ± 0.74	–
12	28.61 ± 0.76	29.72 ± 1.28	30.92 ± 1.26	29.49 ± 0.84	26.08 ± 0.68	–
15	32.22 ± 0.74	32.27 ± 1.98	24.13 ± 1.37	–	–	–
18	31.38 ± 0.66	31.74 ± 1.65	–	–	–	–
21	31.35 ± 0.59	32.03 ± 1.97	–	–	–	–
24	32.34 ± 1.04	32.33 ± 2.05	–	–	–	–
27	31.90 ± 0.66	32.74 ± 5.03	–	–	–	–
30	32.11 ± 0.60	33.98 ± 2.76	–	–	–	–
33	37.77 ± 3.17	35.52 ± 4.38	–	–	–	–
36	33.78 ± 0.00	35.71 ± 4.37	–	–	–	–
39	34.39 ± 0.00	35.95 ± 4.41	–	–	–	–
42	34.47 ± 0.00	36.14 ± 4.43	–	–	–	–
60	37.03 ± 0.00	36.30 ± 4.41	–	–	–	–
120	–	–	–	–	–	–

SEM = Standard error of the mean.

n = 10.

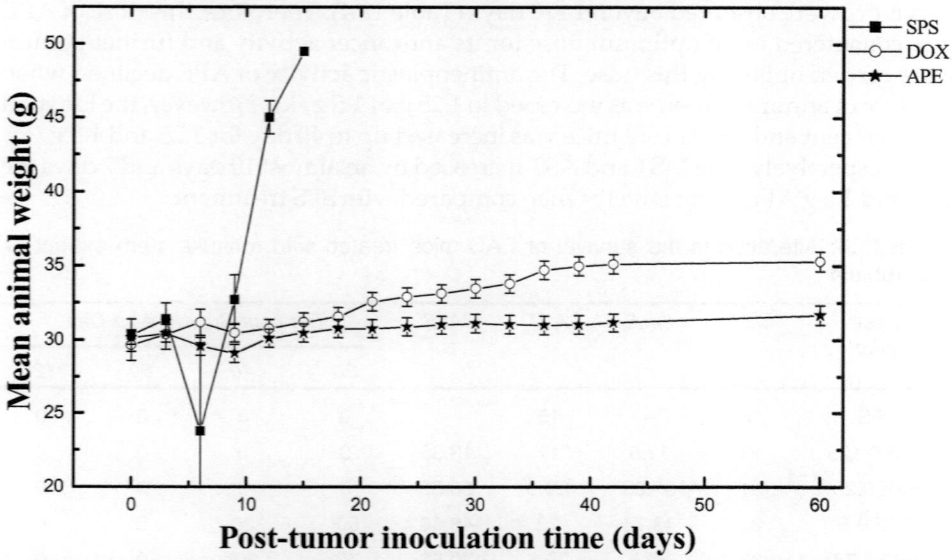

Figure 11.2: Effect of APE (1g/kg) or DOX (1 mg/kg) on the mean body weight of EAC mice.

days, respectively. The median life span (per cent IMLS) increased by 90.3 per cent and increase in average life span (per cent IALS) was by 93.3 per cent (Table 11.4).

Treatment of EAC mice with various doses of APE resulted in a dose dependent enhancement in cell killing effect and remission of tumor up to 1 g/kg body weight, which was reflected in reduced body weight gain (Figures 11.1 and 11.2) and a further increase in the APE dose caused a decline in its antineoplastic activity (Table 11.4). Treatment of EAC mice with 0.125 g/kg body weight APE increased the survival by 2 days when compared to the concurrent CMC treatment. The MST and AST were 18 and 17 days, respectively and showed no protection against tumor growth (Table 11.4). Doubling the dose of APE to 0.25 g/kg body weight had some ameliorating effect and all the animals died within 21 days. The MST and AST increased to 21 and 20.3 days, respectively, when compared to SPS treatment. A further increase in the dose of APE up to 0.5 g/kg or 0.75 g/kg body weight did not show improvement in reducing the tumor growth and weight gain, where almost all animals succumbed to death by 45 days (Table 11.4). The MST and AST were found to be 23 and 22 days, respectively for 0.5 g/kg group and 26 and 25.6 days, respectively for 0.75 g/kg group. The administration of 0.5 and 0.75 g APE increased the life span of EAC mice up to 40 days (Table 11.4). When the APE dose was raised up to 1 g/kg the 40 day survival was almost doubled where, 90 per cent of the animals were alive (Table 11.3). The average weight gain was inhibited by 75 per cent when compared with SPS control group (Table 11.3). The MST and AST also increased to 35 days approximately and the increase in median life span (per cent IMLS) and increase in average life span (per cent IALS) were greater than 120 per cent. APE at 1 g/kg body weight exhibited the best anti-tumor activity amongst all the doses tested as 40 per cent of tumor free survivors were observed beyond 120 days (Table 11.4). Therefore, this dose of APE was considered as an optimum dose for its anticancer activity and further studies were carried out using this dose. The antineoplastic activity of APE declined when the dose of administration was increased to 1.25 g or 1.5 g/kg. However, the life span of 30 per cent and 10 per cent mice was increased up to 40 day for 1.25 and 1.5 g/kg APE, respectively. The MST and AST increased by an almost 10 days and 7 days for 1.25 and 1.5 g APE, respectively when compared with SPS treatment.

Table 11.4: Alteration in the survival of EAC mice treated with rohituka stem extract or doxorubicin.

Dose (g/kg)	*MST*	*IMLS*	*AST*	*IALS*	*Per cent Survivors on Day*			
					30	*40*	*60*	*120*
SPS	16	–	15	–	0	0	0	0
APE 0.125	18	12.5	17	13.33	0	0	0	0
APE 0.25	21	31.25	20.3	35.33	0	0	0	0
APE 0.5	23	43.75	22	46.66	50^{b}	20	0	0
APE 0.75	26	62.5	25.6	70.66	70^{c}	30^{a}	0	0
APE 1	35.5	121.8	35	133.33	100^{d}	90^{d}	40^{b}	40^{b}
APE 1.25	26.5	65.62	25.7	71.33	60^{b}	30^{a}	0	0
APE 1.5	23.5	46.87	22.5	50	30^{c}	10	0	0
DOX 1 mg/kg	29.5	90.32	29	93.37	100^{d}	50^{b}	10	10

p < a = 0.05; b = 0.002; c = 0.001; d = 0.0001 and no symbols = non-significant when compared with SPS.

SPS: Sterile physiological saline; APE: Alcoholic extract of *Aphanamixis polystachya* and DOX: Doxorubicin hydrochloride.

Altered Administration Schedule

The alteration in the nine-day administration schedule of 1 g/kg APE once every day, once every alternate day or once every two alternate days resulted in an elevation in the survival of mice. The greatest effect was observed when a single dose of APE was administered for 9 consecutive days. The MST and AST were found to be 32.5 days each. Administration of 1 g on alternate days and on every two alternate days also resulted in the increase in the MST and AST. However this schedule of administration was not as potent as split dose regimen (Table 11.5). Even though changes in single administration of drug to alternate day and alternate two days reduced the survival when compared to daily adminstration, it was higher than the SPS treated groups (Table 11.5).

Since alteration in single administration did not improve the antineoplastic activity of APE, it was decided to split 1 g/kg APE into two equal doses and test the effect of this schedule on the EAC tumor cell killing. The EAC animals receiving 0.5 g/kg of APE twice a day with a gap of 12 h for nine consecutive days showed a significant tumor remission as evident by retardation in the body weight gain of EAC mice and increase in the life span, where 50 per cent of the animals survived beyond 120 days, a 10 per cent elevation when compared with the single administration regimen (Table 11.5).

Table 11.5: The effect of altered administration schedule on the survival of EAC mice treated with 1000 mg/kg body weight rohituka stem extract.

Dose (g/kg)	*MST*	*IMLS*	*AST*	*IALS*	*Per cent Survivors on Day*			
					30	*40*	*60*	*120*
SPS	16	–	15.5	–	0	0	0	0
APE 1000	35	118.75	34	119.35	100[d]	90[d]	40[b]	40[b]
APE 1000-AD	27	68.75	26.5	70.96	80[d]	60[c]	30[a]	0
APE 1000-A2D	22	37.5	21.75	40.32	60[c]	50[b]	20	0
APE 500+500-D	48	200	47.25	204.8	100[d]	90[d]	60[c]	50[b]

APE administered on AD = alternate days; A2D = alternate two days; SD = two equal split doses of 0.5 g/kg daily.

p< a = 0.05; b = 0.002; c = 0.001; d = 0.0001 and no symbols = non-significant when compared with SPS treatment.

SPS: Sterile physiological saline; APE: Alcoholic extract of *Aphanamixis polystachya* and DOX: Doxorubicin hydrochloride

Stage Specific Evaluation

The stage specific evaluation of the antineoplastic activity of APE was carried out in EAC animals at 1, 3, 6, 9, 12 or 15 days, (stage I, II, III, IV, V and VI, respectively), where two split doses of 0.5 g/kg of APE or a single dose of 1 mg/kg of DOX was administered everyday for nine days, consecutively at stage I, II, III, IV, V and VI (Table 11.6). The APE was effective in increasing the life span of EAC mice up to stage III, where the animals did survive beyond 120 days, which is reflected in the body weight changes. The APE treatment caused higher survival (Table 11.6) and tumor remission during stage I, II and III.

The administration of APE at stage IV also lengthened the life span of 40 per cent animals up to 40 days. The administration of APE at stage V and VI also increased the life span of EAC mice when compared to SPS treatment. The effect of APE dwindled especially after stage III of tumor development (Table 11.6).The treatment of mice with 1 g/kg (two doses of 0.5 g each) APE at various stages of tumor development resulted in an increase in the MST up to 49, 41, 34, 28, 22 and 16.5 days at stage I, II, III, IV, V and VI, respectively, when compared with the DOX treatment (Table 11.6). Similarly, the IMLS and IALS also increased after APE treatment depending on the stage of drug inoculation. The IMLS and IALS declined depending on the treatment stage and the lowest values were observed for stage VI. The IMLS of 206.25, 156.25, 112.5, 75, 37.5 and 3.1 per cent, whereas the IALS of 223.23, 176.66, 120, 75, 43.33 and 3.33 per cent was recorded for stage I, II, III, IV, V and VI, respectively (Table 11.6).

The administration of APE resulted in 50, 40 and 30 per cent survivors by 120 days at stage I, II and III respectively of tumor development. The MST and AST increased by 3, 2.6, 2, 1.7 and 1.4 folds at stage I, II, III, IV and V, respectively after APE treatment when compared with the DOX treatment. The IMLS was 1.9, 1.9, 1.7, 1.6 and 1.7 folds greater for stage I, II, III, IV and V respectively, while the IALS was elevated by 2.1, 2.1, 1.7, 1.4 and 2.2 folds for stage I, II, III, IV and V respectively, when compared with concurrent DOX group (Table 11.6).

Table 11.6: Survival of EAC mice treated with rohituka stem extract or doxorubicin at various stages of tumor development.

Stages of Tumor	*MST*	*IMLS*	*AST*	*IALS*	*Per cent Survivors on Day*			
					30	*40*	*60*	*120*
SPS	16	–	15	–	–	–	–	–
APE I	49	206.25	48.5	223.33	100[d]	90[d]	60[d]	50[d]
APE II	41	156.25	41.5	176.66	100[d]	90[d]	50[d]	40[c]
APE III	34	112.5	34	120	100[d]	70[d]	40[c]	20
APE IV	28	75	26.25	75	60[c]	40[b]	0	0
APE V	22	37.5	21.5	43.33	20	0	0	0
APE VI	16.5	3.125	15.5	3.33	0	0	0	0
DOX I	31	106.66	30	106.89	100[d]	50[c]	10	10
DOX II	29	81.25	27.5	83.33	90[d]	40[b]	10	10
DOX III	26.5	65.62	25.5	70	60[c]	20	10	0
DOX IV	23.5	46.87	23	53.33	30[a]	10	0	0
DOX V	19.5	21.875	18	20	0	0	0	0
DOX VI	17	6.25	16	6.66	0	0	0	0

$p<$ a = 0.05; b = 0.002; c = 0.001; d = 0.0001 and no symbols = non-significant when compared with SPS treatment.

SPS: Sterile physiological saline; APE: Alcoholic extract of *Aphanamixis polystachya* and DOX: Doxorubicin hydrochloride

The tumors were inoculated and allowed to grow for 1, 3, 6, 9, 12 or 15 days before APE treatment and these days have been designated as stage I, II, III, IV, V and VI, respectively.

Biochemical Estimations

The spontaneous values of LOO, GSH, GSHPx, GST, SOD and CAT did not change significantly in EAC cells with assay time (Table 11.7).

Lipid Peroxidation

Treatment of EAC mice with APE caused a time dependent elevation in the LOO and a peak level was reached at 8 h (2 folds; $p<0.05$) that remained unaltered up to 12 h ($p<0.05$) (Figure 11.3a). The LOO declined steadily thereafter without reaching to normal level even up to 24 h PLDA (1.4 folds more when compared with 0 h PLDA). The level of lipid peroxidation was significantly higher up to 16 h PLDA in APE

Table 11.7: Alteration in the antioxidant enzymes and lipid peroxidation in the Ehrlich ascites carcinoma (EAC) cells treated with rohituka stem extract or doxorubicin.

Treatment	*PLDA (h)*	*LPx*	*GSH*	*GSHPx*	*GST*	*CAT*	*SOD*
SPS	0	4.09 ± 0.52	2.63 ± 0.09	8.63 ± 0.67	0.247 ± 0.023	0.37 ± 0.026	2.92 ± 0.051
	1	4.08 ± 0.51	2.62 ± 0.07	8.49 ± 0.52	0.252 ± 0.024	0.368 ± 0.032	2.89 ± 0.053
	2	4.08 ± 0.37	2.63 ± 0.07	8.60 ± 0.56	0.250 ± 0.020	0.352 ± 0.032	2.91 ± 0.041
	4	4.09 ± 0.44	2.62 ± 0.08	8.54 ± 0.61	0.249 ± 0.021	0.36 ± 0.028	2.92 ± 0.05
	8	4.11 ± 0.56	2.62 ± 0.06	8.64 ± 0.24	0.251 ± 0.019	0.37 ± 0.024	2.90 ± 0.042
	12	4.11 ± 0.50	2.62 ± 0.08	8.60 ± 0.58	0.250 ± 0.022	0.36 ± 0.029	2.88 ± 0.043
	16	4.13 ± 0.48	2.63 ± 0.09	8.58 ± 0.42	0.248 ± 0.021	0.377 ± 0.031	2.91 ± 0.040
	24	4.13 ± 0.67	2.62 ± 0.09	8.62 ± 0.49	0.249 ± 0.023	0.36 ± 0.024	2.90 ± 0.049
APE	0	4.09 ± 0.44	2.62 ± 0.08	7.67 ± 0.51	0.196 ± 0.017	0.24 ± 0.022	2.75 ± 0.044
	1	6.17 ± 0.46	2.18 ± 0.06^{a}	5.10 ± 0.45	0.16 ± 0.011	0.203 ± 0.02	2.21 ± 0.047^{b}
	2	7.29 ± 0.41^{a}	1.86 ± 0.05^{b}	4.29 ± 0.50^{a}	0.107 ± 0.01^{a}	0.126 ± 0.017^{a}	2.06 ± 0.043^{b}
	4	7.70 ± 0.42^{a}	1.52 ± 0.05^{b}	4.16 ± 0.43^{a}	0.083 ± 0.01^{a}	0.105 ± 0.019^{b}	1.75 ± 0.041^{c}
	8	8.45 ± 0.46^{a}	1.31 ± 0.06^{b}	4.08 ± 0.42^{b}	0.109 ± 0.013^{a}	0.105 ± 0.018^{a}	1.43 ± 0.048^{c}
	12	8.38 ± 0.36^{a}	1.49 ± 0.05^{b}	4.13 ± 0.46^{a}	0.116 ± 0.014^{a}	0.128 ± 0.02^{a}	1.90 ± 0.040^{c}
	16	7.33 ± 0.37^{a}	1.64 ± 0.04^{b}	4.79 ± 0.51^{a}	0.124 ± 0.015^{a}	0.17 ± 0.021^{a}	2.08 ± 0.040^{b}
	24	6.38 ± 0.41	2.11 ± 0.06^{a}	5.21 ± 0.38^{a}	0.13 ± 0.012^{a}	0.208 ± 0.018^{a}	2.21 ± 0.041^{b}

Contd...

Table 11.7–*Contd...*

Treatment	*PLDA (h)*	*LPx*	*GSH*	*GSHPx*	*GST*	*CAT*	*SOD*
DOX	0	4.09 ± 0.39	2.62 ± 0.08	7.93 ± 0.67	0.20 ± 0.016	0.26 ± 0.021	2.71 ± 0.053
	1	6.12 ± 0.42	2.1 ± 0.07[a]	5.14 ± 0.65	0.162 ± 0.016	0.21 ± 0.021	2.07 ± 0.051[b]
	2	7.18 ± 0.49[a]	1.92 ± 0.05[a]	4.5 ± 0.54[a]	0.117 ± 0.01[a]	0.14 ± 0.02[a]	1.94 ± 0.052[b]
	4	7.41 ± 0.59[a]	1.74 ± 0.05[b]	4.32 ± 0.40[a]	0.101 ± 0.01[a]	0.13 ± 0.019[a]	1.58 ± 0.043[c]
	8	7.37 ± 0.68[a]	1.58 ± 0.04[b]	4.26 ± 0.48[a]	0.115 ± 0.011[a]	0.12 ± 0.017[b]	1.86 ± 0.043[b]
	12	7.30 ± 0.61[a]	1.71 ± 0.05[b]	4.52 ± 0.49[a]	0.13 ± 0.013[a]	0.127 ± 0.021[a]	2.10 ± 0.042[b]
	16	6.23 ± 0.58	2.04 ± 0.05[a]	5.11 ± 0.57[a]	0.149 ± 0.014	0.192 ± 0.021[a]	2.12 ± 0.041[b]
	24	5.87 ± 0.49	2.17 ± 0.06[a]	5.53 ± 0.61	0.16 ± 0.015	0.227 ± 0.023	2.40 ± 0.042[a]

p < a = 0.05; b = 0.01; c = 0.005 and No symbols = non-significant when compared concurrent SPS treated group.

DOX: Doxorubicin 1 mg/kg; APE: Alcoholic extract of *Aphanamixis polystachya* 1g/kg body weight administered intraperitoneally 24 h after tumorization of mouse once daily for 9 consecutive days, and PLDA: Post last drug administration.

LPx = nM TBARS/10^6 EAC cells; GSH = nM/10^6 EAC cells; CAT = nM H_2O_2 decomposed/min/10^6 EAC cells; SOD = U/mg protein/10^6 EAC cells; GSHPx = nM NADPH consumed/min/10^6 EAC cells; GST = mM GSH-CDNB conjugate formed/min/10^6 EAC cells.

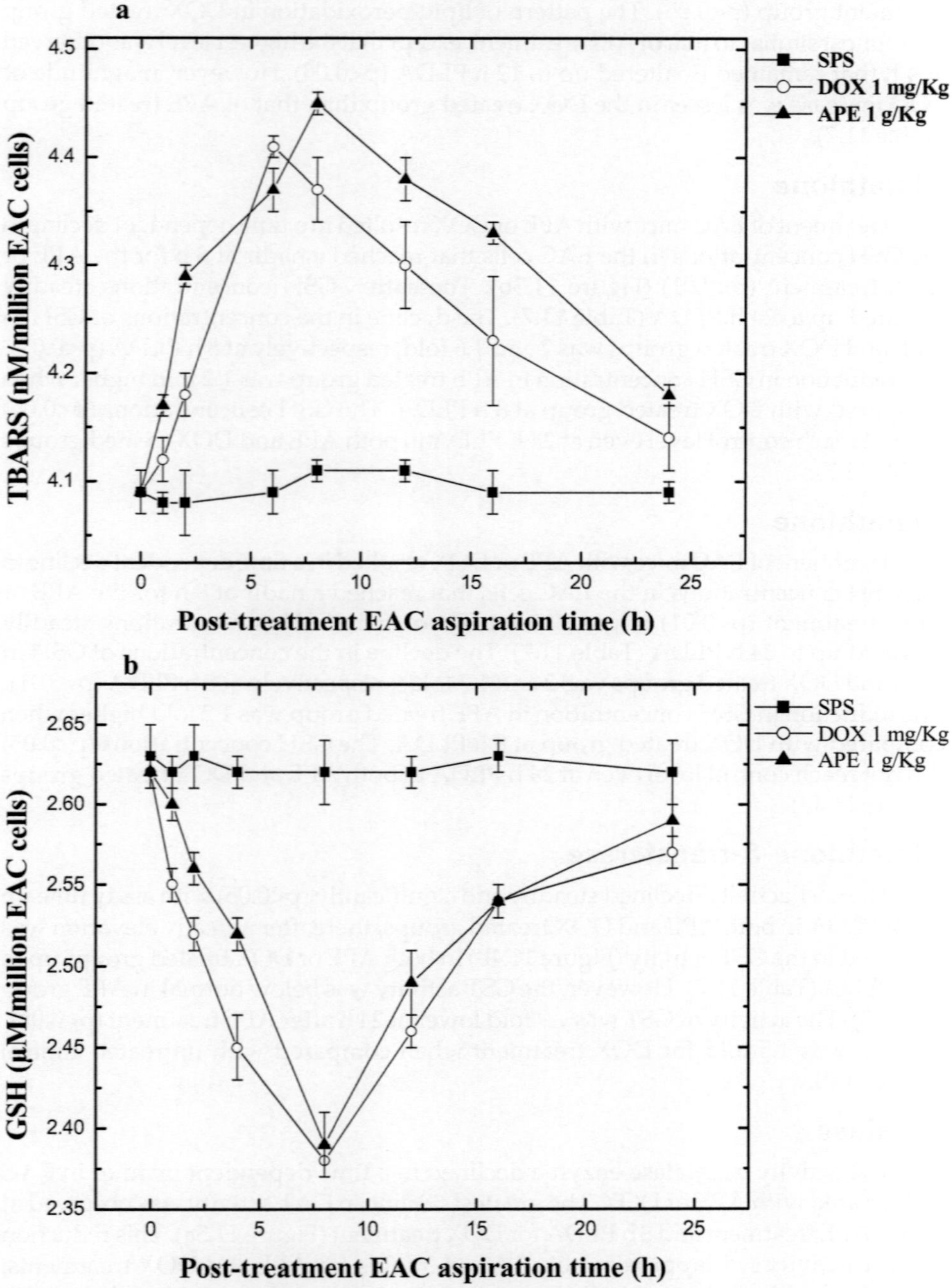

Figure 11.3: Alteration in the Lipid peroxidation and GSH concentration in EAC cells treated with APE or DOX. a, Changes in lipid peroxides as TBARS (nM/million EAC cells); b, Changes in GSH (μM/million EAC cells). Closed squares, saline control; open circles, DOX positive control and closed triangles, APE.

treatment group ($p<0.05$). The pattern of lipid peroxidation in DOX treated group was almost similar to that of APE treatment, except that the highest level was observed at 4 h that remained unaltered up to 12 h PLDA ($p<0.05$). However, magnitude of LOO increase was lesser in the DOX treated group than that of APE treated group (Table 11.7).

Glutathione

Treatment of EAC mice with APE or DOX resulted in a time dependent decline in the GSH concentrations in the EAC cells that reached a nadir at 8 h for the APE or DOX treatment ($p<0.01$) (Figure 11.3b). Thereafter, GSH concentrations steadily elevated up to 24 h PLDA (Table 11.7). The decline in the concentrations of GSH in APE and DOX treated groups was 2 and 1.6 fold, respectively at 8 h PLDA ($p<0.01$). The reduction in GSH concentration in APE treated group was 1.2 fold higher when compared with DOX treated group at 8 h PLDA. The GSH concentrations ($p<0.05$) did not reach control level even at 24 h PLDA in both APE and DOX treated groups (Table 11.7).

Glutathione

Treatment of EAC mice with APE or DOX resulted in a time dependent decline in the GSH concentrations in the EAC cells that reached a nadir at 8 h for the APE or DOX treatment ($p<0.01$) (Figure 11.4a). Thereafter, GSH concentrations steadily elevated up to 24 h PLDA (Table 11.7). The decline in the concentrations of GSH in APE and DOX treated groups was 2 and 1.7 folds, respectively at 8 h PLDA ($p<0.01$). The reduction in GSH concentration in APE treated group was 1.2 fold higher when compared with DOX treated group at 8 h PLDA. The GSH concentrations ($p<0.05$) did not reach control level even at 24 h PLDA in both APE and DOX treated groups (Figure 11.4a).

Glutathione-*S*-transferase

The GST activity declined steadily and significantly ($p<0.05$) with assay time up to 3 h PLDA in both APE and DOX treated groups, thereafter a steady elevation was observed in the GST activity (Figure 11.4b) in both APE or DOX treated groups up to 24 h PLDA (Table 11.7). However, the GST activity was below normal in APE group ($p<0.05$). The activity of GST was 1.9 fold lower at 24 h after APE treatment ($p<0.05$), while it was 1.5 fold for DOX treatment when compared with untreated control (Table 11.7).

Catalase

The activity of catalase enzyme declined in a time dependent manner in EAC cells treated with APE or DOX. The greatest decline in CAT activity was observed at 4 h for APE treatment and 8 h PLDA for DOX treatment (Figure 11.5a). This reduction in CAT activity was approximately 3.4 and 3 folds for APE and DOX treatments, respectively. The reduction in CAT activity was significantly ($p<0.02$)higher at 12 h PLDA in both the groups (Table 11.7). The decline in the CAT activity was greater in APE treatment than the DOX treatment (Figure 11.5a). However, the CAT activity did not reach control level in both DOX and APE groups (Table 11.7). The decline in

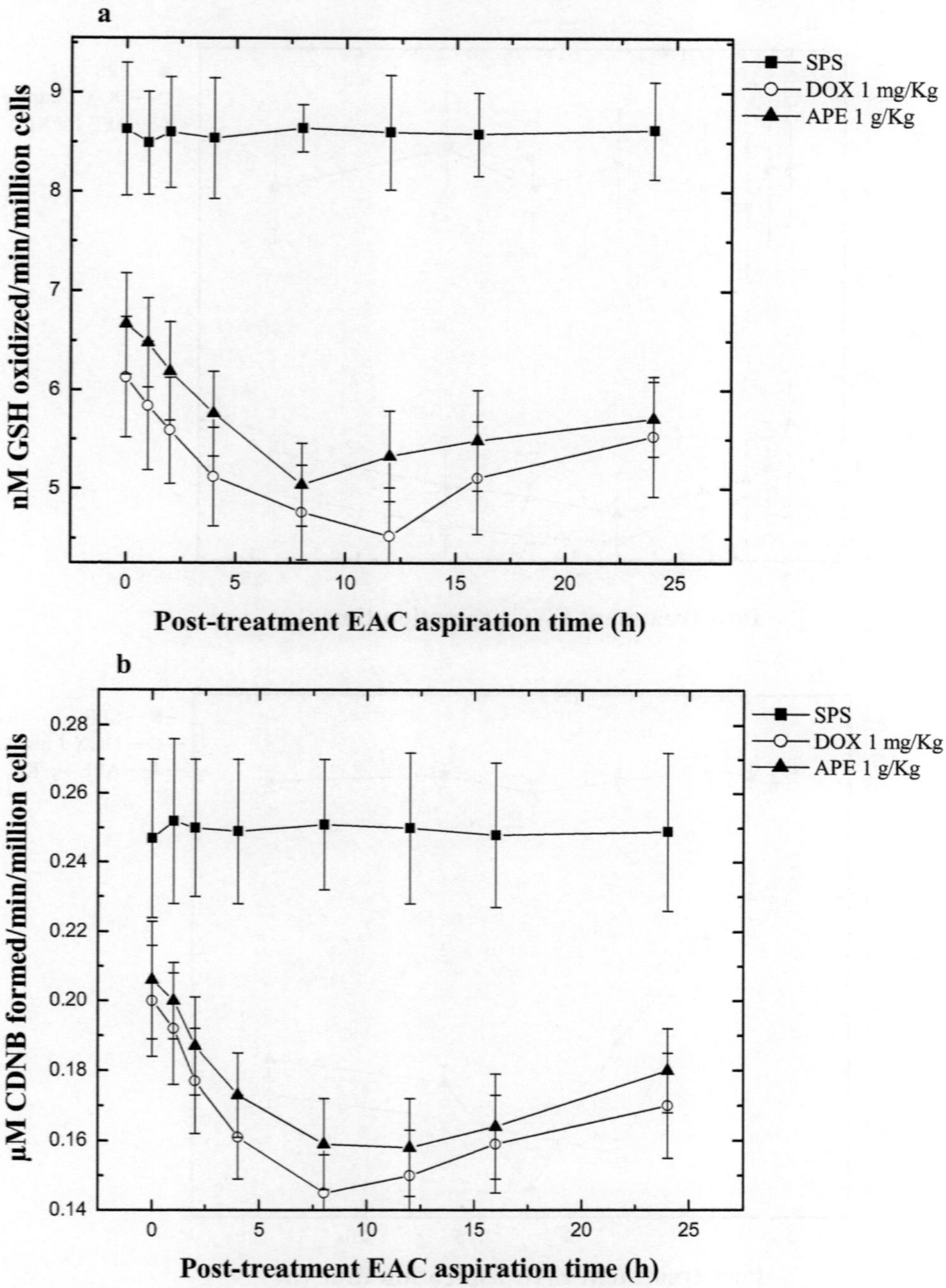

Figure 11.4: Alteration in the GSHPx, and GST activities in EAC cells treated with APE or DOX. a, Changes in GSHPx as nM GSH oxidized/min/million cells; b, Changes in GST as μM CDNB formed/min/million cells. Closed squares, saline control; open circles, DOX positive control and closed triangles, APE.

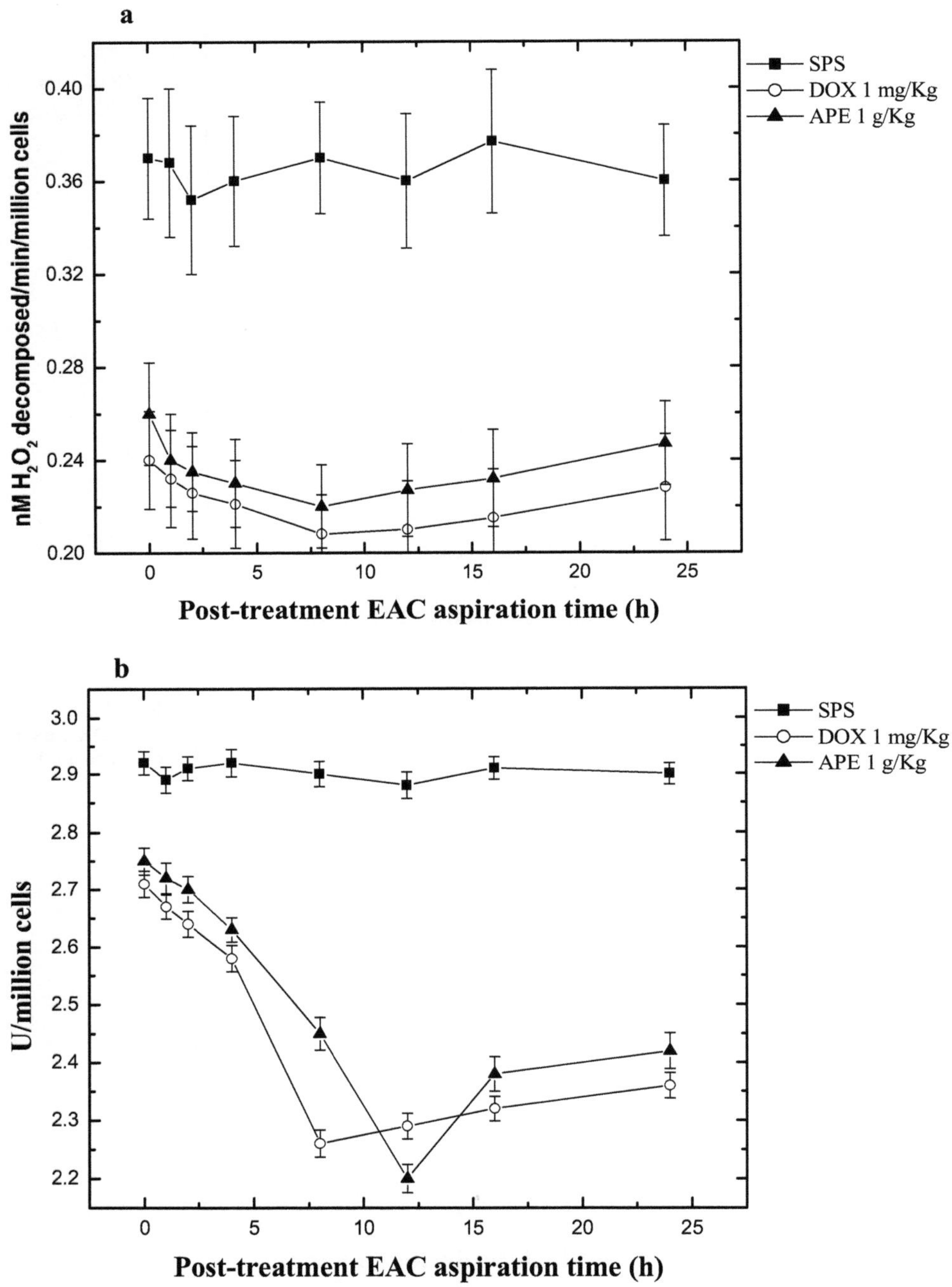

Figure 11.5: Alteration in the CAT, and SOD activities in EAC cells treated with APE or DOX. a, Changes in CAT (nM H_2O_2 decomposed/min/million cells); b, Changes in SOD (Units/ million cells). Closed squares, saline control; open circles, DOX positive control and closed triangles, APE.

catalase activity was almost 1.7 and 1.6 folds greater in the APE ($p<0.05$) and DOX treated groups at 24 h PLDA when compared with the concurrent untreated control (Table 11.7).

Superoxide Dismutase

Administration of mice with APE or DOX resulted in a time dependent decline in the activity of superoxide dismutase in EAC cells up to 8 h PLDA (1.9 folds) and 4 h (1.7 folds) and increased steadily up to 24 h, which was significantly higher in DOX treated group when compared with the 4 h PLDA (Figure 11.5b). The SOD activity in APE or DOX treatment group did not reach control level even by 24 h PLDA (Table 11.7). The decline in SOD activity was 1.2 fold higher in the APE treated group ($p<0.01$) at 24 h PLDA when compared with SOD activity of the concurrent untreated control (Table 11.7).

Discussion

Nature has been generous to bestow a variety of plants and other resources for the benefit of mankind. Plants perform all the living functions including communication and defense. Plants synthesize and make use of chemicals such as terpenes, alkaloids, tannins, phenols etc. for their defense during their survival. That is why different plants contain different types of one or more of these chemicals. Humans have employed plants and herbs for various purposes including health care since the time immemorial, and plant-derived chemicals have played an important role in the treatment of various human diseases including cancer in the modern system of medicine (Newman and Cragg, 2007). The systemic drug discovery and development have established a reasonable armamentarium of useful chemotherapeutic agents during the last six decades (Harvey, 2008). The success with most of the chemotherapeutic agents has been limited due to their high toxicity, lack of tumor selectivity, low efficacy, recurrence and ineffectiveness against drug resistant or multi-drug resistant cancers (Walko and McLeod, 2009). Therefore, the continued commitment to the arduous tasks involved in the discovery of new anticancer therapeutic agents with less toxicity, higher efficacy and better selectivity remains critically important. This target can only be achieved by screening newer paradigms and agents for the management of neoplastic diseases. Therefore, the present study was undertaken to evaluate the antineoplastic activity of *Aphanamixis polystachya* in Ehrlich ascites tumor bearing mice.

Indian medicinal plants like *Alstonia scholaris*, *Ervatamia heyncana*, *Rubia cordifolia*, *Withania somnifera*, *Hygrophila spinosa*, *Podyphyllum hexandrum*, *Tylophora indica* and *Tinospora cordifolia* have been reported to possess antineoplastic activity in different tumor models *in vitro* and *in vivo* (Chitnis *et al.*, 1972; Adwankar *et al.*, 1980; Devi *et al.*, 1992; Mazumdar *et al.*, 1997; Goel *et al.*, 1998; Jagetia *et al.*, 1998, 2005; Jagetia and Baliga, 2005; Jagetia and Rao, 2006). APE treatment caused the regression of EAC tumor in a dose dependent manner up to 1g/kg body weight as is evident by the maximum number of tumor free survivors at the end of 120 days. Increase in APE dose beyond 1 g resulted in a decline in its antineoplastic activity. As far as authors are aware this is probably the first systematic report on the anticancer properties of APE in tumor bearing mice.

Flavopiridol derived from plant alkaloid rohitukin obtained from the leaves and stems of *Aphanamixis polystachya* and *Dyoxylum binectariferum* has been found to be active against various cancers (Senderowicz *et al.*, 1998). The clinical efficacy of an anticancer agent lies in its ability to inhibit the proliferation of tumors not only in early stages but also in the late stages of its development. Therefore, the stage specific antineoplastic activity of 1g/kg APE was evaluated at different stages of the tumor development. The results from the stage specific evaluation show that APE inhibited the increase in the body weight gain in animals due to tumor cell proliferation during various stages effectively and it was very effective until stage III, which may be due to an efficient tumor cell killing. The studies of the anticancer activity of plants at different stages of tumor development are scanty, however, *Alstonia scholaris*, has been reported to inhibit tumor growth in mice when administered in mid stages (Jagetia and Baliga, 2004). APE was effective in increasing the median survival of EAC mice by 18, 12, 7.5, 4.5 and 2.5 days and average survival time by 18.5, 14, 9.5, 3 and 3.5 days at stage I, II, III, IV and V of tumor development when compared with the concurrent positive DOX control. The efficacy of APE lies in that, it even killed the tumor cells, when administered during the late stages of tumor development, where APE treatment lengthened the life span, where 40 and 20 per cent of animals survived beyond 120 days, especially during II and III stages. APE is found to be active even at stage IV and V where a tumor free survivors of 40 per cent beyond 40 days and 20 per cent up to 30 days, respectively was recorded. Ehrlich ascites carcinoma (EAC) is an undifferentiated malignancy. Penetration of EAC to tissues and tumor development is due to the absence of tumor-associated transplantation antigens (Lettre *et al.*, 1972). APE might have inhibited penetration of EAC cells into other organs of EAC mice as the spleens of APE-administered mice were found to be absolutely normal in size and color.

The exact mechanism of action of APE is not known. The cells killing effect of APE may not be due single mechanism but it may be due to the interplay of several mechanisms. APE has reduced glutathione and its related enzymes including GST and GSHpx accompanied by an increase in the lipid peroxidation, which may one of the reasons of its cytotoxic effect on EAC tumor in the present study. This contention is supported by the observation that increased concentration of GSH and GSH related enzymes like GST in the tumor cells have been reported to make the tumor refractory to treatment, while their depletion has been reported to enhance the cell death and apoptosis of the tumor cells along with the loss of essential sulfhydryl groups that result in an alteration of the calcium homeostasis and eventually loss of cell viability (Tew, 1994; Townsend and Tew, 2003). The lipid peroxidation is another important event related to cell death and has been reported to cause severe impairment of membrane function through increased membrane permeability and membrane protein oxidation, DNA damage, cytotoxicity and eventually cell death (Linden *et al.*, 2008). The increase in LOO and reduction in GSH by APE, may have damaged the DNA of EAC cells thereby bringing effective tumor cell killing. APE has also depleted the inventory levels of CAT, and SOD, in EAC cells, and disturbed the redox status which may have made the tumor cells more vulnerable to death. It may have also caused a cell cycle arrest at G_2+M phase and thereby killing the tumor cells. The nuclear factor (NF)-κB has been expressed by tumors and suppression of NF- κB by APE may have been also responsible for the effective cell kill in the present study. APE has been reported to inhibit NF-κB activity in various leukemic cells lines *in vitro* (Lampronti *et al.*, 2005).The cell killing effect of APE may be attributed to the presence of alkaloids including rohitukin, amoorastatin and 12α-hydroxyamoorastatin and limonoids like, polystachin, prieurianin, hispidin C, aphanamixin and aphananin (Connolly *et al.*, 1976; Polonsky *et al.*, 1979; Harmon *et al.*, 1979; Mulholland and Naidoo, 1999).

Natural medicines have gained popularity over synthetic drugs in recent years with the belief that they are much safer and have led to the tremendous growth of phytopharmaceutical usage. While it is known that plant extracts are active against cancer, the standard approach has been to isolate, synthesize and administer the single chemical compound thought to be responsible for this effect. This may also be the reason for higher toxicity and low efficacy of single components. However, different components in a botanical may have synergistic activities and there is also some evidence that the presence of multiple compounds in a plant extract can buffer the toxic effects of a single constituent (Williamson, 2001).

Conclusions

The treatment of APE has killed EAC cells and increased the life span of EAC mice when compared with the control. APE has also been effective in treating the tumor during the different stages of development. However, the effect dwindled with the advanced stages of tumor development. Cytotoxic effect of APE may be due to increased lipid peroxidation, depletion in GSH, GST, GSHpx, CAT and SOD and suppression in the transactivation of NF-κb.

Acknowledgements

The financial assistance from the Council of Scientific and Industrial Research (CSIR), Government of India, New Delhi in the form of senior research fellowship awarded to Dr. Venkatesha VA to carry out this study is gratefully acknowledged. We are thankful to Dr. G.K. Bhat (PoornaPragna College, Udupi, India) for his expert help in identifying *Aphanamixis polystachya*.

References

Abramowitz, M., and Stegun, I.A. (1972). Handbook of Mathematical Functions, Dover Publications Inc., New York.

Adwankar, M.K., Chitnis, M.P., Khandalekar, D.D., and Bhadsavale, C.G. (1980). Anti-cancer activity of the extracts of *Rubia cordifolia* Linn. (NSC b668893). *Indian Journal of Experimental Biology*, **18**: 102.

Aebi, H. (1974). Catalase. In: Methods of Enzymatic Analysis, Hans Ulhrich Bergmeyer (Eds), VerlagChemie International, Florida, pp 673-682.

Ahn, J.W. (1994) Cytotoxiclimonoids from *Melia azedarachta* var. *japonica*. *Phytochemistry*, **36**: 1493-1496.

Chandrasekharan, S., and Chakraborty, T. (1968). Aphanamixol, a diterpene alcohol from *Aphanamixis polystachya* (wall) Parker. *Journal of Indian Chemical Society*, **45**: 208.

Chitnis, M.P., Khandalekar, D.D., Adwankar, M.K., and Sahasrabudhe, M.B. (1972). Anti-cancer activity of the extracts of stem and leaf of *Tylophora indica*. *Indian Journal of Medical Research*, **60**: 359-362.

Chopra, R.N., Nayar, S.L., and Chopra, I.C. (1956). *Glossary of Indian Medicinal Plants*. Council of Scientific and Industrial Research, New Delhi, India.

Connolly, J.D., Okorie, D.A., de Wit, D.L., and Taylor, D.A.H. (1976). Structure of Dregeanin and Rohitukin. *Journal of Chemical Society, Chemical Communication*, **22**: 909-910.

Cragg, G.M., and Newman, D.J. (2005). Plants as a source of anti-cancer agents. *Journal of Ethnopharmacology*, **100**: 2-79.

Devi, P.U., Sharada, A.C., Solomon, F.E., and Kamath, M.S. (1992). *In vivo* growth inhibitory effect of *Withania somnifera* (Ashwagandha) in a transplantable mouse tumor, sarcoma 180. *Indian Journal of Experimental Biology,* **30:** 169-172.

Dhar, M.L., Dhar, M.M., Dhawan, B.N., Mehrotra, B.N., and Ray, C. (1968). Screening of Indian plants for biological activity. Part I. *Indian Journal of Experimental Biology*, **6:** 232-247.

Geran, R.I., Greenberg, N.H., Mac Donald, M.M., Schumacher, A.M., and Abbott, B.J. (1972). Protocols for screening chemical agents and natural products against animal tumors and other biological systems. *Cancer Chemotherapy Reports*, **Part 3**: 25-27.

Ghosh, M.N. (1984) Fundamentals of Experimental Pharmacology. Scientific book Agency, Kolkata, India.

Goel, H.C., Prasad, J., Sharma, A., and Singh, B. (1998). Antitumour and radioprotective action of *Podophyllum hexandrum*. *Indian Journal of Experimental Biology*, **36**: 583-587.

Gole, M.K., and Dasgupta, S. (2002). Role of plant metabolites in toxic liver injury. *Asia Pacific Journal of Clinical Nutriton*, **11(1):** 48-50.

Gunatilaka, A.A.L., BolzaniVda, S., Dagne, E., Hofmann, G.A., Johnson, R.K., McCabe, F.L., Mattern, M.R., and Kingston, D.G. (1998). Limonoids showing selective toxicity to DNA repair-deficient yeast and other constituents of *Trichelia emetica*. *Journal of Natural Products*, **61**: 179-184.

Habig, W.H., Pabst, M.J., and Jacoby, W.S. (1974). Glutathione-*S*-transferases. The first enzymatic step in mercapturic acid formation. *Journal of Biological Chemistry*, **249**: 7130-7139.

Harmon, A.D., Weiss, U., and Silverton, J.V. (1979). The structure of rohitukine, the main alkaloid of *Amoora rohituka* (Syn. *Aphanamixis polystachya*) (Maliaceae). *Tetrahedron Letters*, **8:** 721-724.

Havey, A. (2008). Natural product in drug discovery. *Drug Discovery Today*, **13**: 894-901.

Ho, Y.S., Magnenat, J.L., Roderick, T., Cao, B.J., Gargano, M., Sugawara, M., and Funk, C.D. (1997). Mice deficient in cellular glutathione peroxidase develop normally and show no increased sensitivity to hyperoxia. *Journal of Biological Chemistry*, **272**: 16644–16651.

Jagetia, G.C. (2007). Radioprotective potential of plants and herbs against the effects of ionizing radiations. *Journal Clinical Biochemistry and Nutrition*, **40(2):** 74-81.

Jagetia, G.C., and Baliga, M.S. (2004). Effect of *Alstonia scholaris* in enhancing the anticancer activity of berberine in the Ehrlich ascites carcinoma bearing mice. *Journal of Medicinal Food*, **7(2):** 235-244.

Jagetia, G.C., and Baliga, M.S. (2005).The effect of seasonal variation on the antineoplastic activity of *Alstonia scholaris* R. Br. in HeLa cells. *Journal of Ethnopharmacology*, **96(1-2):** 37-42.

Jagetia, G.C., Nayak, V., and Vidyasagar, M.S. (1998). Evaluation of the antineoplastic activity of guduchi (*Tinospora cordifolia*) in cultured HeLa cells. *Cancer Letters*, **127:** 71-82.

Jagetia, G.C., and Rao, S.K. (2006). Evaluation of cytotoxic effects of dichloromethane extract of Guduchi (*Tinospora cordifolia* Miers ex Hook F and THOMS) on cultured HeLa cells. *Evidenced Based Complementary and Alternative Medicine*, **3:** 267-272.

Jagetia, G.C., and Venkatesha, V.A. (2006). Treatment of mice with stem bark extract of *Aphanamixis polystachya* reduces radiation-induced chromosome damage. *International Journal of Radiation Biology*, **82(3):** .197-209.

Jagetia, G.C., Venkatesh, P., and Baliga, M.S. (2005). *Aegle marmelos* (L.) Correa inhibits the proliferation of transplanted Ehrlich ascites carcinoma in mice. *Biological and Pharmaceutical Bulletin*, **28(1):** 58-64.

Jain, S.A., and Srivastava, S.K. (1985). 8- *C*-methyl-quercetin-3-*O*-b -D-xylopyranoside, a new flavone glycoside from the roots of *Amoora rohituka*. *Journal of Natural Products*, **48:** 299-301.

Jemal, A., Bray, F., Center, M.M., Ferlay, J., Ward, E., and Forman, D. (2011). Global cancer statistics. *CA Cancer Journal for Clinicians*, **61(2):** 69-90.

Kinghorn, A.D., Carcache de Blanco, E.J., Chai, H.B., Orjala, J., Farnsworth, N.R., Soejarto, D.D., Oberlies, N.H., Wani, M.C., Kroll, D.J., Pearce, C.J., Swanson, S.M., Kramer, R.A., Rose, W.C., Fairchild, C.R, Vite, G.D., Emanuel, S., Jarjoura, D., and Cope, F.O. (2009). Discovery of anticancer agents of diverse natural origin. *Pure and Applied Chemistry*, **81(6):** 1051-1063.

Kundu, A.B., Sanghamitra, R., and Chatterjee, A. (1985). Aphananin, a triterpene from *Aphanamixis polystachya*. *Phytochemistry*, **24**: 2123-2125.

Lampronti, I., Khan, M.T., Bianchi, N., Ather, A, Borgatti, M., Vizziello, L., Fabbri, E., and Gambari, R. (2005). Bangladeshi medicinal plant extracts inhibiting molecular interactions between nuclear factors and target DNA sequences mimicking NF-kappa B binding sites. *Medicinal Chemistry*, **1(4)**: 327-333.

Lettre, R., Paweletz, D., Werner, D., and Granzow, C. (1972). Sublines of the Ehrlich-Lettre mouse ascites tumor-a new tool for experimental cell research. *Naturwissenschaften*, **59**: 59-63.

Linden, A., Gülden, M., Martin, H.J., Maser, E., and Seibert, H. (2008). Peroxide-induced cell death and lipid peroxidation in C6 glioma cells. *Toxicology in Vitro*, **22(5)**: 1371-1376.

Lovell, D.P. (1986). Variation in barbiturate sleeping time in mice: 1. Strain and sex differences. *Laboratory Animals*, **20**: 85-90.

Mann, J. (2002). Natural products in cancer chemoptherapy: past, present and future. *Nature Reviews Cancer*, **2**: 143-148.

Mazumdar, U.K., Gupta, M., Maiti, S., and Mukherjee, D. (1997). Antitumor activity of *Hygrophila spinosa* on Ehrlich ascites carcinoma and sarcoma-180 induced mice. *Indian Journal Experimental Biology*, **35**: 473-477.

Misra, H.P., and Fridovich, I. (1972). The role of superoxide anion in the autooxidation of epinephrine and a simple assay for superoxide dismutase. *Journal of Biological Chemistry*, **247**: 3170–3175.

Moron, M.S., Pierre, J.N., and Mannervick, B. (1979). Levels of glutathione, glutathione reductase and glutathione-*S*-transferase activities in rat lung and liver. *Biochimica Biophysica Acta*, **582**: 67–71.

Mulholland, D.A., and Naidoo, N. (1999). Limonoids from *Aphanamixis polystachya*. *Phytochemistry*, **51**: 927-930.

Newman, D.J., and Cragg, G. (2007). Natural Products as Sources of New Drugs over the Last 25 Years. *Journal of Natural Products*, **70**: 461-477.

Nias, A.H.W. (1990). Treatment of Cancer. In: Sikora K, Halnan KE (Eds) Chapman and Hall Medical, London, pp. 53-92.

Ohkawa, H., Ohishi, N., and Yagi, K. (1979). Assay of lipid peroxides in animal tissues by thiobarbituric acid reaction. *Analytical Biochemistry*, **95**: 351–358.

Polonsky, J., Varon, Z., Marazano, C., Arnoux, B., Pettit, G.R., and Schmid, J.M. (1979). The structure of amoorastatone and the cytotoxic limonoid 12-hydroxyamoorastatin. *Experientia*, **35**: 987-989.

Prieur, D.J., Young, D.M., Davis, R.D., Cooney, D.A., Homan, E.R., Dixon, R.L., *et al.* (1972). Procedures for preclinical toxicologic evaluation of cancer chemotherapeutic agents: protocols of the laboratory of toxicology. *Cancer Chemotherapy Reports*, **4**: 1-28.

Rabi, T., and Gupta, R.C. (1995). Antitumor and cytotoxic investigation of *Amoora rohituka*. *International Journal of Pharmacognosy*, **33** : 359-361.

Senderowicz, A.M., Headlee, D., Stinson, S.F., Lush, R.M., Kalil, N., Villalba, L., Hill, K., Steinberg, S.M., Figg, W.D., Tompkins, A., Arbuck, S.G., and Sausville, E.A.

(1998). Phase I trial of continuous infusion flavopiridol, a novel cyclin-dependent kinase inhibitor, in patients with refractory neoplasms. *Journal of Clinical Oncology*, **16:** 2986-2999.

Srivastava, S.K., and Agnihotri, V.K. (1985). New glycosides from the stem bark of *Aphanamixis polystachya*. *Current Science*, **54**: 38-40.

Swerdlow, A.J., Higgins, C.D., Smith, P., Cunningham, D., Hancock, B.W., Horwich, A., Hoskin, P.J., Lister T.A., Radford, J.A., Rohatiner, A.Z., and Linch, D.C. (2011). Second cancer risk after chemotherapyf or Hodgkin's lymphoma: a collaborative British cohort study. *Journal of Clinical Oncology*, **29(31):** 4096-4104.

Talukder, F.A., and Howse, P.E. (1995). Evaluation of *Aphanamixis polystachya* as repellants, antifeedants, toxicants and protectants in storage against *Tribolium castaneum*, (Herbst). *Journal of Stored Product Research*, **31**: 55-61.

Tew, K.D. (1994). Glutathione-associated enzymes in anticancer drug resistance. *Cancer Research*, **54(16):** 4313-4320.

Townsend, D.M., and Tew, K.D. (2003). The role of glutathione-*S*-transferase in anti-cancer drug resistance. *Oncogene*, **22:** 7369-7375.

Walko, C.M., and McLeod, H. (2009). Pharmacogenomic progress in individualized dosing of key drugs for cancer patients. *Nature Clinical Practice Oncology*, **3:** 153–162.

Warier, P.S. (1994). *Indian Medicinal Plants*. Orient Longman Limited, New Delhi, India.

Williamson, E.M. (2001). Synergy and other interactions in phytomedicines. *Phytomedicine*, **8:** 401-409.

World Health Organization (1998). Quality control methods for medicinal plants. In: Determination of arsenic and heavy metals, England pp 62-63.

Medicinal Plants: Phytochemistry, Pharmacology and Therapeutics Vol. 3 (2014) *Pages* **315–322**
Editor-in-Chief: **V.K. Gupta**
Published by: **DAYA PUBLISHING HOUSE, NEW DELHI**

12

Antihyperglycemic and Antihyperlipidemic Profile of *Morinda lucida* Leaf Fractions in Rats

Adejuwon Adewale Adeneye

ABSTRACT

Ethanolic decoction made from the fresh leaves of Morinda lucida Benth. (Rubiaceae) is highly valued in the local management of diabetes mellitus among the Nigerian traditional healers. 50 mg/kg of the different solvent fractions (diethyl ether, chloroform, butanol and the residue)that were prepared from the crude ethanolic leaf extract of Morinda lucida (MLE) were evaluated for their antihyperglycemic and antihyperlipidemic profile in both normal and alloxan-induced hyperglycemic rats for 14 days, with the view of providing a template for drug discovery and development from this African plant in the near future. The effect of each fraction on the fasting blood glucose (FBG) on the 1st, 8th and 15th day post-treatment was evaluated using glucose monitoring system. The effect of the fractions on serum triglyceride and total cholesterol on 15th day post-treatment was also determined using standard procedures. Results showed that in the normal and alloxan-induced diabetic rats, oral treatment with MLE_c and MLE_b resulted in significant ($p<0.05$, $p<0.01$ and $p<0.001$) time-dependent lowering of FBG, serum triglyceride and total cholesterol in the treated rats with MLE_c producing the most significant ($p<0.001$) hypoglycemic and hypolipidemic effects. Thus, the present study shows MLE_c to be the most effective antihyperglycemic and antihyperlipidemic fraction of all the solvent fractions of MLE tested suggesting that potential antidiabetic drug may be resident in MLE_c.

Keywords: Alloxan hyperglycemic rats, Antihyperglycemic, Antihyperlipidemic, Ethanolic leaf fractions, *Morinda lucida* Benth.

1 Department of Pharmacology, Faculty of Basic Medical Sciences, Lagos State University College of Medicine, Ikeja, Lagos, Nigeria

* *Corresponding author*: E-mail: adeneye2001@yahoo.com, aaadeneye@gmail.com

Introduction

Diabetes mellitus (DM) remains the most common endocrine disorder of carbohydrate metabolism with over 170 million adult sufferers estimated to have been affected worldwide and about two-thirds of this figure residing in the developing countries (Wild *et al.*, 2004). This estimate is expected to double by the year 2030 (Wild *et al.*, 2004; Menichini *et al.*, 2011). In view of the growing prevalence of the disease, there is an increasing interest towards developing integrated approaches in the prevention and management of DM by exploring the potentials offered by the indigenous herbal therapies (Marles and Farnsworth, 1996). Herbal drugs are prescribed widely because of their effectiveness, relatively low cost and based on the general belief that they are free of intolerable side-effects (Venkatesh *et al.*, 2003). Also, World Health Organization (2003) has directed further evaluation of the traditional practices and methods (including herbal practices) of managing the disease. In compliance with this directive, many indigenous medicinal plants are currently been screened and evaluated for their antihyperglycemic potentials. In African herbal medicine, there are various native plants that are employed in the local management of DM, one of which cuts across the West African region is *Morinda lucida* Benth.

Morinda lucida Benth., belonging to the family Rubiaceae, is a tropical West Africa rainforest tree with the English name "Brimstone tree". It is known as Sangogo or Bondoukoua longua (in Cote d'Ivoire), Twi, Kònkròmà or Ewe amake (in Ghana), Ewe amake or Atakake(in Togo) and Òruwó or ruwó amongst the Yoruba tribe(South-West Nigeria) (Dalziel, 1937). Amongst the Igbo (South-East Nigeria), it is locally known as "Huka" and "Eze-ogu" (Gill, 1992). Decoctions made from the plant leaves, stem bark and roots are highly valued in the local DM management. The leaves and stem bark of *Morinda lucida* are used as purgative, emetic and diuretic (Gill, 1992). In Nigeria, a mixture of fresh leaves of *Morinda lucida, Momordica charantia, Vernonia amygdalina* and *Dalbergiella welwitschi* are ground together into a fine paste before mixing it with native black soap for bathing a cure for DM (Abimbola-Sodipe, 1986). In South-west Nigeria, fresh leaves of the plant are macerated in fresh palm wine and the filtrate taken orally for blood sugar control in suspected diabetic patients. Earlier, Olajide *et al.* (1999) reported the antidiabetic effect of the aqueous leaf extract of *Morinda lucida* in streptozotocin-induced hyperglycemic rats. We also reported the antihyperglycemic activity of the crude ethanolic leaf extract of *Morinda lucida* in normal and alloxan-induced hyperglycemic rats (Adeneye and Agbaje, 2008). Bearing in mind the fact that identifying and isolating the effective extract fractions and isolating active fractions of a crude extract may prove better therapeutically less toxic and inexpensive than its crude form, the present study aims to evaluate the antihyperglycemic profile of the solvent fractions of the ethanolic leaf extract of *Morinda lucida* Benth in normal and alloxan diabetic rats for 14 days.

Materials and Methods

Plant Collection and Authentication

Two kilograms of fresh leaves of *Morinda lucida* were collected from the same site (as previously reported by Adeneye and Agbaje, 2008) on the outskirt of Low Cost

Housing Estate, Oke-Afa, Isolo, Lagos State, Nigeria in the months of July-August 2009. Plant identification and authentication has earlier been done.

Extract Preparation

500 g of *Morinda lucida* fresh leaves were exhaustively extracted in 1L of 50 per cent ethanol (Aldrich Chemical Co., USA) for 2 hours using Soxhlet extraction procedure. The Soxhlet extractive was filtered using a piece of white cotton cloth and the filtrate obtained was completely dried into an aromatic green-brown solid residue over a water-bath. This procedure was repeated three more times. The residues obtained were pooled and stored in water- and air-tight container and kept refrigerated at -4 °C until required for experiment.

Solvent Fractionation of MLE

50 g of the extract was completely deuterated in 100 ml of distilled water. The deuterated solution was then successively extracted with different partitioning solvents (diethyl ether, chloroform, and butan-1-ol) in the order of their increasing polarity. The fraction obtained with each partitioning solvent was concentrated *in vacuo* using rotary evaporator (BUCHI Rotavapor® Model R-215, Switzerland) with Vacuum Module V-801 EasyVac®, Switzerland) set at a revolution of 70 rpm and a temperature of 35°C. The "extract residue" and the concentrate of each fraction were completely air-dried in an aerated oven preset at 35 °C. The residues left after oven drying were then weighed. This procedure was repeated for 5 more times and each residue was pooled together and stored in clean and dry, water and air-proof containers and preserved in the refrigerator until required for experimentation.

Experimental Animals

Healthy young adult male albino Wistar rats (120-150 g) used in this study were obtained from Zoology Department of the University of Ilorin, Kwara State, Nigeria. The rats were housed in polypropylene cages and handled in accordance with international principles guiding the Use and Handling of experimental animals (United States National Institutes for Health, 1985). Rat feed (Ladokun Feeds, Ibadan, Nigeria) and tap water were provided *ad libitum*. The rats were maintained at an ambient temperature between 25-28 °C, humidity of 56 ± 5 per cent, and 12 h day/night photoperiod.

Oral Treatment of Normal Rats with MLE Fractions

To identify the biologically active fraction(s) of *MLE*, 50 mg/kg of the diethyl ether fraction, chloroform fraction and butanol fraction each was constituted in 10 ml of 5 per cent Tween-80 in distilled water was orally administered to each rat in each treatment group for 14 days.

Thirty-six, young in-bred adult male white albino Wistar rats (120-140 g) were randomly allotted to 6 groups. The rats were then fasted overnight for 12-14 hours but had free access to drinking water. The basal fasting blood glucose of each rat was first determined using One Touch Basic Blood Glucose Monitoring System® (LifeScan Inc., Milpitas, California, U.S.A.). This was then followed by the following treatments:

Group I rats were orally administered 10 ml/kg of 5 per cent Tween-80 in distilled water; Group II rats were orally treated with 20 mg/kg of metformin while Group III-VI rats were orally treated with a single daily dose of 50 mg/kg of diethyl ether (MLE_d), chloroform fraction (MLE_c), and butanol fraction (MLE_b) and the "extract residue" (MLE_r), respectively, for 14 days. The effect of each extract fraction on the blood glucose concentration was then determined on the 15th day after an overnight fasting of the treated rats.

Hyperglycemia Induction in Rats and their Oral Treatment with MLE Fractions

Following a 24-hr fast, rats were made hyperglycemic by injecting each rat with a single intra-peritoneal dose of 120 mg/kg of alloxan monohydrate (Sigma Chemical Company, St. Louis, U.S.A.) dissolved in 3 mM of freshly prepared cold citrate buffer (pH = 4.5).The baseline fasting blood glucose was first determined before alloxan treatment. Six hours after alloxan injection, rats were orally infused with 20 per cent Dextrose (Unique Pharmaceuticals, Sango-Otta, Ogun State, Nigeria) at an oral dose of 10 ml/kg so as to prevent the onset of fatal hypoglycemia which often accompanies administration of alloxan as a result of acute massive pancreatic release of insulin (Prince and Menon, 2001). Gradual onset of hyperglycemia was confirmed on the 3rd day post-induction but a steady hyperglycemic state was achieved by the 5th day post-alloxan treatment. By the 5th day, rats with fasting blood glucose of equal or greater than 200 mg/dl were considered hyperglycemic. Thirty-seven of the alloxan-treated rats had the fasting blood glucose concentration over 200 mg/dl while the remaining three rats had spontaneous resolution of their hyperglycemia.

Six normal and thirty-five alloxan-induced hyperglycemic rats were randomly allocated into 7 treatment groups as follows:

Group I consists of normoglycemic rats which were orally treated with 10 ml/kg of 5 per cent Tween 80 dissolved in distilled water;

Group II consists of alloxan-induced hyperglycemic rats orally treated with 10 ml/kg of 5 per cent Tween 80 in distilled water;

Group III consists of alloxan-induced hyperglycemic rats orally pre-treated with 20 mg/kg of metformin (Glucophage®, Hoechst Marion Roussel Limited, Mumbai, India) dissolved in 5 per cent Tween 80 in distilled water;

Groups IV-VII consist of alloxan-induced hyperglycemic rats that were orally pre-treated with 50 mg/kg of diethyl ether, chloroform, butanol fractions and the "extract residue", respectively. All treatments lasted 14 days after which the fasting blood glucose was determined.

Bioassays

FBG was determined on One Touch Basic Blood Glucose Monitoring System using the glucose oxidase method of Trinder (1969). Serum triglyceride and total cholesterol were determined using standard diagnostic test kits (Randox Laboratories, Crumlin, U.K.) on Automated Clinical System (Sychron ClinicalSystem®, Model: CX5 PRO) (Beckman Coulter Inc., Galway, Ireland).

Data Analysis

Results were expressed as mean ± S.E.M. of six observations. Statistical analysis was done using two-way analysis of variance followed by post-hoc test, Student-Newman-Keuls test on SYSTAT 10.6. Statistical significance were considered at $p<0.05$, $p<0.01$, and $p<0.001$.

Results

Extraction

Crude ethanolic extraction of *Morinda lucida* fresh leaves gave a yield of 15.5 per cent while that of solvent fractionation of MLE with diethyl ether, chloroform and butanol gave yield of 2.1 per cent, 15.6 per cent and 20.0 per cent, respectively, and 62.3 per cent for the "extract residue".

Table 12.1: Effect of *Morinda lucida* leaf fractions on FBG, TG and TC in normal rats.

Treatment Group	FBG (mg/dl) on			TG (mg/dl)	TC (mg/dl)
	Day 1	Day 8	Day 15		
I	58.8 ± 2.8	58.7 ± 4.3	56.0 ± 3.3	91.0 ± 5.7	83.2 ± 4.8
II	60.2 ± 2.2	50.8 ± 1.2^{d}	47.5 ± 2.2^{e}	72.3 ± 5.7$^{\alpha}$	59.8 ± 2.9
III	60.8 ± 1.8	59.5 ± 2.0	58.8 ± 1.5	96.8 ± 4.6	85.5 ± 7.7
IV	59.7 ± 3.0	50.2 ± 1.9^{d}	41.0 ± 1.7^{f}	56.2 ± 3.8$^{\partial}$	49.8 ± 2.3$^{\partial}$
V	59.3 ± 2.3	51.5 ± 2.3^{d}	47.5 ± 5.0^{e}	63.8 ± 12.4$^{\beta}$	55.0 ± 9.7$^{\beta}$
VI	59.0 ± 3.3	57.8 ± 1.8	58.6 ± 1.8	99.5 ± 5.5	88.0 ± 6.7

$^{d, e}$ and ^{f} represent significant decreases at $p<0.05$, $p<0.01$ and $p<0.001$, respectively, when compared to basal value on day 1 and when compared to Group I values on days 1, 8 and 15. α, β and ∂ represent significant decreases at $p<0.05$, $p<0.01$ and $p<0.001$when compared to Group I values.

Group I = 10 ml/kg of 5 per cent Tween-80 in distilled water; Group II = 20 mg/kg of metformin in 5 per cent Tween-80 in distilled water; Group III = single daily oral dose of 50 mg/kg of diethyl ether fraction (MLE_d); Group IV = single daily oral dose of 50 mg/kg of chloroform fraction (MLE_c); Group V = single daily oral dose of 50 mg/kg of butanol fraction (MLE_b); Group VI = single daily oral dose of 50 mg/kg of the "extract residue" (MLE_r)

Effect of MLE Fractions on FBG, Serum Triglyceride and Total Cholesterol in Normal Rats

Daily oral treatment of normal Wistar rats with 50 mg/kg of MLE_c and MLE_b resulted in significant ($p<0.05$, $p<0.01$ and $p<0.001$) time-dependent lowering of FBG in the treated rats with MLE_c inducing the most significant ($p<0.001$) hypoglycemic effect over MLE_b and metformin, particularly,on the 15th day post-treatment (Table 12.1). However, oral treatment with MLE_d and MLE_r did not produce any appreciable changes in the FBG levels (Table 12.1). In a similar pattern, MLE_c and MLE_b produced significant time-dependent ($p<0.05$, $p<0.01$ and $p<0.001$) lowering of the serum total cholesterol and triglyceride in the treated rats, with MLE_c producing the most significant ($p<0.001$) hypolipidemic effect when compared to that of MLE_b and metformin (Table 12.1). Again, MLE_d and MLE_r produced no significant (p>0.05) changes in the serum total cholesterol and triglyceride of the treated rats (Table 12.1).

Effect of MLE Fractions on FBG, Serum Triglyceride and Total Cholesterol in Alloxan-induced Hyperglycemic Rats

Single intraperitoneal injection with 120 mg/kg of alloxan monohydrate resulted in significant ($p<0.001$) hyperglycemia by the 5th day post-induction (Table 12.2). However, oral daily treatment with 20 mg/kg of metformin, 50 mg/kg of MLE_c and MLE_b resulted in significant ($p<0.05$, $p<0.01$ and $p<0.001$) time-dependent hypoglycemic effect in the alloxan-induced hyperglycemic rats, with MLE_c producing the most significant hypoglycemic effect on the 8th and 15th day of treatment (Table 12.2). In a similar pattern, hyperglycemia induction with alloxan was associated with significant ($p<0.001$) elevation in the serum triglyceride and total cholesterol concentrations (Table 12.2). Also, oral treatment with 50 mg/kg of MLE_c caused the most significant ($p<0.001$) reductions in the serum triglyceride and total cholesterol levels compared to those induced by 20 mg/kg metformin and 50 mg/kg MLE_b.

Table 12.2: Effect of *Morinda lucida* leaf fractions on FBG, TG and TC in normal and alloxan-induced hyperglycemic rats.

	Pre-induction FBG (mg/dl)	*Post-induction FBG (mg/dl) on*			*TG (mg/dl)*	*TC (mg/dl)*
		Day 1	*Day 8*	*Day 15*		
I	58.7 ± 2.7	59.3 ± 1.7	58.7 ± 2.1	58.3 ± 3.0	93.7 ± 5.4	80.8 ± 5.1
II	59.3 ± 2.9	250.0 ± 7.3^{c}	259.3 ± 8.2^{c}	276.8 ± 12.2^{c}	204.8 ± 12.2^{c}	179.0 ± 19.3^{c}
III	59.7 ± 3.1	265.7 ± 14.0^{c}	151.2 ± 14.9^{e}	$115.2 \pm 9.2^{a,f}$	116.8 ± 13.5^{e}	105.8 ± 13.5^{e}
IV	59.5 ± 3.5	254.3 ± 5.8^{c}	260.3 ± 11.0^{c}	252.2 ± 11.9^{c}	182.5 ± 6.8^{d}	$148.7 \pm 8.0^{b,d}$
V	59.7 ± 2.7	252.2 ± 2.6^{c}	140.5 ± 8.0^{e}	$107.5 \pm 9.5^{a,f}$	106.8 ± 18.7^{f}	86.2 ± 16.2^{f}
VI	58.8 ± 2.6	250.8 ± 6.3^{c}	207.0 ± 10.7^{d}	$153.2 \pm 14.6^{b,e}$	141.8 ± 13.2^{e}	$128.8 \pm 8.4^{a,e}$
VII	60.17 ± 2.50	253.5 ± 10.5^{c}	237.7 ± 9.1^{c}	233.2 ± 6.8^{c}	173.5 ± 11.8^{d}	$148.2 \pm 8.2^{b,d}$

$^{a, b}$ and ^{c} represent significant increases at $p<0.05$, $p<0.01$ and $p<0.001$, respectively, when compared to pre-induction FBG values and Group I values while $^{d, e}$ and ^{f} represent significant decreases at $p<0.05$, $p<0.01$ and $p<0.001$, respectively, when compared to Group II values on days, 1 8 and 15.

Group I = 10 ml/kg of 5 per cent Tween-80 in distilled water + normal rats; Group II = 10 ml/kg of 5 per cent Tween-80 in distilled water + alloxan-hyperglycemic rats; Group III = 20 mg/kg of metformin in 5 per cent Tween-80 in distilled water + alloxan-hyperglycemic rats; Group IV = single daily oral dose of 50 mg/kg of diethyl ether fraction (MLE_d) + alloxan-hyperglycemic rats; Group V = single daily oral dose of 50 mg/kg of chloroform fraction (MLE_c) + alloxan-hyperglycemic rats; Group VI = single daily oral dose of 50 mg/kg of butanol fraction (MLE_b) + alloxan-hyperglycemic rats; Group VII = single daily oral dose of 50 mg/kg of the "extract residue" (MLE_r) + alloxan-hyperglycemic rats.

Discussion

In the present study, diethyl ether, chloroform, butanol and "extract residue" fractions of the crude ethanolic extract of *Morinda lucida* fresh leaves were evaluated for their antihyperglycemic and antihyperlipidemic profile in normal and alloxan-induced hyperglycemia models. Alloxan, like its counterpart, streptozotocin has been frequently used to induce either type 1 or type 2 diabetes (Frode and Medeiros, 2008).

Alloxan selectively accumulates in the pancreatic β-cells via the GLUT2 glucose transporter, where it targets mitochondrial DNA leading to impairment of mitochondrial signaling function and consequent induction of β-cell apoptosis through several mechanisms including caspases activation and ROS production (Steppel and Horton, 2004; Lenzen, 2008). The ROS mediates the cytotoxic action with the increase in cytosolic Ca^{2+} concentrations, leading to rapid β-cells destruction which results in hypoinsulinemia and eventual hyperglycemia (Szkudelski, 2001). In this study, alloxan-induced DMwas most effectively controlled withrepeated oral treatment with 50 mg/kg of MLE_c, as measured by significant lowering of the FBG, serum triglyceride and total cholesterol levels. These observations suggest that the phytochemicals responsible for the antihyperglycemic activity of ethanolic crude extract of *Morinda* leaf are mostly partitioned into chloroform. Although, the exact antihyperglycemic mechanism(s) of this fraction was not investigated in the present study, hyperinsulinemic mechanism was most unlikely since literature has it that for overt DM to become established using alloxan as an inducing agent, up to 70 per cent pancreatic β-cells would have been destroyed (Jorns *et al.*, 1997; Szkudelski, 2001). Thus, it is most likely that effective glycemic control was achieved via increased action on cellular glucose uptake or intestinal glucose uptake inhibition.Similarly, alloxan is known to induce hyperlipidemia in diabetic animals via increased mobilization of free fatty acids from the peripheral fat deposits (Hussain, 2002). Since MLE_c significantly lowered the serum triglyceride and total cholesterol levels in the treated rats, it may either be inhibiting mobilization of free fatty acids from the peripheral deposits or increasing their deposition into the peripheral tissues. Previous independent studies have shown the crude *Morinda lucida* leaf extract to be rich in flavonoids, alkaloids, tannins and saponin (Olajide *et al.*, 1999; Asuzu and Chineme, 1990). Also, oruwacin (anthraquinones) (Adewunmi and Adesogan, 1984), morindin (glycoside) and tannins have also been isolated from *Morinda lucida* leaves. Literature has equally reported the biological activities of alkaloids and flavonoids to include hypoglycemia, hypolipidemia, hypoazotemia, hypotension among other biological effects (Oladele *et al.*, 1995; Sudheesh *et al.*, 2005). Thus, it is possible that the presence for these components in high concentrations in the chloroform fraction of MLE to be responsible for the antihyperglycemic and antihyperlipidemic activities of this fraction.

In conclusion, the study showed the chloroform fraction of MLE (MLE_c) to be very effective in glucose and lipids homeostasis in both normal and alloxan-induced hyperglycemic rats. Further studies geared towards isolating and characterizing the antihyperglycemic compounds is on-going.

References

Abimbola-Sodipe, R.O.(1986). Traditional treatment for hypertension, stroke, asthma, sickle cell, small pox and diabetes. *In*: The state of medicinal plants research in Nigeria, Ed. By Sofowora, A., Ibadan University Press, Nigeria, pp. 125-133.

Adeneye, A.A., and Agbaje, E.O. (2008). Pharmacological evaluation of oral hypoglycemic and antidiabetic effects of fresh leaves ethanol extract of *Morinda lucida* Benth. in normal and alloxan-induced diabetic rats. *African Journal of Biomedical Research*, **11**: 65-71.

Adewunmi, C.O., and Adesogan, E.K.(1984). Anthraquiones and oruwacin from *Morinda lucida* as possible agents in fasciolasis and schistosomiasis control. *Fitoterapia,* **55**: 259-263.

Asuzu, I.U., and Chineme, C.N. (1990). Effect of *Morinda lucida* leaf extract on *Trypanosoma bruceibrucei* infection in mice. *Journal of Ethnopharmacology,* **30(3)**: 307-312.

Dalziel J.M.(1937). The Useful Plants of West Africa. Crown Agents,London, pp. 403-404.

Gill, L.S.(1992). Ethnomedicinal uses of plants in Nigeria. Uniben Press, Benin City, Nigeria, pp. 166.

Frode, T.S., and Medeiros, Y.S. (2008). Animal models to test drugs with potential antidiabetic activity. *Journal of Ethnopharmacology,* **115**: 173-183.

Hussain, H.E.M.A. (2002). Reversal of diabetic retinopathy in streptozotocin-induced diabetic rats using traditional Indian antidiabetic plant, *A. indica* (L). *Indian Journal of Clinical Biochemistry,* **17(2)**: 115-123.

Jorns, A., Munday, R., Tiedge, M., and Lenzen, S. (1997). Comparative toxicity of alloxan, N-alkyl-alloxans and ninhydrin to isolated pancreatic islets *in vitro*. *Journal of Endocrinology,* **155**: 283-293.

Lenzen, S. (2008). The mechanisms of alloxan- and streptozotocin-induced diabetes. *Diabetologia,* **51**: 216-226.

Marles, F.J., and Farnsworth, N.R. (1996). Antidiabetic plants and their active constituents: an update. *Protocols Journal of Natural Medicine,* **Winter**: 85-111.

Menichini, F., Tundis, R. Loizzo, M.R., Bonesi, M., Liu, B., Jones, P., Persaud, S.J., Mastellone, V., Lombardi, P., Houghton, P.J., Avallone, L., and Menichini, F.(2011). *C. medica*cv Diamante peel chemical composition and influence on glucose homeostasis and metabolic parameters. *Food Chemistry,* **124**: 1083-1089.

Oladele, S.B., Ayo, J.O., and Adaudi, A.O.(1995). Medicinal and physiological properties of flavonoids, coumarin derivatives and anthraquinones of plant origin. *West African Journal of Pharmacology and Drug Research,* **11**: 134-144.

Olajide, O.A., Awe, S.O., Makinde, J. M., and Morebise, O. (1999). Evaluation of the antidiabetic property of *Morinda lucida* leaves in streptozotocin-diabetic rats. *Journal of Pharmacy and Pharmacology,* **51(11)**: 1321-1324

Steppel, J.H., and Horton, E.S. (2004). Beta-cell failure in the pathogenesis of type 2 diabetes mellitus. *Current Diabetes Reports,* **4**: 169-175.

Szkudelski, T. (2001).The mechanism of alloxan and streptozotocin action in β-cells of the rat pancreas. *Physiological Research,* **50**: 537-546.

Trinder, P. (1969). Determination of blood glucose using 4-aminophenzone as oxygen acceptor. *Journal of Clinical Pathology,* **22**: 246-248.

Venkatesh, S., Reddy, G.D., Reddy, B.M., Ramesh, M., and Appa Rao, A.V.N. (2003). Antihyperglycemicactivity of *Caralluma attenuata*. *Fitoterapia,* **74**: 274-279.

Medicinal Plants: Phytochemistry, Pharmacology and Therapeutics Vol. 3 (2013) *Pages* **323–333**
Editor-in-Chief: **V.K. Gupta**
Published by: **DAYA PUBLISHING HOUSE, NEW DELHI**

13

Anti-diabetic Effect of Ethanol Extract of Stem Bark of *Ficus bengalensis* Linn.

Rahul Gupta[1], Samta Johri[2] and A.M. Saxena[3*]

ABSTRACT

The development of effective and safe anti-diabetic drugs is urgent need of today. This could be achieved from natural sources like plants and herbs. The plant Ficus bengalensis Linn. has been used in Indian traditional system of medicine for treatment of ulcers, vomiting, fever, inflammations etc. In traditional claims infusion of bark is supposed to be a powerful tonic and is considered to have specific properties in the treatment of diabetes. So our traditional claims enforced us to evaluate its anti-diabetic activity and probable mechanism of action. Anti-diabetic effect of 95 per cent ethanol extract of stem bark of F. bengalensis Linn., was studied in normal fasted, fed, glucose loaded and streptozotocin induced diabetic male albino rats. Significant blood glucose lowering effect was observed in fasted, fed, glucose loaded and diabetic rats at a single oral dose of 250mg/kg body weight, along with marked degranulation in insulin producing pancreatic β-cells and increase in muscle glycogen store of extract treated rats. Results suggest that stem bark of F. bengalensis produce anti-diabetic effect by stimulating the insulin secretion and glycogenesis.

Keywords: Anti-diabetic, Blood glucose, β-cells, Ethanol extract, Glycogenesis.

Introduction

Ficus bengalensis Linn. (Figure 13.1) of family Moraceae is a large perennial tree, found in all over India and commonly known as Bargad in India (Gupta *et al.*, 2008;

1 Amity Institute of Biotechnology (AIB), Amity University, Lucknow – 226 028, Uttar Pradesh, India

2 Department of Zoology, Mahila P.G. College, Lucknow – 226 018, Uttar Pradesh, India

3 Department of Zoology, University of Lucknow, Lucknow – 226 007, Uttar Pradesh, India

* *Corresponding author*: E-mail: anandmsaxena@rediffmail.com

Figure 13.1: *Ficus bengalensis* Linn.

Mukherjee *et al.*, 2002). All parts of the plant are acrid, sweetish; astringent to the bowels; useful in biliousness, ulcers, erysipelas, vomiting, vaginal complications, fever, and inflammations. An infusion of the bark is supposed to be a powerful tonic and being considered to have specific properties in the treatment of diabetes (Kirtikar and Basu, 1995). Different parts of *F. bengalensis* have been studied for their blood glucose lowering effect (Augusti, 1975; Cherian *et al.*, 1992); however some workers could not get the desirable results (Deshmukh *et al.*, 1960; Joglekar *et al.*, 1963; Vohora and Parasar, 1970). In view of this we tend to investigate the anti-diabetic activity of ethanol extract of stem bark of *F. bengalensis* (collected during fruiting season) in normal and STZ-induced diabetic male albino rats. We were also studied the effect of extract on muscle glycogen content and histology of pancreatic β-cells granularity in normal rats to find out probable mechanism of action.

Materials and Methods

Plant Material and Extraction

Stem bark (Figure 13.2) of *F. bengalensis* was collected fresh during fruiting season in month of November, from Lucknow (latitude 26°52′12″N and longitude

Figure 13.2: Dried stem bark of *F. bengalensis*.

80°51′54″E), India. Plant material was identified with the help of Department of Botany and specimen was deposited as accession number 2/23560/09. Bark was air dried in shade, grinded mechanically and coarse powder extracted in 95 per cent ethanol using soxhlet apparatus (Gupta and Saxena, 2008). The extract was concentrated to a semisolid material free from alcohol (wet stem bark collected-5kg; yield of dry weight/kg wet weight-460g; crude extract yield/100g dry weight-18.0g). All the chemicals and reagents used were of analytical grade from Merk and Sigma Chemicals.

Experimental Models

Charls Foster strain male *albino* rats (120-140g) were used and guidelines for animal ethics were strictly followed. Animals were maintained on Hind Lever pellets diet (Mumbai, India), housed in polypropylene cages at a temperature of 23 ± 2°C and relative humidity 60 ± 5 with 12 h each of dark and light cycle. Water was allowed *ad libitum*. The rats were divided in to experimental and control groups of six rats each. Acute toxicity study (Mutalik *et al.*, 2005) of extract on behavioral aspect was carried out up to dose of 2500mg/kg body weight and found to be safe. Extract was orally fed to experimental group by metal canula at a single dose of 250mg/kg body weight as prepared in 2 per cent gum acacia suspension and control was fed only 2 per cent gum acacia The blood glucose lowering effect of the extract was examined in following four different experimental models (Chandrasekar *et al.*, 1988):

(a) Fasted

Blood was collected (at 0h) from the tail vein of the overnight (18h) fasted animals and extract was fed. Again blood samples were collected at 1, 3 and 4h after extract feeding. Blood glucose concentration was estimated by Nelson's Somogyi method (1944).

(b) Fed

Pellets in access amount were put in the cages on the previous evening, so that some pellets were found left over in the next morning. Blood was collected before (at 0h) and after administration of the extract at 1, 3 and 4h for glucose estimation.

(c) Glucose Loaded

Animals were fasted for 18 h and blood was collected (at 0h) for glucose estimation. Now extract was fed and after half an hour glucose (1.5g/kg b.w. oral) solution was administered and blood samples were collected for sugar estimation at 1/2, 1 and 3 hours.

(d) Diabetic

Rats were made diabetic by injecting streptozotocin (35mg/kg b.w.) dissolved in chilled citrate-phosphate buffer (*pH* 4.5) through tail (Theodorou *et al.*, 1980). At the moment of injection the rats had fasted for 18 hours. After the streptozotocin injection, all rats were returned to their cages and given free access to food and water. Two to three days after injection, diabetes was confirmed by checking urine glucose by Diastex (Saxena *et al.*, 1991). Then the experiments were performed using identical procedure as in the fasted model (*a*).

Study on Muscle Glycogen

Muscle glycogen content of extract treated and control group of well fed normal rats (each of 5 animals) was estimated at the most effective hour of blood glucose lowering. Rats were sacrificed and portion from muscle tissue of diaphragm was taken out quickly for glycogen estimation (Carroll *et al.*, 1955).

Statistical Analysis

Mean and standard error were determined and students't' test was performed between experimental and control group.

Histological Study of Pancreatic β-cells

Pancreas of anesthetized animals of extract treated and control group were taken out at the most effective hours associated with blood glucose lowering and fixed in alcoholic Bouin's fluid. Paraffin sections of pancreas were double stained with haematoxylin and eosine, and granularity of β-cells was compared under microscope (David, 1991).

Results

Effect on Blood Glucose

In the present study very significant blood glucose lowering effect was observed at 3h and 4h ($P<0.001$) in extract treated fasted rats as compared to control (Table 13.1). Maximum lowering calculated in terms of percentage from initial value was higher in extract treated fasted rats (27.49 per cent at 4h) in comparison to 2 per cent

Table 13.1: Effect of 95 per cent ethanolic extract of Stem bark of *Ficus bengalensis*, at a single oral dose of 250mg/kg body weight, on blood glucose level of fasted, fed and STZ-induced diabetic male albino rats.

Group	*Treatment*	*Blood Glucose Level mg/100ml (Mean ± S.E.) at time (hours)*				*Maximum per cent of Blood Glucose Lowering from Initial Value*
		0h	*1h*	*3h*	*4h*	
Fasted	Control	94.93 ± 1.76 (6)	92.48 ± 1.31 (6)	83.59 ± 1.85 (6)	84.66 ± 0.99 (6)	11.95 per cent at 3h
	F. bengalensis Stem bark	92.76 ± 2.00 (6)	90.73 ± 1.31 (6)	70.24 ± 1.89[c] (6)	67.26 ± 2.22[c] (6)	27.49 per cent at 4 h
Fed	Control	110.76 ± 3.72 (6)	105.03 ± 3.91 (6)	99.75 ± 3.83 (6)	101.34 ± 4.63 (6)	9.94 per cent at 3 h
	F. bengalensis Stem bark	104.59 ± 2.93 (6)	98.98 ± 2.92 (6)	83.75 ± 1.55[b] (6)	89.56 ± 1.89[a] (6)	19.93 per cent at 3h
Diabetic	Control	188.27 ± 6.60 (6)	182.75 ± 4.55 (6)	191.92 ± 3.06 (6)	193.98 ± 6.18 (6)	2.93 per cent at 1h
	F. bengalensis Stem bark	184.53 ± 2.79 (6)	185.22 ± 2.21 (6)	172.16 ± 4.05[b] (6)	157.26 ± 6.72[b] (6)	14.78 per cent at 4h

Significance between the control and experimental group: [a]$P<0.05$, [b]$P<0.01$, [c]$P<0.001$.

In parenthesis the number of rats used is given.

Table13.2: Effect of 95 per cent ethanolic extract of Stem bark of *F. bengalensis*, at a single oral dose of 250mg/kg body weight, on glucose tolerance (1.5g/kg b. w. orally) of albino rats.

Group	*Treatment*	*Blood Glucose Level mg/100ml (Mean ± S.E.) at time (hours)*				*Maximum per cent of Blood Glucose Rise from Initial Value*
		0h	*1/2h*	*1h*	*3h*	
Glucose	Control	78.14 ± 2.97 (6)	89.52 ± 2.00 (6)	90.50 ± 1.28 (6)	88.30 ± 2.90 (6)	15.82 per cent at 1h
Loaded	*F. bengalensis* Stem bark	74.36 ± 3.04 (6)	79.78 ± 2.31[b] (6)	84.08 ± 2.36[a] (6)	78.91 ± 4.26 (6)	13.07 per cent at 1h

Significance between the control and experimental group: [a]$P<0.05$, [b]$P<0.01$

In parenthesis the number of rats used is given.

gum acacia (11.95 per cent at 3h). Ethanol extract produced significant ($P<0.01$) blood glucose lowering at 3h and slight but significant ($P<0.05$) lowering at 4h in fed group. Maximum percentage of blood glucose lowering in treated group was 19.93 per cent at 3h, but in control group it was 9.94 per cent.

Significant ($P<0.01$) blood glucose lowering was observed at 3h and 4h in STZ-induced diabetic rats. Maximum lowering was 14.78 per cent at 4h in extract treated group and 2.93 per cent at 1h in control group (2 per cent gum acacia). In glucose loaded model, *F. bengalensis* (stem bark) extract produced significant suppression of peak value at 1/2h ($P<0.01$) and 1h ($P<0.05$) as compared to corresponding control group (Table 13.2). Assessed from maximum percentage of blood glucose rise, it was 13.07 per cent at 1h in treated and 15.82 per cent at 1h in control group. Extract was found to be safe up to 2500mg/kg body weight.

Effect on Muscle Glycogen

Administration of extract significantly ($P<0.05$) increased the muscle glycogen content of normal fed rats at the most effective hour (3h) of corresponding blood glucose lowering. Muscle glycogen content of treated group was 592 ± 20.48mg/100g of tissue, where as in control group it was 480.8 ± 32.50mg/100g of tissue.

Effect on pancreatic β-cells

In histological study extract induced prominent degranulation in pancreatic β-cells of fasted rats was observed at 4h as compared to respective control. In fed model the marked degranulation in β-cells of treated rats was observed at 3h. Whereas, in Glucose loaded model slight degranulation in β-cells of islet of Langerhans was observed at 1/2h (Figures 13.3A-F).

Discussion

Four test models of different backgrounds of blood glucose level have been used to increase the sensitivity of experimental procedure and generate adequate baseline data. The ethanolic extract was orally administered as this route was thought to be most preferable and in tune with the route of administration of 250mg/kg b.w. dose as reported earlier. In the present investigation 95 per cent ethanolic extract of stem bark of *F. bengalensis* was found to produce significant lowering of blood glucose levels in fasted, fed and streptozotocin-induced diabetic male albino rats. Extract was also improved the oral glucose tolerance. Blood glucose lowering effect induced by stem bark extract, persisted for 3 to 4 hours after extract feeding, hence stem bark of *F. bengalensis* definitely lowers the blood glucose in rats. In extract treated normal rats reduction in blood glucose is associated with the increased muscle glycogen content and degranulation in pancreatic β-cells. Degranulation in β-cells indicating that the rate of extract induced insulin release was greater than the rate at which β-cells replenished their insulin store, resulted in blood glucose lowering. We can conclude that 95 per cent ethanol extract of stem bark of *F. bengalensis* definitely having anti-diabetic effect. Therefore, our systematic study supports the claims of Indian traditional system of medicine as mentioned in Vedas.

Figures 13.3(A-F): Microscopic observation of Histological sections of pancreas of *Ficus bengalensis* (Stem bark) extract treated and control (2 per cent gum acacia treated) rats (Magnification-63X).

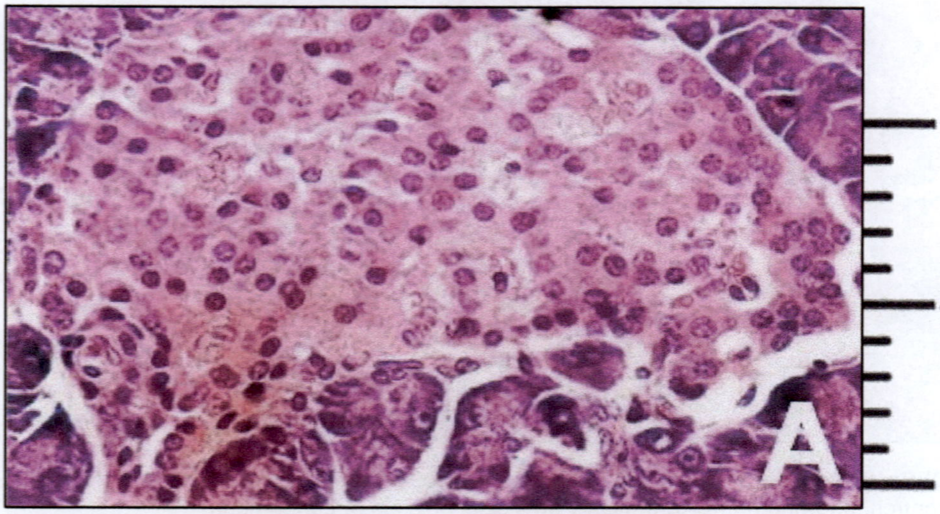

(A) Section of Fasted rat pancreas, sampled at 4 hour post-2 per cent gum acacia treatment (control), showing normal granularity of β-cells of islets of Langerhans.

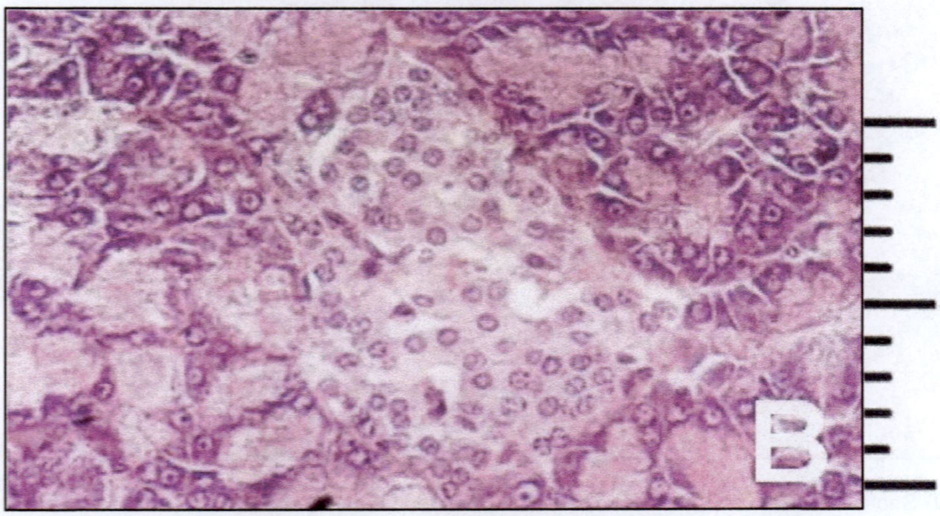

(B) Section of Fasted rat pancreas, sampled at 4 hour post- *F. bengalensis* extract treatment associated with maximum blood glucose lowering, showing degranulated β-cells of islets of Langerhans as compared to control (A).

Contd...

Figures 13.3(A-F)–*Contd...*

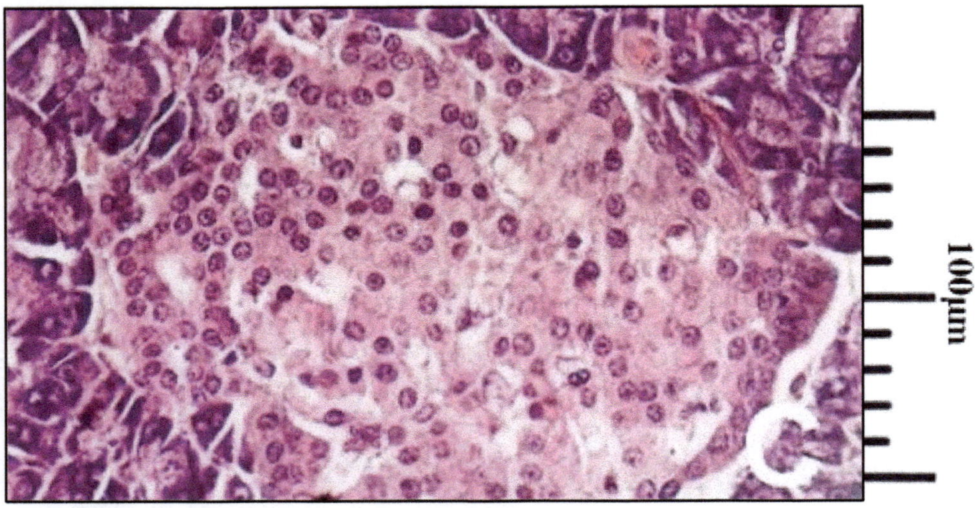

(C) Section of Fed rat pancreas, sampled at 3 hour post-2 per cent gum acacia treatment (control), showing normal granularity of β-cells of islets of Langerhans.

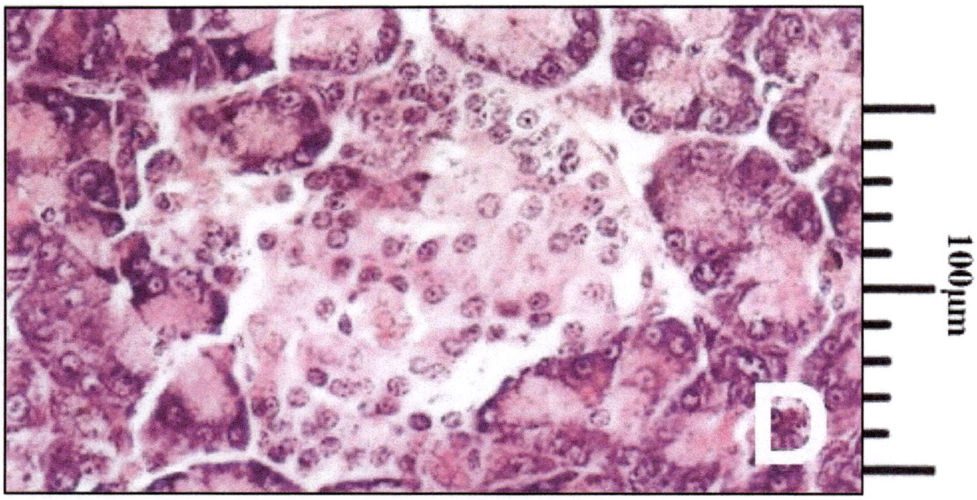

(D) Section of Fed rat pancreas, sampled at 3 hour post- *F. bengalensis* extract treatment associated with maximum blood glucose lowering, showing marked degranulation in β-cells of islets of Langerhans as compared to control (C).

Contd...

Figures 13.3(A-F)–*Contd...*

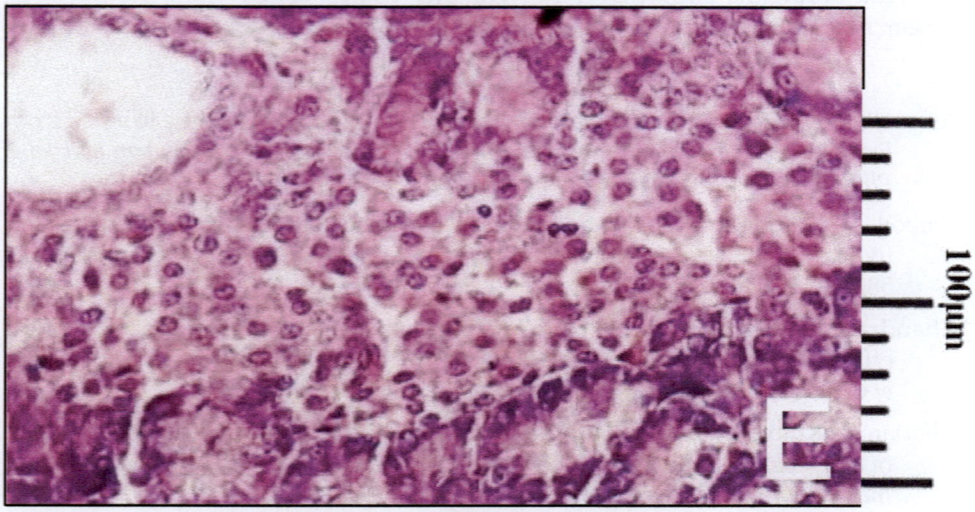

(E) Section of Glucose loaded rat pancreas, sampled at 1/2 hour of glucose feeding post-2 per cent gum acacia treatment (control), showing normal rich granularity of β-cells of islets of Langerhans.

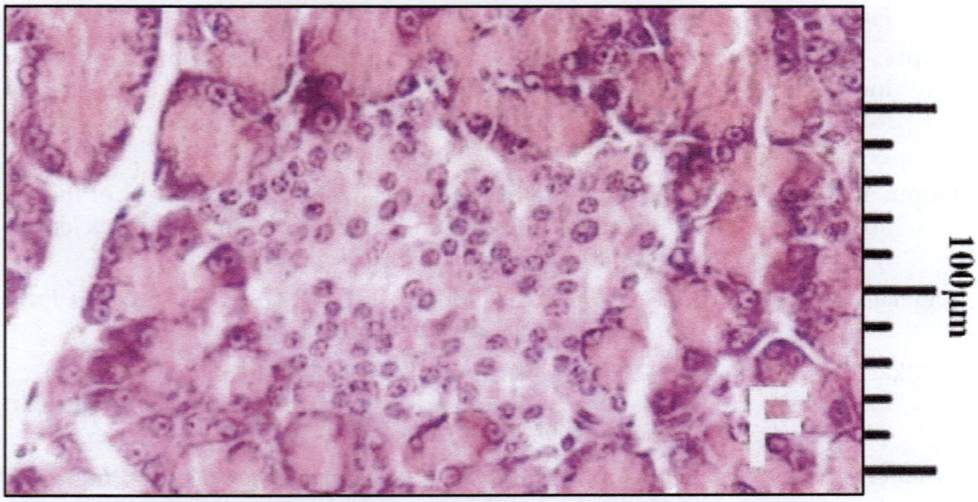

(F) Section of Glucose loaded rat pancreas, sampled at the most effective hour of blood glucose lowering, *i.e.* 1/2 hour of glucose feeding post-*F. bengalensis* extract treatment, showing marked degranulation in β-cells of islets of Langerhans as compared to control (E).

Acknowledgements

Authors are thankful to Dr. P.K. Agnihotri, Toxicology Division, Central Drug Research Institute, Lucknow, India for his help in histological preparation of pancreas.

References

Augusti, K.T. (1975). Hypoglycemic action of bengaloside, a glucoside isolated from *Ficus bengalensis* Linn. in normal and alloxan diabetic rabbits. *Indian Journal of Physiology and Pharmacology*, **19:** 218-220.

Carroll, N.V., Longley, R.W., and Roe, J.H. (1955). The determination of glycogen in liver and muscle by use of Anthrone reagent. *Journal of Biological Chemistry*, **2:** 583-593.

Chandrasekar, B., Mukherjee, B., and Mukherjee, S.K. (1988). Blood sugar lowering effect of *Tricosanthes dioica* Roxb. in experimental rat models. *International Journal of Crude Drug Research*, **26:** 102-106.

Cherian, S., Kumar, R.V., Augusti, K.T., and Kidwai, J.R. (1992). Hypoglycemic effect of a glycoside of pelargonidin isolated from the bark of *Ficus bengalensis* Linn. *Indian Journal of Biochemistry and Biophysics*, **29:** 380-382.

David, S.K. (1991). Handbook of Histology and Histochemical techniques, First Ed., C.S.B. Publishers, Delhi, India, pp. 1-41.

Deshmukh, V.K., Shrotri, D.S., and Aiman, R. (1960). Isolation of hypoglycemic principle from the bark of *Ficus bengalensis* Linn. *Indian Journal of Physiology and Pharmacology*, **4:** 182.

Gupta, R., and Saxena, A.M. (2008). Assessment of Hypoglycemic Activity of Indigenous Herbs. *In:* Phytochemicals: A Therapeutant for Critical Disease Management. (Eds.) D.R. Khanna, A.K. Chopra, G. Prasad, D.S. Malik, R. Bhutiani. Daya Publishing House, Delhi, India, pp. 99-106.

Gupta, R., Bajpai, K.G., Johri, S., and Saxena, A.M. (2008). An overview of Indian novel traditional medicinal plants with anti-diabetic potentials. *African Journal of Traditional Complimentary and Alternative Medicine*, **5:** 1-17.

Joglekar, G.V., Shrotri, D.S., Aiman, R., and Balwani, J.H. (1963). A study on *Ficus bengalensis* Linn. *Journal of Indian Medical Association*, **40:** 11.

Kirtikar, K.R., and Basu, B.D. (1995). *Ficus bengalensis In:* Indian Medicinal Plants, 2nd ED. (1933), Ed. by Blatter E, Caius JF and Mhaskar KS. L.M.B. Publishers, Allahabad, Vol. 3; Reprint 1995, pp. 2312-2313.

Mukherjee, S.K., Saxena, A.M., and Shukla, G. (2002). *Ficus bengalensis In*: Progress of diabetes research in India during 20th century. Ed. by Singh G, Mukherjee TK. National Institute of Science and Communication (CSIR), New Delhi; Reprint 2006, pp. 8-11.

Mutalik, S., Chetana, M., Sulochana, B., Devi, P.V., and Udupa, N. (2005). Effect of Dianex, an herbal formulation on experimentally induced diabetes mellitus. *Phytotherapy Research*, **9:** 409-415.

Nelson, N. (1944). A photometric adaptation of the Somogyi method for the determination of glucose. *Journal of Biological Chemistry,* **153:** 375-381.

Saxena, A.M., Bajpai, M.B., and Mukherjee, S.K. (1991). Swerchirin induced blood sugar lowering of streptozotocin treated hyperglycemic rats. *Indian Journal of Experimental Biology,* **29:** 674-675.

Theodorou, N.A., Helena, V.T., Margaret and Howell, S.L. (1980). Problems in the use of carbonate diffusion chambers for syngeneic pancreatic islet transplantation in rats. *Diabetologia,* **18:** 313-317.

Vohora, S.B., and Parasar, G.C. (1970). Antidiabetic studies on *Ficus bengalensis* Linn. *Indian Journal of Pharmacology,* **32:** 68.

Medicinal Plants: Phytochemistry, Pharmacology and Therapeutics Vol. 3 (2014) *Pages* **335–349**
Editor-in-Chief: **V.K. Gupta**
Published by: **DAYA PUBLISHING HOUSE, NEW DELHI**

14

Review on Phytochemistry and Pharmacological Aspects of *Jatropha gossypifolia* Linn.

Pratibha Singh[1]* and Ajay Singh[1]

ABSTRACT

The antiseptic qualities of aromatic and medicinal plants and their extracts have been recognized since antiquity, while an attempt to characterize these properties in the laboratory dates back to the early 1900s. Medicinal herbs have curative properties due to presence of various complex chemical substance of different composition, which are found as secondary plant metabolites in one or more parts. These plant metabolites according to their composition are grouped as alkaloids, glycosides, corticosteroids, essential oils etc. Jatropha gossypifolia, (Euphorbiaceae) is an herb. This plant originated from Brazil and it is now cultivated widely in tropical countries throughout the world. In Greek word "Jatros" meaning a "Doctor" and "Trophe" meaning "Nutrition", the plant has a reputation for extract of roots, stem, leaves, seeds and fruits widely used in traditional medicine. The plant has been reported as antimicrobial, molluscicidal, insecticidal, anti-malarial, hepato-protective, anti-oxidant, anti-cancerous, anti-inflammatory and anti-coagulant etc.

Keywords: *Jatropha gossypifolia* Linn. Medicinal plants.

Introduction

Medicinal herbs are the local people heritage with global importance. World is endowed with a rich wealth of medicinal herbs, which have made important

1 Natural Product Laboratory, Department of Zoology, D.D.U. Gorakhpur University, Gorakhpur – 273 009, U.P., India

* *Corresponding author*: E-mail: pratibha3gkp@rediffmail.com

contribution to the welfare and quality of life of urban as well as rural communities especially in tropics and sub-tropics (Sofowora, 1993).

During the early years of human existence, many plant materials by instinct, intuition of trial and error were used to combat different ailments. There has ever been an increasing interest investigating various plants for use in folkloric medicine (Ogundare, 2007).

The plant kingdom provides a tremendous reservoir of various chemical substances with potential therapeutic properties. In recent times, plant analyses have been effectively and efficiently carried out with the aid of advanced instruments. This enriched medicinal constituent extract of plants. The use of plant extracts in the treatment of numerous human diseases has been reported by Khan *et al*. (2006).

Jatropha gossypifolia Linn. is a multipurpose plant with many attributes and considerable potential. In Greek word "Jatros" means "a doctor" and "Trophe" means "nutrition" (Damor *et al.*, 2010). *Jatropha gossypifolia* is commonly known as Belly ache bush, locally in Maharashtra, India it is known as Ratanjot.

Origin and Geographic Distribution

All species of *Jatropha* L. are members of the family Euphorbiaceae. *Jatropha gossypifolia* is native to tropical America from Mexico to Paraguay and the Caribbean region and is adapted to a variety of habitats. It grows wildly almost through India. It is intolerant of shade, prefers arid environment. Although it may survive for a season in moderate shade and needs full or nearly full sun for long-term survival and fruiting. This evergreen plant is common in waste places, throughout India, in the southern part; it is cultivated for hedges too (Nayak *et al.*, 2009).

Plant Description

Jatropha gossypifolia is a bushy, gregarious shrub, up to 1.8 m tall, 3-5 lobed, approximately 20 cm long and wide, with leaves having a long petiole, covered with glandular hairs. Petioles are covered with coarse dark brown hairs and the young leaves are sticky. Stem has thin, often greenish bark, which exudes copious amount of watery sap when cut. Flower unisexual, red crimson or purple in corymbs, fruits are fleshy to dry. The seeds are greenish capsule-like seeds, ellipsoid. Seedling with epigeal germination (Kawanga, 2007).

Kingdom:	Plantae
Division:	Magnoliophyta
Class:	Magnoliopsida
Order:	Malpighiales
Family:	Euphorbiaceae
Subfamily:	Crotonoideae
Tribe:	Jatropheae
Genus:	*Jatropha*
Species:	*gossypifolia*

The Yoruba speaking tribe of Nigeria call it Lapalapa (Odebiyi and Sofowora, 1998; Ogundare, 2007). In different languages as well as in different country it has different name *viz.* (Tables 14.1 and 14.2).

Table 14.1: Different names of *Jatropha gossypifolia* in different languages in India (Nayak *et al.*, 2009).

Different Languages	*Common Name*
Hindi	Bagbherenda, Janglia-randi, Safed arand
English	Physic nut, Purging nut
Sanskrit	Kanananaeranda, Parvata-randa
Uriya	Baigoba
Telugu	Adaviyamudama, Kondamudamu
Gujarati	Ratanjota
Malayalam	Kattavanakku
Bengal	Bagbherenda
Bombay	Irandi, Jepal
Tamil	Adalai, Kadalamanakku
Marathi	Mogalierand

Table 14.2: Different names of *Jatropha gossypifolia* in different Countries (Nayak *et al.*, 2009).

Different Country	*Common Name*	*Different Country*	*Common Name*
French	Pourghere, Pignon d' Inde	German	Purgier-nub, Brechnab
Mexico	Pinoncillo	Costa Rica	Coquillo,Tempate
Puerto Rico	Tartago	Brazil	Mundubi-assu
Peru	Pinol	China	Yu-la-Tzu
Netherlands	Purge-ernot	Italian	Fagiala d' India
Guatemala	Pinon	Nepal	Kadam
Thailand	Sabudam	Philippines	Tubangbakod
Indonesia	Jarak budge	Coted' Ivoire	Bagani
Togo	Kpoti	Senegal	Tabanani
Angola	Mupuluka	Nigeria	Butuje
Tanzania	Makaen	South Africa	Purgeerboontjie
Saudi Arabia	Dand barri, Habel melak	Portugal	Purgueira
Myanmar	Kesugi, Thinbankyekku	Cambodia	Lohong kwang
India	Ratanjot		

Uses in Traditional Medicine

Plant-derived drugs remain an important resource, especially in developing countries, to combat serious diseases. Approximately 60-80 per cent of the world population still relies on traditional medicines for the treatment of common illnesses (WHO, 2002; Zhang, 2004). Different parts of *J. gossypifolia* are used in different countries in many ways. The plant is antibiotic, insecticidal and used for toothache and as blood purifier (Balee, 1994). It possesses significant anticancer, hepatoprotective and pesticidal activity (Hartwell, 1969; Chatterjee *et al.*, 1980; Panda *et al.*, 2009a; Panda *et al.*, 2009b). The root, stems, leaves, seeds, and fruits of the plant have been widely used in traditional folk medicine in many part of different country. The young stem of the plant is used as toothbrush as well as to clean the tongue in the treatment thrush (Ogundare, 2007). The stem sap stops bleeding and itching of cuts and scratches (Labadie *et al.*, 1989; Morton, 1968; Morton, 1980). The stem latex has been shown to possess coagulant activity and its mechanism of action as haemostatic agent found to be by precipitation of coagulant factors (Oduola *et al.*, 2005). Grinded paste of tuber is also locally used in the treatment of hemorrhoids (Ogundare, 2007).

The leaves of *J. gossypifolia* are used for intermittent fevers, carbuncles, eczema, itches and sores on the tongues of babies, swollen mammae, stomachache and venereal disease and as blood purifier (Balee, 1994). The leaf decoction is used for bathing wounds (wound healing) (Morton, 1968; Morton, 1981). In Southern Nigeria, the fresh leaf extract applied with crushed leaf by herbalists and local people to stop bleeding from skin and nose (Morton, 1981; Omoregbe *et al.*, 1996). The leaf extract has been used as an anticoagulant for biochemical and haematological analyses (Oduola *et al.*, 2005).

A decoction of the bark is used as an emmenagogue (Omoregbe *et al.*, 1996). The roots are employed against leprosy, as an antidote for snakebite and in urinary complaints (Kirtikar *et al.*, 1996; Banerji *et al.*, 1993; Damor *et al.*, 2010). Seeds are purgative, and used for body pain. The seeds act as an emetic and to expel internal parasites (Oduola *et al.*, 2005). The oil is applied internally as abortifacient and externally as rubefacient to treat rheumatic conditions and a variety of skin infections (Kawanga, 2007).

Use in Modern Medicine

The demand for dietary oils in the world has been increasing. They also constitute an essential part of industrial input. The physico-chemical properties of the seed and seed oil of *Jatropha gossypifolia* were assessed by standard methods. The seed contains per 100 g: 35.8 per cent crude oil of iodine value 107.25, 13.40 per cent protein, 9.25 per cent fibre, 30.32 per cent carbohydrate and 6.0g/kg saponins. The fatty acid composition of the seed oil is: unsaturated fatty acids 74 per cent of which palmitic acid upto 31 per cent and oleic acid and linoleic acid 68-84 per cent and further traces of caprylic, myristic, palmitoleic, palmitic, oleic, stearic, linoleic, vernolic, arachidic, behenic and lignoceric acids were found (Ogbobe *et al.*, 1993). The energy value of the seed oil is 42,000 kJ/kg (Kawanga, 2007).

An alcoholic root extract of *Jatropha gossypifolia* showed significant inhibitory activity in different human cancer cell lines. This finding led to the isolation of the macrocyclic diterpenes jatrophenone (Ravindranath *et al.*, 2003). Evaluation of this activity in the P-388 lymphocytic leukaemia test *in vitro* and *in vivo* and in Eagle's carcinoma of the nasopharynx *in vitro* revealed that most of these components were very active. In addition, it was found that jatrophone had directly inhibitory effects on concentrations of cardic and smooth muscle preparations, which were typically non competitive in nature. The action of jatrophone may involve more than one mechanism; this might be related to alterations in the Ca^{++} handling by cells (Kawanga, 2007).

The diterpenoids jatrophones A and B were isolated from the roots, shown very little biological activity (Kawanga, 2007).

An ethanol extract from the stems caused a significant and dose-dependent reduction of the systolic blood pressure when given orally to rats. An aqueous extract of the stems increased the amplitude between the systolic tone and the diastolic tone of the arterial pressure in guinea pig (Abreu *et al.*, 2003; Kawanga, 2007).

A leaf extract showed significant activity as an anticoagulant for haematological analyses. The anticoagulant effect of the extract was found to be comparable to that of dipotassium ethylenediamine tetraacetic acid. The leaf extract must be purifier to remove interfering substances to make it suitable for biochemical analyses (Oduola *et al.*, 2005).

Phytoactive Constituents

Phytoactive chemicals are present in virtually all plant tissues *e.g.*, leaf, roots stems, fruits etc. Phytochemicals must be concentrated in leaves, stems and roots rather than fruits and flowers. Plant extract is the best source of Phytochemicals, which are classified as secondary metabolites, are produced mainly by the shoot part of the plant, often their function in the plant is unknown but some phytochemical are known to have structural, functional and general defense against plant pathogens.

Phytochemicals have been responsible for various activities, such as cytoprotective, immunodulatory, immunoadjuvant and antimicrobial potential (Gautam *et al.*, 2004a, b; Luc and Arnold, 2005; Abiy *et al.*, 2005). Among the great variety of secondary compounds found in plants, phenolics and terpenoids represent the main antimicrobial agents. Aromatic compound such as phenol, alkaloids and its derivative *e.g.*, flavonoids, coumarinns have been identified as antimicrobial agents (Siddiqui *et al.*, 2009).

Phytoactive chemicals found in tissues of *J. gossypifolia* include two triterpenoids from leaves (Tinto and John, 1992). Three known flavonoids *viz.* apigenin, vitexin and isovitexin isolated from leaves of *J. gossypifolia* (Subramanian *et al.*, 1971). The bark contains the alkaloids jatrophine and a lignan jatrodien is found in stem (Das *et al.*, 1996; Matsuse *et al.*, 1999; Omoregbe *et al.*, 1996; Robineau, 1991). 2, 3-Bis-(hydroxymethyl)-6,7-methylenedioxy-1-(3′,4′-dimethoxphenyl)-naphthalene has been isolated from *Jatropha gossypifolia*.

The latex of *J. gossypifolia* contains the cyclic heptapeptide, cyclogossine A (Horsten *et al.*, 1996) and cyclic octapeptides Cyclogossine B (Auvin-Guette *et al.*, 1997). The aerial parts contain lignin *i.e.*, gossypiline (Das *et al.*, 1998) and gossypifan (Das and Das, 1995).

Jatrophenone, a macrocyclic bioactive diterpene has also been isolated from this plant (Ravindranath *et al.*, 2003). Rahman *et al.* (1990) reported two diterpenes from *J. gossypifolia*, Jatropholone A and Jatrophatrione.

Beside these different chemical compounds such as protease curcain, steroids, saponin, glycosides, carbohydrates, reducing sugars and numerous fatty acids were detected, and then the plant may be helpful for treating different common ailments (Burkill, 1994).

The amino acids *viz.* alanin, arginine, cystine, glycine, leycine, isoleucine, methionone, valine; the fatty acid *viz.* linoleic, oleic, palmitic acids, polyunstarated ester of 12-deoxy-16-hydroxyphorbol and proteins (Seeds). Ceryl and myricyl alcohols, carnavbyl palmitate, patmitone and β-sitosterol. Coumarino-lignans, fraxetin, jatrophone, jatropholones, A and B, 2α and 2β-hydroxy jatrophones, 2β-hydroxy- 5, 6 –isojatrophone (Kumar, 2006).

Pharmacological Activity

Antimicrobial Activity

Kumar *et al.* (2006) reported that 500mg/L crude extract from leaves of *J. gossypifolia* L. completely inhibited eight microorganisms: *Bacillus subtilis, Bacillus pumilus, Bacillus cereus* var. *mycoides, Bordetela bronchiseptica, Staphylococus epidermis, Klebsiella pueumoniae, Streptococcus faecalis* and *Candida albicans*.

Hexane extracts of the fresh fruits showed significant activity against fungi and some bacteria (Pandey *et al.*, 1996). Jatrophenone, a macrocyclic diterpene possess antibacterial activity against the organism, *Staphylococcus aureus*. Its activity was compared to that of the standard compound, penicillin G (Ravindranath *et al.*, 2003).

Jatropha gossypifolia showed antimicrobial activity in a range of 75-1200μg/ml against both bacteria as well as fungal pathogens (Dabur *et al.*, 2007).

Ogundare (2007), obtained the chloroform and methanol leaf extract by using the cold extraction method. These extracts had antimicrobial effect against *Salmonella typhiI* with 12.0 and 12.5 mm as their zones of inhibition and *Staphylococcus aureus* with 6.0 and 18.0 mm as their zones of inhibition, respectively. *Candida albicans* was susceptible to both leaf extracts with zones of inhibition ranging from 30.0 to 35.0 mm. *Pseudomonas aeroginosa* also susceptible to *J. gossypifolia* both leaf extracts with zones of inhibition ranging from 10.0 to 12.0 mm, respectively. Phytochemical screening showed the presence of phlobatannins and tannins in the methanol extract and phlobatannins and alkaloids in the chloroform extract of the *Jatropha gossypifolia*.

Kumar (2006) reported that petroleum ether, methanolic and aqueous leaf extracts of *J. gossypifolia* have antibacterial and antifungal activity. Methanol and aqueous extracts at 25 and 50 mg/ml of concentration were found to be having significant antibacterial activity against *Staphylococcus aureus* and *Streptococcus pyrogene* (gram positive bacteria) and *E. coli* and *Pseudomonas aeroginosa* (gram negative bacteria). All the three extracts at 25 and 50 mg/ml of concentration were found to be having mild to moderate antifungal activity against *Candida albicans* and *Aspergillus flavus*.

Hydroxy fatty acids shows, wide range of antifungal (Sjogren *et al.*, 2003) and antibacterial (Mundt *et al.*, 2003) activities. In *Jatropha gossypifolia*, the oil content of seed containing hydroxyl fatty acids is 33 per cent. *Jatropha gossypifolia* oil appeared preferentially more active with respect to Gram reaction, exerting greater inhibitory activity against Gram-positive organism *Staphylococcus aureus* and among the fungi; it is more active against *Penicillium notatum* and *Aspergillus niger* as compared to *Candida albicans* (Neogi *et al.*, 2008).

Anthelmintic Activity

Kumar (2006), reported that Petroleum ether leaf extact of *Jatropha gossypifolia* at a concentration of 50 mg/ml showed only paralysis but no mortality of the earthworms whereas the methanolic and aqueous extracts showed mild anthelmintic activity as same concentration.

Piscicidal Activity

Plant extracts are referred to as botanicals and when poisonous to fish are called piscicides (Burkill, 1985). Such piscicidal plants contain different active ingredients known as alkaloids such as nicotine, pyrethrum, ryania, rotenone and coumerin, resin, akuammine, tannins, saponins and diosgenin (Wang and Huffman, 1991). However, these alkaloids are toxic to fish and other aquatic organisms at high concentrations and wear off within a short time (Crandall and Goodnight, 1962; Olaifa *et al.*, 1987; Kulakkattolickal, 1987; Adewunmi, 1990; Fafioye, 2005).

Singh and Singh (2002), reported that dilute aqueous solutions of leaf and stem bark were active in causing death of the fishes. The toxic effect of stem bark of *Jatropha gossypifolia* was time as well as dose dependent. There was significant negative correlation between LC_{50} and exposure periods. Thus, the LC_{50} value of stem bark extracts of *Jatropha gossypifolia* were decreases from 4.61 g/l (24 h) > to 4.34 g/l (96h). It has been suggested that these plant products cannot be used directly in freshwater bodies, without their detailed studies on long-term effects on non-target organism as well their structure activity relationship.

Molluscicidal Activity

The stem sap was strongly active against the snail *Lymnaea acuminata* in its aquatic environment. The toxicity of the sap is partly due to inhibitory effects on acetylcholinesterase, and on acid and alkaline phosphatases in the snail (Singh and Agarwal, 1988). The leaf and stem bark extracts contain some neurotoxins, which might be active at neuromuscular system of the exposed animals (Singh and Agarwal, 1990; Tiwari *et al.*, 2004).

The cyclic peptides cyclogossine A, B and C were isolated from the sap. Methanol and n-butanol extracts of unripe seeds showed significant molluscicidal activity against the snails *Lymnaea luteola* and *Indoplanorbis exustus*, the n-butanol extract being more toxic to the egg and adult stages of both snails (Sukumaran *et al.*, 1995; Kawanga, 2007).

Jatrophone from roots of *Jatropha gossypifolia* showed significant molluscicidal activity against snail *Biomphalaria glabrata* (Kawanga, 2007). Adewunmi and Morquis

(1980) also studied the molluscicidal properties of methanolic extracts of the fruit of *Jatropha gossypifolia*.

Stem sap was found to inhibit the growth of *Helminthosporium oryzae* and *Alternaria brassicicola*. The etahnolic stem extract showed significant larvicidal activity against larvae of the tick *Boophilus microplus* (Kawanga, 2007).

Insecticidal Activity

Khumrungsee *et al.* (2009) studied the insecticidal property of leaf extract of *Jatropha gossypifolia* against the beet armyworm *Spodoptera exigua* (Hubner) (Lepidoptera: Noctuidae), a polyphagous pest. The leaf extract showed insecticidal activity with an LC_{50} of 35,000 ppm and 26,500 ppm after 24 and 48 hours of expose. Toxicity of *J. gossypifolia* leaf extract against *S. exigua* larvae seems to be slightly higher than other botanical insecticides. *J. gossypifolia* extract was responsible for biological disturbances in insects.

Anti-malarial Activity

Crude hot water extracts from aerial parts were examined for antimalarial properties against *Plasmodium falciparum in vitro*, and were found to be capable of 100 per cent growth inhibition (Gbeassor *et al.*, 1989).

Hepato-protective Activity

The extract of aerial part of *Jatropha gossypifolia* were screened for its hepato-protective activity in carbon tetrachloride induced liver damaged in Wister albino rats. The results indicate that *Jatropha gossypifolia* Linn. have potent hepato-protective action against carbon tetrachloride induced hepatic damage in rats. Among three extracts *viz.* petroleum ether, aqueous and methanolic extracts; petroleum ether extract shown to possess maximum protectivity and methanolic extract showed the minimum activity (Panda *et al.*, 2009).

Anti-oxidant Activity

The therapeutic potential of medicinal plants as anti-oxidants in reducing free radical induced tissue injury (Pourmorad *et al.*, 2006; Shahwar *et al.*, 2010). The major constituents of biological membranes are lipids and proteins. Reactive oxygen species can easily initiate damage of the cell membrane constituent *i.e.* phospholipids, lipoproteins by propagating a reaction cycle (Raja *et al.*, 2006; Prakash *et al.*, 2009). In the past few years interest in the search of new natural antioxidants has grown because reactive stress is linked to many diseases. It has been mentioned by many authors that anti-oxidant activity of plant is due to their phenolic compounds (Duh *et al.*, 1999; Dragland *et al.*, 2003; Wang, 2003; Wu *et al.*, 2004). Previous reported work on *Jatropha gossypifolia* revealed the presence of phenolic compound *i.e.*, flavonoids with flavonol structure (Sankara *et al.*, 1971).

Anti-cancerous Activity

Some naturally occurring phorbols (family of naturally occurring compounds that can be referred to as tigliane diterpenes; Evans, 1986) are tumor inhibitors, inhibit human immunodeficiency virus (HIV) replication, and possess antileukemic activity.

Jatrophone, a macrocyclic diterpenoids isolated from *Jatropha gossypifolia* showed significant inhibitory activity against cancer cells under *in vitro* and *in vivo* conditions (Kupchan *et al.*, 1970). Jatrophone also possessed significant antileukemic activity against P-388 lymphocytic leukemia at 27 and 12 mg/kg cytotoxicity (ED_{50}) against KB cell culture at 0.17 µg/ml (Goel *et al.*, 2007).

Haemostatic Agent

Oduola *et al.* (2005, 2007), reported that the *Jatropha gossypifolia* stem latex as a haemostatic agent. The reduced clotting and bleeding times recorded in the experiment with stem latex as compared to those without stem latex. This showed that stem latex has coagulant properties and is safe for humans.

Anti-inflammatory Activity

Panda *et al.* (2009), reported the anti-inflammatory and analgesic effect of the methanolic and petroleum ether extracts of *Jatropha gossypifolia*. The methanolic and petroleum ether extracts showed greater anti-inflammatory and analgesic effect comparative to the standard drugs, indomethacin and diclofenac sodium respectively. The methanolic extract of aerial parts exhibited more significant activity than petroleum ether extract in the treatment of pain and inflammation.

Anti-coagulant Activity

Oduola *et al.* (2005), reported that the leaf extract of *Jatropha gossypifolia* is suitable as an anticoagulant for heamatological analysis but must be purified to remove interfering substances for biochemical analysis. The anticoagulant effect of the extract was found to be highest at a concentration of 0.1 ml per ml of blood. This amount of leaf extract is significantly lower than the amount used from conventional anticoagulants such as heparin and warfarin, which are used therapeutically to control thrombosis.

Their Use with Aid of New Scientific Tools

Infectious diseases are the leading cause of death world-wide. Antibiotic resistance has become a global concern. The clinical efficacy of many existing antibiotics is being threatened by the emergence of multidrug-resistant pathogens. Many infectious diseases have been known to be treated with herbal remedies throughout the history of mankind. Natural products, either as pure compounds or as standardized plant extracts, provide unlimited opportunities for new drug leads because of the unmatched availability of chemical diversity (Neogi *et al.*, 2008). India is a varietal emporium of medicinal plants and is one of the richest countries in the world in regard to genetic resources of medicinal plants. It exhibits a wide range in topography and climate, which has a bearing on its vegetation and floristic composition (Parekh and Chandra, 2007). By the dint of availability and greater acceptance among users, there is a continuous as well as urgent need to purify the phyto-active chemicals of the plant containing as secondary metabolites. Futhermore, it is popular due to low costing and its biodegradability.

With the aid of new biotechnological tools, it is needed to explore new sources other than plant, in which they currently found. At last, by the help of these biotechnological tools we have to search new techniques for the purification of the phyto-active chemical from plant and other source.

Conclusions

Recently there has been upsurge of interest in the therapeutic potential of medicinal plants. Green herbs represents a reservoir of effective chemicals, the rapeutants and can provide valuable sources of antimicrobial agents, natural pesticides, molluscicides with additional advantage that they are easily biodegradable, which are less likely to cause environmental contamination. In addition, these plants could represent striking anti-oxidant agents, which provide prophylaxis against various diseases like heart diseases, stroke and cancers. The bioassay-guided fractionation procedure to characterize and isolate the antimicrobial and antioxidant active constituents is needed.

Herbal drugs which are used in various traditional medicines need detailed investigation with phytochemical and pharmacological approach. In the present review it is explored all the detail of the *Jatropha gossypifolia* including information about its botany, habitat, traditional and modern uses, it is commonly found throughout the world and at waste places throughout India.

Acknowledgments

One of the authors (Pratibha Singh) is thankful to University Grants Commission, New Delhi, for the award of the Dr. D.S. Kothari Post Doctoral Fellowship Scheme (No. F.4-2/2006 (BSR)/13-214/2008 (BSR), dated 1st October 2009) for financial assistance.

References

Abiy, Y., Solomon, D., Jacob, D.M., Christine, C.B., Matthias, H., and Martin, G.P. (2005). Antimicrobial flavonoids from the stem of *Erythrina burtiii*. *Fitot.*, **96**: 496-499.

Abreu, I.C., Marinho, A.S., Paes, A.M., Freire, S.M., Olea, R.S., Borges, M.O., and Borges, A.C. (2003). Hypotensive and vasorelaxant effects of ethanolic extract from *Jatropha gosspifolia* L. in rats. *Fitot.*, **74**: 650-657.

Adewumi, C.O. (1990). Plant molluscicides. Potential of Aridan, *Tetrapleura tetraptera* for schistosomiasis control in Nigeria. *The Sci. of Tot. Environ.*, **102**: 21-33.

Adewunmi, C.O., and Morquis, V.O. (1980). Molluscicidal evaluation of some *Jatropha* species. *J. Crude Drug Res.*, **18**: 141-145.

Auvin-Guette, C., Baraguey, C., Blond, A., Pousset, J.L., and Bodo, B. (1997). Cyclogossine B, a cyclic octapeptide from *Jatropha gossypifolia*. *J. Nat. Prod.*, **60**: 1155-1157.

Balee, W. (1994). *Footprints of the forest, ka' apor ethnobotany- the historical ecology of plant utilization by an Amazonian people*. New York: Columbia University Press.

Banerji, J., and Das, B. (1993). MAPA, Dept. of Chemistry, University College of Science, Culcutta, India. **15**: 1002-1017.

Burkill, H.M. (1994). *The useful plants of West Tropical Africa*. Royal Botanic Garden, Kew, UK. **2**: 636.

Burkill, H.N. (1985). *The useful plants of West Tropical Africa*. Royal Botanical Garden, Kew, UK. **1**: 19.

Chatterjee, A., Das, B., Adityachaudhary, N., and Dabkirtaniya, S. (1980). Note on the insecticidal properties of the seeds of *Jatropha gossypifolia* Linn. *Indian J. Agric. Sci.*, **50**: 637-638.

Crandall, C.A., and Goodnight, C.J. (1962). Effects of sublethal concentrations of several toxicants on growth of the common guppy, *Lebistes reticulates*. *Limnol. Oceanogra*, **7**: 233-239.

Dabur, R., Gupta, A., Mandal, T.K., Singh, D.D., Bajpai, V., Gurav, A.M., and Lavekar, G.S. (2007). Antimicrobial activity of some Indian medicinal plants. *Afr. J. Trad.* CAM, **4**: 313-318.

Damor, R., Khatri, P., Samantha, K.C., and Singh, N. (2010). Pharmacognostic and phytochemical investigations of *Jatropha gossypifolia* Linn. Leaves. *Int. J. Pharma. Res. Develop.*, **2**: 1-8.

Das, B., and Banerji, J. (1988). Arylnaphthalene lignin from *Jatropha gossypifolia*. *Phytochemistry*, **27**: 3684-3686.

Das, B., and Das, R. (1995). Gossypifan, a lignin from *Jatropha gossypifolia*. *Phytochemistry*, **40**: 931-932.

Das, B., Rao, S.P., Srinivas, K., and Das, R. (1996). Jatrodien, a lignin from stems of *Jatropha gossypifolia*. *Phytochemistry*, **41**: 985-987.

Das, R., Das, B., and Kashinatham, A. (1998). Gossypiline, a new lignin from *Jatropha gossypifolia*. *Nat. Prod. Sci.*, **4**: 238-240.

Dragland, S., Senoo, H., Wake, K., Holte, K., and Blomhoff, R. (2003). Several culinary and medicinal herbs are important sources of dietary antioxidant. *J. Nut.*, **133**: 1286-1290.

Duh, P.D., Tu, Y.Y., and Yen, G.C. (1999). Free-radical scavenging capacity and antioxidant properties of some selected onions (*Allium cepa* L.) and garlic (*Allium sativum* L.) extracts. *Lebensmwiss Technol.*, **32**: 269-277.

Evans, F.J. (1986). Naturally occurring phorbol esters. CRC Press, *Boca Raton, FL*, pp. 271-288.

Fafioye, O.O. (2005). Plants with piscicidal activities in Southwestern Nigeria. *Turk. J. Fish. Aqua. Sci.*, **5**: 91-97.

Gbeassor, M., Kossou, Y., Amegbo, K., Souza, C., Koumaglo, K., and Denke, A. (1989). Antimalarial effects of eight African medicinal plants. *J. Ethnopharmacol.*, **25**: 115-118.

Goel, G., Makkar, H.P.S., Francis, G., and Becker, K. (2007). Phorbol esters: Structure, biological activity and toxicity in animals. *Int. J. Toxicol.*, **26**: 279-288.

Guatam, M., Diwanay, S., Gairola, S., Shinde, Y.S., Patki, P., and Patwardhan, B. (2004a). Immunoadjuvant potential of *Asparagus racemosus* aqueous extract in experimental system. *J. Ethnopharmacol.*, **91**: 251-255.

Guatam, M., Diwanay, S., Gairola, S., Shinde, Y.S., Jadhay, S.S., and Patwardhan, B.K. (2004b). Immune response modulation to DPT vaccine by aqueous extract of *Withanis somnifers* in experimental system. *Int. J. Immunopharmacol.*, **4**: 841-849.

Hartwell, J.L. (1969). Plants used against cancer, a survey. *Lloydia.*, **32**: 153-205.

Horsten, S.A., Van den Berg, A.J., Kettenes-van den Bosch, J.J., Leeflang, B.R., and Labadie, R.P. (1996). Cyclogossine A: a novel cyclic heptapeptide isolated from latex of *Jatropha gossypifolia*. *Planta Med.*, **62**: 46-50.

Kawanga, V. (2007). *Jatropha gossypiifolia* L. In: Schmelzer, G.H., and Gurib-Fakim, A. (Editors). Prota 11: Medicinal plants/Plantes medicinales 1. [CD- Rom]. PROTA, Wageningen, Netherlands.

Khan, M.R., Omoloso, A.D., and Barewai, Y. (2006). Antimicrobial activity of *Maniltoa schefferi* extracts. *Fitot.*, **77**: 324-326.

Khumrungsee, N., Bullangpoti, V., and Pluempanupat, W. (2009). Efficiency of *Jatropha gossypifolia* L. (Euphorbiaceae) against *Spodoptera exigua* HUbner (Lepidoptera: Noctuidae): Toxicity and its detoxifying enzyme activities. *KKU Sci. J.*, **37**: 50-55.

Kirtikar, K.R., and Basu, B.D. (1996). "*The Indian Medicinal Plants*". 2nd Edition, Allahabad, India. **3**: 2240- 2247.

Kulakkattolickal, A.T. (1987). Piscicidal plants of Nepal primary toxicity screening using carp (*Genopharyngodon idella*) fingerlings. *J. Ethnopharmacol.*, **21**: 1-9.

Kumar, V.P., Chauhan, N.S., Padh, H., and Rajani, M. (2006). Search for antibacterial and antifungal agents from selected Indian medicinal plants. *J. Ethnopharmacol.*, **107**: 182-188.

Kumar, Y. (2006). Evaluation of leaf extracts of *Jatropha gossypifolia* Linn. for wound healing, antimicrobial and anthelmintic activities. Dissertation submitted to Rajiv Gandhi University of Health Sciences, Karnataka, Bangalore, pp. 1-87.

Kupchan, S.M., Sigel, C.W., and Matz, M.J. (1970). Jatrophone, a novel macro-cyclic diterpenoid tumor inhibitor from *Jatropha gossypifolia*. *J. Am. Chem. Soc.*, **92**: 4476-4477.

Labadie, R.P., Nat van der, J.M., Simons, J.M., Kroes, B.H., Kosasi, S., Berg van den, A.J.J, t' Hart, L.A., Sluis van der, W.G., Abeysekera, A., Bamunuarachchi, A., and De Silva, K.T.D. (1989). An ethanopharmacognostic approach to the search for immunomodulators of plant origin. *Planta Med.*, **55**: 339-348.

Luc, P., and Arnold, J.V. (2005). Bioguided isolation of pharmacologically active components still a valuable strategy for finding a new lead compounds. *J. Ethanopharmacol.*, **100**: 57-60.

Matsuse, I.T., Lim, Y.A., Hattori, M., Correa, M., and Gupta, M.P. (1999). A search for anti-viral properties in Panamanian medicinal plants, the effects on HIV and its essential enzymes. *J. Ethanopharmacol.*, **64**: 15-22.

Morton, J.F. (1968). A survey of medicinal plants of Curacao. *Econ. Bot.*, **22**: 87-102.

Morton, J.F. (1980). Caribbean and Latin American folk medicine and its influence in the United States. *Quart. J. Crude Drug Res.*, **18**: 57-75.

Morton, J.F. (1981). *Atlas of medicinal plants of middle America: Bahamas to Yucatan*. Charles C. Thomas, Springfield, USA; 1420.

Mundt, S., Kreitlow, S., and Jansen, R. (2003). Fatty acids with antibacterial activity from the cyanobacterium *Oscillatoria redekei* HUB 051. *J. Appl. Phycol.*, **15**: 263-267.

Nayak, B.S., and Patel, K.N. (2009). A marvel plant- Jatropha: An appraisal. *Int. J. Pharma. Res.*, **1**: 35-39.

Neogi, U., Saumya, R., Mishra, R.K., and Raju, K.C. (2008). Lipid content and *in vitro* antimicrobial activity of oil seeds of some Indian medicinal plants. *Curr. Res. Bacteriol.*, **1**: 1-6.

Odebiyi, O., and Sofowora, E.A. (1998). Phytochemical screening of Nig. medicinal plants II; *Lloyd*, **41**: 234-236.

Oduola, T., Popoola, G.B., Avwioro, O.G., Oduola, T.A., Ademosum, A.A., and Lawal, M.O. (2007). Use of *Jatropha gossypifolia* stem latex as a haemostatic agent: how safe is it? *J. Med. Plants Res.*, **1**: 14-17.

Oduola, T., Adeosun, O.G., Oduola, T.A., Avwioro, O.G., and Oyeniyi, M.A. (2005). Mechanism of action of *Jatropha gossypifolia* stem latex as a haemostatic agent. *Euro. J. Gen. Med.*, **2**: 140-143.

Oduola, T., Avwioro, O.G., and Ayanniyi, T.B. (2005). Suitability of the leaf extract of *Jatropha gossypifolia* as an anticoagulant for biochemical and haematological analyses. *Af. J. Biotechnol.*, **4**: 679-681.

Ogbobe, O., and Akano, V. (1993). The physico-chemical properties of the seed and seed oil of *Jatropha gosspifolia*. *Plant Food. Human Nut.*, **43**: 197-200.

Ogundare, A.O. (2007). Antimicrobial effect of *Tithonia diversifolia* and *Jatropha gossypifolia* Leaf extracts. *Trends Appl. Sci. Res.*, **2**: 145-150.

Olaifa, J.I., Erhan, W., and Akingbohungbe, A.E. (1987). Insecticidal activity of some Nigerian plants. *Insect Sci. Applica.*, **8**: 221-224.

Omoregbe, R.E., Lkuebe, O.M., and Ihimire, I.G. (1996). Antimicrobial activity of some medicinal plants extracts on *Echericia coli*, *Salmonella paratyphi* and *Shigella dysentriae*. *Af. J. Med. Sci.*, **25**: 373-375.

Panda, B.B., Gaur, K., Nema, R.K., Sharma, C.S., Jain, A.K., and Jain, C.P. (2009a). Hepato-protective activity of *Jatropha gossypifolia* against carbon tetrachloride-induced hepatic injury in rats. *Asian J. Pharma. Clinical Res.*, **2**: 50-54.

Panda, B.B., Gaur, K., Kori, M.L., Tyagi, L.K., Nema, R.K., Sharma, C.S., and Jain, A.K. (2009b). Anti-Inflammatory and analgesic activity of *Jatropha gossypifolia* in experimental animals models. *Global J. Pharmacol.*, **3**: 1-5.

Pandey, A., Maity, B.R., and Samaddar, K.R. (1996). Antifungal activity of plant latex towards certain fungal organisms. *J. Mycopathol. Res.*, **34**: 35-40.

Parekh, J., and Chandra, S.V. (2007). *In vitro* antimicrobial activity and phytochemical analysis of some Indian medicinal plants. *Turk. J. Biol.*, **31**: 53-58.

Pourmorad, F., Hosseinimehr, S.J., and Shahabimajd, N. (2006). Antioxidant activity, phenols, flavonoid contents of selected Iranian medicinal plants. *Af. J. Biotech.*, **5**: 1142-1145.

Prakash, V., Mishra, P.K., and Mishra, M. (2009). Screening of medicinal plant extracts for antioxidant activity. *J. Med. Plant. Res.*, **3**: 608-612.

Rahman, M., Ismail, H.M., and Yin, L.T. (1990). Jatropholone A and Jatrophatrione, two diterpenes from *Jatropha gossypifolia*. *Pertanika*, **13**: 405-408.

Raja, S.N., Ahamad, H., and Kumar, V. (2006). Antioxidant activity in some selected Indian medicinal plants. *Cytisus scoparius Link-A natural antioxidant*, **6**: 1-7.

Ravindranath, N., Venkataiah, B., Ramesh, C., Jayaprakash, P., and Das, B. (2003). Jatrophenone, a novel macrocyclic bioactive diterpene from *Jatropha gossypifolia*. *Chem. Pharmacol. Bull.*, **51**: 870-871.

Robineau, L. (1991). *Towards a Caribbean pharmacopeia*, (Tramil 4) Workshop: Scientific Research and Popular Use of Medical plants in the Caribbean. Enda Caribe, Santo Damingo, UNAH, pp. 287-295.

Sankara, S.S., Nagarajan, S., and Sulochana, N. (1971). Flavonoids of the leaves of *Jatropha gossypifolia*. *Phytochemistry*, **10**: 1690.

Shahwar, D., Shafiq-ur- Rehman, Ahmad, N., Ullah, S., and Raza, M.A. (2010). Antioxidant activities of the selected plants from the family Euphorbiaceae, Lauraceae, Malvaceae and Balsaminaceae. *Af. J. Biotech.*, **9**: 1086-1096.

Siddiqui, S., Verma, A., Rather, A.A., Jabeen, F., and Meghvansi, M.K. (2009). Preliminary phytochemicals analysis of some important medicinal and aromatic plants. *Advances Biol. Res.*, **3**: 188-195.

Singh, A., and Agarwal, R.A. (1988). Possibility of using latex of Euphorbiales for snail control. *Sci. Total Environ.*, **77**: 231-236.

Singh, A., and Agarwal, R.A. (1990). Molluscicidal properties of synthetic pyrethroids. *J. Med. Appl. Malacol.*, **2**: 141-144.

Singh, D., and Singh, A. (2002). Piscicidal effect of some common plants of India commonly used in freshwater bodies against target animals. *Chemosphere*, **49**: 45-49.

Sjogren, J., Magnusson, J., Broberg, A., Schnurer, J., and Kenne, L. (2003). Antifungal 3-hydroxy fatty acids from *Lactobacillus plantarum* milab 14. *Appl. Environ. Microbiol.*, **69**: 7554-7557.

Sofowora, E.A. (1993). *Medicinal Plants and Traditional Medicine in Africa*. 2nd Edn. John and Wiley and Sons Ltd., England, pp: 55-62.

Subramanian, S.S., Nagarajan, S., and Sulochana, N. (1971). Flavonoids of the leaves of *Jatropha gossypifolia*. *Phytochemistry*, **10**: 1690.

Sukumaran, D., Parashar, B.D., and Rao, K.M. (1995). Toxicity of extracts of *Jatropha gosspifolia* and *Saponaria saponaria* as molluscicide against freshwater snails *Lymnaea luteola* and *Indoplanorbis exustus*, the vectors of animal schistosomiasis. *Fitot.*, **66**: 393-398.

Tinto, W.F., and John, M.D. (1992). Triterpenoids of *Jatropha gossypifolia*. *J. Nat. Prod.*, **55**: 807-809.

Tiwari, S., Singh, S.K., and Singh, A.(2004). Toxicology and biochemical alterations induced by different fractions of *E. royleana* latex in freshwater harmful vector snail *L. acuminata*. *Indian J. Exp. Biol.*, **42**: 1220-1225.

Wang, S., and Huffman, J.B. (1991). Botanochemicals: Supplements to petrochemicals. *Eco. Bot.*, **35**: 369-382.

Wang, S.Y. (2003). Antioxidant capacity of berry crops, culinary herbs and medicinal herbs. *Acta Hortic.*, **620**: 461-473.

World Health Organization, (2002). *WHO Traditional Medicine Strategy 2002-2005*, Geneva: World Health Organization.

Wu, X., Beecher, G.R., Holden, J.M., Haytowitz, D.B., Gebhardt, S.E., and Prior, R.L. (2004). Lipophilic and hydrophilic antioxidant capacities of common foods in the United States. *J. Agri. Food Chem.*, **52**: 4026-4037.

Zhang, X. (2004). *Traditional medicine: its importance and protection*. In: Protecting International Dimensions. Part 1. The role of Traditional knowledge in healthcare and agriculture – Twarog, S., and P. Kapoor, eds., New York: United Nations. 3-6.

Medicinal Plants: Phytochemistry, Pharmacology and Therapeutics Vol. 3 (2014) *Pages* **351–356**
Editor-in-Chief: **V.K. Gupta**
Published by: **DAYA PUBLISHING HOUSE, NEW DELHI**

15

Essential Oil Constituents of Rhizome Oil of *Alpinia* Species from South India

Archana P. Raina[1]* and Z. Abraham[2]

ABSTRACT

Alpinia species (Family; Zingiberaceae) are used as a food additive, spice and in indigenous system of medicine. In the present study three different species of Alpinia viz; Alpinia galanga, Alpinia calcarata and Alpinia officinarum, collected from South India were studied for essential oil content and its composition. The essential oils from rhizomes were isolated by hydrodistillation. The oil percentage was maximum in A. calcarata (0.73 to 0.96 per cent), followed by A. galanga (0.27-0.62 per cent) and A. officinarum (0.26 per cent). The essential oil compositions were analyzed by capillary GC and GC/MS. Chemical profiles of essential oil of these species showed quantitative and qualitative variations among major compounds. A total of thirty compounds were identified in these oils. 1,8-cineole was present as the major component in all three species. Its percentage varied from 40.92-72.49 per cent in A. galanga, 32.5-46.77 per cent in A. calcarata and 44.17 per cent in A. officinarum. In addition to cineole, a substantially high content of α-fenchyl acetate (19.87 per cent) was observed in A. calcarata followed by 8.91 per cent in A. officinarum whereas it was negligible in A. galanga. Other components found in appreciable amounts in A. calcarata were α-pinene (2.48 per cent), camphene (6.02 per cent), β-pinene (4.14 per cent), camphor (5.27 per cent), α-terpineol (6.44 per cent) and methyl cinnamate (2.68 per cent). Chemical composition of A. officinarum oil showed α-pinene (1.99 per cent), camphene(3.16 per cent), β-pinene (5.68 per cent), camphor (2.51 per cent), α-terpineol (6.35 per cent), α-fenchyl acetate (8.91 per cent) and methyl cinnamate (1.88 per cent). The drugs prepared from the A. galanga, A. officinarum

1 Germplasm Evaluation Division, National Bureau of Plant Genetic Resources, New Delhi – 110 012, India

2 National Bureau of Plant Genetic Resources Regional Station, Kerala Agricultural University, Thrissur -680 656, Kerala, India

* *Corresponding author*: E-mail: aprraina@yahoo.co.in

and A. calcarata due to high 1,8-cineole can be used in the treatment of rheumatism, bronchial catarrh, asthma and in reducing pain. Therefore, there is a promising possibility to utilize these plant species native to South India for Industrial purpose.

Keywords: *Alpinia calcarata, Alpinia galanga, Alpinia officinarum,* Essential oil composition, GC/MS.

Introduction

The genus *Alpinia* belongs to the family Zingiberaceae with *Alpinia galanga* L. Willd. (greater galangal), *Alpinia officinarum* Hance (lesser galangal) and *Alpinia calcarata* Rosc. as the important species. They are found in tropical countries, including Sri Lanka, India, Thailand and Malaysia (Anon 1985; Chopra *et al.,* 1986). The drugs prepared from these *Alpinia* species are used in the treatment of rheumatism, bronchial catarrh, asthma and in reducing pain. It is used to stimulate digestion, purify blood, improve voice and to treat inflammation (Kiritikar and Basu 1935). The dry rhizomes of *A. galanga* are used in flavouring foods, meat dishes and curries in Malaysia (De Pooter *et al.,* 1985). It also yields an essential oil of commercial importance used for preparation of perfumes, pharmaceuticals, room sprays, lotion and cosmetic products.

The essential oil in the rhizome of *A. galanga* has been reported by many researchers from several countries (Charles *et al.,* 1992; De Pooter *et al.,* 1985; Lawrence 2005; Mallavarapu *et al.,* 2002; Mori *et al.,* 1995; Raina *et al.,* 2002; Scheffer *et al.,* 1981). These oils were found to possess compositional differences, suggesting the existence of chemotypes in this species. The oil from Indonesia was made up predominantly of monoterpenoids with pinenes (18.6 per cent) and 1,8-cineole (47.3 per cent) while the Malaysian oil was characterized by sesquiterpenoids with (E)-β-farnesene (18.2 per cent) and β-bisabolene (16.2 per cent) as the major components (Jantan *et al.,* 2004; Mori *et al.,* 1995; Scheffer *et al.,* 1981). While the rhizome oil of *A. officinarum* has been very less studied and reported by Lawrence *et al.*(1969). 1,8-Cineole (eucalyptol) has been found to be the major constituent in the rhizome oil of *A. calcarata* (Arambewela *et al.,* 2005; Kaul *et al.,* 2005; Kong *et al.,* 2000; Rath *et al.,* 1994; Tiwari *et al.,* 1999).

Keeping in view the pharmacological properties of *Alpinia* species, present study was planned to investigate chemical composition of three important species of *Alpinia* namely *A. galanga, A. calcarata and A. officinarum* from India.

Materials and Methods

Plant Material

The plant materials of *A. galanga, A. calcarata* and *A. officinarum* were collected from different locations of South India. These were established in the field gene bank of NBPGR Regional Station, Thrissur. Rhizomes were collected at the time of flowering for two consequent years for chemical analysis.

Extraction of Oil

Freshly harvested rhizome samples (500g each) were cut into pieces, semi-dried and hydrodistilled using a Clevenger- type apparatus for 4 h. The oil samples were stored at 4°C in an airtight glass vial after drying over anhydrous sodium sulphate.

Gas Chromatography and Gas Chromatography-Mass Spectrometry

GC analysis of the volatile constituents in oil samples was performed using Perkin Elmer Autosystem gas chromatograph, fitted with flame ionization detector and a Carbowax 20M capillary column (50m × 0.25 mm; 0.25µm coating). Nitrogen was used as carrier gas at the flow rate of 1mL/min. The oven temperature was programmed from 60°C to 180°C at 3°C/min. Detector temperature was maintained at 250°C. The sample (0.1 µl) was injected neat in split ratio (1: 40) at 250°C. Area percentage reports obtained by GC/FID were directly applied for quantification purpose without any corrections. GS/MS analysis was carried out on Agilent GC/MS model 5975C/7890 GC equipped with HP-5 MS capillary column (30m × 0.25 mm i.d.; film thickness 0.25 µm). Temperature programming was done as mentioned above. Helium was used as the carrier gas at 1 mL/min flow rate. Mass spectra were recorded over 40-400 amu range at 1 scan/s with ionization energy of 70 eV and ion source temperature at 250°C. The split ratio used was 1: 40. The constituents of the oils were identified by comparing the retention times of the peaks with those reported in literature, matching mass spectral data of the peaks with the data of their authentic standards or the data of the NIST library or literature values (Adams, 1990).

Results and Discussion

The essential oil content of *Alpinia galanga* was found to vary from 0.27-0.62 per cent in dry rhizomes. Thirty compounds accounting 92.47 per cent to 97.70 per cent were identified in this oil (Table 15.1). 1,8-cineole was present as the major component and its percentage varied from 40.92-72.49 per cent in *A. galanga*. Other components present were α-terpineol (2.79 per cent), α-pinene (3.64 per cent), terpinen-4-ol (2.83 per cent), γ-terpinene (0.40 per cent), eugenol (0.61 per cent), geranyl acetate (0.55 per cent), caryophyllene oxide (0.92 per cent), β-sesquiphellandrene (2.63 per cent), methyl eugenol (0.59 per cent) and (E)⁻nerolidol (0.53 per cent). The occurrence of 1,8-cineole as major component in the rhizome oil of *A. galanga* in the present investigation is in agreement with the of finding of Dutta and Nath (2003) and Mallavarapu *et al.*(2002). In contrast, methyl cinnamate (45 per cent), α-terpineol (19.1 per cent), *trans* β-farnesene (18.2 per cent) and myrcene (94.5 per cent) five chemotypes of this species are reported. It was concluded that our South Indian *A. galanga* is 1,8-cineole rich chemotype.

The rhizomes of *A. calcarata* on hydrodistillation yielded a colourless to pale yellow volatile oil with camphoraceous and spicy odour. The oil percentage in rhizomes varied from 0.73 per cent to 0.96 per cent. The analysis of oil by GC and GC/MS revealed the presence of 29 components. The volatile constituents identified in the rhizome oils of *A. calcarata* showed that it contained 14 monoterpenes (65.64 per cent), 4 sesquiterpenes (1.25 per cent), 3 monoterpene esters (21.32 per cent) and rest aromatics (4.05 per cent) which accounted for 92.86 per cent of the total oil. The most dominant monoterpene was 1,8-cineole which comprised 37.21 per cent of the total oil, followed by α-fenchyl acetate 19.87 per cent. Other compounds of the oil were α-pinene(4.13 per cent), camphene(6.02 per cent), β-pinene (3.89 per cent), camphor (4.38 per cent), α-terpineol(5.71 per cent) and (E)⁻methyl cinnamate (2.63 per cent). Arambewela *et al.* (2005) reported 1,8-cineole as major compound in rhizome and leaf oils of *A. calcarata* grown in Sri Lanka whereas α-fenchyl acetate on the other hand was the major constituent in its root oil. Chemical composition of essential oil isolated from *A. officinarum* showed 1,8- cineole (44.17 per cent) as the major constituent present followed by α-fenchyl acetate(8.91 per cent), α-pinene(1.99 per cent), camphene (3.16 per cent), β-pinene (5.68 per cent), camphor(2.51 per cent), α-terpineol (6.35 per cent), and (E)- methyl cinnamate(1.88 per cent).

Table 15.1: Essential oil composition (per cent)* of rhizome oil of *Alpinia* species.

Compound	*Alpinia galanga*	*Alpinia calcarata*	*Alpinia officinarum*
α-Thujene	0.02	tr	0.01
α-Pinene	3.64	2.48	1.99
Camphene	0.18	6.02	3.16
β-Pinene	0.82	4.14	5.68
Myrcene	–	0.24	0.90
Limonene	0.14	–	–
1,8- Cineole	72.49	37.21	44.17
γ-Terpinene	0.4	0.01	0.17
p-Cymene	0.68	0.93	0.15
Terpinolene	0.01	–	–
Fenchone	0.22	0.79	0.59
α-Fenchyl acetate	0.23	19.87	8.91
Camphor	0.04	5.27	2.51
Linalool	0.29	0.34	0.2
Bornyl acetate	0.59	0.36	0.79
Terpinen-4-ol	2.83	0.31	4.49
β-Caryophyllene	0.48	0.63	4.32
β-Farnesene	–	0.41	–
α-Terpineol	2.79	6.44	6.35
Borneol	1.11	0.41	0.06
Geranyl acetate	0.55	0.29	–
β-Bisabolene	2.20	0.21	–
Geraniol	0.51	0.17	–
Methyl eugenol	0.59	1.48	2.93
(E)- Nerolidol	0.53	0.54	0.25
(E)- Methyl cinnamate	0.24	2.68	1.88
Caryophyllene oxide	0.92	0.05	0.20
β-Sesquiphellandrene	2.63	–	–
Farnesol	0.27	0.25	0.07
Eugenol	0.61	0.28	0.20

* Data indicates relative area percentage.

Results of present studies showed compositional variation between oils of *A. calcarata* and *A. galanga*. The latter consists of predominantly 1,8-cineole whereas the former comprised of α-fenchyl acetate (19.87 per cent), camphene (6.02 per cent) and camphor(5.27 per cent) in addition to 1,8-cineole(37.21 per cent). Oil composition of *A. officinarum* showed similarity to that of *A. calcarata* due to presence of camphor and α-fenchyl acetate in it. Among all the three species studied, the concentration of 1,8-cineole was maximum in *A. galanga* oil

1,8-Cineole is an important aroma chemical reported to possess expectorant, antiseptic and anesthetic properties and is used widely in pharmaceutical preparations. Since plant collections of *Alpinia* species native to South India were found rich in 1,8-cineole, these can thus be exploited for use in pharmaceutical and agrochemical industry.

References

Adams, R. P. (1990). Identification of Essential Oils by Ion Trap Mass Spectroscopy. Academic Press, San Deigo, CA., USA.

Anonymous. (1985). The Wealth of India: Raw Materials. Vol I, C.S.I.R., New Delhi, India. pp.198-199.

Arambewela, L. S. R., Kumaratunge, Ajith, Arawwawela, Menuka, Owen, Noel L., and Li, Du. (2005). Volatile oils of *Alpinia calcarata* Rosc. Grown in Sri Lanka. *J. Essent. Oil Res.*, **17**: 124-125.

Charles, D.J., Simon, J.E., and Singh, N.K. (1992). The essential oil of *Alpinia galanga* Willd. *J. Essent. Oil Res.*, **4**: 81-82.

Chopra, R.N., Nayar, S.L., and Chopra, I. C. (1986). Glossary of Indian Medicinal Plants, pp.13, Council of Scientific and Industrial Research, New Delhi, India.

De Pooter Herman, L., Muhammad, N.O., Brigitte, A. C., and Niceas, M.S.(1985). The essential oil of greater galanga (*Alpinia galanga*) from Malaysia. *Phytochemistry*, **24**: 93-96.

Dutta, M., and Nath, S.C. (2003). Essential oil components of the rhizome of *Alpinia galanga* Willd. native to North East India.(In): Bioprospecting of commercially important plants. Proc. Nat. Symp., ISAB-JC, pp. 213-216, Borah, R.C., Talukdar, A. Kataky, J C.S., Unni, B.G., Modi, M.K., and Deka, P.C. Eds.

Jantan, I. B., Ahmad, F. B., and Ahmad, A. S. (2004). Constituents of the rhizome and seed oils of greater galanga *Alpinia galanga*(L.) Willd from Malaysia. *J Essent. Oil Res.*, **16**: 174-176.

Kaul, P.N., Rajeswara Rao, B.R., Singh, K., and Bhattacharya, A.K. (2005). Volatile constituents of essential oils isolated from different parts of *Alpinia calcarata* Rosc. J. *Essent. Oil Res.*, **17**: 7-9.

Kirtikar, K.R., and Basu, B. D. (1935). Indian Medicinal Plants, Vol IV, pp. 2445-2446, IInd edn, M/s. Bishen Singh Mahendra Pal Singh, New Delhi.

Kong, L. Y., Qin, M. J., and Niwa, M. (2000). Diterpenoids from the rhizomes of *Alpinia calcarata*. *J. Nat. Prod.*, **63**: 939-942.

Lawrence, B.M. (2005). Galangal Oil. In: Progress in Essential Oils. *Perf. and Flav.*, **30**: 66-72.

Lawrence, B.M., Hogg, J.W., and Terhune, S.J. (1969). Essential oils and their constituents. Part 2. The oil of *Alpinia officinarium* Hance. *Perfum. Essent. Oil Rec.*, **60**: 88-96.

Mallavarapu, G.R., Rao, L., Ramesh, S., Dimri, B.P, Rajeswara Rao, B.R., Kaul, P.N., and Bhattacharya, A.K. (2002). Composition of the volatile oils of *Alpinia galanga* rhizomes and leaves from India. *J. Essent. Oil Res.*, **14**: 397-399.

Mori, H., Kubota, K., and Kobayashi, A. (1995). Potent aroma components of rhizomes from *Alpinia galanga* Willd. L. *Nippon Shokuhin Kagaku Kogaku Kaishi*, **42**: 989-995.

Raina,V.K., Srivastava, S.K., and Syamasundur, K.V. (2002). The essential oil of greater galangal (*Alpinia galanga* L. Willd) from the lower Himalayan region of India. *Flav. Fragn. J.*, **17**: 358-360.

Rath, S.P., Sahoo, S.B., and Srinivasulu, C. (1994). Analysis of cultivated *Alpinia calcarata*. *Indian J. Nat Prod.*, **10**: 12-13.

Scheffer, C.J.J., Gani, A., and Baerheim-Svendsen, A. (1981). Monoterpenes in the essential rhizome oil of *Alpinia galanga* L. Willd. *Sci. Pharm.*, **49**: 337-346.

Tiwari, A., Pant, A.K., Mathela, C.S., Mengi, N. E., Kohl, E., and Bestmann, H.J. (1999). Volatile constituents of *Alpinia calcarata*. *J. Essent. Oil Res.*, **11**: 739-741.

Medicinal Plants: Phytochemistry, Pharmacology and Therapeutics Vol. 3 (2013) *Pages* **357–368**
Editor-in-Chief: **V.K. Gupta**
Published by: **DAYA PUBLISHING HOUSE, NEW DELHI**

16

Phytochemical Studies on *Gracilaria pygmaea* using TLC, FT-IR and HPLC

M. Johnson[1]*, E. Petchiammal[1], N. Janakiraman[1] and A. Babu[1]

ABSTRACT

The present study was aimed to reveal the phytochemical properties of Gracilaria pygmaea. To identify the functional constituents present in the crude extracts, the spectroscopic and chromatographic analysis were performed. The different extracts of G. pygmaea showed the presence of phenolic compounds, saponins and flavonoids with varied degree. TLC profile of G. pygmaea showed three distinct phenolic spots in the methanolic extract with different Rf values 0.172, 0.534 and 0.810. Steroids TLC profile of G. pygmaea methanolic extract displayed only one distinct spot with the Rf value 0.068. The HPLC chromatogram of G. pygmaea chloroform extracts showed four compounds (peaks) with the different retention time 1.713, 2.160, 2.320 and 3.047 min respectively. Benzene extracts of G. pygmaea illustrated nine compounds at different retention times viz., 1.970, 2.147, 2.320, 2.430, 2.627, 2.937, 3.050, 3.340 and 3.600 min respectively. The results of FTIR analysis showed different peaks at 3315.41, 2358.78, 1668.31, 1402.15, 1114.78, 750.26, 653.82 and 607.54 respectively. The peak values of G. pygmaea confirmed the presence of functional groups such as amides, phosphorus compounds, alkenes, secondary alcohols, benzene ring and halogen compounds. The results of the present study confirmed that G. pygmaea may be rich sources of phytoconstituents which can be isolated and further screened for various biological activities. In addition, these can be used as biochemical markers in the pharmaceutical industries to identify the authentic G. pygmaea and differentiate from its adulterants.

Keywords: FTIR, Functional group, *Gracilaria pygmaea,* HPLC, Phytochemical, TLC.

1 Centre for Plant Biotechnology, Department of Plant Biology and Plant Biotechnology, St. Xavier's College (Autonomous), Palayamkottai – 627 002, Tamil Nadu, India

* *Corresponding author*: E-mail: ptcjohnson@gmail.com

Introduction

Seaweeds are one of the commercially important living marine resources that belong to the primitive groups of non flowering plants (Bermudes and Margulis, 1985). They have been used as food stuff in Asia for centuries as it contains carotenoids, dietary fibres, proteins, essential fatty acids, vitamins and minerals. Fresh and dry seaweeds are extensively consumed by people especially living in the coastal areas. Edible seaweeds contain a significant amount of proteins, vitamins and minerals which are essential nutrition for human (Mohamed Fayaz *et al.*, 2005). Bio-stimulant properties of seaweeds are explored for use in agriculture and the antimicrobial activities for the development of novel antibiotics. Seaweeds have some valuable medicinal components such as antibiotics, laxatives, anti-coagulants, anti-ulcer products and suspending agents in radiological preparations. Seaweeds have recently received significant attention for their potential as natural antioxidants. Many metabolites isolated from marine algae have bioactive efforts (Gopala *et al.*, 2007; Oh *et al.*, 2008). Many of these compounds are bioactive and have been extensively studied by using bioassays and pharmacological assays (Paul and Fenical, 1987). As a consequence of an increasing demand for biodiversity in the screening programs seeking therapeutic drugs from natural products, there is now a greater interest in marine organisms, especially algae. Screening of seaweeds for bioactivity and bioactive constituents is quite imperative. It is well known that many drugs can be prepared from marine source which could profitably be used in pharmaceutical industries. There is a growing demand and need for new bioactive drugs to control many bacterial and fungal diseases of plants and animals (McGill and Hardy, 1992).

Gracilaria genus (Gracilariales, Rhodophyta) is a macroalgae group with more than 300 species of which 160 have been accepted taxonomically. These are usually red, green or greenish brown with a three-phase cycle and can be found in tropical and subtropical seas (Skriptsova *et al.*, 2001). *Gracilaria* are important for the industrial and biotechnological uses because they have phycocolloids, the main source of agar α-(1,4)-3,6-anhydro-L-galactose and β-(1,3)-D-galactose with little esterification in cell wall (Kain and Destombe, 1995). Among the carbohydrates, agar and other polysaccharides are present in *G. confervoides* (Misawa, 1977), *G. dura* (Murano *et al.*, 1992), *G. chilensi* and *G. secundata* (Brasch *et al.*, 1983; Hemmingson *et al.*, 1996). These algae also produce important bioactive metabolites like the primary compound with antibiotic activity acrylic acid (Glombitza, 1979), and the eicosanoids which are derivatives C20 polyunsaturated fatty acid (PUFA) metabolism through oxidative pathways that originate mainly from arachidonic acid and eicosapentaenoic acids, the precursors of prostaglandins (Glickman, 1987; Imbs *et al.*, 2001). Species such as *G. asiatica* and *G. lichenoids* contain PGE2 (Chapman and Chapman, 1980; Sajiki, 1997). PGF2 and 15-keto-PGE2 were respectively isolated from *G. lichenoids* and *G. asiatica* (Imbs *et al.*, 2001); *G. verrucosa* contains PGA2 that appears to be responsible for a gastrointestinal disorder, known as "ogonori" poisoning in Japan (Fusetani and Hashimoto, 1984). Although a number of phytochemical and bio-efficacy studies on seaweeds were carried out at global level, there is no report on the phytochemical studies of *Gracilaria pygmaea* to the best of our knowledge. Therefore, the present study was aimed to analyse the phytochemical properties of *G. pygmaea* using TLC, HPLC and FTIR.

Materials and Methods

Gracilaria pygmaea was collected by handpicking from the coast of Rasthacaud, Kanyakumari District, Tamil Nadu, India (Lat N 08°08′308″ E77°32′80″). The collected seaweeds were cleaned well with sea water to remove all the extraneous matter such as epiphytes, sand particles, pebbles and shells and brought to the laboratory in plastic bags. The collected seaweeds were then thoroughly washed with tap water followed by distilled water. For drying, washed seaweeds were blotted on the blotting paper and spread out at room temperature in shade. The shade dried seaweeds were grounded to fine powder using tissue blender. The powdered samples were then stored in refrigerator for further use.

Preparation of Extract

10 g of air dried powder of *G. pygmaea* was extracted with 60 mL of solvents *viz.*, petroleum ether, benzene, chloroform, acetone, methanolic and aqueous. The sample was kept in dark for 72 h with intermittent shaking. After incubation, the solution was filtered through filter paper and the filtrate was collected (crude extracts).

Fluorescence Analysis

The powdered materials were treated with various reagents such as 50 per cent nitric acid, acetone, ethanol, 50 per cent sulphuric acid, 1N HCl, 1N NaOH. The crude extracts (solvents extracted and acids/base treated extracts) were examined under visible and UV light and changes in colour were recorded (Indian Pharmacopoeia, 1996).

Phytochemical Screening

The different extracts were tested for steroids, alkaloids, phenolic compounds, saponins, flavonoids and anthroquinones. Phytochemical screening of the extracts was carried out according to the standard method described by Harborne (Harborne, 1998).

TLC Analysis

TLC was carried out on 10 × 20 cm silica gel plates (Merck, Germany). The phenolic and steroids compounds present in *G. pygmaea* were qualitatively detected by TLC. Chloroform and methanol (9: 1 ratio) was served as mobile phase. Folin ciocalteau reagent was served as spraying agent to detect the phenolic compound present in *G. pygmaea*. The occurrence of blue colour spot in the TLC chromatogram indicated the presence of phenolic compounds in *G. pygmaea* extracts. For steroids, benzene and methanol (9: 1 ratio) was used as mobile phase. 5 per cent alcoholic sulphuric acid was exercised as the spraying agent to detect steroids presence in *G. pygmaea*. The appearance of bluish green colour spot in the TLC chromatogram indicated the presence of steroidal compounds in *G. pygmaea*.

HPLC Analysis

The different extracts of *G. pygmaea* were centrifuged at 3000 rpm for 10 min and then filtered through Whatmann No.1 filter paper using high pressure vacuum pump. The sample is diluted to 1: 10 with the same solvents. HPLC method was performed on a Shimadzu LC-10AT VP HPLC system, equipped with a model LC-10AT pump, UV-Vis detector SPD-10AT, Rheodyne injector fitted with a 20 µL loop and auto injector SIL-10AT. A Hypersil BDS C-18 column (4.6 × 250 mm, 5 µm size) with a C-18 guard column was used. The elution was carried out with gradient solvent systems with a flow rate of 1 mL min-1 at ambient temperature (25-28°C). The mobile phase was consisted of 0.1 per cent v/v methanol (solvent A) and water (solvent B). The mobile phase was prepared daily, filtered through a 0.45 µm and sonicated before use. Total running time was 15 min. The sample injection volume was 20 µL while the wavelength of the UV-Vis detector was set at 254 nm (Sharanabasappa *et al.*, 2007; Mallikharjuna *et al.*, 2007).

FTIR Analysis

FTIR analysis was performed using Perkin Elmer Spectrophotometer system, which was used to detect the characteristic peaks and their functional groups. The peak values of the FTIR spectrum were recorded. Each and every analysis was repeated twice and confirmed the spectrum (Janakiraman *et al.*, 2011).

Results

The fluorescence analysis of the seaweed *G. pygmaea* treated with various reagents was recorded in Table 16.1. Preliminary phytochemical screening of six different chemical compounds (steroids, alkaloids, phenolic groups, saponins, flavonoids and anthroquinone) were tested in six different extracts of *G. pygmaea* and the results were illustrated in Table 16.2. Thus out of (1 x 6 x 6 = 36) 36 tests for the presence or absence of the above compounds, 7 tests gave positive results and the remaining 29 gave negative results. The 7 positive results show the presence of phenolic groups, saponins and flavonoids. Alkaloids, steroids and anthraquinone did not show any positive result for their presence in any of the tested six extracts of *G. pygmaea*. Among the six different extracts tested, aqueous extract failed to show the presence of any of the compounds. *G. pygmaea* showed the maximum presence of flavonoids and saponins in three different extracts. Phenolics were present only in methanolic extract. Among the six different extracts, benzene and chloroform shows the presence of maximum number (2/6) of metabolites. Next to that petroleum ether, methanol and acetone extracts showed the presence of only one metabolite (Table 16.2).

Phenolics and steroids present in methanolic extracts of *G. pygmaea* were separated using TLC. Three distinct phenolic spots were detected in the methanolic extract of *G. pygmaea* with different Rf values 0.172, 0.534 and 0.810. Steroids profile of *G. pygmaea* methanolic extract displayed only one distinct spot with the Rf value 0.068.

The qualitative HPLC fingerprint profile of chloroform and benzene extracts of *G. pygmaea* were selected at a wavelength of 254 nm due to sharpness of the peaks and proper baseline. Extracts prepared by cold extraction was subjected to HPLC for the isolation and identification of constituents present in the *G. pygmaea.* The HPLC chromatogram of *G. pygmaea* chloroform extracts showed four compounds (peaks) with the different retention time 1.713, 2.160, 2.320 and 3.047 min respectively. The profile of *G. pygmaea* chloroform extracts displayed one prominent peak at a retention time of 2.320 min and some moderate peaks at a retention time of 1.713, 2.160 and

3.047 min respectively (Figure 16.1). Benzene extracts of *G. pygmaea* illustrated nine compounds with different retention times *viz.*, 1.970, 2.147, 2.320, 2.430, 2.627, 2.937, 3.050, 3.340 and 3.600 min respectively. The profile displayed one prominent peak at a retention time of 2.627 min and some moderate peaks were also observed at a retention time of 1.970, 2.147, 2.320, 2.430, 2.937, 3.050, 3.340 and 3.600 respectively (Figure 16.2).

Table 16.1: Fluorescence analysis color change with the presence of acid.

Light	*50 per cent Sulphuric Acid*	*1N HCl*	*Acetone*	*NaOH*	*Nitric Acid*
Flourescent	Blackish Brown	Dark green	Light green	Brown	Yellowish brown
UV	Blackish Brown	–	–	Brown	Yellow

Table 16.2: Preliminary phytochemical analysis of crude extracts of *G. pygmaea*

Name of the Extract	*Phenolics*	*Saponins*	*Flavonoids*
Aqueous	–	–	–
Methanol	+	–	–
Acetone	–	–	+
Benzene	–	+	+
Chloroform	–	+	+
Pet. Ether	–	+	–

The FTIR spectrum was used to identify the functional group of the active components based on the peak value in the region of infrared radiation. The crude powder of *G. pygmaea* was passed into the FTIR and the functional groups of the components were separated based on its peak ratio. The results of FTIR analysis of crude powder of *G. pygmaea* was shown in Table 16.3. This showed different peaks at 3315.41, 2358.78, 1668.31, 1402.15, 1114.78, 750.26, 653.82 and 607.54 respectively (Figure 16.3). It confirmed the presence of functional groups such as amides, phosphorus compounds, alkenes, secondary alcohols, benzene ring and halogen compounds.

Discussion

It is a real fact that the importance of marine organisms as a source of new substances is growing. With marine species comprising approximately a half of the total global biodiversity, the seaweeds offer an enormous resource for novel compounds and it has been classified as the largest remaining reservoir of natural molecules to be evaluated for drug activity. Recent findings revealed that seaweeds contained antibacterial (Tuney *et al.*, 2007; Karthikaidevi *et al.*, 2009), antiviral (Serkedjieva, 2004), antifungal (Tang *et al.*, 2002), cytotoxic (Smith *et al.*, 1994) and larvicidal potentials (Thangam *et al.*, 1991). Phenolic compounds are commonly found in plants, including seaweeds, and have been reported to have a wide range of biological

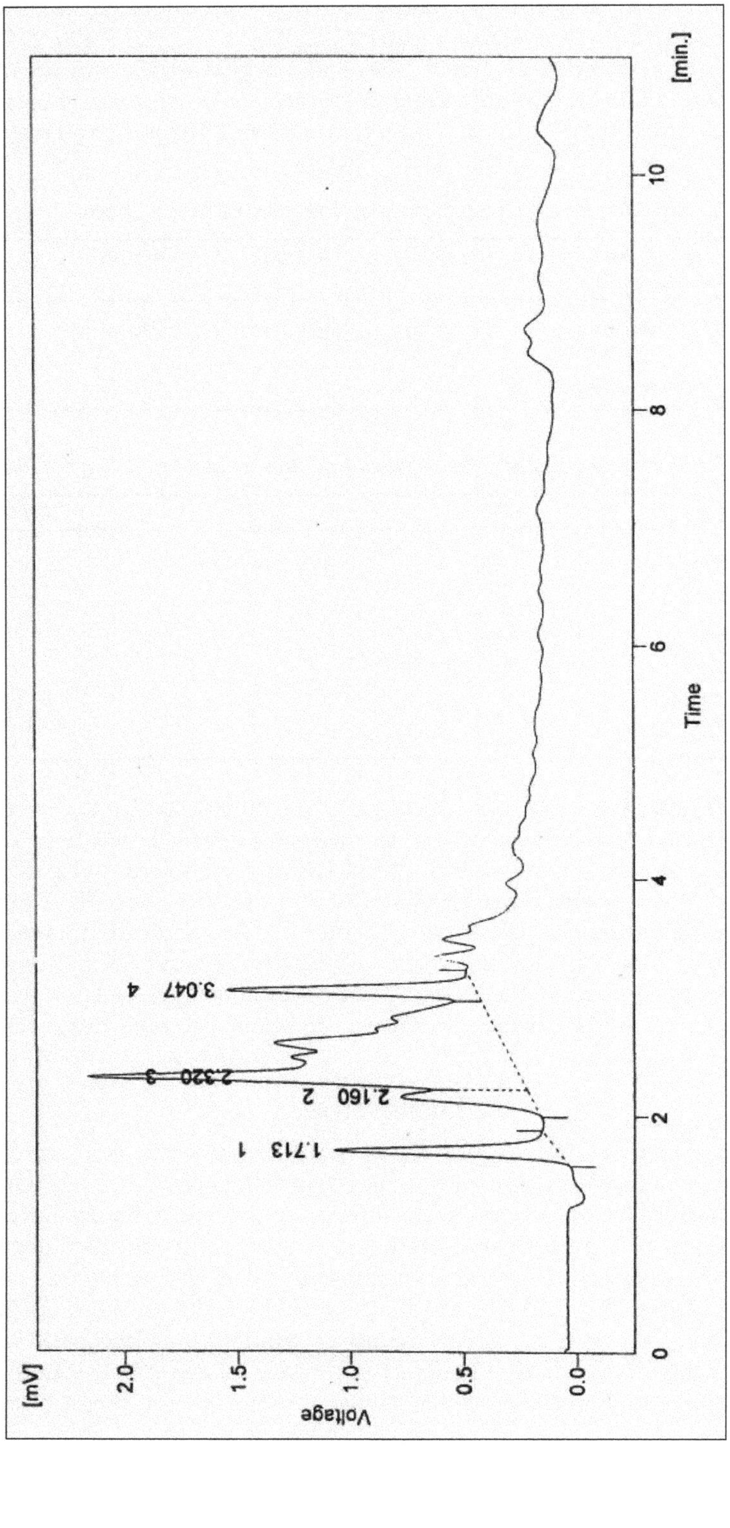

Figure 16.1: HPLC Chromatogram of the Chloroform extracts of *G. pygmaea.*

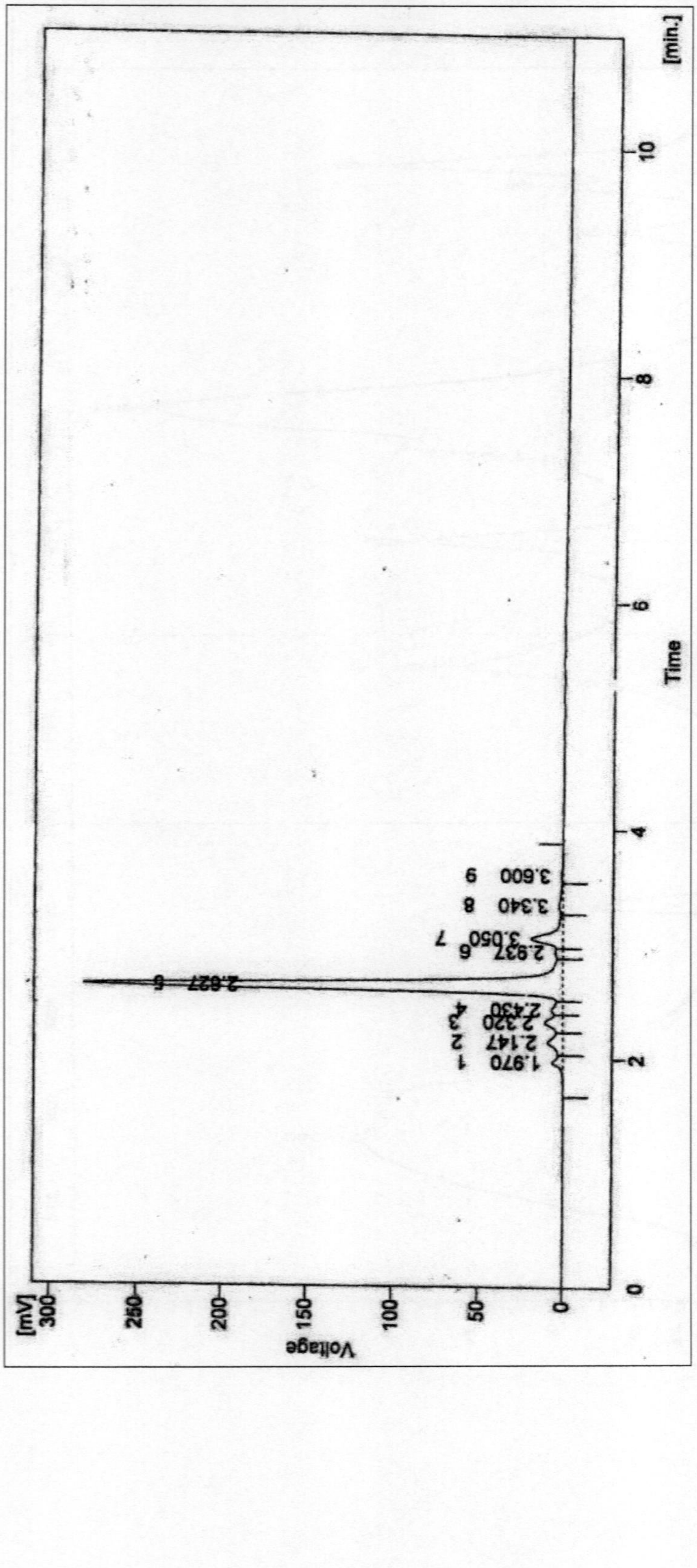

Figure 16.2: HPLC Chromatogram of the Benzene extracts of *G. pygmaea*.

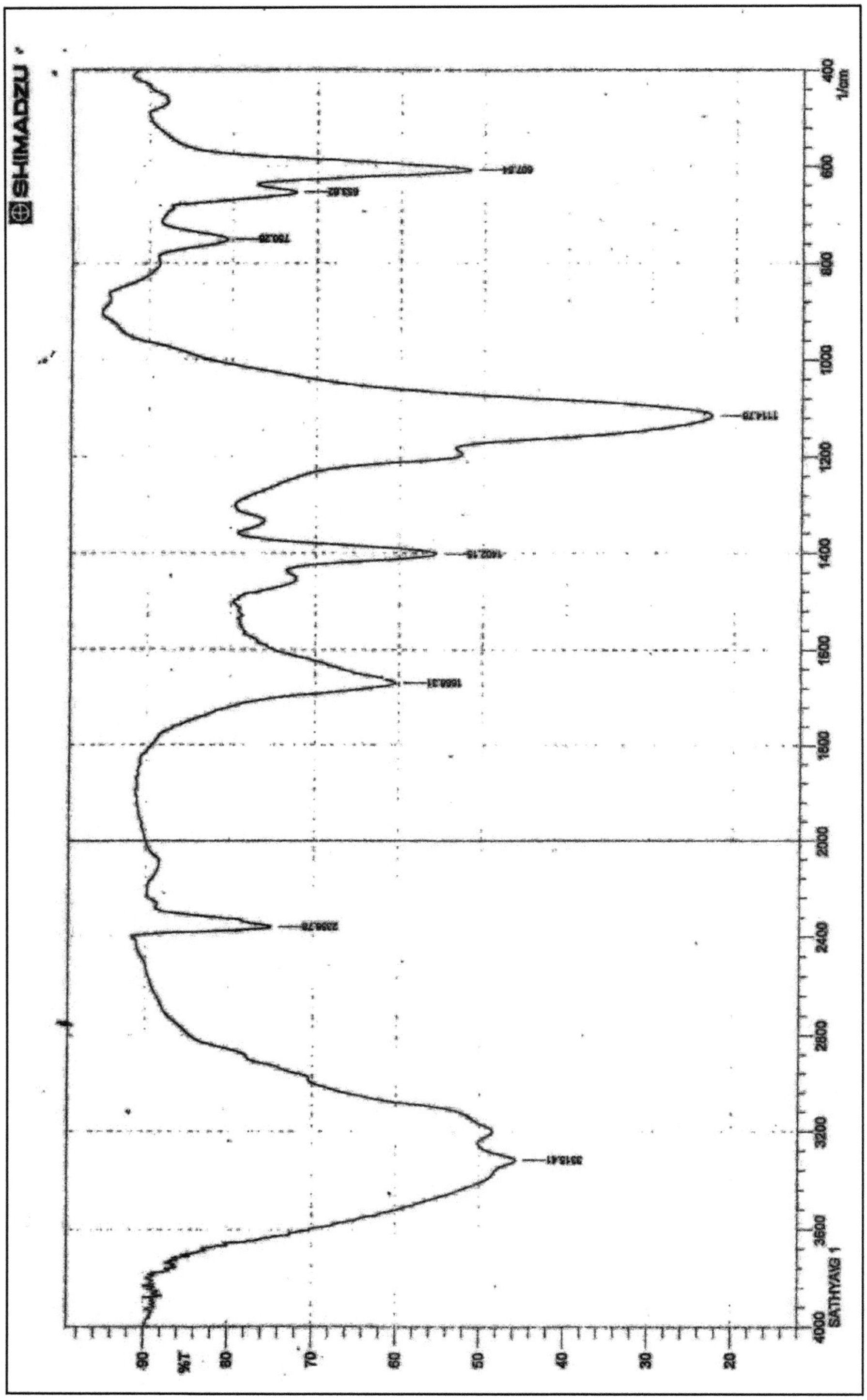

Figure 16.3: FT-IR Spectrum of crude powder of *G. pygmaea*.

Table 16.3: FTIR peak values on crude powder of *G. pygmaea*.

Peak Values	*Compounds*
3315.41	Amides-bonded NH
2358.78	Phosphorus compound
1668.31	Cross-Conj. Dienones
1402.15	Alkenes
1114.78	Secondary alcohols
750.26	Benzene ring with three adjacent hydrogen atoms
653.82	Halogen compounds–C-Br axial
607.54	Halogen compounds-Other Monobromides

activities including antioxidant properties (Kuda *et al.*, 2007; Wang *et al.*, 2009). In the present study also the phenolics presence was confirmed by the qualitative analysis in the methanolic extracts of *G. pygmaea*. Flavonoids, the largest groups of phenolic compounds are known to contain a broad spectrum of chemical and biological activities including antioxidant and free radical scavenging properties (Kahkonen *et al.*, 1999). Flavonoids include flavonols, flavones, catechins, proanthocyanidins, anthocyanidins and isoflavonoids (Ndhlala *et al.*, 2007). Flavonoids are important in plant defense mechanisms against invading bacteria and other types of environmental stress, such as wounding and excessive light or ultraviolet (UV) radiation (Harborne, 1994). In the present study also, flavonoids were observed in acetone, benzene and chloroform extracts of *G. pygmaea*. Saponins possess specific physical, chemical and biological activities that make them useful as drugs. Some of these biological properties include antimicrobial, anti-inflammatory, anti-feedent and hemolytic effects (George *et al.*, 2002). In the present investigation, saponins showed its presence in benzene, chloroform and petroleum ether extracts of *G. pygmaea*.

The fluorescence analysis is adequately sensitive and enables the precise and accurate determination over a satisfactory concentration range without several time-consuming dilution steps prior to analysis of pharmaceutical samples (Pimenta *et al.*, 2006). TLC studies employed the Rf values to distinguish the seaweeds from other species and adulterant (Amit Joshi *et al.*, 2011). In the present study also we developed the phenolics and steroids profile of *G. pygmaea* using TLC to distinguish between each other. HPLC identification test are required to confirm the presence of the active constituents and potential adulterant in ayurvedic drugs. In the present study, HPLC profile for chloroform and benzene extracts of *G. pygmaea* were evolved. These studies on *G. pygmaea* exhibited novel markers in standardization as useful analytical tools to check not only the quality of the powder but also the presence of adulterants in ayurvedic drugs. The FTIR spectrum was used to identify the functional group of the active components based on the peak value in the region of infrared radiation. The results of FTIR analysis showed different peaks which confirmed the presence of functional groups such as amides, phosphorus compounds, alkenes, secondary alcohols, benzene ring and halogen compounds.

Seaweeds known as medicinal are rich in secondary metabolites which include alkaloids, glycosides, flavonoids, saponins, tannins, steroids and related active metabolites. These are of great medicinal value and have been extensively used in the drug and pharmaceutical industry (Balandrin and Klocke, 1988). Recently, a number of studies have been reported on the phytochemistry of seaweeds across the world (Wang *et al.*, 2009; Cox *et al.*, 2010; Manivannan *et al.*, 2011). In the present investigation also, *G. pygmaea* have been selected from Rasthacaud coastal areas, Tamil Nadu, India for phytochemical screening and it confirmed the presence of phenolics, saponins and flavonoids with varied degree. In addition fluorescence, TLC, HPLC and FTIR profiles can be used as biochemical markers in the pharmaceutical industries to identify the authentic mother plants and differentiate from its adulterants. These studies will also help the manufacturers for identification and selection of raw materials for drug production.

References

Amit Joshi, Vijay Baghel, Pathak, A.K., and Tailang, M. (2011). Phytochemical studies of whole plant of *Cyathocline lyrata*–CASS. *International Journal of Pharmaceutical Studies and Research*, 59-64.

Balandrin, M.J., and Klocke, J.A. (1988). Medicinal, aromatic and industrial materials from plants. Ed. By Bajaj, Y.P.S., Springer-Verlag, Berlin, Heidelberg, pp. 1-36.

Bermudes, D., and Margulis, L. (1985). Symbiosis as a mechanism of evolution: Status of the symbiosis theory. *Symbiosis*, **1**: 101-124.

Brasch, D.J., Chuah, C.T., and Melton, L.D. (1983). Marine algal polysaccharides, Part 2. The agar-type polysaccharide from the red alga *Gracilaria secundata*. *Carbohydr. Res.*, **115**: 191-198.

Chapman, V.J., and Chapman, D.J. (1980). Seaweeds and Their Uses; Springer-Verlag: Berlin, Germany, **62**: 112-122.

Cox, S., Abu-Ghannam, N., and Gupta, S. (2010). An assessment of the antioxidant and antimicrobial activity of six species of edible Irish seaweeds. *Int. Food Res. J.*, **17**: 205-220.

Fusetani, N., and Hashimoto, K. (1984). Prostaglandin E2: A candidate for causative agent of "Ogonori" poisoning. Bull. *Jpn. Soc. Sci. Fish.*, **50**: 465-469.

George, F., Zohar Kerem, Harinder, P.S.M., and Klaus Becker. (2002). The biological action of saponins in animal systems: A review. *British Journal of Nutrition*, **88**(6): 587-605.

Glickman, M. (1987). Utilisation of seaweed hydrocolloids in the food industry. *Hydrobiology*, **151**: 31-47.

Glombitza, K.W., Hoppe, H.A., and Levring, T. (1979). Marine Algae in Pharmaceutical Science. Ed. By Walter de Gruyter, New York, USA, pp. 303-342.

Gopala, S.V.K., Panchagnula Aditya, L., Gottumukkala, V., and Subbaraju. (2007). Synthesis, structural revision, and biological activities of 4-chloroaurone, a metabolite of marine brown alga *Spatoglossum variabile*. *Tetrahedron*, **63**(29): 6909-6914.

Harborne, J.B. (1994). The Flavonoids: Advances in Research Since 1986. London, UK.

Harborne, J.B. (1998). Phytochemical Methods, A guide to modern techniques of plant analysis, 3rd Edn. Chapman and Hall, New York, pp. 1-150.

Hemmingson, J.A., Furneaux, R.X., and Murray-brown, V.H. (1996). Biosynthesis of agar polysaccharides in *Gracilaria chilensis* bird. *Carbohydr. Res.*, **287**: 101-115.

Imbs, A.B., Vologodskaya, A.V., Nevshupova, N.V., Khotimchenko, S.V., and Titlyanoy, E.A. (2001). Response of prostaglandin content in the red alga *Gracilaria verrucosa* to season and solar irradiance. *Phytochemistry*, **58**: 1067-1072.

Indian Pharmacopoeia (1996). Controller of Publication, Government of India

Janakiraman, N., Sahaya Sathish, S., and Johnson, M. (2011). UV-VIS and FTIR spectroscopic studies on *Peristrophe bicalyculata* (Retz.) Nees. *Asian Journal of Pharmaceutical and Clinical Research*, **4**(4): 125-129.

Kahkonen, M.P., Hopia, A.I., Vuorela, H.J., Rauha, J.P., Pihlaja, K., Kujala, T.S., and Heinonen, M. (1999). Antioxidant activity of plant extracts containing phenolic compounds. *Journal of Agricultural and Food Chemistry*, **47**: 3954-3962.

Kain, J.M., and Destombe, C. (1995). A review of the life history, reproduction and phenology of *Gracilaria*. *J. Appl. Phycol*, **7**: 69-281.

Karthikaidevi, G., Manivannan, K., Thirumaran, G., Anantharaman, P., and Balasubaramanian, T. (2009). Antibacterial properties of selected green seaweeds from Vedalai coastal waters; Gulf of Mannar marine biosphere reserve. *Global Journal of Pharmacology*, **3**(2): 107-112.

Kuda, T., Kunii, T., Goto, H., Suzuki, T., and Yano, T. (2007). Varieties of antioxidant and antibacterial properties of *Ecklonia stolonifera* and *Ecklonia kurome* products harvested and processed in the Noto peninsula, Japan. *Food Chemistry*, **103**: 900-905.

Mallikharjuna, P.B., Rajanna, L.N., Seetharam, Y.N., and Sharanabasappa, G.K. (2007). Phytochemical studies of *Strychnos potatorum* L. f.–A medicinal plant. *E-Journal of Chemistry*, **4**(4): 510-518.

Manivannan, K., Karthikai Devi, G., Anantharaman, P., and Balasubramanian, T. (2010). Antimicrobial potential of selected brown seaweeds from Vedalai coastal. *Middle-East Journal of Scientific Research*, **4**(2): 100-112.

McGill, A.S., and Hardy, R. (1992). Review of the methods used in the determination of antibiotics and their metabolites in farmed fish. *In*: Chemotheraphy in aquaculture: From Theory to Reality, Ed. By Michel, C., and Alderma, D.J. Office International Des Espizooties, Paris, pp. 343-380.

Misawa, M. (1977). Production of natural substances by plant cell cultures described in Japanese patents. *In:* Plant Tissue Culture Its Bio-Technol, Ed. By Barz, W., Reinhard, E., Zenk, M.H., Springer-Verlag, Berlin, Germany, 17-26.

Mohamed Fayaz, K.K., Namitha, K.N., Chidambara Murthy, M., Mahadeva Swamy, R., Sarada Salma Khanam, P.V., and Subbarao. (2005). Chemical composition, iron bioavailability and antioxidant activity of *Kappaphycus alvarezii* (Doty). *J Agri Food Chem*, **53**: 792-797.

Murano, E., Toffanin, R., Paoletti, S., and Rizzo, R. (1992). Pyruvate-rich agarose from the red alga *Gracilaria dura*. *Planta Med.*, **58**: A588–A589.

Ndhala, A.R., Kasiyamhuru, A., Mupure, C., Chitindingu, K., Benhura, M.A., and Muchuweti, M. (2007). Phenolic composition of *Flacourtia indica, Opuntia megacantha* and *Sclerocarya birrea*. *Food Chemistry*, **103**: 82-87.

Oh, K.B., Lee, J.H., Chung, S.C., Shin, J., Shin, H.J., and Kim, H.K. (2008). Antimicrobial activities of the bromophenols from the red alga *Odonthalia corymbifera* and some synthetic derivatives. *Bioorganic and Med Chem Lett*, **18**: 104-108.

Paul, V.J., and Fenical, W. (1987). Natural Products Chemistry and Chemical Defense in Tropical Marine Algae of the Phytum Chlorophyta. *In*: Bioorganic Marine Chemistry, Ed. By. Scheuer, P.J., Springer-Verlag, Berlin, pp. 1-29.

Pimenta, A.M., Montenegro, M.C., Ara Ujo, A.N., and Martinez, J.C. Application of sequential injections analysis to pharmaceutical analysis. *Journal of Pharm. Biomed. Annals*, **40**: 16-34.

Sajiki, J. (1997). Effect of acetic acid treatment on the concentrations of arachidonic acid and prostaglandin E2 in the red algae, *Gracilaria asiatica* and *G. rhodocaudata*. *Fish Sci.*, **63**: 128-131.

Serkedjieva, J. (2004). Antiviral activity of the red marine algae *Ceramium rubrum*. *Phytotherapy Research*, **18**: 480-483.

Sharanabasappa, G.K., Santosh, M.K., Shaila, D., Seetharam, Y.N., and Sanjeevarao, I. (2007). Phytochemical Studies on *Bauhinia racemosa* Lam. *Bauhinia purpurea* Linn. and *Hardwickia binata* Roxb. *E-Journal of Chemistry*, **4**(1): 21-31.

Skriptsova, A.V., Titlyanova, T.V., and Titlyanov, E.A. (2001). Red algae of the genus *Gracilaria* in south of the Russian far east. Russ. *J. Mar. Biol*, **27**: S38–S52.

Smith, P., Hiney, M.P., and Samuelsen, O.B. (1994). Bacterial resistance to antimicrobial agents used in fish farming: a critical evaluation of method and meaning. *Ann. Rev. Fish. Dis.*, **4**: 273-313.

Tang, H.F., Yi, Y.H., Yao, X.S., Xu, Q.Z., Zhang, S.Y., and Lin, H.W. (2002). Bioactive steroids from the brown algae *Sargassum carpophyllum*. *Journal of Asian Natural Product Research*, **4**: 95-105.

Thangam, T.S., and Kathiresan, K. (1991). Mosquito larvicidal activity of marine plant extracts with synthetic insecticides. *Botanica Marina*, **34**: 537-539.

Tuney, I., Cadirci, B.H., Unal, D., and Sukatar, A. (2007). In antimicrobial activities of crude extracts of marine algae from the coast of Izmir (Turkey). *Fres. Environ. Bull.*, **16**: 428-434.

Wang, B.G., Zhang, W.W., Duan, X.J., and Li, X.M. (2009). *In vitro* antioxidative activities of extract and semi-purified fractions of the marine red alga, *Rhodomela confervoides* (Rhodomelaceae). *Food Chem.*, **113**: 1101-1105.

Medicinal Plants: Phytochemistry, Pharmacology and Therapeutics Vol. 3 (2014) *Pages* **369–382**
Editor-in-Chief: **V.K. Gupta**
Published by: **DAYA PUBLISHING HOUSE, NEW DELHI**

17

Linseed: A Review on Phytochemistry, Pharmacology and Therapatic Potentials

Deepak Mahadev Kasote[1]*

ABSTRACT

Linseed (Linum usitatissimum L) is an oil bearing crop, recently becoming more popular as medicinal food plant owing its wide range therapeutic potentials. The aim of this article is to provide comprehensive information on the botany, traditional uses, phytochemistry, pharmacological and therapeutic potential of linseed. So far, linseed reported to have potent biological and pharmacological activities such as antioxidant, hepatoprotective, immunomodulatory, anti-cancerous, anti-diabetes, cardioprotective, etc. Omega-3 fatty, alpha-linolenic acid (ALA) and lignan, secoisolariciresinol diglucoside (SDG) are major phyto-constituents of linseed responsible for most of these activities. However, further studies are warranted in the context of their mode of actions, bioavailability and pharmacokinetics. Moreover, there is also need to further substantiate promising activities of linseed by in detail clinical trials.

Keywords : Lignan, Linseed, *Linum usitatissimum* L., Omega-3-fatty acid, Pharmacology, Phytochemistry, Therapeutic potential.

Introduction

'Linseed' or 'flaxseed (*Linum usitatissimum* L.) belongs to the family Linaceae and known as *'Uma'* in Sanskrit, *'Alsi'* and *'Tisi'* in Hindi and *'Javas'* in Marathi. It is cultivated throughout the world especially for oil and fiber. Archaeological findings suggest that linseed has been used by human for about 10,000 years. There is a record of linseed being used in bread in Jordan and Greece 3,000 years ago (Vaisey-Genser

1 MACS–Agharkar Research Institute, G.G. Agarkar Road, Pune – 411 004, M.S., India
* *Corresponding author*: E-mail: deepak_kasote@yahoo.com

and Morries, 2003; Johnsson, 2009). Moreover, linseed has long history of use in India as both foodstuff and medicine. In Ayurvedic System of Medicine, it has been used for constipation (Tarpila, 2005). It also used as a component in some herbal preparations, generally used in treatment of inflammatory disorders, gynecological practices, treatment of anemia, bronchial asthma, stomach cancer, gastritis, diabetes mellitus, hyperthyroidism, prostatic hypertrophy, urethritis, nephrosclerosis etc. (Zanwar *et al.*, 2010a).

Now days, linseed is becoming more popular as medicinal or functional food due their diverse range of health promoting properties. In some parts of India, it is consumed as *chutney*. In Northern America and Europe linseed is received at low levels as components in specially breads, as a seed dressing on buns and various other bakery products (Shakir and Madhusudhan, 2007). Accumulating evidences suggest that omega-3 fatty, alpha-linolenic acid (ALA) and lignan secoisolariciresinol diglucoside (SDG) are responsible for most of the health promoting potentials of linseed.

The aim of this article is to provide comprehensive information on the botany, phytochemistry, pharmacological and therapeutic potential of linseed.

Botanical Description

Linseed is a blue colored crop adapted to both warm and cold temperate climates. It is an erect annual plant growing to 120 cm tall and 3 mm broad. The flowers are pure pale blue, 1.5-2.5 cm diameter, with five petals. The fruit is a round, dry capsule 5-9 mm diameter, containing several 4-7 mm long flat and oval with a pointed tip, and varies in color from dark brown to yellow (Figure 17.1). However, both brown and yellow linseed verities have similar values. Today, it is cultivated in more than 50 countries, the majority of them in the northern hemisphere. World linseed production area is about 2,339,210 ha. Canada is the main linseed producer, followed by China, United States and India (Rubilar *et al.*, 2010). Share of India in world linseed production is about 13.8 per cent (Nagaraj, 2009). In India, it is cultivated in Rabi season; and matures in 120-170 days after sowing. It can be grown in well drained, fertile, medium/heavy soils like sandly loam, clay loam or silty clays with pH 5-7. (Nagaraj, 2009).

Phytochemistry

Oil

Linseed contains 32-45 per cent of oil. Out of this about 73 per cent is of polyunsaturated fatty acids (PUFAs). Further, PUFAs constitutes 51–55 per cent omega-3 fatty acid, ALA, and 15–18 per cent linoleic acid (Hettiarachchy *et al.*, 1990; Oomah and Mazza, 1993)

Protein

The amino acid pattern of linseed protein is similar to that of soybean protein. Linseed reported to contain about 22 per cent of protein. Major proteins are globulins, which make up 18.6 per cent of the total protein, while albumin content represents 17.7 per cent of the total protein (Chung *et al.*, 2005; Rubilar *et al.*, 2010).

Figure 17.1: Linseed flowering and seeds.

Carbohydrates

Linseed is low in carbohydrates (sugars and starches), providing only 1 per cent. For this reason, it contributes little to total carbohydrate intake (Morris, 2007).

Fiber

Linseed contains about 35-45 per cent of fiber, generally classified as either dietary fiber or functional fiber. Out of the total fiber, two-thirds is insoluble and one-third soluble. Insoluble fiber consists of cellulose, hemicellulose and lignin. Soluble fiber is in the form of mucilaginous material composed of polysaccharides (Hosseinian, 2006; Tarpila *et al.*, 2005).

Phenols

Lignans

Secoisolariciresinol diglucoside (SDG) is major lignan present in the linseed. The aglycone of SDG is often referred to as secoisolariciresinol or SECO, (Mazur *et al.*, 1996). The content of SDG varies between 6-29 g/kg (Beejmohun *et al.*, 2007; Charlet *et al.*, 2002; Eliasson *et al.*, 2003; Johnsson *et al.*, 2000; Johnsson *et al.*, 2002). In addition to SDG, smaller quantities other type lignans such as matairesinol, isolariciresinol, lariciresinol, demethoxy-secoisolariciresinol and pinoresinol have also been reported in linseed (Meagher *et al.*, 1999; Sicilia *et al.*, 2003). SDG has been shown to be the precursor for the mammalian lignans, enterolactone (EL) and enterodiol (ED), which are produced in the colon by the action of bacterial flora (Axelson *et al.*, 1982) (Figure 17.2).

Phenolic Acids

Linseed contains 8-10 g/kg of phenolic acids, out of which about 5 g/kg is of esterified phenolic acids and 3-5 g/kg of etherified phenolic acid (Oomah *et al.*, 1995). In linseed, most of the phenolic acids such as p-hydroxybenzoic, trans-ferulic and trans-p-coumaric acids are ester bound. However, phenolic acids like trans and cis-sinapic, *o*-coumaric, *p*-droxybenzoic, trans-*p*-coumaric, vanillic acids and caffiec acid are in free state (Babrowski and Soulski, 1984; Kozlowska *et al.*, 1983) (Figure 17.2).

Phenylpropanoid Glucoside

In linseed, phenylpropanoid glucoside, linusitamarin is present together with other phenylpropanoid glucosides, linocinamarin and daucosterol (Luyengi, *et al.*, 1993) (Figure 17.2).

Flavonoids

It contain about 0.3-0.75 g/kg of flavonoids. They are present in the form of their glucoside such as herbacetin 3, 8-O-diglucopynanoside, herbacetin 3, 7-O-dimethyl ether, and kaempferol 3, 7-O-diglucopyranoside (Qui *et al.*, 1999) (Figure 17.2).

Tannins

Tannins are reported in n-butanol fraction of linseed along with phenolic acid and their glycosides (Kasote *et al.*, 2011a; Kasote *et al.*, 2011b).

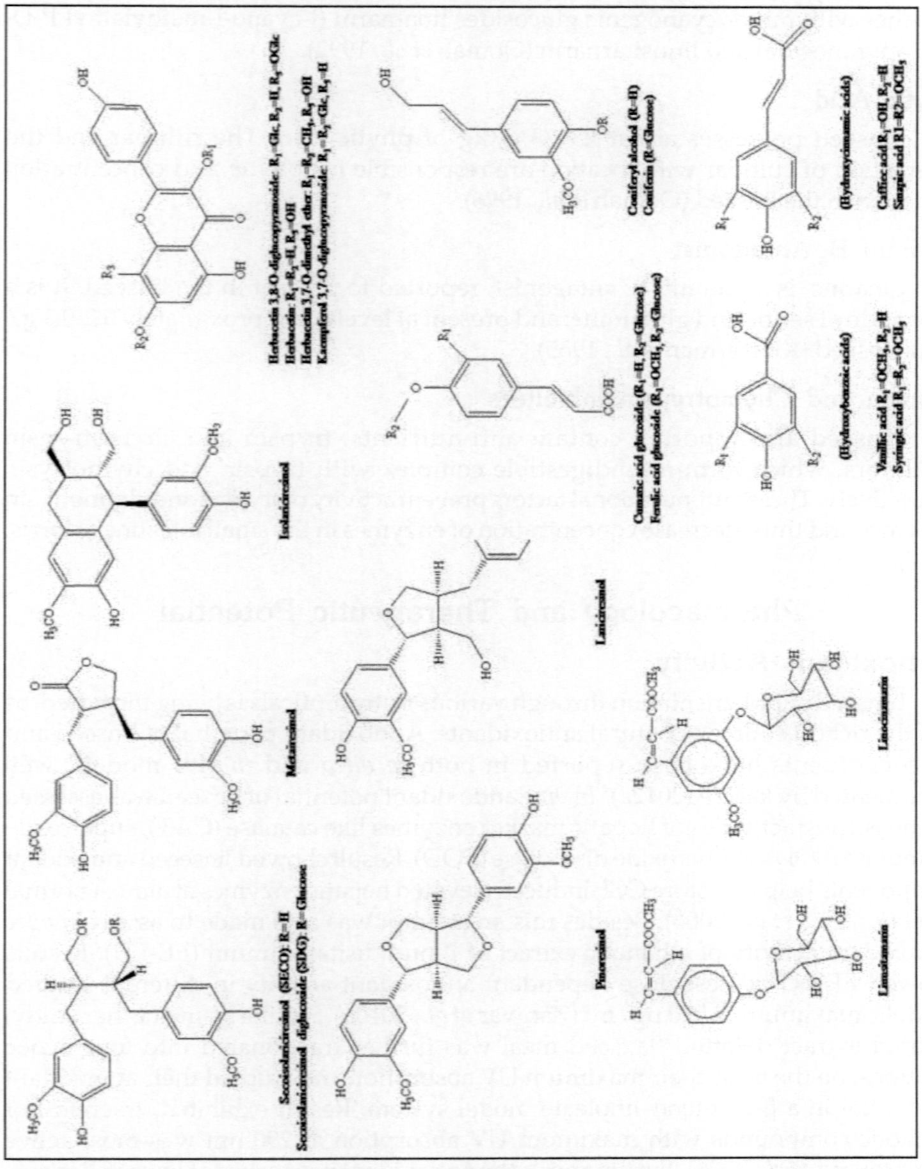

Figure 17.2: Linseed phenolics (Johnsson, 2009; Meagher *et al.*, 1999; Qui, 1999; Riihinen, 2005; Westcott *et al.*, 2002).

Antinutrients

Cyanogenic Glucosides

In linseed, linustatin (β- gentiobioside of acetone cyanohydrin) and neolinustatin (methyl ethyl ketone cyanohydrin,) are the major cyanogenic glucosides, present together with minor cyanogenic glucosides linamarin (l-cyano-l-methylethyl P-D-glucopyranoside) and linusitarnarin (Oomah *et al.*, 1992).

Phytic Acid

Linseed possesses about 23–33 g/kg of phytic acid. The cultivar and the interaction of cultivar with location are responsible for phytic acid concentration variation in the linseed (Oomah *et al.*, 1996).

Vitamin B_6 Antagonist

Linatine, is a vitamin B_6 antagonist, reported to present in the linseed. It is a dipeptide of serine and glutamine; and present at levels of approximately 0.2-0.1 g/kg of linseed (Klosterman *et al.*, 1955).

Trypsin and Chymotrypsin Inhibitors

Linseed also reported contain anti-nutrients, trypsin and chymotrypsin inhibitors, which form an indigestible complex with trypsin and chymotrysin respectively. These anti-nutritional factors prevent activity of trypsin and chymotrysin enzymes and thus, decrease concentration of enzymes in the small intestine (Morris, 2007).

Pharmacology and Therapeutic Potential

Antioxidant Activity

Linseed supplementation through various nutraceuticals is being increased, as it is the richest source of natural antioxidants. Antioxidant potential of linseed and its constituents have been reported in both *in vitro* and *in vivo* models, well documented by kasote (2012a). *In vivo* antioxidant potential of linseed was assessed by measuring activities of hepatic marker enzymes like catalase (CAT), superoxide dismutase (SOD) and peroxide dismutase (POD). Result showed linseeed antioxidant components help to restore CCl_4 induced elevated hepatic enzymes at almost normal level (Rajesha *et al.*, 2006). Besides this an attempt was also made to assess *in vitro* antioxidant activity of ethanolic extract of Linum usitatissimum (EE-LU). Results showed EE-LU possess dose dependant antioxidant activity in different studied models, maximum at 500 µg/ml (Zanwar *et al.*, 2010a). Similarly, in another study, ethanol extract defatted flaxseed meal was further fractionated into four major fractions, on the basis their maximum UV absorption and studied their antioxidant pontential in a β-carotene–linoleate model system. Result exhibited, fraction I of phenolic compounds with maximum UV absorption at 290 nm was most active (Amarowicz *et al.*, 2006). Furthermore, the antioxidant properties of linseed lignans, SECO and SDG and their major oxidative compounds using 2,2'-azobis(2-amidinopropane) dihydrochloride (AAPH) in *in vitro* model of lipid peroxidation was assessed. Results showed that the antioxidant mechanisms of both SECO and

SDG would be based on hydrogen donation ability of SECO and SDG to trap AP free radicals and consequently inhibiting radical chain propagation (Hosseinian, 2006). In addition to SDG, antioxidant potential of mammalian lignans enterodiol and enterolactine of SDG have been also reported (Kitts *et al.*, 1999). Phenolic acids accumulated at high concentrations such as p-coumaric acid and ferulic acid glucosides were found to possess potent antioxidant properties (Yuan *et al.*, 2008). DPPH radical scavenging potential has been also attributed to linseed flavonoids like herbacetin 3, 8-*O*-diglucopynanoside, herbacetin 3, 7-*O*-dimethyl ether, kaempferol 3, 7-*O*-diglucopyranoside (Qui *et al.*, 1999). Ether insoluble phenolic components such as tannins, phenolic acids, etc. of n-butanol fraction of linseed also exhibited significant *in vitro* lipid peroxidation inhibitory property and reducing power close to the standard antioxidants (Kasote *et al.*, 2011a; Kasote *et al.*, 2011b).

Hepatoprotective Potential

Linseed found to possess hepatoprotective activity, which could be due to its antioxidant potential. It has been found that rat feeding with linseed chutney (15 per cent, w/w) in a supplemented diet leads to depletion of carbon tetrachloride (CCl_4) intoxicated increased levels of serum hepatic markers enzymes glutamate oxaloacetate transminase (GOT), glutamate pyruvate transaminase (GPT) and alkaline phosphatase (ALP). Authors concluded that the observed hepatoprotective property could due to the antioxidant potential of linseed (Shakir and madhusudhan, 2007). Similarly, it was also found that feeding of linseed containing (5 per cent to 10 per cent) helps to restoration CCl_4 intoxicated increased activities of hepatic enzymes such as CAT, SOD and POD (Rajesha *et al.*, 2006). Moreover, hepatoprotective potential was also reported to hull fraction of linseed. Authors concluded that SDG content of linseed hull could be responsible this hepatoprotective potential (Rajesha *et al.*, 2010). In addition to whole linseed supplementation, hepatoprotective potential of its ether insoluble phenolic components of n-butanol fraction (EPC-BF) was also studied. Results of this study outlined that EPC-BF have potent hepatoprotective potential close to standard drug silymarin (Kasote *et al.*, 2012b).

Immunomodulatory Activity

Linseed oil is unique in its high content of omega-3 fatty acid, ALA. Which known influence on the several indices of immune status. By now, effect of linseed feeding on modulation of immune response has been studied in various animals including cattle caws, rodent, human and birds such as chicken. It has been observed that supplementation of linseed diet to human tend to suppress the cell-mediated immunity without affecting the humoral immunity (Kelley *et al.*, 1991). Similarly, it has also been reported that n-3 PUFA from linseed oil had no effect on antibody production in rabbits and rats (Kelley *et al.*, 1988; Kim and Lee, 1992). It has observed that feeding of linseed diet to postpartum dairy cows, affects on their immune response (Lessard *et al.*, 2003). It has also found that supplementation of high level heat treated linseed to chicken, affects antibody response but, not delayed hypersensitivity reaction (Ahmed *et al.*, 2009). Moreover, immunomodulatory potential of ether insoluble phenolic components of n-butanol fraction (EPC-BF) of linseed was assessed in rat. Result of this study showed that EPC-BF modulate immune response by lower cell mediated immune response. However, it does not affect on neutrophil adhesion and humoral immune response (Kasote *et al.*, 2012c).

Anti-cancerous Property

Accumulating evidences suggest that linseed consumption has protective effect in breast and prostate cancer. The effect of linseed supplementation in treatment of breast cancer (including mammary gland cancer) was studied in rats and mice; showed a reduction in tumor development, tumor size, tumor growth and metastasis (Chen *et al.*, 2002; Chen *et al.*, 2003; Chen *et al.*, 2004; Jungestrom *et al.*, 2007; Wang *et al.*, 2005). This protective effect was further confirmed the in human intervention study (Thompson *et al.*, 1996). It was also found that diet supplemented with linseed meal inhibits estradiol-induced growth, angiogenesis, and secretions of vascular endothelial growth factor in human breast cancer xenografts *in vivo* (Jungestrom *et al.*, 2007). Moreover, linseed inhibited the growth and development of prostatic carcinoma in mice (Lin *et al.*, 2002). Linseed has been shown to reduce the aberrant crypts (early marker of colon cancer); and prevent colon tumor development in rat (Bommareddy *et al.*, 2006; Jenab and Thompson, 1996; Serraino and Thompson, 1992). Several experimental and clinical studies have demonstrated that ALA helps to reduces colon cancer (Dwivedi *et al.*, 2005).

Prevention of Cardiovascular Diseases

So far, cardioprotective potential of linseed was attributed to its ALA and lignan content. Cardioprotective effects of linseed enterolignans have been studied in human and animals. Zanwar *et al.* (2011) reported the cardioprotective potential of linseed lignan concentrates in rat. Whole linseed modestly reduces total cholesterol (TC) and low density lipoprotein cholesterol (LDL-C) with known mechanisms (Pellizzon *et al.*, 2007). Moreover, linseed has been also shown to be effective in reducing hypercholesterolemic atherosclerosis without lowering serum cholesterol (Prasad, 1997).

Antidiabetic Activity

Linseed oil has been proved its importance in treatment of diabetes (Sekine *et al.*, 2008). Ghule *et al.* (2012) showed that antihyperglycemic potential to ethanolic extract of *L. usitatissimum* (EE-LU), which mediated through inhibition of reactive oxygen species (ROS) and preservation of endogenous antioxidant enzymes. Moreover, SDG known to prevent or delay the development of type I and II diabetes in diabetes-prone rats, probably by decreasing oxidative stress in body tissues (Prasad, 2000). Further, it was also found that consumption of linseed mucilage in combination with other mucilage may offer added benefit for controlling Type II diabetes mellitus (Thakur *et al.*, 2009). Amylase inhibitors, used in the treatment of Type II diabetes were reported in the linseed isopropanol, acetone and methyl-tertiary-butyl ether extracts (Sudha *et al.*, 2011).

Other Pharmacological and Therapeutic Potentials

Linseed oil known to have laxative property and proved its importance in prevention of oxidative lung damage, inflammation and fibrosis (Lee *et al.*, 2009). It has reported that dietary linseed may modestly lower serum level of sex steroid hormones *i.e.* estrogens and androgens, especially in overweight/obese women due its lignan (Sturgeon *et al.*, 2008). Moreover, the phyto-constituents of linseed may reduce skin lesions due to hypersensitivity (O'Neill *et al.*, 2002). Cyclolinopeptide A (CLA), is a cyclic nonapeptide from isolated from linseed and reported to have strong immunosuppressive and antimalarial activity along with the ability to inhibit cholate uptake into hepatocytes (Picur *et al.*, 2006). Further, it is also found that linseed supplementation may be an effective alternative therapy to alleviate postmenopausal symptoms and may protect cardiovascular system by altering lipid profile favorably, also may improve bone-turnover with reduction in bone-fracture risk (Osman *et al.*, 2011). Linseed possesses fungistatic activity and could be used as a multifunctional food ingredient (Xu *et al.*, 2008). Linseed oil proved its usefulness in the treatment of burns, anxiety, neurotoxicity and kidney disease (Sankaran *et al.*, 2006; Moneim *et al.*, 2011).

Linseed Toxicity

Sometimes, linseed may exhibits toxicity on consumption, mainly due to the presence of the phytoestrogen, lignan, cyanogenic compounds and cadmium content (Madhusudhan, 2009). In one study, fetotoxic effect was attributed to ethanolic extract of linseed meal (Zanwar *et al.*, 2010b). However, at present there are not strong evidences regarding cyanogenic and cadmium content toxicity of linseed (Madhusudhan, 2009).

Conclusions

In recent times, linseed is becoming more popular as medicinal food owing to its wide range of therapeutics potentials. Among all plants, linseed is rich in omega-3 fatty acid, ALA and lignan, SDG; and most the reported pharmacological and therapeutics potentials of linseed are due to these two constituents. Findings regarding beneficial effects of linseed in treatment of cancer, diabetes and cardiovascular diseases are really fascinating. However, promising results in this context should be further substantiated by in detail clinical trials. Furthermore, there is a need to study mode of actions and bioavailability of functional linseed phytoconstituents in sufficient detail. Efforts are also essential to investigate the safety and efficacy of linseed and its product consumption.

References

Ahmed, A.S., Al-Yousif, Y.M., and Najib, H. (2009). Effect of dietary high levels of heat treated locally produced flaxseeds on body weight gain and immune response of pre-layer local chicken. *Egyptian Poultry Science*, **29:** 821-835.

Amarowicz, R., Wanasundara, P.K.J.P.D., and Shahidi, F. (2006). Chromatographic separation of flaxseed phenolics. *Nahrung*, **38:** 520-526.

Axelson, M., Sjovall, J., Gustafsson, B.E., and Setchel, K.D.R. (1982). Origin of lignans in mammals and identification of a precursor from plants. *Nature*, **298:** 659–660.

Babrowski, K.J., and Soulski, F.W. (1984). Composition of free and hydrolysable phenolic acids in defatted flours of ten oilseeds. *Journal of Agricultural and Food Chemistry*, **32**: 128-130.

Beejmohun, V., Fliniaux, O., Grand, E.´, Lamblin, F., Bensaddek, L., Christen, P., Kovensky, J., Fliniaux, M.A., and Mesnard, F. (2007). Microwave-assisted extraction of the main phenolic compounds in flaxseed. *Phytochemical Analysis*, **18:** 275–282.

Bommareddy, A., Arasada, B.L., Mathees, D.P., and Dwivedi, C. (2006). Chemopreventive effects of dietary flaxseed on colon tumor development. *Nutrition and Cancer*, **54:** 216-222.

Charlet, S., Bensaddek, L., Raynaud, S., Gillet, F., Mesnard, F., and Fliniaux, M.A. (2002). An HPLC procedure for the quantification of anhydrosecoisolariciresinol application to the evaluation of flax lignan content. *Plant Physiology and Biochemistry*, **40:** 225–229.

Chen, J.M., Stavro, P.M., and Thompson, L.U. (2002). Dietary flaxseed inhibits human breast cancer growth and metastasis and downregulates expression of insulin-like growth factor and epidermal growth factor receptor. *Nutrition and Cancer*, **43:** 187-192.

Chen, J.M., Tan, K.P., Ward, W.E., and Thompson, L.U. (2003). Exposure to flaxseed or its purified lignan during suckling inhibits chemically induced rat mammary tumorigenesis. *Experimental Biology and Medicine*, **228:** 951-958.

Chen, J.M., Hui, E., Ip, T., and Thompson, L.U. (2004). Dietary flaxseed enhances the inhibitory effect of tamoxifen on the growth of estrogen-dependent human breast cancer (MCF-7) in nude mice. *Clinical Cancer Research*, **10:** 7703-7711.

Chung, M., Lei, B., and Li-Chan, E. (2005). Isolation and structural characterization of the major protein fraction from NorMan flaxseed (*Linum usitatissimum* L.). *Food Chemistry*, **90:** 271–279.

Dwivedi, C., Natarajan, K., and Matthees, D. (2005). Chemopreventive effects of dietary flaxseed oil on colon tumor development. *Nutrition and Cancer*, **51:** 52-58.

Eliasson, C., Kamal-Eldin, A., Andersson, R., and Åman, P. (2003). High performance liquid chromatographic analysis of secoisolariciresinol diglucoside and hydroxycinnamic acid glucosides in flaxseed by alkaline extraction. *Journal of Chromatography A*, **1012:** 151–159.

Ghule, A.E., Jadhav, S.S., and Bodhankar, S.L. (2012). Effect of ethanolic extract of seeds of *Linum usitatissimum* (Linn.) in hyperglycaemia associated ROS production in PBMNCs and pancreatic tissue of alloxan induced diabetic rats. *Asian Pacific Journal of Tropical Biomedicine*, 1-3.

Hettiarachchy, N.S., Hareland, G.A., Ostenson, A., and Bladner-Shank, G. (1990). Chemical composition of 11 flaxseed varieties grown in North Dakota. *Proceeding of Flax Institute*, **53**: 36.

Hosseinian, F. (2006). Antioxidant properties of flaxseed lignans using *in vitro* model systems, Ph.D. thesis, University of Saskatchewan, Canada, pp. 1.

Jenab, M., and Thompson, L.U. (1996). The influence of flaxseed and lignans on colon carcinogenesis and beta-glucuronidase activity. *Carcinogenesis,* **17:** 1343-1348.

Johnsson, P. (2009). Bioactive phytochemicals in flaxseed with particular emphasis on the secoisolariciresinol oligomer, Ph.D. thesis, Swedish University of Agricultural Sciences, Uppsala, pp 13.

Johnsson, P., Kamal-Eldin, A., Lundgren, L.N., and Åman, P. (2000). HPLC method for analysis of secoisolariciresinol diglucoside in flaxseeds. *Journal of Agricultural and Food Chemistry,* **48:** 5216–5219.

Johnsson, P., Peerlkamp, N., Kamal-Eldin, A., Andersson, R.E., Ersson, R. Lundgren, L.N., and Åman, P. (2002). Polymeric fractions containing phenol glucosides in flaxseed. *Food Chemistry,* **76:** 207–212.

Jungestrom, M.B., Thompson, L.U., and Dabrosin, C. (2007). Flaxseed and its lignans inhibit estradiol-induced growth, angiogenesis, and secretion of vascular endothelial growth factor in human breast cancer xenografts *in vivo. Clinical Cancer Research,* **13:** 1061-1067.

Kasote, D.M., Hegde, M.V., and Deshmukh, K.K. (2011a). Antioxidant activity of phenolic components from n-butanol fraction (PC-BF) of defattedflaxseed meal. *American Journal of Food Technology,* **6:** 604-612.

Kasote, D.M., Hegde, M.V., and Deshmukh, K.K. (2011b). HPTLC and LC-ESI-MS characterization of ether insoluble phenolic components from n-butanol fraction (EPC-BF) of defatted flaxseed meal. *International Journal of Drug Development and Research,* **3:** 126-130.

Kasote, D.M. (2012a). Flaxseed phenolics as natural antioxidants: A review. *International Food Research Journal* (Accepted).

Kasote, D.M., Badhe, Y.S., Zanwar, A.A., Hegde, M.V., and Deshmukh, K.K. (2012b). Hepatoprotective potential of ether insoluble phenolic components from n-butanol fraction of defatted flaxseed meal against CCl_4 intoxication in rats. *Journal of Pharmacy and Bioallied Sciences* (In press).

Kasote, D.M., Zanwar, A.A., Devkar, S.T., Hegde, M.V., and Deshmukh, K.K. (2012c). Immunomodulatory activity of ether insoluble phenolic components of n-butanol fraction (EPC-BF) flaxseed in rat. *Asian Pacific Journal of Tropical Biomedicine* (In press).

Kelley, D.S., Nelson, G.J., Serrato, C.M., Schmidt, P.C., and Branch, L.B. (1988). Effects of type of dietary fat on indices of immune status of rabbits. *Journal of Nutrition,* **18:** 1376-84.

Kelley, D.S., Branch, L.B., Love, J.E., Taylor, P.C., Rivera, Y.M., and Iacono, J.M. (1991). Dietary α-linolenic acid and immunocompetence in humans. *American Journal of Clinical Nutrition,* **53:** 40–46.

Kim, S.H., and Lee, L. (1992). Effects of n-6, n-3 fatty acids and vitamin A levels on humoral immunity in rats. *Korean Journal of Nutrition,* **25:** 468–475.

Kitts, D.D., Yuan, Y.V., Wijewicreme, A.N., and Thompson, L.U. (1999). Antioxidant activity of the flaxseed lignan secoisolariciresinol diglycoside and its mammalian lignan metabolites enterodial and enterolactone. *Molecular and Cellular Biochemistry,* **202:** 91-100.

Klosterman, H.J., Smith, F., and Clagett, C.O. (1955). The constitution of linocinnamarin. *Journal of the American Chemical Society,* **77:** 420–421.

Kozlowska, H., Zadernowski, R., and Sosulski, F.W. (1983). Phenolic acids in oilseed flour. *Nahrung,* **27:** 449-453.

Lee, J.C., Krochak, R., Blouin, A., Kanterakis, S., Chatterjee, S., Arguiri, E., Vachani, A., Solomides, C.C., Cengel, K.A., and Christofidou-Solomidou, M. (2009). Dietary flaxseed prevents radiation-induced oxidative lung damage, inflammation and fibrosis in a mouse model of thoracic radiation injury. *Cancer Biology and Therapy,* **8:** 47-53.

Lessard, M., Gagnon, N., and Petit, H.V. (2003). Immune response of postpartum dairy cows fed flaxseed. *Journal of Dairy Science,* **86:** 2647–2657.

Luyengi, L., Pezzuto, J.M., waller, D.P., Beecher, C.W.W., and fong, H.H.S. (1993). Linusitamarin, new phenylpropanoid glucoside from *Linum usitatissimum. Journal of Natural products,* **56:** 2012-2015.

Madhusudhan, B. (2009). Potential benefits of flaxseed in health and disease: A perspective. *Agriculturae Conspectus Scientificus,* **74:** 67-72.

Mazur, W., Fotsis, T., Wähälä, K., Ojala, S., Salakka, A., and Adlercreutz, H. (1996). Isotope dilution gas chromatographic-mass spectrometric method for the determination of isoflavonoids, coumestrol, and lignans in food samples. *Analytical Biochemistry,* **233:** 169-180.

Meagher, L.P., Beecher, G.R., Flanagan, V.P., and Li, B.W. (1999). Isolation and characterization of the lignans, isolariciresinol and pinoresinol, in flaxseed meal. *Journal of Agricultural and Food Chemistry,* **47:** 3173-3180.

Moneim, A.E., Dkhil M.A., and Al-quraishy, S. (2011). Effect of flaxseed oil on lead acetate –induced neurotoxicity in rats. *Biology of Trace Elements Research,* DOI 10.1007/s12011-011-9055-4.

Morris, D.H. (2007). Flax- A health and nutrition primer, 4th ed., Flax Council of Canada, Canada.

Nagaraj, G. (2009). Linseed. *In:* Oilseeds properties, processing, products and procedures, New Delhi Publishing Agency, New Delhi, pp. 123.

O'Neill, W., McKee, S., and Clarke, A.F. (2002). Flaxseed (*Linum usitatissimum*) supplementation associated with reduced skin test lesional area in horses with Culicoides hypersensitivity. *Canadian Journal of Veterinary Research,* **66:** 272–277.

Oomah, D.B., Mama, G., and Kenaschuk, E.O. (1992). Cyanogenic compounds in flaxseed. *Journal of Agricultural and Food Chemistry,* **40:** 1346-1348.

Oomah, B.D., and Mazza, G. (1993). Flaxseed proteins: A review. *Food Chemistry*, **48:** 109-114.

Oomah, B.D., Kenaschuk, E.O., and Mazza, G. (1995). Phenolic acids in flaxseed. *Journal of Agriculture and Food Chemistry*, **43:** 2016-2019.

Oomah, B.D., Mazza, G., and Kenaschuck, E.O. (1996). Flavonoid content of flaxseed: Influence of cultivar and environment. *Euphytica*, **90:** 163-167.

Osman, H.F., Yousef Ayad S.K., and El-Mahdy, A.A. (2011). The potential effect of flaxseed on female postmenopausal rats. *Nature and Science*, **9:** 1-8.

Pellizzon, M.A., Billheimer, J.T., Bloedon, L.T., Szapary, P. O., Daniel J., and Rader D.J. (2007). Flaxseed reduces plasma cholesterol levels in hypercholesterolemic mouse models. *Journal of the American College of Nutrition*, **26:** 66–75

Picur, B., Cebrat, M., Zabrocki, J., and Siemion, I.Z. (2006). Cyclopeptides of *Linum usitatissimum*. *Journal of Peptide Science*, **12:** 569-74.

Prasad, K. (1997). Dietary flaxseed in the prevention of hypercholesterolemic atherosclerosis. *Atherosclerosis*, **132:** 69–76.

Prasad, K., Mantha, S.V., Muir, A.D., and Westcott, N.D. (1998). Reduction of hypercholesterolemic atherosclerosis by CDC-flaxseed with very low alpha-linolenic acid. *Atherosclerosis*, **136:** 367–375.

Prasad, K. (2000). Oxidative stress as a mechanism of diabetes in diabetic BB prone rats: Effect of secoisolariciresinol diglucoside (SDG). *Molecular and CellularBiochemistry*, **209:** 89-96.

Qiu, S., Lu, Z., Luyengi, L., Lee, S.K., Pezzuto, J.M., Farnsworth, N.R., Thompson, L.U., and Fong, H.S. (1999). Isolation and characterization of flaxseed (*Linum usitatissimum*) constituents. *Pharmaceutical Biology*, **37:** 1–7.

Rajesha, J., Murthy, C.K.N., Kumar, M.K., Madhusudhan, B., and Ravishankar, G. (2006). Antioxidant potentials of flaxseed by *in vivo* model. *Journal of Agricultural and Food Chemistry*, **54:** 3794-3799.

Rajesha, J., Ranga Rao, A., Karuna Kumar, M., and Ravishankar, G.A. (2010). Hepato-protective potential of hull fraction from Indian flaxseed cultivar. *Asian Journal of Medical Sciences*, **1:** 20-25.

Rubilar, M., Gutiérrez, C., Verdugo, M., Shene, C., and Sineiro, J. (2010). Flaxseed as a source of functional ingredients. *Journal of soil science and plant nutrition*, **10:** 373 – 377.

Sankaran, D., Bankovic-Calic, N., Peng, C.Y., Ogborn, M.R., and Aukema, H.M. (2006). Dietary flax oil during pregnancy and lactation retards disease progression in rat offspring with inherited kidney disease. *Pediatric Research*, **60:** 729-33.

Serraino, M., and Thompson, L.U. (1992). Flaxseed supplementation and early markers of colon carcinogenesis. *Cancer Letter*, **63:** 159-165.

Shakir, K.A.F., and Madhusudhan, B. (2007). Hypocholesterolemic and hepatoprotective effects of flaxseed chutney: Evidence from animal studies. *Indian Journal of Clinical Biochemistry*, **22:** 117-12.

Sicilia, T., Niemeyer, H.B., Honig, D.M., and Metzler, M. (2003). Identification and stereochemical characterization of lignans in flaxseed and pumpkin seeds. *Journal of Agricultural and Food Chemistry,* **51:** 1181-1188.

Sturgeon, S.R., Heersink, J.L., Volpe, S.L., Bertone-Johnson, E.R., Puleo, E., Stanczyk, F.Z., Sabelawski, S., Wahala, K., Kurzer, M.S., and Bigelow, C. (2008). Effect of dietary flaxseed on serum levels of estrogens and androgens in postmenopausal women. *Nutrition and Cancer,* **60:** 612-8.

Sudha, P., Zinjarde, S.S., Bhargava, S.Y., and Kumar, A.R. (2011). Potent alpha-amylase inhibitory activity of Indian Ayurvedic medicinal plants. *BMC Complementary and Alternative Medicine,* **11:** doi: 10.1186/1472-6882-115.

Tarpila, A., Wennberg, T., and Tarpila, S. (2005). Flaxseed as a functional food. *Current Topics in Nutraceutical Research,* **3:** 167-188.

Thakur, G., Mitra, A., Pal, K., and Rousseau, D. (2009). Effect of flaxseed gum on reduction of blood glucose and cholesterol in Type 2 diabetic patients. *International Journal of Food Science and Nutrition,* **60:** 126-136.

Thompson, L.U., Rickard, S.E., Orcheson, L.J., and Seidl, M.M. (1996). Flaxseed and its lignan and oil components reduce mammary tumor growth at a late stage of carcinogenesis. *Carcinogenesis,* **17:** 1373-1376.

Vaisey-Genser, M., and Morris, D. H. (2003). Introduction: History of the cultivation and uses of flaxseed. *In:* Flax, The genus Linum, Muir, A.D., and Westcott N.D., Ed., Taylor and Francis, London, pp. 1-21.

Wang, L., Chen, J.M., and Thompson, L.U. (2005). The inhibitory effect of flaxseed on the growth and metastasis of estrogen receptor negative human breast cancer xenografts is attributed to both its lignan and oil components. *International Journal of Cancer,* **116:** 793-798.

Xu, Y., Hall, C., Wolf-Hall, C., and Manthey, F. (2008). Fungistatic activity of flaxseed in potato dextrose agar and a fresh noodle system. *International Journal of Food Microbiology,* **121:** 262-7.

Yuan, J.P., Li, X., Xu, S., Wang, J., and Liu, X. (2008). Hydrolysis kinetics of secoisolariciresinol diglucoside oligomers from flaxseed. *Journal of Agricultural and Food Chemistry,* **56:** 10041-10047.

Zanwar, A.A., Hegde, M.V., and Bodhankar, S.L. (2010a). *In vitro* antioxidant activity of ethanolic extract of *Linum usitatissimum*. *Pharmacologyonline,* **1:** 683-696.

Zanwar, A.A., Aswar, U.M., Hegde, M.V., and Bodhankar S.L. (2010b). Estrogenic and embryo-fetotoxic effects of ethanol extract of *Linum usitatissimum* in rats. *Journal of Complementary and Integrative Medicine,* 7: 1-15.

Zanwar, A.A., Hegde, M.V., and Bodhankar S.L. (2011). Cardioprotective activity of flax lignan concentrate extracted from seeds of *Linum usitatissimum* in isoprenalin induced myocardial necrosis in rats. *Interdisciplinary Toxicology,* **4:** 90–97.

Medicinal Plants: Phytochemistry, Pharmacology and Therapeutics Vol. 3 (2014) *Pages* **383–411**
Editor-in-Chief: **V.K. Gupta**
Published by: **DAYA PUBLISHING HOUSE, NEW DELHI**

18

Herbal Products as Adjuvants or Bio-modulators in Diabetic Therapy: Evidence of Potentiation of Therapeutic Profiles, Increased Bioavailability and Improvement in Vital Biochemical Markers

Jaywant U. Deshmukh[1], Vikas B. Gawali[1], Deepa K. Ingawale[2], Rahul S. Somani[1] and Suresh R. Naik[2*]

ABSTARCT

Many herbal drugs and phytochemicals proved to have great potential in the treatment of diabetes and diabetic complications. In the present study, polyherbal formulations, Dolabi (HF) and Karnim (HFK) largely prescribed in India, Pakistan and other South Asian countries were investigated for antidiabetic activity on alloxan induced diabetic rat model. Development of alloxan diabetes in rats showed a significant elevation of plasma glucose, lipid profile, glycosylated haemoglobin, depleted liver glycogen and alteration in the histoarchitecture of pancreatic β-islets. Such clinical picture found to be similar to human type 2 diabetes mellitus. Administration of alloxan diabetic rats with HF (140 mg/kg) and HFK (200 mg/kg), metformin (MF) (200 mg/kg) and glibenclamide (GC) (4 mg/kg) orally daily for 28 days significantly decreased plasma glucose, glycosylated haemoglobin, plasma cholesterol, triglyceride, increased liver glycogen and improved histoarchitecture of β-islets of pancreas (regeneration,

1 Sinhgad College of Pharmacy, S. No. 44/1, Vadgaon (Bk.), Off Sinhgad Road, Pune – 411 041, Maharashtra, India

2 Sinhgad Institute of Pharmaceutical Sciences, S. No. 309/310, Off Mumbai-Pune Expressway, Kusgaon (Bk.), Lonavala, Pune – 410 401, Maharashtra, India

* *Corresponding author*: E-mail: srnaik5@rediffmail.com

hyperplasia, restoration of the cellular size and number). Furthermore, HF and HFK co-administered individually with MF or GC orally daily for 28 days improved pharmacokinetic profile such as area under curve (AUC), concentration maximum (C_{max}), peak concentration time (T_{max}), elimination half-life ($t_{1/2}$), clearance/elimination rate (K_e), kinetic constant (K_a) and volume of distribution (V_d) of MF and GC significantly.

The present experimental findings demonstrate HF and HFK perse treatment showed hypoglycemic activity and prevent the progression of diabetes mellitus. The anti-diabetic activity of HF and HFK may be due to the presence of antioxidant phytochemicals like flavonoids (catechin, quercetin and rutin), triterpenes, saponins (oleanolic acid and gymnamic acid). However, the improvement in pharmacokinetic profile (bioavailability) of MF and GC is not known, but it is believed to be through the inhibition of microsomal metabolising enzymes. The present findings of PK/PD studies demonstrated that such herbomineral formulation can be a used as biomodulators or adjuvants along with clinically effective drugs in the treatment of diabetes mellitus.

Keywords: Herbomineral formulations, Antidiabetic activity, Pharmacokinetic profiles, Serum lipids, Histoarchitecture of pancreatic β cells.

Introduction

The term diabetes mellitus largely described as "a metabolic disorders of multiple aetiology characterised by chronic hyperglycaemia with disturbances of carbohydrate, fat and protein metabolism resulting from endocrine defects in insulin secretion, insulin action or both, and affect a total population of 220 million worldwide in the year 2020 (King *et al.*, 1998; Amos *et al.*, 2010). There are 2 common forms of diabetes: Type 1 or insulin-dependent diabetes mellitus (IDDM) and Type 2 or non-insulin-dependent diabetes mellitus (NIDDM) (Xie *et al.*, 2003). Type 1 diabetes is caused by the autoimmune destruction of pancreatic β-islets, leading to a near total deficiency in insulin secretion. The type 2 diabetes is most common form of diabetes, accounting > 90 per cent of patients. It is caused by the resistance to the action of insulin and/or with a deficiency in insulin secretion (DeFronzo, 1997). Both the types of diabetes have distinct causes, 2 distinct common features: hyperglycemia and various complications (retinopathy, nephropathy, neuropathy, amputations, and increased risk of coronary artery disease and stroke) (Taylor, 1999). Therefore, an effective control of blood glucose is the key factor for prevention or reversion of diabetic complications and improving the quality of life for both Type 1 and Type 2 diabetic patients (The Diabetes Control and Complications Trial Research Group, 1993).

Oral hypoglycemic agents are useful in the treatment of diabetic mellitus, but their use is restricted by their pharmacokinetic properties, secondary failure and side effects. Hence, World Health Organisation (WHO) expert committee on diabetes has listed in its recommendations that traditional methods of treatment for diabetes should be further studied (Naggar, 2005).

Many Indian plants as well as herbal formulations have been studied on diabetes and its associated complications are: *Cajanus cajan* (Esposito Avella *et al.*, 1991), *Corinia indica* (Singh *et al.*, 1985), *Allium cepa* (Augusti, 1973), *Allium sativum* (Jain and

Vyas, 1975), *Aloe vera* (Ajabnoor, 1990), *Caesalpinia bonducella* (Rao *et al.*, 1994), *Ficus bengalenesis* (Achrekar *et al.*, 1991), *Nelumbo nucifera* (Mukherjee *et al.*, 1997) *Gymmema sylveatre* (Chattopadhyay, 1999), *Momordica charantia* (Ali *et al.*, 1993, Fernades *et al.*, 2007), *Ocimum sanctum* (Rai *et al.*, 1997), *Pterocarpus marsupium* (Manickam *et al.*, 1997), *Swertia chirayita* (Saxena *et al.*, 1991), *Syzigium cumini* (Coimbra *et al.*, 1992), *Tinospora cordifolia* (Stanely *et al.*, 2000), *Trigonella foenumgraecum* (Khosla *et al.*, 1995). *Picrorhiza kurroa* (Joy and Kuttan, 1999; Mandlik *et al.*, 2008), *Zingiber officinale* (Shanmugam *et al.*, 2011) (Mishra *et al.*, 2012; Mandlik *et al.*, 2008).

Many herbs and their phytoconstituents are reported to excert immunomodulatory activity, which help to improve the depressed immune functions in diabetes due to excessive consumption of proteins for energy production (Mishra *et al.*, 2012). Furthermore, such ingredients also help to ameliorate incidences of severe infections, a common feature of diabetes due to elevated blood glucose. Impaired liver function due excessive consumption of glycogen and reduced gluconeogenesis are also a major contributing biochemical event in the development of diabetic complications. Many herbs improve the functioning of liver and thereby maintain the metabolic balance of the body (Marchesini *et al.*, 2001). It is established that the oxidative stress also play a major role in the genesis of diabetes (Fatmah *et al.*, 2012). Many herbs are known to ameliorate the oxidative stress by their anti-stress property, (Bhattacharya *et al.*, 1997, Rai *et al.*, 2003, Gupta *et al.*, 2007) and thereby able to prevent the stress related complications in diabetes (Jeanette *et al.*, 2005; King and Loeken, 2004; Ceriello, 2006).

Pharmacokinetic interaction occurs when herbs able to affect/alter the process of the absorption, distribution, metabolism, protein binding or excretion of a drug that can result in altered bioavailability of drug or its metabolites. In addition, most of the pharmacokinetic drug interactions involves inhibition or stimulation of drug metabolising microsomal enzymes and drug transporters (Boullata *et al.*, 2005) results in altered blood level.

The herbomineral products, HF and HFK are widely prescribed by Ayurvedic practioners, in India and South East Asian countries including Pakistan for diabetic patients. The various constituents of HF and HFK are outlined in (Table 18.1).

Keeping in mind the possible antidiabetic/hypoglycaemic activity as well as the wide prescription of HF and HFK (herbomineral products) by ayurvedic practioners, it was interesting to study the effect perse of HF and HFK on alloxan diabetic rats as well as to establish their possible clinical efficacy experimentally as diabetic modulators or as an adjuvant along with clinically used antidiabetic drugs (MF and GC). In addition, the effect of HF and HFK on PK profile of MF and GC were also studied.

Materials and Methods

Experimental Animals

Wistar rats of either sex, weighing 140-160g, were purchased from Intox, Pune, India and housed under standard environmental conditions (temperature 24 ± 1°C, light/dark cycle: 14h/10h). Rats were fed with standard pellet diet (Amrut feed,

Pune, India) and water *ad libitum*. Animals described as fasted were deprived of food for 10h but had free access to water.

Table 18.1: Contents of various herbs in Herbomineral Formulation *Dolabi* (HF) and *Karnim* (HFK).

Contents of various herbs in Herbomineral Formulation Dolabi (HF)		
Unani Name	*Botanical Name and Parts of the Plant Used for preparation*	*Each 600 mg HF Tablet Contains (mg)*
Aqaqiya	*Acacia arabica* (Bark and gum)	166.6
Banslochen	*Bambusa arundinaceae* (Leaves and bark)	132.0
Tukhm hammaz	*Rumex vesicarius* (Leaves)	83.3
Gurmar booti	*Gymnema sylvestre* (Leaves)	27.7
Maghz jamun	*Eugenia jambolana* Lans (Seeds)	27.7
Kushta baiz murgh	Egg gallius domestics (oxide of egg shell)	13.8
Kushata khabsul hadeed	Iron (oxide of iron rust)	13.8
Kushta jast	Zinc oxide	41.6
Contents of various herbs in Herbomineral Formulation Karnim (HFK)		
Common Name	*Botanical Name and Parts of the Plant Used for preparation*	*Each 400 mg HFK Capsule Contains (mg)*
Karela	*Momordica charantia* (Fruit)	150
Neem	*Azadirachta indica* (Leaves)	75
Tulsi	*Ocimum sanctum* (Entire plant)	75
Kutki	*Picrorrhiza kurroa* (Root)	35
Sounth	*Zingiber officinale* (Rhizome)	30
Shudha guggul	*Pure commiphera mukul*	35

The institutional animal house registered with Government of India and the registration number is (SCOP/IAEC/Approval/2009-10/08). Experimental protocols were reviewed and approved by the institutional animal ethics committee (IAEC) and conforms to Indian National Science Academy Guidelines for the use and care of experimental animal's research.

Drugs and Chemicals

Alloxan monohydrate was purchased from Loba chemie, Mumbai, India. Bulk technical MF and GC powder were obtained as gift samples from Aventis Pharma India, and Franco-India Pharmaceuticals Ltd, respectively. Acetonitrile, phosphate buffer, methanol, sodium hydroxide, glacial acetic acid and HPLC grade water were purchased from Merck chemicals Pvt Ltd, India. HF (tablets) and HFK (capsules) were purchased from local chemist shop. Various biochemical assays were performed by using readymade reagents kits supplied by Trans Asia Biomedicals Ltd. India.

Preparation of MF, GC and Herbal Products HF and HFK

The tablets of HF herbal products were crushed into fine powder and content of HFK capsule were removed and prepared a homogenous suspension in normal saline using mortar and pestle prior to use. The antidiabetic drugs, MF and GC were also prepared as suspension in 1 per cent (w/v) carboxy methyl cellulose (CMC).

Induction of Diabetes in Rats

Diabetes was induced in rats by intraperitonial administration (i.p.) of alloxan monohydrate (150 mg/kg). Four days later, blood samples were drawn and blood glucose was assayed to confirm the establishment of diabetes. The rats showing blood glucose > 250 mg per cent was selected for antidiabetic activity studies.

Pharmacodynamic Studies

The alloxan diabetes rats were randomly divided into 10 groups (6/group) and received the different treatments daily for 28 days mentioned below. Blood samples were withdrawn on different days viz 0, 3rd, 7th, 14th, 21st and 28th (24 hrs after previous dose) from retro orbital vein under light ether anesthesia for glucose determination. On 29th day, overnight fasted rats were anesthetised with diethyl ether and blood was collected directly by cardiac puncture for various biochemical studies. The pancreas were also separated, washed with cold saline and preserved at -20ºC until use.

Group I- Normal control

Group II- Diabetic control (CMC 1 per cent 10 ml/kg, B/W, Orally)

Group III- HF (140 mg/kg, B/W, Orally)

Group IV- HFK (200 mg/kg, B/W, Orally)

Group V- MF (200 mg/kg, B/W, Orally)

Group VI- GC (4 mg/kg, B/W, Orally)

Group VII- MF (200 mg/kg, B/W, Orally) + HF (140 mg/kg, B/W, Orally)

Group VIII- MF (200 mg/kg, B/W, Orally) + HFK (200 mg/kg, B/W, Orally)

Group IX- GC (4 mg/kg, B/W, Orally) + HF (140 mg/kg, B/W, Orally)

Group X- GC (4 mg/kg, B/W, Orally) + HFK (200 mg/kg, B/W, Orally)

Biochemical Studies

Blood glucose was determined by glucose-oxidase/peroxidase method (Trinder, 1969) using visible spectrophotometer. Glycosylated haemoglobin (GHb) was determined in heparinised whole blood by the method of Godkar (1994) and serum lipids were assayed by the methods described by Mc Grown *et al.* (1983) using readymade reagents kits.

Measurement of Body Weight

The body weights of diabetic rats were recorded on 0, 7, 14, 21 and 28th day during the treatment period by using digital balance.

Histopathological Studies of Pancreas

The pancreas were isolated, immediately after sacrifice, washed with cold normal saline, and preserved in 10 per cent v/v formalin. After dehydration the pieces of pancreas were embedded in paraffin wax, cut into 4-6 μm thick sections and stained using haematoxylin and eosin. The sections were observed under light microscope for the alterations in histoarchitecture of pancreas.

Pharmacokinetic Studies

The alloxan diabetic rats were randomly divided into 7 groups (6/group) and received the respective treatments daily for 28 days mentioned below.

Group I MF (200 mg/kg, B/W, Orally)

Group II GC (4 mg/kg, B/W, Orally)

Group III- MF (200 mg/kg, B/W, Orally) + HF (140 mg/kg, B/W, Orally)

Group IV- MF (200 mg/kg, B/W, Orally) + HFK (200 mg/kg, B/W, Orally)

Group V- GC (4 mg/kg, B/W, Orally) + HF (140 mg/kg, B/W, Orally)

Group VI- GC (4 mg/kg, B/W, Orally) + HFK (200 mg/kg, B/W, Orally)

Group VII- Diabetic rats received CMC 1 per cent (200 mg/kg, B/W, Orally)

Extraction of MF and GC from Plasma and HPLC Assay

After the last dose on 28th day, rats were sacrificed under light ether anaesthesia. Blood samples were collected directly from heart at 0.5, 1.0, 2.0, 4.0, 8.0 and 24.0 hrs time intervals in heparinised eppendrof tubes. The blood samples were centrifuged at 3000 x g (Remi RM-12C India) to separate plasma. 1.0 ml of acetonitrile was added to 0.5 ml plasma and vortexed for 30 sec. The mixture was centrifuged at 5000 x g for 30 min. The upper layer 100 μl was separated and filtered through nylon membrane filter (0.22 μ, 13 mm), 20 μl of filtrate was used for the assay of MF and GC by HPLC (JASCO, Japan) following the method described by Wanjari *et al.* (2008) and Rajendran *et al.* (2007) respectively.

HPLC conditions for MF and GC assay

Parameters	*Metformin Assay*	*Glibenclamide Assay*
Stationary Phase C-18 Column	Nucleosil-Si-C18 (250 X 4.6 mm) C-18 Column	Nucleosil-Si-C18 (250 X 4.6 mm) C-18 Column
Mobile Phase	Acetonitrile: Ammonium acetate buffer (pH 4.2-4.6) [90: 10 v/v]	Acetonitrile: Phosphate buffer (pH 3.5) [60: 40 v/v]
Sample	MF 10 mg/ml [10 μL]	GC 10 mg/ml [10 μL]
Retention time	2.8 min	7.8 min
Flow rate	1 ml/min	0.6 ml/min
Detection	232 and 236 nm	253 nm
Time cycle	15 min	15 min

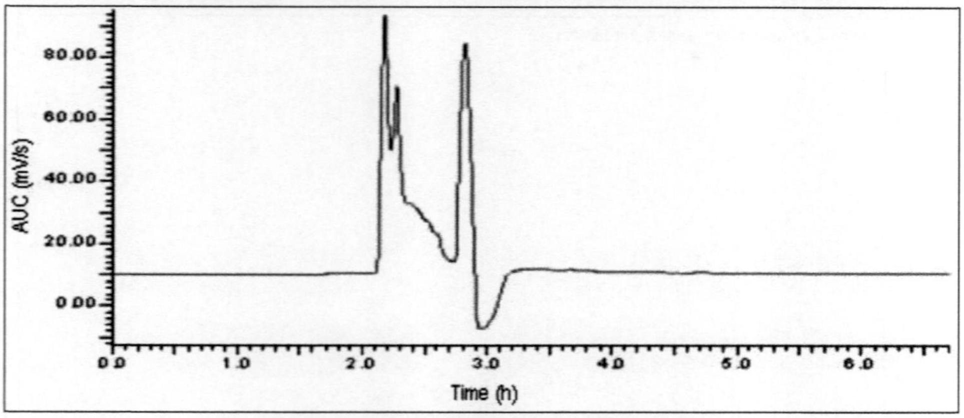

(a) Rat plasma spiked with metformin

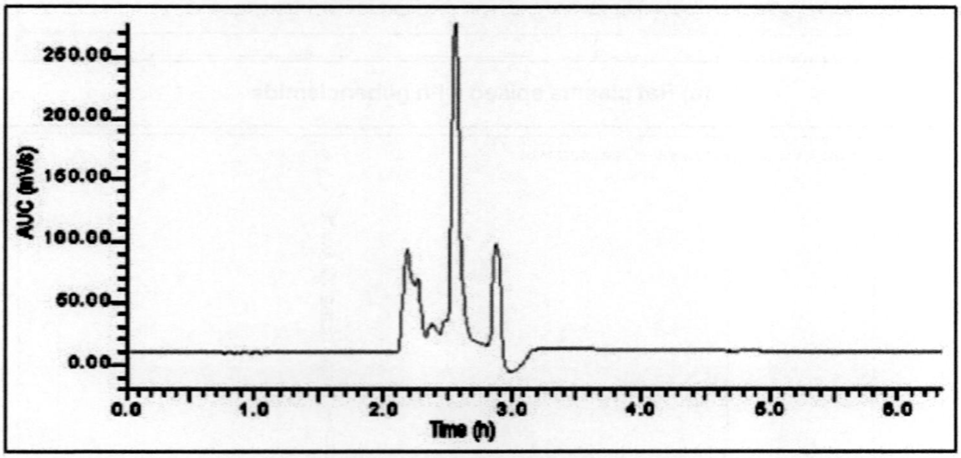

(b) Rat plasma metformin

Figure 18.1a and b: Comparative HPLC chromatogram of rat plasma samples of metformin.

Caliberation curve was established by spiking rat plasma with known amount of MF and GC 5, 10, 15, 20, 25 and 50µg/ml. The caliberation curve for MF and GC was generated by plotting the peak area versus concentration and used for determining analytes. The determination of MF and GC from plasma was validated for its selectivity, limit of detection and limit of quantitation, linearity, precision, recovery and stability in plasma.

Pharmacokinetic Studies and Data Analysis

The plasma samples obtained at various time intervals from rat were analysed by HPLC method. Concentrations of plasma drugs (MF and GC) were calculated from the standard curve. The plasma concentration profile obtained was fed into S-inverse (S-INV), a computer software on BASIC® (Version 1.12) to determine

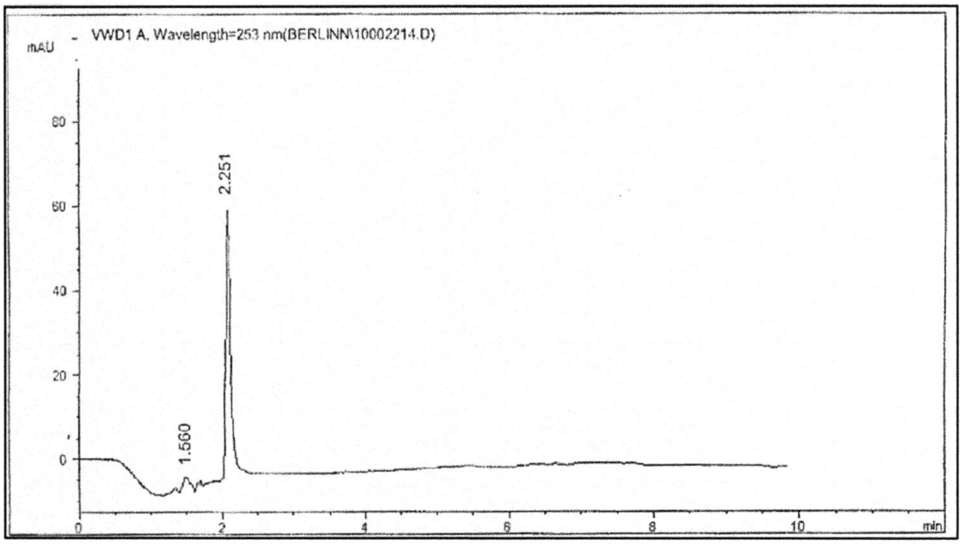

(a) Rat plasma spiked with glibenclamide

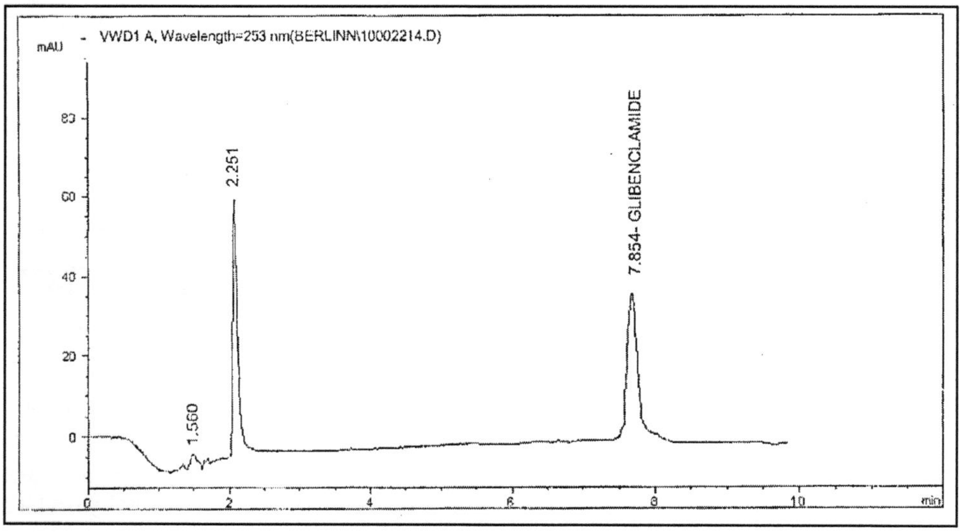

(b) Rat plasma glibenclamide

Figures 18.2a and b: Comparative HPLC chromatogram of rat plasma samples of glibenclamide.

pharmacokinetic parameters which include: concentration maximum (C_{max}) and the corresponding peak Time (T_{max}), elimination half-life ($t_{1/2}$), clearance rate (K_e), kinetic constant (K_a) and volume of distribution (V_d) (Boltan, 1997; CDER, 2001).

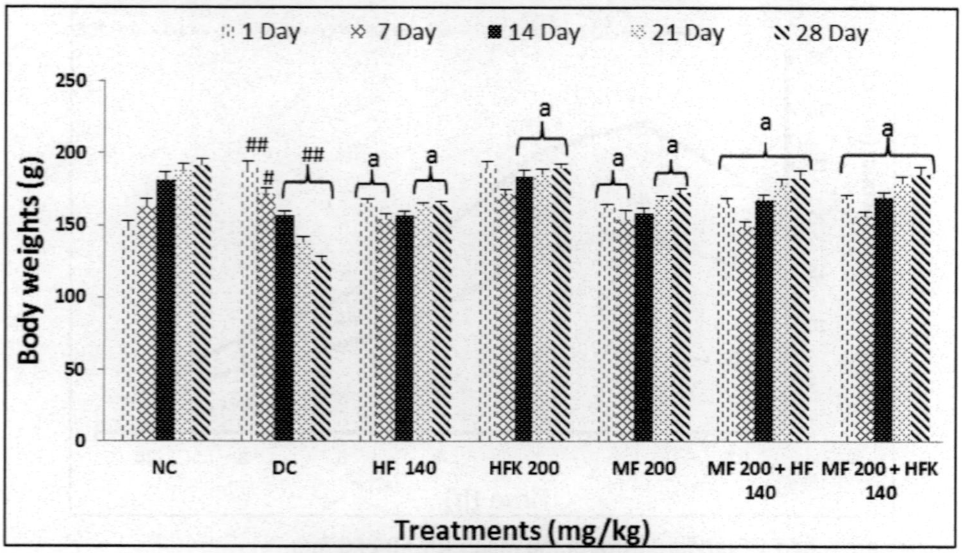

(a) Metformin with HF and HFK

Values are mean ± Standard Error of the Mean (SEM) of 6 rats.

P<0.01; compared with normal control; a = ** P<0.01 compared with diabetic control.

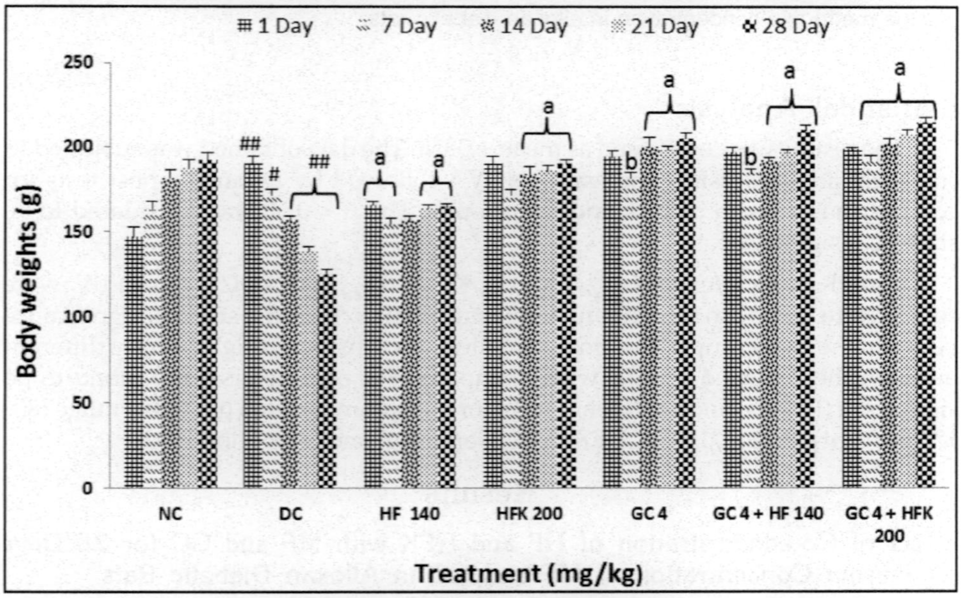

(b) Glibenclamide with HF and HFK

Values are mean ± Standard Error of the Mean (SEM) of 6 rats.

P<0.01; compared with normal control; a = ** P<0.01 and b= * P<0.05; compared with diabetic control.

Figure 18.3: Effect of 28 days perse treatment of HF, HFK, MF, GC and their combination on body weight in alloxan diabetic rats (Figures a and b).

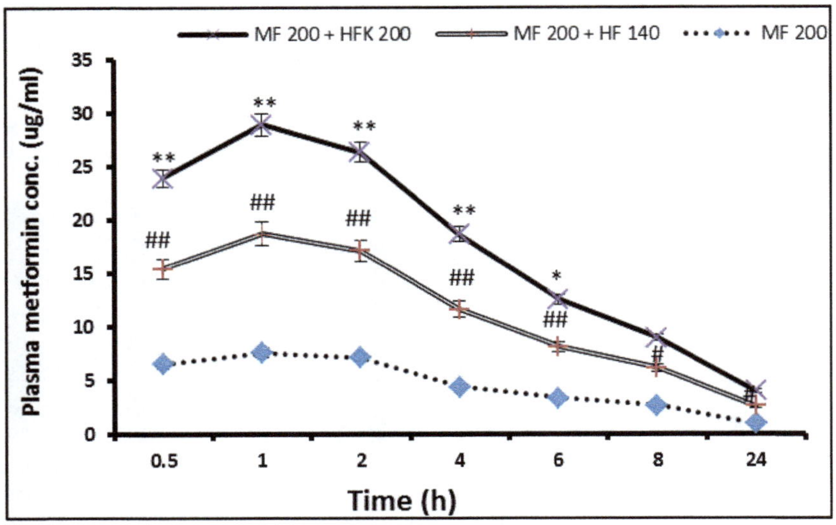

Values are mean ± Standard Error of the Mean (SEM) of 6 rats.

##P<0.01; #P<0.05; **P<0.01 and *P<0.05 compared with metformin 200 mg/kg

Figure 18.4: Effect of co-administration of metformin with HF and HFK (for 28 days) on plasma metformin concentration in alloxan diabetic rats.

Statistical Analysis

The results were expressed as mean ± SEM. The data obtained was subjected to statistical analysis using one way ANOVA followed by Dunnett's post tests for comparison between control and test groups. The $P < 0.05$ was considered to be statistical significant.

The PK parameters of C_{max}, T_{max}, $t_{1/2}$, K_e, AUC $_{0\text{-}1}$ and AUC $_{0\text{-}¥}$, K_a and V_d were subjected to Wilcoxon signed rank test using SPSS to check the stastical significance between the two groups. The bioavailability was considered significantly different between the 2 treated groups when compared PK parameters were found to be significant (P<0.05) and confidence level for the parameters fall out of the range (80-125 per cent) (Bolton, 1997; CDER, 2001; Derendorf and Meibohm, 1999).

Results

Effect of Co-administration of HF and HFK with MF and GC for 28 Days on Plasma Concentration of MF and GC in Alloxan Diabetic Rats

The plasma concentration-time profile of MF (200 mg/kg) alone, MF (200 mg/kg) + HF (140 mg/kg) and MF (200 mg/kg) + HFK (200 mg/kg) in alloxan diabetic rats is shown in Figure 18.4. Similarly, plasma concentration of GC (4 mg/kg) alone, GC (4 mg/kg) + HF (140 mg/kg) and GC (4 mg/kg) + HFK (200 mg/kg) is shown in Figure 18.5.

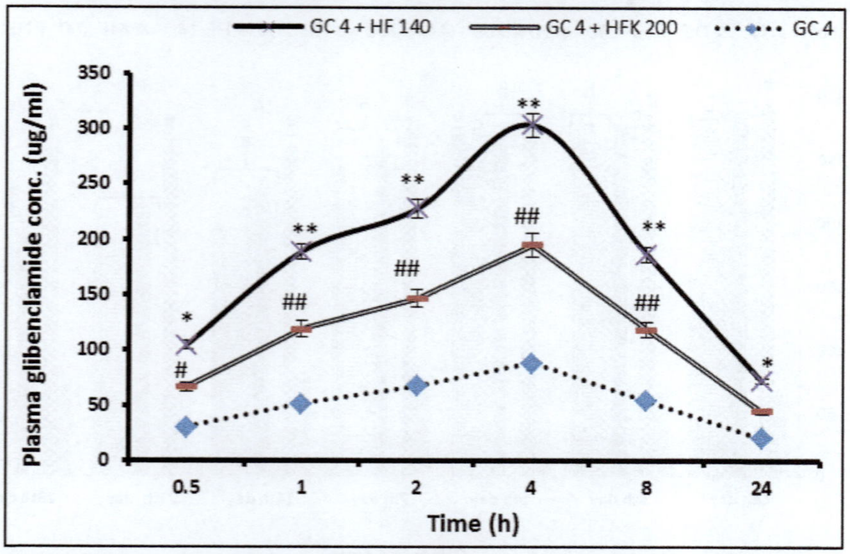

Values are mean ± Standard Error of the Mean (SEM) of 6 rats.
##P<0.01; #P<0.05; **P<0.01 and *P<0.05 compared with glibenclamide 4 mg/kg

Figure 18.5: Effect of co-administration of glibenclamide with HF and HFK (for 28 days) on plasma glibenclamide concentration in alloxan diabetic rats.

It is observed that co-administration of HF and HFK elevated plasma concentration of MF and GC significantly at various time intervals studied. The maximum concentration was observed at 2h and 4h for MF and GC respectively. The blood levels of these drugs remained significantly high till 8h (Figures 18.4 and 18.5).

Effect of 28 Days Treatment of MF and GC alone and their Combination with HF and HFK on Pharmacokinetic Profile of these Drugs

Co-administration of HF and HFK with MF or GC improved significantly the PK parameters *viz.* AUC, V_d, C_{max}, T_{max}, $t_{1/2}$, K_e and K_a as compared to MF or GC alone treated group (Table 18.4)

Effect of Co-administration of Herbal Products (HF and HFK) on MF and GC Plasma Glucose in Alloxan Diabetic Rats

Both MF and GC perse pretreatment for 28 days reduced the plasma glucose of alloxan induced diabetic rats. Co–administration of herbal products (HF and HFK) with MF or GC showed a greater reduction in plasma glucose as compared to their perse treatment *i.e.* from 3rd day onwards and upto 29 days. It was observed that reduction of plasma glucose was greater in HF + MF as well as HF + GC (Figures 18.6a and b).

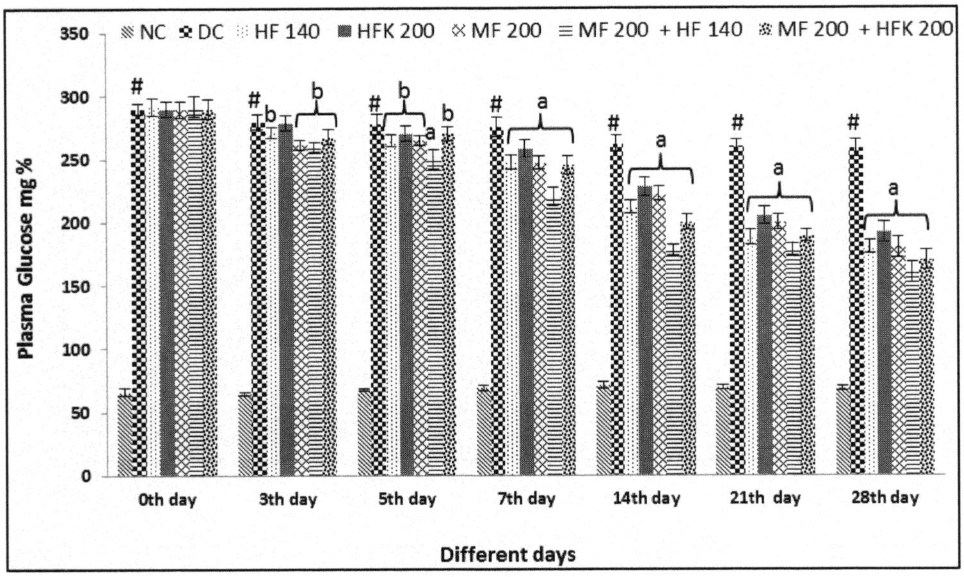

(a) Metformin with HF and HFK

Values are mean ± Standard Error of the Mean (SEM) of 6 rats.

# P<0.01; compared with normal control; a P<0.01 and b P<0.05 compared with diabetic control.

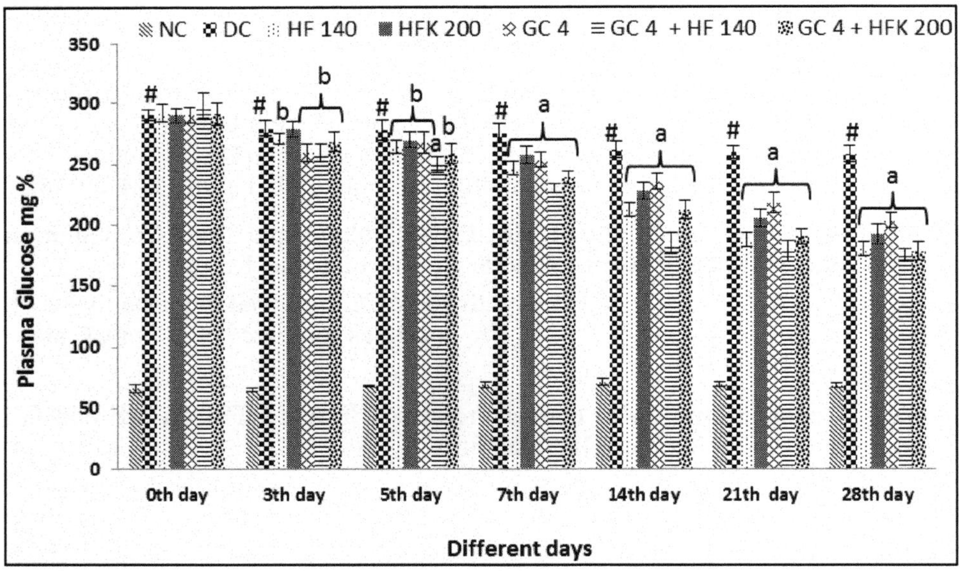

(b) Glibenclamide with HF and HFK

Values are mean ± Standard Error of the Mean (SEM) of 6 rats.

# P<0.01; compared with normal control; a P<0.01 and b P<0.05 compared with diabetic control.

Figure 18.6: Effect of 28 days treatment of metformin, glibenclamide alone and their combination with HF and HFK on plasma glucose in alloxan diabetic rats (a and b).

Table 18.2: Effect of 28 days perse treatment of HF, HFK, metformin (MF), glibenclamide (GC) and their combination on glycosylated haemoglobin (GH), cholesterol (CH) and triglyceride (TG) in alloxan diabetic rats.

Treatment (Dose mg/kg)	*Liver Glycogen (mg/g)*	*Glycosylated Haemoglobin (per cent)*	*Cholesterol (mg per cent)*	*Triglyceride (mg per cent)*
Normal control (NC)	2.316 ± 0.15	4.00 ± 0.22	68.0 ± 3.0	98.4 ± 14.0
Diabetic control (DC)	0.6655 ± 0.07 ##	9.00 ± 0.25 ##	112.2 ± 4.1 ##	168 ± 16.12 ##
HF (140)	1.17 ± 0.06*	6.9 ± 0.3	94.0 ± 2.7**	148 ± 18.4**
HFK (200)	1.145 ± 0.10*	7.1 ± 0.3	93 ± 2.8**	152 ± 16.0**
MF (200)	1.28 ± 0.05**	6.34 ± 0.3**	89.8 ± 4.0**	130 ± 12.0**
MF (200) + HF (140)	1.87 ± 0.05**	5.5 ± 0.2**	74.2 ± 4.1**	110 ± 11.8**
MF (200) + HFK (200)	2.15 ± 0.04**	5.8 ± 0.3**	80.2 ± 5.0**	126 ± 10.8**
GC (4)	1.371 ± 0.09**	6.2 ± 0.1**	84.8 ± 6.3**	136 ± 9.8**
GC (4) + HF (140)	1.620 ± 0.30**	5.2 ± 0.3**	60.0 ± 5.8**	124 ± 8.4**
GC (4) + HFK (200)	1.827 ± 0.10**	5.8 ± 0.2**	74.0 ± 6.3**	131 ± 6.8**

Values are mean ± Standard Error of the Mean (SEM) of 6 rats.

P<0.01; compared with normal control.

a = ** P<0.01 and b= * P<0.05; compared with diabetic control.

Table 18.3: Effect of co-administration of metformin, glibenclamide with HF and HFK for 28 days on plasma concentration of metformin and glibenclamide in alloxan diabetic rats.

Treatments (mg/kg) for 28 Days	*Plasma Concentration (µg/ml) of Metformin at Different Time Intervals (h)*					
	0.5	*1.0*	*2.0*	*4.0*	*8.0*	*24.0*
MF (200)	6.5 ± 0.4	7.6 ± 0.5	7.1 ± 0.3	4.4 ± 0.2	3.3 ± 0.06	2.6 ± 0.2
MF (200)+ HF (140)	8.9 ± 0.5 **	11.1 ± 0.8**	10.0 ± 0.6**	7.2 ± 0.01**	4.8 ± 0.1**	3.5 ± 0.1##
MF (200) + HFK (200)	8.5 ± 0.4 **	10.2 ± 0.6**	9.2 ± 0.5**	7.0 ± 0.2**	4.4 ± 0.1*	2.8 ± 0.2
	Plasma Concentration (µg/ml) of Glibenclamide at Different Time Intervals (h)					
GC (4)	30.0 ± 1.0	51.4 ± 2.0	68.0 ± 1.9	89.0 ± 2.1	54.5 ± 1.0	20.4 ± 1.0
GC (4)+ HFK (200)	36.8 ± 0.6##	68.2 ± 1.2**	79.0 ± 1.0*	106.5 ± 2.4**	64.6 ± 2.0*	25.0 ± 1.0
GC (4) + HF (140)	38.0 ± 0.5##	69.6 ± 1.0**	81 ± 1.2**	108.4 ± 1.9**	68.0 ± 2.2*	27.4 ± 1.8##

Values are mean ± Standard Error of the Mean (SEM) of 6 rats.

P<0.01; compared with normal control.

a = ** P<0.01 and b= * P<0.05; compared with diabetic control.

Table 18.4: Effect of 28 days treatment of HF, HFK, metformin, glibenclamide and their combination on pharmacokinetic parameters.

Parameters	*Treatment (dose) mg/kg for 28 Days*					
	MF 200	*MF (200) + HF (140)*	*MF (200) + HFK (200)*	*GC 4*	*GC (4) + HF (140)*	*GC (4) + HFK (200)*
Cmax	7.60 ± 0.58	11.10 ± 1.38**	10.2 ± 1.23**	89.00 ± 1.95	106.5 ± 0.62**	108.40 ± 0.614**
Tmax	1.00 ± 0.58	1.00 ± 0.58	1.00 ± 0.46	4.00 ± 0.00	4.00 ± 0.00	4.00 ± 0.00
$t_{1/2}$	8.51 ± 0.01	12.34 ± 1.77**	13.153 ± 0.84**	5.885 ± 0.002	6.34 ± 0.002*	6.6 ± 0.173*
Ke	0.08 ± 0.01	0.09 ± 0.01	0.093 ± 0.02	0.120 ± 0.01	0.105 ± 0.00	0.105 ± 0.00
Ka	0.08 ± 0.02	0.11 ± 0.01*	0.035 ± 0.001	0.052 ± 0.004	0.044 ± 0.001	0.045 ± 0.001
AUC	62.37 ± 5.48	84.12 ± 2.90**	164.10 ± 4.8**	1363 ± 16.25	1610 ± 8.8**	1613 ± 8.76**
V_d	0.042 ± 0.002	0.094 ± 0.01**	0.074 ± 0.02**	0.029 ± 0.002	0.034 ± 0.002*	0.026 ± 0.002

Values are mean ± Standard Error of the Mean (SEM) of 6 rats.

a = ** P<0.01 and b= * P<0.05; compared with diabetic control.

Effect of 28 Days alone Treatment of HF, HFK, MF, GC and their Combinations on Glycosylated Haemoglobin, Liver Glycogen, Plasma Cholesterol (CH) and Triglycerides (TG) in Alloxan Diabetic Rats

MF and GC alone treatment significantly reduced the elevated glycosylated haemoglobin of alloxan diabetic rats. The herbal products, HF and HFK alone treatment also reduced significantly the elevated glycosylated haemoglobin of alloxan diabetic rats. Co-administration of the herbal products, HF and HFK with MF or GC showed greater reduction in glycosylated haemoglobin of alloxan diabetic rats (Table 18.2).

The treatment with MF or GC alone reduced significantly the plasma CH and TG in alloxan diabetic rats. Co-administration of herbal products HF and HFK with MF or GC showed greater reduction in the plasma CH and TG of alloxan diabetic rats. HF and HFK alone treatment also reduced the plasma CH and TG of alloxan diabetic rats, but the reduction was comparatively less than that of MF or GC treated rats. Hypocholesterolemia effect of HF + GC > HFK + GC and HF + MF > HFK + MF treated rats (Table 18.2).

Effect on Histoarchitecture of Pancreas of Alloxan Diabetic Rats Treated with MF, GC, HF, HFK and their Combinations

The normal rats showed normal cellular population in the pancreatic β-islets of Langerhans in pancreatic histology (Figures 18.7A and 18.8A). In alloxan diabetic rats, the β-islets of Langerhans showed diffused necrotic changes, decreased cellular population, extensive damage, reduction in size and absence of hyperplasia (Figures 18.7B and 18.8B). The alloxan diabetic rats treated with HF, MF, HFK and GC alone showed mild to moderate degree of diffused necrotic changes, mild reduction in the size and number of the β-islets of the pancreas (Figures 18.7C, 18.8C, 18.7D and 18.8D). The combined treatment of MF + HF, MF + HFK, GC + HF and GC + HFK restored to a great extent, the numbers and the cellular size of β-islets towards normal with hyperplasia (Figures 18.7E, 18.8E, 18.7F and 18.8F).

Discussion

The integrated pharmacodynamic/pharmacokinetic (PK/PD) concepts provide a highly promising tool and can also help to extrapolate from the preclinical data, simulation of drug responses as well as clinical effectivity. Thus, PK/PD concepts are playing a crucial role in the drug develpemnet process for the pharmaceutical industry (Derendorf and Melbohm, 1999). Based on the present PK/PD findings of MF, GC and their combination with herbomineral products (HF and HFK), it was possible to derive a good correlation between PD effect as well as PK profiles.

In PK study, co-administration of MF or GC with HF and HFK increased significantly bioavailability of these drugs. The increased bioavailability was reflected fully in all the PK paradigms *viz.* AUC, C_{max}, $t_{1/2}$, V_d, K_e and K_a studied. The process of absorption of drugs is generally faster in the small intestine than stomach region (Rawland and Tozer, 1996). The rapid and complete absorption of drugs from intestine is related to surface area, permeability and blood flow. Therefore, one of the reasons for increased bioavailability may be due to HF and HFK effects on various

Figure 18.7: Histoarchitecture of Pancreas with Metformin and its combination with HF and HFK in alloxan induced diabetic rats.

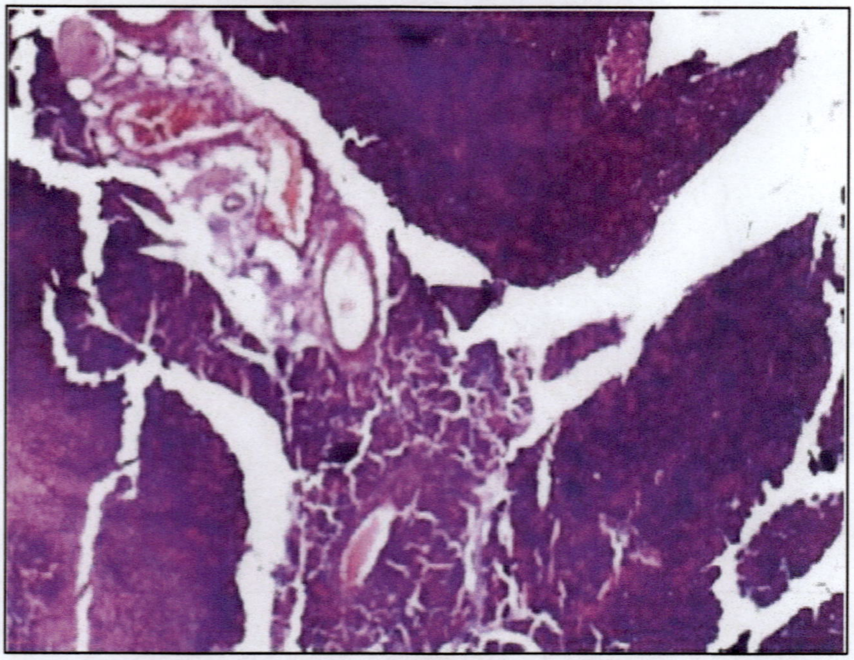

(A) NC

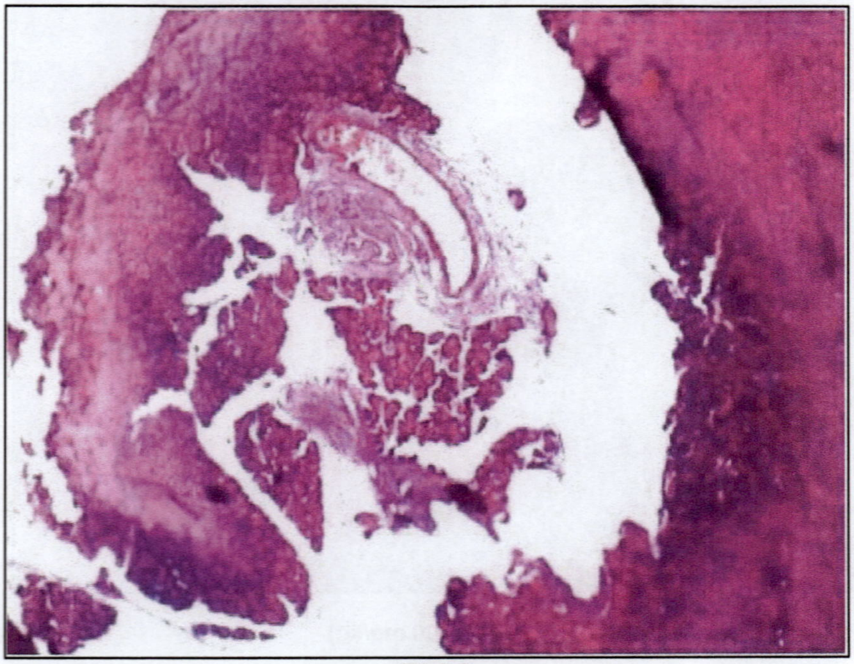

(B) DC

Contd...

Figure 18.7–*Contd...*

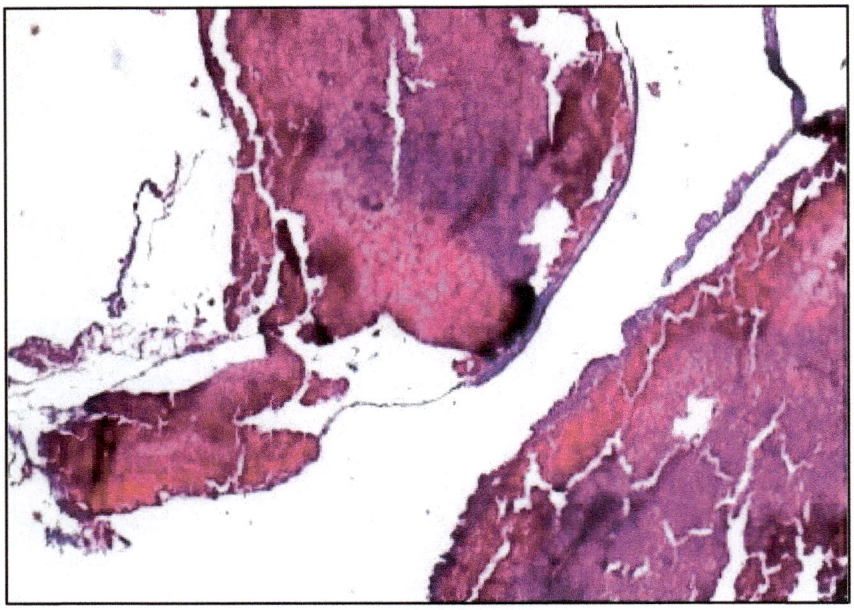

(C) HF (140 mg/kg)

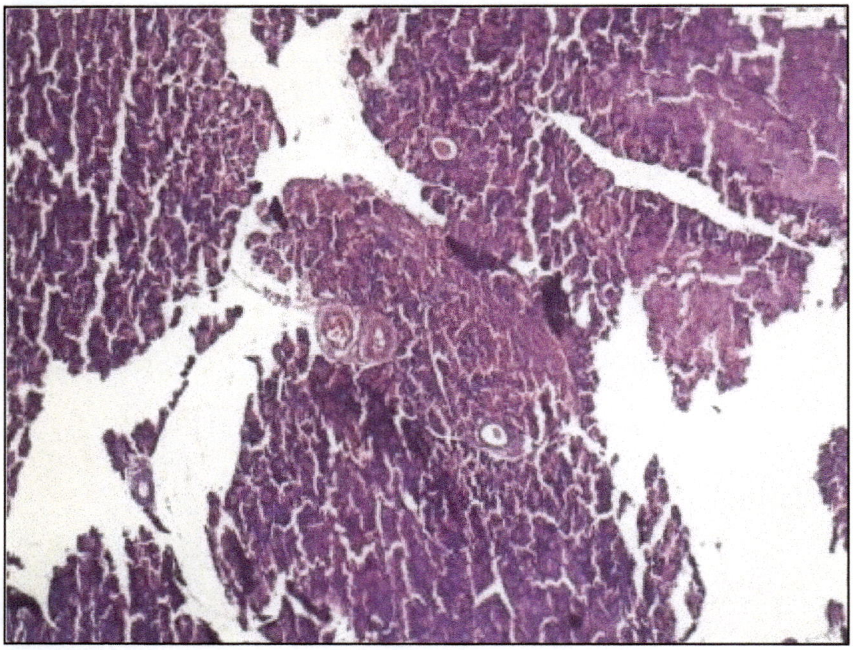

(D) MF (200 mg/kg)

Contd...

Figure 18.7–*Contd...*

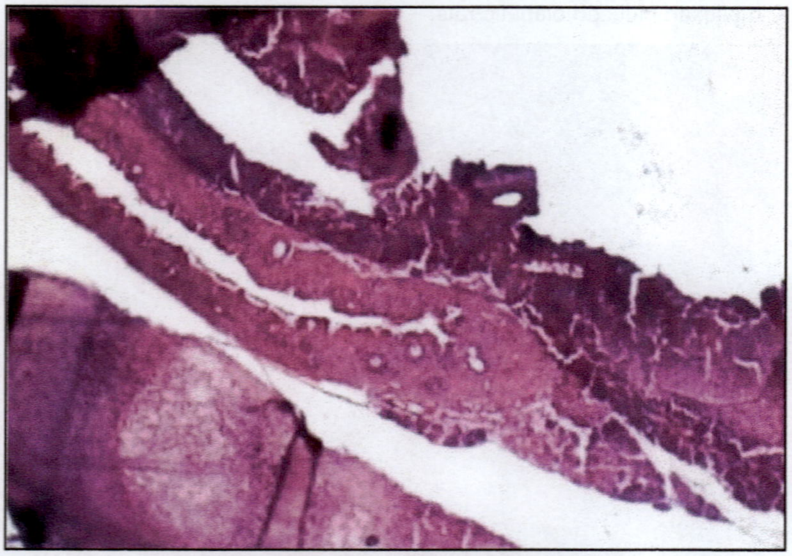

(E) MF (200 mg/kg) + HF (140 mg/kg)

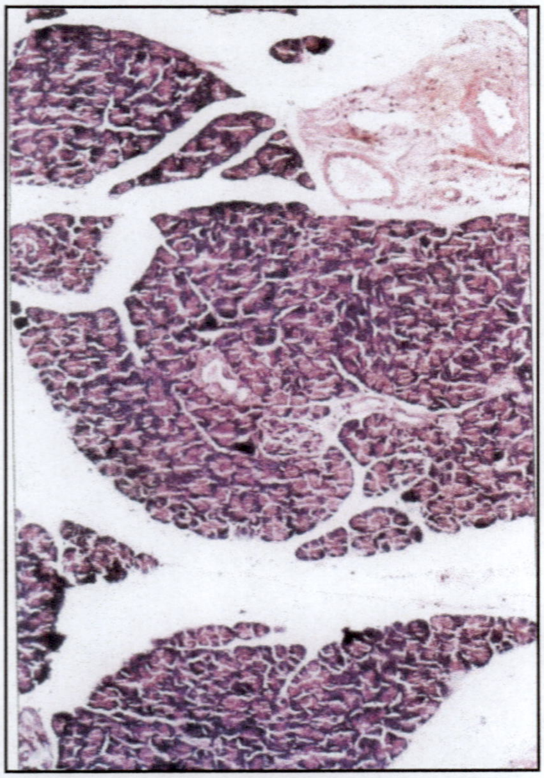

(F) MF (200 mg/kg) + HFK (200 mg/kg) at 400 X

Figure 18.8: Histoarchitecture of Pancreas with Glibenclamide and its combination with HF and HFK in alloxan induced diabetic rats.

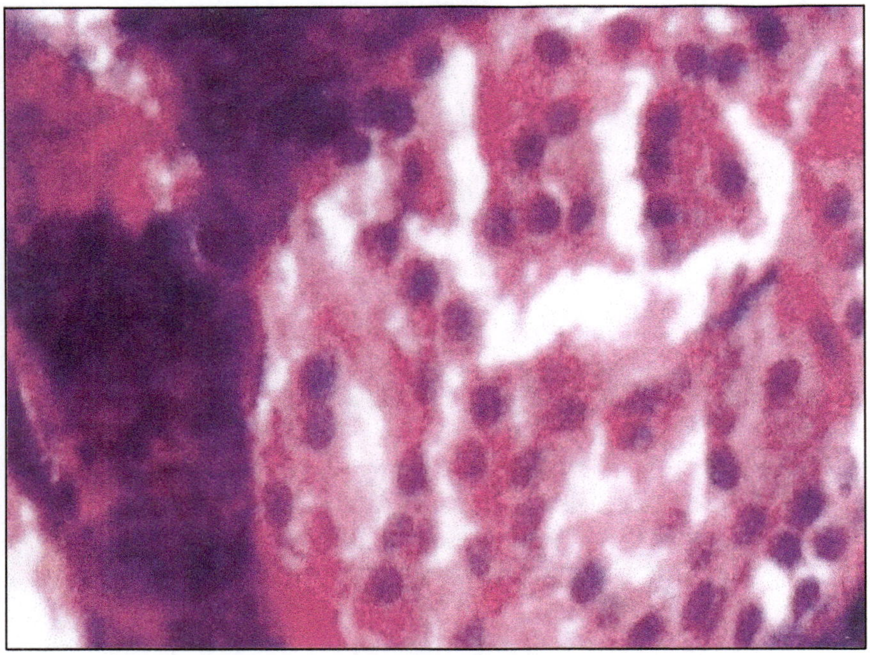

(A) NC

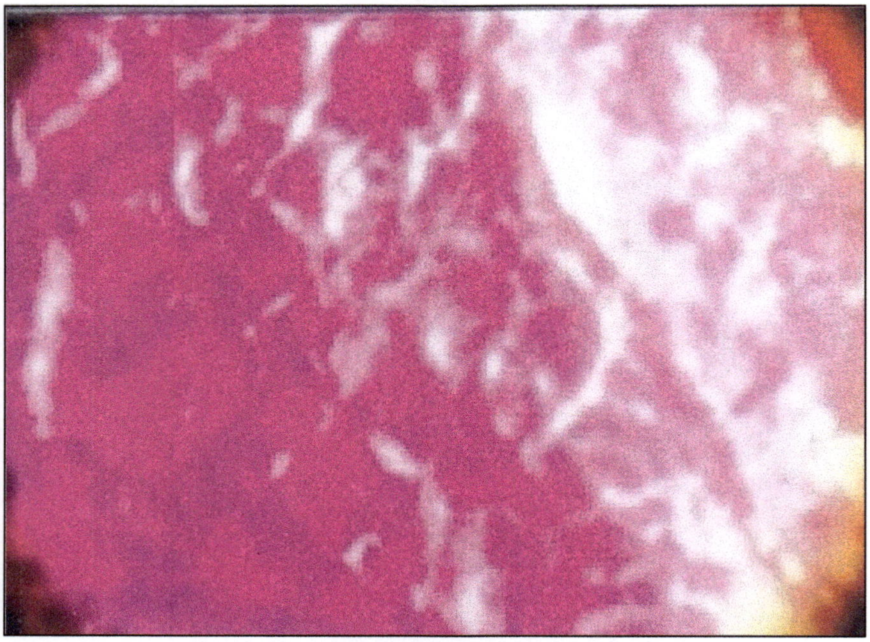

(B) DC

Contd...

Figure 18.8–*Contd...*

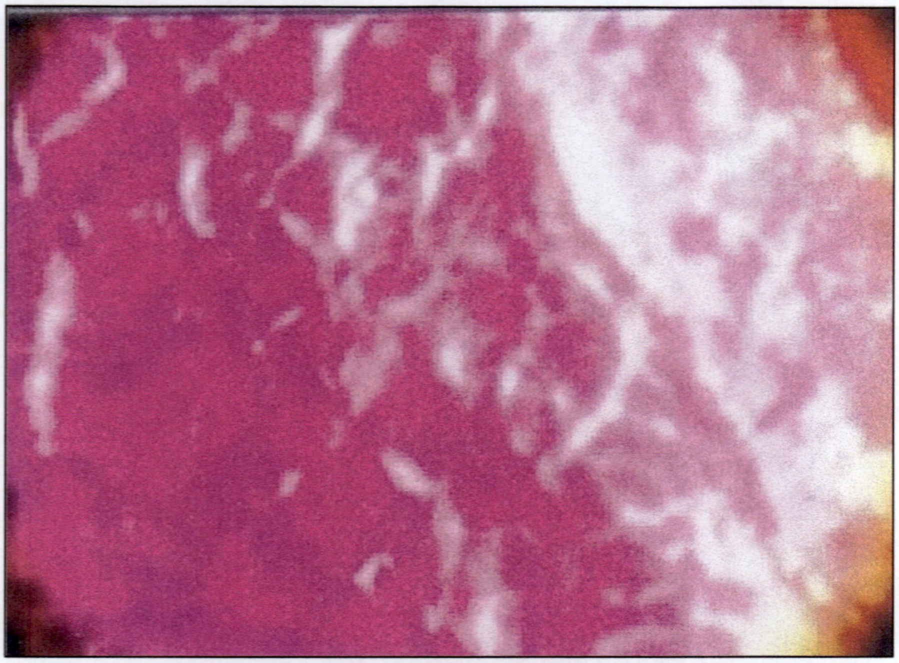

(C) HFK (200 mg/kg)

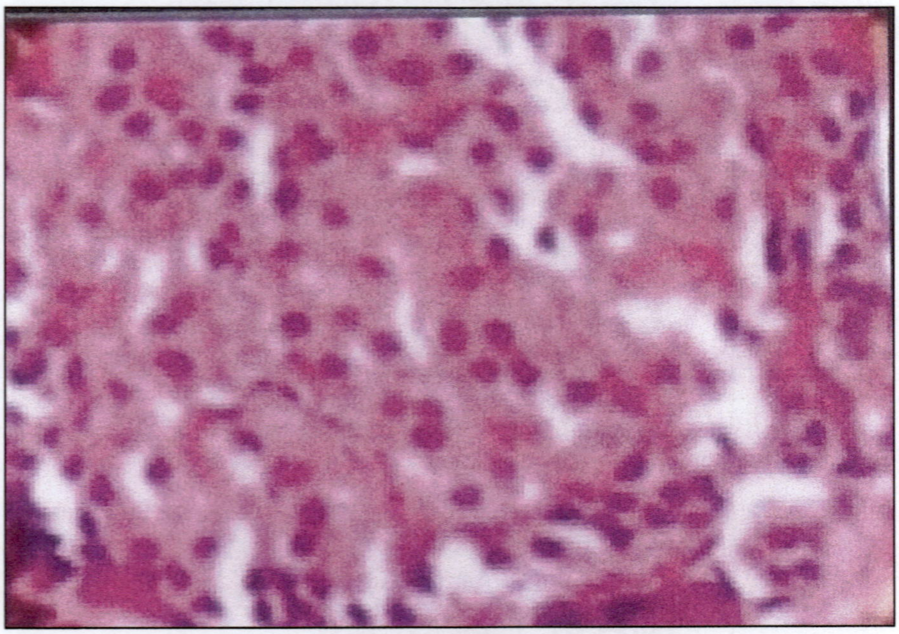

(D) GC (4mg/kg)

Contd...

Figure 18.8–*Contd...*

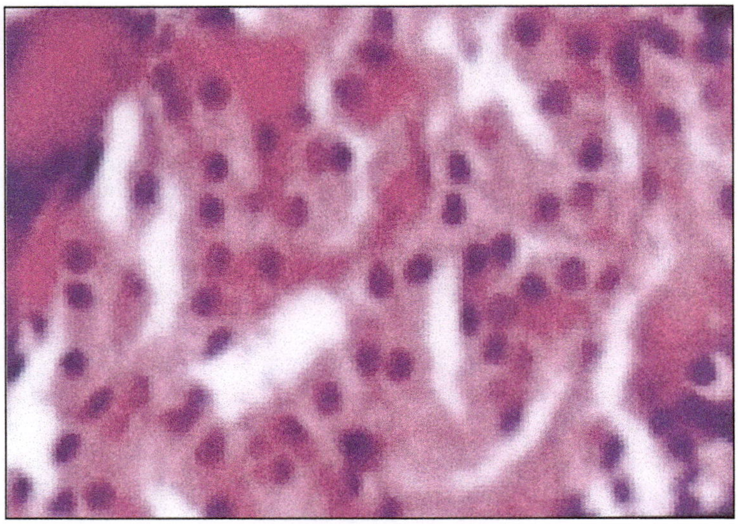

(E) GC (4 mg/kg) + HFK (200 mg/kg)

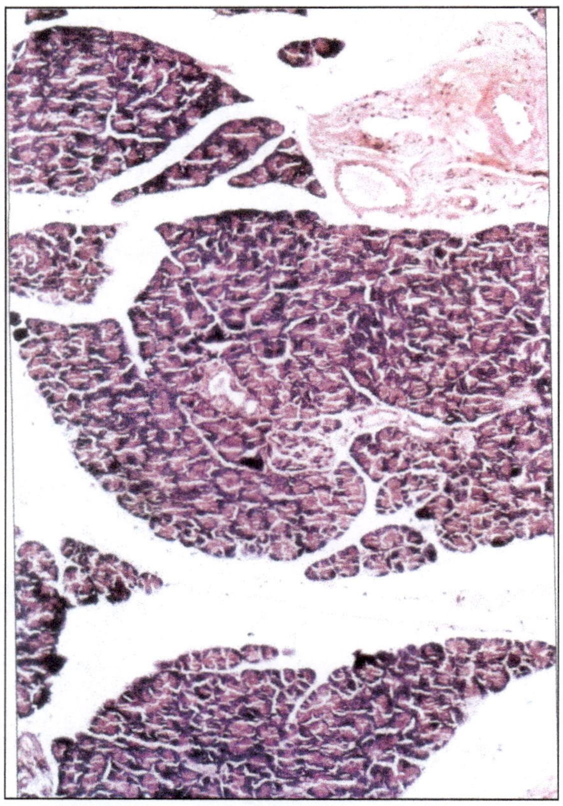

(F) GC (4 mg/kg) + HF (140 mg/kg) at 400 X

physicochemical factors like, lipid solubility, non-ionised drug molecules permeability and drug diffusion across the biological membrane etc. in the intestinal milieu (Baggot, 1978).

The other major contributing factor is the metabolism of drugs, which is largely influenced by microsomal metabolising enzymes. The documented reports indicate that GC is metabolised in the liver by the microsomal enzymes *viz.* CYP_{3A4}, CYP_{2C9} and CYP_{2C29} (Van Giersbergar *et al.*, 2002; Naritomi *et al.*, 2004; Dixit *et al.*, 2008). However, it was subsequently confirmed that CYP_{3A4} plays a major role (96 per cent) in the metabolism of GC (Naritomi *et al.*, 2004). Similarly, the involvement of microsomal enzymes CYP2C11, CYP201 and CYP3A has been implicated in MF metabolism (Choi *et al.*, 2007). Therefore, it is likely, that the presence of phytoconstituents of HF and HFK may be inhibiting or modifying these microsomal metabolising enzymes. Furthermore, the presence of flavonoids, alkaloids, saponins and minerals may influence directly or indirectly physicochemical process like membrane permeability, diffusion across biological membranes, causing an improvement in the absorption of drugs in the intestinal milieu. The various ingredients of HFK, especially *Picrorhiza kurroa, Zingiber officinale* and *Ocimum sanctum* are anti-hyperglycaemic in nature, hence may participate in reducing the elevated blood glucose via stimulation of insulin secretion from pancreatic β-islets in diabetic condition (Vats *et al.*, 2004). Furthermore, herbs *Picrorhiza kurroa, Zingiber officinale* and *Ocimum sanctum* are reported to enhance the glucose uptake process by muscles (Bangar *et al.*, 2009). The other plausible explanations one can put forward for enhanced hypoglycemic activity of MF or GC by herbomineral products are: a) stimulation of insulin like substances or inhibiting insulinase activity (Chang *et al.*, 1980), b) stimulation of β-islets of pancreas to secrete increased insulin (Chakravarty *et al.*, 1980) and c) facilitating regeneration process of pancreatic β-islets (Bopanna *et al.*, 1997). Other indirect beneficial effects that can be derived from phytoconstituents like *Azadirachta indica* and *Picrorhiza kurroa* is the activation of diminished immune system during diabetes condition by reducing the consumption of proteins for energy derived physiological process, prevention of the incidences of severe infection, an effective control of hyperglycemia and restoration of liver dysfunctioning by preventing the excessive break down of glycogen and improving gluconeogenesis. Precipitation of hyperglycaemia due to oxidative stress is a common occurring salient feature in diabetes mellitus. Such oxidative stress can be prevented and/or ameliorated to maintain metabolic balance of the body by the presence of the herb, *Ocimum sanctum*, a well-documented anti-stress agent (Vats *et al.*, 2004). Similarly, *Acacia arabica, Bambus arundinaceae, Gymnema sylvestre* and *Eujenia jambolana* can also show to establish wide spectrum of hypoglycemic, hypolipidemic and antioxidant activities in animal experiments (Baig *et al.*, 2012). Similarly, *Acacia arabica, Bambus arundinaceae, Gymnema sylvestre* and *Eujenia jambolana* have direct or indirect effect on carbohydrate metabolism like inhibition of polyl pathways which triggers severe metabolic changes leading to activation of protein kinase causing the release of insulin from remaining β-islets of pancrea (Derubertis and Craven, 1994). *Eugenia jambolana* elicits hypoglycemic activity presumably through the stimulation of surviving β-islets of Langerhans to release more insulin. In addition, it elevates glucose-6-phosphate content of liver indicating overall increase in glucose influx which augments the utilisation of glucose and increasing insulin secretion through the increased activity of cathepsin (Sagrawat, 2006). *Gymnemea sylvestre* contains gymneic acid, which is reported to inhibit the hyperglycemic response to adrenaline and growth factor (Anturkar *et al.*, 1995). Zinc is a well-known antioxidant which prevents the onset of diabetes mellitus by blocking NF-jB (a ROS-sensitive transcription factor that regulates immune responses) activation in the pancreas. (Zargar *et al.*, 2002). Hence, it is believed the presence of phytoconstituents in HF and HFK might be beneficial in such wider perspective point of view and to exert hypoglycemic effect as well as the enhancement of the hypoglycemic activity of MF and GC.

Thus, by co-administration of HFK and HF with MF or GC, absorption processes were either augmented or depressed the metabolism or both the events by the presence of phytoconstituents leading to greater bioavailability of MF and GC. This is also reflected in improved PD effects *viz.* hypoglycemia, hypolipidemia, inhibition of glycosylated haemoglobin, increased body weights and improved histology of β-islets of pancreas. Considering the PD and PK profiles of MF and GC with concomitant treatment of herbal products (HF and HFK), it is suggested that these herbal products may be an excellent biomodulators or adjuvants and may be allowed to use along with antidiabetic drugs in clinical medicine after the clinical studies on human subjects.

Conclusions

Our PK/PD model studies of MF and GC along with herbal products (HF and HFK) demonstrated the beneficial effects on both PD and PK profile indicating its potentiation effect. Such beneficial therapeutic multi-target effects of herbal products on carbohydrate as well as lipid metabolism may be responsible for improved PK/PD parameters. There are clear trends to show that mainstream in pharmaceutical research is moving away from single molecule or single target approach to combination of molecule as well as multiple target approaches (Wermuth, 2004). It may be essential to adapt a golden triangle consisting of traditional knowledge; modern medicine and science with systemic orientation which would pave a way to evolve innovative approaches for new, safe, affordable an effective therapeutic agents (Mashalkar, 2005). Alternatively such herbal products can also be used as an adjuvant along with antidiabetic drugs in clinical medicine. It is essential that some of the proved therapeutic effects of established herbal drugs/products needs an extensive studies to explore their combination along with clinically effective drugs to ascertain their beneficial effects in various diseases like cardiovascular, hepatic, diabetes and immune associated diseases.

Acknowledgements

The authors wish to thank Prof. M. N. Navale, Founder President, Sinhgad Technical Education Society, Pune and Dr. K. S. Jain, Principal, Sinhgad College of Pharmacy, Vadgaon, Pune for providing facilities to carry out this work.

References

Achrekar, S., Kaklij, G.S., Pote, M.S., and Kelkar, S.M. (1991). Hypoglycemic activity of *Eugenia jambolana* and *Ficus bengalenesis*: Mechanism of action. *In Vivo*, **5**: 143-147.

Ajabnoor, M.A. (1990). Effect of aloes on blood glucose levels in normal and alloxan diabetic mice. *Journal of Ethnopharmacology*, **28**: 215-220.

Ali, L., Khan, A.K., Mamun, M. I., Mosihuzzaman, M., Nahar, N., Nur-e-Alam, M., and Rokeya, B. (1993). Studies on hypoglycemic effects of fruit pulp, seed and whole plant of *Momordica charantia* on normal and diabetic model rats. *Planta Medica*, **59**: 408-412.

Amos, A.F., McCarty, D.J., and Zimmet, P. (2010). The rising global burden of diabetes and its complications: estimates and projections to the year 2010. *Diabetic Medicine*, **14**: 1-85.

Anturkar, S.D., Gopumadhavan, B.J., and Mitra, S.K. (1995). Effect of D-400, a herbal formulation, on blood sugar of normal and alloxan-induced diabetic rats. *Indian Journal of Physiology and Pharmacology*, **39**: 95-100.

Augusti, K.T. (1973). Studies on the effects of a hypoglycemic principle from *Allium Cepa* Linn. *Indian Journal of Medical Research*, **61**: 1066-1071.

Baggot, J.D. (1978). Some aspects of chemical pharmacokinetics in veterinary medicine. *Journal of Veterinary Pharmacology and Therapeutics*, **1**: 5.

Baig, M. A., Gawali, V. B., Patil, R.R., and Naik, S.R. (2012). Protective effect of herbomineral formulation (Dolabi) on early diabetic nephropathy in streptozotocin-induced diabetic rats. *Journal of Natural Medicines*, **66**: 500-509.

Bangar, O., Jarald, E., Asghar, S., and Ahmad, S. (2009). Antidiabetic activity of a polyherbal formulation (Karnim Plus). *International Journal of Green Pharmacy*, **3**: 211-214.

Bhattacharya, S.K., Satyan, K.S., and Chakrabarti, A. (1997). Effect of Trasina, an Ayurvedic herbal formulation, on pancreatic islet superoxide dismutase activity in hyperglycaemic rats. *Indian Journal of Experimental Biology*, **35**: 297-299.

Boltan, S. (1997). Experimental design in clinical trials. In: Pharmaceutical statistic, Practical and Clinical Applications, 3rd ed. *Marcel Dekker Inc*, **80**: 384-443.

Bopanna, K.N., Kannan, J., Sushma, G., Balaraman, R., and Rathod, S.P. (1997). Antidiabetic and antihyperlipidemic effect of neem karnel powder on alloxan diabetic rats. *Indian Journal of Pharmacology*, **29**: 162-167.

Boullata, J. (2005). Natural health products interactions with medication. *Nutrition in Clinical Practice*, **20**: 33-51.

Center for Drug Evaluation and Research (CDER), (2001). Guidance for Industry Bioanalytical Method Validation. Department of Health and Human Services Food and Drug administration. Rockville, Maryland.

Ceriello, A. (2006). Oxidative stress and diabetes-associated complications. *Endocrine Practice*. **12**: 60-62.

Chakravarthy, B.K., Gupta, S., Gambhir, S.S., and Gode, K. D. (1980). Pancreatic beta cell regeneration-A novel anti-diabetic mechanism of *Petrocarpus marsupium*. Roxb. *Indian Journal of Pharmacology*, **12**: 123-127.

Chang, M.L.W., and Johnson, M.A. (1980). Effect of garlic on carbohydrate metabolism and lipid synthesis in rats. *Journal of Nutrition*, **110**: 931-936.

Chattopadhyay, R. R. (1999). A comparative evaluation of some blood sugar lowering agents of plant origin. *Journal of Ethnopharmacology*, **67**: 367-372.

Choi, Y.H., Inchul, L., and Myung, G. (2007). Effects of bacterial lipopolysaccharide on the pharmacokinetics of metformin in rats Young H. *International Journal of Pharmaceutics*, **337**: 194-201.

Coimbra, T.C., Danni, F.F., Blotta, R.M., da Periara, C.A., Guedes, M.D., and Graf, R.G. (1992). Plants employed in the treatment of diabetes mellitus; results of an ethnopharmacological survey in Porto Alegre, Brazil. *Fitoterapia*, **63**: 320-322.

DeFronzo, R.A. (1997). Pathogenesis of type 2 diabetes: metabolic and molecular implications for identifying diabetes genes. *Diabetes Reviews*, **5**: 177-269.

Derendorf, H., and Meibohm, B. (1999). Modelling of pharmacokinetic/ pharmacodynamic (PK/PD) relationships: *Concepts and Perspectives Pharmaceutical Research*, **16**: 176-185.

Derubertis, F.R., and Craven, P.A. (1994). Activation of protein kinase C in glomerular cells in diabetes: Mechanisms and potential links to the pathogenesis of diabetic glomerulopathy. *Diabetes*, **43**: 1-8.

Diabetes Control Complications Trial Research Group. (1993). The effect of intensive treatment of diabetes on the development and progression of long-term complications in insulin-dependent diabetes mellitus. *New England Journal of Medicine*, **329**: 977-986.

Dixit, P.V., Umathe, S.N., Virendra, K., Bansod, K.U., and Wanjari, M.M. (2008). Quercetin pretreatment increases the bioavailability of pioglitazone in rats: Involvement of CYP_{3A} inhibition. *Biochemical Pharmacology*, **75**: 1670-1676.

Esposito, A. M., Diaz, A., De Gracia, I., De Tello, R., and Gupta, M.P. (1991). Evaluation of traditional medicine: Effects of *Cajanus cajan* L. and of *Cassia fistula* L. on carbohydrate metabolism in mice. *Revista Medica de Panama*, **16**: 39-45.

Fatmah, A. M., Siti, B. B., Zariyantey, A. H., Nasar, A., and Jamaludin, M. (2012). The Role of oxidative stress and antioxidants in diabetic complications. *Sultan Qaboos University Medical Journal*. **12**: 5-18.

Fernandes, N.P.C., Lagishetty, C.V., Panda, V.S., and Naik, S.R. (2007). An experimental evaluation of the antidiabetic and antilipidemic properties of a standardized *Momordica charantia* fruit extract. *BMC Complementary Alternative Medicine*. **7**: 29.

Godkar, P. B. (1994). Clinical biochemistry: Principles and Practice, 3rd Edition. Bhalani Publishing House, Bombay, 115-116.

Gupta, P., Yadav, D.K., Siripurapu, K.B., Palit, G., and Maurya, R. (2007). Constituents of *Ocimum sanctum* with antistress activity. *Journal of Natural Products*, **70**: 1410-1416.

Jain, R.C., and Vyas, C.R. (1975). Garlic in alloxan-induced diabetic rabbits. *American Journal of Clinical Nutrition*, **28**: 684-685.

Jeanette, S. J., Alex, K. H., David, J. R., and Adviye, E. (2005). Oxidative stress and the use of antioxidants in diabetes: Linking basic science to clinical practice. *Cardiovascular Diabetology*, **4**: 5.

Joy, K.L., and Kuttan, R. (1999). Anti-diabetic activity of *Picrorrhiza kurroa* extract. *Journal of Ethnopharmacology*, **67**: 143-148.

Khosla, P., Gupta, D.D., and Nagpal, R.K. (1995). Effect of *Trigonella foenumgraecum* (Fenugreek) on blood glucose in normal and diabetic rats. *Indian Journal of Physiology and Pharmacology*, **39**: 173-174.

King, G.L., and Loeken, M.R. (2004). Hyperglycemia-induced oxidative stress in diabetic complications. *Histochemistry and Cell Biology*, **122**: 333-338.

King, H., Aubert, R.E., and Herman, W.H. (1998). Global burden of diabetes, 1995-2025: Prevalence, numerical estimates, and projections. *Diabetes Care*, **21**: 1414-1431.

Kondeti, R. S., Korivi, M., Kesireddy N., Chia H., and Kesireddy, S. R. (2011). Protective effect of dietary ginger on antioxidant enzymes and oxidative damage in experimental diabetic rat tissues. *Food Chemistry*, **124**: 1436-1442.

Mandlik, R.V., Desai, S.K., Naik, S.R., Sharma, G., and Kohli, R.K. (2008). Antidiabetic activity of a polyherbal formulation (DRF/AY/5001). *Indian Journal of Experimental Biology*, **46**: 599-606.

Manickam, M., Ramanathan, M., Jahromi, M.A., Chansouria, J.P., and Ray, A.B. (1997). Antihyperglycemic activity of phenolics from *Pterocarpus marsupium*. *Journal of Natural Products*, **60**: 609-610.

Marchesini, G., Brizi, M., and Blanchi, G. (2001). Nonalcoholic fatty liver disease: A feature of the metabolic syndrome. *Diabetes*, **50**: 1844-1850.

Mashelkar, R. A. (2005). Global voices of science: India's R and D: reaching for the top. *Science*, **307**: 1415-1417.

Mcgrown, M.W., Joseph, D.A., Strandberg, D.R., and Zak, B. A. (1983). Peroxidase coupled method for the colorimetric determination of serum triglyceride. *Clinical chemistry*, **29**: 538-542.

Mishra, N., Tandon, V. L., and Gupta, R. (2012). Immunomodulation by Hibiscus *rosa-sinensis*: Effect on the humoral and cellular immune response of *Mus musculus*. *Pakistan Journal of Biological Sciences*, **15**: 277-283.

Mukherjee, P.K., Saha, K., Pal, M., and Saha, B.P. (1997). Effect of *Nelumbo nucifera* rhizome extract on blood sugar level in rats. *Journal of Ethnopharmacology*, **58**: 207-213.

Naggar, E. M. (2005). Anti-diabetic effect of *Cleome droserifolia* aerial parts: Lipid peroxidation induced oxidative stress in diabetic rats. *Acta Veteriniaria Brno*, **74**: 347.

Naritomi, Y., Terashita, S., and Kagayama, A. (2004). Identification and relative constitutions of human CYP_{450} isoforms involved in the metabolism of glibenclamide and lansoprazole: evaluation of an approach based on the *in vitro* substrate disappearance rate. *Xenobiotica*, **34**: 415-428.

Rai, D., Bhatia, G., Sen, T., and Palit, G. (2003). Anti-stress effects of *Ginkgo biloba* and *Panax ginseng* a comparative study. *Journal of Pharmacological Sciences*, **93**: 458-464.

Rai, V., Iyer, U., and Mani, U.V. (1997). Effect of Tulsi (*Ocimum sanctum*) leaf powder supplementation on blood sugar levels, serum lipids and tissue lipids in diabetic rats. *Plant Foods and Human Nutrition*, **50**: 9-16.

Rajendran, S.D., Philip, B.K., Gopinath, R., and Suresh, B. (2007). RPHPLC method for the estimation of glibenclamide in human serum. *Indian Journal of Pharmaceutical Sciences*, **69**: 796-799.

Rao, V.V., Dwivedi, S.K., and Swarup, D. (1994). Hypoglycemic effect of *Caselpinia bonducella* in rabbits. *Fitoterpia*, **65**: 245- 247.

Rawland, M., and Tozer, T. N. (1996). In: Clinical Pharmacokinetics: Concepts and Applications, 3rd Edition, Lea and Febiger, Philadelphian USA, 287-293.

Sagrawat, H., Mann, A.S., and Kharya, M.D. (2006). Pharmacological potential of *Eugenia jambolana*: A review. *Pharmacognosy Magazine*, **2**: 96-105.

Saxena, A.M., Bajpai, M.B., and Mukherjee, S.K. (1991). Swerchirin induced blood sugar lowering of streptozotocin treated hyperglycemic rats. *Indian Journal of Experimental Biology*, **29**: 674-675.

Singh, N., Singh, S.P., Vrat, S., Misra, N., Dixit, K.S., and Kohli, R.P. (1985). A study on the anti-diabetic activity of *Coccinia indica* in dogs. *Indian Journal of Medical Science*, **39**: 27-29.

Stanely, P., Prince, M., and Menon, V.P. (2000). Hypoglycemic and other related actions of *Tinospora cordifolia* roots in alloxan-induced diabetic rats. *Journal of Ethnopharmacology*, **70**: 9-15.

Taylor, S.I. (1999). Deconstructing type 2 diabetes. *Cell*, **97**: 9-12.

Trinder, P. (1969). Determination of blood glucose using an oxidase peroxidase system with a non-carcinogenic chromogen. *Annals of Clinical Biochemistry*, **6**: 24-30.

Van Giersbergen, P.L., Treiber, A., Clozel, M., Bodin, F., and Dingemanse, J. (2002). *In vivo* and *in vitro* studies exploring the pharmacokinetic interaction between bosentan, a dual endothelin receptor antagonist and glyburide. *Clinical Pharmacology and Therapeutics*, **71**: 253-262.

Vats, V., Grover, J.K., and Rathi, S.S. (2002). Evaluation of anti-hyperglycemic and hypoglycemic effect of *Trigonella foenum-graecum* Linn, *Ocimum sanctum* Linn and *Pterocarpus marsupium* Linn in normal and alloxanized diabetic rats. *Journal of Ethnopharmacology*, **79**: 95-100.

Vats, V., Yadav, S.P., and Grover, J. K. (2004). Ethanolic extract of *Ocimum sanctum* leaves partially attenuates streptozotocin-induced alterations in glycogen content and carbohydrate metabolism in rats. *Journal of Ethnopharmacology*, **90**: 155-160.

Wanjari, M. M., There, A.W., Tajne, M. R., Chopde, C. T., and Umathe, S. N. (2008). Rapid and simple RPHPLC method for the estimation of metformin in rat plasma. *Journal of Pharmaceutical Sciences*, **70**: 198-202.

Wermuth, C.G. (2004). Multitargeted drugs: The end of the one target one disease philosophy. *Drug Discovery Today*, **9**: 826-827.

Xie, J.T., Wang, A., Mehendale, S., Wu, J., Aung, H.H., Dey, L., Qiu, S., and Yuan, C.S. (2003). Anti-diabetic effects of *Gymnema yunnanense* extract. *Pharmacological Research*, **47**: 323-329.

Zargar, A.H., Bashir, M.I., and Masoodi, S.R. (2002). Copper, zinc and magnesium levels in type-1 diabetes mellitus. *Saudi Medical Journal*, **23**: 539-542.

Medicinal Plants: Phytochemistry, Pharmacology and Therapeutics Vol. 3 (2014) *Pages* **413–421**
Editor-in-Chief: **V.K. Gupta**
Published by: **DAYA PUBLISHING HOUSE, NEW DELHI**

19

Phytochemical and Histochemical Studies on Different Species of *Selaginella*

A. Sivaraman[1] and M. Johnson[1*]

ABSTRACT

The present study was aimed to reveal the phytochemical composition of various Selaginella species using qualitative tests and histochemical analysis. Different ferns viz., Selaginella involvens (Sw.) Spring, Selaginella intermedia (Bl.) Spring, Selaginella inaequalifolia (Hook. and Grev.) Spring and Selaginella tenera (Hook. and Grev.) Spring were collected from Western Ghats, South India. The powdered materials were extracted by cold extraction with various solvents viz., petroleum ether, chloroform, acetone and ethanol. Phytochemical screening of the extracts was carried out according to the method described by Harborne. Histochemical tests were made on the fresh sections of the stem treated with the following reagents to identify the presence or absence of metabolites phenol, lignin and tannin. Phytochemical analysis revealed the presence of steroids, phenolics, tannins, saponins, flavonoids, alkaloids, aminoacids and anthroquinone in various species of Selaginella. Histochemical studies revealed the presence of phenols, tannin, and lignin in the different parts of the stipes of selected four Selaginella species with varied percentage. Hence this study offers a base of using Selaginella species as herbal alternative for the synthesis of active compounds.

Keywords: Phytochemistry, Histochemistry, *Selaginella,* Pteridophytes, Stipes.

1 Centre for Plant Biotechnology, Department of Plant Biology and Plant Biotechnology, St. Xavier's College (Autonomous), Palayamkottai – 627 002, Tamil Nadu, India

* *Corresponding author*: E-mail: ptcjohnson@gmail.com

Introduction

Medicinal plants, since time immemorial, have been used in virtually all cultures as a source of medicine. The widespread use of herbal remedies and healthcare preparations, as those described in ancient texts such as Bible and the Vedas, and obtained from commonly used traditional herbs and medicinal plants, has been traced to the occurrence of natural products with medicinal properties. Medicinal plants are staging a comeback and 'renaissance' is happening all over the globe (Joy *et al.*, 2001). Ferns and their fern allies have little economic significance to humankind as compared to other groups, especially angiosperms; hence these are poorly understood, overlooked or often neglected by the society. None the less, being the major constituent of biodiversity their role in the system of nature cannot be overlooked or neglected in the global changing environmental scenario (Sharpe, 2011). Despite the large number of ferns and their allied species, a considerable percentage of them are rare and threatened. With increasing utilization of land and natural resources, it is feared that many of these threatened taxa will become yet rarer, more vulnerable and endangered, and in several cases may finally become extinct, as any disturbance or imbalance in their narrowly confined ecosystems is liable to lead to their extermination (Chandra *et al.*, 2008).

There are about 1200 species of Pteridophytes in India. Like flowering plants, pteridophytes are also used by tribals in India and other countries. *Selaginella,* with about 700 species, is world-wide in distribution, but most of the species are tropical and many of them grow in damp forests (Pichi-Sermolli, 1977). Alston (1945) enumerated 58 species of *Selaginella* from India and Sri Lanka. Of these, 45 species has been reported from India. According to Dixit (1984), 18 species are in South India (Tamil Nadu and Kerala) and 12 species have been reported from the Western Ghats, South India (Manickam and Irudayaraj, 1992). Considerable numbers of pteridophytes are used in Chinese medicines. The Chinese herb *Selaginella doederleinii* is used as an anticancer drug (Lee and Lin, 1988; Huang, 1992). The medicinal use of *Selaginella* in India goes back to the period of 'Ramayana' (Sah, 2008). According to the principles of traditional Chinese medicine, *Selaginella* has sweet, spicy, bitter and cold properties, and is associated with the liver, lung and stomach meridians. Its main functions are to clear heat, reduce toxicity and drain damp heat. More recent research has shown that *Selaginella* may be effective against both acute and chronic hepatitis. *Selaginella* may also be used externally to help stop bleeding and promote wound healing (Tinde van Andel *et al.*, 2007). Prepared *Selaginella* can be found in pill, powder and tablet forms. *Selaginella* should be taken with caution by patients diagnosed with cold deficiency. Taking large amounts of *Selaginella* may result in loss of appetite and abdominal discomfort. There are no known drug interactions associated with *Selaginella* (Lee *et al.*, 1999). *Selaginella doederleinii,* a popular anticancer herb, may contain an as yet unidentified substance that contributes to reversible bone marrow suppression (Duraiswamy *et al.*, 2010). Histochemical studies can be used to study the chemical composition of the cells resulting from growth and developments and these data can be used to make inferences about the importance of secondary metabolites in growth and development of plants. With this knowledge, the present study was aimed to reveal the phytochemical composition of various *Selaginella* species using qualitative tests and histochemical analysis.

Materials and Methods

Plant Collection

Healthy, disease free fern allies of *Selaginella involvens* (Sw.) Spring, *Selaginella intermedia* (Bl.) Spring, *Selaginella inaequalifolia* (Hook. and Grev.) Spring and *Selaginella tenera* (Hook. and Grev.) Spring were collected from Kakachi stream, Tirunelveli Hills (1000 m) in the Western Ghats. The fern allies were identified by Dr. V. Irudayaraj, St. Xavier's College, Palayamkottai.

Plant Extraction and Preliminary Phytochemical Screening

The fresh materials were washed in running tap water for 5 min and dried using blotting papers. The washed plant materials were air and shade dried for two weeks and pulverized to powder using tissue blender. The powdered materials (10 g) were extracted by cold extraction with 60 mL of various solvents *viz.*, petroleum ether, chloroform, acetone and ethanol. The samples were kept in dark for 72 h with intermittent shaking. The extracts were filtered using Whattman filter paper (No.1) and then concentrated in vacuum at 40°C using rotary evaporator. The residues obtained were stored in a freezer -70°C until further tests. Phytochemical screening of the extracts was carried out according to the method described by Harborne (1998).

Histochemical Analysis

Fresh free hand sections were made from the fronds of *S. involvens, S. intermedia, S. inaequalifolia* and *S. tenera*. Histochemical tests were made on the fresh sections of the stem treated with the following reagents to identify the presence or absence of metabolites phenol, lignin and tannin. Lugol's Iodine is used to detect tannins and lignins (Haridass and Suresh Kumar, 1985; Chamberlin, 1924), 10 per cent sodium nitrite, 20 per cent urea and 10 per cent acetic acid were used to detect phenolics (Reeve, 1951). The stained sections were observed under Motic trinocular microscope (Japan). They were photographed at different magnifications and at different views. The results were registered on photomicrograph. Based on the photographs taken, anatomical description and localization of tested chemicals were done.

Results

In the present study, the phytochemical screening of petroleum ether, chloroform, acetone and ethanolic extracts of *S. involvens, S. intermedia, S. inaequalifolia* and *S. tenera* were carried out and the results were tabulated in Table 19.1. Steroids showed their presence in all the tested 16 extracts followed by phenolics and tannins in 8 extracts. Next to that, 7 extracts were illustrated their existence of saponins. Flavonoids demonstrated their occurrence in 6 different extracts. Alkaloids, aminoacids and anthraquinone showed their presence only in ethanolic extracts of *S. involvens, S. intermedia, S. inaequalifolia* and *S. tenera*.

Among the different extracts of *S. involvens,* steroids showed their presence in all the four extracts followed by flavonoids in 3 different extracts. Phenolics, tannins and saponins demonstrated their existence in 2 different extracts. In *S. intermedia,* steroids demonstrated their occurrence in all the four extracts followed by phenolics,

Table 19.1: Phytochemical analysis on different species of *Selaginella.*

Metabolites	*S. involvens*				*S. intermedia*				*S. inaequalifolia*				*S. tenera*			
	P	*C*	*A*	*E*	*P*	*C*	*A*	*E*	*P*	*C*	*A*	*E*	*P*	*C*	*A*	*E*
Steroids	+	+	+	+	+	+	+	+	+	+	+	+	+	+	+	+
Alkaloids	–	–	–	+	–	–	–	+	–	–	–	+	–	–	–	+
Phenolics	–	–	+	+	–	–	+	+	–	–	+	+	–	–	+	+
Saponins	–	+	+	–	–	–	+	+	–	–	+	–	–	+	+	–
Tannins	–	–	+	+	–	–	+	+	–	–	+	+	–	–	+	+
Anthraquinone	–	–	–	+	–	–	–	+	–	–	–	+	–	–	–	+
Amino acids	–	–	–	+	–	–	–	+	–	–	–	+	–	–	–	+
Flavonoids	–	+	+	+	–	–	–	+	–	–	–	+	–	–	–	+

Table 19.2: Histochemical analysis on different species of *Selaginella*.

Parts	*S. involvens*			*S. intermedia*			*S. inaequalifolia*			*S. tenera*		
	P	*T*	*L*	*P*	*T*	*L*	*P*	*T*	*L*	*P*	*T*	*L*
Epidermis	–	3^{+}	3^{+}	–	+	3^{+}	3^{+}	3^{+}	3^{+}	3^{+}	2^{+}	2^{+}
Hypodermis	–	2^{+}	2^{+}	–	–	2^{+}	2^{+}	2^{+}	3^{+}	3^{+}	2^{+}	2^{+}
Ground tissue	2^{+}	–	–	2^{+}	–	–	2^{+}	–	2^{+}	2^{+}	–	–
Endodermis	3^{+}	–	2^{+}	3^{+}	–	–	2+	–	–	2^{+}	–	–
Xylem	2^{+}	–	3_{+}	2^{+}	–	3+	2^{+}	2^{+}	3^{+}	3^{+}	–	2^{+}
Phloem	–	–	+	3^{+}	–	+	2+	2^{+}	2^{+}	–	–	–

P: Phenol; T: Tannin; L: Lignin.

+: Mild, 2^{+}: Average, 3^{+}: high, –: Data not available.

saponins and tannins in two different extracts. Among various extracts of *S. inaequalifolia,* steroids illustrated their existence in all the four extracts whereas phenolics and tannins were present in 2 extracts only. In *S. tenera* also, steroids showed their presence in all the four extracts whereas phenolics, saponins and tannins were present in 2 extracts only.

The occurrence and distribution of various chemicals (phenols, lignin and tannins) in different tissues of the stipe was recorded in Table 19.2 and Figure 19.1. Based on the intensity of the resulting colour (metabolites intensity) it has been scored into three categories of concentrations such as: Low (+), Medium (2+) and High (3+).

Histochemical studies revealed the presence of phenols, tannin and lignin in the selected four *Selaginella* species with varied percentage. Phenols are mostly present in all the four selected species. *S. involvens* shows higher concentration of phenols in the endodermis (Figure 19.1A). Minimum concentration of phenols was present in the ground tissue and xylem of *S. involvens* (Figure 19.1A). In *S. intermedia,* higher concentration of phenols was present in the endodermis and phloem (Figure 19.1D). Lower concentration of phenols was present in the ground tissue and xylem of *S. intermedia* (Figure 19.1E). *S. inaequalifolia* expresses maximum concentration of phenols only in the epidermis (Figure 19.1I). Minimum concentrations of phenols were illustrated in all other parts of *S. inaequalifolia* (Figure 19.1I). *S. tenera* shows high concentration of phenols in the epidermis, hypodermis and xylem (Figure 19.1M). In ground tissue and endodermis of *S. tenera,* minimum concentrations of phenols were present (Figure 19.1I).

Tannins are mostly present in all the four selected species. *S. involvens* shows higher concentration of tannin in the epidermis (Figure 19.1B). Less concentration of tannin were present in the hypodermis of *S. involvens* (Figure 19.1B). *S. intermedia* expresses maximum concentration of tannin in the epidermis only (Figure 19.1F). Tannin is absent in the all other parts of the *S. intermedia.* In *S. inaequalifolia,* high concentrations of tannin were observed in the epidermis (Figure 19.1J). Minimum amount of tannin was present in the hypodermis, xylem and phloem (Figure 19.1J). In *S. tenera,* tannin showed minimum concentration in epidermis and hypodermis (Figure 19.1N). Tannin is absent in all other parts of the *S. tenera* (Figure 19.1N).

Lignin is present at high concentration in the epidermis and xylem of *S. involven* (Figure 19.1C). Lignin showed lesser amount in the hypodermis, endodermis and phloem of *S. involvens* (Figure 19.1C). *S. intermedia* expresses maximum concentration of lignin in the epidermis and xylem (Figure 19.1G). Lignin is minimum in the hypodermis and phloem (Figure 19.1H). Lignin is present at high concentration in the epidermis, hypodermis and xylem of *S. inaequalifolia* (Figure 19.1L). Minimum concentration of lignin is present in the ground tissue and phloem of *S. inaequalifolia* (Figure 19.1K). *S. tenera* express minimum concentration of lignin in the epidermis, hypodermis and xylem (Figure 19.1O).

Discussion

The beneficial medicinal effects of plant materials typically result from the secondary metabolites present in the plant although, it is usually not attributed to a single compound but a combination of the metabolites. The medicinal actions of

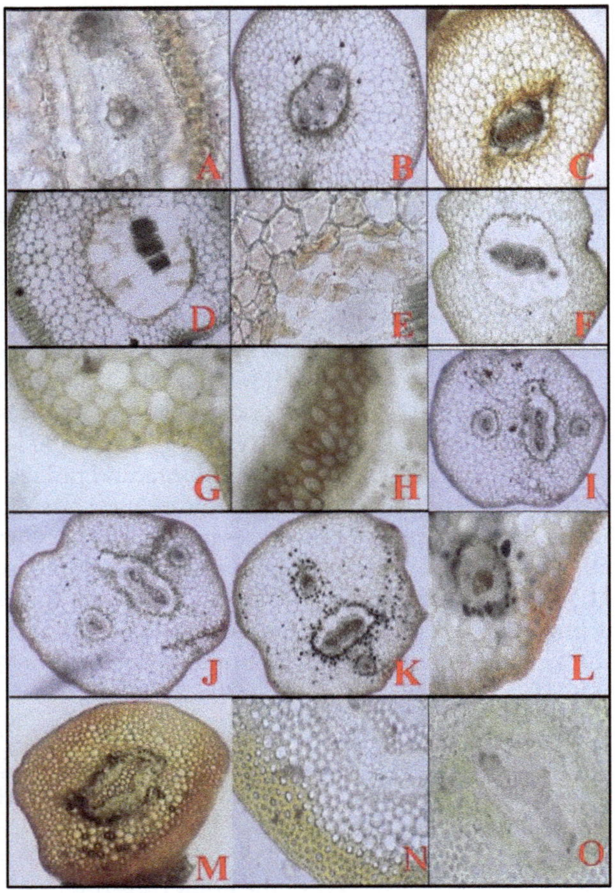

Figure 19.1

A: Phenol presence in ground tissue, endodermis and xylem of *S. involvens;* B: T.S of stipe- A portion enlarged in *S. involvens* showing the presence of tannin in epidermis and hypodermis; C: T.S of stipe – A portion enlarged in *S. involvens* showing the presence of lignin in epidermis, hypodermis, endodermis and xylem; D: T. S of stipe – A portion enlarged in *S. intermedia* showing the presence of phenol in endodermis and phloem; E: Phenol presence in ground tissue, endodermis, xylem and phloem of *S. intermedia;* F: T.S of stipe showing tannin presence in *S. intermedia* in the epidermis; G: Lignin presence in epidermis, hypodermis and of *S. intermedia;* H: Lignin presence in xylem of *S. intermedia;* I: Entire view of stipe *S. inaequalifolia* showing phenol in epidermis, hypodermis, endodermis xylem and phloem; J: Entire view of stipe *S. inaequalifolia* showing tannin in epidermis, hypodermis, xylem and phloem; K: T.S of stipe – presence of lignin in *S. inaequalifolia* in the epidermis, hypodermis, xylem and phloem; L: Lignin presence in epidermis, hypodermis, ground tissue, xylem and phloem of *S. inaequalifolia;* M: T.S of stipe – entire view of *S. tenera* showing phenol present in epidermis, hypodermis,endodermis and xylem; N: Tannin presence in epidermis and hypodermis of *S. tenera;* O: *S. tenera* showing presence of lignin in epidermis, hypodermis and xylem.

plants are distinctive to a particular plant species or group, reliable with the concept that the combination of secondary metabolites in a particular plant is taxonomically distinct (Parekh *et al.*, 2005). Steroids may serve as an intermediate for the biosynthesis of downstream secondary natural products and it is believed to be a biosynthetic precursor for cardenolides in plants (Daly, 1998). The results of the present study determined the presence of steroids in all the four extracts of studied *Selaginella* species. Tannins have been found to possess antiviral, antibacterial, antiparasitic effects, antiinflammatory, antiulcer and antioxidant property for possible therapeutic applications (Lu *et al.*, 2004; Kolodziej *et al.*, 2005). In the present study we revealed the tannins presence in acetone and ethanolic extracts of all the four species of *Selaginella*. Tannins also showed their presence in different parts of the all the studied *Selaginella* species. Saponins possess specific physical, chemical and biological activities that make them useful as drugs. Some of these biological properties include antimicrobial, anti-inflammatory, anti-feedent and hemolytic effects (George *et al.*, 2002). The results of the present study showed the presence of saponins in different extracts of *Selaginella* species except petroleum ether. Alkaloid, which is one of the largest phytochemical groups in plants, has amazing effect on humans and this has led to the development of powerful pain killer medications (Staerk *et al.*, 2002). In the present study, alkaloids presence was determined only in the presence of ethanolic extracts of all the four species. Phenolics may serve in plant defense mechanisms to counteract reactive oxygen species (ROS) in order to survive and prevent molecular damage and damage by microorganisms, insects, and herbivores (Vaya *et al.*, 1997). Different types of phenolic compounds have different antioxidant activities, which is dependent on their structure (Bajpai *et al.*, 2005). In the present study, we revealed the phenolics presence in acetone and ethanolic extracts and also in different parts of all the four species of *Selaginella*. Flavonoids, the major group of phenolic compounds were reported for their antimicrobial, antiviral and spasmolytic activity. They are able to scavenge hydroxyl radicals, superoxide anion radicals and lipid peroxy radicals, which highlights many of the flavonoid health-promoting functions (Havsteen, 2002). The present study results illustrated the presence of flavonoids in different extracts of *Selaginella* species except petroleum ether. Lignin is found in the cell walls of plants. Lignins impart strength to cell walls, facilitate water transport, and impede the degradation of wall polysaccharides, thus acting as a major line of defense against pathogens, insects and other herbivores (Sederoff *et al.*, 1999). In the present study, lignins illustrated their presence in all the different parts of stipes of various *Selaginella* species. Thus the present investigation on phytochemical and histochemical studies on different *Selaginella* species confirmed the presence of phenolics, tannins, alkaloids, saponins, steroids and lignin, which are of great medicinal value and have been extensively used in the drug and pharmaceutical industry. Hence this study offers a base of using *Selaginella* species as herbal alternative for the synthesis of active compounds.

Acknowledgements

The authors are thankful to the authorities of St. Xavier's College (Autonomous), Palayamkottai, Tamil Nadu, India for the infrastructure facilities. The authors are thankful to DR. V. Irudayaraj for his valuable suggestion and constant support.

References

Alston, A.H.G. (1945). An enumeration of the Indian species of *Selaginella. Proc. Natl. Inst. Sci. India.* **11B**: 211-235.

Bajpai, M., Pande, A., Tewari, S.K., and Prakash, D. (2005). Phenolic contents and antioxidant activity of some food and medicinal plants. *International Journal of Food Sciences and Nutrition*, **56(4):** 287-291.

Chamberlin, C.J. (1924). Methods in plant histology. Univ. Chicago Press, Chicago.

Chandra, S., Fraser-Jenkins, C.R., Kumari, A., and Srivastava, A. (2008). A summary of the status of threatened pteridophytes of India. *Taiwania,* **53(2):** 170-209.

Daly, J.W. (1998). Thirty years of discovering arthropod alkaloids amphibian skin. *Nat. Prod.,* **61**: 162-172.

Dixit, R.D. (1984). *A census of Indian Pteridophytes*. Howrah: Botanical Survey of India, Dept. Environment, pp. 1-177.

Duraiswamy, H., Nallaiyan,S., Nelson, J., Rathinasamy, P., Johnson, M., and Irudayaraj, V. (2010). *Asian Pacific Journal of Tropical Medicine,* 412-420.

George, F., Zohar Kerem, Harinder, P.S.M., and Klaus Becker. (2002). The biological action of saponins in animal systems: A review. *Brit J Nutr.*, **88(6):** 587-605.

Harborne, J.B. (1973). Phytochemical Methods, Chapman and Hall, London.

Haridass, E.T., and Suresh Kumar, N. (1985). Some techniques in the study of insect-host plant interactions. In: *Dynamics of Insect Plant interactions*. Ananthakrishnan, Ed., Entomology Research Institute, Loyola College, Madras.

Havsteen, B.H. (2002). The biochemistry and medical significance of flavonoids. *Pharmacol and Therapeutics*, **96**: 67-202.

Huang, K.C. (1992). *The pharmacology of Chinese herbs*. CRC publishers, pp. 388.

Joy, P.P., Thomas, J., Mathew, S., and Skaria, B.P. (2001). Medicinal plants. *Tropical Horticulture,* **2**: 449-632.

Kolodziej, H., and Kiderlen, A.F. (2005). Antileishmanial activity and immune modulatory effects of tannins and related compounds on Leishmania parasitised RAW 264.7 cells. *Phytochemistry*, **66(17):** 2056-2071.

Lee, H., and Lin, J.Y. (1988). Antimutagenic activity of extracts from anticancer drugs in Chinese medicine. *Mutat Res.,* **204(2):** 229-234.

Lee, I.S., Nishikawa, A., Furukawa, F., Kasahara, K., and Kim, S. U. (1999). *Cancer Lett.,* **44(1):** 93-99.

Lu, L., Liu, S.W., Jiang, S.B., and Wu, S.G. (2004). Tannin inhibits HIV-1 entry by targeting gp41. *Acta Pharmacol Sin*, **25(2):** 213-218.

Manickam, V.S., and Irudayaraj, V. (1992). *Pteridophyte Flora of the Western Ghats, South India.* B.I. Publications, Pvt. Ltd. New Delhi.

Parekh, J., Jadeja, D., and Chanda, S. (2005). Efficacy of aqueous and methanol extracts of some medicinal plants for potential antibacterial activity. *Turk J Biol*, **29**: 203-210.

Pichi-Sermolli, R.E.G. (1977). Tentamen Pteridophytorum genera in Taxonomicum ordinem reoigendi. *Webbia*, **31:** 313-512.

Reeve, R.M. Histochemical tests for polyphenols in plant tissues. *Stain Technol.* **1951**; 26: 91-96.

Sah, P. (2008). *The Journal of American Science*, **4(3):** 65-67.

Sederoff, R.R., MacKay, J.J., Ralph, J., and Hatfield, R.D. (1999). Unexpected variation in lignin. *Curr Opin Plant Biol.*, **2**: 145-152.

Sharpe, J.M. (2011). Ferns and fern-allies of Sikkim, a pictorial handbook part I. *Am Fern J*, **101(2):** 129-130.

Staerk, D., Lykkeberg, A.K., Christensen, J., Budnik, B.A., Abe, F., and Jaroszewki, J.W. (2002). *In vitro* cytotoxic activity of phenanthroindolizidine alkaloids from *Cynanchum vincetoxicum* and *Tylophora tanake* against drug-sensitive and multidrugresistant cancer cells. *J Nat Prod*, **65(9):** 1299-1302.

Tinde van Andel, Behari-Ramdas, J., Havinga, R., and Groenendijk, S. (2007). *Ethnobotany Research and Applications*, **5**: 351-372.

Vaya, J., Belinky, P.A., and Aviram, M. (1997). Antioxidant constituents from licorice roots: Isolation, structure elucidation and antioxidative capacity toward LDL oxidation. *Free Radical Biol. Med.*, **23(2):** 302-313.

Medicinal Plants: Phytochemistry, Pharmacology and Therapeutics Vol. 3 (2014) *Pages* **423–439**
Editor-in-Chief: **V.K. Gupta**
Published by: **DAYA PUBLISHING HOUSE, NEW DELHI**

20

Amazon Piperaceae with Potential Insecticide Use

M. Fazolin[1], J.L.V. Estrela[1], K.K.L. Yamaguchi[2], F.A. Pieri[2,3] and V.F. Veiga-Junior[2*]

ABSTRACT

The Amazon rainforest is a potential source of essential oils, such as those found in the family Piperaceae, to which the species known as long pepper and pepper jack belong. The genus Piper comprises around 700 species, among the 140-300 species that are part of the rainforest flora of the Amazon region. The literature describes the composition of its essential oils, with various biological activities such as larvicide, antimicrobial, and antioxidant activities, among others. The use of essential oils as a future alternative to insecticides is a field that is growing, and this chapter presents a literature review of studies focusing on Amazonian Piperaceae essential oils that have potential insecticide use. The species Piper aduncum L., P. callosum, P. divaricatum, P. hispidinervum, P. hostimannianum, P. humaytanum, P. marginatum, P. nigrum L., and P. tuberculatum, have shown excellent results in studies to evaluate their potential as plants with biological activities that can be used to control pests that cause damage to agricultural crops, or to human health. The essential oil of P. aduncum is the one that has been most widely studied for its potential as an insecticide, showing effectiveness in the control of various species of agricultural pests in Brazil e.g., Cerotoma tingomarianus Bechyné, Tenebrio molitor L., Solenopsis saevissima. The Reports in the literature on the essential oils of Piperaceae demonstrate their ability to inhibit or delay insect maturation, reducing reproductive capacity, and causing death by starvation or direct toxicity. The main constituents present in the composition of the essential oils are phenylpropanoids,

1 Embrapa ACRE, Brazilian Agricultural Research Corporation, Rio Branco, Acre, Brasil

2 Institute of Exact Sciences, Federal University of Amazonas, Manaus, Brazil

3 Leônidas and Maria Deane Institute, Oswaldo Cruz Foundation, Manaus, Brazil

* *Corresponding author*: E-mail: valdirveiga@ufam.edu.br

monoterpenes and sesquiterpenes, such as eugenol, safrole, α-humulene, β-caryophyllene, β-farnesene, and α-bisabolol. The wide biological diversity, largely unexplored, especially in tropical regions like the Amazon rainforest, holds great potential for research into new products that could replace conventional insecticides, or be used as part of synergistic formulations in the efforts to control insect resistance to these products.

Keywords: Amazon rainforest, *Piperaceae*, Essential oils, Insecticide use.

Introduction

Angiosperms first appeared 90 million to 130 million years ago, with approximately 500,000 species (Castro *et al.*, 2004). In Brazil, there are thought to be around 200,000 species (Labandeira *et al.*, 1994). Of this total, at least 80,000 species grow in the Amazon rainforest (Pletsch and Santana, 1995).

The diversity of the flora is directly related to the structural diversity of secondary metabolites produced, which are currently estimated more than 200,000 compounds (phenolics in general, isoprenoids, alkaloids, and polyketides, among others) (Dixon, 2001).

The family Piperaceae is thought one of the most primitive of the angiosperms, and can be considered a "fossil plant" (Taylor and Hickey, 1992). A biogeographical analysis of the distribution of species of *Piper* on the North and South American continents revealed three distinct regions of occurrence: Amazon, Central America and Mexico, and the Atlantic forest (Jamarillo and Marquis, 2004).

The secondary metabolism of this family is presented as one of the most versatile of the botanical families known. The metabolites accumulated by species of Piperaceae are characterized by being derived from mixed biosynthesis (shikimate/acetate), resulting in the production of aromatic amides or aromatic compounds, essentially phenylpropanoids, of the lignan and neolignan types, and the occurrence of terpenoids, flavonoids and other classes of natural products with biological activity (Alécio *et al.*, 1997; Gottlieb *et al.*, 1995; Parmar *et al.*, 1997).

Estimates suggest that under normal development, 20 per cent of the carbon fixed by the plants flows to the shikimic acid pathway (Sangwan *et al.*, 2001), hence the importance, to the plant, of the compounds produced in this metabolic pathway.

The Piperaceae family comprises 12 genera, with about 2000 species. Of these, the bioactivity of the compounds has only been evaluated for only ten percent (Andrade *et al.*, 2009). In Brazil, 700 species have been reported for the genus *Piper*, of which between 140 and 300 are presumed to be part of the Amazon flora (Yuncker, 1972, 1973; Jaramillo and Manos, 2001).

The diversity of *Piper* can be explained by its unique adaptation to the natural habitat. The highest number of species is found in tropical rainforests. They are pioneer species, and then used in disturbed areas for their regeneration (Andrade *et al.*, 2009).

The aim of this study was to point out, by means of the currently available knowledge, the potential of Piperaceae species that occur naturally in the Amazon, as a source of promising compounds for pest control.

Materials and Methods

We performed a literature search using the databases SCOPUS, ISI, PubMed, SCIELO, and search tools available on the Internet. Search keywords such as "Piperaceae", "Piper oil" and "insecticide" were used, plus the names of the main chemical classes present in *Piper*, in various combinations or alone.

The constituents of essential oils and plant extracts from the Piperaceae family presented in the study "Chemical Variability in essential oils of Piper species in the Amazon" (Andrade *et al.*, 2009) were used as reference. The study was conducted with 230 specimens of Piperaceae, representing 53 plant species collected in the Brazilian states of the Amazon region, including Acre, Amapá, Amazonas, Maranhão, Mato Grosso, Pará, Roraima and Tocantins.

To indicate the potential of the species, the research started with the premise that secondary metabolites, such as most monoterpenes, sesquiterpenes, phenylpropanoids and some amides, present biological interference on insects. The metabolites considered were obtained from essential oils and methanol, ethanol and dichloromethanic extracts, pondering the accumulated knowledge for each type of product.

The prevalence of the chemical compounds monoterpenes, phenylpropanoids and sesquiterpenes for all species was defined by the greater percentage of occurrence in the oils, and in some cases, these were attributed to more than one compound, since that the presence of those compounds in the sample differ from minor values than 50 per cent of highest value observed.

Results and Discussion

The family Piperaceae comprises one of the most important plant families in the Amazon rainforest. Most species of this genus are used in traditional and popular medicine, prompting interest in their chemical constitutions, especially the minor components contained in the plants. The phenylpropanoid compounds produced by plants of the genus *Piper* are largely responsible for the biological activity of some species of the Amazon, such as apiol, dillapiole, myristicin, elemicin, eugenol, methyl eugenol, and safrole ethyl piperonyl ketone, all of which have been found in essential oils of species previously studied (Dyer and Palmer, 2004).

The studies are subdivided into the major classes found in the chemical composition of essential oils and the activities of these insecticides.

Essential Oils

The formation of essential oils, and their secretory structures, are primitive traits in plants, followed by the trend to replace the oil by other metabolites (Gottlieb and Salatino, 1987). These anatomical structures have evolved from oleiferous cells, secretory cavities and channels and glandular hairs. These structures characterize evolutionary lineages of angiosperms, and their occurrence, along with the storage and volatility of the oil, are important characteristics that define an essential oil. Volatilized plant substances can penetrate insects when the plant is injured, altering the hormonal balance of these insects resulting in a more sophisticated defense mechanism (Gottlieb and Salatino, 1987).

The constituents of essential oils are classified as terpenoids (most constituencies are synthesized in the chloroplast and cytoplasm) and phenylpropanoids (synthesized via shikimate). Although phenylpropanoids are not very common constituents in essential oils, some species, such as *Piper* genus, contain significant proportions of these compounds (Deschamps and Biasi, 2009).

Phenylpropanoids

The deamination of phenylalanine to cinnamic acid, and it transformation into lignoids (flavonoids in condensed tannins, cinnamyl alcohols in lignins), was the crucial determining biochemical phenomenon in the colonization of the land by plants. Phenylpropanoids exert pronounced biological actions in animals (Gottlieb and Salatino, 1987). The primary structures of phenylpropanoids found in Piperaceae species are shown in Figure 20.1.

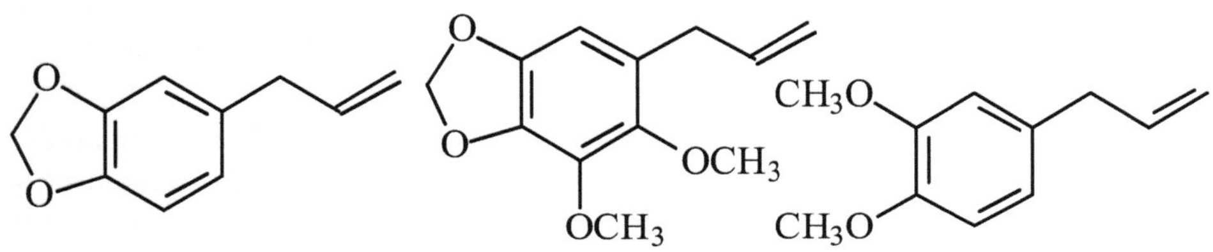

Figure 20.1: Examples of phenylpropanoids found in Piperaceae species.

A differentiated group of phenylpropanoids is lignans, which encompass a broad spectrum of structural models and molecular sizes. They are found in various parts of the plant (stem, rhizome, roots, seeds, oils, resins, flowers, leaves and bark) and their amounts vary depending, on the tissues and the species (Davin and Lewis, 1998).

The functions of lignans are related primarily to plant defense, and their formation is constitutive or stress-induced; their deposits contribute to the durability, color and quality of plant tissue (Burlat *et al.*, 2001). This component is also attributed with antioxidant, antiherbivore, bactericide, fungicide, ad antiviral functions, and phytotoxicity to other plant species (Chu *et al.*, 1993). There is evidence that the accumulation of lignans in damaged sites inhibit the enzymes secreted by fungi, preventing the lignan from breaking down (Ward, 1997).

Lignins and lignans are accumulated in the same tissue; lignin give rigidity to the tissue, while lignan protects the plant against microorganisms (Davin Lewis, 1998).

The binding of lignans to the methylenedioxyphenyl group is a characteristic of Piperaceae, and they are considered important cytochrome P450-dependent monooxygenase inhibitors, used as synergists of natural insecticides (Murkerjee *et al.*, 1979; Bernard *et al.*, 1990). Synthetic derivatives of the chemical group showed synergistic effect for carbamate insecticides (Wilkinson *et al.*, 1966), and also act alone as an insecticide (Star, 2005).

Terpenoids

Terpenes have an important ecological role by acting as internal and external messengers, functioning as allelopathic agents, insect repellent, or to insect attractant, to promote plant pollination (Harrewijan *et al.*, 2001). There is evidence that its biosynthesis is induced by insect feeding, probably through oral secretion from the individuals (Paré and Tumlinson, 1997).

Terpenes can be classified according to the number of C5 units: monoterpenes, C10, sesquiterpenes, C15; diterpenes, C20; sesquiterpenes, C25, triterpenes, C30, and tetraterpenes, C40 (Castro *et al.*, 2004).

The insecticidal action of mono and sesquiterpenes appears to be related to the inhibition of acetylcholinesterase. The majority of studies report that the higher terpenoids act by inhibiting or retarding growth, damaging the insect maturation, reducing its the reproductive capacity, and suppressing the appetite, which may lead to death of the insect by starvation or direct toxicity (Viegas Júnior, 2003).

The majority of monoterpenes are volatile, and are the basic constituent of aromatic oils (essential oils or essences). Some are precursors of a special class of substances, iridoids and secoiridoids, others as structural units of some types of complex alkaloids (Castro *et al.*, 2004).

Some sesquiterpenes are present in various essential oils, such as α-humulene, β-caryophyllene, β-farnesene and α-bisabolol. The chemical structure of some constituents of the essential oils is found in Figure 20.1. Other more complex and more functionalized sesquiterpenes have an ecological function, or are active constituents of some medicinal plants such as sesquiterpene lactones (Castro *et al.*, 2004).

Amides

Based on the report of Parmar *et al.* (1997), 145 different amides had been isolated from plants of the Piperaceae family, this number can now be assumed to be over 300, though no references have been found to support this.

Piperine (Figure 20.3) was the first amide to be isolated from the fruit of the *Piper* species, and its chemical constituents have widely investigated, including the unsaturated lipophilic amides, considered as precursor metabolites in the synthesis of various other amide analogs. Besides being the major group of plant metabolites, these are primarily responsible for the insecticidal activity, Parmar *et al.* (1997).

The insecticidal action of the amide piperine, derived from *Piper nigrum* L, was proven in 1924 (Su, 1977). It is considered a metabolite precursor of the synthesis of

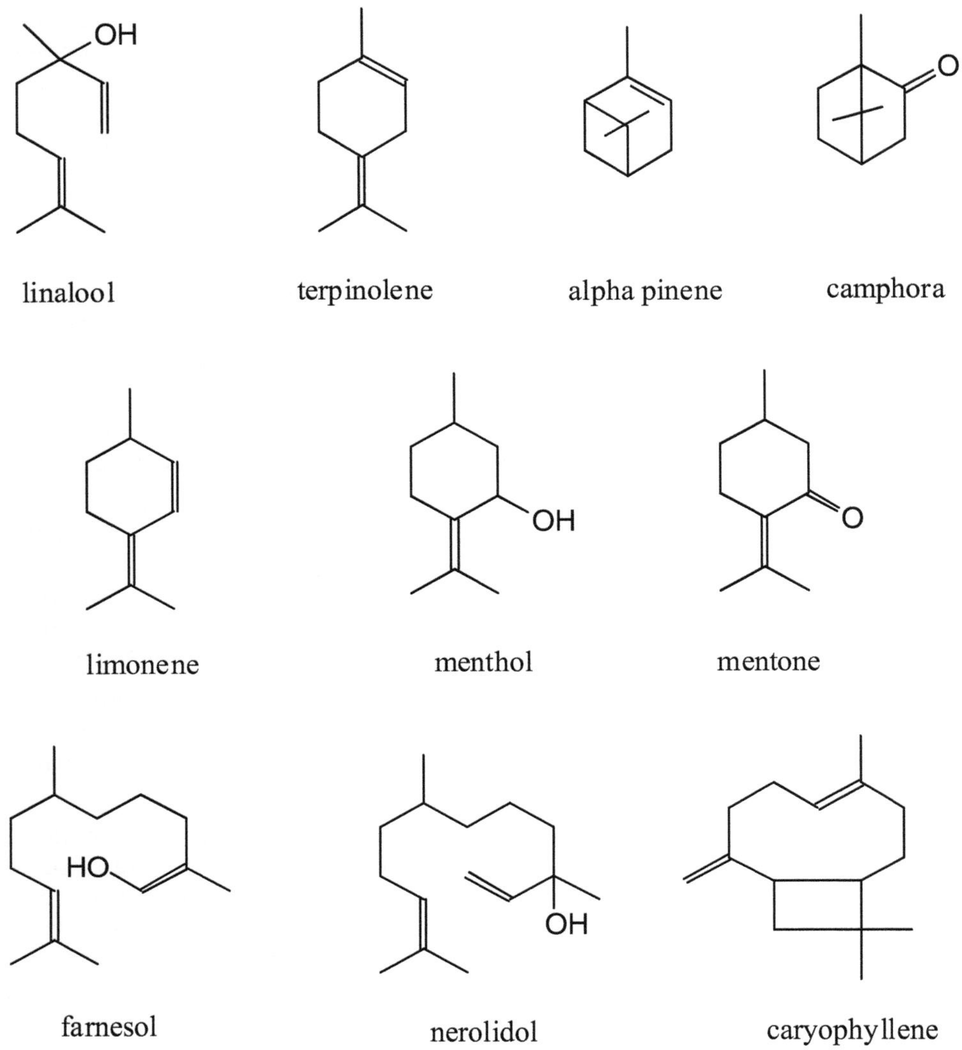

Figure 20.2: Examples of monoterpenes and sesquiterpenes.

several other amide analogs. In general, amides have insecticidal neurophysiological action (Scott *et al.*, 2008).

Plants of the Piperaceae family constitute a source of long chain unsaturated isobutylamides with insecticidal properties, like piperine (Strunz and Finlay, 1994). Estrela *et al.* (2005) investigated the piperine amide analogs with the N-hexyl, N-isopropyl, and N-isopentyl groups bound to isopentyl (3,4-methylenedioxyphenyl) amide. They observed that these amides caused high toxicity to the caterpillar *S. frugiperda*, causing mortality and deformities involved in its vital activities.

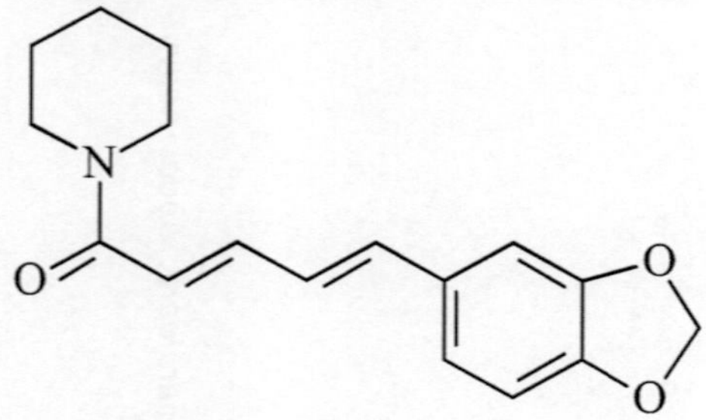

piperine

Figure 20.3: Isolated amide from species of the genus *Piper*.

It can be inferred that the presence of the methylenedioxyphenyl bound to some amide groups confers stability to the molecule, and removing this radical, according in Elliott *et al.* (1987), practically nullifies the insecticidal action of the amide.

This was evidenced by the work of Scott *et al.* (2002), who demonstrated that piperamides are bifunctional when combined with the molecule methylenedioxyphenyl, acting as neurotoxin and cytochrome P450 enzyme inhibitor. This chemical characteristic is common in plants of *Piper*, and it is considered a defense strategy of the plant, against herbivores (Navickiene *et al.*, 2006).

Of the 53 Piperaceae species evaluated, 45 (84.9 per cent) presented sesquiterpenes as the predominant compounds, followed by 17 others (32.7 per cent) with a predominance of monoterpenes, and of these, 15 species presented sesquiterpenes as the second most predominant compounds. Only five species (9.4 per cent) were predominant in the phenylpropanoid in essential oil, and only in *P. marginatum* and *P. millegranum* were predominant together with monoterpenes, sesquiterpenes and phenylpropanoids (Table 20.1).

The effectiveness of essential oils for the control of at least one species of insect that is harmful to crops or human health was demonstrated for only seven Piper species (13.2 per cent): *P. aduncum* L., *P. callosum*, *P. divaricatum*, *P. hispidinervum*, *P. hostimannianum*, *P. humaytanun* and *P. marginatum*.

The essential oil of *P. aduncum* is the most widely studied oil in terms of insecticidal properties, and has efficacy in the control of the crop pests: *C. tingomarianus* Bechyné, *Tenebrio molitor* L, 1758 (Fazolin *et al.*, 2005 and Fazolin *et al.*, 2007), *Sitophilus zeamais* Mots., 1865 (Star *et al.*, 2006), *Solenopsis saevissima* Smith 1855 (Souto 2005), all of which are present in Brazil. *Choristoneura rosaceana* (Harris) an important polyphagous pest in North America and Europe was also controlled by *P. aduncum* oil (Laroque *et al.*, 1999). Moreover, it was also found to controls insect vectors of human diseases: *Anopheles marajoara* (Galvão and Damasceno, 1942) and *Aedes aegypti* L (Souto, 2005). Its insecticidal action is related to the wealth of lignans associated with dillapiole, a phenylpropanoid that in this association, inhibits reactions of the cytochrome P450 monooxygenase (Murkerjee *et al.*, 1979; Bernard *et al.*, 1990). Bernard *et al.* (1995) argue that this compound alters the detoxification capacity of insects,

Table 20.1: Piper species occurring in the Amazon and quantification of the compounds of essential oils with potential insecticide use.

Piper Species of Amazon	*Number of Specimens*	*Monoterpenes (Per cent)*	*Sesquiter-penes (Per cent)*	*Phenylpro-panoids (Per cent)*	*Species of Insects Controlled*
Piperaduncum L.	21			97.0	*C. tingomarianus; T. molitor; S. zeamais; Anopheles marajoara; A. aegypti; S. saevissima; Choristoneura rosaceana*
Piper alatipentiolatum Yuncker	3		99.8		
Piper amapaense Yuncker	1		96.9		
Piper anonifolium Kunth	8*	81.8	75.5		
Piper aramanum C.DC.	2		84.7		
Piper arboreum Aubl. Vr. *Arboreum*	8		99.8		
Piper baccans (Miq.) C.DC.	2		82.9		
Piper bartlingianum (Miq.) C.DC.	2		98.3		
Piper brachypetiolatum Yuncker	1		79.5		
Piper brasiliense C.DC.	1		75.3		
Piper brevesanum Yuncker	1		96.4		
Piper callosum Ruiz and Pav.	8			83.9	*Anopheles marajoara; Aedes aegypti; S. saevissima*
Piper carniconnectivum C.DC.	6		97.7		
Piper cernnum Vell.	1	39.6		45.0	
Piper colubrinum (Link ex Kunth) Link ex C.DC.	3*	71.1	96.6		
Piper crassinervium Kunth	1		70.2		
Piper cyrtopodon (Miq.) C.DC.	7		81.1		
Piper dactylostigmum Kunth	1		87.9		
Piper demeraenum (Miq.) C.DC.	4	96			

Contd...

Table 20.1–*Contd...*

Piper Species of Amazon	*Number of Specimens*	*Monoterpenes (Per cent)*	*Sesquiter-penes (Per cent)*	*Phenylpro-panoids (Per cent)*	*Species of Insects Controlled*
Piper dilatatum Rich.	10		94.8		
Piper divaricatum G. Mey	11			92.1	*Anopheles marajoara; A. aegypti; S. saevissima*
Piper duckei C.DC.	2		98.8		
Piper durilignum	1		98.6		
Piper erectipilum	1		67.9		
Piper glandulosissimum	7		84.1		
Piper gurupanum	2		100.0		
Piper hispidinervum C.DC.	12			97.3	*T. molitor; S. zeamais*
Piper hispidum Sw.	12*	83.6	97.9		
Piper hostmannianum (Miq.) C.DC.	8*	62.1	78.8		*A. aegypti*
Piper humaytanum Yuncker	5		87.4		*A. aegypti*
Piper kegelianum (Miq.) C.DC.	1	86.2	83.0		
Piper klotzschianum (Kunth) C.DC.	1	82.6			
Piper krukoffii Yuncker	1			83.3	
Piper macedoi Yuncker	2*	53.2	33.9		
Piper manausense Yuncker	4		77.0		
Piper marginatum sensu lato	26*	58.9	87.9	91.4	*A. aegypti; Anopheles marajoara; S. saevissima*
Piper marstersianum C.DC.	1		91.8		
Piper millegranum Yuncker	4*	43.8	73.3	45.7	

Contd...

Table 20.1–*Contd...*

Piper Species of Amazon	*Number of Specimens*	*Monoterpenes (Per cent)*	*Sesquiter-penes (Per cent)*	*Phenylpro-panoids (Per cent)*	*Species of Insects Controlled*
Piper montealegreanum Yuncker	1	31.8	67.8		
Piper nigrispicum C.DC.	1		96.7		
Piper ottonoides Yuncker	1		94.8		
Piper ovatilimbum C.DC.	1		89.2		
Piper pellitum C.DC.	1	25.7	61.3		
Piper peltatum (L.) Miq.	5		96.2		
Piper piresii Yuncker	1		97.6		
Piper plurinervosum Yuncker	1	42.6	54.6		
Piper reticulatum L.	2		60.5		
Piper schwackei C.DC.	8		98.9		
Piper trichocarpum C.DC.	2		72.7		
Piper tuberculatum Jacq. Var. *Tuberculatum*	11	83.8	69.9		
Piper utinganum Yuncker	1		95.2		
Piper vitaceum Yuncker	1	49.6	43.6		
Piper wachenheimii Trel.	1	60.5	36.8		

Adapted from Andrade *et al.* (2009).

–* Predominant compounds in two different specimens within the same species.

which are poisoned with compounds present in the food that would normally be eliminated gradually.

This phenomenon is of interest due the fact that in coevolution between herbivores and plants, the selection of individuals of both plant and animal species occurs, depending on the ability to adapt and survive in this chemical struggle. Thus, the mode of action of dillapiole is altering the capacity of the insect, through the loss of chemical adaptation, developed over time, this capacity returning to the primitive stages of protection.

In addition to the insect species mentioned above, the essential oils of *P. hostimannianum*, *P. humaytanun* and *P. marginatum* are also effective against *A. aegypty*, an important vector of the dengue fever virus (Mitchell *et al.*, 2007).

Of the 53 species occurring in the Amazon, only *P. tuberculatum* had isolated that amides that showed insecticidal action; pelitorin and 4.5 dihydro piperlonguminine (Table 20.2). These amides were effective in the control of important pests of soybean *Anticarsia gemmatalis* Hueb., 1818 and sugar cane, *Diatraea sacharallis* (Fabr, 1794) (Navickiene, 2007). The extracts of this Piperaceae also presented insecticide activity against *A. aegypti* (Pohlit *et al.*, 2004) (methanol and aqueous extracts), *Diatraea sacharallis* (methanol, ethanol and dichloromethanol extracts) (Soberon *et al.*, 2006) and *Ostrinia nubilalis* (Hübner) (ethanol extract) (Bernard *et al.*, 1995), the latter being an important polyphagous pest distributed throughout Europe, Asia, Africa, North America (Pohlit *et al.*, 2004).

Aqueous extracts of *P. aduncum* also presented efficient control of *Aetalion reticulatum* (L., 1767) (Smith *et al.*, 2007), known as the "cricket of the orchards," and a widespread pest in Brazil.

Based on the above, it is clear that there is need for further research with Amazonian Piperaceae, in order to evaluate its potential insecticide use. Considering an estimated incidence of approximately 250 species of Piperaceae in the Amazon, only 27.2 per cent (53 species) have been investigated for their chemical constituents, and only 6 (2.4 per cent) of them have had some kind of assessment for their insecticide potential, whether through the use of essential oils, extracts or isolation products, or synthesis of amides.

Considering the wealth of insecticidal compounds of these Amazonian *Piper* species, as demonstrated by Andrade *et al.* (2009) (Tables 20.1 and 20.2), mainly present in essential oils rich in mono- and sesquiterpenes and phenylpropanoids, it can be concluded that there is much still to be investigated.

If the objective assessment of the insecticide potential is made on the basis of the predominance of phenylpropanoids, at least three species should be further investigated: *P. cernnum*, *P. Kukroffii* and *P. millegranum*, whose specimens showed 45.0 per cent, 83.3 per cent and 45.7 per cent of that compound, respectively (Table 20.1). It is noteworthy that in the case of occurrence of levels near 45 per cent phenylpropanoids, this is ordinarily associated with the occurrence of terpenoids in high proportions, which can increase the potential for insecticide use these species of Piperaceae.

Table 20.2: Piper species that occur in the Amazon and quantification of extracts or isolated compounds with potential insecticide use.

Piper Species of Amazon	*Isolated Amides with Insecticide Activity*	*Species of Insects Controlled*	*Types of Extracts Evaluated as Insecticide*	*Species of Insects Controlled*
Piper aduncum L.			1-Methanol; 2- Aqueous; 3- Ethanol	*A. aegypti (1 and 2); Aetalion* sp. (*2*); *O. nubilalis* (*3*)
Piper tuberculatum Jacq. Var. *tuberculatum*	Pelitorin and 4,5-dihydro piperlonguminine	*A.gemmatalis; D. sacharallis*	1-Methanol; 2- Aqueous; 3- Ethanol;4- Dichlorome-thanol	*A.aegypti* (*1 and 2*); *O. nubilalis.* (*3*); *Diatraea sacharallis* (*1,3, and 4*)

Adapted from Andrade *et al.* (2009).

Also, considering the wealth of phenylpropanoids in the essential oils of eight species (Table 20.1), there is a need to evaluate the potential use as a synergistic when combined with conventional insecticides. This may attract the interest of the agrochemical industry to promote a more economically advantageous substitute for piperonyl butoxide, which is produced from safrole (*P. hispidinervum*) and is currently the most widely used synergistic in insecticides worldwide.

Of the 53 species of *Piper* evaluated by Andrade *et al.* (2009), 45 (84.9 per cent) species, rich in terpenoids (Table 20.1), have not been evaluated for their insecticide potential. This opens a vast prospect of bioprospecting essential oils that can be used for insect pest control.

With the rapid growth in demand for products of organic origin, and that are compatible for use in subsistence farming, the potential insecticide use of essential oils and extracts of Piperaceae may justify greater investment in research with plants of the Amazon.

There is a visible lack of information on the insecticidal activity of the different extraction methods and amides isolated from 53 species evaluated in this study, with only two species–*P. aduncum* and *P. tuberculatum* presenting results that indicate the potential use of these substances (Table 20.1). Regarding the isolation of amides with insecticidal activity, *P. tuberculatum* was the only species studied, resulting in the isolation of pelitorine and 4.5 diidropiperlonguminine with marked insecticidal activity. Bernard *et al.* (1995) considers that the isobutylamides from *Piper*, because they contain nitrogen atoms with simple and low molecular weight, could theoretically be biosynthesized with low production costs.

In terms of the simplicity of obtaining some types of extracts, mainly aqueous and alcoholic, these have higher possibilities for use, but the limitation to their successful use is the lack of knowledge of their components and guarantee of homogeneity of the product.

Besides furthering phytochemical knowledge of Piperaceae, there is still much to be done to domesticate species of commercial interest, as well as improving and adapting the industrial process of hydrodistillation. In relation to the spray application of essential oils of *Piper* in particular, evaluations of the phytotoxic effect of lethal concentrations to the pests in certain plant species should be observed.

Consumer concerns over the environment, food quality and side effects of synthetic insecticides have led researchers to test alternative forms of pest control. Research into alternatives to synthetic chemical insecticides is very important, not only for controlling insect pest infestations, but also for safeguarding consumers' health against toxic residues (Sriranjini and Rajendran, 2008).

Recent works have reported high insecticide capacity of oils of the *Piper* species. Coitinho *et al.* (2010) evaluated the persistence of essential oils and of the compound eugenol in corn seeds infested with maize weevil, *Sitophilus zeamais*. The results of this study indicate that of the seven essential oils tested, the essential oil of *P. marginatum* was more persistent on maize grain for 120 days of storage, with better results than the standard eugenol.

Essential oil of *P. hispidinervum* showed insecticidal activity against *Macrosiphum euphorbiae* (Thomas) (Hemiptera: Aphididae) on rose bushes (Smith *et al.*, 2012).

Conclusion

Based on this literature review, it is concluded that several compounds present in *Piper* species have bioactivity in insects that cause diseases in plants and animals, enabling the use of the isolates and extracts or oils from these plants as alternatives to synthetic chemical insecticides for the control of pest infestations, minimizing the issue of toxic residues and lowering production costs. However, further research is needed to evaluate the chemical composition of oils and extracts of a wider range of Amazonian *Piper* species, and to evaluate their activity and isolated compounds in the control of insects that are harmful to agriculture or human and animal health.

Acknowledgement

The authors thank the Brazilian National Council for Scientific and Technological Development (CNPq) and the Brazilian Coordination for the Improvement of Higher Education Personnel (CAPES) for financial support.

References

Alécio, A.A., Bolzani, V.S., Young, M.C.M., Kato, M.J., and Furlan, M. (1997). Antifungal amides from *Piper hispididum*. *Journal Natural Products*, **61**: 637-639.

Andrade, E.H., Guimarães, E.F., and Maia, J.G.S. (2009). Variabilidade química em óleos essenciais de espécies de *Piper* da Amazônia. FEQ/UFPA, Belém.

Bernard, C.B., Arnason, J.T., Philogene, B.J.R., Lam, J., and Waddel, T. (1990). *In vivo* effect of mixtures of allelochemicals on the life cycle of the European corn borer, *Ostrinia nubilalis*. *Entomologia experimentalis et applicata*, **57**: 17-22.

Bernard, C.B., Krishinamurty, H.G., Chauret, D., Durst, T., Philogene, B.J.R., Sanchés-Vindas, P., Hasbaun, C., Poveda, L., Roman, L.S., and Arnason, J.T. (1995). Insecticidal defenses of piperaceae from the neotropics. *Journal Chemical Ecology*, **21**: 801-814.

Burlat, V., Kwon, M., Davin, L.B., and Lewis, N.G. (2001). Dirigent proteins and dirigent sites in lignifying tissues. *Phytochemistry*, **57**: 883-897.

Castro, H.G., Ferreira, F.A., Silva, D.J.H., and Mosquim, P.R. (2004). Contribuição ao estudo das plantas medicinais: Metabólitos secundários. Gráfica Suprema e Editora, 113p, Visconde do Rio Branco.

Chu, A., Dinkova, A., Davin, L.B., Bedgar, D.L., and Lewis, N.G. (1993). Stereospecificity (+)- piroresinol and (+) – lauriciresinol reductases from *Forsythia intermedia*. *Journal Biologycal Chemistry*, **268:** 27026- 27033.

Coitinho, R.L.B.C., Oliveira, J.V., Gondim Júnior, M.G.C., and Câmara, C.A.G. (2010). Persistência de óleos essenciais em milho armazenado, submetido à infestação de gorgulho do milho. *Ciência Rural*, **40:** 1492-1496.

Dixon, R.A. (2001). Natural products and plants disease resistance. *Nature*, **411**: 843-847.

Dyer, A., and Palmer, D.N. (2004). Piper: A model genus for studies of photochemistry, ecology and evolution. Kluwer Academic Plenum Publishers, New York.

Elliott, M., Farnham, A.W., Janes, N.F., Johnson, D.M., and Pulman, D.A. (1987). Synthesis and insecticidal activity of lipophilic amides. Part 4: The effect of substituents on the phenyl group of 6-phenylhexa-2,4-dienamides. *Pesticide Science*. **18**: 223-228.

Estrela, J.L.V., Guedes, R.N.C., Maltha, C.R.A., Magalhães, L.C., and Fazolin, M. (2005). Toxicidade de amidas análogas à piperina para *Spodoptera frugiperda* (J. E. Smith, 1797) (Lepidoptera: Noctuidae). *Magistra*, **17**: 69-75.

Estrela, J.L.V., Fazolin, M., Catani, V., Alécio, M.R., and Lima, M.S. (2006). Toxicidade de óleos essenciais de *Piper aduncum* e *Piper hispidinervum* em *Sitophilus zeamais*. *Pesquisa Agropecuária Brasileira*, **41**: 217-222.

Fazolin, M., Estrela, J.L.V., Catani, V., Lima, M.S., and Alécio, M.R. (2005). Toxicidade do Óleo de *Piper aduncum* L. a adultos de *Cerotoma tingomarianus* Bechyné (Coleoptera: Chrysomelidae). *Neotropical Entomology*, **34**: 485-489.

Fazolin, M., Estrela, J.L.V., Catani, V., Alécio, M.R., and Lima, M.S. (2007). Propriedade inseticida dos óleos essenciais de *Piper hispidinervum, Piper aduncum e Tanaecium nocturnum* sobre *Tenebrio molitor*. *Ciência e Agrotecnologia*, **31**: 113-120.

Gottlieb, O.R., and Salatino, A. (1987). Função e evolução de óleos essenciais e de suas estruturas secretoras. *Ciência e Cultura*, **39**: 707-716.

Gottlieb, O.R., Birin, M.R., and Kaplan, M.A.C. (1995). Biosynthetic interdependence of lignins and secondary metabolites in angiosperms. *Phytochemistry*, **40**: 99.

Harreewijan, P., Van Osten, A.M., and Piron, P.G.M. (2001). Natural Terpenoids as Messengers: A Multidisciplinary Study of their Production, Biological Functions and Practical Applications, Kluwer Academic Publishers, London.

Jaramillo, A.M., and Manos, S.P. (2001). Phylogeny and patterns of floral diversity in the genus *Piper* (Piperaceae). *American Journal of Botany*, **88**: 706-716.

Jaramillo, A.M., and Marquis, R. (2004). Current Perspectives on the Classiûcation and Phylogentics of the Genus *Piper* L. *In*: DYER A, PALMER DN (eds.). *Piper:* A model genus for studies of phytochemistry, ecology, and evolution. Kluwer Academic Plenum Publishers, New York. 179-198 p.

Labandeira, C.C., Dilcher, D.L., Davis, D.R., and Wagner, D.L. (1994). Ninety- seven million yearsof angiosperm-insect association: paleobilogical insight into the meanig coevolution. *Procccedings of the National Academy of Ciences of the United States of America*, **91**: 12278-12282.

Laroque, N., Vincent, C., Bélanger, A., and Bourassa, J.P. (1999). Effects of tansy essential oil from *Tanacetum vulgare* on biology of oblique-banded leafroller, *Choristoneura rosaceana*. *Jounal of Chemical Ecology*, **25**: 1319-1330.

Lewis, N.G., and Davin, L.B. (1998). Biochemical control of monolignol coupling and structure during lignin and lignan biosynthesis. *ACS Symposium Series*, **679**: 334- 361.

Morais, S.M., Facundo, V.A., Bertini, L.M., Cavalcanti, E.S.B., Anjos Júnior, J.F., Ferreira, A.S., Brito, E.S., and Souza Neto, M.A. (2007). Chemical composition and larvicidal activity of essencial oils from *Piper* species. *Biochemical Systematics and Ecology*, **35**: 670-675.

Mukerjee, S.K., Saxena, V.S., and Tomar, S.S. (1979). New methylenedioxyphenyl synergists for pyrethrins. *Journal of the Science and Food Agriculture*, **27**: 1209-1211.

Navickiene, H.M.D., Morandim, A.A., Alécio, A.C., Regasini, L.O., Bergamo, D.C.B., Telascrea, M., Cavalheiro, A.J., Lopes, M.N., Bolzani, V., Furlan, M., Marques, M.O.M., Young, M.C.M., and Kato, M.J. (2006). Composition and antifungal activity of essential oils from *Piper aduncum*, *Piper arboreum*and, *Piper tuberculatum*. *Quimica Nova*, **29**: 467-470.

Paré, W., and Tumlinson, J.H. (1997). Biosynthesis of volatiles induced by insect herbivory in cotton plants. *Plant Physiology*, **114**: 1161-1167.

Parmar, V.S., Jain, S.C., Bsht, K.S., Jain, R., Taneja, P.A., Tyagi, O.D., Prasad, A.K., Wengel, J., and Olsen, C.E.P. (1997). Phytochemistry of the genus *Piper*. *Phytochemistry*, **46**: 597-673.

Pohlit, A.M., Quignard, L.J., Nunomura, S.M., Tadei, W.P., Hidalgo, A.F., Pinto, A.C., Santos, E.V.M., Morais, S.K.R., Saraiva, R.C.G., Ming, L.C., Alecrim, A.M., Ferraz, A.B., Pedroso, A.C.S., Diniz, E.V., Finney, E.K., Gomes, E.O., Dias, H.B., Souza, K.S., Oliveira, L.C.P., Don, L.C., Queiroz, M.M.A., Henrique, M.C., Santos, M., Lacerda Júnior, O.S., Pinto, P.S., Silva, S.G., and Graça, Y.R. (2004). Screening of plants found in the State of Amazonas, Brazil for larvicidal activity against *Aedes aegypti* larvae. *Acta Amazonica*, **34**: 97-105.

Pletsch, M., and Sant´ana, A.*E.G.* (1995). Secondary compound accumulation in plants- The application of plant biotechnology to plant improvement. *Chemistry of Amazonian*, **5**: 51-64.

Rajendran, S., and Sriranjini, V. (2008). Plant products as fumigants for stored product insect control. *Journal of Stored Products Research*, **44**: 126-135.

Sangwan, N.S., Farooqi, A.H.A., Shabid, F., and Sangwan, R.S. (2001). Regulation of essential oil production in plants. *Plant Growt Regulation*, **34**: 3-21.

Scott, I.M., Puniani, E., Durst, T., Phelps, D., Merali, S., Assabgui, R.A., Sanchez-Vindas, P., Poveda, L., Philogé, B.J.R., and Arnason, J.T. (2002). Insecticidal activity of *Piper tuberculatum* Jacq. extracts: synergistic interaction of piperamides. *Agricultural and Forest Entomology*, **4**: 137-144.

Scott, I.M., Jensen, H.R., Philogéne, B.J.R., and Arnason, J.T. (2008). A review of *Piper* spp. (Piperaceae) phytochemistry, insecticidal activity and mode of action. *Phytochemistry Review*, **7**: 65–75.

Soares, C.S.A., Silva, M., Costa, M.B., Bezerra, C.E., Carvalho, L.M., and Soares, A.H.V. (2012). Atividade inseticida de óleos essenciais sobre o pulgão da roseira *Macrosiphum euphorbiae* (Hemiptera: Aphididae). *Revista Brasileira de Agroecologia*, **7**: 169-175.

Soberón, V.G., Rojas, C., Saavedra, J., Massuo, J.K., and Delgado, G.E. (2006). Acción biocida de plantas de *Piper tuberculatum* Jacq. sobre *Diatraea saccharalis* (Lepidoptera, Pyralidae). *Revista Peruana de Biologia*, **13**: 107-112.

Souto, R.N.P. (2005). Avaliação das atividades repelentes, larvicida e inseticida de oleos essenciais de *Piper* da Amazônia contra *Anopheles marajoara* Galvão and Damasceno, *Sregomyia aegypti* (L. e *Solenopsis saevissima* (F. Smith,1855). Thesis (Doutor degree in Zoology), Federal University of Pará, Belém.

Strunz, G.M., and Finlay, H. (1994). Concise, efficient new synthesis of pipercide, an insecticidal unsaturated amide from *Piper nigrum*, and related compounds. *Tetrahedron*, **50**: 11113-11122.

Su, H.C.F. (1977). Inseticidal properties of black pipper to rice weevil and cowpea weevils. *Journal of Economic Entomology*, **70**: 19-22.

Taylor, D.W., and Hickey, L.J. (1992). Philogenetic evidence for the herbaceous origin of angiospermae. *Plant Systematics and Evolution*, **180**: 177-178.

Viegas Júnior, C. (2003). Terpenos com atividade inseticida: uma alternativa para o controle químico de insetos. *Química Nova*. 26: 57-76.

Yuncker, T.G. (1972). New Piperaceae of Brazil I: Piper – Group, I, II, III, IV. *Hoehnea*. **2**: 19-366.

Ward, R.S. (1997). Lignans, neolignans and related compounds. *Natural Product Reports*.**14**: 43-74.

Wilkinson, C., Metcalf, R.L., and Fukuto, T.R. (1996). Some structural requirements of methylenodioxyphenyl derivates as synergists of carbamate insecticides. *Journal of Agricultural and Food Chemistry*, **14**: 73-79.

Medicinal Plants: Phytochemistry, Pharmacology and Therapeutics Vol. 3 (2014) *Pages* **441–457**
Editor-in-Chief: **V.K. Gupta**
Published by: **DAYA PUBLISHING HOUSE, NEW DELHI**

21

Cannabis and Cannabinoids: Phytochemistry, Pharmacology and Therapeutics–A Review

Leena Parihar[1]* and Priyanka Kapoor[2]

ABSTRACT

Cannabis is a commonly used illicit drug that has physiological and psychological effect, and has various intake routes (intravenous, smoke, inhalation and oral). Cannabis contains sixty one cannabinoids, of which the major active ingredient, δ9-Tetrahydrocannabinol (THC), appears to be primarily responsible for its psychoactive, immunomodulatory effects and is also seen to have adverse effects in reproductive and respiratory system.

Keywords: THC, Cannabinoids, Lymphocyte, Neurotransmitter, Psychoactive, Immunomodulatory.

Introduction

Cannabis sativa (cannabis) is among the earliest plants cultivated by human being. The first evidence of the use of cannabis was found in China, where archaeological and historical findings indicate that that plant was cultivated for fibers since 4.000 B.C., with the fibers obtained from the cannabis stems, the Chinese manufactured strings, ropes, textiles, and even paper (Li and lin,1974). The Chinese also used cannabis fruits as food which are small (3 to 5 mm), elliptic, smooth, with a hard shell, and contain one single seed within (Touwn,1981).

1 Division of Research and Development, Lovely Professional University, Punjab, India

2 Department of Biotechnology and Biosciences, Lovely Professional University, Punjab, India

* *Corresponding author*: E-mail: leena.parihar@gmail.com

Cannabis and its Constituents

1. Cannabinoids: 61 known (Table 21.1)
2. Nitrogenous compounds: 20 known
3. Amino acids: 18 known
4. Proteins, glycoproteins, and enzymes: 9 known
5. Sugars and related compounds: 34 known
6. Hydrocarbons: 50 known
7. Simple alcohols: 7 known
8. Simple aldehydes: 12 known
9. Simple ketones: 13 known
10. Simple acids: 20 known
11. Fatty acids: 12 known
12. Simple esters and lactones: 13 known
13. Steroids: 11 known
14. Terpenes: 103 known
15. Noncannabinoid phenols: 16 known
16. Flavanoid glycosides: 19 known
17. Vitamins: 1 known
18. Pigments: 2 known (Turner *et al.*, 1980)

Cannabinoid content generally changes with change in environmental conditions. It has been seen that increase in cannabinoid content with increase of temperature (32° C. vs. 22° C.) (Boucher *et al.*, 1974).

Adverse Effects of Cannabis

Cannabis preparations are largely derived from the female plant of *Cannabis sativa.* The primary psychoactive constituent is delta-9-tetrahydrocannabinol (THC) and its content is highest in the flowering tops, declining in the leaves, lower leaves, stems, and seeds of the plant. Marijuana (THC content 0·5–5·0 per cent) is prepared from the dried flowering tops and leaves; hashish (THC content 2–20 per cent) consists of dried cannabis resin and compressed flowers; and hashish oil may contain between 15 per cent and 50 per cent THC. Sinsemilla and Netherwood varieties of cannabis may have a THC content of up to 20 per cent. Cannabinoids act on a specific receptor that is widely distributed in the brain regions involved in cognition, memory reward, pain perception, and motor coordination. These receptors respond to an endogenous ligand, anandamide, which is much less potent and has a shorter duration than THC. The identification of a specific cannabinoid antagonist promises to improve our understanding of the role of cannabinoids in normal brain function (Adams and Martin, 1996).

Table 21.1: Cannabinoids: 61 known.

Phytocannabinoid	*Known Variant*	*Structure*	*Mechanism*	*Assay*	*Properties*
CBD (Thomas *et al.*, 2007; Pertwee,1997)	7	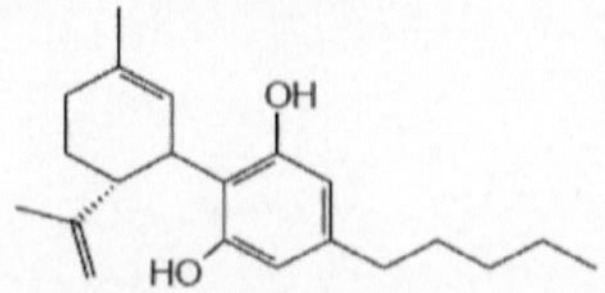	Antagonist of CB1/CB2 agonists	[35S]GTPgS binding to mouse brain membranes (CB1) and to hCB2-CHO cell membranes	CBD antagonises cannabinoid induced antispasmodic effect in the isolated vas deferens as well as the *in vivo* responses to D9-THC in animals and humans.
(Thomas *et al.*, 2007)			CB2 inverse agonist	[35S]GTPgS binding to hCB2-CHO cell membranes	To be determined. Potential role in CBD-induced anti-inflammatory effects
(Ligresti *et al.*, 2006; Capasso *et al.*, 2008)			FAAH inhibition	Measurement of [14C]ethanolamine released from [14C]anandamide by membranes prepared from N18TG2 cells	CBD reduces FAAH expression in the inflamed intestine and probably via this mechanism, reduces inflammation-induced intestinal hypermotility in mice
(Ligresti *et al.*, 2006)			Anandamide reuptake inhibitor	[14C]anandamide uptake by basophilic leukaemia or MDA-MB-231 cells	To be determined
(Ryberg *et al.*, 2007)			GPR55 antagonist	Antagonism of CP55970-induced activation of [35S] GTPgS binding to transfected HEK293S cells	To be determined
(Ahrens *et al.*, 2009)			Positive allosteric modulator at a1 and a1b glycine receptors.	Measurement of the current response to glycine in HEK 293 cells expressing a1 or a1b receptors	To be determined. In the dorsal horn of the spinal cord, glycine acts as an inhibitory postsynaptic neurotransmitter, with a role in chronic pain after inflammation or nerve injury

Contd...

Table 21.1–*Contd...*

Phytocannabinoid	*Known Variant*	*Structure*	*Mechanism*	*Assay*	*Properties*
			Opioid receptor ligand	Inhibition of [3DNM] (m opioid receptor ligand) binding to rat brain synaptosomal membranes	To be determined. CBD could potentially enhance the effects of opiates
			Positive Allosteric modulator at m and d opioid receptors	H3-DAMGO and H3-naltrindole (m and d opioid receptor ligand) binding to rat cerebral cortical membranes	The effect occurs at very high concentrations and cannot be expected to contribute to the *in vivo* action of CBD
(De Petrocellis *et al.*, 2008)			TRPA1 agonist	Increase of $[Ca^{2+}]i$ in TRPA1-HEK-293 cells	To be determined. Potential role in CBD analgesic effects
			TRPM8 antagonist	Antagonism of icilin- or menthol-induced increase in $[Ca^{2+}]$ in TRPM8-HEK-293 cells	To be determined. Potential role in CBD analgesic effects. Potential role in prostate Carcinoma
(De Petrocellis *et al.*, 2008)			TRPV1 agonist	Increase of $[Ca2^{+}]i$ in TRPV1-HEK-293 cells	To be determined. TRPV1 is involved in CBD anti-psychotic and analgesic effects
(Qin *et al.*, 2008)			TRPV2 agonist	Ca^{2+} mobilization in TRPV2-HEK-293 cells	The effect is shared b yD9-THC and CBN TRPV2 activation by CBD may mediate CGRP release from cultured rat dorsal root ganglion neurons
(Carrier, 2006)			Adenosine uptake competitive inhibitor	[3H]adenosine uptake in murine microglia and macrophages	CBD decreases TNF-a in wildtype but not in A2A receptor deficient mice. Its anti-inflammatory effects in

Contd...

Table 21.1–*Contd...*

Phytocannabinoid	*Known Variant*	*Structure*	*Mechanism*	*Assay*	*Properties*
					the retina are linked to the inhibition of adenosine uptake
Sullivan *et al.*, 2009)			PPARg agonist	Reporter gene assay, competition-binding assay and adipogenesis assay	CBD induces vasorelaxation and stimulation of fibroblasts into adipocytes via PPARg activation
			5-HT1A agonist	Displacement of [3H]8-OH-DPATin CHO cells transfected with5-HT1A receptors; [35S]GTPgS binding to transfected CHO cells	5-HT1A is involved in CBD-induced antischemic and anxiolytic properties
(Petwee, 2008)			Antagonist of the putative abnormal-CBD receptor	Antagonism of the vasodilator response of abnormal-CBD	CBD attenuates the vasodilator response to anandamide
(Drysdale *et al.*, 2006)			Regulator of intracellular[Ca^{2+}]	Ca^{2+} imaging experiments in hippocampal cultures	To be determined. Potential basis for the neuroprotective and antiepileptic properties of CBD
(Ross *et al.*, 2008)			T-type Ca^{2+} channel inhibitor	Electrophysiological recordings in transfected HEK293 cells and sensory neurons	To be determined. Potential role in CBD-induced nociception and antiepileptic effects
(Jenny *et al.*, 2009)			Suppressor of tryptophan degradation	Measurements in human peripheral blood mononuclear cells	To be determined. Tryptophan is a precursor of 5-HT. Potential role in pain, inflammation and depression

Contd...

Table 21.1–*Contd...*

Phytocannabinoid	*Known Variant*	*Structure*	*Mechanism*	*Assay*	*Properties*
(Takeda *et al.*, 2009)			5-Lipoxygenase inhibitor	Enzymatic assay in a cell-free system	The effect is observed at very high concentrations. However, the 5-lipoxygenase pathway may be involved in CBD-induced antimitotic effect in glioma cells. CBD decreases 5-lipoxygenase in tumour tissues *in vivo*
(Takeda *et al.*, 2009)			15-Lipoxygenase inhibitor	Enzymatic assay in a cell-free system	To be determined. 15-Lipoxygenase is involved in developing atherosclerosis
(Evans *et al.*, 1987)			Phospholipase A2 modulator c	Enzymatic assay in a cell-free system	CBD exerts a biphasic stimulation of PGE2 release in human synovial cells CBD exerts anti-inflammatory effects in rodents
CBD (Anxiolytic, Antipsychotopic, Antiepileptic, Neuroprotective, Vasorelaxant, Antispasmodic, Anti-ischemic, Antiproliferative, Anticancer, Antidiabetic, Antipsoriatic, Analgesic, anti-inflammatory, immunosuppressant					
D9-THCV Ma *et al.*, 2008)		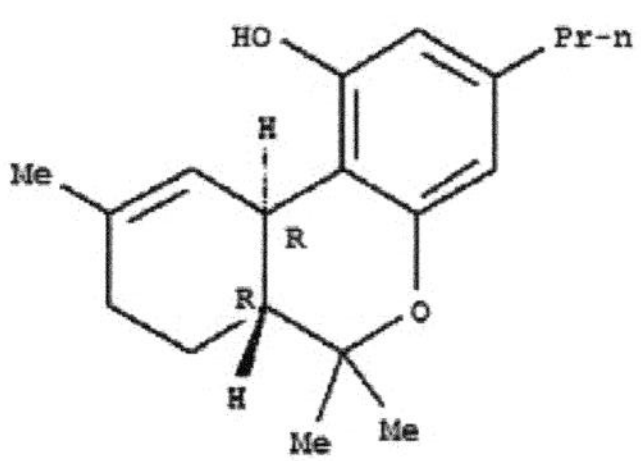	CB1 antagonist	Antagonism of cannabinoid agonist-induced [35S]GTPgS binding to mouse whole brain, cerebellar and piriform cortical membranes	D9-THCV increases central inhibitory neurotransmission with a therapeutic potential in epilepsy- and decreases food intakethrough CB1 antagonism. D9-THCV attenuates D9-THC-induced hypothermia and antinociception *in vivo*

Contd...

Table 21.1–*Contd...*

Phytocannabinoid	*Known Variant*	*Structure*	*Mechanism*	*Assay*	*Properties*
(Petwee, 2008)			CB2 partial agonist	Inhibition of forskolin-induced stimulation of cAMP productionby hCB2-CHO cells.	D9-THCV stimulates mesenchymal stem cells via CB2 receptors
D9-THCV(Anorectic, Bone Stimulant, Antiepileptic)					
Cannabigerol (CBG)	6	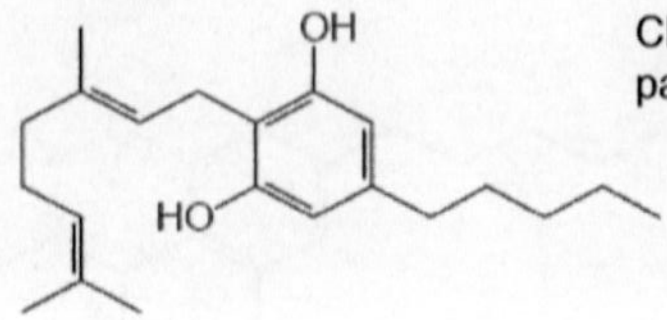	CB1 and CB2 partial agonist	Displacement of [3H] CP55, 940 from mouse brain membranes of hCB2-CHO cell membranes. [35S]GTPgS binding to mouse brain membranes (CB1) and to hCB2-CHO cell membranes	Non-psychotomimetic. It acts as an α2-adrenergic receptor agonist, 5-HT1A receptor antagonist, and CB1 receptor antagonist
(Ligresti *et al.*, 2006)			Anandamide re-uptake inhibitor	[14C]anandamide uptake by basophilic leukaemia or MDA-MB-231 cells	To be determined. Potential applications similar to those of inhibitors of endocannabinoid degradation
(De Petrocellis *et al.*, 2008)			TRPA1 aganist	Increase of $[Ca^{2+}]i$ in TRPA1-HEK-293 cells	To be determined. Potential role in analgesia
			TRPV1 agonist	Increase of $[Ca^{2+}]i$ in TRPV1-HEK-293 cells	To be determined. Potential role in analgesia
(De Petrocellis *et al.*, 2008)			TRPM8 antagonist	Antagonism of icilin- or menthol-induced increase in $[Ca^{2+}]i$ in TRPM8-HEK-293cells	To be determined. Potential role in analgesia and in the treatment of prostate carcinoma.

Contd...

Table 21.1–*Contd...*

Phytocannabinoid	*Known Variant*	*Structure*	*Mechanism*	*Assay*	*Properties*
(Evans *et al.*, 1987)			Phospholipase A2 modulatorc	Enzymatic assay in cell-free system	CBG reduces PGE2 release inhuman synovial cells
CBG (Antiproliferative, Anti Bacterial, Bone stimulant)					
Cannabichromene (CBC)	4	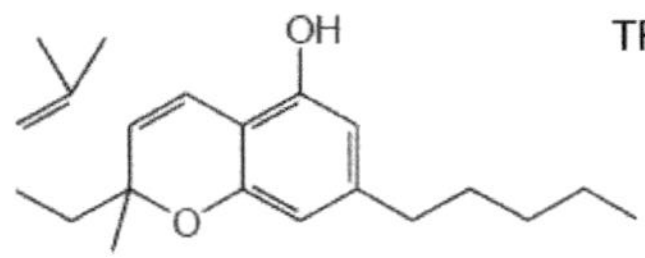	TRPA1 agonist	Increase of $[Ca^{2+}]i$ in TRPA1-HEK-293 cells	To be determined. Potential role in analgesia
(Ligresti *et al.*, 2006)			Anandamide re-uptake inhibitor	[14C]anandamide uptake by basophilic leukaemia or MDA-MB-231 cells	To be determined. Potential applications similar to those of inhibitors of endocannabinoid degradation
CBC(Anti-inflammatory, Analgesic, bone stimulant, Antimicrobial, Antiproliferative)					
D-9-Tetrahydro-cannabinol (D-9-THC)	9	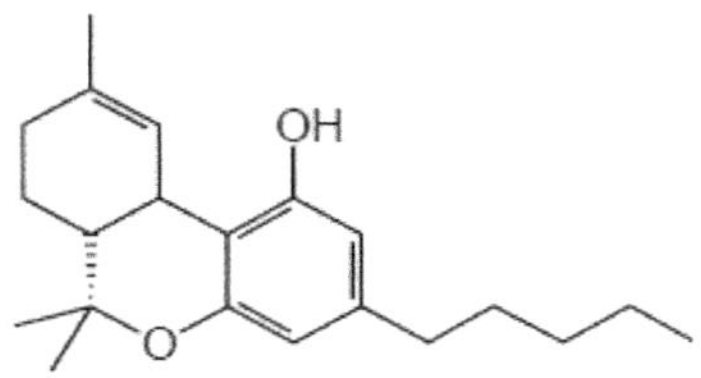			Main Psychoactive component
D-8-Tetrahydro-cannabinol(A-8-THC)	2	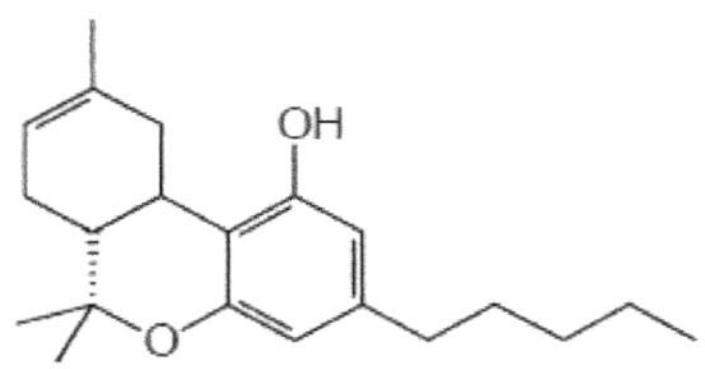			

Contd...

Table 21.1–*Contd...*

Phytocannabinoid	*Known Variant*	*Structure*	*Mechanism*	*Assay*	*Properties*
D9-THCA			TRPA1 partial agonist	Increase of $[Ca^{2+}]$I in TRPA1-HEK-293 cells	To be determined. Potential role in analgesia
(De Petrocellis *et al.*, 2008)			TRPM8 antagonist	Antagonism of icilin- or menthol-induced increase in $[Ca^{2+}]$i in TRPA1-HEK-293 cells	To be determined. Potential role in analgesia and in the treatment of prostate carcinoma.
D-9 THCA (Antispasmodic, Antiproliferative)					
CBDA			TRPA1 partial agonist	Increase of $[Ca^{2+}]$I in TRPA1-HEK-293 cells	To be determined. Potential role in analgesia
(Ligresti *et al.*, 2006)			TRPV1 agonist	Increase of $[Ca^{2+}]$i in TRPV1-HEK-293 cells	To be determined. Potential role in analgesia
(De Petrocellis *et al.*, 2008)			TRPM8 antagonist	Antagonism of icilin- or menthol-induced increase in$[Ca^{2+}]$i in TRPA1-HEK-293 cells	To be determined. Potential role in analgesia and in the treatment of prostate carcinoma.
(Takeda *et al.*, 2008)			COX-2 inhibitor	Enzymatic assay	To be determined. The effect is not shared by D9-THC or CBD; D9-THCA weakly activeat 100 mM. Potential role in inflammation
Cannabicyclol (CBL)	3	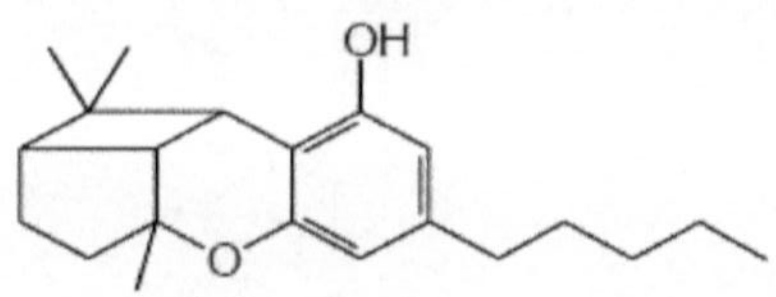			Non-psychotomimetic cannabinoid

Contd...

Table 21.1–*Contd...*

Phytocannabinoid	*Known Variant*	*Structure*	*Mechanism*	*Assay*	*Properties*
Cannabielsoin(CBE)	3	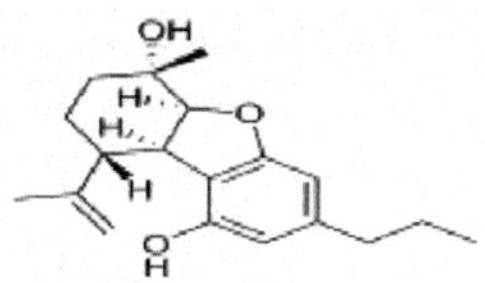			
Cannabinol (CBN)	6	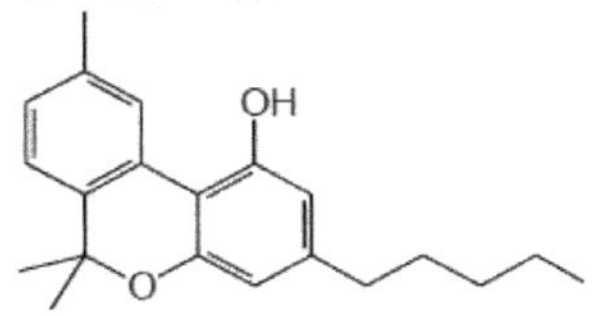		Mildly psychoactive. Its affinity to the CB2 receptor is higher than for the CB1 receptor (ii)	
Cannabinodiol (CBND)	2	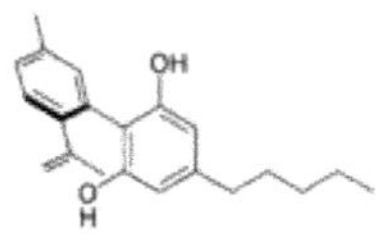			
Cannabitriol (CBT)	6	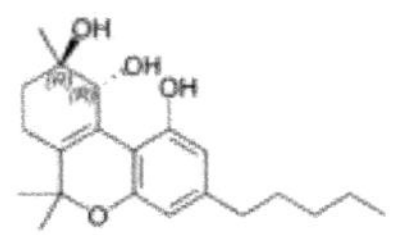			
Miscellaneous types	9				
Other cannabinoids	4				

NR = not reported.

Abbreviations: CBD: cannabidiol; D9-THCV: D9-tetrahydrocannabivarin; CBC: cannabichromene; CBG: cannabigerol; D9-THCA: D9-tetrahydrocannabinolic acid;: CBDA: cannabidiolic acid; D9-THC: D9-tetrahydrocannabinol; CBN: cannabinol; TRPV1: transient receptor potential vanilloid type 1; TRPV2: transient receptor potential vanilloid,type 2; TRPA1: transient receptor potential ankyrin type 1; TRPM8: transient receptor potential melastatin type 8; COX-2: cycloxygenase-2; 5-HT1A: 5-hydroxytryptamine. receptor subtype 1A; FAAH: fatty acid amide hydrolase. adenotes that the effect occurs via the equilibrate nucleoside transporter. bdenotes that the effect occurs via mitochondrial uptake and release or via L-type voltage gated [Ca2+] channel. cdenotes activation at low concentrations: inhibition at higher concentrations.

Effect of Cannabis on Respiratory system

Chronic heavy cannabis smoking is associated with increased symptoms of chronic bronchitis, such as coughing, production of sputum, and wheezing. Lung function is significantly poorer and there are significantly greater abnormalities in the large airways of marijuana smokers than in non-smokers. Carcinogenic effect may be related to Cannabis smoking and it is mutagenic both *in vitro* and *in vivo* (Leuchtenberger,1983), thus having adverse effect on human immunity. Cannabinoids impair cell-mediated and humoral immunity in rodents, decreasing resistance to infection, and non-cannabinoids in cannabis smoke impair alveolar macrophages (Munson and Fehr,1983). The relevance of these findings to human health is uncertain because the doses of THC used in animal studies have been very high, and tolerance may develop to the effects on immunity in human beings (Hollister,1992). Tashkin and colleagues have reported evidence of an additive effect of marijuana and tobacco smoking on histopathological abnormalities in lung tissue (Tashkin *et al.*, 1990). Tashkin also proved that long-term cannabis smoking may also increase the risks of respiratory cancer (Tashkin,1993).

Effect of Cannabis on Reproductive system

Chronic administration of high doses of THC to animals lowers testosterone secretion, impairs sperm production, motility, and viability, and disrupts the ovulatory cycle (Bloch,1983). Cannabis administration during pregnancy reduces birth weight in animals (Abel,1985).

Immunomodulatory Effect

The THC molecule main component of cannabinoid is a neutral, lipid soluble substance which can cross alveolar membranes when cannabis is smoked, resulting in near instantaneous appearance in the blood stream and distribution to tissues.

Alveoli macrophages are prime immune-effector cells in the lung which protect the lungs against infectious microorganisms, inhaled foreign substances, and tumour cells. During tissue inflammation, the number of alveolar macrophages increases and has been seen that in habitual cannabis smokers, alveolar macrophages are twice in number as compared to non-smokers, and smokers of both cannabis and tobacco had twice as many again. Cannabis smokers are less efficient to kill fungi, pathogenic bacteria, and tumour target cells due to reduced ability of alveoli macrophages. It behaves as an Immunosuppressant by decrease production of proinflammatory cytokines, which are important regulators of macrophage function (Barbers *et al.*, 1987).

The use of Cannabinoids in autoimmune disorders are discussed by various researchers, the effect of cannabidiol (CBD) (cannabis constituent) in collagen-induced arthritis in mice (malfait *et al.*, 2000), a psychoactive constituent (THC) effect on appetite, weight and nausea examined in AIDS patient, has also been used to treat anorexia nervosa in patients with acquired immunodeficiency syndrome. THC used for MS patient where it decreases EAE inflammation leading to reduced neurological effects and delayed MS effects (Lyman *et al.*, 1989), in other study it has been seen that the suppression of EAE by cannabinoids may be related to their effect on corticosterone secretion within experimental animals (Wirguin *et al.*, 1989). It is used externally as a poultice for corns, sores, varicose veins, gout and rheumatism. It is useful in neuralgia, gout, rheumatism, delirium tremens, insanity, infantile convulsions, insomnia, gonorrhea, menorrhagia, chronic cystitis and urinary affections (Razak, 2012). There is no conclusive evidence that consumption of cannabinoids impairs human immune function, as measured by numbers of T lymphocytes, B lymphocytes, or macrophages, or immunoglobulin concentrations as the consumption is not very high as compared in other experimental animals where cannabinoids impair cell-mediated and humoral immunity in rodents, decreasing resistance to infection, and non-cannabinoids in cannabis smoke impair alveolar macrophages (Hollister,1992).

Cannabis use during pregnancy reduces birth weight in animals, even some studies showed the increased risk of nonlymphoblastic leukaemia, rhabdomyosarcoma, and astrocytoma in children whose mothers reported using cannabis during their pregnancies (Robison *et al.*, 1989; Grufferman *et al.*, 1993; Kuitjen *et al.*, 1952)

Nabilone appears to be a beneficial, well-tolerated treatment option for fibromyalgia patients, with significant benefits in pain relief and functional improvement (Lynn and Herkenham,1994).

Cannabinoids exert specific receptor-mediated actions on the immune system of rats. At high concentrations, cannabinoids exert membrane effects at sites where they are sequestered nonspecifically. Binding to cannabinoid receptors in the immune system are confined to B lymphocyte-enriched areas *i.e.*, the marginal zone of the spleen, cortex of the lymph nodes and nodular corona of Peyer's patches and absent in T-lymphocyte (thymus and periarteriolar lymphatic sheaths of the spleen) or macrophage- enriched areas (liver and lung) (Facci *et al.*, 1995).

Psychoactive Effects

Psychosis is a generic term used to describe severe mental illness characterized by the presence of delusions, hallucinations, and other associated cognitive and behavioral impairments that interfere with the ability to meet the ordinary demands of life. It is measured using standardized diagnostic criteria for psychotic conditions documented in standard literature such as the DSM-IV (*Diagnostic and statistical manual of mental disorders*, fourth edition) (Mackenzie, 2010).

Various studies have determined that cannabis users are predisposed to an increased risk of psychosis by a factor of 1.77 to 11 times the general population (Fergusson *et al.*, 2006). It is further evident that the incidence of psychosis is higher in subjects consuming a larger amount of cannabis (Andreasson *et al.*, 1987; Arseneault *et al.*, 2002; Caspi *et al.*, 2005; Fergusson *et al.*, 2005; Henquet *et al.*, 2005). Certain factors have been found to increase the susceptibility to cannabis associated psychosis such as a past diagnosis of a psychotic disorder (Van *et al.*, 2002; degenhardt *et al.*, 2003), psychotic or paranoid symptoms at baseline(Henquet *et al.*, 2005), a family history of psychotic disorder (Miller *et al.*, 2001), and the Val/Val variant of the Catechol-O-methyltransferase (COMT) gene (Caspi *et al.*, 2005). This gene has a role in regulating dopamine concentrations and has been implicated in the development of schizophrenia.

δ9-THC has been used as a psychoactive agent for thousands of years other than as analgesia, anti-inflammation, immunosuppression, anticonvulsion, alleviation of intraocular pressure in glaucoma, and attenuation of vomiting (Dewey, 1986). Cannabinoids are a group of C21 compounds occurring in the glandular hairs of *Cannabis sativa* (Indian hemp) and consequently in hashish and marijuana, the well known drug of abuse (Samuelsson, 1999; Norcerino *et al.*, 2000). The most important representative cannabinoid, δ9-Tetrahydrocannabinol (δ9-THC), has psychoactive properties due to their interaction with the cannabinoid CB1 receptors within the brain. CB1 receptors are found mainly at the terminals of central and peripheral neurons where they usually mediate inhibition of neurotransmitter release (pertwee,1997). Good quality hashish contains 4–10 per cent and marijuana 0.1–2.7 per cent δ9-THC (Samuelsson, 1999). Cannabidiol, the main component of the glandular hairs (up to 15 per cent), is a non- psychoactive cannabinoid; which exerts a plethora of pharmacological effects, including anti-convulsive, sedative, hypnotic, anti-psychotic, anti-nausea and anti-inflammatory actions (Mechoulam *et al.*, 2002). Cannabidiol is a potent antioxidant compound and it has been recently proposed to have a neuroprotective role during ischemic damage (Hampson *et al.*, 1998; Hampson *et al.*, 2000).

In animal mode, cannabinoid receptor agonists target CB2 (peripheral) cannabinoid receptors (activation)- inhibits acute, inflammatory and neuropathic pain responses. (Malan *et al.*, 2003)

Conclusions

Cannabis is a commonly used illicit drug that has physiological and psychological effect, and has various intake routes (intravenous, smoke, inhalation and oral). Cannabis contains over sixty cannabinoids, of which the major active ingredient, δ9-Tetrahydrocannabinol (THC), appears to be primarily responsible for its immunomodulatory and psychoactive effects.

References

Abel, E,L. (1985). Effects of prenatal exposure to cannabinoids. In: Pinkert TM, ed. Current research on the consequences of maternal drug abuse. NIDA Research Monograph no 59. Washington: Department of Health and Human Services.

Adams, I.B., and Martin, B.R.(1996). Cannabis: pharmacology and toxicology in animals and humans. *Addiction,* **91**: 1585–1614.

Andreasson, S., Allebeck, P., Engstrom, A., and Rydberg, U. (1987). Cannabis and schizophrenia: A longitudinal study of Swedish conscripts. *Lancet*, 1483-1486.

Ahrens, J., Demir, R., Leuwer, M., de la Roche, J., Krampfl, K., Foadi, N., Karst, M., and Haeseler, G. (2009). The nonpsychotropic cannabinoid cannabidiol modulates and directly activates alpha-1 and alpha-1-Beta glycine receptor function. *Pharmacology,* **83:** 217–222.

Arseneault, L., Cannon, M., Poulton, R., Murray, R., Caspi, A., and Moffitt, T.E. (2002). Cannabis use in adolescence and risk for adult psychosis: Longitudinal prospective study. *BMJ.*, **325**: 1212-1213.

Barbers, R.G., Gong, H.J., Tashkin, D.P., Oishi, J., and Wallace, J. M. (1987). Differential examination of bronchoalveolar ravage cells in tobacco cigarette and marijuana smokers. *American Review of Respiratory Disease*, **135**: 1271-1275.

Bloch, E. (1983). Effects of marijuana and cannabinoids on reproduction, endocrine function, development and chromosomes. In: Fehr KO, Kalant H, eds. Cannabis and health hazards. Toronto: Addiction Research Foundation.

Boucher, F., Cosson, L., Unger, J., and Paris, M.R. (1974). Le *Cannabis sativa* L.; races chemiques ou varietes. *Pl. Med. Phytotherap.*, **8**: 20-31.

Capasso, R., Borrelli, F., Aviello, G., Romano, B., Scalisi, C., Capasso, F. and M.R. Izzo, A.A. (2008). Cannabidiol, extracted from *Cannabis sativa*, selectively inhibits inflammatory hypermotility in mice. *Br. J. Pharmacol.*, **154**: 1001–1002.

Carrier, E.J. (2006). Inhibition of an equilibrative nucleoside transporter by cannabidiol: a mechanism of cannabinoid immunosuppression. *Proc. Natl. Acad. Sci. USA.*, **103**: 7895–7900.

Caspi, A., Moffitt, T., Cannon, M., McClay, J., Murray, R., and Harrington, H. (2005). Moderation of the effect of adolescent-onset cannabis use on adult psychosis by a functional polymorphism in the catechol-o-methyltransferase gene: Longitudinal evidence of a gene x environment interaction. *Biol Psychiatry*, **57**: 1117-1127.

Degenhardt, L., Hall, W., and Lynskey, M.(2003). Testing hypotheses about the relationship between cannabis use and psychosis. *Drug Alcohol Depend.*, **71**: 37-48.

Dewey, W. L. (1986). Cannabinoid pharmacology. *Pharmac. Rev.*, **38**: 151–178.

De Petrocellis, L., Vellani, V., Schiano-Moriello, A., Marini, P., Magherini, P.C., Orlando, P., and Di Marzo, V. (2008). Plant-derived cannabinoids modulate the activity of transient receptor potential channels of ankyrin type-1 and melastatin type-8. *J. Pharmacol. Exp. Ther.*, **325**: 1007–1015.

Drysdale, A.J., Ryan, D., Pertwee, R.G., and Platt, B. (2006). Cannabidiol-induced intracellular Ca2+ elevations in hippocampal cells. *Neuropharmacology*, **50**: 621–631.

Evans, A.T., Formukong, E., and Evans, F.J. (1987). Activation of phospholipase A2 by cannabinoids. Lack of correlation with CNS effects. *FEBS Lett.*, **211**: 119–122.

Facci, L., Dal Toso, R., Romanello, S., Buriani, A., Skaper, S. D., and Leon, A. (1995). Mast cells express a peripheral cannabinoid receptor with differential sensitivity to anandamide and palmitoylethanolamide. *PNAS.*, **92 (8)**: 3376-3380.

Fergusson, D.M., Poulton, R., Smith, P.F., and Boden J.M. (2006). Cannabis and psychosis. *BMJ.*, **332(7534)**: 172–175.

Fergusson, D.M., Horwood, L.J., and Ridder, E.M. (2005). Tests of causal linkages between cannabis use and psychotic symptoms. *Addiction*, **100**: 354-366.

Grufferman, S., Schwartz, A.G., Ruymann, F.B., and Mauer, H.M.(1989). Parent's use of cocaine and marijuana and increased risk of rhabdomyosarcoma in their children. *Cancer, Causes and Control*, **4:** 217–224.

Hampson, A. J., Grimaldi, M., Axelrod, J., and Wink, D. (1998). Cannabidiol and (-) D9-tetrahydrocannabinol are neuroprotective antioxidants. *Proc. Natl Acad. Sci. USA.*, **95**: 8268–8273.

Hampson, A. J., Grimaldi, M., Lolic, M., Wink, D., Rosenthal, R., and Axelrod, J. (2000). Neuroprotective antioxidants from marijuana. *Ann. NY Acad. Sci.*, **899:** 274–282.

Henquet, C., Krabbendam, L., Spauwen, J., Kaplan, C., Lieb, R., and Wittchen, H.U. (2005). Prospective cohort study of cannabis use, predisposition for psychosis, and psychotic symptoms in young people. **330**: 11.

Hollister, L.E. (1992). Marijuana and immunity. *J Psychoactive Drugs*, **24**: 159–164.

Jenny, M., Santer, E., Pirich, E., Schennach, H., and Fuchs, D. (2009). Delta-9-tetrahydrocannabinol and cannabidiol modulate mitogen-induced tryptophan degradation and neopterin formation in peripheral blood mononuclear cells *in vitro*. J.*Neuroimmunol.*, **207**: 75–82.

Kuitjen, R.R., Bunin, G.R., Nass, C.C., and Meadows, A.T. (1952). Parental occupation and childhood astrocytoma. *Cancer Res.*, **52**: 782–86.

Li, H.L., and Lin, H. (1974). An archaeological and historical account of *Cannabis* in China. *Econ Bot.*, **28(4):** 437-447.

Leuchtenberger, C. (1983). Effects of marihuana (cannabis) smoke on cellular biochemistry of *in vitro* test systems. In: Fehr KO, Kalant H, eds. Cannabis and health hazards. Toronto: Addiction Research Foundation.

Ligresti, A., Moriello, A.S., Starowicz, K., Matias, I., Pisanti, S., De Petrocellis, L., Laezza, C., Portella, G., Bifulco, M., and Di Marzo, V. (2006). Antitumor activity of plant cannabinoids with emphasis on the effect of cannabidiol on human breast carcinoma. *J.Pharmacol. Exp. Ther.*, **318**: 1375–1387.

Lynn, A. B., and Herkenham, M. (1994). localization of cannabinoid receptors and nonsaturable high-density cannabinoid binding sites in peripheral tissues of the rat: implications for receptor-mediated immune modulation by cannabinoids. *JPET.*, **268(3)**: 1612-1623.

Lyman, W.D., Sonett, Jr., Brosnan, C.F., Elkin, R., and Bornstein, M.B. (1989). Delta-9tetrahydrocannabinol- a novel treatment for experimental autoimmune encephalomyelitis. *J.Neuroimmunol.*, **23**: 73-81.

Ma, Y.L., Weston, S.E., Whalley, B.J., and Stephens, G.J. (2008). The phytocannabinoid Delta(9)-tetrahydrocannabivarin modulates inhibitory neurotransmission in the cerebellum. *Br. J. Pharmacol.*, **154:** 204–215

Mackenzie, T,B. (2010). Psyche Primer, Scientific Foundation Course University of Minnesota, Medical School.

Malan, T., Philipr Mohab, M., Ibrahim, Josephine Lai., Todd, W., Vanderah, Alexandros Makriyannis, and Frank, Porreca. (2003). CB_2 cannabinoid receptor agonists: pain relief without psychoactive effects? *Current Opinion in Pharmacology*, **3(1):** 62-67.

Malfait, A.M., Gallily, R., and Sumariwalla, P.F. (2000). The non pyschoactive cannabis constituent canabidiol is an oral anti-arthritic Therapeutic in murien collagen-induced arthiritis. *Proc Natl Acad Sci USA.*, **97:** 9561-9566.

Massi, P., Valenti, M., Vaccani, A., Gasperi, V., Perletti, G., Marras, E., Fezza, F., Maccarrone, M., and Parolaro, D. (2008). 5-Lipoxygenase and anandamide hydrolase (FAAH) mediate the antitumor activity of cannabidiol, a nonpsychoactive cannabinoid. *J. Neurochem.*, **104**: 1091–1100.

Mechoulam, R., Parker, L. A., and Gallily, R. (2002). Cannabidiol: an overview of some pharmacological aspects. *J. Clin. Pharmacol.*, **42**: 11S–19S.

Miller, P., *et al.* (2001).Genetic liability, illicit drug use, life stress and psychotic symptoms: preliminary findings from the Edinburgh study of people at high risk for schizophrenia. *Soc Psychiatry Psychiatric Epidemiol.*, **36**: 338-42.

Munson, A.E., and Fehr, K.O. (1983). Immunological effects of cannabis. In: Fehr KO, Kalant H, eds. Cannabis and health hazards. Toronto: Addiction Research Foundation.

Nocerino, E., Amato, M., and Izzo, A. A. (2000). Cannabis and cannabinoid receptors. *Fitoterapia*, **71:** S6–S12.

Pertwee, R. G. (1997). Pharmacology of cannabinoid CB1 and CB2 receptors. *Pharmacol. Ther.*, **74**: 129–180.

Pertwee, R.G. (2008). The diverse CB1 and CB2 receptor pharmacology of three plant cannabinoids: delta9-tetrahydrocannabinol, cannabidiol and delta9-tetrahydrocannabivarin. *Br. J. Pharmacol.*, **153:** 199–215.

Qin, N., Neeper, M.P., Liu, Y., Hutchinson, T.L., and Lubin M.L., Flores, C. M. (2008). TRPV2 is activated by cannabidiol and mediates CGRP release in cultured rat dorsal root ganglion neurons. *J. Neurosci.*, **28**: 6231–6238.

Razak, N. (2012). Shodhganga.inflibnet.ac.in/bitstream/10603/./10_chapter per cent 205.pdf

Robison, L.I., Buckley, J.D., Daigle, A.E., *et al.* (1989). Maternal drug use and the risk of childhood nonlympholastic leukemia among offspring: an epidemiologic investigation implicating marijuana. *Cancer*, **63**: 1904–1911.

Ross, H.R., Napier, I., and Connor, M. (2008). Inhibition of recombinant human T-type calcium channels by Delta9-tetrahydrocannabinol and cannabidiol. *J. Biol. Chem.*, **283**: 16124–16134.

Ryberg, E., Larsson, N., Sjögren, S., Hjorth, S., Hermansson, N.O., Leonova, J., Elebring, T., Nilsson, K., Drmota, T., and Greasley, P.J. (2007). The orphan receptor GPR55 is a novel cannabinoid receptor. *Br. J. Pharmacol.*, **152**: 1092–1101.

Samuelsson, G. (1999). Drugs of Natural Origin. Swedish Pharmaceutical Press, Stockholm, Sweden.

Scutt, A., and Williamson, E.M. (2007). Cannabinoids stimulate fibroblastic colony formation by bone marrow cells indirectly via CB2 receptors. *Calcif. Tissue Int.*, **80**: 50–59.

Sullivan, S.E., Sun, Y., Bennett, A.J., Randall, M.D., and Kendall, D.A. (2009). Time-dependent vascular actions of cannabidiol in the rat aorta. *Eur. J. Pharmacol.*, **612**: 61–68.

Takeda, S., Usami, N., Yamamoto, I., and Watanabe, K. (2009). Cannabidiol-20,60-dimethyl ether, a cannabidiol derivative, is a highly potent and selective 15-lipoxygenase inhibitor. *Drug. Metab. Dispos.*, **37:** 1733–1737.

Takeda, S., Misawa, K., Yamamoto, I., and Watanabe, K. (2008). Cannabidiolic acid as a selective cyclooxygenase-2 inhibitory component in cannabis. *Drug Metab. Dispos.*, **36:** 1917–1921.

Touwn, M. (1981). The religious and medicinal uses of Cannabis in China, India and Tibet. *J Psychoactive Drugs*, **13(1):** 23-34.

Tashkin, D.P., Fligiel, S., and Wu, T.C. (1990). Effects of habitual use of marijuana and/or cocaine on the lung. In: Chiang CN, Hawks RL, eds. Research findings on smoking of abused substances. National Institute on Drug Abuse Research Monograph 99. Rockville, Maryland: National Institute on Drug Abuse.

Tashkin, D.P. (1993). Is frequent marijuana smoking harmful to health? *Western J Med.*, **158**: 635–637.

Thomas, A., Baillie,G.L., Phillips, A.M., Razdan, R.K., Ross, R. A., and Pertwee, R.G. (2007). Cannabidiol displays unexpectedly high potency as an antagonist of CB1 and CB2 receptor agonists *in vitro*. *Br. J. Pharmacol.*, **150**: 613–623.

Turner, C.E., ElSohly, M.A., and Boeren, *E.G.* (1980). Constituents of *Cannabis sativa* L. XVII. A review of the natural constituents. *Journal of Natural Products*, **43**: 169-234.

Van Os, J., Bak, M., Hanssen, M., Bijl, R., de Graaf, R., and Verdoux, H. (2002). Cannabis use and psychosis: a longitudinal population-based study. *Am J Epidemiol.*, **156**: 319-27.

Wirguin, I., Mechoulam, R., Breuer, A., Schezen, E., Weidenfeld, J., and Brenner, T. (1989). Supression of experimental autoimmune encephalomyelitis by cannabinoids. *Immunopharmacology*, **28:** 209-214.

Medicinal Plants: Phytochemistry, Pharmacology and Therapeutics Vol. 3 (2014) *Pages* **459–474**
Editor-in-Chief: **V.K. Gupta**
Published by: **DAYA PUBLISHING HOUSE, NEW DELHI**

22

Amazonian Medicinal Plants: Current Applications in Dentistry

M.C.C. Mussi[1], F.A. Pieri[2,3] and V.F. Veiga Junior[3]*

ABSTRACT

Tropical rainforests, particularly those located in Brazil, concentrate the highest biodiversity in the world. However, knowledge of the potential chemical and pharmacological properties of plants from the Brazilian rainforests, including the Amazon rainforest, is poorly described. Further studies are needed to establish large-scale commercialization of products from the Amazon, thereby promoting the sustainability of the rainforest and creating a means of subsistence for the local population. The Amazon has a wide diversity of plants, which have been used by indigenous communities for thousands of years, due to their medicinal properties. In dentistry, there are several plants with potential uses ranging from plants with antimicrobial effects to those that present tumor inhibitory activity. This chapter gives an extensive literature review of studies conducted with Amazonian plants in the field of dentistry. It describes plants or their compounds with possible inclusion in the dental field, as confirmed by laboratory tests or clinical trials. Reports of isolated cases will not be presented in this survey, as it does not deal with scientific evidence. Recent studies have provided some important plants with desirable therapeutic activities in the dental field. Oils and extracts from Copaifera sp. trees have been suggested by several studies, due to their antimicrobial, anti-inflammatory, healing and anti-cancer properties, among others. Copaiba oil has been linked to antibacterial activity, especially against bacteria of the oral microbiota. Extract of Psidium guajava has been evaluated for its antifungal properties, especially against Candida albicans, obtaining positive results in some tests. The formulation of natural dyes, such Euterpe oleracea, as dental plaque disclosing

---

1 Dentistry School, University of São Paulo

2 Leônidas and Maria Deane Institute, Oswaldo Cruz Foundation, Manaus, Brazil

3 Institute of Exact Sciences, Federal University of Amazonas, Manaus, Brazil

* *Corresponding author*: E-mail: valdirveiga@ufam.edu.br

agents has shown superior efficacy to some synthetic dyes used for this purpose. Another Amazonian plant with potential properties for use in dentistry is Caesalpinia ferrea, which has been indicated for its analgesic, anti-inflammatory, antibacterial and antiulcerogenic properties. Besides these, many other Amazonian plants are presented for their medicinal properties relevant to dentistry, as demonstrated in several studies that will be presented in this chapter. Some products are already commercially available, containing active compounds of some of these plants. These include endodontic cements, restorative materials, mouthwashes, toothpastes, gels, prophylactic pastes, and disinfectants, among others. One Amazonian plant, known locally as Jambu, contains a compound called spilantol that causes a tingling sensation. This chapter presents an overview of the current knowledge on the use of Amazonian medicinal plants in dentistry, serving as a guide in the selection of plants with potential use in dentistry for future studies, thereby encouraging possible therapeutic applications of these products.

Keywords: Medicinal plants, Amazon, Dentistry, Analgesic, Antimicrobial, Anti-inflammatory, Healing.

Introduction

The most common oral diseases are dental caries, periodontal disease, oral cancer, oral infectious diseases, trauma wounds and hereditary diseases. Worldwide, 60-90 per cent of school children and almost 100 per cent of most adults have dental caries (World Health Organization, 2003). The prevalence of periodontal disease, measured by the Community Periodontal Index (CPI), with a more severe score or signs of periodontal disease (CPI score 4) ranges from 10 per cent to 15 per cent in adult population worldwide. The most prevalent score worldwide is CPI score 2, clinically described as gingival bleeding and calculus. However it is noted that oral diseases are more common in children and adults among the poorest and more disadvantaged sectors of society (Petersen and Ogawa, 2005).

Some of these diseases have been treated with herbal medicines. Copaiba and Sacaca, among others, have been used for gingivitis, mouth abscess, inflammation and ulcers.

Therefore, most research on the use of plants in dentistry is focused on the treatment of diseases, through experiments involving microbiological, anti-inflammatory, healing, analgesic, and antitumor properties.

Brazil, particularly the Amazon rainforest, has a rich flora containing one of the most complete ecosystems, with numerous plants presenting potential medicinal uses (Di Stasi and Hiruma-Lima, 2002). The Amazon is the largest tropical rainforest ecosystem in the world, with a rich diversity of vegetation, which includes species as yet unexplored and unknown (Amorozo *et al.*, 1988).

The World Health Organization (1996) defines the concept of medicinal plants as "plants (wild or cultivated) used for medicinal purposes". The use of medicinal plants in the treatment of diseases, which for many people might seem like mysticism, is currently the object of scientific research with proven legitimacy.

This rich plant diversity of the Amazon rainforest provides great potential for the application of herbal medicine, including in the field of dentistry. The rich cultural diversity associated with the use of plants in this field, combined with scientific validation of this knowledge, are areas that need to be explored further.

Despite the undoubted potential phytochemical of the Amazon rainforest, the use of plants to treat oral diseases, and systemic diseases with oral manifestations, has been little explored. However, this area has an important role as a source of raw materials with application in dentistry.

There are some plant-based products that are already used in dentistry, such as eugenol, which is a phenolic compound obtained from the Indian clove (*Syzygium aromaticum* L.), used in endodontic treatment (Mathew and Hegde, 2008; Bhuiyan *et al.*, 2010). Nevertheless, few plant-based products have been validated in relation to their different substances for use in oral diseases.

Research in dentistry, particularly in the area of microbiology, has some limitations, one of which is the use of medicinal plants in oral biofilms. Thus, many microbiological studies are testing medicinal plants for use in dentistry with planktonic cells (microorganisms that do not organize communities and are dispersed in conventional liquid culture). This feature does not reflect the real conditions found in the oral cavity, since the bacteria arranged as biofilm express distinct properties than that found in planktonic cells, including increased resistance to antimicrobial agents (Lewis, 2001; Jefferson, 2004).

Numerous studies have evaluated the use of plant species of the Amazon in dentistry, such as *Copaifera* sp, *Euterpe olerace, Psidium guajava*, among others. The market stimulates research to develop new medicines based on plants, as well as encouraging the scientific discoveries that serve to legitimize the creation of a "plant in fashion", however we must pay careful attention to the side effects. In the case of dentistry, plants rich in polyphenols are not recommended for daily use because they disrupt the oral microbiota due to their antibacterial effect (Fukai *et al.*, 1991; Smullen *et al.*, 2007). Polyphenols also cause tooth staining, as it is believed that some salivary proteins, such as proline-rich protein, can mediate increased staining of the tooth enamel by polyphenols (Proctor *et al.*, 2005).

Due to the extensive use of medicinal plants in different therapeutic applications, this chapter aimed to review the literature on Amazonian species indicated for the treatment of diseases of the oral cavity, in order to contribute to the screening and direction of new research involving effective and validated herbal medicines for oral diseases.

Material and Methods

A literature review was carried out using the databases SCOPUS, ISI, PubMed, SCIELO, and search tools available in the Internet. The search terms used included "Amazonian plants", "herbal", "medicinal plants", "oral bacteria", "Dentistry", "Antimicrobial", and "dental", alone and in various combinations in Portuguese and English.

Results and Discussion

As a result of this research we found 141 references including 4 books, 97 articles, 3 doctoral thesis, 16 master's degree dissertations, 3 course conclusion dissertations, 16 abstracts and 2 extended abstracts, which formed the basis of this literature review.

Of the 141 references, 74 were related to antibacterial effect, 31 tested the antifungal effects and 11 evaluated the efficacy of herbal plants in the treatment of periodontal disease. The antineoplastic effect, the formation of dentin, pulp repair, prevention of caries, endodontic drugs, healing effect and anxiolytic capacity were also studied. Only 12 studies have evaluated plants in Dentistry through clinical trials (Figure 22.1).

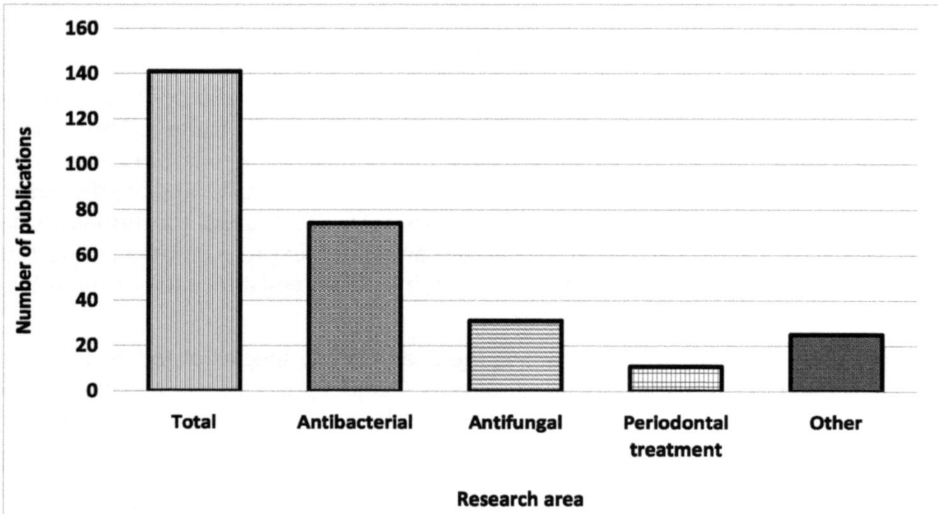

Figure 22.1: Most frequently researched activities related to the use of Amazonian medicinal plants in dentistry.

The essential oil of *Croton cajucara* Benth (popularly known as sacaca), rich in linalool, was tested against standard strains, and against biofilms formed by these strains and wild strains. The oil suspensions inhibited the growth of standard strains of *Candida albicans*, *Staphylococcus aureus*, *Streptococcus sobrinus*, *Streptococcus mutans* and *Porphyromonas gingivalis* (Alviano *et al.*, 2005). These micro-organisms are associated with diseases of the oral cavity such as caries, periodontal disease and candidiasis. Oral *Streptococcus* is involved in numerous human diseases, such as dental caries, purulent infections in various parts of the body, and infective endocarditis, among others. (Hardie and Whiley, 2006). *Candida albicans* is a commensal fungus responsible for oropharyngeal and disseminated candidiasis (Conti *et al.*, 2009).

The linalool purified fraction was inhibitory against *C. albicans*. Micro-organisms obtained from saliva samples from human subjects with fixed orthodontic appliances, as well as reference strains, were used to construct an artificial plaque, which was exposed to essential oil or the purified fraction. In the form of a microbial suspension, the essential oil was toxic to all micro-organisms, while the fraction mainly inhibited the growth of *C. albicans*. The effects of linalool on *C. albicans* cell biology were evaluated by electron microscopy, demonstrating that it induced a reduction in cell size (Alviano *et al.*, 2005).

Park *et al.* (2012) tested the effects of antibacterial linalool against cariogenic and periodontal bacteria and their cytotoxicity in human cells of the oral tissue. The objective was to test whether this compound can be used as a candidate for pre-clinical and clinical trials in future oral hygiene products. However, the authors concluded that the concentration of linalool should be kept below 0.4 mg/ml. Additionally, other compounds with antimicrobial activity against bacteria and periodontal cariogenic can be used in combination.

Croton cajucara oil also showed anti-inflammatory and antinociceptive effects that can be applied in dentistry. The essential oil obtained a significant anti-inflammatory effect in rodents models of acute and chronic inflammation when administered at lower doses (100µg/mL). This effect was attributed to inhibition of cyclooxygenase, but more study is needed to determine the course of action responsible for the anti-inflammatory effect. Regarding the antinociceptive effect, the oil was effective only when administered at the highest dose (1000µg/mL). This effect seems related to the interaction with the opioid system (Bighetti *et al.*, 1999).

These experimental findings indicate that *Croton cajucara* Benth is a plant with potential for tests against periodontal disease, due to its antibacterial effect against periodontopathic bacteria, as well as for its anti-inflammatory potential. However, care should be taken to use lower concentrations due to its cytotoxicity (Graim *et al.*, 2008).

Some medicinal plants have also been tested for their potential as plaque disclosing agents. Emmi and Barroso (2005) developed two dental plaque disclosing agents; the first based on anthocyanins (present in açaí, *Euterpe oleracea*) and the other based on bixin/norbixin (present in urucum, *Bixa orellana*). These products were compared *in vivo* with dental plaque disclosing based of synthetic dyes on the market, for the presence or absence of dental plaque disclosed by the dye in 42 subjects. The researchers concluded that the natural pigments anthocyanin, bixin and norbixin were effective in disclosing plaque, but that anthocyanin was more effective than the mixture of bixin and norbixin and the synthetic dyes. There are numerous benefits of the use of açai as a plaque disclosing agent, including the absence of side effects.

The research of the group yielded a patent (U.S. 1,182,935 B2) for a plaque disclosing product based of natural dyes, such as *Euterpe oleracea* and *Bixa orellana* associated with vehicles used by the pharmaceutical area (Nazareth *et al.*, 2007).

Psidium guajava is a small tree, whose leaves extract has several described properties such as analgesic, anti-inflammatory and antimicrobial activity. Metwally *et al.* (2010) isolated five flavonoid compounds (quercetin, quercetin-3-O-α-L-arabinofuranoside, quercetin-3-O-β-D-arabinopyranoside, quercetin-3-O-β-D-glucoside and quercetin-3-O-β-D-galactoside) from *P. guajava* extract and evaluated its antimicrobial activities. The results showed good activities for the extracts and the isolated compounds. Quercetin and its glycosides have potent antimicrobial activity against *S. aureus*, *E. coli*, *P. aeruginosa*, and *C. albicans*.

Caesalpinia ferrea fruits are used as an antimicrobial agent in oral infections. Its extract had effect against *Candida albicans*, *Streptococcus mutans*, *Streptococcus salivarius*, *Streptococcus oralis* and *Lactobacillus casei*. *Caesalpinia ferrea* fruit extract inhibited *in vitro* growth of oral pathogens in planktonic and biofilm models supporting its use as medicine for oral infections (Sampaio *et al.*, 2009).

Acmella oleracea, commonly known as Jambu, is a plant that has been used traditionally as a medication for toothache, arthritis and fever. One of the main properties of this plant is its anesthetic activity. This property was assessed by pain experiences of patient during the nonsurgical treatment of periodontitis using a test gel based on *Acmella oleracea* and compared to a placebo gel used as an adjunct to scaling and root and planning. The results showed that the gel containing the test *Acmella oleracea* was effective in reducing the patient's perception of pain, compared to the placebo gel used as an adjunct for proposed periodontal treatment (Girish *et al.*, 2011).

This plant also showed antimicrobial property against bacteria of the oral cavity such as *Pseudomonas aeruginosa*, *Streptococcus salivarius*, *Streptococcus viridans* and *Streptococcus mutans*. The methodology used to evaluate antimicrobial effect was agar diffusion. Different parts of *Acmella oleracea*, as root, stem, leaves and inflorescence were selected for testing and characterization of antibacterial secondary metabolites. As pure solvents, the authors used acetone, chloroform, ethanol, methanol or water, however, the water solutions had no effect on the microorganisms. The researchers did not describe having performed solvent controls. Despite this, they suggest that the results showed that *Acmella oleracea* is a drug capable of inhibiting the bacteria growth found in dental caries, as it is an abundant source of secondary metabolites (Thompson *et al.*, 2012). The plant has existing applications in the pharmaceutical industry, including patents already registered as formulations for relief of dental pain, swelling and gingival infections, periodontitis and mouthwash (Pandey *et al.*, 2007; Adler, 2006, Shimada and Gomi, 1995).

Holetz *et al.* (2002) tested the hydroalcoholic extract against *Escherichia coli* (ATCC 25922), *Pseudomonas aeruginosa* (ATCC 15442), *Bacillus subtilis* (ATCC 6623), *Staphylococcus aureus* (ATCC 25923), *Candida albicans*, *C. krusei*, *C. parapsilosis* and *C. tropicalis*. *Acmella oleracea* was weak against all tested microorganisms.

The studies showed a wide results discrepancy, therefore, more studies are needed in order to evaluate the real efficacy of *Acmella oleracea* against the oral microorganisms with both planktonic and biofilm cells. Part of the discrepancy between the results can be attributed to the different methods used, the composition of the plant, and the solvents used.

Lippia species are used in folk medicine to treat colds, flu, bronchitis and cough. In many cases, the leaves and flowers are used in the form of an oral infusion or administered as a poultice (Di Stasi and Hiruma-Lima, 2002). In dentistry, the oil of *Lippia sidoides* Cham, as well as their constituents carvacrol and thymol, were tested against various micro-organisms as cariogenic *Streptococcus* species and against

Candida albicans, using the broth dilution and disk diffusion tests. The essential oils thymol and carvacrol exhibited potent antimicrobial activity against the microorganisms, with minimal inhibitory concentrations ranging from 0.625 to 10.0 mg/mL (Botelho *et al.*, 2007). The most sensitive microorganisms were *Candida albicans* and *Streptococcus mutans*.

Botelho *et al.* (2009) evaluated the safety and efficacy of a *Lippia sidoides* mouthwash on plaque and gingivitis in fifty-five subjects. The non-alcoholic solution was prepared from essential oil extracted from the leaves at a concentration of 1 per cent. The participants were randomly assigned to use the mouth rinse solution with *L. sidoides* for a period of seven days. The control group used chlorhexidine 0.12 per cent. The results demonstrated a decrease in plaque levels, gingival inflammation scores, and levels of gingival bleeding at the end of the seven days, but there was no statistically significant difference ($p > 0.05$) between the test group and the control group in any of the clinical parameters assessed. The results showed that rinse with *L. sidoides* was safe and effective in reducing plaque and gingivitis. Side effects exhibited by patients who used the *L.sidoides* mouthwash were a mild sensation of burning in the mouth, changes in the sense of taste, and epithelial desquamation.

According Fontenelle *et al.* (2007), oral administration of *L. sidoides* oil (117.95 mg/kg/day) for thirty days did not induce any significant histopathological, hematological or biochemical changes.

Protium heptaphyllum is a plant that is widely throughout Brazil, including the Amazon region. The species *Protium heptaphyllom* March exudes a resin from its trunk, which has anti-inflammatory properties (Siani *et al.*, 1999).

A pentacyclic triterpene isolated from *Protium heptaphyllum*, α,β-amyrin, was tested in a rat model of orofacial pain induced by formalin or capsaicin. The authors reported that α,β-amyrin attenuates orofacial pain, in part, through a peripheral opioid mechanism (Holanda Pinto *et al.*, 2008a).

This same substance, obtained from *Protium heptaphyllum*, had anti-inflammatory potential at acute phase of ligature-induced periodontitis in rat. The results showed that α,β-amyrin modulates acute periodontal inflammation, reducing the neutrophil infiltration, oxidative stress, and the production of the TNF-α pro-inflammatory cytokine. These characteristics suggest that this substance may be useful as a therapeutic agent for the treatment of gingivitis and for delaying the progression of periodontal disease (Holanda Pinto *et al.*, 2008b).

Despite the anti-inflammatory and analgesic properties described, no significant antimicrobial effect has been described for α,β-amyrin. However, its association with other substances, or medicinal plants with antimicrobial properties for topical use in periodontal disease, is a potential source of study.

Pothomorphe umbellata is a shrubby or herbaceous plant known as "caapeba", "capeba" or "pariparoba" and has been used as a medicinal plant (Di Stasi and Hiruma-Lima, 2002) because of its various properties such as antimicrobial (Pimenta, 2000), and anti-inflammatory properties (Perazzo *et al.*, 2005), among others.

Ropke *et al.* (2006) determined the effect of ethanolic root extract of *P. umbellata* in metalloproteinases (MMP-2 and MMP-9). Matrix metalloproteinases (MMPs) have been related to an important role in invasion and metastasis of squamous cell carcinoma (Walk and Weed, 2011). *P. umbellata* was able to inhibit the activity of MMP-9 constituent *in vivo* in rats sacrificed 2 hours after UVB irradiation, however MMP-2 was not altered *in vivo* (Ropke *et al.*, 2006).

Although the Ropke *et al.* (2006) study was not directed to oral carcinoma, the data contribute to knowledge of the photoaging prevention mechanism by *P. umbellata.* This fact may provide a basis for the use of this plant in prevention of cancer caused by sun exposure, such as cancer of the lip.

Pimenta (2008) evaluated *in vitro* the antimicrobial effectiveness of *Pothomorphe umbellata* extract against *Candida albicans* (ATCC 90028), *Candida parapsilosis* (ATCC 22019), *Candida krusei* (ATCC 6258), *Candida glabrata* (ATCC 22019), *Staphylococcus aureus* (ATCC 25923), *Escherichia coli* (ATCC 25922) and *Enterococcus faecalis* (ATCC 29212). The authors reported that hydro alcoholic extract has antifungal and antibacterial activity mainly against *Staphylococcus aureus*.

Pothomorphe umbellata was bactericidal by the methods of antibiogram and broth microdilution against *Enterococcus faecalis*, including standard and hospital strains of bacteria (Atindehou *et al.*, 2002). *Enterococcus faecalis* is a Gram-positive bacteria with an important role in hospital-acquired (nosocomial) infections. It is resistant to most frequently used antibiotics, and can still be found in root canals (Devriese *et al.*, 2006), where is resistance to intra-root antimicrobials. Therefore, it is often found in cases of failure of conventional endodontic therapy and persistent periapical lesions. This fact has led to the development of research to evaluate the effectiveness of the antimicrobial action of intra-root medications against this microorganism (Peciuliene *et al.*, 2000).

Sponchiado-Junior (2006) evaluated the *in vitro* antimicrobial activity of a intracanal medication with *P. umbellata* for control of *Enterococcus faecalis*. They investigated the antimicrobial activities of hexane, butanol and ethyl acetate fractions by agar diffusion method. Based on the results, the authors performed the bioautography assay to detect constituents that exhibited antimicrobial activity. The results of the bioautography showed that the ethyl acetate fraction and its sub-fractions showed at least two active antimicrobials, with chemical structures of different polarities. Thus, the ethyl acetate fraction from *P. umbellata* presented activity capable of inhibiting the growth of *E. faecalis*, a fact that gives great potential for its use as a constituent of intra-root medication.

Marques *et al.* (2011) tested the biocompatibility of a paste with *Pothomorphe umbellata* and calcium hydroxide, using propylene glycol as a carrier. The material was implanted together with a polyethylene tube in the subcutaneous tissue of rats. Histological analysis after 7, 21 and 42 days showed that the inflammatory response caused was higher than the control at both initial times. After 42 days, the control and the tested substance showed mild inflammatory infiltrate.

All these features described make *Pothomorphe umbellata* a plant with potential use in dentistry, especially in the field of endodontics. However, more studies need to be carried out to confirm its antimicrobial activity and cytotoxicity effects, and to demonstrate features that are important for an endodontic material, such as ease of manipulation, lack of contraction after insertion, the fact that it does not stain the tooth structure, etc.

Theobroma grandiflorum, commonly known as cupuaçu, is an Amazon fruit that contains polyphenols (Pugliese, 2010). This substance may enhance the properties of collagen, increase collagen synthesis, and accelerate the conversion of insoluble to soluble collagen during development (Cetta *et al.*, 1977; Rao *et al.*, 1983). Shinohara (2011) evaluated the remineralization effect of cupuaçu extracts in artificial root caries by confocal laser scanning, and observed that cupuaçu seed extract has a positive effect on the remineralization of artificial root caries. Despite this property, there have been no significant work on the antimicrobial property of this material.

Cecropia peltata L. is a plant found in the Amazon and Atlantic forest. It is used in traditional medicine to treat diabetes mellitus (Izquierdo *et al.*, 2011). In dentistry, *C. peltata* has described activity, by the modified agar diffusion test, against *E. coli*, *B. cereus* and *S. aureus*. However, strains of *S.* β*-hemolytic*, *P. aureginosa* and *C. albicans* were more resistant to the plant extract. It is believed that the presence of steroids and amino acids in *C. peltata* could correspond to its high antimicrobial activity against *E. coli* (Rojas *et al.*, 2006).

Bacteria such as *Escherichia coli*, *Pseudomonas aeruginosa*, and *Staphylococcus aureus* have been widely studied for their role in nasocomial pneumonia. Nosocomial pneumonia, particularly when associated with mechanical ventilation, is a common infection in ICUs. The main etiologic factors include opportunistic and colonizing bacteria of the oral cavity (Scannapieco and Mylotte, 1996).

Although *C. peltata* has antimicrobial activity against some bacteria responsible for nosocomial pneumonia, few studies have explored the potential of the plant. There are some reports of its wound-healing activity (Nayak, 2006), however, there are no studies that extrapolate these findings to dentistry, for example in the progression of periodontal disease, or intraoral wound repair.

Momordica charantia is a plant found in the Amazon with hypoglycemic function (Baldwa *et al.*, 1977). Arayapisit *et al.* (2004) tested the inhibition capacity of *Momordica charantia* in a spinocellular carcinoma cell line of the mouth (HSC-4). After 24 hours of treatment, the levels of growth inhibition were analyzed by the MTT assay, and DNA fragmentation was determined by TUNEL method. The results showed that *Momordica charantia* inhibited cell proliferation, with IC50 of 0.27 per cent (w/v) compared with 3.125 per cent (v/v) obtained in the positive control. Apoptotic cell properties such as cell reduction, formation of membrane bubbles and chromatin condensation were also observed after treatment with the plant.

The copaiba tree (*Copaifera* sp.) is a large tree that exudes an oil-resin and is one of the most commonly medicinal plants by the Amazonian population (Lopez *et al.*, 2008). Several attributes are described in the literature for Copaiba oil, and it has potential for use in medicine and dentistry, due to its anti-inflammatory (Veiga Junior *et al.*, 2007) and antimicrobial (Pieri *et al.*, 2010a; Pieri *et al.*, 2010b; Pieri *et al.*, 2012) actions, among others.

Among many effects of copaiba oil, we highlight its antibacterial action. Mussi *et al.* (2010) tested the activity of *Copaifera officinalis* oil against *Porphyromonas gingivalis* and *Fusobacterium nucleatum*, establishing the Minimum Inhibitory Concentration (MIC) and Minimum Bactericidal Concentration (MBC) of bacteria. The authors also performed tested to determine the Subinhibitory Concentration (SC) and the auto-aggregation and co-aggregation ability of the bacteria exposed to concentrations of the tested solution (SC). The results obtained as MIC against *Porphyromonas gingivalis* and *Fusobacterium nucleatum* were 2 µl/ml and 100µL/mL respectively. The solution containing copaiba was bactericidal at all the concentrations tested against *F.nucleatum*, with an MBC value of 100µL/mL (MBC = CIM), *for P. gingivalis* copaiba balsam was bacteriostatic at all the tested concentrations. The solution of copaiba oil inhibited the auto-aggregation ability and unlike chlorhexidine digluconate, was effective in inhibiting the process of and co-aggregation between *P. gingivalis* and *F.nucleatum.*

Copaiba oil also presents biocompatibility capacity related to dentistry. Lima *et al.* (2011) evaluated histological repair after pulpotomy and pulp covering with *Copaifera langsdorffi* oil resin, green propolis extract, fibrin sponge and iodoform based paste. For all the experimental periods, an inflammatory response was observed, consisting predominantly of neutrophils. The lowest intensity found was in the *Copaifera langsdorffii* group and the highest intensity was in the fibrin sponge group, which exhibited periapical microabscesses. Necrosis was observed in all the groups, and their length increased with time, except for the teeth protected with *Copaifera*. Formation of a mineralized tissue barrier in the area of exposure of the pulp occurred only in *Copaifera langsdorffii* group. Therefore, the resin extracted from the copaiba oil showed biocompatible behavior, stimulating the formation of dentin.

The oil of *Copaifera multijuga* was compared for the sealing of root canals with other commercial sealers (Sealer 26®, Endofill® and AH plus®). The results showed that Sealer 26® and AH plus® had a longer setting time. With regard to solubility and disintegration, only Endofill ® did not obtain a satisfactory performance. Sealer 26® showed the biggest dimensional change. The authors noted that the experimental cement based in *Copaifera multijuga* showed satisfactory physiochemical properties (Garrido *et al.*, 2010).

In future, these desirable characteristics may facilitate its use as a chemotherapeutic agent in the treatment and prevention of periodontal disease (Pieri *et al.*, 2009), as well as a restorative material. Recent studies have tried to prove the described activities in laboratory tests, and demonstrate the antimicrobial action of copaiba oil on some plaque-forming bacteria (Pieri *et al.*, 2009; Pieri *et al.*, 2010a).

Chicaro (2005) evaluated the possible antineoplastic action of two plants (Copaiba and Andiroba oils) and a synthetic medicine (dexamethasone) in a squamous carcinoma lineage of the oropharynx (FADU), and analyzed whether there was a correlation between NFkB protein expression (transcription factor of genes involved in the process of carcinogenesis and inflammation) and inhibition of cell proliferation. The copaiba oil showed a potent action inhibition of proliferation and induction of apoptosis, the latter verified by the Tunel positive test for apoptotic cells. The andiroba oil and dexamethasone inhibited the proliferation of these cells, but did not induce apoptosis, the action of he latter drug being time and concentration dependent. The copaiba and andiroba and dexamethasone, natural and synthetic anti-inflammatory, respectively, promoted action of inhibition in FADU cell proliferation line, having a relationship with decreased levels of NFkB protein which is responsible for the regulation of genes involved in the process of cell proliferation and survival. According to Pedreira (2007), copaiba oil also has chemopreventive behavior in the development of cancer induced by 9.10-Dimethyl-1.2 benzanthracene (DMBA) in hamsters, based on parameters such as lesion diameter and macroscopic and microscopic features.

In addition to the described properties, copaiba balsam appears to have only mild side effects, so if its properties and low cytotoxicity are proven, it has potential for use. Studies indicate that the copaiba oil at the administered doses (0.3ml/kg to 0.9ml/kg) did not cause maternal toxicity or teratogenicity in mice (Lawrence *et al.*, 2009). Therefore, this oil can be used for prolonged periods, or added to the composition of toothpastes, mouthwashes and dental materials. However, for this, more research needs to be conducted, especially concerning its toxicity, in order to facilitate its possible use.

Conclusions

This review showed that in the last decade, interest in studying the medicinal properties of plants in the Amazon for use in Dentistry had increased greatly. In relation to microbial properties, we noted that most of the studies do not use biofilm, only bacterial planktonic cells. This makes it difficult to predict the behavior of the studied plant in the oral cavity, as the bacteria arranged in biofilm have greater resistance to microbial agents.

Many studies do not identify the chemical composition of the studied plant, which makes it difficult to compare the results obtained in different studies. Another factor is the discrepancy in methods used, since there is no standardization of the technique according to the studied substance.

We also noted that few studies with plants used in dentistry have been carried out with clinical trials, which increase the credibility of results due to reproduction of the micro-environment. But for this to occur, it is first necessary to establish the cytotoxicity of medicinal plants and their metabolites.

Acknowledgement

The authors thank CNPq-National Council for Scientific and Technological Development for financial support.

References

Adler, R.J. (2006). Compositions for the acute and/or long term treatment of periodontal diseases using herb extracts. WO Pat. 2006059196.

Alviano, W.S., Mendonça-Filho, R.R., Alviano, D.S., Bizzo, H.R., Souto-Padrón, T., Rodrigues, M.L., Bolognese, A.M., Alviano, C.S., and Souza, M.M. (2005). Antimicrobial activity of *Croton cajucara* Benth linalool-rich essential oil on artificial biofilms and planktonic microorganisms. *Oral Microbiology and Immunolog*, **20**(2): 101-105.

Amorozo, M.C.M., and Gély, A. (1988). Uso de plantas medicinais por caboclos do baixo Amazonas Barcarena, PA, Brasil. *Boletim do Museu Paraense Emílio Goeldi*, **4**: 47.

Arayapisit, T., Boonanatanasarn, K., Suppakpatana, P., Janebodin, K., Sripairojthikoon, W., and Chunhabundit, P. (2004). Effect of Five Thai Herbs on Oral Squamous Cell Carcinoma. The Annual Scientific Meeting, 19th International Association for Dental Research-Southeast Asia Division and 13th Southeast Asia Association for Dental Education.

Atindehou, K.K., Koné, M., Terreaux, C., Traore, D., Hostettmann, K., and Dosso, M. (2002). Evaluation of the antimicrobial potential of medicinal plants from the Ivory Coast. *Phytotherapy Research*, **16**: 497-502.

Baldwa, V.S., Bhandra, C.M., Pangaria A., and Goyal, R. K. (1977). Clinical trials in patients with diabetes mellitus of an insulin-like compound obtained from plant source. *Upsala Journal of Medical Sciences*, **82**: 39-41.

Bhuiyan, N.I, Begum, J., Nandi, N.C., and Akter F. (2010).Constituents of the essential oil from leaves and budsnof clove (*Syzigium caryophyllatum* (L.) Alston). *African Journal of Plant Science*, **4**: 451-454.

Bighetti, E.J., Hiruma-Lima, C.A., Gracioso, J.S., and Brito, A.R. (1999). Anti-inflammatory and antinociceptive effects in rodents of the essential oil of *Croton cajucara* Benth. *Journal of Pharmacy and Pharmacology*, **51**: 1447-1453.

Botelho, M.A., dos Santos, R.A., Martins, J.G., Carvalho, C.O., Paz, M.C., Azenha, C., Ruela, R.S., Queiroz, D.B, Ruela, W.S., Marinho, G., and Ruela, F.I. (2009). Comparative effect of an essential oil mouthrinse on plaque, gingivitis and salivary *Streptococcus mutans* levels: a double blind randomized study. *Phytotherapy Research*, **23**: 1214-1219.

Botelho, M.A., Nogueira, N.A., Bastos, G.M., Fonseca, S.G., Lemos, T.L., Matos, F.J., Montenegro, D., Heukelbach, J., Rao, V.S., and Brito, G.A. (2007). Antimicrobial activity of the essential oil from *Lippia sidoides*, carvacrol and thymol against oral pathogens. *Brazilian Journal of Medical and Biological Research*, **40**: 349-356.

Cetta, G., Pallavicini, G., Tenni, R., and Bisi, C. (1977). Influence of flavonoid-copper complexes on cross linking in elastin. *Italian Journal of Biochemistry*, **26**: 317–327.

Chicaro, C. F. (2009). Análise da expressão da proteína nf-kappaB antes e depois do tratamento com dexametasona e os óleos de copaíba e andiroba em cultura de células de carcinoma epidermóide bucal. Thesis of University in São Paulo.

Conti, H.R., Shen, F., Nayyar, N., Stocum, E., Sun, J.N., Lindemann, M.J., Ho, A.W., Hai, J.H., Yu, J.J., Jung, J.W., Filler, S.G., Masso-Welch, P., Edgerton, M., and Gaffen, S.L. (2009). Th17 cells and IL-17 receptor signaling are essential for mucosal host defense against oral candidiasis. *The Journal of Experimental Medicine*, **206**: 299-311.

Devriese, L., Baele, M., and Butaye, P. (2006). The Genus Enterococcus: Taxonomy. *In:* Prokaryotes, Vol. 4, Ed. By Dworkin, M., Falkow, S., Rosenberg, E., Schleifer K.H., and Stackebrandt, E., Springer Science+Business Media, USA, pp. 163.

Di Stasi, L.C., and Hiruma-Lima, C.A. (2002). Plantas Medicinais na Amazônia e na Mata Atlântica. Editora UNESP, São Paulo.

Emmi, D.T., and Barroso, R.F.F. (2005). A biodiversidade amazônica na promoção da saúde bucal: elaboração de evidenciador de placa dental utilizando os corantes do açaí e urucum e a análise comparativa de sua eficácia em relação aos corantes sintéticos. *UFPA*. **2**: 49.

Fontenelle, R.O., Morais, S.M., Brito, E.H., Kerntopf, M.R., Brilhante, R.S., Cordeiro, R.A., Tomé, A.R., Queiroz, M.G., Nascimento, N.R., Sidrim, J.J., and Rocha, M.F. (2007). Chemical composition, toxicological aspects and antifungal activity of essential oil from *Lippia sidoides* Cham. *Journal of Antimicrobial Chemotherapy*, **59**: 934-940.

Fukai, K., Ishigami, T., and Hara, Y.(1991). Antibacterial activity of tea polyphenols against phytopathogenic bacteria. *Agricultural and Biological Chemistry*, **55**: 1895-1897.

Garrido, A.D., Lia, R.C., França, S.C., da Silva, J.F., Astolfi-Filho, S., and Sousa-Neto, M.D. (2010) Laboratory evaluation of the physicochemical properties of a new root canal sealer based on *Copaifera multijuga* oil-resin. *International Endodontic Journal*, **43**: 283-291.

Girish, S., Srinivas, K., Dr., and Sunil, J. (2011). An *in vivo* evaluation of the pain perception using an herbal gel (*Acmella oleracea*) as an adjunct to scaling and root planing. *Journal of PEARLDENT*, **2**: 52-54.

Graim, J.F.S., Lopes Filho, G.J., Brito, M.V.H., and Matos, L.T.M.B. (2008). Histologic evaluation of rats' liver after *Croton cajucara* Benth (sacaca) administration. *Acta Cirúrgica Brasileira*, **23**: 130-4.

Hardie, J.M., and Whiley, R.A. (2006). The Genus Streptococcus Oral. *In:* The prokaryotes. 3rd ed. Vol 4, Ed. By Dworkin, M., Falkow, S., Rosenberg, E., Schleifer, K.H., and Stackebrandt E. Springer Science+Business Media: Singapura.

Holanda Pinto, S.A., Pinto, L.M., Cunha, G.M., Chaves, M.H., Santos, F.A., and Rao, V.S. (2008b) Anti-inflammatory effect of alpha, beta-amyrin, a pentacyclic triterpene from Protium heptaphyllum in rat model of acute periodontitis. *Inflammopharmacology*, **16**: 48-52.

Holanda Pinto, S.A., Pinto, L.M., Guedes, M.A., Cunha, G.M., Chaves, M.H., Santos, F.A., and Rao, V.S. (2008a). Antinoceptive effect of triterpenoid alpha,beta-amyrin in rats on orofacial pain induced by formalin and capsaicin. *Phytomedicine*, **15**: 630-634.

Holetz, F.B., Pessini, G.L., Sanches, N.R., Cortez, D.A.G, Nakamura, C.V., and Dias Filho, B.P. (2002). Screening of some plants used in Brazilian folk medicine for the treatment of infectious diseases. *Memórias do Instituto Oswaldo*, **97**: 1027-1031.

Izquierdo, A.M, Torres, M.P N., Jiménez, G.S., and Sosa, F.C. (2011). Changes in biomass allocation and phenolic compounds accumulation due to the effect of light and nitrate supply in *Cecropia peltata* plants. *Acta Physiologiae Plantarum*, **33**: 2135-2147.

Jefferson, K.K. (2004). What drives bacteria to produce a biofilm? *FEMS Microbiology Letters*, **236**: 163-173.

Lewis, K. (2001). Riddle of biofilm resistance. *Antimicrobial Agents and Chemotherapy*, **45**: 999-1007.

Lima, R.V., Esmeraldo, M.R., de Carvalho, M.G., de Oliveira, P.T., de Carvalho, R.A., da Silva, F.L. Jr., and de Brito Costa, E.M. (2011). Pulp repair after pulpotomy using different pulp capping agents: a comparative histologic analysis. *Pediatric Dental Journal*, **33**: 14-18.

Lopez, C., Shanley, P., Fantini, AC., and Cronkleton, M.C. (2008). Riquezas da floresta: frutas, plantas medicinais e artesanato na América Latina. CIFOR, Indonésia.

Lourenço, A.C.S., Miguel, L.K., Guarido, K.L., Sensiate, L.A., and Salles, M.J.S. (2009). Copaiba oil (*Copaifera langsdorfii* Desf.) on mouse reproductive patterns and embryonic or fetal development. *Revista Brasileira de Plantas Medicinais*, **11**: 407-413.

Marques, A.A., Sponchiado, E.C. Jr, Garcia, L.F., Garrido, A.D., França, S.C., and Lia, R.C. (2011). Morphological analysis of tissue reaction caused by a new endodontic paste in subcutaneous tissue of rats. *Journal of Conservative Dentistry*, **14**: 309-13.

Mathew, S.T., and Hegde, M.N. (2008). Shear bond strength of eugenol- and non-eugenol–based endodontic sealers to gutta-percha and dentin: An *in vitro* study. *Journal of Conservative Dentistry*, **11**: 30-36.

Metwally, A.M., Omar, A.A., Harraz, F.M., and El Sohafy, S.M. (2010). Phytochemical investigation and antimicrobial activity of *Psidium guajava* L. leaves. *Pharmacognosy Magazine*, **6**: 212-8.

Mussi, M.C.M. (2011). Análise da atividade antimicrobiana dos óleos de copaíba (*Copaifera officinalis*) e de melaleuca (*Melaleuca alternifolia*) sobre *Fusobacterium nucleatum* e *Porphyromonas gingivalis*: determinação das concentrações inibitórias e bactericidas mínimas e efeito de concentrações subinibitórias sobre a agregação. Thesis of University of São Paulo.

Nayak BS. (2006). *Cecropia peltata* L(Cecropiaceae) has wound-healing potential: a preclinical study in a Sprague Dawley rat model. *The International Journal of Lower Extremity Wounds*, **5**: 20-26.

Nazaré, R.F.E., Emmi, D.T., Barroso, R.F.F., and Rocha, P.O.R. (2007). Bacterial plaque evidencing composition based on natural colorants. US 1,182,935 B2, US Patent.

Pandey, H.K., Rawut, P.S., Kumar, N., and Verma, G.S. (2007). A herbal formulation for toothache and related disorders and a process for preparation thereof. IN Patent. 2004DE00260.

Park, S.N., Lim, Y.K., Freire, M.O., Cho, E., Jin, D., and Kook, J.K. (2012). Antimicrobial effect of linalool and α-terpineol against periodontopathic and cariogenic bacteria. *Anaerobe*, **18**: 369-372.

Peciuliene, V., Balciuniene, I., Eriksen, H.M., and Haapasalo M. (2000). Isolation of *Enterococcus faecalis* in previously root-filled canals in a lithuanian population. *Journal of Endodontics*, **26**: 593-595.

Pedreira, E.N. (2007). Avaliação do efeito inibidor tumoral do óleo resinina de copaíba *in natura* (*Copaifera reticulata*) e manipulado artesanalmente no modelo de carcinogênese bucal experimental DMBA induzida. Thesis of University in São Paulo.

Perazzo, F.F., Souza, G.H., Lopes, W., Cardoso, L.G., Carvalho, J.C., Nanayakkara, N.P., and Bastos, J.K. (2005). Anti-inflammatory and analgesic properties of water-ethanolic extract from *Pothomorphe umbellata* (Piperaceae) aerial parts. *Journal of Ethnopharmacology*, **99**: 215-220.

Petersen, P.E., and Ogawa H. (2005). Strengthening the prevention of periodontal disease: the WHO approach. *Journal of Periodontology*, **76**: 2187-2193.

Pieri, F.A., José, R.M., Galvão, N.N., Nero, L.A., and Moreira, M.A.S. (2010b). Antimicrobial activity of autoclaved and non autoclaved copaiba oil on *Listeria monocytogenes*. *Ciência Rural*, **40**: 1797-1801.

Pieri, F.A., Mussi, M.C., Fiorini, J.E., and Schneedorf, J.M. (2010a). Efeitos clínicos e microbiológicos do óleo de copaíba (*Copaifera officinalis*) sobre bactérias formadoras de placa dental em cães. *Arquivo Brasileiro de Medicina Veterinária e Zootecnia*, **62**: 578-585.

Pieri, F.A., Mussi, M.C.M., and Moreira, M.A.S. (2009). Óleo de copaíba (*Copaifera* sp.): histórico, extração, aplicações industriais e propriedades medicinais. *Revista Brasileira de Plantas Medicinais*, **11**: 465-472.

Pieri, F.A., Mussi, M.C.M., Fiorini, J.E., and Moreira, M.A.S., Schneedor, J.M. (2012). Bacteriostatic effect of copaiba oil (*Copaifera officinalis*) against *Streptococcus mutans*. *Brazilian Dental Journal*, **23**: 36-38.

Pimenta, D.O.M.M. (2008). Avaliação *in vitro* da atividade antifúngica e antibacteriana de extratos de *Pothomorphe umbellata* e *Struthantus* sp. e análise da expressão proteica. Thesis of Ribeirão Preto University.

Proctor, G.B., Pramanik, R., Carpenter, G.H., and Ree G.D. (2005). Salivary Proteins interact with dietary constituents to modulate tooth staining. *Journal of Dental Research*, **84**: 73-78.

Pugliese, G.A. (2010). Compostos fenólicos do cupuaçu (*Theobroma grandiflorum*) e do cupulate: composição e possíveis benefícios. Thesis of University of São Paulo.

Rao, C.N., Rao, V.H., and Steinmann, B. (1983). Bioflavonoid-mediated stabilization of collagen in adjuvant-induced arthritis. *Scandinavian Journal of Rheumatology*, **12**: 39–42.

Rojas, J.J, Ochoa, V.J., Ocampo, S.A., and Muñoz, J.F. (2006). Screening for antimicrobial activity of ten medicinal plants used in Colombian folkloric medicine: A possible alternative in the treatment of non-nosocomial infections. *BMC Complementary and Alternative Medicine*, **6**: 2.

Ropke, C.D., da Silva, V.V., Kera, C.Z., Miranda, D.V., de Almeida, R.L., Sawada, T.C., and Barros, S.B. (2006). *In vitro* and *in vivo* inhibition of skin matrix metalloproteinases by *Pothomorphe umbellata* root extract. *Photochemistry and Photobiology*, **82**: 439-442.

Sampaio, F.C., Pereira, M.S., Dias, C.S., Costa, V.C., Conde, N.C., and Buzalaf, M.A. (2009). *In vitro* antimicrobial activity of *Caesalpinia ferrea* Martius fruits against oral pathogens. *Journal of Ethnopharmacology*, **124**: 289-94.

Scannapieco, F.A., and Mylotte J.M. (1996). Relationships between periodontal disease and bacterial pneumonia. *Journal of Periodontology*, **67**: 1114-1122.

Shimada, T., and Gomi, T. (1995). Spilanthol-rich essential oils for manufacturing toothpastes or other oral compositions. JP Pat. 07090294.

Shinohara, M.S. (2011). Remineralization effect of cupuaçu seed extract on artificial root caries. IADR General Session and Exhibition.

Siani, A.C., Ramos, M.F.S., Menezes-De-Lima Jr., O., Ribeiro-Dos-Santos, R., Fernadez-Ferreira, E., Soares, R.O.A., Rosas, E.C., Susunaga, G.S., Guimarães, A.C., Zoghbi,M.G., and Henriques M.G.M.O. (1999). Evaluation of anti-inflammatory-related activity of essential oils from the leaves and resin of species of Protium. *Journal of Ethnopharmacology*, **66**: 57-69.

Smullen, J., Koutsou, G.A., Foster, H.A., Zumbé, A., and Storey, D.M. (2007). The antibacterial activity of plant extracts containing polyphenols against *Streptococcus mutans*. *Caries Research*, **41**: 342-349.

Sponchiado-Junior, E.C. (2006). Atividade antibacteriana contra o *Enterococcus faecalis* de uma medicação intracanal contendo ativos fitoterápicos de *Pothomorphe umbellata*. Thesis of Federal University of Amazonas.

Thompson, T.S., Sukesh, K., and Singh, D. (2012). A study on the antimicrobial effect of *Acmella oleraceae* against dental caries bacteria. *International Journal of Pharmaceutical Sciences and Research*, **3**: 1194-1197.

Veiga Junior, V.F., Rosas, E.C., Carvalho, M.V., Henriques, M.G., and Pinto, A.C. (2007). Chemical composition and anti-inflammatory activity of copaiba oils from *Copaifera cearensis* Huber ex Ducke, *Copaifera reticulata* Ducke and *Copaifera multijuga* Hayne: A comparative study. *Journal of Ethnopharmacology*, **112**: 248-254.

Walk, E.L., and Weed, S.A. (2011). Recently identified biomarkers that promote lymph node metastasis in head and neck squamous cell carcinoma. *Cancers*, **3**: 747-772.

World Health Organization. (1996). Expert Committee on Specifications for Pharmaceutical Preparations. Thirty-fourth report. World Health Organization, Geneva.

World Health Organization. (2003). The World Oral Health Report 2003. World Health Organization, Geneva.

Medicinal Plants: Phytochemistry, Pharmacology and Therapeutics Vol. 3 (2014) *Pages* **475–487**
Editor-in-Chief: **V.K. Gupta**
Published by: **DAYA PUBLISHING HOUSE, NEW DELHI**

23

Radical Scavenging Effects of Zingiberaceae Plants– *Alpinia galanga* and *Curcuma zedoaria*

Manu Babu[1], Kavitha Sankar P.C.[1] and Beena Joy[1*]

ABSTRACT

In the present study the antioxidant activity and the composition of the active fraction of the rhizomes of two varieties of Zingiberacea family–Alpinia galanga and Curcuma zedoaria were investigated. The plant rhizomes were subjected to successive extraction using different solvents such as hexane, ethyl acetate and methanol. The extraction efficiency was evaluated in terms of total phenolic content. The compounds present in the active ethyl acetate extract of Alpinia galanga and Curcuma zedoaria were identified by GC- MS analysis. The TPC for A. galanga and C. zedoaria were found to be high for the ethyl acetate fraction with values 102.3 ± 1.26 mg/g and 93.1 ± 1.26 mg/g, respectively. The ethyl acetate extract of Alpinia galanga and Curcuma zedoaria showed potential radical scavenging activity as revealed by DPPH assay with IC_{50} value of 119.29 ± 1.46µg/ml and 82.32 ± 1.60µg/ml, with ABTS assay with IC_{50} value of 17.38 ± 1.10µg/ml and 33.9 ± 1.24µg/ml respectively. The ethyl acetate fraction of the plants showed high reducing power in accordance with high total phenolic content determined. The GC-MS data revealed that 2, 4-dimethyl benzoic acid (65.2 per cent) as the most prominent compound present in the Alpinia galanga extract. Besides this compound, eugenol (6.55 per cent) and β-farnesene (6.37 per cent) were also identified. Isocurcumenol

1 Agroprocessing and Natural Product Division, National Institute for Interdisciplinary Science and Technology, Thiruvananthapuram – 695 019, Kerala, India

* *Corresponding author*: E-mail: bjoy1571@gmail.com

(25.24 per cent) and methyl sterolate (24.94 per cent) were the major compounds identified in Curcuma zedoaria extract. The results suggest Alpinia galanga to be a more promising plant than Curcuma zedoaria with respect to its antioxidant potential.

Keywords: *Alpinia galanga, Curcuma zedoaria,* DPPH, ABTS, GC-MS analysis, IC_{50} value.

Introduction

Phytochemicals have potent bioactive properties and have provided multiple active compounds in the past (Aggarwal *et al.,* 2003). Recently a large number of natural products have been isolated from different plants. Yet the broad spectrum of activities of these natural products are not much explored for the development as therapeutics. Plants are an excellent source of an amazing diversity of secondary metabolites which are not essential for normal bodily functions but are still biologically active and are of medicinal value. Natural products from plant sources have distinct advantages over synthetic products in some respects. First, the natural products due to their structural and functional features can coordinate with biologically active proteins in a better way than synthetic compounds. Second, the ability of these drugs to locate multiple targets simultaneously has the potential to minimize the likelihood of treatment resistance. Antioxidants are deployed to prevent generation of free radicals and ROS or to scavenge those formed. The phenolic compounds, nitrogen compounds, vitamins, terpenoids, and some other endogenous metabolites found in medicinal plants are known to possess antioxidant activity. Deficiency of anti oxidative defense mechanism may lead to oxidative stress and associated disorders such as coronary heart diseases, neural disorders, diabetes, arthritis and cancer which can be significantly reduced by consuming natural antioxidant rich medicinal plants.

The purpose of the present study were to evaluate the antioxidant activity of two genera of Zingiberaceae by DPPH, ABTS assays after sequential extraction of the rhizome and GC-MS profiling of the major phytochemical constituents in the active fraction of plant extracts. The total phenolic content of the samples are also measured. Many studies have successfully correlated the phenolic content with antioxidant activity. In the study done by Emami (2007), the antioxidant activity due to presence of phytochemicals in natural products is correlated to the plant secondary metabolites such as flavonoids, alkaloids, tannins, phenolic compounds.

The plants belonging to Zingiberaceae family are considered safe for human consumption as these species are excellent candidates for development of novel chemotherapeutics. Both the plants chosen for the study are official in many pharmacopoeias and are reported to have several therapeutic activities (Jain *et al.,* 2007). *Curcuma zedoaria* is a constituent of many Ayurvedic formulations and Chinese medicines. *Curcuma zedoaria* rhizomes are reported to possess anti-inflammatory activity (Jang *et al.,* 2001), antifungal, antiulcer and hepatoprotective activity. *Alpinia galanga* commonly known as galangal is an important cultivated plant in India. Besides the use of plant in culinary and cosmetics (Mohd *et al.,* 2003), it is mainly used in traditional medicines (Yang and Eilerman, 1999). It has also been reported to possess a variety of biological activities, including antioxidant (Ly *et al.,* 2003), antimicrobial (Chit *et al.*, 2001), antiulcer, antifungal (Ficker *et al.*, 2003), anti-HIV (Tewtrakul *et al.*, 2003) and antitumor properties.

Materials and Methods

Materials

Alpinia galanga and *Curcuma zedoaria* rhizomes were collected from the Central Tuber Research Institute (ICAR), Thiruvananthapuram during April 2009, identified and authenticated by the taxonomist and kept at the herbarium of NIIST, Thiruvananthapuram. The rhizomes were cut into small pieces and dried overnight in a tray in RRL-NC dryer at 40°C. They were then ground with a blender to make powder.

Reagents

Gallic acid, 2, 2-Diphenyl-1-Picrylhydrazyl (DPPH), 2, 2′-azino-bis (3-ethylbenzothiazoline–6-sulphonic acid (ABTS), Trolox, were purchased from Sigma Aldrich (St. Louis, MO, US). Disodium hydrogen phosphate, Dipotassium hydrogen phosphate, sodium dihydrogen phosphate, Tris HCl buffer, Folin-Ciocalteu reagent (FCR) were purchased from Sisco Research laboratory (*SRL*) (Mumbai, India). Sodium hydroxide, Trichloroacetic acid (TCA), Potassium hydroxide was purchased from Ranbaxy chemicals. All other chemicals and solvents used were of analytical grade

Extraction Procedure

The air-dried powdered rhizome (200g) of *Alpinia galanga* and *Curcuma zedoaria* was extracted sequentially with different solvents in the increasing order of polarity *viz.* hexane, ethyl acetate and methanol in a soxhlet apparatus (Mukherjee pulok, 2002). These solvent extracts were filtered through Whatman No. 1 filter paper and the solvents were evaporated in vacuum rotary evaporator to dryness. The stock solution prepared by making up each of the extracts up to a definite volume and stored below 4°C in a refrigerator. The concentrates were dissolved in dimethyl sulfoxide (DMSO) and used for the study.

Evaluation of Antioxidant Activity

In vitro antioxidant activity of the extract was determined by DPPH radical scavenging activity, total phenolic content and total reducing power using GEAC (gallic acid equivalent antioxidant capacity) and ABTS radical scavenging activity using TEAC (trolox equivalent antioxidant activity) and IC_{50} value determined.

Total Phenolic Content (TPC)

The Total Phenolic Content was determined using Folin- Ciocalteu reagent (Shingh *et al.*, 2009). Appropriately diluted extracts and standard gallic acid were made up to 3.5 ml using distilled water in a series of test tubes. The plant extracts and standard were treated with 0.5 ml Folin- Ciocalteu reagent and incubated for 3 minutes at room temperature. The reaction was then neutralized with the addition of 1ml of 20 per cent Na_2CO_3. The reaction mixture was incubated at room temperature for 90 minutes and the absorbance of the blue color developed was read at 760nm using a

Spectrophotometer (Shimadzu UV Vis–Spectrophotometer, model 2450). The Total Phenolic Content was expressed as gallic acids equivalent in milligram per gram of sample, using a standard curve generated with gallic acid.

DPPH Radical Scavenging Capacity

2, 2-Diphenyl-1-Picrylhydrazyl (DPPH) radical scavenging capacity was estimated by the method of Wang (2002). To 2.8 ml of 0.1mM DPPH (3mg DPPH in 100ml methanol) taken in 3ml cuvette, appropriately diluted standard (Gallic acid) and extracts in 0.2ml solution was added and absorbance read using Shimadzu UV-Vis Spectrophotometer, model 2450. Data acquisition was immediately started and continued for 20 minutes. The decrease in absorbance at 517nm was noted continuously from time 0 to 20 minutes. From this percentage radical scavenging activity was calculated.

$$\text{Radical Scavenging Activity (RSA)} = [(A_0 - A_1)/A_0 \times 100]$$

where,

A_0 is the absorbance of the control and A_1 is the absorbance of the sample.

From the calibration curve obtained with different concentrations of various extracts of *Alpinia galanga* and *Curcuma zedoaria*, the Inhibitory concentration required for 50 per cent quenching of free radicals (IC_{50}) value was calculated. The IC_{50} was defined as the concentration of an antioxidant extracts which was required to quench 50 per cent of the initial DPPH under the experimental conditions given.

Total Reducing Power

The reducing power of the various extracts of *Alpinia galanga and Curcuma zedoaria* was measured according to Oyaizu method (1986). Different concentrations of various extracts and standard Gallic acid were made up to 1ml with distilled water and add 2.5ml Phosphate buffer (0.2mM, pH 6.6) and Potassium ferricyanide. The reaction mixture was incubated at 50°C for 20 minutes. After the incubation 2.5ml Trichloroacetic acid (TCA) (20 per cent) was added and mixed well. Then 2.5ml of the solution was taken and mixed with 2.2ml of distilled water and added 0.5ml of Ferric chloride. The absorbance of the reaction mixtures was read at 700nm used a UV visible spectrophotometer. Higher absorbance of the reaction mixture indicated greater the reducing power.

ABTS Cation Decolorisation Capacity

The experiment was carried out using an improved 2,2'-azino-bis(3-ethylbenzthiazoline-6-sulphonic acid (ABTS) decolorisation assay (Re *et al.*, 1999) and involved the generation of $ABTS^+$ chromophore by the oxidization of ABTS with potassium persulphate. It is applicable for both hydrophilic and lipophylic compounds. The ABTS radical cation ($ABTS^+$) was produced by reacting 7mM stock solution ABTS with 2.45mM Potassium Persulphate. The mixture was allowed to stand in the dark for at least 6 hours at room temperature before use. The $ABTS^+$ solution was diluted to an absorbance 0.7 ± 0.05 at 736nm. Absorbance was measured 7 minutes after the initial mixing of different concentrations of the extracts. The $ABTS^+$

decolorisation capacity of the extracts were compared with the standard Trolox. A standard curve was prepared by measuring the reduction in the absorbance of $ABTS^+$ solution at different concentration of Trolox over a period of 7 minutes.

Chemical Composition of Ethyl Acetate Fraction by GC-MS Analysis

The compounds present in the extracts were identified by GC-MS (GC-MS Shimadzu QP 2010). Column DB-5 was used for the analysis. For GC–MS detection, an electron ionization system with ionization energy of 70 eV was used. Helium gas was used as a carrier gas at a constant flow rate of 1 ml/min. Oven temperature were programmed as follows: 80-200°C at the rate of 5°C per min and held at 200°C for 25 min. The column temperatures are programmed from 50°C to 280°C at the rate of 100 C/min. The injection port temperature was 200°C and interface temperature 200°C, spilt ratio 1: 75.The percentage composition of oil was calculated automatically from peak areas without any correction. The retention indices of compounds were determined relative to the retention times of a series of n- alkanes with linear interpolation. Identification of the components was done by comparison of their mass spectra with Wiley GC-MS library as well as by comparing them with those reported in the literature. The identification of each compound was confirmed by comparison of its retention index either with those of authentic compounds or from literature.

Statistical Analysis

Experimental data are expressed as mean ± SD. Student's t test was applied for expressing the significance and P value less than 0.05 was considered as significant.

Results and Discussion

Sequential extraction leads to an initial fractionation of constituents based mainly on their nature of polarity. The extraction efficiency was evaluated in terms of total phenolic content. Among the hexane, ethyl acetate and methanol fraction, the TPC was found to be high for the polar extracts, highest for the ethyl acetate fraction of both the plant extract species and hence chosen for further assays and GC-MS analysis. Higher activity for the polar extracts indicates that the polar phytochemicals are responsible for the antioxidant activity.

Total Phenolic Content

The total phenolic content was observed to be high in ethyl acetate fraction (102.3 ± 1.26 mg/g for *Alpinia galanga* and 93.1 ± 1.26 mg/g for *Curcuma zedoaria*). Gallic acid was used as standard. Poly phenolic compounds are products of secondary metabolism in plants, and antioxidant activity is mainly due to their redox properties and chemical structure which aid in chelating to metals and scavenging free radicals. It is well known that these phenolic compounds are viewed as powerful antioxidants due to their ability to donate hydrogen and to form stable radicals. The antioxidant activity was lesser in the case of *Curcuma zedoaria* as shown in the results (Table 23.1). However it is obvious that the TPC measured by the Folin- Ciocalteu procedure do not give a full picture of the quantity or quality of the phenolic constituents in the extracts (Katsube *et al.*, 2004).

Table 23.1: Total phenolic content of extracts of *Alpinia galanga* and *Curcuma zedoaria*.

Sl.No.	Plant	Extract	TPC (mg/g of extract)
1	*Alpinia galanga*	Hexane	34.8 ± 1.20
		Ethyl acetate	102.3 ± 1.26
		Methanol	49.3 ± 1.20
2	*Curcuma zedoaria*	Hexane	46.8 ± 1.62
		Ethyl acetate	93.1 ± 1.26
		Methanol	49.4 ± 1.36

Values are mean of ± S.D, n= 5.

DPPH Scavenging Activity

DPPH is one of the powerful free radical which is used to evaluate the electron donating capacity of antioxidants. Antioxidant compounds either transfer electrons or hydrogen atoms to DPPH radical and neutralizes the free radical character which is visualized by the colour change from purple to yellow. Due to reduction of DPPH there will be a decrease in absorbance value. The decrease in absorbance depends on the amount of antioxidant compounds. The DPPH radical strongly absorbs at 518 nm and upon reduction the absorption decreases. The IC_{50} value for the ethyl acetate fraction of *A. galanga* and *C. zedoaria* were found to be low compared to the other extracts (IC_{50} values are 119.29 ± 1.46 µg/mL and 82.32 ± 1.60µg/mL respectively) (Table 23.2). Among the extracts of the two plants studied, the ethyl acetate fraction was found to have high total phenolic content (Table 23.1). The polar phenolics in the plant extracts are mainly responsible for the free radical scavenging of natural antioxidants (Chen and Ho, 1995). Several mechanisms reveal phenolic groups as radical scavengers by oxidation of diphenols to quinines which further degrade to radical scavenging degradation products. All the extracts were able to reduce the stable purple colored DPPH free radical to yellow colored 2, 2 -diphenyl -1-picrylhydrazine. As the concentration of extract increases, the radical scavenging activity also increases. Polar extracts exhibited stronger activity than non-polar extracts. Zaeoung *et al.* (2005) had reported high 1, 1-diphenyl-2-picrylhydrazyl (DPPH) radical scavenging activity of methanol and water extracts of rhizomes of *A. galanga.*

The phenolic antioxidants present in *Curcuma zedoaria* have high affinity towards DPPH radical as revealed from IC_{50} value (Table 23.2) and the activity probably depend on side chain structures and substitution patterns of the benzene ring.

Total Reducing Power

Reducing power is also widely used in evaluating antioxidant activity of plant polyphenols. In this assay, the presence of reductants in the antioxidant sample cause the reduction of the Fe^{3+}/fericyanide complex to the Fe^{2+}/ferrous form, so the

Table 23.2: DPPH evaluation of Gallic acid and extracts of *Alpinia galanga* and *Curcuma zedoaria.*

Sl.No.	*Plant*	*Extract/Standard*	*IC $_{50}$ (µg/ml)*
1.	–	Gallic acid	1.57 ± 0.98
		Hexane extract	345.63 ± 2.80
2.	*Alpinia galanga*	Ethyl acetate extract	119.29 ± 1.46
		Methanol extract	210.79 ± 1.30
3.	*Curcuma zedoaria*	Hexane extract	806.2 ± 3.10
		Ethyl acetate extract	82.32 ± 1.60
		Methanol extract	158.96 ± 1.24

Values are mean of ± S.D, n= 5.

reducing power of the sample can be monitored by measuring the formation of Perl's Prussian blue at 700nm. In the case of reducing power experiment as the absorbance value increases the antioxidant activity also increases. The experiments for the total reducing power calculation showed that ethyl acetate extract of both the plants showed higher reducing power as in the case of total phenolic content.The antioxidant activity of phenolic compounds in the extract is due to their redox properties, which enable to absorb and neutralize free radicals and quench singlet and triplet oxygen species. The reducing property depends on reductones in the extract, which exert antioxidant action by donating a hydrogen atom to the free radical. The phenolic antioxidants present in the extract of both the plants in the present study have high reducing power equivalents in ethyl acetate extract even though it was less in comparison with the standard gallic acid. The activity of the extracts could possibly be increased either by fractionation or by active principle isolation (Figures 23.1 and 23.2).

ABTS Cation Decolorisation Capacity

ABTS radical assays, can be used in both organic and aqueous solvent systems, therefore often used in evaluating total antioxidant power of single compounds and complex mixtures of various plants. The inhibitory concentration for 50 per cent decolourisation was observed to be less in the ethyl acetate fraction of *Alpinia galanga* (17.38 ± 1.10µg/ml) and *Curcuma zedoaria* (33.9 ± 1.24µg/ml). Unlike DPPH scavenging effect, the ABTS decolorisation capacity was more promising to *Alpinia galanga* than to *Curcuma zedoaria* (Table 23.3).

Analysis of the Chemical Components in the Active Ethyl Acetate Fraction of *Alpinia galangal* and *Curcuma zedoaria*

Gas Chromatography Mass Spectrophotometry analysis revealed that 2, 4-dimethyl benzoic acid with 65.2 per cent was the most prominent compound present in the active ethyl acetate extract of *Alpinia galanga*. Percent composition of major constituents in the extract are given in Table 23.4a. Apart from 2, 4-dimethyl benzoic acid, twenty two other compounds were also identified in the ethyl acetate extract. Eugenol with 6.55 per cent and β-farnesene with 6.37 per cent were the next more

Table 23.3: Evaluation of ABTS Cation Decolorisation Capacity of trolox and extracts of *Alpinia galanga* and *Curcuma zedoaria.*

Sl.No.	*Plant*	*Extract/Standard*	*IC $_{50}$ (μg/ml)*
1.	–	Trolox	2.066 ± 0.78
		Hexane extract	56.68 ± 1.42
2.	*Alpinia galanga*	Ethyl acetate extract	17.38 ± 1.10
		Methanol extract	38.26 ± 1.20
3.	*Curcuma zedoaria*	Hexane extract	205.5 ± 1.46
		Ethyl acetate extract	33.9 ± 1.24
		Methanol extract	43.64 ± 1.42

Values are mean of ± S.D, n= 5.

Table 23.4a: Chemical composition of the active Ethyl acetate fraction of *Alpinia galanga* by GC-MS.

Sl.No.	*Compound*	*Percentage*	*Retention Time*
1.	1,8-Cineole	0.42	3.287
2.	Ciscarenyl acetate	0.23	7.095
3.	α-Terpinyl acetate	0.84	7.183
4.	Eugenol	6.55	7.318
5.	α-Cubene	0.83	7.540
6.	Geramyl acetate	0.58	7.633
7.	Dodecane	0.17	7.733
8.	Caryophyllene	0.80	8.135
9.	β-farnesene	6.37	8.473
10.	α-humulene	0.25	8.583
11.	Penta decane	4.84	8.972
12.	β-bisabolene	0.38	9.270
13.	δ-Cadinene	0.15	9.431
14.	β-sesquiphellandrene	0.26	9.477
15.	Caryophyllene oxide	0.30	10.227
16.	Germacrone	0.18	10.350
17.	2,4-dimethyl benzoic acid	65.20	11.035
18.	1-pentadecane	2.57	11.103
19.	8-heptadecane	0.23	11.208
20.	Heptadecane	0.27	11.332
21.	Cis-farnesol	1.82	13.015
22.	Hexa decanoic acid	1.06	14.964
23.	Linoleic acid	0.33	19.01

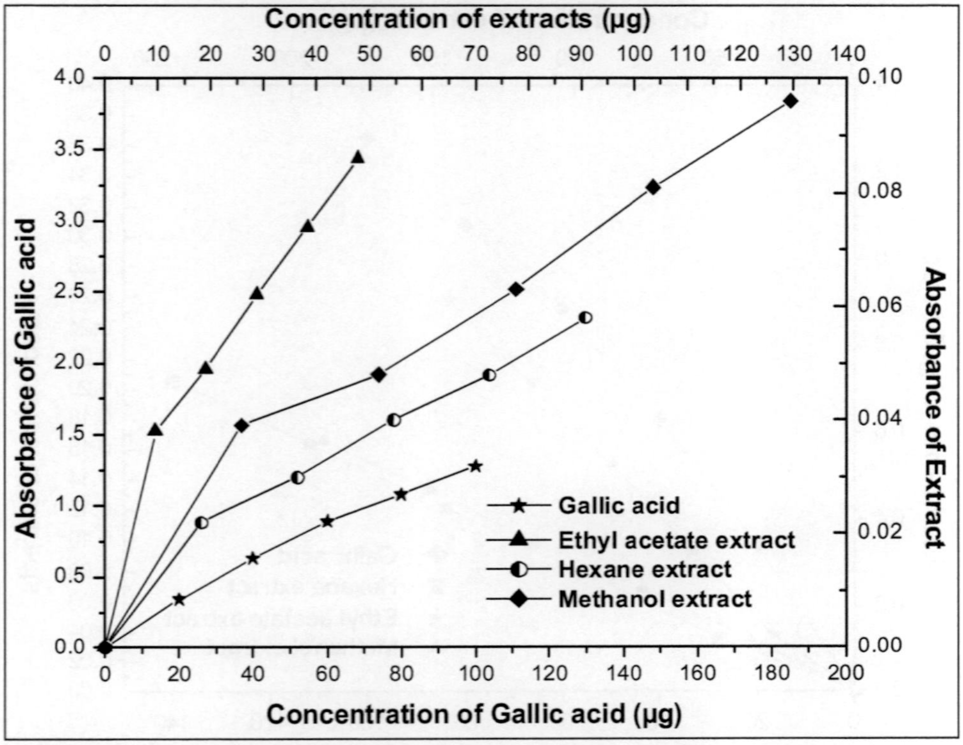

Figure 23.1: Total reducing power evaluation of *Alpinia galanga.*

prominent compounds present in the extract. Pentadecane, cis-farnesol, caryophyllene, α-terpinyl acetate, α-cubene, 1, 8-cineole, β-bisbolene are the other minor constituents in the ethyl acetate fraction. 1, 8-cineole is reported to be an oxygenated monoterpene and the hydroxyl derivatives of 1, 8-cineole are primary aromatic components in the plant. The sesquiterpene hydrocarbon identified were β-bisbolene (0.38 per cent) and α-humelene (0.25 per cent). Caryophyllene (0.80 per cent) is a sesquiterpene and cis-farnesol is the major sesequiterpene oxygenated compound identified. Sookkongwaree *et al.* (2006) reported that the rhizome extract contain 1,8-cineole, β-bisabolene, β-selinene as main compounds and α-selinene, farnesene, 1,2-dicarboxylic acid, germacrene, pentadecane as minor constituents. 1, 8-cineole, α-fenchyl acetate are reported to be present in the rhizome extract by Mallavarapu *et al.* (2002). The above reports are in agreement with the study conducted and differ in the percentage composition. But needless to mention that the variation in composition for a particular species depend on seasonal, geographical conditions as well as on the method opted for extraction (Burt, 2004). As *Curcuma zedoaria* contains bioactive principles, the constituents of *zedoary* have been investigated extensively and it is recognized to be a rich source of terpenoids. Chemical investigations of the *Curcuma zedoaria* rhizomes reveal the presence of several sesquiterpenoids namely

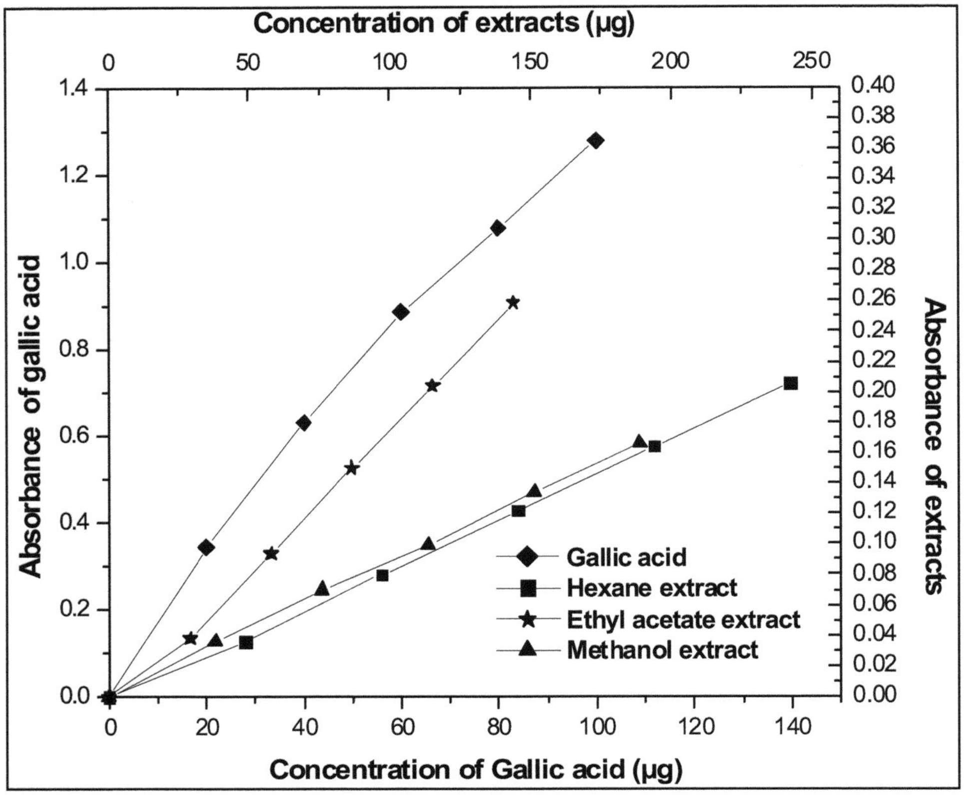

Figure 23.2: Total reducing power evaluation of *Curcuma zedoaria*.

zederone, curcolone, curcumenol, furanodiene, curzerene, curzerenone, epicurzerenone, procurcumenol, isocurcumenol, curcumadiol, curcumol, curdione, dehydrocurdione, zingiberene, β-turmerone and ar-turmerone (Purkayastha, 2006).

Our studies on the *Curcuma zedoaria* rhizome extract native to Kerala showed isocurcumenol (25.24 per cent) and methyl sterolate (24.94 per cent) as the major compounds. Percentage composition of major constituents in the extract is given in Table 23.4b. Some sesquiterpenoids like curcumenone, curcarabranols A and B, 4-epicurcumenol, neocurcumenol have been reported from C. zedoaria (Matsuda *et al.*, 2001a,2001b; Jang *et al.*, 2001). The major sesquiterpenes, including furanodiene, germacrone, curdione, neocurdione, curcumenol, isocurcumenol, aerogidiol, zedoarondiol, curcumenone and curcumin were found to show protective effect against Dgalactosamine/lipopolysaccharide induced acute liver injury in mice (Matsuda *et al.*, 1998). Germacrone, isocurcumenol, β-eudesmol and β-dictyopterol show potent vasorelaxant effects (Matsuda *et al.*, 2001a). The phytochemical investigations in different *Curcuma zedoaria* from China and Malaysia reported the isolation of six sesquiterpenes including isocurcumenol (Aspollah *et al.*, 2010). The same compound was found to be the major constituent in the rhizome extract from

Kerala, India. The present finding could indicate the existence of different chemotypes and regional variation for *C. zedoaria*. Studies by Lakshmi *et al.* (2011) has reported the significant nontoxic nature and antitumour effects of isocurcumenol, a sesquiterpenoid compound isolated from *Curcuma zedoaria* on the human lung, nasopharyngeal, and leukemic cells as well as the murine lymphoma cells.

Table 23.4b: Chemical composition of the active fraction of *Curcuma zedoaria* by GC- MS.

Sl.No.	*Compound*	*Retention Time*	*Percentage*
1.	Eucalyptol	9.786	2.04
2.	Camphor	12.44	1.58
3.	β-selinene	21.008	1.03
4.	Azulene	21.225	0.87
5.	4,5,6,6A-tetrahydro-2-[1H] pentalenone	23.752	14.49
6.	Germacrone	25.783	1.87
7.	Isocurcumenol	26.623	25.24
8.	9-octadecynoic acid methyl ester	26.948	1.36
9.	Dihydro neoclovene	27.018	3.48
10.	Methyl sterolate	27.738	24.94
11.	Isovelleral	27.886	5.92
12.	Eremanthron	28.127	2.89
13.	Elemene	28.883	3.75
14.	Octahydroanthracene	31.650	1.25
15.	Isolongifolene	31.961	9.31

Conclusions

In vitro antioxidant activities of different polarity based extracts of *Alpinia galanga* and *Curcuma zedoaria* were compared by four different methods. The Ethyl acetate extract of *Alpinia galanga* and *Curcuma zedoaria* possessed significant antioxidant activity compared to the other extracts. In comparison with the *Curcuma zedoaria*, the Ethyl acetate fraction of *Alpinia galanga* was more antioxidant in nature. This study revealed that the total phenolic content was high in ethyl acetate extract and hence the scavenging activity as well as the total reducing power. The GC-MS analysis revealed that 2, 4-dimethyl benzoic acid with 65.2 per cent was the most prominent compound present in the active fraction of *Alpinia galanga. Curcuma zedoaria* showed isocurcumenol and methyl sterolate as the major compounds. The present study reveal that isocurcumenol rich *Curcuma zedoaria* variety from Kerala can be exploited to develop antitumour drugs. The potential effects of *Alpinia galanga* could be attributed to the chemical moieties present in it. This is the first report on the comparison of antioxidant activity in terms of chemical potential of these two plants. The results of the present study further support the authenticity of *Alpinia galanga* by ancient people for several medicinal purposes.

Acknowledgements

The authors are thankful to the Director, National Institute for Interdisciplinary Science and Technology for providing all the facilities and CSIR for the financial support to carry out the investigations and Central Tuber Crops Research Institute for the plant material.

References

Aggarwal, B.B., Kumar, A., and Bharti A.C. (2003). Anticancer potential of curcumin: preclinical and clinical studies. *Anticancer Research*, **23 (1A):** 363–398.

Aspollah, S.M., Wah, T.S., and Saad, S.M. (2010). Bioactive sesquiterpenes from *Curcuma ochrorhiza* and *Curcuma heyneana*. *Natur Prod. Res.*, **24(9):** 838–845.

Burt, S. (2004). Essential oils: Their antibacterial properties and potential applications in foods–a review. *International Journal of Food Microbiology*, **94:** 223–253.

Chen, C.W., and Ho, C.T. (1995). Antioxidant properties of polyphenols extracted from green and black tea. *Journal of Food Lipids*, **2:** 35-46.

Chit, K., Myint, W., Thein, K., Maw, W.W., Myint, M.M., Than, A., and Khin, M. (2001). Cyclic AMP Phosphodiesterase inhibitory activity and chemical screening of four medicinal plants. *Pharmaceutical Biol.*, **39:** 181-183.

Emami, S. (2007). Antioxidant activity of leaves and fruits of Iranian conifers. *eCAM.*, **3**: 313-19.

Ficker, C.E., Smith, M.L., Susiarti, S., Leaman, D.J., Irawati, C., and Arnason, J.T. (2003). Inhibition of human pathogenic fungi by members of *Zingiberaceae* used by the Kenyah(Indonesia Borneo). J *Ethnopharmacol.*, **85(2-3):** 289-293.

Jain, S., Shrivastava, S., Nayal, S., and Sumbhate, S. (2007). Plant review trends in *Curcuma longa* Linn. *Pharmacog Rev.*, **1 (1):** 119-128.

Jang, M.K., Sohn, D.H., and Ryu, J.H. (2001). A curcuminoid and sesquiterpenes as inhibitors of macrophage TNF-alpha release from *Curcuma zedoaria*. *Planta med.*, **67(6):** 550-552.

Katsube, T., Tabata, H., Ohta, Y., Yamasaki, Y., Anuurad, E., Shiwaku, K., and Yamane, Y. (2004). Screening for antioxidant activity in edible plant products: Comparison of low-density lipoprotein oxidation assay, DPPH radical scavenging assay and Folin-Ciocalteu assay. *J Agri Food Chem.*, **52:** 2391–2396.

Lakshmi, S., Padmaja, G., and Remani, P. (2011). Antitumor effects of isocurcumenol isolated from *Curcuma zedoaria* rhizome on human and murine cancer cells. *Int. J. Med. Chem.*, doi: 10.1155/2011/253962.

Ly, T.N., Shimoyamada, M., Kato, K., and Yamauchi, R. (2003). The health benefits of traditional Chinese plant medicines: Weighing the scientific evidence. *J Agric. Food Chem.*, **51:** 4924- 4929.

Mallavarapu, G.R., Rao, L., Ramesh, S., Dimri, B.P., Rao, B.R., Kaul, R., and Bhattacharya, A.K. (2002). Composition of the volatile oil of *Alpinia galanga* rhizome and leaves from India. *J Essen Oil Res.*, **14:** 397-399.

Matsuda, H., Ninomiya, K., Morikawa, T., and Yoshikawa, M. (1998). Inhibitory effect and action mechanism of sesquiterpenes from *Zedoariae rhizoma* on D-galactosamine/lipopolysaccharide -induced liver injury. *Bioorg. Med. Chem. Lett.*, **8:** 339-344.

Matsuda, H., Morikawa, T., Ninomiya, K., and Yoshikawa, M. (2001a). Hepatoprotective constituents from *Zedoariae rhizoma*: absolute stereostructures of three new carabrane-type sesquiterpenes, curcumenolactones A, B, and C. *Bioorg. Med. Chem.*, **9:** 909-916.

Matsuda, H., Morikawa, T., Ninomiya, K., and Yoshikawa, M. (2001b). Absolute stereostructure of carabrane-type sesquiterpene and vasorelaxantactive sesquiterpenes from *Zedoariae rhizoma*. *Tetrahedron*, **57:** 8443-8453.

Mohd, M.S., Chin, C.B., Chen, L.L., and Sim, N.L. (2003). Antifungal activity of the essential oils of nine *Zingiberaceae* species. *Pharmaceutical Biol.*, **41(5):** 302-307.

Mukherjee pulok, K. (2002). Quality control of herbal drugs. Edn 1, Nirali publications, Mumbai.

Oyaizu, M. (1986). Studies on products of browning reaction: antioxidative activity of products of browning reaction prepared from glucosamine. *Japanese Journal of Nutrition*, **44:** 307–315.

Purkayastha, J., Nath, S.C., and Klinkby, N. (2006). Essential oil of the rhizome of *Curcuma zedoaria (Christm.) rosc*. Native to Northeast India. *J Essential Oil Res.*, **18(2):** 154–155.

Re, R., Pellegrini, N., Proteggente, A., Pannala, A., Yang, M., and Rice-Evans, C. (1999). Antioxidant activity applying an improved ABTS radicalcation decolorization assay. *Free Radical Biology and Medicine*, **26:** 1231–1237.

Shingh, J., Upadhyay, A.K., Singh, S., and Rai, M. (2009). Total phenolics content and free radical scavenging activity of brassica vegetables. *J Food Science Technology*, **46:** 595–597.

Sookkongwaree, K., Geitmann, M., Roengsumran, S., Petsom, A., and Danielson, U.H. (2006). Inhibition of viral proteases by Zingiberaceae extracts and flavones isolated from *Kaempferia parviflora*. *Pharmazie.*, **61(8):** 717-721.

Tewtrakul, S., Subhadhirarakul, S., Kummu, S., and Landcarin, S. (2003). HIV-1 protease inhibitor effect of medicinal plants used as self medication by AIDS patients. *J Sci Techno.*, **25(2):** 239-243.

Wang, S.Y., Kuo, Y.H., Chang, H.N., Kang, P.L., Tsay, H.S., and Lin, K.F. (2002). Profiling and characterization antioxidant activities in *Anoectochilus formosanus* Hayata. *J Agric Food Chem.*, **50:** 1859-1865.

Yang, X., and Eilerman, R.G. (1999). Pungent principles of *Alpinia galanga* (L.) Swartz and its applications. *J Agri Food Chem.*, **47:** 1657-1662.

Zaeoung, S., Plubrukarn, A., and Keawpradub, N. (2005). Cytotoxic and free radical scavenging activities of Zingiberaceous rhizomes. *Songklanakarin J Sci Techno.*, **27(4):** 799-812.

Medicinal Plants: Phytochemistry, Pharmacology and Therapeutics Vol. 3 (2014) *Pages* **489–498**
Editor-in-Chief: **V.K. Gupta**
Published by: **DAYA PUBLISHING HOUSE, NEW DELHI**

24

In vitro Effects of *Curcuma xanthorrhiza* Extracts and Constituents on Glutathione Transferase Activity

Nurul Afifah Mohd Salleh[1], Mohd Rohaimi Ab Halim[1] and Sabariah Ismail[1]*

ABSTRACT

Herbal supplements are commonly used as alternative medicines to treat various diseases which are often administered in combination with therapeutic drugs, raising the potential of herb-drug interactions. The interaction may involve having an herb component causing an increase or decrease in the amount of drug in the blood through induction or inhibition of drug metabolizing enzymes. Curcuma xanthorrhiza is a native Indonesian plant and traditionally utilized for a range of illness including liver damage, hypertension, diabetes and cancer. Glutathione transferase (GST) is a phase II enzyme which is involved in detoxification and metabolism of anticancer drugs such as chlorambucil. Induction or inhibition of GST enzyme activity by Curcuma xanthorrhiza may affect the metabolism of anticancer drugs that are highly metabolized by GST enzyme. In the present study, the effects of ethanolic, hexane and aqueous extracts of Curcuma xanthorrhiza and its three main pure compounds, namely curcumin, curcumene and xanthorrhizol on glutathione transferase (GST) activity in vitro were investigated. The ability of these plant extracts and their pure compounds to inhibit GST activity was examined on rat liver cytosolic fraction and was analyzed using ultraviolet (UV) absorbance at 340 nm. In this study, 1–chloro-2,4- dinitrobenzene (CDNB) was employed as a substrate to monitor GST activity. Tannic acid which was used as the positive inhibitor showed an IC_{50} value of 16.73 ± 0.79 µg mL^{-1}. The ethanolic extract of Curcuma xanthorrhiza demonstrated inhibition at the highest concentration used with IC_{50}

1 Centre for Drug Research, Universiti Sains Malaysia, 11800 Penang, Malaysia
* *Corresponding author*: E-mail: sabaris@usm.my

value of 233.80 ± 3.71 μg mL^{-1}. The IC_{50} values for hexane and aqueous extracts of Curcuma xanthorrhiza and their pure compounds were not determined due to the percentage of inhibition being less than 50 per cent. These findings suggest that Curcuma xanthorhhiza shows weak inhibitory potential towards GST activity.

Keywords: Glutathione-S-transferase, *Curcuma xanthorrhiza,* Enzyme inhibition, 1-chloro-2,4-dinitrobenzene, Tannic acid, Xanthorrhizol, Curcumin.

Introduction

Curcuma xanthorrhiza or commonly known as " temu lawak" or Javanese turmeric is a member of the ginger family (Zingeberacea) which is a native Indonesian plant. It is well grown in Thailand, Philippines, Sri Lanka and Malaysia. *Curcuma xanthorrhiza* is a low growing plant with a root (rhizome) that looks like ginger (Sears, 2005). Traditionally, this plant is popularly used as an ingredient in health supplements known as "jamu" and "maajun", or individually used to remedy certain health problems including hepatitis, liver complaints, diabetes, rheumatism, anticancer, hypertension and heart disorders (Ruslay *et al.*, 2007, Sears, 2005). *Curcuma xanthorrhiza* has also shown antidiuretic, anti-inflammatory, anti-oxidant, anti-hypertensive, anti-hepatotoxic, anti-bacterial and antifungal effects (Sears, 2005).

Xanthorrhizol, the major component of the essential oil of *Curcuma xanthorrhiza*, is a bisabolane-type sesquiterpenoid. This compound makes up nearly 46.3 per cent of the total components of the essential oil obtained through hydrodistillation (Yew *et al.*, 2009). Xanthorrhizol is a special marker for *Curcuma xanthorrhiza* and thus its presence differentiate this plant from other *Curcuma* species. Xanthorrhizol has been reported to encompass a wide range of biological activities such as antibacterial, antiseptic and antibiotic (Nur, 2006).

Curcumin belongs to the phenol group and comprises almost 1 to 2 per cent in ethyl acetate fraction of *Curcuma xanthorrhiza*. It is the second essential active compound found in this plant (Sharin *et al.*, 2007). Recently, curcumin has garnered interest as a potential anticancer agent, for both chemopreventative and chemotherapeutic purposes (Aggarwal *et al.*, 2003; Aggarwal and Shishodia, 2006).

The glutathione transferase (GST) plays a significant role in detoxification across multiple kingdoms and phyla. Three distinct superfamilies of GST are recognized: the cytosolic GST (which form the largest superfamily), the mitochontrial, and the microsomal GST (Oakley, 2011). In humans, most classes of cytosolic GST catalyze conjugation reactions with drugs and other xenobiotics (Hayes *et al.*, 2005). For instance, GST participates in the detoxification of many anticancer drugs including chlorambucil (Hayes, 1995). Since interest in herbal supplements has grown rapidly among patients with cancer, the risk for herb drug interactions is a growing concern. To date, there is paucity of studies on the effect of *Curcuma xanthorrhiza* on GST activity. In 2011, Tan *et al.* reported that aqueous extract of *Curcuma xanthorrhiza* showed negligible inhibition towards GST activity. Therefore in this study, the potential of *Curcuma xanthorrhiza* extracts and its constituents to affect the activity of glutathione transferase was evaluated using rat liver cytosolic fraction as a model system.

Materials and Methods

Chemicals

1–chloro-2,4- dinitrobenzene (CDNB), reduced glutathione (GSH), cupric sulphate ($CuSO_4$), Folin-Ciocalteau's phenol reagent, sodium potassium tartrate sodium carbonate (Na_2CO_3) and curcumin were obtained from Sigma Chemicals Company (USA). Tannic acid that was used as the positive inhibitor was purchased from HmbG Chemical (German). The powder of dipotassium hydrogen phosphate was obtained from Riedel-de Haen (Germany). Potassium chloride (KCl) was acquired from BDH Chemical (UK). All the reagents and organic solvents used throughout this project were of analytical grade.

Plant Materials

Standardized ethanolic, hexane and aqueous extracts of *Curcuma xanthorrhiza* were prepared in our laboratory. The purity of the extracts was verified by comparing the physical and spectroscopic data.

Animal

Six Sprague Dawley rats weighing from 150 to 200 g provided by Animal Holding Facility of Universiti Sains Malaysia were used in this study. The rats were housed in a temperature controlled room at 22±2°C under a 12 h light/ 12 h dark cycle for one week prior to sacrifice. The animals were fed with water and food *ad libitium*.

Preparation of Rat Cytosolic Fraction

Rat livers were removed immediately after sacrificed, rinsed with ice-cooled distilled water followed by ice-cooled 67 mM potassium phosphate buffer (pH 7.4), blotted dry and weighed. Isolated rats liver samples were homogenized in 3 times volumes of 67 mM potassium phosphate buffer (pH 7.4) containing 1.15 per cent potassium chloride, using a Potter–Elvehjem homogenizer. After centrifugation of the homogenate fraction at 12,500× *g* for 20 minutes at 4°C, the resultant supernatant was decanted to ultracentrifuge tubes (Optiseal™) and centrifuge at 100,000× *g* for 60 min in a Optima™ TLX refrigerated ultracentrifuge (Beckman Coulter, Inc., USA). The supernatant obtained was the cytosolic fractions and its protein concentration was determined by the Lowry method (Promory, 2008). The cytosolic fractions were stored at "80 °C until used.

Determination of GST Activity on *Curcuma xanthorrhiza* Extracts and Pure Compounds

The inhibition studies using cytosolic fraction as GST source was carried out according to a previous method (Habig *et al.*, 1974) with slight modifications (Azizi *et al.*, 2010) using 1-chloro-2,4- dinitrobenzene (CDNB) as substrate. Incubation mixtures (300 ìL) contained 0.1 M potassium phosphate buffer pH 6.5, 30 mM CDNB, 30 mM GSH, and GST enzymes (0.125 mg/mL rat liver cytosolic fraction). The concentrations of plant extracts and tannic acid ranging from 0.01-500 µg mL^{-1} were used. The concentrations of pure compounds used in this study ranged from 0.01-500 µM. The reason for not determining the activity above 500 µg mL^{-1} and 500 µM was because maximum absorbance was reached at concentration higher than that. GST mediated conjugation of 1-chloro-2,4- dinitrobenzene CDNB to glutathione (GSH) was measured using a PlateCHAMELEON™ multitechnology plate reader, 425-106 (Hidex Oy, Finland) at the wavelength of 340 nm for 5 minutes.

Statistical Analysis

All experiments were carried out in 5 replicates and expressed as the Mean±SEM. All the computation analysis was carried out using GraphPad Prism® 5 software.

Results

Results are shown in Figures 24.1 and 24.2 and Table 24.1.

Discussion

In this study, *Curcuma xanthorrhiza* extracts namely ethanolic, hexane and aqueous were investigated for their effects on glutathione transferase (GST) activity. The effects of the three main compounds in this plant namely curcumin, curcumene and xanthorrhizol on GST activity were also evaluated. The conventional reaction catalyzed by GST is the formation of a thioether between GSH and an electrophilic substrate. The reaction can be exemplified by using the probe substrate 1-chloro-2,4-dinitrobenzene (CDNB), which can react enzymatically or nonenzymatically with nucleophiles. Hence, the kinetics of thioether formation can be monitored over time and the activity can be calculated (Azizi *et al.*, 2010).

Table 24.1: IC_{50} values for *Curcuma xanthorrhiza* ethanolic extract and tannic acid.

Plant	*IC_{50} Value (µg mL^{-1})*
Curcuma xanthorrhiza extracts	
Ethanolic	233.80 ± 3.71
Hexane	ND
Aqueous	ND
Curcuma xanthorrhiza constituetnts	
Curcumin	ND
Curcumene	ND
Xanthorrhizol	ND
Tannic acid (positive inhibitor)	16.73 ± 0.79

Values are Mean ± SEM for n = 5 determinations.

ND: Not determined due to percentage of inhibition being less 50 per cent.

For GST inhibition assay, our *in vitro* results indicate that the addition of *Curcuma xanthorrhiza* extracts into the reaction mixture lead to concentration dependent inhibition of cytosolic GST. The percentage inhibition of GST specific activity for

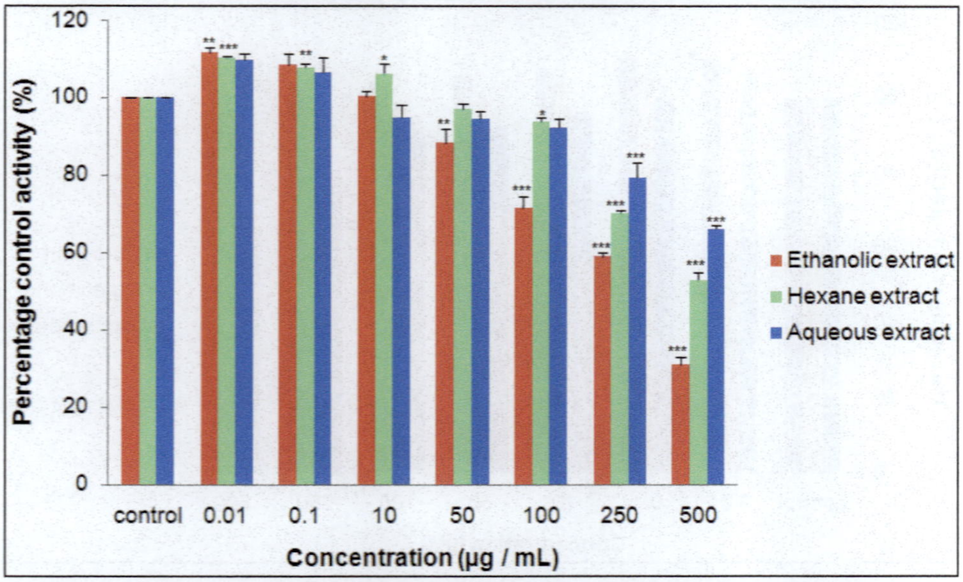

Figure 24.1: Effect of ethanolic, hexane and aqueous extracts of *Curcuma xanthorrhiza* on GST activity. Each bar represents the mean value ± SEM of percentage of GSTs specific activity compared with control for five determinations (n = 5). Statistical analysis was conducted using one way ANOVA and Dunnet test. * $p < 0.05$ *vs.* control, ** $p < 0.01$ *vs.* control, *** $p < 0.001$ *vs.* control.

ethanolic extract varied from 11-68 per cent, 2-47 per cent for hexane extract and 4-33 per cent for aqueous extract. Curcumin had a very low significant inhibition towards GST activity with inhibition ranging from 4-37 per cent, whereas there was no significant inhibition observed for curcumene and xanthorrhizol (Figure 24.2). Earlier studies have shown that curcumin itself is a potent inhibitor of the human GSTs A1-1, M1-1, and P1-1, using CDNB as a substrate, with IC_{50} values of 18.8 µM, 0.3 µM and 15.1 µM respectively. The IC_{50} of curcumin itself towards rat liver GSTs was 4.2 µM (Appiah-Opong *et al.*, 2009).

The IC_{50} values were determined for the extracts that demonstrated more than 50 per cent inhibition (Appiah-Opong *et al.*, 2007). The IC_{50} values were calculated using the GraphPad Prism® 5 (Version 5.01, GraphPad Software, Inc., USA) by plotting the percentage control activity versus log concentration of the plant extracts and pure compounds. Based on Figure 24.1, only *Curcuma xanthorrhiza* ethanolic extract showed inhibition at the highest concentration used with an IC_{50} value of 233.80 ± 3.71 µg mL^{-1}. For hexane and aqueous extracts, an accurate IC_{50} value could not be determined because percentages of inhibition were less than 50 per cent. Previous study by Tan *et al.* in 2011 had also reported that an aqueous extract of *Curcuma xanthorrhiza* did not show any significant inhibition towards GST activity. To date, there are no published reports for *in vitro* GST inhibition by *Curcuma xanthorrhiza* ethanolic and hexane extracts.

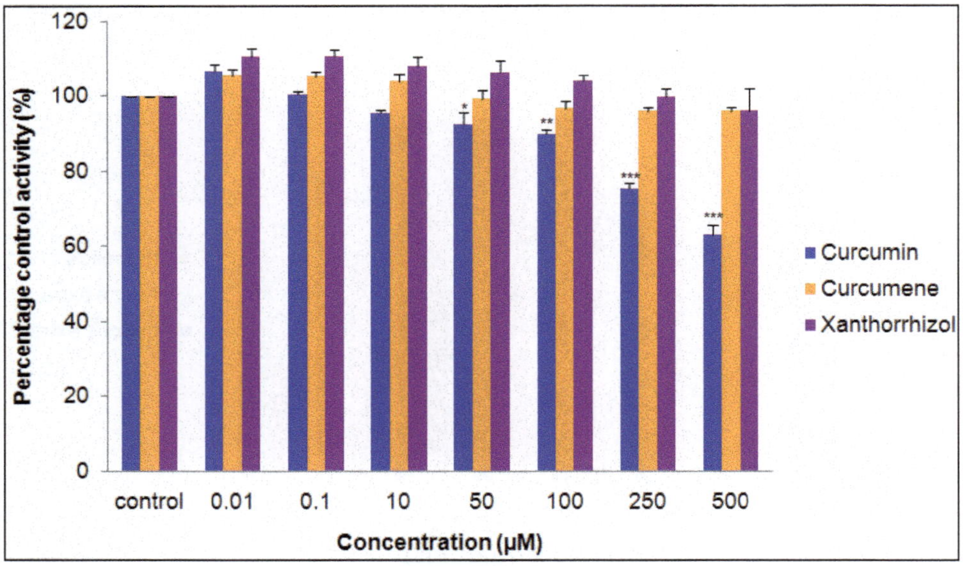

Figure 24.2: Effect of curcumin, curcumene and xanthorrhizol on GST activity. Each bar represents the mean value ± SEM of percentage of GSTs specific activity compared with control for five determinations ($n = 5$). Statistical analysis was conducted using one way ANOVA and Dunnet test. * $p < 0.05$ *vs.* control, ** $p < 0.01$ *vs.* control, *** $p < 0.001$ *vs.* control.

In this study, tannic acid which is the known inhibitor for GST enzyme showed an IC_{50} value of 16.73 ± 0.79 µg mL^{-1}. This value is higher than previous report of an IC_{50} value of 1.78 µg mL^{-1} (Zhang and Das, 1994). This is expected since IC_{50} is a relative value and its magnitude depends upon the concentration of chemicals used in the assay (Burlingham and Widlanski, 2003). Tannic acid (hydrolyzable tannin) is a polyphenol ubiquitously present in plants and easily accesible to animals and human through their diets (Das *et al.*, 1992). Zhang and Das (1994) reported that tannic acid showed concentration-dependent inhibition of rat liver GSTs with CDNB as substrate. The authors also proved that tannic acid is a potent inhibitor amongst the polyphenols tested.

Ethanolic extract of *Curcuma xanthorrhiza* showed 14 times higher IC_{50} value compared to tannic acid. These findings indicated that *Curcuma xanthorrhiza* extracts have weak inhibitory potential towards GST activity. In addition, only curcumin has low significant inhibition for concentration from 250 to 500 µM while the other two main compounds of this plant, curcumene and xanthorrhizol did not show any significant inhibition when added into GST reaction mixture. Phenolic substances, particularly those present in dietary and medicinal plants, have been reported to possess substantial anticarcinogenic and antimutagenic activities (Koo *et al.*, 2004).

GST activity has been shown to be modulated by natural plant products (Wortelboer *et al.*, 2003; Zhang *et al.*, 2003). Previous studies had reported that plant

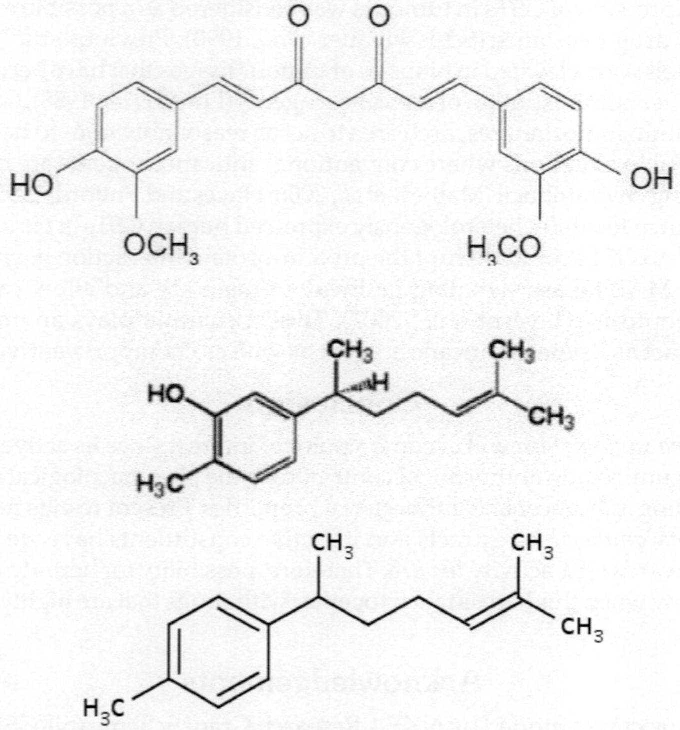

Figure 24.3: The structures for *Curcuma xanthorrhiza* constituents, curcumin (A), xanthorrhizol (B) and curcumene (C).

phenolic compounds such as tannic acid, ellagic acid, silybin, and curcumin were responsible for the inhibition of GSTs *in vitro* (Zhang and Das, 1993; Bartholomaeus *et al.*, 1994; Hayashi *et al.*, 2007). Among the three main compounds found in *Curcuma xanthorrhiza*, curcumin has been the subject of numerous studies over the past decades due to its pharmacological properties including antioxidant, antiinflammatory, anticancer, cholesterol lowering effects and a powerful immune system booster (Koo *et al.*, 2004; Mague *et al.*, 2004).

There are many reports on herb-drug interactions, adverse effect of herbs and their ability to influence the therapeutic effect of prescribed drugs (Ernst, 2003; Low and Tan, 2007; Mills *et al.*, 2005). It is possible that one drug may affect the metabolism of the other drug when both of them are metabolized by the same GST enzyme. Much less is known about the specific drugs that are metabolized by GST enzyme, except that the *in vitro* detoxification of some of the anticancer drugs [*e.g.*, 1,3-bis(2-chloroethyl)-1-nitrosourea (BCNU), busulfan, chlorambucil, melphalan, and thio-TEPA] is catalyzed by individual GST enzymes.

Drugs that are known to be *in vivo* substrates for human GST enzymes include acetaminophen, valproic acid and busulfan (Chang, 2010).

Overexpression of GSTs in tumours was considered as a possible mechanism of tumour cell drug resistance (Schisselbauer *et al.*, 1990). Previous study had shown that GST levels were elevated in biopsies of tumour tissues that have become resistant to therapy after administration of anticancer agents (Lewis *et al.*, 1988). GST inhibitors have become an important research area to act as reasonably non- toxic modulatory agents, useful in situations where conventional anticancer agents are detoxified by GST including chlorambucil (Mathew *et al.*, 2006; Hayes and Pulford, 1995). Curcumin has been shown to inhibit heterologously expressed human GSTs *in vitro*. The binding of curcumin to GSTs could disrupt the protein-protein interaction *in vivo* and hence disrupt the MAP kinase signaling pathway in the body and allow cancer cells to undergo apoptosis (Hayeshi *et al.*, 2007). Thus, curcumin plays an important role which may act as a potent anticancer agent as well as chemopreventive agent.

Conclusion

Curcuma xanthorrhiza will become a point of interest since its active constituents namely curcumin and xanthorrhizol contribute to the pharmacological effects of this plant including anticancer and antibacterial properties. Present results have indicated that *Curcuma xanthorrhiza* extracts and its active constituents have weak inhibitory potential towards GST activity *in vitro*. Therefore, possibility for herb-drug interaction to occur is low when this herb is taken together with drugs that are highly metabolized by GST.

Acknowledgements

This project was funded by NKEA Research Grant Scheme (NRGS), Ministry of Agriculture and Agro-Based Industry Malaysia. The authors wish to acknowledge Centre for Drug Research, Universiti Sains Malaysia for technical support for this research project.

References

Aggarwal, B., Kumar, A., and Bharti, A. (2003). Anticancer potential of curcumin: Preclinical and clinical studies. *Anticancer Res.* **23**(1A): 363–398.

Aggarwal, B., and Shishodia, S. (2006). Molecular targets of dietary agents for prevention and therapy of cancer. *Biochem Pharmacol.*, **71**(10): 1397–1421.

Ahokas, J. T., Nicholls, F. A., Ravenscroft, P. J and Emmerson, B. T. (1985). Inhibtion of purified rat liver glutathione S-transferase isozymes by diuretic rats. *Biochemical Pharmacolgy*, **34**(12): 2157-2161.

Appiah-Opong, R., Commandeur, J. N. M., Istyastonol, E., Bogaards, J.J., and Vermeulen, N.P.E. (2009). Inhibition of human glutathione *S*-transferases by curcumin and analogues. *Xenobiotica*, **39**(4): 302–311.

Appiah-Opong, R., Commandeur, J. N. M., Vugt-Lussenburg, B., and Vermeulen, N. P. E. (2007). Inhibition of human recombinant cytochrome P450s by curcumin and curcumin decomposition products. *J.Toxicol.*, **235** 83.

Azizi, J., Ismail, S., Mordi, M. N., Ramanathan, S., Said, M. I., and Mansor, S. M. (2010). *In vitro* and *in vivo* effects of three different *Mitragyna speciosa korth* leaf extracts on phase II drug metabolizing enzymes-glutathione transferases (GSTs). *Molecules*, **15**: 432-441.

Bartholomaeus, A. R., Bolton, R., and Ahokas, J.T. (1994). Inhibition of rat liver cytosolic glutathione S-transferase by silybin. *Xenobiotica*, **24**(1) : 17-24.

Boyer, T.D. (1989). The glutathione S-transferase: An update. *Hepatology*, **9** : 486–496.

Burlingham and Widlanski (2003). An intuitive look at the relationship of Ki and IC50 : A more general use for the dixon plot. *Journal of Chemical Education*, **80** : 214-218.

Chang, T. K. H. (2010). Drug metabolizing enzymes. *In:* Handbook of drug-nutrient interaction, Ed. by Boullata, J. I., and Armenti, V. I., Humana Press, New York, USA, pp. 85-117.

Ernst, E. (2002). The risk-benefit profile of commonly used herbal therapies: Ginkgo, St. John's Wort, Ginseng, Echinecea, Saw Palmetto, and Kava. *Ann. Intern. Med.*, **136**: 42-53.

Habig, W. H., Pabst, M. J., and Lakoby, W. B. (1974). Glutathione-S-transferase the first enzymatic step in mercapturic acid formation. *J. Biol. Chem.*, **249**: 7130-7139.

Hayes, J. D., and Pulford, D. J. (1995). The glutathione-S-transferase supergene family: Regulation of GST and the contribution of the isoenzymes to cancer chemoprotection and drug resistance. *Crit. Rev. Biochem. Mol. Biol.*, **30**: 445-600.

Hayes, J. D., Flanagan, J. U., and Jowsey, I. R. (2005). Glutathione transferases. *Ann. Rev. Pharmacol. Toxicol.*, **45** : 51–88.

Hayeshi, R., Mutingwende, I., Mavengere, W., Masiyanise, V., and Mukanganyama, S. (2007). The inhibition of human glutathione-S-transferases activity by plant phenolic compounds ellagic acid and curcumin. *Food and Chemical Toxicology*, **45**: 286-295.

Koo, J.Y., Kim, H.J., Jung, K-O., and Park, K-Y. (2004). Curcumin inhibits the growth of AGS human gastric carcinoma cells *in vitro* and shows synergism with 5-flourouracil. *J. Med. Food*, **7**: 117–121.

Lewis, A., Hayes, J., and Wolf, C. (1988). Glutathione and glutathione dependent enzymes in ovarian adenocarcinoma cell lines derived from a patient before and after the onset of drug resistance: Intrinsic differences and cell cycle effects. *Carcinogenesis*, **9**: 1283-1287.

Low, W. Y., and Tan, H. M. (2007). Asian traditional medicine for erectile dysfunction. *Journal of Men's Health and Gender*, **4**: 245-50.

Mague, J.T., Alworth, W.L., and Payton, F.L. (2004). Curcumin and derivatives. *Acta. Crystallogr.*, **60**: 608–610.

Mathew, N., Kalyanasundram, M., and Balaraman, K. (2006). Glutathione-S-transferase (GST) inhibitors. *Expert Opin. Ther. Paterns*, **16**(4): 431-444.

Mills, E., Cooper, C., Seely, D., and Kanfer, I. (2005). African herbal medicines in the treatment of HIV: Hypoxis and sutherlandia: An overview of evidence and pharmacology. *Nutrition Journal*, **4**: 19.

Nur, S. W. (2006). Perbandingan sistem ekstraksi dan validasi penentuan xanthorrhizol dari temulawak secara kromatografi cair kinerja tinggi. *Skripsi Bogor: Falkutas Matematika dan Ilmu Pengetahuan Alam, Institut Pertanian Bogor.*

Oakley, A. (2011). Glutathione transferases : A structural perspective. *Drug Metabolism Reviews*, **43**(2) : 138-151.

Pomory, C.M. (2008). Color development time of the lowry protein assay. *Anal. Biochem.*, **378** : 216-217.

Ruslay, S., Abas, F., Shaari, K., Zainal, Z., Maulidiani., Sirat, H., Israf, D. A., and H. Lajis, N. (2007). Characterization of the components in the active fractions of health gingers (*Curcuma xanthorrhiza* and *Zinger zerumbet*) by HPLC-DAD-ESIMS. *Food Chemistry*, **104**: 1183-1191.

Schisselbauer, J. C., Silber, R., Papadopoulos, E., Abrams, K., Lacreta, F. P., and Tew K. D.(1990). Characterization of glutathione-*S*-transferase expression in lymphocytes from chronic lymphocytic leukemia patients. *Cancer Res.*, **50**: 3562-3568.

Sears, B. (2005). In the anti-inflammation zone: Reversing the silent epidemic that's destroying our health. HarperCollins Publishers, New York, USA. pp. 75-87.

Sparreboom, A., C.Cox, M., R. Acharya, M., and D. Figg, W. (2003). Herbal remedies in the United States : Potential adverse interactions with anticancer agents. *J. Clin. Oncol.*, **22** : 2489-2503.

Tan, M. S. M. Z., Ab Halim, M. R., Mustaffa, F., Mohd Ali, N. I., Ismail, S., and Mahmud, R. (2011). Inhibitory effect of selected Malaysian herbal plants on glutathione-S-transferase activity. *International Journal of Pharmacology*, **7**: 349–355.

Wortelboer, H.M., Usta, M., van der Velde, A.E., Boersma, M.G., Spenkelink, B., van Zanden, J.J., Rietjens, I.M.C.M., van Bladeren, P.J., and Cnubben, N.H.P. (2003). Interplay between MRP inhibition and metabolism of MRP inhibitors: The case of curcumin. *Chem. Res. Toxicol.*, **16**: 1642–1651.

Yew, H.C., Fariza, J.N., Rozie, S., Thiam, T.T., Hawariah, L.P.A., Hasnah, M.S., Badrul, A.A.R., Noor, R.A., and Zakiah, I. (2009). Combined xanthorrhizol-curcumin exhibits synergistic growth inhibitory activity *via* apoptosis induction in human breast cancer cells MDA-MB-231. *Cancer Cell Int.*, **9(1)**.

Yugarani, T., Tan, B. K. H and Das, N.P. (1992). Effects of polyphenols natural products on the lipid profiles of rats fed high fat diets. *Lipids*, **27**(3): 181-186.

Zhang, K., and Das, N.P. (1994). Inhibitory effects of plant polyphenols on rat liver glutathione-S-transferases. *Biochem. Pharmacol.*, **47**: 2063–2068.

Zhang, K., Wong, K.P., and Chow, P.(2003). Conjugation of chlorambucil with GSH by GST purified from human colon adenocarcinoma cells and its inhibition by plant polyphenols. *Life Sci.*, **72**: 2629–2640.

Zhang, K., and Wong, K. M. (1997). Glutathione conjugation of chlorambucil: Measurement and modulation by plant polyphenols. *Biochem. J.*, **325**: 417-422.

Previous Volumes Contents

— Volume 1 —

2010, xiii+464p., col. plts., figs., tabls., ind., 25 cm Rs. 1800
ISBN 978-81-7035-627-1

— Volume 2 —

2012, xi+429p., col. plts., figs., tabls., ind., 25 cm Rs. 2400

ISBN 978-81-7035-768-1

Index

A

B

N

Q

R

S

www.ingramcontent.com/pod-product-compliance
Ingram Content Group UK Ltd.
Pitfield, Milton Keynes, MK11 3LW, UK
UKHW052005051026
15624UKWH00003B/581